Advances in Nonradiative Processes in Solids

NATO ASI Series

Advanced Science Institutes Series

A series presenting the results of activities sponsored by the NATO Science Committee, which aims at the dissemination of advanced scientific and technological knowledge, with a view to strengthening links between scientific communities.

The series is published by an international board of publishers in conjunction with the NATO Scientific Affairs Division

A	**Life Sciences**	Plenum Publishing Corporation
B	**Physics**	New York and London
C	**Mathematical and Physical Sciences**	Kluwer Academic Publishers
D	**Behavioral and Social Sciences**	Dordrecht, Boston, and London
E	**Applied Sciences**	
F	**Computer and Systems Sciences**	Springer-Verlag
G	**Ecological Sciences**	Berlin, Heidelberg, New York, London,
H	**Cell Biology**	Paris, Tokyo, Hong Kong, and Barcelona
I	**Global Environmental Change**	

Recent Volumes in this Series

Volume 248—Lower-Dimensional Systems and Molecular Electronics
edited by R. M. Metzger, P. Day, and G. C. Payavassiliou

Volume 249—Advances in Nonradiative Processes in Solids
edited by Baldassare Di Bartolo

Volume 250—The Application of Charge Density Research to Chemistry and Drug Design
edited by George A. Jeffrey and Juan F. Piniella

Volume 251—Granular Nanoelectronics
edited by David K. Ferry, John R. Barker, and Carlo Jacoboni

Volume 252—Laser Systems for Photobiology and Photomedicine
edited by A. N. Chester, S. Martellucci, and A. M. Scheggi

Volume 253—Condensed Systems of Low Dimensionality
edited by J. L. Beeby, P. K. Bhattacharya, P. Ch. Gravelle, F. Koch, and D. J. Lockwood

Volume 254—Quantum Coherence in Mesoscopic Systems
edited by B. Kramer

Series B: Physics

Advances in Nonradiative Processes in Solids

Edited by

Baldassare Di Bartolo

Boston College
Chestnut Hill, Massachusetts

Assistant Editor

Xuesheng Chen

Boston College
Chestnut Hill, Massachusetts

Plenum Press
New York and London
Published in cooperation with NATO Scientific Affairs Division

Proceedings of a NATO Advanced Study Institute on
Advances in Nonradiative Processes in Solids,
held June 15-19, 1989,
in Erice, Italy

Library of Congress Cataloging-in-Publication Data

NATO Advanced Study Institute on Advances in Nonradiative Processes in Solids (1989 : Erice, Italy)
Advances in nonradiative processes in solids / edited by Baldassare Di Bartolo.
p. cm. -- (NATO ASI series. Series B, Physics ; vol. 249)
"Proceedings of a NATO Advanced Study Institute on Advances in Nonradiative Processes in Solids, held June 15-19, 1989, in Erice, Italy"--T.p. verso.
"Published in cooperation with NATO Scientific Affairs Division."
Includes bibliographical references and index.
ISBN 0-306-43838-0
1. Solid state physics--Congresses. I. Di Bartolo, Baldassare.
II. Title. III. Title: Nonradiative processes in solids.
IV. Series: NATO ASI series. Series B, Physics ; v. 249.
QC176.A1N318 1989
530.4'1--dc20 91-11747
CIP

A Division of Plenum Publishing Corporation
233 Spring Street, New York, N.Y. 10013

Printed in the United States of America

Happy are those who can tie together
in their thoughts, the past, the present,
and the future.

—Alexis de Toqueville

PREFACE

This book presents an account of the course "Advances in Nonradiative Processes in Solids" held in Erice, Italy, from June 15 to 29, 1989. This meeting was organized by the International School of Atomic and Molecular Spectroscopy of the "Ettore Majorana" Centre for Scientific Culture.

An area of solid state research that continues to attract the attention of experimental and theoretical physicists is that of nonradiative relaxation processes of excited solids. The interest in these processes stems from their technological relevance, and from the difficulty in the quantitative characterization and differentiation of their various pathways. The decay channels leading to the ground state include the conversion of electronic excitation energy into phonon energy, nonradiative transfer of excitation energy, upconversion processes, etc.

Considerable advances have been achieved in understanding and modeling the <u>radiative</u> process that follow the electronic excitations of solids; the progress in this field has been instrumental in the development of new solid-state devices and laser materials. On the other hand, these advances have underscored the inadequacy in the understanding of the <u>nonradiative</u> relaxation processes.

This course dealt with the advances in physical modeling, mathematical formalisms and experimental techniques relevant to the quantitative characterization of the various pathways of nonradiative relaxation of solids in excited electronic states.

Nonradiative processes are generally considered drawbacks in such systems as phosphors and laser materials; however, some positive features of these processes are intricately connected with the negative ones and the design of good and efficient systems may require a careful compromise between these two aspects. In accordance with these developments, the course reviewed the theoretical and experimental advances in the field and examined the role of nonradiative processes in the performance of efficient solid-state devices.

The initial lectures treated the fundamental aspects of the theory and were followed by specific treatments of various systems. Each lecturer developed a coherent aspect of the program starting at a somewhat fundamental level and ultimately reaching the frontier of knowledge in the field. The formal lectures were complemented by seminars and discussions on specific topics and applications and by problem sessions.

The course was addressed to workers in the field of solid-state physics from universities and industry. Senior scientists in related fields and graduate students were encouraged to participate.

A total of eighty participants came from forty-three laboratories and eighteen different countries (Algeria, China, France, Germany, Greece, India, Ireland, Israel, Italy, the Netherlands, Poland, Portugal, the Soviet Union, Spain, Switzerland, Turkey, the United Kingdom, and the United States).

The secretaries of the course were John Di Bartolo and Clyfe Beckwith.

Thirty-six lectures divided into eleven series were given. In addition twelve (one- or two-hour) "long" seminars and twenty-six "short" seminars were presented. Several problem sessions were organized; the purpose of these sessions was to create a closer interaction between participants and lecturers. During a problem session a lecturer did not produce any new material, but explained in greater detail some points of the theory already presented, answered more question, presented examples, etc.

Three topics of general interest to physicists, "Particles and Elementary Excitations", "Cold Fusion" and "X-ray Astrophysics", were treated in special sessions.

Two round-table discussions were held. The first round-table discussion took place the fourth day of the meeting and had as subjects an evaluation of the work done in the first days of the course and the consideration of suggestions and proposals regarding the format of the lectures, the level of the presentations, etc. This round-table discussion favorably affected the progress of the course. The second round-table discussion was held at the end of the course: the attendees had the opportunity of evaluating the work done during the entire meeting and to discuss various proposals for the next course of the International School of Atomic and Molecular Spectroscopy.

During the final session Professor Imbusch presented a summary of the meeting and identified the major themes which appeared in various degrees throughout the school.

I wish to express my sincere gratitude to Drs. Gabriele and Savalli, to Mr. Pilarski and to all the personnel of the "Ettore Majorana" Centre who contributed so much to create a congenial atmosphere for our meeting. I wish to acknowledge the sponsorship of the meeting by the NATO Scientific Affairs Division, the ENEA Organization, Boston College, the National Science Foundation, the European Physical Society, the Italian Ministry of Public Education, the Italian Ministry of Scientific and Technological Research, and the Sicilian Regional Government.

I would like to thank the members of the Organizing Committee (Doctor Baldacchini, Professors Curie and Klingshirn), Professor Imbusch, Doctor Auzel, the secretaries of the course (John Di Bartolo and Clyfe Beckwith), Giovanni Galfano, and Daniel Di Bartolo, Mr. Michele Strazzera and Avv. Giuseppe Luppino.

It was my privilege to be the director of this course.

B. Di Bartolo
Director of the International
School of Atomic and Molecular
Spectroscopy of the "Ettore
Majorana" Centre

Erice, June 29, 1989

CONTENTS

SPECIAL TOPICS (Abstracts)

SHORT SEMINARS (Abstracts)

NON-RADIATIVE RELAXATION OF SOLIDS:

DIFFERENT PATHWAYS TO THE GROUND STATE

B. Canny and D. Curie

Laboratoire de Physique des Milieux Condensés
Tour 13, étage 4e
Université Paris VI
75252 Paris Cedex 05 France

ABSTRACT

Non-radiative processes are a very general item, and they may occur equally well in gases, liquids, or solids. Here we are specifically interested in electronic radiationless processes in solids. The most important example is radiationless deactivation of the excited state, but phonon absorption for instance can also lead to an upper electronic state. In the simplest case, radiative and non-radiative transitions compete and occur in the same ion or more generally within the same luminescent center ("one-center processes"). As pointed out by Struck and Fonger, the case of a low phonon coupling typically leads to the so-called "exponential energy gap" law for multiphonon transitions, while the case of a strong phonon coupling often leads to a Mott-Seitz type law in the temperature range of interest i.e. an activation energy followed by a cascade phonon emission process. However, these authors realized that even when this law fits correctly the experimental results, the activation energy involved in the Mott-Seitz formula is not necessarily equal to the height of the crossing point of the configurational curves. The reason for this discrepancy is that the conventional Mott-Seitz process actually occurs only in the limiting case of very high temperatures, while at lower temperatures tunneling from low vibrational levels in the excited state to the ground state may occur.

In addition, even if radiationless transitions can be predicted to occur between the emitting excited state and the ground state, often the quenching processes involve excitation via one or several other excited levels (for instance this is the case in thermal quenching of ruby).

Both the number of emitted (or absorbed) phonons, and the strength of the coupling constant, are to be considered as well. Kun Huang has shown that the pathway to the ground state involves, in some cases the highest energy phonons, in other cases the largest coupling constant or the mean value of the phonon distribution. As a matter of fact, any specific case must be studied separately.

More intricate pathways take place, in which radiationless transitions occur via some kind of energy transfer from the excited center to a "Killer center". The mechanism of impurity poisoning in conventional II - VI or III - V phosphors is not yet clear, but important progresses have been made in the recent years. Transfers have been shown to occur towards a center which emits I.R. photons for absorbed U.V. or visible exciting radiation.

Some attention is also given to trapping processes and Auger effects in semi-conducting luminescent materials. Characteristic examples of these items are discussed shortly.

I. INTRODUCTION

Historically, two kinds of phenomena have been described first in some details:

- Thermal quenching of luminescence
- Impurity poisoning ("Killer effects")

I.A. Thermal Quenching of Luminescence

The typical relationship between the fluorescent lifetime τ and the efficiency η (fig. 1) has been first described by F. Perrin and coworkers on the case of uranyl salts, solutions and glasses [1,2].

A

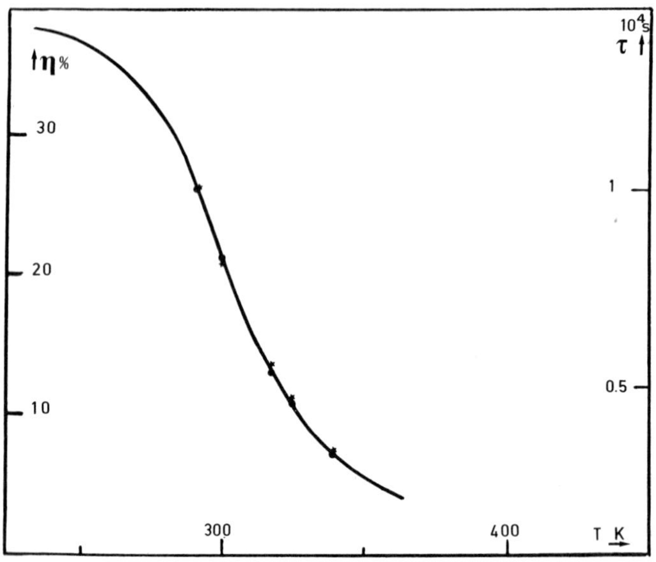

B

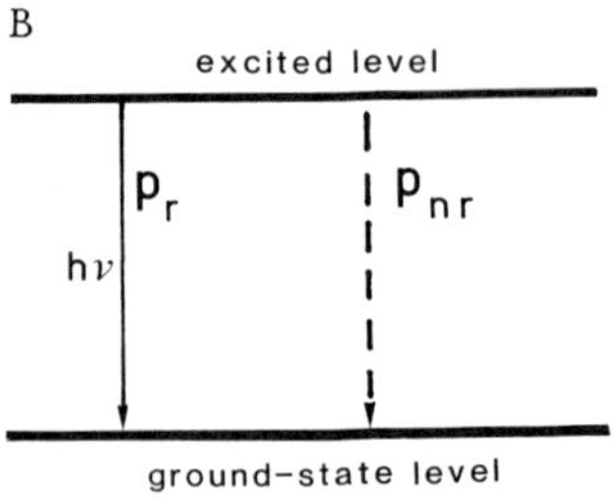

Fig.1.A Proportionality between τ and η (see text). The data are relevant to uranyl sulfate dissolved in sulfuric acid.

Fig.1.B The simplest energy level scheme for taking into account this relationship.

Let us denote by:

P_r the radiative transition probability.
P_{nr} the non radiative transition probability.
The two following expressions are straightforward:

$$1/\tau = P_r + P_{nr} \qquad (1)$$

$$\eta = \frac{P_r}{P_r + P_{nr}} \qquad (2)$$

In many cases, $P_{nr}(T=0) = 0$.
Hence

$$\frac{1}{\tau(T=0)} = P_r \qquad \eta(T=0)) = 1 \qquad (3)$$

P_r is assumed not to change versus temperature T, while P_{nr} increases with T. Hence

$$\eta(T) = \frac{\tau(T)}{\tau(T=0)} \qquad (4)$$

But even if $P_{nr}(T=0) \neq 0$, the proportionality relation holds:

$$\frac{\eta(T)}{\eta(T=0)} = \frac{\tau(T)}{\tau(T=0)} \qquad (5)$$

We shall now make some remarks about the formulas (4) and (5). From a priori consideration, we may conclude as follows:

-Typically, they work for a 2-levels system (see below, however).
-They may work even if the well-known Mott-Seitz formula:

$$\eta(T) = \frac{1}{1 + C \exp(-\Delta E/kT)} \qquad (6)$$

is not valid.

Now, what actually happens:

- Formulas (4), (5) and (6) are valid much more generally than we should expect. However some typical breakdowns occur, which are sometimes not well understood.

We shall give as an example the behaviour of F-centers in alkali-halides [3].

A three-channel decay takes place for the excited state (see fig.2).

The observed mean life is thus:

$$\frac{1}{\tau} = \frac{1}{\tau_r} + \frac{1}{\tau_{nr}} + \frac{1}{\tau_0} \exp(-E_i/kT) \qquad (7)$$

$\tau_0 \sim 10^{-12}$ s.

A

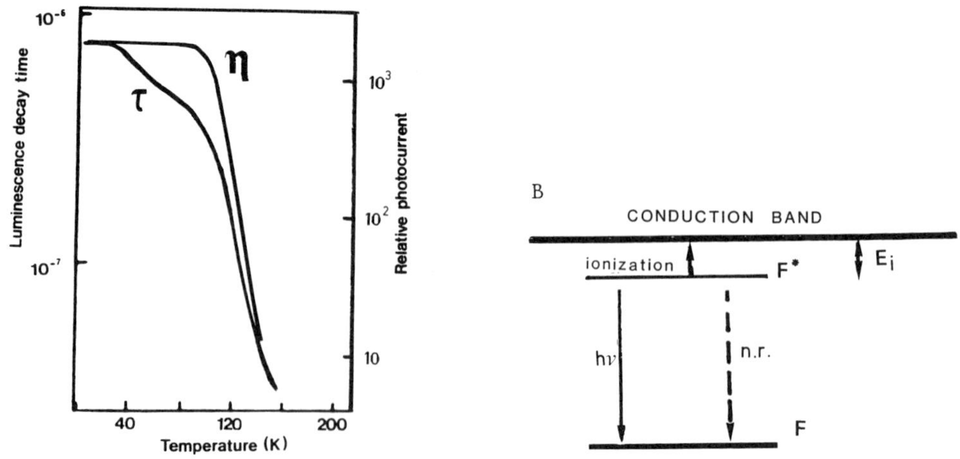

Fig.2A Ionization process, radiative and non-radiative transitions for F-centers in KCl [4].

Fig.2B Experimental behaviour of τ and η.

At low temperature, $1/\tau_{nr}$ is assumed to be negligible and we turn to a Mott-Seitz type law (with a completely different mechanism).

$$\tau(T) = \frac{\tau_r}{1 + \frac{\tau_r}{\tau_0}\exp(-E_i/kT)} \tag{8}$$

$$\eta(T) = \frac{1}{1 + \frac{\tau_r}{\tau_0}\exp(-E_i/kT)} \tag{9}$$

Experimentally (fig.2B):

-quenching occurs for $T \geq 60$ K and formula (9) for η is valid.

-the proportionality relationship between $\tau(T)$ and $\eta(T)$ does not hold in a wide range of temperatures.

Many explanations have been suggested for this discrepancy, for instance, mixing of 2p and 2s states.

The behaviour of τ and η when pressure is changed instead of temperature is reported by Drickaman et al. in [5].

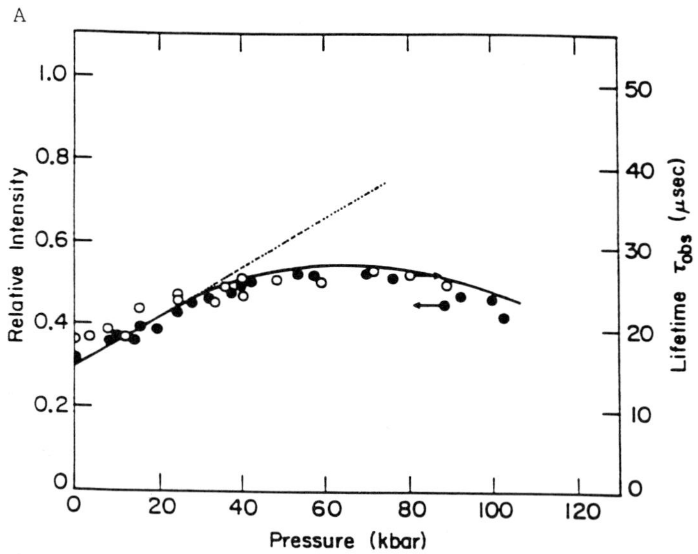

Fig.3A Effect of pressure on intensities (•) and lifetimes (o) in $CaMoO_4$ [6].

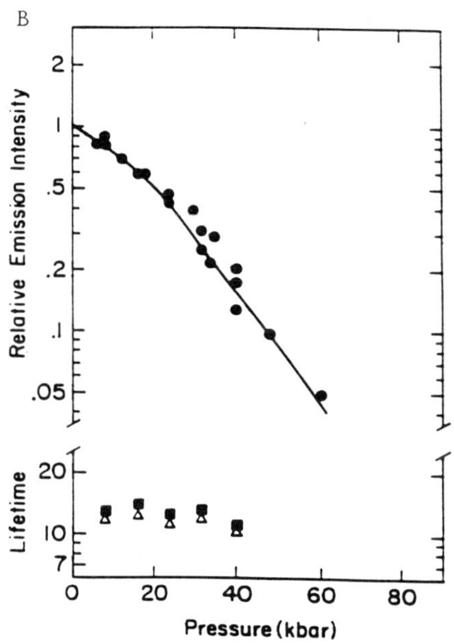

Fig.3B Luminescence intensities (•) and lifetimes versus pressure for ZnS:Cu,Al.

Tyner and Drickamer [6] have shown that in a material as $CaMoO_4$, in which excitation and emission processes can be described by means of a single configurational coordinate model, the efficiency η and the mean life τ show a similar behaviour when pressure changes. On the other hand, in a ZnS:Cu,Al phosphor where "delocalized excitation" occurs, η and τ exhibit a totally different behaviour [7].

The actual decay is complex but fits with two exponentials law with lifetimes of the order 10^{-6}(■) and 10^{-4} s.(Δ).

These lifetimes are approximately independent of pressure [7].

In a similar way, when concentrational quenching occurs, frequently the F.Perrin relationship does not hold, because p_r cannot be considered as a constant when concentration is increased.

I.B. "Killer Centers" in Luminescence

Since the very beginning of the story of conventional Lenard phosphors (i.e. CaS, ZnS, ZnCdS etc...), it has been observed that accidental contamination by very small amounts of iron (or Ni or Co) results in a drastic decrease in the luminescent output of the material. Transfers to "Killer centers" have been assumed to occur. However, many problems arise about this "Killer effect".

First, Shionoya and coworkers [8] have shown that transfers to isolated Fe ions cannot be the mechanism for the high efficiency of deactivation. A highly important point towards the solution of this puzzling problem was performed by Godlewski, Cavenett and co-workers [9, 10], who have detected by ODMR experiments the formation of Fe-Cu pairs.

Fe^{2+} in ZnS produces 3 infra-red emission bands:

0.98 μm	$^3T_1(^3H) \dashrightarrow {}^5E(^5D)$
1.45 μm	$^3T_1(^3H) \dashrightarrow {}^5T_2(^5D)$
3.4 μm	$^5T_2(^5D) \dashrightarrow {}^5E(^5D)$

In ZnS:Cu,Fe:

a) when $n_{Cu} > n_{Fe}$, one observes

- 2 visible bands (blue and green emission) due to D-A type centers containing copper.

- the I.R. band from isolated Cu^{2+} ions ($^2E \dashrightarrow {}^2T_2$ emission with ZPL 0.858 eV).

- the 3 I.R. emission bands from Fe^{2+}.

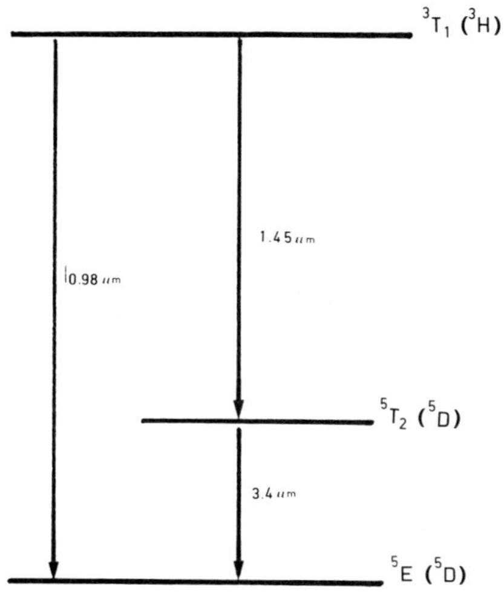

Fig.4 The I.R. emission spectrum of Fe^{2+}.

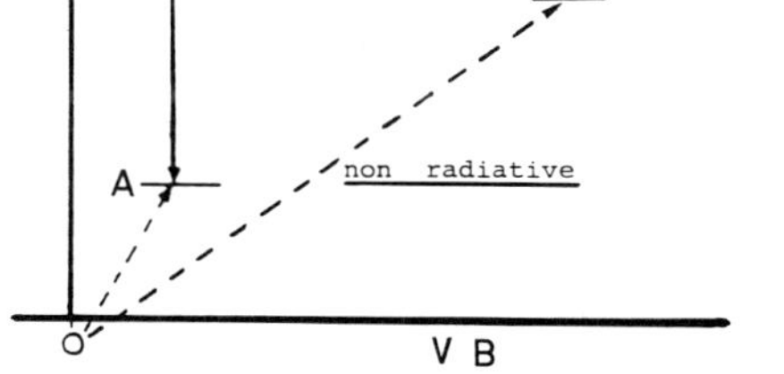

Fig.5 Energy transfer and infra-red transitions in ZnS:Cu,Fe [10].

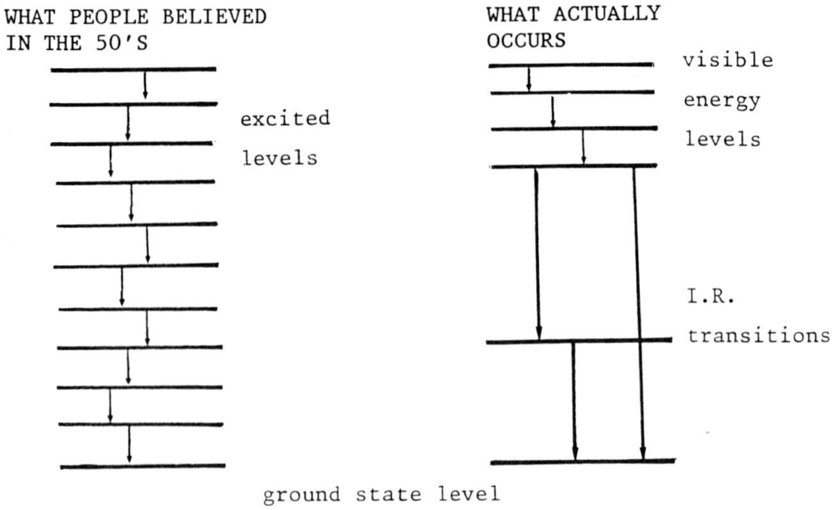

Fig.6 The scheme of Fe^{++} levels and the Killer effect mechanism of Iron in Zn S.

b) for $n_{Fe} > n_{Cu}$, visible and I.R. emission from copper are suppressed, while one observes the 0.98 μm and 1.45 emission bands from Fe^{2+} shifted only slightly by the copper interaction in the pairs.

Fe^{3+}-Cu pairs capture non-radiatively an electron from the conduction band (or from an excited D-A pair) whence the I.R. emission from the Fe^{2+}-Cu pair occurs.

However, the problem of the quantum efficiency of these I.R. emissions remains unsolved (i.e. we don't know to what extent these I.R. photons compete with non-radiative transitions). Up to now, nobody has proven that the number of emitted I.R. photons is equal to the number of missing visible photons.

II. THE CASE OF INTERNAL TRANSITIONS WITHIN THE SAME CENTER

This case is conventionally treated by using the configurational curves diagram (see the article by Prof. Di Bartolo in this book for the definition of these diagrams). One-dimensional configurational coordinates are generally used, but this approximation theoretically needs the interaction between the optical system and one specific normal mode of vibration to be preponderating. This is often the situation with the breathing mode. Now, a correct treatment of radiative and non-radiative transitions would involve 3N configuration coordinates, where N is the number of atoms in the crystal.

Pioneering references on configurational curves diagrams are A. Von Hippel [11], F. Seitz [12], N.F. Mott [13,14].

II.A. The Case of Broad-Band Phosphors

Pioneering work in fact was dealing with this case: strong linear coupling, i.e. large change in the equilibrium position of the ions between ground state and excited state.

According to the Mott-Seitz model for non-radiative transitions (fig.7):

High temperature rate (i) involves a thermal activation energy ΔE (the pathway CX) followed by a cascade process (XA). Hence the well-known Moss-Seitz formula:

$$\eta(T) = \frac{1}{1 + C\,\exp(-\Delta E/kT)} \tag{10}$$

The low temperature rate (ii) can be visualized as a tunnel effect CE leading from the excited state configurational curve to the ground state curve.

In recent years, it has been realized that, even if it may happen that formula (10) gives the gross behaviour (*) of $\eta(T)$ in the quenching temperature range, the activation energy ΔE in this formula is as a matter of fact smaller than the height of the crossing point X [15]. For downward transitions:

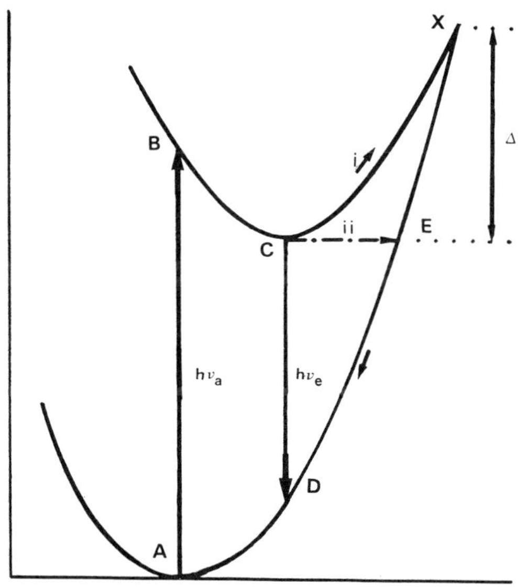

Fig.7 High temperature processes (Mott-Seitz mechanism, i) and low temperature processes (tunnel effect, ii) on the one-dimensional configurational curve diagram.

(*) See, however, F.G. Imbusch's article in this book.

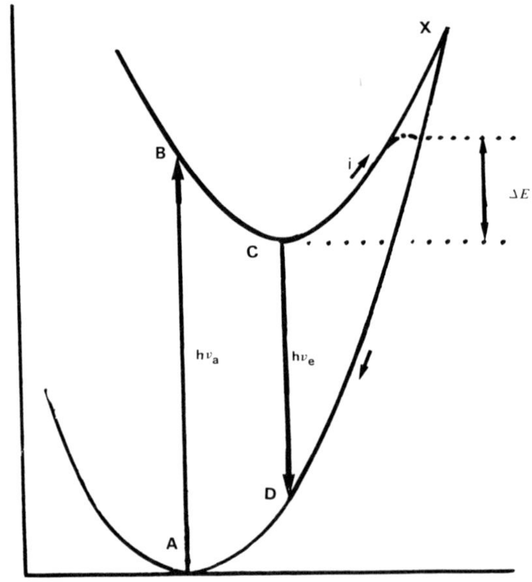

Fig.8 At intermediate temperatures, a phonon-assisted tunnel effect may occur, activation energy where activation energy ΔE is smaller than according to the conventional Mott-Seitz mechanism. Note, however, that this representation is merely a simple way to describe the actual phenomenon, which includes multiphonon transitions occuring at any level between C and X (see text).

$$\Delta E < E_X - E_C$$

The simple Mott-Seitz theory works better for upward transions (i.e. the path A X C):

ΔE only slightly $< E_X - E_A$.

This can be easily understood from the unified model developed by Struck and Fonger for the case of one dimensional configurational coordinate diagrams [16, 17, 18, 19, 20]. (*)

Except at OK, the transfer from the excited state configurational curve to the ground state curve may take place at the height of any vibrational level. (**)

The transfer is particularly efficient near the height of the crossing point X, but a summation upon all vibrational levels must be performed for a convenient treatment of the non-radiative transition rate.

(*) See also the articles by B. Di Bartolo, F. Auzel, G. Blasse, C. Struck, etc... in this book.

(**) See for instance Paragraph V of B. Di Bartolo's article, or Fig.3 of C. Struck's article.

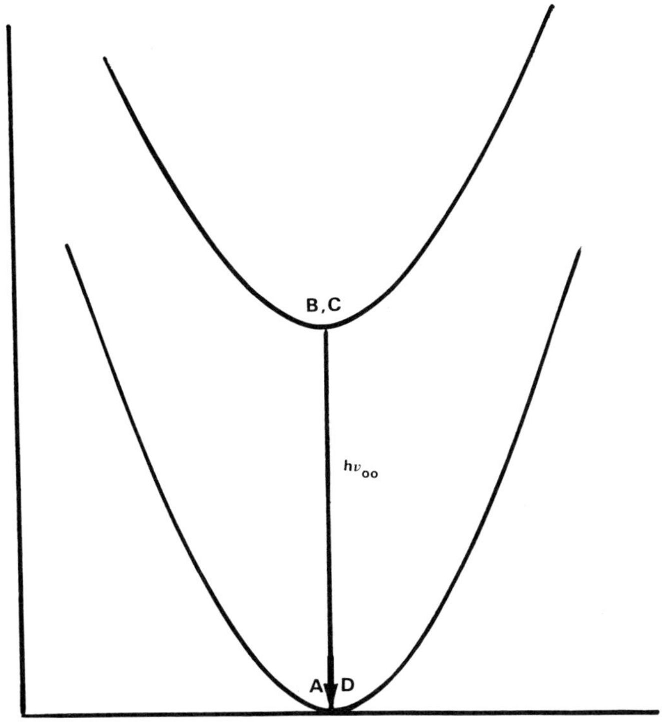

Fig.9 The case of a purely quadratic coupling: No shift between the equilibrium position of the ions.

II.B. The Case of Narrow-Line Phosphors

Such is the case for most trivalent rare earth phosphors, and also for the R-lines in ruby etc...

The Huang-Rhys coupling constant is near zero (typically $s \leq 0.1$), hence the change in the equilibrium position of the ions between ground state and excited state is zero or can be neglected.

The main aspect of the emission spectrum is a zero phonon line $h\nu_{00}$ with a very high intensity, accompanied by very small phonon replicas.

The crossing point X, if any, is extremely high and non-radiative transitions according to the Mott-Seitz mechanism cannot take place.

The non-radiative decay rate depends in an exponential way on the net number of emitted phonons.

As a result of first order perturbation calculations,

$$P_{nr} = \epsilon^{p} \qquad \epsilon \ll 1$$

$$p = \Delta E/\hbar\omega_{eff} \tag{11}$$

At low temperatures, (11) results in the exponential energy gap law:

$$P_{nr}(\Delta E) = p_{nr}(0) \exp(-\alpha \Delta E) \tag{12}$$

Where:

$$\alpha = \frac{\text{a constant } \gamma}{\hbar\omega_{eff}} \tag{13}$$

The behaviour for increasing temperatures is simply given by:

$$P_{nr}(T) = P_{nr}(T=0)\ (\bar{n} + 1)^p \tag{14}$$

while the calculation in the limiting case of very high temperatures leads to an activation energy law:

$$P_{nr} \sim p_0 \exp(-W/kT) \tag{15}$$

but W of course is no more related with the crossing point X (which should be infinitely high if the curves for ground state an excited state have the same curvature).

More generally, if S is small but not zero:

$$P_{nr} \sim \frac{e^{-S}}{p!} (\bar{n} + 1)^p\ e^{-2\bar{n}S} \tag{16}$$

II.C. F-Centers Luminescence in NaCl, KCl etc...

This case is characterized by a considerably large Stokes shift $h\nu_a \sim$ visible or UV, $h\nu_e \sim$ I.R. and therefore a highly efficient quenching of fluorescence. In the best cases, I.R. luminescence from F-centers is observed only at low temperatures, typically below liquid nitrogen, in other cases it is not observed at all, even at temperatures about a few Kelvins. The explanation for this peculiar behaviour has been given by Dexter [21]. Here we follow more specifically ref. [22]. (*)

The following assumptions will be made for the sake of simplicity

- single mode configurational diagram
- $\hbar\omega_{\text{excited state}} = \hbar\omega_{\text{ground state}}$

Optical absorption leads from point A to point B (Fig.10), emission comes back from C to D, but if the crossing point X is situated below point B, non-radiative transitions will take place instead of emission. Hence the Bartram-Stoneham criterion for the occurrence of F-center luminescence (at low temperatures, of course):

(*) For a more detailed treatment, look at the article by G. Baldacchini in this book.

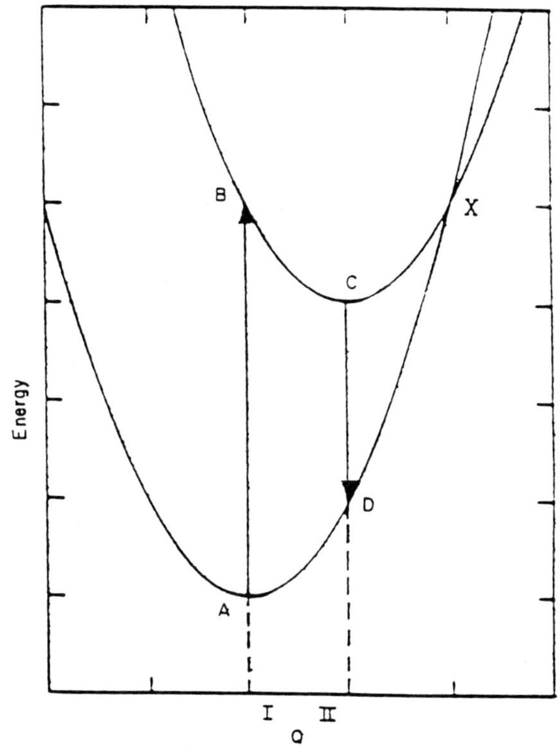

Fig.10 The configuration diagram relevant to F-centers (R.H. Bartram and A.M. Stoneham [22]).

<u>Point X situated above point B</u>

Introducing the following parameter

$$\Lambda = \frac{S\ \hbar\omega_{phonons}}{E_{absorbed}} \tag{17}$$

this criterion results in $\Lambda \leq 1/4$ (18)

In spite of the over-simplifications in the model, this criterion works rather well:

<u>No luminescence</u> (both predicted and observed)
LiI $\Lambda \sim 0.8$

NaBr, NaI
LiCl, LiBr, LiF } $\Lambda = 0.32 - 0.38$

Luminescence (at low temperatures)

NaCl	$\Lambda = 0.26$
NaF KCl, KBr, KI RbCl, RbBr, RbI RbF	$\Lambda \sim 0.16 - 0.24$

CsF, CsCl, CsBr, CsI $\Lambda \sim 0.009$-0.19

The parameter

$$\Lambda' = 1/2\ (1 - E_{emitted}/E_{absorbed}) \quad (19)$$

can be used as well (Λ' should be equal to Λ in the simple model). As a matter of fact Λ' is slightly larger than Λ.

More recently [23], the same authors developed a multicoordinate configurational curve model, in which they consider both a promoting mode and an accepting mode, and also a set of lattice modes which ultimately absorb the excess vibrational energy.

II.D. Different Paths of Decay of Self-Trapped Excitons in Solid Rare Gases

The spectra due to radiative decay of excitons in solid xenon or argon exhibit [24]:

- line spectra from: $^3P_1$, $^1P_1$ and metastable $^3P_2$ level of the free atom;

- band spectra from the exciton dimer Ar-Ar* or Xe-Xe*.

- but in addition photo-induced desorption has been observed for excitons trapped at the surface [25].

In Fig.11, it can be seen that the minimum in the ground state configurational curve is so flat that decomposition of the dimer Ar-Ar* may occur, either after vibrational relaxation or radiative relaxation from the excited S.T. state.

II.E. Three (or more) Parabolae Models

Two cases of a special interest will be discussed here:

1. Quenching of the Ruby R - Lines via the $^4T_2$ State [27]. (Fig.12).

The ruby R-line emission is due to the transition $^2E \rightarrow {}^4A_2$, with a very small Huang-Rhys coupling constant S. Hence thermal activation energy according to the Mott-Seitz mechanism leading directly

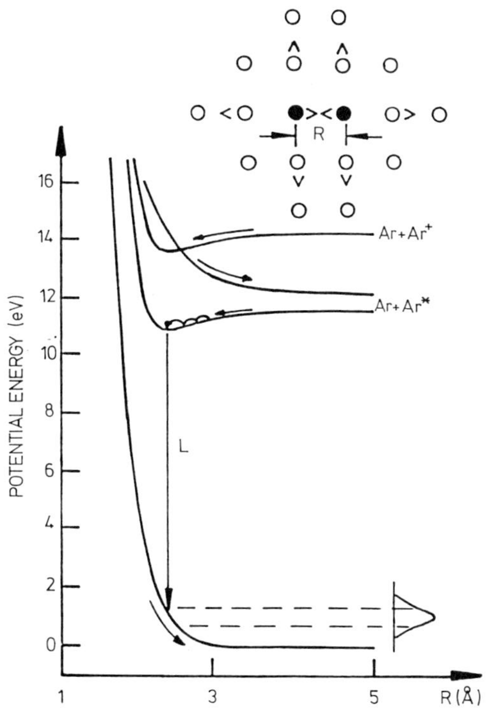

Fig.11 Potential energy curves for S.T. exciton in solid argon (T. Kloiber and G. Zimmerer [26]). The relaxation energy in the ground state is sufficient to overcome the cohesive energy. R means the internuclear distance.

to the $^4A_2$ ground state cannot occur. But Fig.12 shows clearly that a transfer involving the intermediate state $^4T_2$ is quite possible.

Other intermediate levels, such as for instance $^4T_1$, can also be considered a priori, but G. Blasse and co-workers have clearly shown that consideration of $^4T_1$ instead of $^4T_2$ would not lead to a convenient fit of the theoretical quenching curve versus the experimental one.

2. Quenching of Rare-Earth Luminescence via a Charge Transfer State. The important case of Eu^{3+} doped phosphors is given here as an example (Fig 13).

The dotted arrow denotes quenching via the charge transfer state (C.T.). It needs a thermal activation except if the CT state is at very low energies. (If it happens to be the case, the luminescence is totally quenched). This process competes with multiphonon transition processes, involving high-energy vibrations. Energy transfer towards other ions may also occur. The relative importance of the charge transfer process

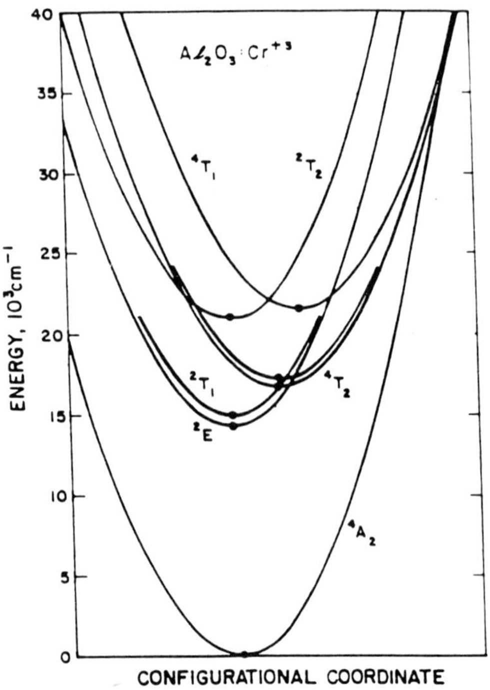

Fig.12. S.C.C. diagram of the Cr^{3+} states in ruby [27].

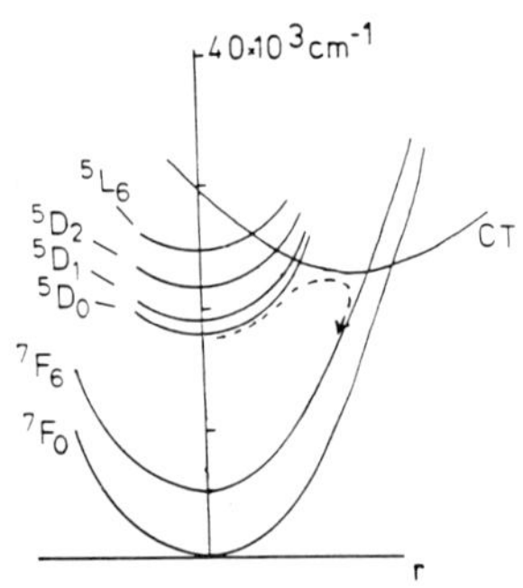

Fig.13 Quenching of Eu^{3+} luminescence via the charge transfer state (Fonger and Struck [29]), (Blasse and Sabbatini [30]).

depends on the height of the CT level compared with the Eu^{3+} ion levels. High quantum efficiencies for CT excitation are only to be expected if the CT state is at high energy (typically, 40,000 cm^{-1}), [30].

II.F. Experimental Examples of the Exponential Energy Gap Law [12]

This law fits most experimental results very nicely, and such a good fit is rather suprising in view of the numerous approximations made in the perturbation calculus [31,32].

Fig.14, which can be found in any paper on this topic, is relevant to several R.E. ions in the same crystalline laser material (Y Al 0_3).

Fig.15 shows that the exponential law works equally well in glasses (GLS = germanium, lanthanum, sulphur glass, ALS = aluminum lanthanum, sulpher).

The exponential dependance on p, the net number of emitted phonons, works equally well, but it has recently received some more elaborated formulations (*).

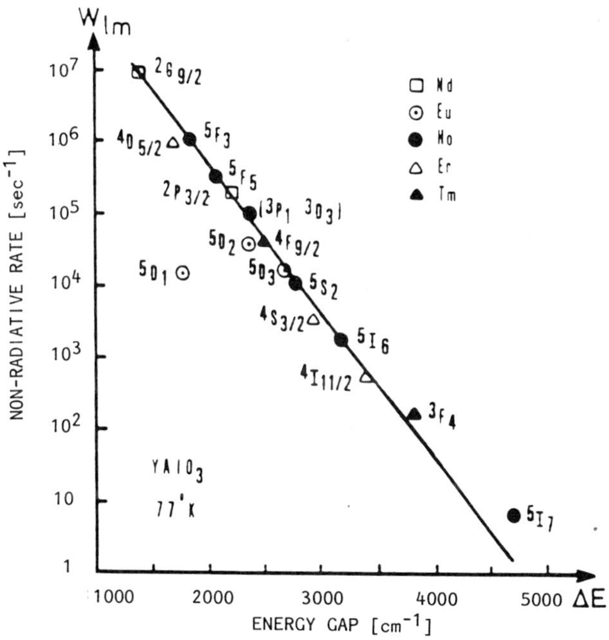

Fig.14 M.J.Weber [33]. The exponential energy gap law in laser crystals.

(*) See especially the paper by C.J. Donnelly in this book.

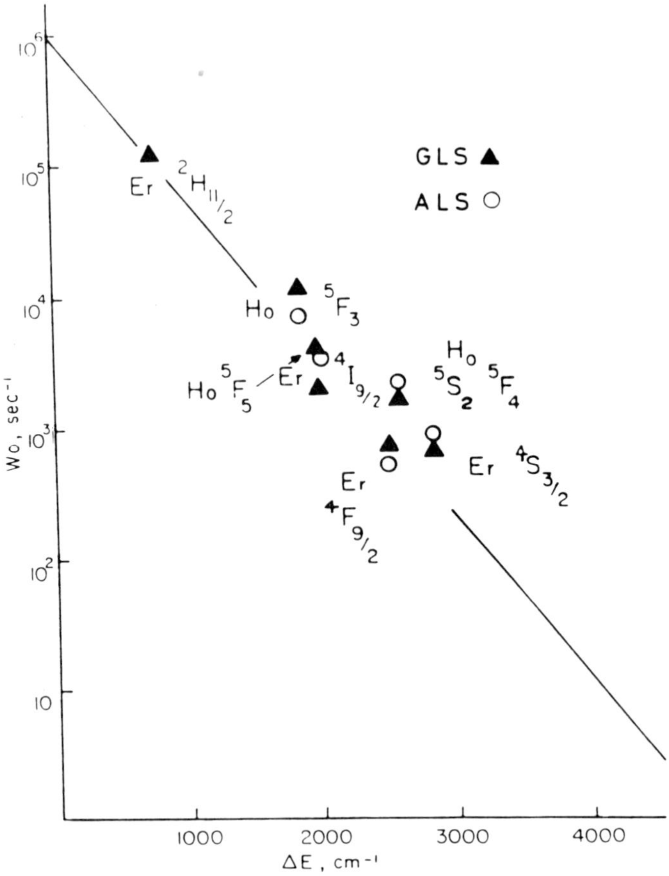

Fig.15 from R. Reisfeld [34]. The exponential energy gap law in glasses.

II.G. The Choice of the "Effective Phonon Energy" $\hbar\omega_{eff}$ in Formula (13)

Kun Huang [35] has theoretically studied the case of a statistical distribution of phonons emitted in multiphonon transitions.

We quote here his main conclusion: even in the absence of any selection rule which could forbid the interaction with some kind of vibrational mode, "the relative roles of the different energy phonons vary enormously with transition energy and coupling strength". Hence it is very hard to give any general tendency, the proper choice of the effective phonon energy must be studied separately in any specific case. However, it seems reasonable to accept Moss' general idea [36], that, in the case of very weak coupling, the highest energy phonons may predominate. But, in this school for instance, F. Auzel describes examples in which $\hbar\omega_{eff}$ approaches the mean value of the phonon distribution. In other cases, if a highly strong peak is the main feature in the phonon distribution, it may happen that $\hbar\omega_{eff}$ will coincide with the energy of this peak.

A very peculiar example has been described by Jia and Yen [37]: the luminescence of Mn^{2+} ions in

$Cs\ Mn\ Cl_3,\ 2\ H_2O$

$Cs\ Mn\ Cl_3,\ 2\ D_2O$

They found that the replacement of H by D reduced the decay rate by a factor more than 10; therefore the main component in the decay is a multiphonon process, in which H (or D) plays a dominant role.

However, most modes coupled to radiative transitions have relatively low energies, about 0.2 eV.

The energy to be dissipated via localized vibrations is ~ 2eV, whence the importance of high frequency local modes induced by H or D. These modes are weakly coupled to the radiative transitions in the luminescence centers, but become predominent in non-radiative transitions.

III. NONRADIATIVE TRANSITIONS IN SEMICONDUCTORS

III.A. Internal Transition within the Luminescent Center

In semi-conductors, radiative and non-radiative processes can take place within the same center, exactly as they do for rare earths or transition elements incorporated at low concentrations into insulating materials. Competition takes place between p_r and p_{nr} and the theoretical treatment in such a case is similar, it involves the configurational curve diagram.

A typical example is the treatment of infra-red optical processes in III-V compounds doped by vanadium, chromium, iron, nickel or cobalt [38]. In II-VI compounds, similar considerations would be valid as well. Specifically, H.J. Schulz and co-workers study the following ions:

V^{3+} (d^2)	$Fe^{2+}(d^6)$
Cr^{3+} (d^3)	$Co^{2+}(d^7)$
	Ni^+ (d^9)

embedded into GaAs, GaAlAs, GaP, InP.

Let τ_0 the mean life for radiative processes ($p_r = 1/\tau_0$), the transition probability for non-radiative processes is assumed to be (in Schulz's notation)

$$P_{nr} = \frac{1}{\tau_{th}} \exp(-E_a/kT) \qquad (20)$$

E_a the thermal activation energy. Hence the observation mean life at temperature T is

$$\tau(T) = \frac{\tau_0}{1 + \frac{\tau_0}{\tau_{th}} \exp(-E_a/kT)} \qquad (21)$$

	$GaAs:V^{3+}$	$GaAs:Cr^{3+}$	$GaAs:Fe^{2+}$	$GaAs:Ni^{+},Sn$	$GaP:Ni^{+}$
τ_0 (μs)	86	0.9	10	4.2	9
E_a cm^{-1}	200	1600	400	120	1290
τ_0/τ_{th}	25	5.10^4	180	15	3.10^4

while the quantum efficiency $\eta(T)$ should be proportional to $\tau(T)$. Experimental determination of $\tau(T)$ allows to determine all the parameters of the configurational diagram. The preceding table gives only τ_0, τ_{th} and E_a [38]:

However, we should like to bring the reader's attention upon the general aspects of the conclusions derived by Crasemann and Schulz:

- "the formulas can be understood from a conventional configurational diagram model".

- "the magnitude of the transition probabilities depends on covalency", but "one should refrain, however, from far-reaching conclusions, as the tendencies may just reflect features inherent in the models". We emphasize that this call for carefulness seems particularly adequate in all covalency problems.

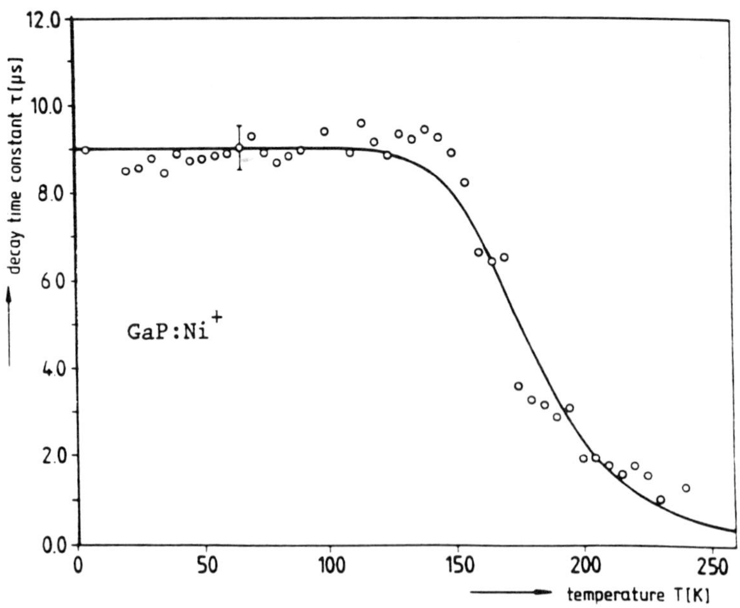

Fig.16. Quenching of the decay time constant in $GaP:Ni^{+}$ [38]. Observe the usual Mott-Seitz appearance of this curve.

III.B. Specific Processes for Semi-Conductors

Consulting the book by M. Janos "Deep Levels in Semi-Conductors" [39] is of a particular importance here. We follow his description of the phenomena.

There are 4 processes concerning recombination at deep traps:
- optical recombination
- cascade mechanism
- multiphonon transition
- Auger effect

1. Cascade Mechanism for Nonradiative Trapping Processes. Such a mechanism is connected with the problem of "giant traps" discussed by M. Lax about thirty years ago [40, 41].

Doubly or triply charged impurities (Nickel, Copper, Gold, etc...) in silicon and germanium produce traps with enormously high cross-sections extending in the range 10^{-15} cm^2 to 10^{-13} cm^2. Direct multiphonon capture into the ground state of the trap cannot explain these cross-sections.

The main process involved in the capture mechanism is the capture into an highly excited state, followed by a cascade transition into the ground state.

A recent theoretical treatment has been given in [42], the capture cross-section turns out to be:

$$\sigma_c = \frac{4\pi}{3 l_0} \frac{e^2 Z}{\epsilon kT} \left[\frac{e^2 Z}{\epsilon (kT + 2.74\ ms^2)} \right]^2 \tag{22}$$

ϵ the dielectric constant
s the sound velocity
eZ the charge of the impurity center
$l_0 = v\tau$ where τ is the electron relaxation time.

The hydrogen model is accepted to be valid for the excited states. Formula (22) leads to asymptotic dependences as follows:

$$\sigma_c \sim T^{-1} \quad \text{for } kT \ll ms^2$$
$$\sigma_c \sim T^{-3} \quad \text{for } kT \gg ms^2$$

in good agreement with experiments. The above theory suits for charged centers. Clearly enough, "neutral centers exhibit large separation of the excited states, the number of states is too small to suit the cascade theory" [39].

2. Capture Cross-Section Due to Multiphonon Recombination. Probably R. Kubo [43] was the first who used a tentative configuration diagram model, in which one of the two curves is relevant to the bottom of the free carrier continuum (fig.17).

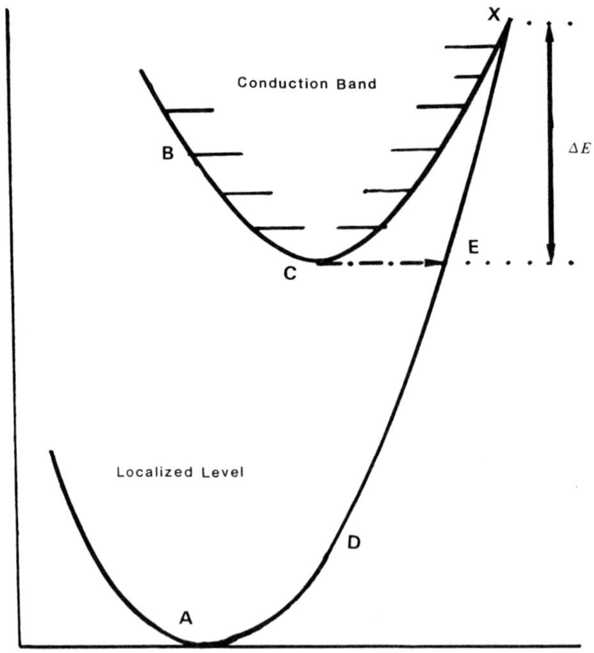

Fig.17 Configuration diagram for trapping of an electron from the conduction band into a deep localized level. Note that the physical meaning of the configurational coordinate in such a case raises problems!

At high temperatures, one joins the activation energy process:

$$\sigma = \sigma_\infty \exp(-\Delta E/kT) \tag{23}$$

$\sigma_\infty \sim 10^{-15}\ cm^2$, while at low temperatures the low-coupling theory of multiphonon transitions works:

$$\sigma = \sigma_0 (\bar{n} + 1)^p e^{-2\bar{n}S} \tag{24}$$

$p\hbar\omega$ is the energy difference between the electronic levels of the ground state and the free state.

$\bar{n}$ is as usual the mean number of phonons at temperature T.

$$\bar{n} = \frac{1}{e^{\hbar\omega/kT} - 1} \tag{25}$$

It may be seen, $\sigma \to \sigma_0$ for $T \to 0K$.

According to [39], it happens in many cases that "it is difficult to reconcile the values of S derived from low- and high-temperature parts of the spectra".

A more elaborate theory, involving several configurational coordinates, has been described in [44].

3. Auger Recombination. Hundreds of papers are published every year, dealing with Auger recombination in solids. However we must emphasize the pioneering role of Landsberg, who especially pointed out that Auger effect is just the inverse process of impact ionization [45]. Therefore we shall bring attention on the following papers by J.W. Allen and co-workers [46,47].

These papers are dealing with Auger effects in electroluminescent materials such as ZnS:Mn and ZnSe:Mn, in which high field electroluminescence occurs by means of impact ionization.

Let $\tau_r = 1/p_r$ be the mean life due to radiative processes, $\tau_{nr} = 1/p_{nr}$ the mean life due to non-radiative processes (which will be considered as being independent on T because clearly we exclude the thermal quenching temperature range), γ the probability of the Auger effect, and n the concentration of free carriers.

The observed mean life τ is then given by:

$$\frac{1}{\tau} = \underbrace{\frac{1}{\tau_r} + \frac{1}{\tau_{nr}}}_{\text{let this be } 1/\tau_0} + \gamma n \tag{26}$$

We may write $$\frac{1}{\tau} = \frac{1}{\tau_0} + \gamma n \tag{27}$$

If the charge carriers come predominantly from a level with activation energy E:

$$n = \beta T^{3/2} \exp(-E/kT) \tag{28}$$

whence:

$$\frac{1}{\tau} = \frac{1}{\tau_0} + \gamma \beta T^{3/2} \exp(-E/kT) \tag{29}$$

$$\ln\left[T^{-3/2} \left(\frac{1}{\tau} - \frac{1}{\tau_0}\right)\right] = \ln(\gamma \beta) - E/kT \tag{30}$$

ln of the quantity between square brackets is thus predicted to be a linear function of E/kT.

This prediction works quite well (fig.18).

For several Mn^{2+} -doped materials, in which E ~ 0.2 eV, the following values of the Auger coefficient γ are found:

ZnSe:Mn	τ_0 ~ 0.2 ms	γ ~ 5.10^{-10} cm^3 s^{-1}
ZnS:Mn	τ_0 ~ 1.8 ms	γ ~ $4\text{-}20.10^{-10}$ cm^3 s^{-1}
CdF_2:Mn	cubic form τ_0 ~ 60 ms	γ ~ 2.10^{-15} cm^3 s^{-1}

In all the above cases, we are dealing with the same transition $^4T_1 \rightarrow {}^6A_1$ of the Mn^{2+} ion.

We see that the γ coefficient is about 10^5 times larger in ZnS:Mn and ZnSe:Mn, which are electroluminescent materials, than in CdF_2:Mn, which is not.

IV. MISCELLANEOUS

Finally we call attention upon some recent work upon processes of a great interest.

IV.A. A Puzzling Problem: The Effect of Pressure on the Ruby Fluorescence Lifetime

L.D. Merkle, R. Powell and co-workers [48] have shown that the observed mean life τ of $^2E \rightarrow {}^4A_2$ transition in ruby increases linearly with applied hydrostatic pressure.

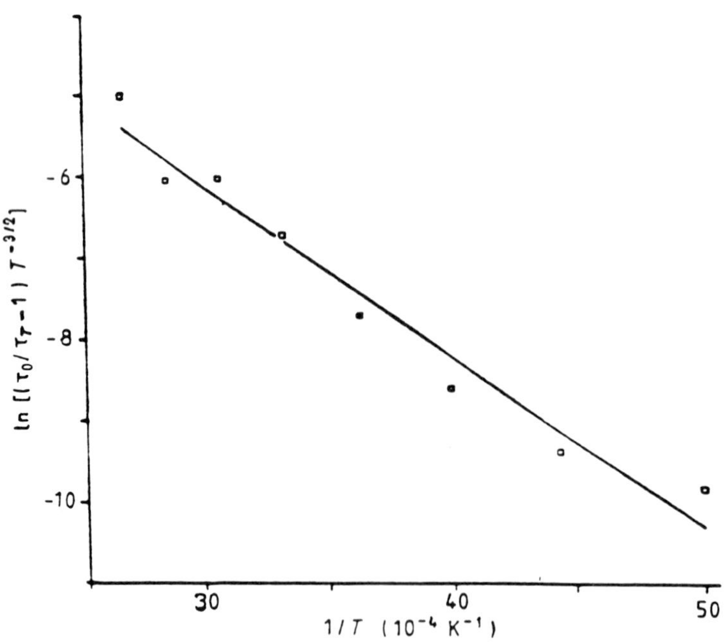

Fig.18 Auger effect in Zn Se: Mn [47].

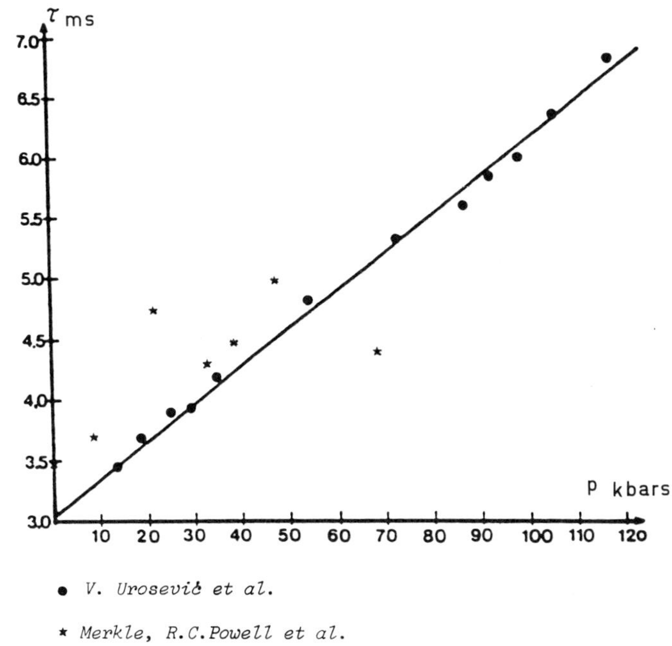

Fig.19 Linear increase of the mean life τ of ruby R. line versus pressure [48], 49].

This result has been confirmed and extended to a larger range of applied pressures, by V. Urosevic and co-workers [49].

According to G.F. Imbusch (private communication during this school), pressure-induced changes in the interaction energy between $^4T_2$ and ^2E levels can explain about 50% of the shift. But changes in non-radiative recombination probability as well as energy transfer between chromium ions may also explain part of the effect.

IV.B. Interface Recombination in Quantum Wells

In GaAs - GaAlAs multiple quantum well lasers, operating at room temperature, D. Bimberg et al. [50] have studied the recombination rate for different amounts of doping.

Let

$$R(n) = An + Bn^2 \qquad (31)$$

be this recombination rate. An is a monomolecular term (partly excitonic). This term is partly radiative, partly non radiative:

$$A = \frac{1}{\tau_r} + \frac{1}{\tau_{nr}} \qquad (32)$$

The bimolecular term Bn^2 is radiative. Therefore we may separate R(n) into 2 parts:

$$R(n) = R_r(n) + R_{nr}(n) \tag{33}$$

with

$$R_r(n) = \frac{n}{\tau_r} + \beta n^2 \tag{34}$$

$$R_{nr}(n) = \frac{n}{\tau_{nr}} \tag{35}$$

The efficiency is, as usual:

$$\eta = \frac{R_r(n)}{R_r(n) + R_{nr}(n)} \tag{36}$$

Typically at $n \sim 10^{18}$ cm^{-3}, $\eta \sim 0.7 - 0.8$
$\tau_r \sim 8$ ns $\qquad \tau_{nr} \sim 12$ ns

But the most interesting result is that B in quantum wells is found to be about 5 times larger than B in the bulk GaAs material:

$B_{Quantum\ Well} \sim 1.5\ 10^{-10}\ cm^3\ s^{-1}$ at R.T.

$B_{Bulk} \sim 0.34\ 10^{-10}\ cm^3\ s^{-1}$

These results can be correlated with those obtained by Sermage, Meichenin et al. [51].

These authors performed direct measurements of the interface recombination velocity (Fig.20): its value (denoted by S on Fig.20) can be about 5 times larger then the bulk value S_0.

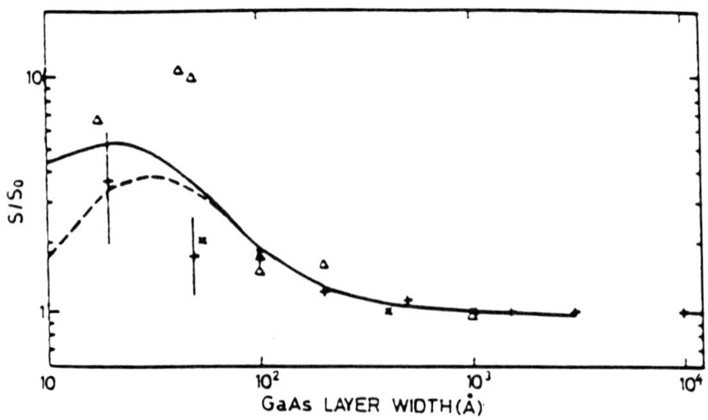

Fig.20 Ratio of the interface recombination velocity in GaAs quantum wells to its value in thick layers for the three series: first MBE series (+), second MBE series (Δ) and MOCVD series (X). Calculated models for barrier (—) and interface (---) recombinations.

Of course, in this introductory paper, we did not develop highly important items such as up-conversion, energy transfer etc. which will be treated by the other contributors to this book.

As a kind of conclusion, we shall give an example of a situation in which non-radiative recombinations have been deliberately used in a luminescence problem.

The problem was the emission spectrum of ZnS samples doped by rare earths. Unavoidable contamination by copper produces a broad band emission spectrum, which creates a real problem for studying green or blue emission lines. A very simple way to avoid this problem has been used by M. Bancie-Grillot and P. Bourtayre [52]. These authors deliberately added some Co^{2+}, which acts as a killer of copper luminescence:

$$Zn:5.10^{-3}\ R.E. + (?)Cu + 10^{-6}Co^{2+}$$

Thus the rare earth emission could be studied more easily.

REFERENCES

1. F. Perrin, Thesis, Ann. de Phys., 12, 169 (1929).
2. R. Delorme and F. Perrin, J. Physique, 10, 177 (1929).
3. B. Henderson, Optical Spectroscopy of Colour Centers, in Spectroscopy of Solid State Laser Type Materials, B. Di Bartolo ed., Plenum Press, 1987, p.112-113.
4. R.K. Swank and F.C. Brown, Phys. Rev. 130, 34 (1963).
5. H.G. Drickamer, G.B. Schuster, D.S. Mitchell, in Radiationless Transitions, S.H. Lin ed., Academic Press 1980, p.289.
6. C.E. Tyner and H.G. Drickamer, J. Chem. Phys., 67, 4103 (1977).
7. G.L. House and H.G. Drickamer, J. Chem. Phys., 67, 3321 (1977).
8. M. Tabei, S. Shionoya, H. Ohmatsu, J. Appl. Phys., 14, 240 (1975).
9. M. Godlewski, W.E. Lamb, B.C. Cavenett, J. Phys. C., 15, 3925 (1982).
10. M. Godlewski and M. Scowronski, Phys. Rev., B 32, 4007 (1985).
11. A. Von Hippel, Zeitschrift für Phys., 101, 680 (1936).
12. F. Seitz, Trans. Faraday. Soc., 35, 74 (1939).
13. N.F. Mott, Proc. Roy. Soc., A, 167, 384 (1938).
14. N.F. Mott, Electronic Processes in Ionic Crystals, Clarendon Press, Oxford (1940).
15. W.H. Fonger and C.W. Struck, in Radiationless Processes, B. Di Bartolo ed., Plenum Press 1980, p.475.
16. C.W. Struck and W.H. Fonger, J. of Luminescence, 10, 1, (1975).
17. W.H. Fonger and C.W. Struck, J. of Luminescence, 17, 241, (1978).
18. C.W. Struck and W.H. Fonger, J. of Luminescence, 18-19, 101 (1979).
19. C.W. Struck and W.H. Fonger, Phys. Rev., B 19, 4400 (1979).
20. C.W. Struck and W.H. Fonger, J. Chem. Phys., 60, 1988 (1974).
21. D.L. Decter, Solid State Phys., 6, 355 (1958).
22. R.H. Bartram and A.M. Stoneham, Solid State Comm., 17, 1593 (1975).
23. A.M. Stoneham and R.H. Bartram, Solid State Electron, 21, 1325 (1978).
24. T. Kloiber, H.J. Kmeicik, M. Kruse, M. Schreiber and G.Zimmerer, J. of Luminescence, 40-41, 593 (1988).
25. T. Kloiber, W. Laasch, G. Zimmerer, F. Coletti and J.M. Debever, Europhys. Lett. 7, 77 (1988).

26. T. Kloiber and G. Zimmerer, Radiation Effects and Defects in Solids, 1989 (in press).
27. W.H. Fonger and C.W. Struck, Phys. Rev. B 11. 3215 (1975).
28. G. Blasse, in Radiationless Processes, B. Di Bartolo ed., Plenum Press 1980, p.316.
29. C.W. Struck and W.H. Fonger, J. Chem. Phys. 64, 1784 (1976).
30. G. Blasse and N. Sabbatini, Materials Chem. and Phys., 16, 237 (1987).
31. F. Auzel, in Radiationless Processes, B. Di Bartolo ed., Plenum Press 1980, p.213.
32. L.A. Risberg, in Radiationless Processes, B. Di Bartolo ed., Plenum Press 1980, p.369.
33. M.J. Weber, Phys. Rev. B8, 54 (1973).
34. R. Reisfeld, in Radiationless Processes, B. Di Bartolo ed., Plenum Press 1980, p.489.
35. K. Huang, J. of Luminescence, 31-32, 738 (1984).
36. H.W. Moos, J. of Luminescence, 1-2, 106 (1970).
37. W. Jia and W.M Yen, Acta Phys. Sin., 32, 346 (1983).
38. J.H. Crasemann and H.J. Schulz, in Defects in Semi-Conductors (H.J. Von Bardeleben ed.), Materials Science Forum, 10-12, 693 (1986).
39. M. Jaros, Deep Levels in Semi-conductors, Adam Hilger Ltd., Bristol 1982, pp.180-190.
40. M. Lax, J. Chem. Phys., 20, 1752 (1952).
41. M. Lax, Phys. Rev. 119, 1502 (1960).
42. V.N. Abakumov, V.I. Perel, I.N. Yassievich, Sov. Phys. Semi-Cond., 12, 1 (1978).
43. R.Kubo, Phys. Rev. 86, 929 (1952).
44. H. Sumi, J. of Luminescence, 40-41, 76 (1988).
45. P.T. Landsberg and D.J. Robbins, Solid State Electr., 21, 1289 (1978).
46. N.T. Gordon and J.W. Allen, J. Phys. D., Appl. Phys. 18, 951 (1985).
47. S.G. Ayling and J.W. Allen, J. Phys. C., 20, 4251 (1987).
48. L.D. Merkle, I.L. Spain and R.C. Powell, J. Phys. 14, 2027 (1981).
49. V. Urosevic, B. Panic, B. Jovanic, L.J. Zekovic, P. Savic, Chem. Phys. Lett. 155, 325 (1989).
50. E.H. Bottcher, K. Ketterer, D. Bimberg, G. Weiman, W. Schapp, Appl. Phys. Lett., 50, 1074 (1987).
51. B. Sermage, M.F. Pereira, F. Alexandre, T. Beerens, R. Azoulay, C. Tallot, A.M. Jean-Louis, D. Meichenin, J. Physique, Colloque C5, 135 (1987).
52. M. Bancie-Grillot and P. Bourtayre, J. Physique, Colloque C2, 116 (1968).

THE ROLE OF NONRADIATIVE PROCESSES IN THE SPECTROSCOPY OF OPTICALLY ACTIVE CENTERS IN SOLIDS

B. Di Bartolo

Department of Physics
Boston College
Chestnut Hill, MA 02167, USA

ABSTRACT

This article deals with the basic mechanisms responsible for the radiative and nonradiative relaxations of optically active centers in solids. In the first section the systems and the basic interactions affecting them are defined. The second section deals with the quantum theory of molecular systems: the adiabatic (Born-Oppenheimer) approximation is introduced and its implications are discussed. The subject of the third section is the "phonon" system, namely the system of vibrations of the solid; the two pictures of harmonic oscillators and, when applicable, of phonons, are considered. The fourth section deals with the Franck-Condon principle and the basic processes of absorption, induced emission, and spontaneous emission; a simple model based on a single normal coordinate is used in order to derive the Pekar formula and the Huang and Rhys formula. The fifth section introduces the basic mechanism for the nonradiative relaxation, due to deviations from the adiabatic approximation, and treats the nonradiative transitions in terms of the normal coordinates of the solid. Finally the last section deals with the phenomenon of energy transfer among optically active centers in solids, a process that may provide an additional nonradiative decay path to an excited optical center.

I. INTRODUCTION

We shall be dealing with ionic, laser-type solids doped with transition metal ions or rare earth ions. The dopant ions are responsible for the spectral properties of these solids both in absorption and emission and, consequently, are referred to as "optically active" ions or centers.

The basic interactions are related to the spectral characteristics of these systems: i) the ion-photon interaction that produces the phenomena of absorption, induced emission, and spontaneous emission, and ii) the ion-phonon interaction that may produce nonradiative processes.

Advances in Nonradiative Processes in Solids
Edited by B. Di Bartolo, Plenum Press, New York, 1991

II. ENERGY LEVELS

II.A. The Hamiltonian of a Molecular System

The Hamiltonian of a molecular system, when spin orbit and other less important interactions are neglected, may be written as follows:

$$H = T_r + T_R + V(\vec{r}_i, \vec{R}_s) \tag{1}$$

where

$$T_r = \sum_{s=1}^{n} \frac{(\vec{p}_i)^2}{2m} = \text{kinetic energy of the electrons} \tag{2}$$

$$T_R = \sum_{s=1}^{N} \frac{(\vec{p}_s)^2}{2M} = \text{kinetic energy of the nuclei} \tag{3}$$

$$V(\vec{r}_i, \vec{R}_s) = V_{ee} + V_{nn} + V_{ne} \tag{4}$$

with

$$V_{ee} = 1/2 \sum_{\substack{i=1 \\ i \neq j}}^{n} \sum_{j=1}^{n} \frac{e^2}{|\vec{r}_i - \vec{r}_j|} \tag{5}$$

$$V_{nn} = 1/2 \sum_{\substack{s=1 \\ s \neq t}}^{N} \sum_{t=1}^{N} \frac{e^2 Z_s Z_t}{|\vec{R}_s - \vec{R}_t|} \tag{6}$$

$$V_{ne} = - \sum_{i=1}^{n} \sum_{s=1}^{N} \frac{e^2 Z_s}{|\vec{R}_s - \vec{r}_i|} \tag{7}$$

The Schroedinger equation is given by

$$H\Psi(\vec{r}_i, \vec{R}_s) = E\Psi(\vec{r}_i, \vec{R}_s) \tag{8}$$

or

$$T_r\Psi + T_R\Psi + V\Psi = E\Psi \tag{9}$$

We assume at this point that the eigenfunctions Ψ of H have the following form, called a Born-Oppenheimer product [1]

$$\Psi(\vec{r}, \vec{R}) = \phi(\vec{R})\psi(\vec{r}, \vec{R}) \tag{10}$$

where we have dropped the subscripts of the coordinates. Replacing (10) into (9) we obtain:

$$\phi(\vec{R})T_r\psi(\vec{r}, \vec{R}) + T_R\phi(\vec{R})\psi(\vec{r}, \vec{R}) + V(\vec{r}, \vec{R})\phi(\vec{R})\psi(\vec{r}, \vec{R}) = E\phi(\vec{R})\psi(\vec{r}, \vec{R}) \tag{11}$$

But

$$T_r\phi(\vec{R})\psi(\vec{r},\vec{R}) = \phi(\vec{R})T_R\psi(\vec{r},\vec{R}) + \psi(\vec{r},\vec{R})T_R\,\phi(\vec{R}) - \sum_s \frac{\hbar^2}{M_s}\vec{\nabla}_s\,\phi(\vec{R})\cdot\vec{\nabla}_s\,\psi(\vec{r},\vec{R}) \tag{12}$$

Therefore (11) becomes

$$\left\{-\sum\frac{\hbar^2}{M_s}\vec{\nabla}_s\phi(\vec{R})\cdot\vec{\nabla}_s\psi(\vec{r},\vec{R}) + \phi(\vec{R})T_R\phi(\vec{r},\vec{R})\right\} \tag{13}$$

$$+\ \phi(\vec{r},\vec{R})T_R\phi(\vec{R}) + \phi(\vec{R})T_r\psi(\vec{r},\vec{R}) + V(\vec{r},\vec{R})\phi(\vec{R})\psi(\vec{r},\vec{R}) = E\phi(\vec{R})\psi(\vec{r},\vec{R})$$

At this point, we shall neglect the terms in { } brackets; we shall say more about the validity of this (so-called <u>adiabatic</u>) approximation later. Eq. (13) then becomes

$$\frac{\psi}{\phi}\,T_R\phi + \left[T_r + V\right]\psi = E\psi \tag{14}$$

The terms in the [] brackets above represent the Hamiltonian of the system if the nuclei are kept fixed in space ($T_R = 0$); we call this Hamiltonian H_e

$$H_e = T_r + V(\vec{r},\vec{R}) \tag{15}$$

The functions $\psi(\vec{r},\vec{R})$ that enter the Born Oppenheimer products are chosen to be eigenfunctions of H_e

$$T_r\psi(\vec{r},\vec{R}) + V(\vec{r},\vec{R})\psi(\vec{r},\vec{R}) = \epsilon(\vec{R})\psi(\vec{r},\vec{R}) \tag{16}$$

We note that the energy eigenvalues depend parametrically on $\vec{R}$. Using (16), eq. (14) becomes:

$$T_R\phi(\vec{R}) + \epsilon(\vec{R})\phi(\vec{R}) = E\phi(\vec{R}) \tag{17}$$

Eq. (17) determines the choice of the functions $\phi(\vec{R})$ that enter the Born-Oppenheimer products.

The possibility of using such products as representative of the eigenstates of the system is based on the fact that the terms in the { } brackets of eq. (13) are negligible; this implies that

$$\left|-\sum_s\frac{\hbar^2}{M_s}\vec{\nabla}_s\phi\cdot\vec{\nabla}_s\psi + \phi T_R\psi\right| \ll \left|\psi\,T_R\phi\right| \tag{18}$$

We shall examine later the conditions under which this inequality holds.

For the moment, assuming the adiabatic approximation legitimate, we can say that the solution of the Schroedinger equation for the entire system reduces to the solution of the two equations

$$T_r\psi_k(\vec{r},\vec{R}) + V(\vec{r},\vec{R})\psi_k(\vec{r},\vec{R}) = \epsilon_k(\vec{R})\psi_k(\vec{r},\vec{R}) \tag{19}$$

$$T_R\phi_{k\ell}(\vec{R}) + \epsilon_k(\vec{R})\phi_{k\ell}(\vec{R}) = E_{k\ell}\phi_{k\ell}(\vec{R}) \tag{20}$$

where we have attached proper subscripts to the eigenvalues and eigenfunctions. A stationary state of the system is then given by

$$\Psi_n(\vec{r},\vec{R}) = \psi_k(\vec{r},\vec{R})\phi_{k\ell}(\vec{R}) \tag{21}$$

where ψ_k and $\phi_{k\ell}$ are eigenfunctions of the "electronic" Hamiltonian

$$H_e = T_r + V(\vec{r},\vec{R}) \tag{22}$$

and of the "vibrational" Hamiltonian

$$H_v = T_R + \epsilon_k(\vec{R}) \tag{23}$$

respectively.

II.B. The Meaning of the Adiabatic Approximation

The use of the adiabatic approximation has led us to the two eigenvalue equations (19) and (20) that we want now to consider again. First eq. (19) can be written

$$H_e\psi_k(\vec{r},\vec{R}) = \epsilon_k(\vec{R})\psi_k(\vec{r},\vec{R}) \tag{24}$$

The eigenfunctions H_e represent the electron system when the nuclei are in fixed positions $\vec{R}$: the eigenfunctions and the energy eigenvalues depend parametrically on $\vec{R}$. This means that, as the positions $\vec{R}$ change in time, both ϵ_k and ψ_k follow this change "adiabatically", while the system continues to stay in the state represented by the quantum number k.

The validity of the approximation derives from the fact that, on account of their smaller mass, the electrons go through their orbits many times before the nuclei shift their positions by any considerable amount. This is expressed by the inequality (18) which is valid when

$$|\vec{\nabla}_s\psi(\vec{r}_i,\vec{R}_s)| \ll |\vec{\nabla}_s\phi(\vec{R}_s)| \tag{25}$$

This inequality implies that the functions $\psi(\vec{r},\vec{R})$which represents the motion of the electrons varies slowly with the nuclear coordinates $\vec{R}_s$. From the physical point of view, if the conditions for the validity of the adiabatic approximation are present, the electrons in a certain state k move in a potential $V(\vec{r},\vec{R})$ and follow the lattice motion (i.e., change of $\vec{R}$) without making any transition to any other electronic state m ($m \neq k$).

The eigenvalue equation (20) can be written

$$H_v \phi_{k\ell}(\vec{R}) = E_{k\ell} \phi_{k\ell}(\vec{R}) \tag{26}$$

The eigenfunctions $\phi_{k\ell}$ represent the motion of the nuclei in the potential set by $\epsilon_k(\vec{R})$. The eigenvalues of eq. (24) play here the role of a potential energy. We note that $\phi_{k\ell}$ depends on k, i.e., the electrons' quantum state; k is, however, only a subscript for ϕ and it does not represent a good quantum number. If $k \neq k'$, in general

$$\int \phi_{k\ell}(\vec{R})^* \phi_{k'\ell'}(\vec{R}) d^3\vec{R} \neq \delta_{\ell\ell'} \tag{27}$$

III. THE "PHONON" SYSTEM

III.A. The Nuclear Vibrations Equation

The equation that represents the nuclear vibrations of a molecular system can be written

$$H_v \phi_{k\ell}(\vec{R}) = E_{k\ell}\, \phi_{k\ell}(\vec{R}) \tag{28}$$

where

$$H_v = T_R + \epsilon_k(\vec{R}) \tag{29}$$

We shall assume that the problems of translational and rotational motions have been solved, and the results of these motions have been eliminated from the above equation by a proper choice of coordinates.

We shall now summarize the results that we obtain in the case that the system under consideration is a solid.

III.B. Vibrations of Solids

In dealing with the radiative processes of ions in solids, we have to consider the fact that these ions are part of a structure which undergoes vibrational (thermal) motion. We may at this point make the following observations:

1. The "electronic" energy eigenvalue $\epsilon_k(\vec{R})$ plays the role of a potential in which the atoms of the solid perform their vibrational motions.

2. These motions can be thought of as a superposition of normal modes of vibration which are represented by normal coordinates q_i ($i = 1,2,\ldots\ldots, 3N$), where N = number of atoms in the solid. (More precisely since the solid has also three translational and three rotational degrees of freedom, the number of normal modes of vibration is 3N-6).

3. A normal mode is a pattern of motion in which, in general, all the atoms of the solid participate. There may be, however, normal modes, called "localized", which are related to the motion of a relatively small number of atoms.

4. In treating the vibrations of solids, the "harmonic approximation" is used. According to this approximation, no exchange of energy takes place between the normal modes, and the "potential" energy $\epsilon_k(\vec{R})$ and the kinetic energy are both sums of independent quadratic ($\sim q^2$, $\sim \dot{q}^2$) terms, one for each normal mode.

5. Each normal mode is equivalent to a harmonic oscillator. The vibrations of a solid are then equivalent to a collection of 3N harmonic oscillators. If an oscillator of frequency ω is in its nth excited state, this fact is also expressed by saying that n "phonons" of energy $\hbar\omega$ are present in the solid.

6. The different normal modes are not completely isolated, but are, rather, in speaking terms due to their anharmonicity: this provides the mechanism for reaching thermal equilibrium.

7. When considering the thermal vibrations of a solid, it may be interesting to have an idea of how many phonons may be present in a solid at, say, room temperature. The number of phonons in a frequency interval $(\omega, \omega + d\omega)$ is given by $\bar{n}(\omega)\rho(\omega)d\omega$ where: $\bar{n}(\omega) = (e^{\hbar\omega/kT} - 1)^{-1}$ and $\rho(\omega)$ = density of phonon states $\propto \omega^2/c_s^3$. It is the value of c_s = velocity of sound in solids $\sim 5 \times 10^5$ cm/sec (versus c = velocity of light = 3×10^{10} cm/sec) that makes the number of phonons extremely large. For the sake of comparison, the total number of photons/cm^3 in a typical laser medium (λ = 6300 A, 1W/cm^2) is 10^8, but the number of phonons/cm^3 in a solid (with a Debye temperature T_D = 1000 K) at T = 300 K is $\sim 3.5 \times 10^{23}$! The sheer number of phonons may give us an idea of their importance in affecting the spectral characteristics of ions in solids.

8. The electronic energy $\omega_k(\vec{R})$ can be represented as a surface in a (3N+1)-fold space. If the system of normal coordinates q_i is used, a potential "curve" will correspond to each coordinate q_i as in Fig. 1. This figure represents the potential curve diagrams of a diatomic molecule: for such a system, only one vibrational mode is present (q_i = R).

9. The question now arises: is it legitimate, when considering the potential curves in correspondence to a certain normal coordinate q_i, to put the potential curves for the ground state and for an excited electronic state in the same diagram as in Fig. 1? Electronic excitation corresponds always to a rearrangement of the electron charges in the internuclear spaces: this may in turn produce changes in the molecular architecture, i.e., the actual positions of the nuclei. In a diatomic molecule, the electronic excitation produces a change in the relative distance of the two

nuclei, and in the force constant that controls the vibrational motion; however, since we have only one vibrational coordinate, it is legitimate to represent all the potential curves in the same diagram. This is not the case for a complex molecule, where, in general, angular distortions and changes in the force constants may result from an electronic excitation. This means that, to be precise, a change of electronic state produces also a change in the system of normal coordinates: this change has been called pictorially by Professor Rebane [2] the "scrambling of normal coordinates."

10. According to the adiabatic approximation that we introduced in section II.A. of this article, the quantum states of a solid are represented by the Born-Oppenheimer products

$$\Psi_i(\vec{r},\vec{R}) = \psi_k(\vec{r},\vec{R})\phi_{k\ell}(\vec{R}) \quad (30)$$

We note here that, due to the presence of nonadiabatic terms in the Hamiltonian (these terms are in {} in eq. 13), the states exemplified in (30) are not truly stationary. The non-adiabaticity may furnish the mechanism by which a system moves nonradiatively from a state ψ_i, given by (30) above, to a state

$$\Psi_f(\vec{r},\vec{R}) = \psi_m(\vec{r},\vec{R})\phi_{mn}(\vec{R}) \quad (31)$$

11. Finally, we need to mention the so-called Jahn-Teller effect [3]: complexes with degenerate electronic states are unstable to symmetry - lowering distortions that take place in the way of vibrational modes. This fact was established by Jahn and Teller on the basis of symmetry considerations which do not furnish any quantitative information.

IV. HOW TO DEAL WITH RADIATIVE TRANSITIONS

IV.A. The Franck-Condon Principle

The process of absorption or emission of ultraviolet light by molecules is accompanied by a change of their electronic state. Stated very simply, the Franck-Condon Principle [4,5] recognizes the fact that the change in the electronic structure occurring in an electronic transition is much more rapid than the possible changes in the internuclear distances occurring during the same transition.

By using purely classical arguments, the Franck-Condon principle is stated as follows:

> "During an electronic transition, the electronic state changes so fast that 1) the nuclei do not move and 2) the nuclei do not change their momenta."

This is best illustrated considering a diatomic molecule. Potential curve diagrams for such a system are given in Fig. 1; in it R is the distance between the two nuclei. The condition 1) above means that during an electronic transition, the internuclear distance R must remain constant; the condition 2) implies that the kinetic energy must also remain constant and this means that if the molecule's initial state is

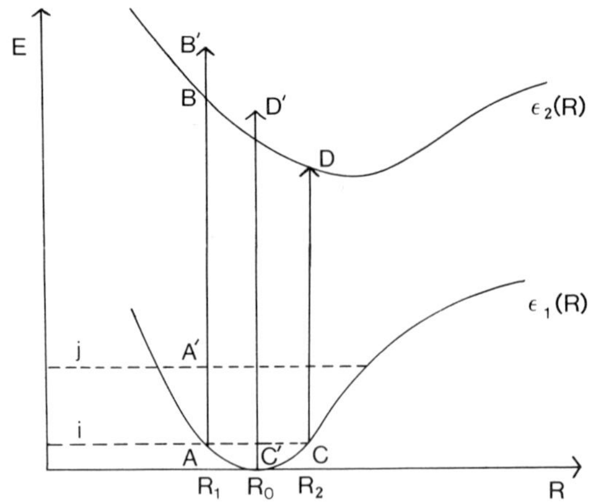

Fig.1. Diagram Illustrating the Franck-Condon Principle.

given by the point A' the molecule after the electronic transition will be found at B' with AA' = BB'. A third condition may be added considering that the molecule spends most of its time at its turning points at which the kinetic energy is zero. This condition 3) states:

> "At the beginning of an electronic transition a molecule is found at the turning points."

Therefore, according to this condition, the following transitions are allowed:

AB, CD

and the following tansitions are forbidden

A'B', C'D'.

In its <u>semiclassical</u> formulation the Franck-Condon principle consists of the three conditions:

1) R = const, 2) p = cost and 3) an electronic transition can take place for any value of R with a probability W(R), as given by quantum mechanics.

If the system is in a particular vibrational quantum state, say the ith vibrational state, then

$$W(R) = |\phi_i(R)|^2 \tag{32}$$

If the system is in thermal equilibrium at a temperature T

$$W_T(R) = \frac{\sum_i e^{-E_i/kT} |\phi_i(R)|^2}{\sum_i e^{-E_i/kT}} \tag{33}$$

For most diatomic molecules at ordinary temperature $\hbar\omega$ is so large (of the order 0.1 eV) that

$$\hbar\omega >> kT \tag{34}$$

and the vibrational states other than the ground state are practically unoccupied; under these conditions, the relevant probability W(R) may well be given by eq. (32) with i = 0.

The above considerations can be generalized by treating a molecule with many vibrational degrees of freedom. In these more complex cases, the relevant probability is given by

$$W_T = \prod_K W_{KT}(q_K) \tag{35}$$

where W_{KT} is of the type (33) with $\omega = \omega_K$ and $R = q_K$ (vibrational coordinate); in the limit of low temperatures or high ω, each W_{KT} is simply of the type (32).

Considering now again the diagram in Fig. 1, we note that the semiclassical Franck-Condon principle allows the transition AB, CD, A′B′ and C′D′ which have probabilities proportional to

$$|\phi_i(R_i)|^2,\ |\phi_i(R_2)|^2,\ |\phi_j(R_1)|^2, \text{ and } |\phi_i(R_0)|^2$$

respectively; in particular, if i = 0, the probability will have its maximum in correspondence to R_0 and transitions such as C′D′ have the greatest importance.

Finally, the quantum mechanical formulation of the Franck-Condon principle is perhaps the most illuminating. The transition probability for a radiative transition in the dipole approximation, is proportional to the square of the matrix element

$$M_{fi} = \langle \vec{\pi}^{\sigma}_{\alpha} \cdot \vec{M} \rangle_{fi} = \langle \vec{\pi}^{\sigma}_{\alpha} \cdot \sum_t q_t \vec{r}_t \rangle_{fi} \tag{36}$$

where $\vec{\pi}^{\sigma}_{\alpha}$ is the unit vector in the direction of the light polarization, and q_t and $\vec{r}_t$ are the charge and the position of the tth charged particle, respectively. The initial state of the system is given in the adiabatic approximation by

$$\Psi_i(\vec{r},\vec{R}) = \psi_k(\vec{r},\vec{R})\phi_{k\ell}(\vec{R}) \tag{37}$$

and the final state by

$$\Psi_f(\vec{r},\vec{R}) = \psi_m(\vec{r},\vec{R})\phi_{mn}(\vec{R}) \tag{38}$$

The electric dipole operator can be written as follows:

$$\vec{M} = -e \sum_i \vec{r}_i + e \sum_s Z_s \vec{R}_s \tag{39}$$

where the sums over i and s extend to the electrons and the nuclei, respectively. We can write

$$\vec{\pi} \cdot \vec{M} = - e \sum_i \vec{r}_i \cdot \vec{\pi} + e \sum_s Z_s \vec{R}_s \cdot \vec{\pi}$$

$$= D_e(\vec{r}) + D_n(\vec{R}) \tag{40}$$

The relevant matrix element is now given by

$$M_{fi} = M_{mn,k\ell}$$

$$= \iint d^3\vec{r} d^3\vec{R} \; \Psi_f(\vec{r},\vec{R})^* \left[D_e(\vec{r}) + D_n(\vec{R})\right] \Psi_i(\vec{r},\vec{R})$$

$$= \iint d^3\vec{r} d^3\vec{R} \; \psi_m(\vec{r},\vec{R})^* \; \phi_{mn}(\vec{R})^* \left[D_e(\vec{r}) + D_n(\vec{R})\right] \psi_k(\vec{r},\vec{R})\phi_{k\ell}(\vec{R})$$

$$= \int d^3\vec{R} \; \phi_{mn}(\vec{R})^* \left[\int d^3\vec{r} \; \psi_m(\vec{r},\vec{R})^* D_e(\vec{r})\psi_k(\vec{r},\vec{R})\right]\phi_{k\ell}(\vec{R})$$

$$+ \int d^3\vec{R} \; \phi_{mn}(\vec{R})^* D_n(\vec{R})\phi_{k\ell}(\vec{R})\left[\int d^3\vec{r} \; \psi_m(\vec{r},\vec{R})^* \psi_k(\vec{r},\vec{R})\right]$$

$$= \int d^3\vec{r} \; \phi_{mn}(\vec{R})^* D_{mk}(\vec{R})\phi_{k\ell}(\vec{R}) + \int d^3\vec{R} \; \phi_{mn}(\vec{R}) D_n(\vec{R})\phi_{k\ell}(\vec{R})\delta_{mk} \tag{41}$$

where

$$D_{mk}(\vec{R}) = \int d^3\vec{r} \; \psi_m(\vec{r},\vec{R})^* D_e(\vec{r})\psi_k(\vec{r},\vec{R}) \tag{42}$$

The following observations can be made:

1) The matrix element M_{fi} consists of two terms. The second term contributes to the matrix element only when $m = k$, namely when the electronic state does not change in the transition and gives rise to infrared absorption by the vibrations.

2) The first term contains the dipole moment of the electrons and corresponds to transitions between different electronic states.

3) If $m = k$ the first term becomes

$$\int d^3\vec{R} \phi_{mn}(\vec{R})^* D_{mm}(\vec{R})\phi_{m\ell}(\vec{R}) \tag{43}$$

where

$$D_{mm}(\vec{R}) = \int d^3\vec{r} |\psi_m(\vec{r},\vec{R})|^2 \; D_e(\vec{r}) \tag{44}$$

Since $D_e(\vec{r})$ is an odd operator, $D_{mm}(\vec{R})$ is different from zero only if the molecule lacks inversion symmetry. If this is the case, the first term in M_{fi} also contributes to the infrared transition.

For transitions between different electronic states, the relevant matrix element is given by

$$M_{fi} = M_{mn;k\ell}$$

$$= \int \phi_{mn}(\vec{R})^* D_{mk}(\vec{R}) \phi_{k\ell}(\vec{R}) R^2 \sin\theta \, dR d\theta d\phi \qquad (45)$$

where $D_{mk}(\vec{R})$ is given by (42).

We shall consider now for simplicity the case of a diatomic molecule; for such a system, the eigenfunctions appearing in the integral (45) are of the type:

$$\phi(R,\theta,\phi) = P_{NK}(R)\theta_{KM}(\theta)\Phi_M(\phi) \qquad (46)$$

The equation for P(R) is

$$\frac{1}{R^2}\frac{d}{dR}\left(R^2 \frac{dP}{dR}\right) + \left\{ -\frac{K(K+1)}{R^2} + \frac{2\mu}{\hbar^2}\left[E - \epsilon(R)\right] \right\} P = 0 \qquad (47)$$

where μ = reduced mass of the molecule [6]. If we set

$$P(R) = \frac{S(R)}{R} \qquad (48)$$

(47) reduces to

$$\frac{d^2S}{dR^2} + \left\{ -\frac{K(K+1)}{R^2} + \frac{8\pi^2\mu}{\hbar^2}\left[E - \epsilon(R)\right] \right\} S = 0 \qquad (49)$$

In order to find S(R) we need to know $\epsilon(R)$, the form of the adiabatic potential. If it is assumed tht the force between the atoms is proportional to the displacement of the internuclear distance from its equilibrium value, then the functions S(R) take the form of harmonic oscillator eigenfunctions. In any case, regardless of the form of the adiabatic potential $\epsilon(R)$, the matrix element will be proportional to the radial integral

$$\int S_{mn}(R)^* D_{mk}(R) S_{k\ell}(R) dR \qquad (50)$$

If we assume that $D_{mk}(R)$ does not depend strongly on the internuclear distance, we may expand it in terms of the displacement from its equilibrium value. According to the Franck-Condon principle, we retain only the first (constant) term in this expansion:

$$D_{mk}(R) = D^o_{mk} \qquad (51)$$

Then for electronic transitions

$$|M_{fi}|^2 \sim |D^o_{mk}|^2 \left| \int S_{mn}(R)^* S_{k\ell}(R) dR \right|^2 \qquad (52)$$

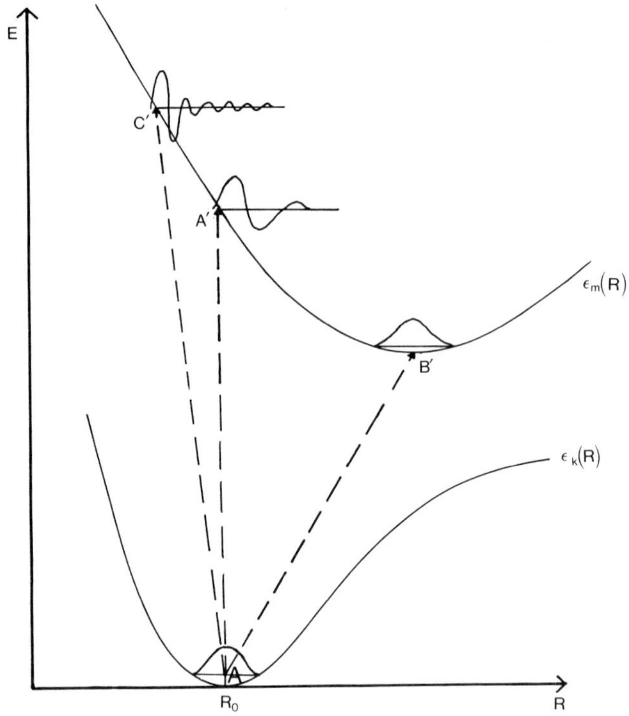

Fig. 2. Diagram Illustrating the Quantum Mechanical Franck-Condon Principle.

The transition probability is then proportional to the square of the overlap integral of the vibrational eigenfunctions of the initial and final states. This overlap controls (apart from D^o) the strength of the transition probabilities. It is to be noted here that $S_{mn}(R)$ and $S_{k\ell}(R)$ are solutions of different Schroedinger equations: they actually belong to different sets of orthonormal eigenfunctions and the subscripts m and k do not play the role of good quantum numbers. Therefore if $m \neq k$, the overlap integral in general may not go to zero even if $n \neq \ell$.

The diagram in Fig. 2 illustrates the role played by the overlap integral in determining the strength of a transition. It is evident from this diagram that the "vertical" transition AA' is the one of highest probability and that the "sloped" transitions AB' and AC' have much lower probabilities; in particular in the former case (AB'), there is practically no overlap, and in the latter case (AC'), the overlap is small because of the oscillating nature of the upper vibrational eigenfunction.

The following considerations can be made:

1) The quantum mechanical Franck-Condon principle replaces the requirements R = const., p = const. of the semiclassical formulation with the notion that transitions for which such conditions are not fulfilled are very unlikely.

2) The Franck-Condon principle in its quantum mechanical formulation is not a selection rule in the conventional sense. It does not say that certain transitions cannot occur but, rather, that they are highly improbable. Selection rules are generally derived from the matrix element of an operator taken between two eigenfunctions of the same orthonormal set, contrary to the present situation where the relevant entity is an overlap integral of two eigenfunctions belonging to different orthonormal sets.

3) If the potential curves of the two electronic states have the same shape, i.e., if

$$\epsilon_k(R) = \epsilon_m(R) + \text{const.} \tag{53}$$

then the two functions S_{mn} and S_k belong to the same set and

$$\int S_{mn}(R)^* S_{k\ell}(R)\,dR = \delta_{n\ell} \tag{54}$$

In this case, the Franck-Condon principle can be expressed as the selection rule: "If the adiabatic potentials of two electronic states are the same, no transition can take place in which the molecule changes its vibrational state." This selection rule is independent from the nature (electric dipole, magnetic dipole, or electric quadrupole) of the transition.

IV.B. Generalization of the Franck-Condon Approximation

In the general case of a solid, the Franck-Condon approximation consists of the following:

1) We neglect the dependence of the matrix element D_{mk} on the position of the nuclei.

2) We set D_{mk} equal to the value it takes in correspondence to the equilibrium configuration of hte nuclei.

The transition probability between two quantum states ψ_i and ψ_f given by (37) and (38), respectively, is proportional to the squared matrix element.

$$|M_{fi}|^2 = |D^o_{mk}|^2 \left|\int \phi_{mn}(\vec{R})^* \phi_{k\ell}(\vec{R})\,d^3\vec{R}\right|^2 \tag{55}$$

Neglecting the scrambling of normal coordinates, we can write

$$\phi_{k\ell}(\vec{R}) = \phi_{kn_1''}(q_1)\phi_{kn_2''}(q_2) \quad \ldots\ldots \tag{56}$$

$$\phi_{mn}(\vec{R}) = \phi_{mn_1'}(q_1)\phi_{mn_2'}(q_2) \quad \ldots\ldots \tag{57}$$

Taking the last two relations into account the square of the overlap integral in (55) becomes

$$|\int \phi_{mn}(\vec{R})^* \phi_{k\ell}(\vec{R}) d^3\vec{R}|^2 = \pi_i |\int \phi_{mn'_i}(q_i)^* \phi_{kn''_i}(q_i) dq_i|^2 \tag{58}$$

We note that the integral which appears in eq. (55) exists even if the scrambling of the normal coordinates is not negligible; in this case, however, the integral does not appear in the form π_i .....

The radiative and non-radiative processes of ions in solids are summarized in Fig. 3 in which we have established a similarity of a sort between the atom-phonon and the atom-phonon interaction. The arrows indicate the direction in which the energy moves. Vibronic processes are a result of the simultaneous interaction of the atomic system with the radiation field and the phonon field.

Other effects like natural width and Lamb shift with their equivalent thermal broadening and thermal shift of sharp lines are also reported to call the attention of the reader on these important processes.

The direct interaction of the radiation field with the thermal vibrations produces the infrared and Raman spectra of solids. Other, more extensive treatments should be consulted [7] for an account of these effects.

IV.C. Radiative Transitions: Radiated and Absorbed Power

The transition probability per unit time in the electric dipole approximation for one-photon processes ($\vec{k} = \vec{k}_\alpha$, polarization $\vec{\pi}^\sigma_\alpha$) is given by [8]:

$$p^\sigma_\alpha d\Omega_\alpha = \begin{cases} \dfrac{4\pi^2 I(\omega_\alpha;\sigma)}{\hbar^2 c} |M_{fi}|^2 \, d\Omega_\alpha \\[2ex] \left\{ \dfrac{4\pi^2 I(\omega_\alpha;\sigma)}{\hbar^2 c} + \dfrac{\omega^3_\alpha}{hc^3} \right\} |M_{fi}|^2 \, d\Omega_\alpha \end{cases} \tag{59}$$

where the upper (lower) row corresponds to absorption (emission),

$d\Omega_\alpha$ = infinitesimal solid angle,

$I_\alpha(\omega_\alpha;\sigma)$ = intensity of radiation per unit angular frequency range, angular frequency ω_α, and polarization σ,

and

$$|M_{fi}|^2 = |\langle \vec{\pi}^\sigma_\alpha \cdot \vec{M} \rangle|^2 = |\langle \vec{\pi}^\sigma_\alpha \cdot \sum_t q_t \vec{r}_t \rangle|^2 \tag{60}$$

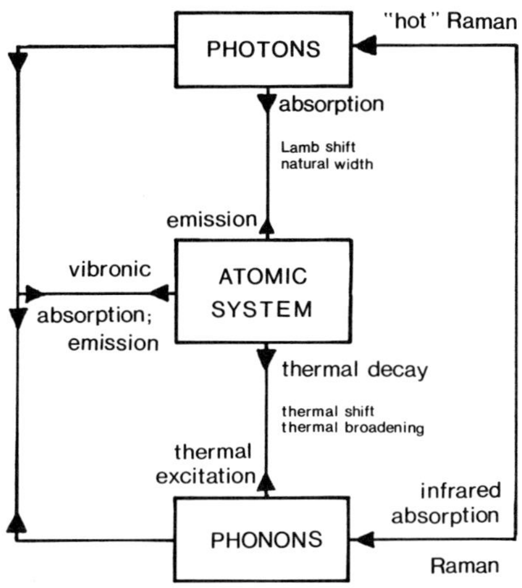

Fig. 3. Diagram Illustrating Various Processes Which are Present in Solids

The probability of spontaneous emission is given by

$$P^{\sigma}_{\alpha}(sp)d\Omega_{\alpha} = \frac{\omega_{\alpha}^{3}}{hc^{3}} |M_{fi}|^{2} d\Omega_{\alpha} \tag{61}$$

and the radiated power due to spontaneous emission by

$$P^{\sigma}_{\alpha}(sp)d\Omega_{\alpha} \hbar\omega_{\alpha} = \frac{\omega_{\alpha}^{4}}{2\pi c^{3}} |M_{fi}|^{2} d\Omega_{\alpha} \tag{62}$$

If several pairs of initial and final states are connected by transitions corresponding to the same frequency, wave vector, and polarization, the power radiated per unit energy range is

$$\frac{\omega_{\alpha}^{4}}{2\pi c^{3}} \sum_{i} N_{i} \sum_{f} |M_{fi}|^{2} \delta(E_{f}-E_{i} + \hbar\omega_{\alpha})d\Omega_{\alpha}$$

$$= A_{+} \sum_{i} N_{i} \sum_{f} |M_{fi}|^{2} \delta(E_{f}-E_{i} + \hbar\omega_{\alpha})d\Omega_{\alpha} \tag{63}$$

where N_i = number of molecules occupying the state i and

$$A_{+} = \frac{\omega_{\alpha}^{4}}{2\pi c^{3}} \tag{64}$$

The expression for the absorption and induced emission transition probability is given by

$$P^{\lambda}_{\alpha}(abs;emi)d\Omega_{\alpha} = \frac{4\pi^{2}I(\omega_{\alpha};\sigma)}{\hbar^{2}c} |M_{fi}|^{2} d\Omega_{\alpha} \tag{65}$$

Correspondingly, the power absorbed is

$$P^{\sigma}_{\alpha}d\Omega_{\alpha}\hbar\omega_{\alpha} = \frac{4\pi^{2}\omega_{\alpha}I(\omega_{\alpha};\sigma)}{\hbar c} |M_{fi}|^{2} d\Omega_{\alpha} \tag{66}$$

and if several pairs of levels are connected by transitions of the same frequency ω_α, wave vector, and polarization, the power absorbed per unit energy range is given by

$$\frac{4\pi^{2}\omega_{\alpha}I(\omega_{\alpha};\sigma)}{\hbar c} \sum_{i} N_{i} \sum_{f} |M_{fi}|^{2} \delta(E_{f}-E_{i}-\hbar\omega_{\alpha})d\Omega_{\alpha}$$

$$A_{-} \sum_{i} N_{i} \sum_{f} |M_{fi}|^{2} \delta(E_{f}-E_{i}-\hbar\omega_{\alpha})d\Omega_{\alpha} \tag{67}$$

where

$$A_{-} = \frac{4\pi^{2}\omega_{\alpha}I(\omega_{\alpha};\sigma)}{\hbar c} \tag{68}$$

We can write (63) and (67) concisely by using the expression

$$A_{\pm} \sum_i N_i \sum_f |M_{fi}|^2 \, \delta(E_f - E_i \pm \hbar\omega_\alpha) d\Omega_\alpha \tag{69}$$

where the plus and minus signs stand for spontaneous emission and absorption, respectively, and A_+ and A_- are given by (64) and (68), respectively.

IV.D. Characteristics of the Entire Spectral Band

We call zero-phonon line a spectral line corresponding to a transition in which the vibrational state of the system does not change.

An absorption or emission zero-phonon lines is accompanied by other electronic transition in which the vibrational state of the system undergoes a change; these transitions are called vibrational-electronic or vibronic transitions. The zero-phonon transition and the accompanying vibronic transitions produce a spectral band.

In order to calculate the integrated intensity of such a band, we have to use the formula

$$A \sum_i N_i \sum_i |M_{fi}|^2 \tag{70}$$

where i is the initial state, f is the final state, and M_{fi} is given by (60). With the notation

i: 0 (electronic), n_i (vibrational)

f: 1 (electronic); n_f (vibrational)

the integrated intensity of the band is given by

$$A \sum_i N_i \sum_f \left| \int d^3\vec{R} \phi_{1n_f}(\vec{R})^* D_{10}(\vec{R}) \phi_{0n_1}(\vec{R}) \right|^2 \tag{71}$$

where

$$D_{10}(\vec{R}) = \int d^3r \, \psi_1(\vec{r},\vec{R})^* \left(-e\sum_t \vec{r}_t \cdot \vec{\pi} \right) \psi_0(\vec{r},\vec{R}) \tag{72}$$

Then

$$A \sum_i N_i \sum_f |M_{fi}|^2$$

$$= A \sum_i N_i \sum_f \int d^3\vec{R} \phi_{1n_f}(\vec{R})^* D_{10}(\vec{R}) \phi_{0n_1}(\vec{R}) \int d^3\vec{R}' \phi_{1n_f}(\vec{R}') D_{10}(\vec{R}')^* \phi_{0n_f}(\vec{R}')^*$$

$$= A \sum_i N_i \iint d^3\vec{R} \, d^3\vec{R}' \left[\sum_f \phi_{1n_f}(\vec{R})^* \phi_{1n_f}(\vec{R}') \right] \left[D_{10}(\vec{R}) D_{10}(\vec{R}')^* \right] \phi_{0n_i}(\vec{R}) \phi_{0n_i}(\vec{R}')^*$$

$$= A \sum_i N_i \iint d^3\vec{R} \, d^3\vec{R}' \, \delta(\vec{R}-\vec{R}') D_{10}(\vec{R}) D_{10}(\vec{R}')^* \phi_{0n_i}(\vec{R}) \phi_{0n_i}(\vec{R})^*$$

$$= A \sum_i N_i \int d^3\vec{R} |D_{10}(\vec{R})|^2 |\phi_{0n_i}(\vec{R})|^2 \qquad (73)$$

This result was derived with the assumption of the validity of the adiabatic approximation, but with no assumption regarding the Franck-Condon approximation. If the latter approximation is considered valid, then

$$A \sum_i N_i \sum_f |M_{fi}|^2 = A \sum_i N_i |D^o_{10}|^2 \int d^3\vec{R} |\phi_{0ni}(\vec{R})|^2$$

$$= A|D^o_{10}|^2 \sum_i N_i = A|D^o_{10}|^2 N \qquad (74)$$

where N = total number of molecules.

The above result indicates that the overall integrated intensity of the <u>entire</u> vibronic band, consisting of the zero-phonon line and the accompanying vibronic lines, is independent of temperature within the limits of the present model:

a. the coefficient A is constant, independent of frequency,

b. the adiabatic approximation is valid, and

c. the Franck-Condon approximation is valid.

IV.E. Radiative Transitions in the Presence of a Localized Vibration

We shall examine this problem under the following assumptions:

a) ω, the frequency of a localized vibrational mode, is such that $\hbar\omega >> kT$, which implies that only the ground vibrational level is populated and

b) the frequency ω is the same in the ground and excited electronic states of the system.

Under these conditions, the intensity of a vibronic (absorption) line of frequency ω corresponding to a transition by which the system undergoes a change of both its electronic and its vibrational states from 0 to n is given by

$$A \sum_i N_i \sum_f |M_{if}|^2 \delta(E_i - E_f + \hbar\omega) = AN|M_{fi}|^2 = AN|D^o_{10}|^2 \left|\int \phi_{1n}(q)\phi_{00}(q)dq\right|^2 \qquad (75)$$

This transition is represented in Fig.4. The quantity

$$L = \frac{\omega^2 q_o^2}{2} \qquad (76)$$

where q_o = displacement of the two adiabatic potential curves, is the <u>Stokes loss</u> and the quantity

$$\ell = \frac{\omega^2 q_o^2}{2} \Big/ (\hbar\omega) \qquad (77)$$

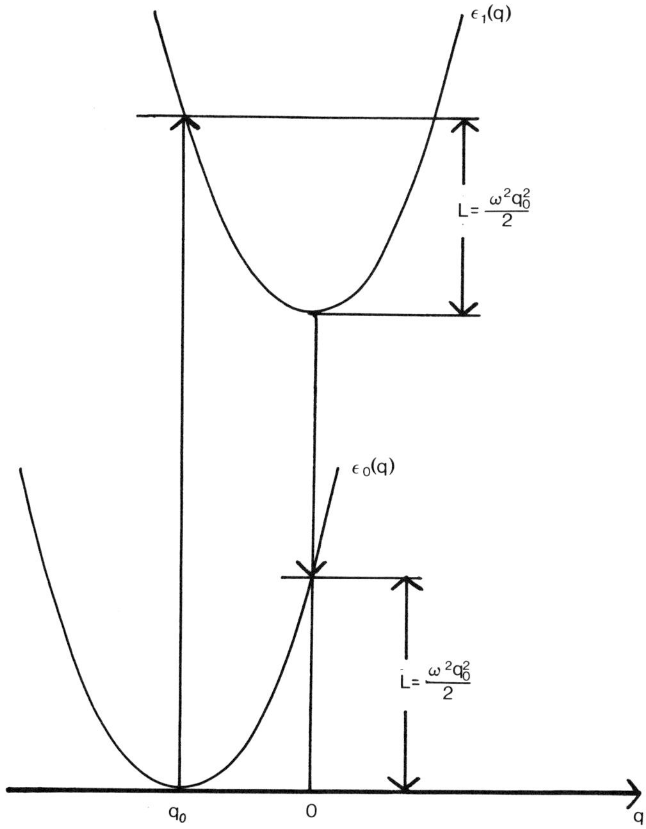

Fig.4. Adiabatic Potential Curves for a Localized Vibrational Mode.

is the Stokes loss in number of vibrational quanta.

The overlap integral of interest is now

$$\int \phi_{1n}(q)\phi_{00}(q)dq \tag{78}$$

The functions ϕ above are harmonic oscillator eigenfunctions. These functions are given by the formula

$$\phi_n(q) = \frac{1}{\sqrt{\bar{q}}} \frac{(-1)^n}{\sqrt{2^n n!\ \pi}} e^{y^2/2} \frac{d^n}{dy^n} e^{-y^2} \tag{79}$$

where

$$y = \frac{q}{\sqrt{\bar{q}}} = - \frac{q}{\sqrt{\hbar/\omega}} = \text{pure number} \tag{80}$$

$$\bar{q} = \sqrt{\frac{\hbar}{\omega}} = \text{zero point energy amplitude} \tag{81}$$

In particular

$$\phi_0(q) = \frac{1}{\sqrt{\bar{q}}} \frac{1}{\sqrt[4]{\pi}} e^{-\frac{1}{2}y^2} \tag{82}$$

Therefore

$$\phi_{00}(q) = \frac{1}{\sqrt{\bar{q}}} \frac{1}{\sqrt[4]{\pi}} e^{-1/2(y+y_o)^2} = \frac{1}{\sqrt{\bar{q}}} \frac{1}{\sqrt[4]{\pi}} e^{-\frac{y^2}{2} - \frac{y_o^2}{2} - yy_o} \tag{83}$$

where

$$y_o = \frac{q_o}{\sqrt{\hbar/\omega}} = \frac{q_o}{\bar{q}} \tag{84}$$

Also

$$\phi_{1n}(q) = \frac{1}{\sqrt{\bar{q}}} \frac{(-1)^n}{\sqrt{2^n n!\sqrt{\pi}}} e^{y^2/2} \frac{d^n}{dy^n} e^{-y^2} \tag{85}$$

and

$$\int_{-\infty}^{+\infty} \phi_{1n}(q)\phi_{00}(q)dq$$

$$= \frac{1}{\bar{q}} \frac{(-1)^n}{\sqrt{2^n n!\ \pi}} e^{\frac{y_o^2}{2}} \int_{-\infty}^{+\infty} e^{-yy_o} \frac{d^n}{dy^n} e^{-y^2}\, dy$$

$$= \frac{(-1)^n}{\sqrt{2^n n!\ \pi}} e^{-\frac{y_o^2}{2}} \int_{-\infty}^{+\infty} e^{-yy_o} \frac{d^n}{dy^n} e^{-y^2}\, dy \tag{86}$$

By successive integrations by part

$$\int_{-\infty}^{+\infty} e^{-yy_o} \frac{d^n}{dy^n} e^{-y^2}\, dy = \int_{-\infty}^{+\infty} e^{-yy_o} \frac{d}{dy}\left(\frac{d^{n-1}}{dy^{n-1}} e^{-y^2}\right)dy$$

$$= \left[e^{-yy_o} \frac{d^{n-1}}{dy^{n-1}} e^{-y^2}\right]_{-\infty}^{+\infty} + y_o \int_{-\infty}^{+\infty} e^{-yy_o} \left(\frac{d^{n-1}}{dy^{n-1}} e^{-y^2}\right)dy$$

$$= y_o \int_{-\infty}^{+\infty} \left(\frac{d^{n-1}}{dy^{n-1}} e^{-y^2} \right) dy = y_o^{\,n} \int_{-\infty}^{+\infty} e^{-y^2 - yy_o}\, dy \tag{87}$$

We solve this integral by completing the square in the exponent

$$\int_{-\infty}^{+\infty} e^{-y^2 - yy_o}\, dy = e^{\frac{y_o^2}{4}} \int_{-\infty}^{+\infty} e^{-\left(\frac{y_o}{2} + y\right)^2} dy \tag{88}$$

We set

$$\frac{y_o}{2} + y = z \tag{89}$$

Then

$$\int_{-\infty}^{+\infty} e^{-y^2 - yy_o}\, dy = e^{\frac{y_o^2}{4}} \int_{-\infty}^{+\infty} e^{-z^2}\, dz = e^{\frac{y_o^2}{4}} \sqrt{\pi} \tag{90}$$

and

$$\int_{-\infty}^{+\infty} \phi_{1n}(q)\phi_{00}(q)\,dq = \frac{(-1)^n}{\sqrt{2^n n!\pi}} e^{-\frac{y_o^2}{2}} y_o^{\,n} e^{\frac{y_o^2}{4}} \sqrt{\pi}$$

$$= \frac{(-y_o)^n}{\sqrt{2^n n!}} e^{-\frac{y_o^2}{4}} \tag{91}$$

Also

$$\left| \int_{-\infty}^{+\infty} \phi_{1n}(q)\phi_{00}(q)\,dq \right|^2 = e^{-\frac{y_o^2}{2}} \left(\frac{y_o^2}{2}\right)^n \frac{1}{n!} \tag{92}$$

But

$$\frac{y_o^2}{2} = \frac{q_o^2}{2\bar{q}} = \frac{q_o^2\omega}{2\hbar} = \frac{q_o^2\omega^2}{2} \Big/ (\hbar\omega)$$

$= \ell =$ Stokes loss in number of vibrational quanta. (93)

Therefore

$$\left| \int_{-\infty}^{+\infty} \phi_{1n}(q)\phi_{00}(q)dq \right|^2 = e^{-\ell} \frac{\ell^n}{n!} . \tag{94}$$

This formula is known as the <u>Pekarian Formula</u>. The function (94) is represented in Fig. 5 for different values of the Stokes loss ℓ. We note here that these patterns are similar to those observed in the electronic spectra of diatomic molecules, apart from the effects caused by the rotation of the molecules, which produces additional fine structure to each vibronic line.

We can now make the following observations:

1) The Pekarian formula expresses the Poisson distribution

$$P(n) = e^{-\bar{n}} \frac{(\bar{n})^n}{n!}, \tag{95}$$

if we replace ℓ with $\bar{n}$.

2) For large $\ell(\bar{n})$, the Pekarian (Poisson) distribution goes over to a Gaussian distribution.

3) If n = 0, we obtain

$$\left| \int_{-\infty}^{+\infty} \phi_{10}(q)\phi_{00}(q)dq \right|^2 = e^{-\ell}. \tag{96}$$

The intensity of the n = 0 line is the smaller, the larger is ℓ.

4) If n = 1, we get

$$\left| \int_{-\infty}^{+\infty} \phi_{11}(q)\phi_{00}(q)dq \right|^2 = e^{-\ell}\,\ell \simeq \ell, \tag{97}$$

for small ℓ.

5) If $\ell = 0$, the expression (94) becomes zero for $n \neq 0$. This means: no vibronic lines without Stokes shift.

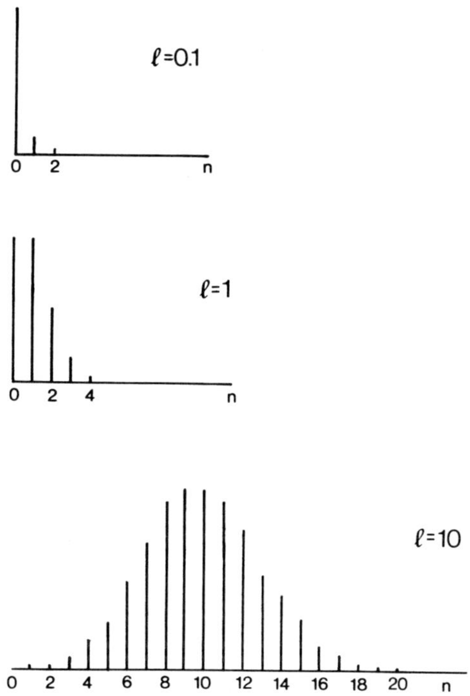

Fig.5. Typical Vibronic Transitions with Creation of Vibrational Quanta

IV.F The Formula of Huang and Rhys

We shall treat now a case similar to the one just examined. As previously, the vibrational frequency ω will be considered to be the same in the ground and in the excited electronic states of the "molecule". However, the restriction on the temperature ($\hbar\omega >> kT$) is removed.

The most general (absorption) transition initiates in the nth vibrational level of the ground electronic state and ends up in the mth vibrational level of an excited electronic state.

In this process, n vibrational quanta are destroyed and m vibrational quanta are created. The net number of vibrational quanta created is

$$p = m - n \tag{98}$$

We shall now obtain the formula of Huang and Rhys by following the derivation made by Curie [9].

The probability of creating a quantum of vibration may be taken to be

$$\pi_e \sim \ell(U+1) \tag{99}$$

where ℓ = Stokes loss in number of vibrational quanta, measure of the strength of the electron-vibration coupling and

$$U = \frac{1}{e^{\hbar\omega/kT} - 1} \tag{100}$$

U represents the average degree of excitation of the quantum vibrational oscillator at temperature T.

Similarly, the probability of destroying a quantum of vibration may be taken to be

$$\pi_a \sim \ell U. \tag{101}$$

Note that if $T = 0$, $U = 0$, $\pi_a = 0$ and $\pi_e \sim \ell$

The probability of a process in which m vibrational quanta are created and n vibrational quanta are destroyed is then

$$\sim \frac{\pi_e^m}{m!} \frac{\pi_a^n}{n!} \tag{102}$$

Normalizing

$$\left[e^{-\pi_e} \frac{\pi_e^m}{m!}\right]\left[e^{-\pi_a} \frac{\pi_a^n}{n!}\right] = e^{-(\pi_e+\pi_a)} \frac{\pi_e^m}{m!} \frac{\pi_a^n}{n!}$$

$$= e^{-\ell(2U+1)} \frac{[\ell(U+1)]^m}{m!} \frac{(\ell U)^n}{n!} = e^{-\ell(2U+1)} \frac{\ell^{n+m}}{n!m!} U^n(U+1)^m \tag{103}$$

The probability that a net number of $p = m-n$ vibrational quanta are created is given by

$$W(p,\ell) = \sum_{n=0}^{\infty} e^{-\ell(2U+1)} \frac{\ell^{2n+p}}{n!(n+p)!} U^n (U+1)^{n+p}$$

$$= e^{-\ell(2U+1)} \sum_{n=0}^{\infty} \frac{\ell^{2n+p}}{n!(n+p)!} \frac{U^{n+p/2}}{U^{p/2}} (U+1)^{n+p/2} (U+1)^{p/2}$$

$$= e^{-\ell(2U+1)} \left(\frac{U+1}{U}\right)^{p/2} \sum_{n=0}^{\infty} \frac{\left[\ell\sqrt{U(U+1)}\right]^{2n+p}}{n!(n+p)!} \qquad (104)$$

or

$$W(p,\ell) = e^{-\ell(U+1)} \left(\frac{U+1}{U}\right)^{p/2} I_p\left[2\ell\sqrt{U(U+1)}\right] \qquad (105)$$

where

$$I_p(x) = \sum_{n=0}^{\infty} \frac{\left(\frac{x}{2}\right)^{2n+p}}{n!(n+p)!} = (i)^{-p} J_p(ix) \qquad (106)$$

= modified Bessel function of the first kind, of order p.

The result (104) above is known as the formula of Huang and Rhys. We can write

$$A \sum_i N_i \sum_f |M_{fi}|^2 \, \delta(E_f - E_i - \hbar\omega) = A\, N\, |D_{10}^o|^2 \, W(p,\ell) \qquad (107)$$

where $W(p,\ell)$ is given by (105) and N is the total number of molecules.

We can now make the following observations:

1) Since

$$\sum_{p=-\infty}^{+\infty} W(p,\ell) = 1 \qquad (108)$$

we find, in agreement with (74),

$$A \sum_i N_i \sum_f |M_{fi}|^2 = A\, N\, |D_{10}^o|^2 \qquad (109)$$

2) If $\ell = 0$ all I_p are zero except I_o; namely, only zero-phonon lines are allowed.

3) At low T only the n = 0 term is relevant. In this case, m = p and

$$W(p,\ell) \to e^{-\ell}\left(\frac{1}{U}\right)^{p/2} \frac{\left(\ell\sqrt{U}\right)^p}{p!} = e^{-\ell} \frac{\ell}{p!} \qquad (110)$$

We reobtain in this case the Pekarian formula.

4) For $n = 0$, $p > 0$ and small, we get

$$W(p,\ell) \to \left(\frac{U+1}{U}\right)^{p/2} \frac{\left[\ell \sqrt{U(U+1)}\right]^p}{p!} = \frac{\ell^p}{p!} (1+U)^p \tag{111}$$

V. NONRADIATIVE DEACTIVATION OF OPTICALLY ACTIVE CENTERS

V.A. Deviations from the Adiabatic Approximation

The "exact" Hamiltonian of a molecular system is given as in Section II.A by

$$H = T_r + T_r + V(\vec{r},\vec{R}) \tag{112}$$

where

T_r = kinetic energy of the electrons

T_R = kinetic energy of the nuclei

$V(\vec{r},\vec{R}) = V_{ee} + V_{nm} + V_{ne}$

In the adiabatic approximation, the eigenstates of the system are given by

$$\Psi_i(\vec{r},\vec{R}) = \psi_k(\vec{r},\vec{R})\phi_{k\ell}(\vec{R}) \tag{113}$$

where

$$\left[T_r + V(\vec{r},\vec{R})\right] \psi_k(\vec{r},\vec{R}) = \epsilon_k(\vec{R})\psi_k(\vec{r},\vec{R}) \tag{114}$$

$$\left[T_R + \epsilon_k(\vec{R})\right]\phi_{k\ell}(\vec{R}) = E_{k\ell}\phi_{k\ell}(\vec{R}) \tag{115}$$

If we operate with H on Ψ_i we find

$$\begin{aligned} H\Psi_i(\vec{r},\vec{R}) &= \left[T_r + T_R + V(\vec{r},\vec{R})\right]\psi_k(\vec{r},\vec{R})\ \phi_{k\ell}(\vec{R}) \\ &= \left[T_r + V(\vec{r},\vec{R})\right]\psi_k(\vec{r},\vec{R})\ \phi_{k\ell}(\vec{R}) + T_R\psi_k(\vec{r},\vec{R})\ \phi_{k\ell}(\vec{R}) \\ &= \epsilon_k(\vec{R})\psi_k(\vec{r},\vec{R})\ \phi_{k\ell}(\vec{R}) + T_R\psi_k(\vec{r},\vec{R})\ \phi_{k\ell}(\vec{R}) \end{aligned} \tag{116}$$

where we have taken advantage of (114). From (115), we obtain

$$\psi_k(\vec{r},\vec{R})T_R\phi_{k\ell}(\vec{R}) + \epsilon_k(\vec{R})\psi_k(\vec{r},\vec{R})\phi_{k\ell}(\vec{R}) = E_{k\ell}\psi_k(\vec{r},\vec{R})\phi_{k\ell}(\vec{R}) \tag{117}$$

and

$$\epsilon_k(\vec{R})\psi_k(\vec{r},\vec{R})\phi_{k\ell}(\vec{R}) = E_{k\ell}\psi_k(\vec{r},\vec{R})\phi_{k\ell}(\vec{R}) - \psi_k(\vec{r},\vec{R})T_R\phi_{k\ell}(\vec{R}) \tag{118}$$

Therefore (116) becomes

$$H\Psi_i(\vec{r},\vec{R}) = E_{k\ell}\psi_k(\vec{r},\vec{R})\phi_{k\ell}(\vec{R}) - \psi_k(\vec{r},\vec{R})T_R\phi_{k\ell}(\vec{R}) + T_R\psi_k(\vec{r},\vec{R})\phi_{k\ell}(\vec{R})$$

$$= E_{k\ell}\psi_k(\vec{r},\vec{R})\phi_{k\ell}(\vec{R}) + A\psi_k(\vec{r},\vec{R})\phi_{k\ell}(\vec{R}) \tag{119}$$

where A, the nonadiabatic operator, is defined by

$$A\psi_k(\vec{r},\vec{R})\phi_{k\ell}(\vec{R}) = T_R\psi_k(\vec{r},\vec{R})\phi_{k\ell}(\vec{R}) - \psi_k(\vec{r},\vec{R})T_R\phi_{k\ell}(\vec{R}) \tag{120}$$

Let us follow up

$$A\psi_k\phi_{k\ell} = T_R\psi_k\phi_{k\ell} - \psi_k\, T_R\phi_{k\ell}$$

$$= \psi_k T_R\phi_{k\ell} + \phi_{k\ell}T_R\psi_k - \sum_s \frac{\hbar^2}{M_s}\vec{\nabla}_s\psi_k \cdot \vec{\nabla}_s\phi_{k\ell} - \psi_k T_r\phi_{k\ell}$$

$$= \phi_{k\ell}T_R\psi_k - \sum_s \frac{\hbar^2}{M_s}\vec{\nabla}_s\psi_k \cdot \vec{\nabla}_s\phi_{k\ell} \tag{121}$$

The first term of the last expression is negligible with respect to the second term in the assumption that

$$|\vec{\nabla}_s\psi_k(\vec{r},\vec{R})| << |\vec{\nabla}_s\phi_{k\ell}(\vec{R}) \tag{122}$$

Therefore

$$A\psi_k(\vec{r},\vec{R})\phi_{k\ell}(\vec{R}) = -\sum_s \frac{\hbar^2}{M_s}\vec{\nabla}_s\psi_k(\vec{r},\vec{R}) \cdot \vec{\nabla}_s\phi_{k\ell}(\vec{R}) \tag{123}$$

V.B. Mechanism for Radiationless Transitions

The previous section's treatment led us to the following result: when we operate on a Born-Oppenheimer product state function of the type (113), we obtain

$$H\Psi_i(\vec{r},\vec{R}) = H\psi_k(\vec{r},\vec{R})\phi_{k\ell}(\vec{R}) = E_i\psi_i(\vec{r},\vec{R}) + A\psi_i(\vec{r},\vec{R}) \tag{124}$$

where $E_i = E_{k\ell}$.

The general solution of the time dependent Schroedinger equation

$$H\Psi(t) = i\hbar\frac{\partial\Psi(t)}{\partial t} \tag{125}$$

can be expanded in terms of the eigenfunctions $\psi_i(\vec{r},\vec{R})$ as follows:

$$\Psi(t) = \sum_i c_i(t)\,\Psi_i(\vec{r},\vec{R})\exp\left(-i\frac{E_i}{\hbar}t\right) = \sum_i c_i(t)\Psi_i(t), \tag{126}$$

where the wave functions

$$\Psi_i(t) = \psi_k(\vec{r},\vec{R})\phi_{k\ell}(\vec{R}) \exp\left(-i \frac{E_{k\ell}}{\hbar} t\right)$$

represent states of definite energy $E_{k\ell} = E_i$. Replacing the expression (126) of $\Psi(t)$ into (125) we obtain

$$H \sum_i c_i(t)\Psi_i(t) = i\hbar\left[\sum_i c_i(t) \frac{\partial\Psi_i(t)}{\partial i} + \sum_i \frac{\partial c_i(t)}{\partial t} \Psi_i(t)\right] \tag{127}$$

Then

$$\sum_i c_i(t)\, H\, \Psi_i(t) = \sum_i c_i(t)E_i\Psi_i(t) + i\hbar \sum_i \frac{\partial c_i(t)}{\partial t} \Psi_i(t) \tag{128}$$

Taking (124) into account, (128) becomes

$$\sum_i c_i(t)E_i\Psi_i(t) + \sum_i c_i(t)A\Psi_i(t) = \sum_i c_i(t)E_i\Psi_i(t) + i\hbar \sum_i \frac{\partial c_i(t)}{\partial t} \Psi_i(t)$$

or

$$\sum_i c_i(t)A\Psi_i(t) = i\hbar \sum_i \frac{\partial c_i(t)}{\partial t} \tag{129}$$

Multiplying by $\Psi_j^*(t)$ and integrating over spatial coordinates, we obtain

$$i\hbar \frac{\partial c_j(t)}{\partial t} = \sum_i c_i(t) < \Psi_j(t)|A|\Psi_i(t) >$$

$$= \sum_i c_i(t) \exp\left(i \frac{E_j - E_i}{\hbar} t\right) < \Psi_j(0)|A|\psi_i(0) > \tag{130}$$

The matrix element of importance is

$$< \Psi_j(0)|A|\psi_i(0) > = < \psi_m(\vec{r},\vec{R})\phi_{mn}(\vec{R})|A|\psi_k(\vec{r},\vec{R})\phi_{k\ell}(\vec{R}) > \tag{131}$$

V.C. Radiationless Transitions

We consider now radiationless transitions between two quantum states defined by

$$|k\ell> = \psi_k(\vec{r},\vec{R})\phi_{k\ell}(\vec{R}) \tag{132}$$

and

$$|mn> = \psi_m(\vec{r},\vec{R})\phi_{mn}(\vec{R}) \tag{133}$$

The probability per unit time of a radiationless transition between these two states is given, according to the Fermi golden rule, by

$$w(k\ell \to mn) = \frac{2\pi}{\hbar} |< mn|A|k\ell>|^2 \quad \rho(E_{mn} = E_{k\ell}) \tag{134}$$

where $\rho\ (E_{mn} = E_{k\ell})$ = density of final states. We know that

$$A\psi_k(\vec{r},\vec{R})\phi_{k\ell}(\vec{R}) = -\sum_s \frac{\hbar^2}{M_s} \vec{\nabla}_s\psi_k(\vec{r},\vec{R}) \cdot \vec{\nabla}_s\phi_{k\ell}(\vec{R}) \tag{135}$$

or, in terms of the vibrational normal coordinates,

$$A\psi_k(\vec{r},q)\ \phi_{k\ell}(q) = \sum_i \frac{\partial\psi_k(\vec{r},q)}{\partial q_i} \frac{\partial\phi(q)}{\partial q_i} \tag{136}$$

where q stands for q_1, q_2 ... On the other hand, the functions $\phi_{k\ell}$ and ϕ_{mn} are expressed as the products of (56) and (57) respectively. By taking these relations into account, we may write

$$< mn|Ak\ell > = <\psi_m\phi_{mn}|A\psi_k\phi_{k\ell} >$$

$$= < \psi_m(\vec{r},q)\ \phi_{mn_1'}(q_1)\phi_{mn_2'}(q_2)\ \ldots\ \phi_{mn_1'}(q_i)\ \ldots\ |$$

$$|\ -\sum_i \hbar^2 \frac{\partial\psi_k(\vec{r},q)}{\partial q_i}\ \phi_{kn_1''}(q_1)\phi_{kn_2''}(q_2)\ \ldots\ \frac{\partial\phi_{kn_i''}(q_i)}{\partial q_i}\ \ldots >$$

$$= -\sum_i \hbar^2 < \psi_m(\vec{r},q)\frac{\partial\psi_k(\vec{r},q)}{\partial q_i} > < \phi_{mn_i'}(q_i)|\frac{\partial\phi_{kn_i''}(q_i)}{\partial q_i}> \prod_{s\neq i} < \phi_{mn_s'}(q_s)|\phi_{kn_s''}(q_s) >$$

$$\simeq -\hbar^2 \sum_i R_i(mk) < \phi_{mn_i'}(q_i)|\ \frac{\partial\phi_{kn_i''}(q_i)}{\partial q_i} > \prod_{s\neq i} < \phi_{mn_s'}(q_s)|\phi_{kn_s''}(q_s) > \tag{137}$$

where

$$R_i(mk) = <\psi_m(\vec{r},q)|\ \frac{\partial\psi_k(\vec{r},q)}{\partial q_i} > \tag{138}$$

The modes q_i responsible for the electronic transition m → k are called <u>promoting modes</u>, the other modes q_s are called <u>accepting modes</u>.

Finally

$$w(k\rightarrow m) = \frac{2\pi}{\hbar} \sum_{\ell n} P_\ell|<\psi_m\phi_{mn}|A|\psi_k\phi_{k\ell}>|^2\delta(E_{mn}-E_{k\ell}) \tag{139}$$

where

$$P_\ell = \frac{e^{-E_{k\ell}/kT}}{\sum_\ell e^{-E_{k\ell}/kT}} \tag{140}$$

is the probability of occupancy of the vibrational state ℓ.

This result is the starting point of more elaborate treatments [10-14].

VI. ENERGY TRANSFER PROCESSES

VI.A. Energy Transfer Among Optically Active Centers in Solids

Energy transfer processes are very important in solid-state luminescent systems, because they may provide an enhancement of the luminescence emission. This is usually achieved by the introduction in the host material, in addition to the ion responsible for the desired emission, called "activator" or "acceptor", of an ion of different type, called "sensitizer" or "donor". The donor ion provides additional spectral bands that absorb the pumping energy of the source. The energy absorbed results in the excitation of the donor ion; in turn this excitation energy is transferred to the acceptor [15].

However, it should be noted that the transfer of energy from an excited ion to an acceptor center that does not emit, a "killer" center, may result in the quenching of the luminescence emission; this phenomenon is considered responsible for the fact that the intensity of the luminescence actually decreases when the concentration of luminescent centers exceeds a certain value.

A typical sequence of events generally consists of the following steps: 1) Excitation of a donor (D) by the absorption of a photon, 2) decay of the donor to the ground state and transition of the acceptor (A) from the ground state to an excited state, and 3) decay of the activator to its ground state. Of course, if the actrivator's decay is radiative, an enhancement of the emission may result; if it is nonradiative, a quenching of the emission will be the consequence.

We note the following:

1) If D = A, i.e. if only ions of one type are present in the solid, the energy may transfer from an ion of a certain type to another ion of the same type. The emitted radiation may be trapped in the solid, leading to a longer decay time. This effect depends on the size and shape of the sample. If the concentration of D ions is large, the transfer can make it easy for the energy to reach killer sites, and may produce a "concentration quenching" of the luminescence.

2) Two different types of transfer are actually possible:
   i) Radiative transfer: the photon emitted by D is absorbed by A. This process consists of two first order sequential processes, the emission of a photon and the reabsorption of the same photon. The lifetime of D is unaffected by such transfer of energy and no additional dacay path is provided to D.
   ii) Nonradiative transfer: the decay of D and the excitation of A take place at the same time via a second order interaction. The lifetime of D is shortened, giving evidence of an additional decay path provided to D.

Since the nonradiative energy transfer is the one that may provide an ion with an additional nonradiative path to the ground state, we shall consider solely this case in what follows. The mechanism that makes such a process possible is the van der Waals interaction between two atomic centers. This interaction is responsible for the lowering of the ground state energy of a two-atom system, resulting in the so-called van der Waals bonding; it is also responsible for the time evolution of a two-atom excited system, i.e. it leads to energy transfer from one (excited) atom to the other (unexcited) atom.

Im many cases of interest the D and A centers in the solid are interacting weakly; quantitatively this means that the shift of an energy level due to the D-A interaction is much smaller than the width of the level. In such cases the absorption bands of D and A are identifiable, in the sense that the absorption spectrum of a sample containing both D and A is simply the superposition of the spectra of the individual ions.

VI.B Relevant Parameters

The rate of the energy transfer between a donor ion (D) and an acceptor ion (A), w_{DA}, depends on the separation between the ions, R. This dependence can be expressed in a multipolar expansion as follows:

$$w_{DA}(R) = C^{(6)}/R^6 + C^{(8)}/R^8 + C^{(10)}/R^{10} + \ldots \tag{141}$$

where the first three terms correspond to dipole-dipole, dipole-quadrupole, and quadrupole-quadrupole interactions, respectively. If there is a dominant multipolar interaction, then the transfer rate takes the simpler form

$$w_{DA}(R) = C^{(n)}/R^n \tag{142}$$

and we can define an "energy transfer" radius, R_0, as follows:

$$w_{DA}(R) = C^{(n)}/R^n = \tau_0^{-1}(R_0/R)^n \tag{143}$$

In the expression, τ_0 is the effective decay time of D in the absence of A and R_0 is the separation between ions at which the energy transfer rate is equal to the decay rate of the donor. In the dipole-dipole case R_0 can be expressed in terms of the overlap integral [15-17].

$$R_0^6 = \epsilon f_A \left[3\pi e^2 c^3 \hbar^5/(2m)\right] \int g_D(E) g_A(E) \frac{dE}{E^4} \tag{144}$$

where ϵ is the quantum efficiency of the donor luminescence (in the absence of the acceptor),

$$\epsilon = \frac{\text{probability of radiative decay}}{\begin{array}{c}\text{(probability of radiative decay}\\ \text{+ probability of nonradiative decay)}\end{array}} = \frac{\tau_{0\,rad}^{-1}}{\tau_0^{-1}} \tag{145}$$

(τ_{0rad} is the radiative decay time of D in the absence of A). The overlap integral involves $g_D(E)$ and $g_A(E)$, the normalized spectral functions of D and A respectively. The f-number or oscillator strength,

f_A, is derived from the absorption data as

$$f_A = \left[mcn/\left(2\pi^2 e^2 C_A\right)\right]\int \alpha_A(\omega)\ d\omega. \tag{146}$$

Here n is the index of refraction, C_A is concentration of ions A, and α_A (ω) the absorption coefficient of A. The parameter $C^{(n)}$ in eq. (142), given by

$$C^{(n)} = R_0^n/\tau_0 \tag{147}$$

is often called the microscopic interaction parameter; it is equal to the probability per unit time of energy transfer when D and A are at the distance of 1 cm.

The energy of an excited donor may be expended by a radiative transition, by transfer of excitation to an acceptor ion or to another donor ion. This latter process may be treated as a diffusion with the probability of donor diffusion dependent on the donor concentration, C_D. The time dependence of the decay of the donor excitation is given by $\rho(t)$, which is a statistical average of the decay rates corresponding to all the processes contributing to the decay of a single donor. We can distinguish three different regimes of donor luminescence depending upon the extent of donor diffusion.

VI.C. Regimes of Donor Decay

1. No Diffusion. In the first regime, where the donor concentration is small, the effects of diffusion are negligible. In the extreme case of an isolated donor, which is far away from all recipients of excitation, the D decay is then expressed by the exponential

$$\rho(t) = \exp(-t/\tau_0) \tag{148}$$

On the other hand, if the donor has an acceptor ion as a nearest neighbor, the exponential decay is modified by the probability of energy transfer as $\rho(t)$ can be expressed as follows:

$$\tau(t) = \exp\left[-t/\tau_0 - t\ w_{DA}(R_m)\right] \tag{149}$$

where R_m is the minimum D-A distance. If this probability is extended to include transfer to any of the other acceptors in the crystal the time-dependent decay will follow the formula [18]

$$\bar{\rho}(t) = \exp\left[-t/\tau_0 - C_A/C_0\Gamma(1-3/n)(t/\tau_0)^{3/n}\right] \tag{150}$$

where $C_0^{-1} = 4\pi R_0^3/3$ and n is either 6 (dipole-dipole), 8 (dipole-quadrupole), or 10 (quadrupole-quadrupole).

The following general observations can be made regarding eq. (150). At early times the decay is faster because the donors close to the acceptors decay first. After a while other donors, which are farther away from acceptors, and remain excited, start transferring energy. At very long times donors that are very far from acceptors will finally decay with their own decay time. The greater C_A/C_0, the longer one has to wait for $\rho(t)$ to decay at the same rate as the one for $C_A/C_0 = 0$.

2. Diffusion-Limited Decay. For larger values of C_D the effects of diffusion begin to appear. The rate of diffusion is determined by the diffusivity D. Assuming a dipole-dipole mechanism, n = 6, for $t \ll [C^{(6)}]^{1/2}/D^{3/2}$, $\rho(t)$ is the same as for the case of no migration, and for $t \gg [C^{(6)}]^{1/2}/D^{3/2}$ the decay rate is given by

$$\rho(t) \; \alpha \; \exp\left[-(t/\tau_0) - K_D t\right] \tag{151}$$

where

$$K_D = 4\pi D C_A R_D$$
$$R_D \simeq \left[C^{(6)}/D\right]^{1/4} \tag{152}$$

The time, $t^* = \left[C^{(6)}\right]^{1/2}/D^{3/2}$ represents a boundary between the "transfer without migration" region and the "transfer with migration" region. At later times the only donors that are still excited are those far away from any acceptor. They now transfer their energy by first diffusing it, i.e. by sending it to a donor near an acceptor. As migration becomes more rapid, the boundary between the two regions shifts: $t^* = \left[C^{(6)}\right]^{1/2}/D^{3/2}$ shifts to shorter times, until, for sufficiently fast migration, the decay appears to be entirely exponential.

There may be a physical situation such that the system is in an intermediate regime between 1 and 2 above. This may be the case when the diffusion process is about as fast as the energy transfer process. In such a case the formula for the decay curve of D is given by

$$\rho(t) = \exp\left[-t/\tau_0 - K_D t - C_A/C_0 \Gamma(1-3/n)(t/\tau_0)^{3/n}\right] \tag{153}$$

3. Fast Diffusion. An increase in the concentration of donors produces a faster migration of the energy, i.e. a larger D, and a larger K_D. The characteristic time t* becomes shorter and shorter until it reaches the value $w_{DA}^{-1}(R_m)$. Any additional increase of C_D has effect on D, but no effect of K_D which takes a "saturated" value. In these conditions the decay of the donor luminescence is purely exponential, since the fast diffusion averages out all the different donor environments.

ACKNOWLEDGEMENTS

The author would like to acknowledge the benefit of discussions with Drs. Charles Struck, Martin Buoncristiani, and Guzin Armagan, and several participants in the 1989 Erice NATO Advanced Study Institute on "Advances in Nonradiative Processes in Solids".

REFERENCES

1. M. Born and J.R. Oppenheimer, Ann. Phys. 84, 457 (1927).
2. K.K. Rebane, Impurity Spectra of Solids, Plenum Press, New York and London, 1970, p.13.
3. M.D. Sturge, in Solid State Physics, Vol.20, F. Seitz, D. Turnbull and H. Ehrenreich, eds., Academic Press, New York and London, 1967, p.91.
4. J. Franck, Trans. Faraday Soc. 21, 536 (1925).
5. E.U. Condon, Phys. Rev. 32, 858 (1928).
6. L. Pauling and E.B. Wilson, Introduction to Quantum Mechanics, McGraw Hill, New York and London, 1935, p.265.
7. B. Di Bartolo and R.C. Powell, Phonons and Resonances in Solids, John Wiley and Sons, Inc., New York, 1976.
8. B. Di Bartolo, in Spectroscopy of Solid-State-Laser-Type Materials, B. Di Bartolo, ed., Plenum Press, New York and London, 1987, p.53.
9. D. Curie in Optical Properties of Ions in Solids, B. Di Bartolo, ed., Plenum Press, New York and London, 1975, p.84.
10. K. Huang and A. Rhys, Proc. Roy. Soc. A204, 406 (1950).
11. M. Lax, J. Chem. Phys. 20, 1752 (1952).
12. R. Kubo and Y. Toyozawa, Progr. Theoret. Phys. 13, 160 (1955).
13. T. Miyakawa and D.L. Dexter, Phys. Rev. B1, 2961 (1970).
14. W.H. Fonger and C.W. Struck, in Radiationless Processes, B. Di Bartolo, ed., Plenum Press, New York and London, 1980, p. 471.
15. B. Di Bartolo, in Energy Transfer Processes in Condensed Matter, B. Di Bartolo, ed., Plenum Press, New York and London, 1984, p. 103.
16. D.L. Dexter, J. Chem. Phys. 21, 836 (1953).
17. B. Di Bartolo, Optical Interactions in Solids, John Wiley and Sons, Inc., New York, 1968.
18. M. Inokuti and F. Hirayama, J. Chem. Phys. 43, 1978 (1963).

RADIATIVE AND NONRADIATIVE RATES IN LUMINESCENCE CENTERS.

C. W. Struck
GTE Laboratories Inc.
and
W. H. Fonger
174 Guyot Avenue
Princeton, NJ 08540

ABSTRACT

The W_p function and related functions for the derivative and the z operators are defined and evaluated. These functions are used to explain the luminescence optical bands and the efficiencies of Eu^{3+}, Tm^{3+}, and Yb^{3+} in oxysulfide hosts. Their usefulness in understanding luminescence centers in general is illustrated in model examples.

I. INTRODUCTION

This paper summarizes some of the work contained in many papers on models of luminescence centers by the authors[1-21].

II. COMPENDIUM OF USEFUL EQUATIONS

II.A. THE WAVEFUNCTIONS

The harmonic oscillator wavefunctions are defined in terms of Hermite polynomials,

$$v_m = H_m(z)\, e^{-z^2/2}(m!2^m\pi^{1/2}x_v)^{-1/2}, \tag{1}$$

which are given by their recursion formula, initiated with the first two values,

$$-2kH_{k-1} + 2zH_k - H_{k+1} = 0,$$

$$H_0 = 1,$$

$$H_1 = 2z. \tag{2}$$

The z and d/dz relations are

Advances in Nonradiative Processes in Solids
Edited by B. Di Bartolo, Plenum Press, New York, 1991

$$\sqrt{2}\, z v_k = \sqrt{k}\, v_{k-1} + \sqrt{k+1}\, v_{k+1}, \tag{3}$$

$$\sqrt{2}\frac{dv_k}{dz} = \sqrt{k}\, v_{k-1} - \sqrt{k+1}\, v_{k+1}. \tag{4}$$

The Schrődinger equation, in dimensioned units and in dimensionless units are, respectively,

$$\left[\frac{\hbar^2}{2M}\frac{d^2}{dx^2} + E_{vm} - \tfrac{1}{2}k_v(x\text{-}a)^2\right]\Psi_m(x,k_v,a) = 0, \tag{5}$$

$$\frac{d^2}{dz^2}v_k - z^2 v_k + (2k+1)v_k = 0. \tag{6}$$

The dimensionless variables are defined through

$$E_{vm} = (m+\tfrac{1}{2})\hbar\omega_v, \qquad \omega_v = \left(\frac{k_v}{M}\right)^{1/2},$$

$$x_v^2 = \frac{\hbar}{\sqrt{Mk_v}}, \qquad z_v = (x\text{-}a)/x_v,$$

$$\tfrac{1}{2}k_v x_v^2 = \tfrac{1}{2}\hbar\omega_v. \qquad \tfrac{1}{2}k_v(x\text{-}a)^2 = z_v^2\left(\frac{\hbar\omega_v}{2}\right),$$

$$\frac{\hbar^2}{2Mx_v^2} = \frac{\hbar\omega_v}{2}. \tag{7}$$

II.B. THE MANNEBACK RECURSION FORMULAS

The Manneback recursion formulas for unequal force constants are

$$k_+\sqrt{n}\,A_{n\text{-}1,m} = \sqrt{m+1}\,A_{n,m+1} + k_+S_u^{1/2}A_{nm} + k\sqrt{m}\,A_{n,m\text{-}1},$$

$$k_+\sqrt{m}\,A_{n,m\text{-}1} = \sqrt{n+1}\,A_{n+1,m} - k_+S_v^{1/2}A_{nm} - k\sqrt{n}\,A_{n\text{-}1,m}, \tag{8}$$

where

$$k = \cos 2\theta, \qquad k_+ = \sin 2\theta,$$

$$S_u^{1/2} = \frac{a_{uv}\cos\theta}{\sqrt{2}}, \qquad S_v^{1/2} = \frac{a_{uv}\sin\theta}{\sqrt{2}},$$

$$\left(\frac{S_v}{S_u}\right)^{1/2} = \left(\frac{\hbar\omega_v}{\hbar\omega_u}\right)^{1/2} = \tan\theta, \qquad a_{uv} = \{2(S_u + S_v)\}^{1/2}. \tag{9}$$

The equal force constants equations are

$$\sqrt{n}\, A_{n-1,m} = \sqrt{m+1}\, A_{n,m+1} + S^{1/2} A_{nm},$$

$$\sqrt{m}\, A_{n,m-1} = \sqrt{n+1}\, A_{n+1,m} - S^{1/2} A_{nm}. \tag{10}$$

The equal-force-constants z_v and z_u operators have the relationships

$$<u_n|z_v|v_m> = A_{n,m,z_v} = \sqrt{\frac{m}{2}}\, A_{n,m-1} + \sqrt{\frac{m+1}{2}}\, A_{n,m+1},$$

$$<v_m|z_u|u_n> = A'_{m,n,z_u} = \sqrt{\frac{n}{2}}\, A_{n-1,m} - \sqrt{\frac{n+1}{2}}\, A_{n+1,m}. \tag{11}$$

The derivative operator expressions for the $A_{nm,d/dz}$ are

$$<u_n|\frac{d}{dz}|v_m> = A_{n,m,d/dz_v} = \sqrt{\frac{m}{2}}\, A_{n,m-1} - \sqrt{\frac{m+1}{2}}\, A_{n,m+1},$$

$$<v_m|\frac{d}{dz}|u_n> = A'_{m,n,d/dz_u} = \sqrt{\frac{n}{2}}\, A_{n-1,m} - \sqrt{\frac{n+1}{2}}\, A_{n+1,m}. \tag{12}$$

II.C. THE EQUAL-FORCE CONSTANTS W_p AND RELATED FUNCTIONS IN ONE DIMENSION

The W_p function is defined by

$$W_p = \sum_m (1-r)\, r^m <u_{p+m}|v_m>^2 = \sum_m (1-r)\, r^m A^2_{p+m,m}, \tag{13}$$

and has the explicit expressions

$$W_p = \exp(-S<2m+1>) \sum_{j=\max(0,-p)}^{\infty} \frac{(S<m>)^j (S<1+m>)^{p+j}}{j!(p+j)!}$$

$$= \left[\frac{<1+m>}{<m>}\right]^{p/2} I_p\left(2S(<m><1+m>)^{1/2}\right),$$

$$<m> = \frac{r}{1-r}, \qquad r = \exp(-\frac{\hbar\omega}{kT}). \tag{14}$$

The W_p satisfies the recursion relationship,

$$S <m> W_{p+1} + p W_p - S<1+m> W_{p-1} = 0. \tag{15}$$

At $T = 0$, W_p is the Poisson distribution,

$$W_p = e^{-S} \frac{S^p}{p!}. \tag{16}$$

When $S(<m><1+m>)^{1/2}/p$ is small, then W_p has the approximate form

$$W_p \approx B_p \equiv e^{-S<2m+1>} \frac{(S<1+m>)^p}{p!}, \qquad p \geq 0,$$

$$\equiv e^{-S<2m+1>} \frac{(S<m>)^p}{p!}, \qquad p<0. \tag{17}$$

When S is large, then W_p is approximately the Gaussian function,

$$G(p) = \frac{1}{\sqrt{2\pi\sigma^2}} \exp\left[-(p\text{-}S)^2/2\sigma^2\right], \tag{18}$$

with $\sigma^2 = S<2m+1>$.

W_p has also the approximate form, based on the asymptotic expansion of the modified Bessel I_p function,

$$W_p \approx W_p^{\cdot} \equiv e^{-S<2m+1>} \left(\frac{<1+m>}{<m>}\right)^{p/2} \frac{e^{y_p}}{\sqrt{2\pi y_p}} \left(\frac{x}{p+y_p}\right)^p,$$

$$x = 2S(<m><1+m>)^{1/2},$$

$$y_p = \sqrt{p^2+x^2} \geq 1. \tag{19}$$

$W_{p,z}$ can be expressed in terms of W_p

$$W_{p,z} = \frac{(p\text{-}S)^2}{S} W_p. \tag{20}$$

$W_{p,d/dz}$ is expressed in terms of $W_{p,z}$ and other W_p functions as

$$W_{p,d/dz} = W_{p,z} - 2<m><1+m>S\,(W_{p+1} - 2W_p + W_{p-1}) \tag{21}$$

Both $W_{p,z}$ and $W_{p,d/dz}$ have 5-W_p expressions,

$$W_{p,O_\pm} = L_p + S \sum_{i=-2}^{2} \alpha_{i,\pm} W_{p+i}.$$

$$\alpha_{-2,z} = \frac{<1+m>^2}{2}, \qquad \alpha_{-2,d/dz} = \frac{<1+m>^2}{2},$$

$$\alpha_{-1,z} = -<1+m>, \qquad \alpha_{-1,d/dz} = -<1+m><2m+1>,$$

$$\alpha_{0,z} = \frac{1}{2} - <m><1+m>, \qquad \alpha_{0,d/dz} = \frac{1}{2} + 3<m><1+m>,$$

$$\alpha_{1,z} = <m>, \qquad \alpha_{1,d/dz} = -<m><2m+1>,$$

$$\alpha_{2,z} = \frac{<m>^2}{2}, \qquad \alpha_{2,d/dz} = \frac{<m>^2}{2}. \tag{22}$$

The average m through which a transition occurs has the definition and the expression

$$<m>_p = \frac{\sum_{m=m_0}^{\infty} m(1-r)r^m A^2_{p+m,m}}{\sum_{m=m_0}^{\infty} (1-r)r^m A^2_{p+m,m}} = \left[\frac{1}{W_p}\right] \sum_{m=m_0}^{\infty} m(1-r)r^m A^2_{p+m,m} \tag{23}$$

and

$$<m>_p W_p = S<m><1+m>\{W_{p+1} - 2W_p + W_{p-1}\} + <m> W_p. \tag{24}$$

II.D. THE UNEQUAL-FORCE-CONSTANTS EXPRESSIONS

For unequal force constants, the analogous expressions are

$$V_{P_V} = \sum_n (1-r_u)\, r_u^n\, V_{P_V,n}, \qquad U_{P_U} = \sum_m (1-r_v)\, r_v^m\, U_{P_U,m},$$

$$V_{P_{V,z}} = \sum_n (1-r_u)\, r_u^n\, V_{P_V,n,z}, \qquad U_{P_{U,z}} = \sum_m (1-r_v)\, r_v^m\, U_{P_U,m,z} \tag{25}$$

where

$$r_u = \exp(-\hbar\omega_u/kT), \qquad r_v = \exp(-\hbar\omega_v/kT),$$

$$V_{P_V,n} = <v_{p_j}|u_n>^2(p_V+1-p_j) + <v_{p_j-1}|u_n>^2(p_j-p_V),$$

$$U_{P_U,m} = <u_{p_i}|v_m>^2(p_U+1-p_i) + <u_{p_i-1}|v_m>^2(p_i-p_U),$$

$$V_{P_V,n,z} = <v_{p_j}|z_u|u_n>^2(p_V+1-p_j) + <v_{p_j-1}|z_u|u_n>^2(p_j-p_V),$$

$$U_{p_U,m,z} = <u_{p_i}|z_v|v_m>^2(p_U+1-p_i) + <u_{p_i-1}|z_v|v_m>^2(p_i-p_U). \tag{26}$$

The indices p_i and p_j depend on m and n, respectively. They are the smallest indices such that

$$p_{j(n)} - n\hbar\omega_u > p_V\hbar\omega_v, \qquad p_{i(m)} - m\hbar\omega_v > p_U\hbar\omega_u. \tag{27}$$

II.E THE MOMENTS

The equal-force-constants moments are

$$<(p\text{-}S)^0> = <1> = 1,$$

$$<(p\text{-}S)^1> = 0,$$

$$<(p\text{-}S)^2> = S<2m+1>,$$

$$<(p\text{-}S)^3> = S$$

$$<(p\text{-}S)^4> = S<2m+1> + 3[S<2m+1>]^2. \tag{28}$$

They are obtained from the moment recursion formula

$$<(p\text{-}S)^\alpha> = \sum_{j=0}^{\alpha-2}\binom{\alpha-1}{j}\beta_{\alpha j}<(p\text{-}S)^j>$$

$$\beta_{ij} = S<2m+1>, \qquad \alpha-j \text{ even},$$

$$= S, \qquad \alpha-j \text{ odd}, \tag{29}$$

using the first two moments to initiate the sequence of calculations.

The first three rigorous unequal-force-constants $v\rightarrow u$ emission p moments are

$$<p_u>_{uv} = S_u + k_\theta<2m+1>_v,$$

$$<(p_u - <p_u>_{uv})^2>_{uv} = (\frac{\omega_u}{\omega_v})S_u<2m+1>_v + 2k_\theta^2<2m+1>_v^2,$$

$$<(p_u - <p_u>_{uv})^3>_{uv} = S_u + 6k_\theta(\frac{\omega_u}{\omega_v})S_u<2m+1>_v^2$$

$$+ 4k_\theta^2(\frac{\omega_v}{\omega_u})<2m+1>_v + 8k_\theta^3<2m+1>_v^3.$$

$$k = \cos 2\theta, \qquad k_{+} = \sin 2\theta, \qquad k_{\theta} = \left(\frac{k}{k_{+}^{2}}\right) = \frac{\omega_u^2 - \omega_v^2}{4\omega_v\omega_u}. \tag{30}$$

The $u \rightarrow v$ absorption p moments are

$$<p_v>_{vu} = = S_v - k_{\theta} <2n+1>_u,$$

$$<(p_v - <p_v>_{vu})^2>_{vu} = (\frac{\omega_v}{\omega_u})S_v <2n+1>_u + 2k_{\theta}^2 <2n+1>_u^2,$$

$$<(p_v - <p_v>_{vu})^3>_{vu} = S_v - 6k_{\theta}(\frac{\omega_v}{\omega_u})S_v <2n+1>_u^2$$

$$+ 4k_{\theta}^2(\frac{\omega_u}{\omega_v}) <2n+1>_u - 8k_{\theta}^3 <2n+1>_u^3. \tag{31}$$

The k'th energy moment is the k'th p moment multiplied by the final-state phonon energy, here $\hbar\omega_v$ in absorption and $\hbar\omega_u$ in emission, raised to the k'th power.

II.F. MULTIPLE COORDINATE MODELS OF A LUMINESCENCE CENTER

The Thermal-Condon factor for the Einstein-Huang-Rhys-Pekar model of N_{Av} oscillators having a common frequency $\hbar\,\omega$ and therefore a common $<m>$ is itself a W_p function with parameters

$$W_p(S, <m>) = \sum_{p_1,\ldots,p_N} W_{p_1} W_{p_2} \ldots W_{p_N} = W_p\left(\sum S_i, <m>\right),$$

$$\sum p_i = p. \tag{32}$$

The z and d/dz operator distributions are given by

$$A^2 M_{p,O_{\pm}}$$

$$\equiv \sum_{\vec{p},\vec{m}} \left| <u_{p_1+m_1} u_{p_2+m_2} \ldots u_{p_N+m_N} \mid A_1 O_{1\pm} + \ldots A_N O_{N\pm} \mid v_{p_1+m_1} v_{p_2+m_2} \ldots v_{p_N+m_N}> \right|^2$$

$$\cdot (1-r)^N r^{m_1+\ldots+m_N}$$

$$= A^2[(1-\gamma)W_{p,O_{\pm}} + \gamma L_p], \tag{33}$$

where

$$A^2 = \sum_{k=1}^{N} |A_k|^2,$$

$$S = \sum_{k=1}^{N} S_k,$$

$$\gamma = 1 - \frac{\left[\sum_{k=1}^{N} A_k S_k^{1/2}\right]^2}{A^2 S},$$

$$0 \leq \gamma \leq 1,$$

$$L_p = \frac{1}{2}(<1+m> W_{p-1} + <m> W_{p+1}), \tag{34}$$

where $W_{p,O_\pm}$, L_p, and the W_p functions within the L_p expression all have the arguments $(S, <m>)$. The $W_{p,O_\pm}$ themselves are given by Eq. (22).

The multiple frequency model expressions for the Condon operator are defined by

$$D_p^K = A^{-2} G_p^K, \tag{35}$$

where the G_p^K are defined by

$$G_p^K(\vec{S}, \vec{\omega_K}, kT) = A^2 \sum_{p_1, p_2, \ldots, p_K} W_{p_1}(S_1, \omega_1, kT)\ W_{p_2}(S_2, \omega_2, kT) \ldots W_{p_K}(S_K, \omega_K, kT), \tag{36}$$

where the summation indices are constrained by

$$\sum_{k=1}^{K} k p_k = p, \tag{37}$$

and

$$S_i = \sum_{j=1}^{g_i} S_{i,j},$$

$$\omega_k = k\omega_1. \tag{38}$$

The W_p have the normal variables $(S, <m>)$, and $<m>$ must be determined for each ω_k through

$$r = \exp(-\hbar\omega/kT), \qquad <m> = \frac{r}{1-r}. \tag{39}$$

This Condon-operator distribution is most readily obtained by the recursion algebra

$$D_q^{K,l} = \sum_{p_1} D_{q-lp_1}^{K,l-1} W_{p_1}(S_l,\omega_l,kT),$$

$$D_q^{K,1} = W_q(S_1,\omega_1,kT) \equiv W_q(S_1,<m_1>),$$

$$2 \leq l \leq K$$

$$D_{q=p}^{K,K} = D_p^K. \tag{40}$$

For the $O_\pm$ operators, the distributions are a sum of two terms,

$$G_{p,O_\pm}^K = \Psi_{p,O_\pm}^K + \Phi_{p,O_\pm}^K, \tag{41}$$

In Eq. (41), $\Psi_{p,O_\pm}^K$ is given by the recursion algebra

$$\Psi_{q,O_\pm}^{K,1} = A_1^2 M_{q,O_\pm}(S_1,\omega_1,kT),$$

$$\Psi_{q,O_\pm}^{K,l} = \sum_{p_1}\left[\Psi_{q-lp_1,O_\pm}^{K,l-1} W_{p_1}(S_l,\omega_l,kT) + D_{q-lp_1}^{K,l-1} A_l^2 M_{p_1,O_\pm}(S_l,\omega_l,kT)\right],$$

$$\Psi_{q,O_\pm}^{K,K} = \Psi_{q,O_\pm}^K \tag{42}$$

The $\Phi_{p,O_\pm}^K$ are obtained by the recursion algebra

$$\Theta_{q,O_\pm}^{K,l} = \sum_{p_1}\left[\Theta_{q,O_\pm}^{K,l-1} W_{p_1}(S_l,\omega_l,kT) + D_{q-lp_1}^{K,l-1}\Gamma_l M_{p_1,O_\pm}((S_l,\omega_l,kT)\right],$$

$$\Theta_{q,O_\pm}^{K,1} = \Gamma_1 X_{q,O_\pm}(S_1,\omega_1,kT),$$

$$\Phi_{q,O_\pm}^{K,l} = \sum_{p_1}\left[\Phi_{q-lp_l,O_\pm}^{K,l-1} W_{p_1}(S_l,\omega_l,kT) + \Theta_{q-lp_1}^{K,l-1}\Gamma_l^* X_{p_1,O_\pm}(S_l,\omega_l,kT)\right.$$

$$+\left(\Theta^{K,l-1}_{q-lp_l,O_\pm}\right)^{*}\Gamma_l X_{p_1,O_\pm}(S_l,\omega_l,kT)\Big],$$

$$X_{p,O_\pm} = <m>(W_p - W_{p+1}) \pm <1+m>(W_{p-1} - W_p),$$

$$\Phi^{K,1}_{q,O_\pm} = 0$$

$$\Phi^{K,K}_{q=p,O_\pm} = \Phi^{K}_{p,O_\pm}. \tag{43}$$

The first of Eq. (43) is the recursion relation which, when initiated with the second of this equation, allows sequential generation of the $\Theta^{K,l}_{q,O_\pm}$ for $2 \leq l \leq K$.

In turn, the third of this equation is the recursion relation for $\Phi^{K,l}_{q,O_\pm}$, which, when initialized as in the fourth equation, allows sequential generation of the $\Phi^{K,l}_{q,O_\pm}$ for $2 \leq l \leq K$ and leads to the desired sum in the last of this Eq. (41).

We adopt the Discretized Debye Equal S and A model through

$$\bar{g}_1 = \frac{27}{8K^3},\quad \bar{g}_k = \frac{12k^2+1}{4K^3},\quad \bar{g}_K = \frac{12K^2-6K+1}{8K^3},\quad \bar{g}_l = g_l/3N_0,$$

$$S\hbar\omega_K = \sum_{k=1}^{K} S_k\hbar\omega_k = \frac{S_0}{K}\sum_{k=1}^{K} kg_k,$$

$$S_k = g_k S_0 = \bar{g}_k\left\{\frac{8K^4}{6K^4+K^2+1}\right\},$$

$$\Gamma_k \equiv \frac{1}{\sqrt{2}}\sum A_{k,l}S^{1/2}_{k,l} = \frac{1}{\sqrt{2}}(A^2_k S_k)^{1/2},$$

$$A^2_k = \sum_l^{g_l} A^2_{kl} = g_l A^2_0. \tag{44}$$

II.G. ENERGY TRANSFER

The transition rate from an energy donor to an energy acceptor has the thermal Franck-Condon weight given by

$$W_{p_{da}} = \sum_{m,l,n,k=0}^{\infty}(1-r_0)r_0^m(1-r_0)r_0^l<d_n|d_m^+>^2<a_k|a_l^+>^2, \tag{45}$$

where the m,n,k,l indices are constrained to follow

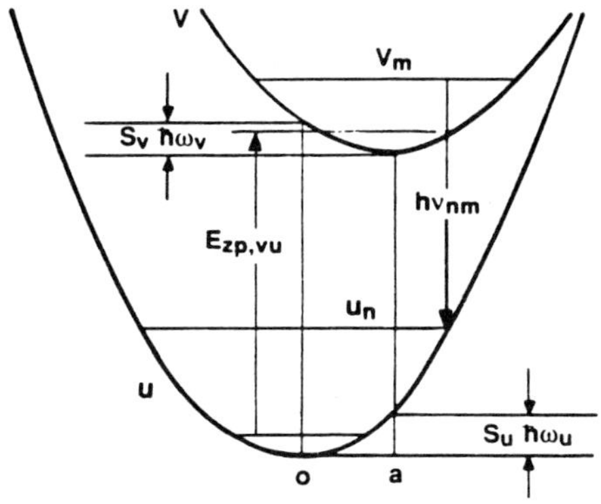

Figure 1. The Single-Configurational-Coordinate Model

$$p_{da}\hbar\omega = [(n-m)+(k-l)]\hbar\omega = h\nu_{zp,d^+d} - h\nu_{zp,a^+a} \tag{46}$$

This nuclear factor is recognized to be

$$W_{p_{da}} = \sum_{p_d=-\infty}^{\infty} W_{p_d}(S_{d'} <m>) W_{p_a}(S_{a'} <m>), \tag{47}$$

where the indices are constrained by

$$p_d + p_a = p_{da}, \tag{48}$$

and is further recognized as itself a W_p function, namely,

$$W_{p_{da}}(S_d + S_{a'} <m>) = \sum_{p_d=-\infty}^{\infty} W_{p_d}(S_{d'} <m>) W_{p_a}(S_{a'} <m>). \tag{49}$$

III. ILLUSTRATIVE NONRADIATIVE TRANSITIONS

III.A. THE SIMPLEST MODEL: ONE COORDINATE AND EQUAL FORCE CONSTANTS

1. The Optical Band Shapes. Figure 1 shows the simplest model of a luminescence center, the single-configurational-coordinate (SCC) model, namely two parabolas representing the ground and the excited states of the center. The two parabolas are displaced along one vibrational coordinate, q, the ground state minimum being at $q=0$ and the upper state minimum at $q=a$. A natural measure of the coordinate offset between the two parabolas are the energies $S_u\hbar\omega_u$ and $S_v\hbar\omega_v$ which are the energy differences in the parabolas between the two minima q values, namely, 0 and a, in units of the phonon energy of the parabola. The two parabolas are also displaced in the energy coordinate. A reasonable measure of this offset is the zero-phonon energy $h\nu_{zp,vu}$.

The curves in Figure 1 satisfy in general the equations,

$$E_v = E_{zp,vu} - \frac{1}{2}(\hbar\omega_v - \hbar\omega_u) + S_v\hbar\omega_v\left(\frac{x}{a} - 1\right)^2,$$

$$E_u = S_u\hbar\omega_u\left(\frac{x}{a}\right)^2, \tag{50}$$

and, for the special case of equal force constants, to which our attention will be limited in this chapter,

$$E_v = E_{zp,vu} + S_0\hbar\omega\left(\frac{x}{a} - 1\right)^2,$$

$$E_u = S_0\hbar\omega\left(\frac{x}{a}\right)^2, \tag{51}$$

where $\hbar\omega_u = \hbar\omega_v \equiv \hbar\omega$ and $S_u = S_v \equiv S_0$. The two equal-force-constants parabolas have their crossover at

$$\frac{x_c}{a} = \frac{1}{2}\left(\frac{E_{zp,vu} + S_0\hbar\omega}{S_0\hbar\omega}\right),$$

$$E_c = \frac{(E_{zp,vu} + S_0\hbar\omega)^2}{4S_0\hbar\omega}. \tag{52}$$

We call the ground state vibrational wavefunctions u_n and the offset state wavefunctions v_m.

Several attributes of the performance of this luminescence center are qualitatively evident in the figure. First, neglecting small corrections due to differences in the zero-point energies, there will be absorption centered at $h\nu_{abs} = h\nu_{zp,vu} + S_v\hbar\omega_v$ and emission centered at lower energy at $h\nu_{emis} = h\nu_{abs} - S_u\hbar\omega_u - S_v\hbar\omega_v$. Second, the shape of these bands will not change if the upper parabola is shifted to higher or lower energy without changing the offset. Third, the optical bands will be broad if the offset is considerable and narrow with the zero-phonon line dominant if the offset is very small. Fourth, as temperature increases and higher vibrational states are occupied, the band widths both in absorption and in emission will widen.

For this simple equal-force-constants center, the bandshapes are given most usefully as a distribution over p, the number of phonons of energy $\hbar\omega$ created by the transition having photon energy $h\nu_{vu,p}$ in absorption, $h\nu_{uv,p}$ in emission, according to

$$p\hbar\omega \equiv (m-n)\hbar\omega = h\nu_{vu,p} - h\nu_{zp,vu},$$

$$p\hbar\omega \equiv (n-m)\hbar\omega = hv_{zp,vu} - hv_{uv,p}, \tag{53}$$

in absorption and in emission respectively. All transitions $u_n \rightarrow v_m$ in absorption and $v_m \rightarrow u_n$ in emission with their indices following the Eq. (4) relationship share a common photon energy. The p index runs from $-\infty$ to ∞, with a positive value implying creating phonons. The maximum of this distribution will be near $p = S_0$.

The normalized bandshapes themselves will be given by Boltzmann-weighted Franck-Condon [22] factors, i.e., squared overlap integrals. This shape function will be

$$W_p = \sum_{n=max(0,-p)}^{\infty} (1-r)\, r^n\, |<v_{p+n}|u_n>|^2 = \sum_{n=max(0,-p)}^{\infty} (1-r)\, r^n A^2_{p+n,n},$$

$$W_p = \sum_{m=max(0,-p)}^{\infty} (1-r)\, r^m\, |<u_{p+m}|v_m>|^2 = \sum_{m=max(0,-p)}^{\infty} (1-r)\, r^m A^2_{p+m,m}, \tag{54}$$

in absorption and in emission, respectively. Here,

$$r = \exp(-\hbar\omega/kT) \tag{55}$$

is the Boltzmann factor and is related to Planck's measure of the temperature by

$$<m> \equiv \frac{\sum_m m r^m}{\sum_m r^m} = r(1-r)\frac{d}{dr}\left[\sum_m r^m\right] = r(1-r)\frac{d}{dr}\left(\frac{1}{1-r}\right)$$

$$= \frac{r}{1-r} = \frac{1}{\exp(\hbar\omega/kT) - 1}. \tag{56}$$

The $<m>$ will be extensively used here as the temperature parameter.

We will use Eq. (54) in this introduction to calculate W_p. However, explicit expressions for W_p as functions of S_0 and $<m>$, and that W_p is a normalized distribution, and recursion relations between W_p's with nearby indices are all derived in Chapter 4. Finally, the calculation of the A_{nm} matrix elements, using the Manneback recursion formulas which themselves are functions of S_0 and $<m>$, is described in Chapter 4. Figure 2 shows the general behavior of these W_p distributions for small ($S_0 = 1/9$), intermediate ($S_0 = 1$), and large ($S_0 = 9$) offsets at a low ($<m> = 1/9$, shown in dots) and a moderate ($<m> = 1$, shown as full curves) temperature.

At low temperature, for small offset the W_p distribution shows a strong zero-phonon line and small sidebands decreasing rapidly with phonon

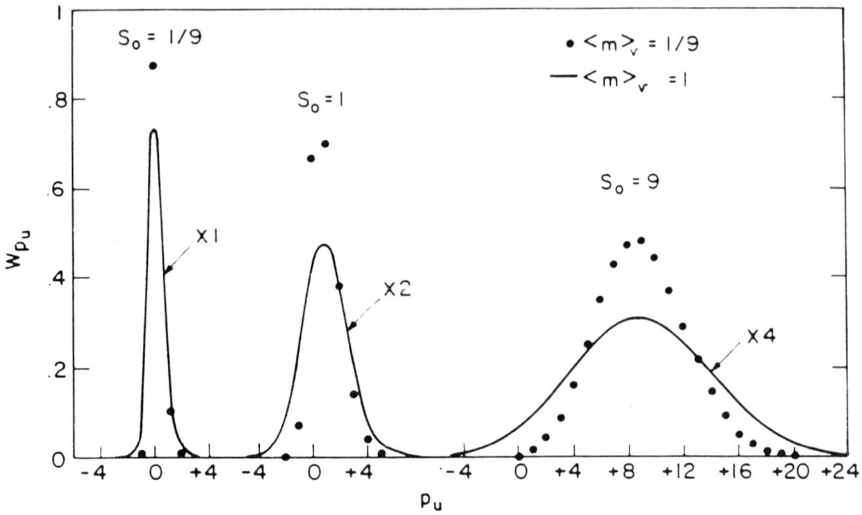

Figure 2. W_p band shapes

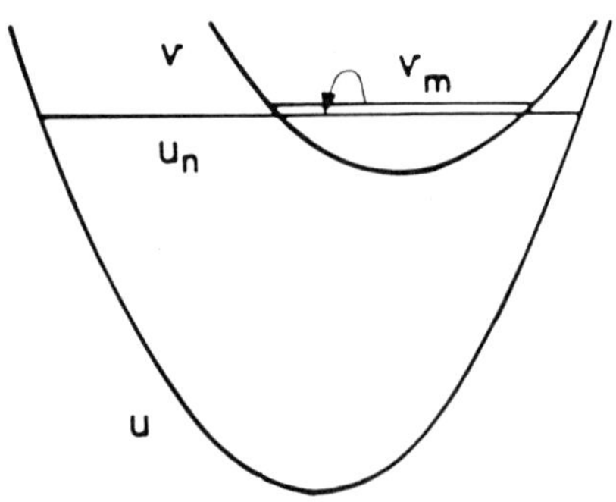

Figure 3. Typical Component of a Nonradiative Transition

number; for large offset the band is broad and somewhat Gaussian; for intermediate offset, one can see distinct assymetry between positive and negative p values favoring the positive (phonon generating) transitions as expected. At high temperature, at all offsets the bands broaden and lose significantly their assymetry.

2. The Nonradiative Rate. From Figure 1 one might also suspect that some nonradiative losses will occur after the excited state is occupied. The loss rate will be strongest when the parabolas intersect at the upper-state minimum. It might be expected that for the case pictured, the $v \rightarrow u$ nonradiative rate will increase with temperature and, as Mott[23] suggested, will be proportional to the Boltzmann factor for the availability of the crossover. The true picture is that the transition goes at some rate from each vibrational state m to that final state $n = p + m$ with closely the same energy, as pictured in Figure 3.

It will be found that the same sum of thermally weighted Franck-Condon factors,

$$R_{uv,p} = A_{uv} W_p \tag{57}$$

with the same W_p as in Eq. (54) and with $A_{uv} \approx 10^{12} - 10^{14}$, will give a very reasonable picture of the nonradiative $v \rightarrow u$ transition rate, where p is determined via

$$p\hbar\omega = h\nu_{zp,vu}, \tag{58}$$

i.e., by the analogue of Eq. (53) with zero photon energy. The average m through which the transition occurs, namely,

$$<m>_p = \sum_m m(1-r)r^m <u_{p+m}|v_m>^2. \tag{59}$$

is at only a small fraction of the crossover energy. We will derive expressions for $<m>_p$ in Chapter 4.

Of course, W_p and thus $R_{uv,p}$ are defined only for integer p values, i.e., only for certain relative placements of the two parabolas along the energy axis. Our practice is to interpolate logarithmically between integer p values for any intermediate placements. This practice assumes that the low frequency vibrations which must play the role of forcing energy conservation have only this interpolative effect on the nonradiative rate.

3. Six Typical Thermal Quenching Behaviors. The nonradiative rate $A_{uv}W_p$ is found to be exceedingly sensitive to the offset S_0, to the energy offset $h\nu_{zp,vu}$, to the temperature, and to differences in the curvatures of the two parabolas. We give here six typical behaviors of thermal quenching for equal curvatures; our discussion of the influence of different curvatures on quenching behavior is postponed to the experimental examples.

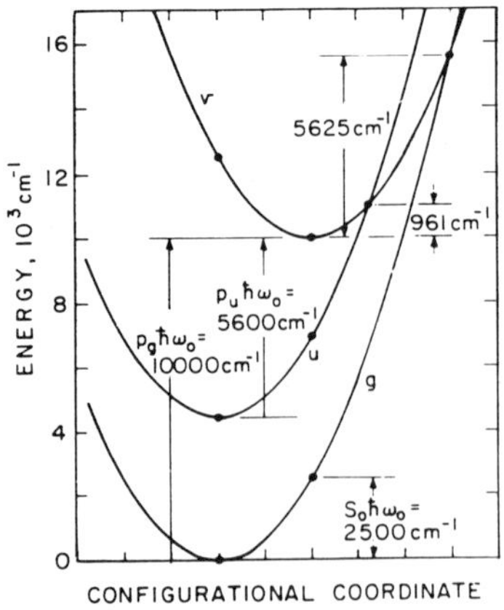

Figure 4. Bottom and Outside Crossovers

In all cases in this chapter we calculate the nonradiative rate using Eq. (57) and show the terms in W_p according to Eq. (54). We will also calculate $<m>_p$ in Eq. (59).

a. Fast Bottom Crossover. Figure 4 shows a center with a fast bottom crossover, where the crossover energy is $\approx 10^3 cm^{-1}$ or less, namely, the $v \rightarrow u$ transition. There will be no observable emission from the v state in such a center even at the lowest temperatures because the nonradiative $v \rightarrow u$ feeding overwhelms the emission rate.

Table 1 gives the description of this bottom crossover rate at four temperatures. The parameter values used in Eqs. (62), (59), and (57) are $\hbar\omega = 400 cm^{-1}$, $S_0 = 6.25$, $p = 14$, $A_{uv} = 10^{13} s^{-1}$.

The transition rate for the bottom crossover is seen from the W_p and $<m>_p$ values in Table 1 to be nearly temperature independent and fast. One can see in the four columns that the higher vibrational levels contribute more with increasing temperature. However, the crossover is at $m = 2.8$, and the transition actually takes place on the average at the $<m>_p$ indicated, below the $m = 2$ level even at moderately high temperatures.

The bottom crossover case describes qualitatively the fast Eu^{3+} CTS → ^5D feeding in oxysulfides[1,2,12] and fast Cr^{3+} $^4T_2$ → ^2E feeding in Al_2O_3 [24,10]

b. Outside Crossover. Table 2 gives the description of a typical outside crossover, the $v \rightarrow g$ transition in Figure 4. The parameter values here are $\hbar\omega = 400 cm^{-1}$, $S_0 = 6.25$, $p = 25$, $A_{uv} = 10^{13} s^{-1}$.

The outside crossover rate, as seen in Table 2, is strongly temperature dependent, rising by a factor of ~190 over the temperature range shown. The participation of higher vibrational levels with increasing temperature is clearly evident, the maxima in the contributions to the total rate being at $m = 0,1,2,5$, at 0K, 259K, 347K, and 547K, respectively.

TABLE 1
Description of a Fast Bottom Crossover

m	$A^2_{14+m,m}$	$(1-r)r^m A^2_{14+m,m}$			
		0K	259K	374K	547K
0	0.00307	0.00307	0.00274	0.00241	0.00200
1	0.01569		0.00151	0.00265	0.00357
2	0.04002		0.00042	0.00144	0.00317
3	0.06654		0.00007	0.00052	0.00184
4	0.07842		0.00001	0.00013	0.00076
5	0.06588			0.00002	0.00022
6	0.03617				0.00005
7	0.00875				0.00001
	W_{14}	0.00307	0.00475	0.00717	0.01162
	$<m>_{14}$		0.547	1.075	1.716
	$<m>$		0.122	0.273	0.536
	$kT/\hbar\omega$		0.45	0.65	0.95
	r		0.1084	0.2147	0.3490

TABLE 2
Description of an Outside Crossover

m	$A^2_{25+m,m}$	$10^7 \cdot (1-r)r^m A^2_{25+m,m}$			
		0K	259K	374K	547K
0	9.818-9	0.0982	0.0876	0.077	0.06
1	1.473-9		0.1423	0.248	0.34
2	1.139-6		0.1193	0.412	0.90
3	6.040-6		0.0686	0.296	1.67
4	2.468-5		0.0303	0.182	2.38
5	8.271-5		0.0111	0.098	2.79
6	2.365-4		0.0034	0.047	2.78
7	5.922-4		0.0009	0.020	2.43
8	1.324-3		0.0002	0.008	1.90
9	2.678-3			0.003	1.34
10	4.957-3			0.001	0.87
11	8.461-3				0.52
12	1.340-2				0.29
13	1.979-2				0.15
14	2.732-2				0.07
15	3.536-2				0.03
16	4.292-2				0.01
	$10^7 W_{25}$	0.0982	0.4637	2.274	18.53
	$<m>_{25}$		1.707	3.548	6.10
	$<m>$		0.122	0.273	0.54
	$kT/\hbar\omega$		0.45	0.65	0.95
	r		0.1084	0.2147	0.3490

The crossover energy is here near $m=14$, but the measure $<m>_{25}$ of the average m level used in the transition is only ~ 6 even at the highest temperature, where a center with a radiative rate as fast as $10^7 s^{-1}$ would be quenched.

The quantum efficiency of the outside crossover transition is

$$\eta_{gv} = \frac{R_{gv}}{R_{gv} + A_{gv} W_p} \tag{60}$$

and the transition lifetime is

$$\tau_v = \frac{1}{R_{gv} + A_{gv} W_p} = \frac{1}{R_{gv}} \eta_{gv}. \tag{61}$$

Since R_{gv} is usually nearly independent of temperature, Eq. (61) shows that η_{gv} and τ_v have the same temperature dependences. For a $10^4 s^{-1}$ radiative rate, the quenching will be almost complete even at 0K, while for a $10^7 s^{-1}$ radiative rate, the quenching will be to $\eta \approx 0.70$ at 259K, to $\eta \approx 0.30$ at 374K, and to $\eta \approx 0.05$ at 547K.

Radiative lifetimes range from $10^{-1} - 10^{-8}$ s. There is some correlation linking large offsets with shorter radiative lifetimes, because large offsets frequently imply significant changes in the electronic wavefunctions and therefore large parity-allowed electronic transition integrals.

This case describes qualitatively the competition between Cr^{3+} $^4T_2 \rightarrow {}^4A_2$ radiative and nonradiative transitions in Al_2O_3 [24,25,10]

c. Small-Offset Multiphonon Emission. Figure 5 shows a typical small-offset center.

Here we expect narrow line emission and absorption, with a dominant zero-phonon line and weak phonon sidebands. We anticipate from the early multiphonon-emission studies that there will be some weak nonradiative transitions. These early studies emerged independently of the treatment of nonradiative rates using the W_p function and came to an approximate form of this function.

Table 3 gives the description of the nonradiative transition for a typical small offset case, the $v \rightarrow u$ transition in Figure 5. The parameter values here are $\hbar\omega = 400 cm^{-1}$, $S_0 = 0.09$, $p = 6$, $A_{uv} = 0.8 \cdot 10^{12} s^{-1}$.

The small-offset transition has the same qualitative features as the outside crossover, in that the contributions to the total rate favor the higher vibrational states with increasing temperature. The difference is quantitative. The small-offset nonradiative rate has only a modest temperature dependence, only a factor of 12 from 0K to 547K vs. the factor of ~ 190 for the outside crossover transition of Table 2. In the small offset case, the crossover is above $m=100$, the transition occurs on average through levels lower than $m=4$, and the appeal of the Mott activation energy to the crossover is totally absent.

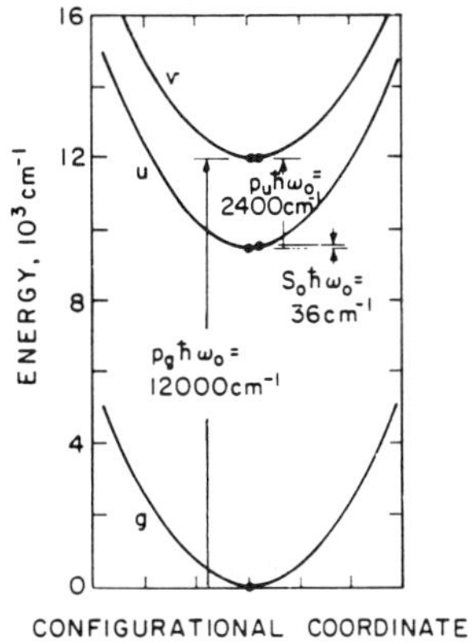

Figure 5. Small-Offset Case

This case describes the competitions between many 4f → 4f narrow line emissions and 4f → 4f nonradiative transitions[26-29]. Indeed, the first appeal for describing these nonradiative transitions was to the Kiel-Moos-Weber multiphonon emission rate alluded to above. We shall not take up here the analytic approximations which link Eq. (54) to this Kiel-Moos-Weber expression, beyond citing Eq. (17).

d. Two-Step Quenching With a Fast Second Step. Figure 6 shows a typical two-step quenching with a fast second step, namely, the $u \rightarrow v \rightarrow t$ sequence. This pattern of electronic states is frequently encountered and will be illustrated in Chapter 5 in oxysulfides activated with rare earths.

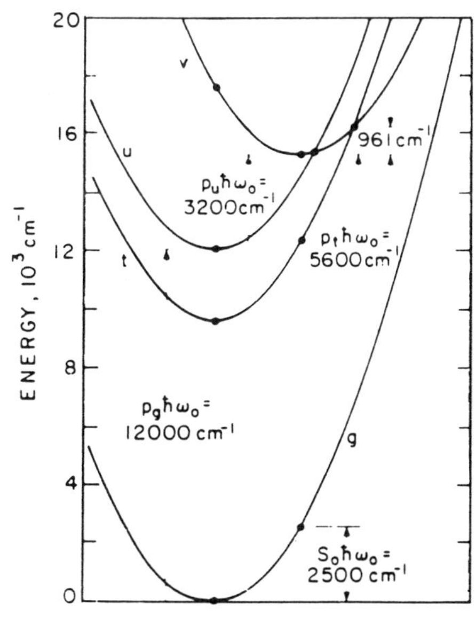

Figure 6. Two-Step Quenching through a Higher Offset State

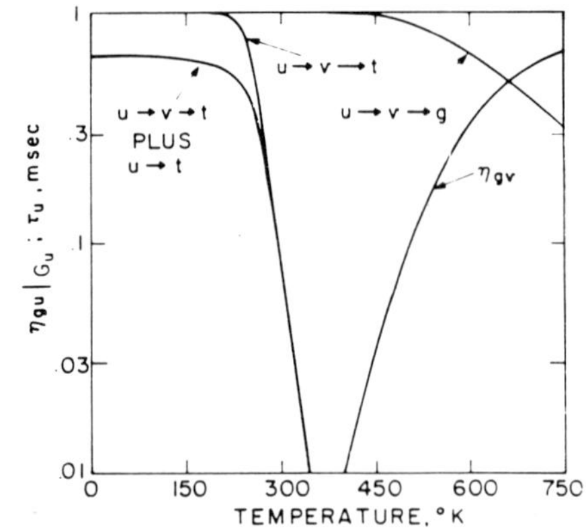

Figure 7. Quantum Efficiencies in Two-Step Quenching Processes with Fast and Slow Second Steps

The center shown has narrow-line absorptions and emissions at 9600 and 12000 cm^{-1} and a broad band absorption near 17700 cm^{-1} The narrow line emission at 12000 cm^{-1} will be precipitously quenched by the two-step process, as shown in Figure 7.

We will give now a notation for the needed rates and a derivation of the efficiency expression used to understand this temperature behavior. The notation $(i \rightarrow f)$ will be used for the rate of the transition $i \rightarrow f$ for unit population in i, and the notation $(i \rightarrow f_1, f_2, \ldots)$ will be used for the sum of the rates of the transitions $i \rightarrow f_1, i \rightarrow f_2 \ldots$

TABLE 3

Description of a Small-Offset Transition

m	$A^2_{6+m,m}$	$0.8 \cdot 10^9 (1-r)\, r^m A^2_{6+m,m}$			
		0K	259K	374K	547K
0	6.746-10	0.540	0.481	0.424	0.35
1	4.601-9		0.356	0.621	0.84
2	1.794-8		0.150	0.520	1.14
3	5.243-8		0.048	0.326	1.16
4	1.277-7		0.013	0.171	0.99
5	2.737-7		0.003	0.079	0.74
6	5.333-7		0.001	0.033	0.50
7	9.649-7			0.013	0.32
8	1.645-6			0.005	0.19
9	2.671-6			0.002	0.11
10	4.162-6				0.06
11	6.265-6				0.03
12	9.154-6				0.02
13	1.303-5				0.01
	$8 \cdot 10^9\, W_6$	0.540	1.0527	2.194	6.46
	$\langle m \rangle_6$		0.830	1.853	3.608
	$\langle m \rangle$		0.122	0.273	0.536
	$kT/\hbar\omega$		0.45	0.65	0.95
	r		0.1084	0.2147	0.3490

For the equal-force constants case, it will be shown in Chapter 4 that inverse rates are related by $(u \to v)/(v \to u) = r^p$. One also knows qualitatively that the bottom crossovers $v \to u,t$ are very fast.

Since excitation leaves the u,v system only by transitions to g and t, the quantum efficiency for $u \to g$ emission is always

$$\eta_{gu} = \frac{(u \to g)n_u}{(u \to g,t)n_u + (v \to g,t)n_v} = \left[1 + \frac{(u \to t)}{(u \to g)} + \frac{(v \to g,t)}{(u \to g)}\frac{n_v}{n_u}\right]^{-1}. \tag{62}$$

In Eq. (62), the populations n_u and n_v satisfy the steady-state rate equations

$$(u \to g,t,v)n_u - (v \to u)n_v = G_u,$$

$$-(u \to v)n_u + (v \to g,t,u)n_v = G_v, \tag{63}$$

where G_u and G_v are the excitation rates into u and v, respectively.

For narrow line G_u excitation ($G_v = 0$) and broad band G_v excitation ($G_u = 0$), the population ratios are found

$$\left.\frac{n_v}{n_u}\right|_{G_u} = \frac{(u \to v)}{(v \to g,t,u)} = \frac{(v \to u)}{(v \to g,t,u)}r^p,$$

$$\left.\frac{n_v}{n_u}\right|_{G_v} = \frac{(u \to g,t,v)}{(v \to u)} = \frac{(u \to g,t)}{(v \to u)} + r^p, \tag{64}$$

respectively. The efficiencies in Eq. (62) are found to be

$$\eta_{gu}|_{G_u} = \left[1 + \frac{(u \to t)}{(u \to g)} + \frac{(v \to g,t)(u \to v)}{(u \to g)(v \to g,t,u)}\right]^{-1}$$

$$= \left[1 + \frac{(u \to t)}{(u \to g)} + \frac{(v \to g,t)(v \to u)}{(u \to g)(v \to g,t,u)}r^p\right]^{-1}.$$

$$\eta_{gu}|_{G_v} = \frac{(v \to u)}{(v \to g,t,u)}\eta_{gu}|_{G_u} \tag{65}$$

The second of Eq. (65) requires after substituting the second of Eq. (64) into Eq. (62) the identity

$$\left[\frac{(v \to u)}{(v \to g,t,u)}\right]\left[1 + \frac{(u \to t)}{(u \to g)} + \frac{(v \to g,t)}{(u \to g)}\cdot\frac{(u \to g,t,v)}{(v \to u)}\right]$$

$$= \frac{(v \to u)}{(v \to g,t,u)} + \frac{(u \to t)}{(u \to g)}\cdot\frac{(v \to u)}{(v \to g,t,u)} + \frac{(v \to g,t)}{(u \to g)}\cdot\frac{(u \to g,t,v)}{(v \to g,t,u)}$$

$$= \frac{[(v\to u)\,(u\to g,t) + (v\to g,t)\,(u\to g,t)]}{(u\to g)\,(v\to g,t,u)} + \frac{(v\to g,t)\,(u\to v)}{(u\to g)\,(v\to g,t,u)}$$

$$= 1 + \frac{(u\to t)}{(u\to g)} + \frac{(v\to g,t)\,(u\to v)}{(u\to g)\,(v\to g,t,u)}, \tag{66}$$

or it may be directly derived by noting that $(v\to u)/(v\to g,t,u)$ is the fraction of G_v excitations that directly feed u and thus are equivalent to G_u excitations.

The lifetime of the u state, defined for G_u excitation, is, using the first expression in Eq. (62) in the second step,

$$\tau_u = \left.\frac{n_u + n_v}{(u\to g,t)n_u + (v\to g,t)n_v}\right|_{G_u} = \frac{n_u + n_v}{(u\to g)n_u}\eta_{gu}|_{G_u} = \frac{1}{(u\to g)}\left(1+\frac{n_v}{n_u}\right)\eta_{gu}|_{G_u} \tag{67}$$

Since n_v/n_u is always small in the examples here,

$$\tau_u \approx \frac{1}{(u\to g)}\eta_{gu}|_{G_u}, \tag{68}$$

and therefore τ_u and $\eta_{gu}|_{G_v}$ are both multiples of $\eta_{gu}|_{G_u}$ and it is sufficient to study $\eta_{gu}|_{G_u}$ for its temperature dependence.

In the first of Eq. (65) the second term in the denominator involving $(u\to t)$ is from the multiphonon emission $u\to t$ process. The third term in the denominator is from the two step $u\to v\to g,t$ processes. While one may use the W_p formula with negative p for the $(u\to v)$ rate, the expression of this term as $(v\to u)r^p$ allows a further useful approximation, namely that the factor $(v\to g,t)\,(v\to u)/(v\to g,t,u)$ is approximately the smaller of $(v\to g,t)$ and $(v\to u)$. For the example in Figure 6, where the v,t crossover is higher than the v,u crossover, $(v\to g,t)$ is smaller and is further approximated as $(v\to t)$ since the bottom crossover $v\to t$ is much faster than the outside crossover $v\to g$. Then Eq. (65) reduces to

$$\eta_{gu}|_{G_u} \approx \left[1 + \frac{(u\to t)}{(u\to g)} + \frac{(v\to t)}{(u\to g)}r^p\right]^{-1}. \tag{69}$$

Figure 7 shows the precipitous quenching caused by the two step $u\to v\to g,t$ processes. We have used here the illustrative values for the $u\to g$ narrow line emission $R_{gu} = 10^3 s^{-1}$, $\hbar\omega = 400 cm^{-1}$. Those for the $v\to t$ transition are $S_0\hbar\omega = 2500 cm^{-1}$, $h\nu_{zp,vt} = p_{vt}\hbar\omega = 5600 cm^{-1}$, $A_{tv} = 10^{13} s^{-1}$, $p_t = 14$, $S_0 = 6.25$. Those for the $u\to t$ multiphonon emission transition are $S_0\hbar\omega = 36 cm^{-1}$, $\hbar\omega = 400 cm^{-1}$, $h\nu_{zp,ut} = p_{ut}\hbar\omega = 2400 cm^{-1}$, $A_{tu} = 0.8\cdot 10^{12} s^{-1}$, $p_{ut} = 6$, $S_0 = 0.09$. These transitions have already been described in Tables 2 and 3. In Figure 7, the precipitous quenching is labeled $u \rightarrow v \rightarrow t$ *plus* $u \rightarrow t$. The quenching induced by $u \rightarrow v \rightarrow t$ itself is also shown. The $u \rightarrow t$ process produces non-unit quantum efficiency even at the lowest temperatures.

This particular case, namely, two-step quenching through a higher offset state, involves negative p values in the $u \rightarrow v$ step and is the most favorable for the Mott-activation-energy approximate form. Indeed, one can show in general that for equal force constants the rate $A_{uv}W_p$ can be approximated by the exponential expression $A\exp(-E_x/kT)$, where

$$E_x = \langle m \rangle_p - \langle m \rangle$$

$$A = A_{uv} A^2_{\langle m \rangle_p}. \tag{70}$$

Eq. (70) is derived by forcing the values and the first derivatives with respect to $(1/kT)$ to match, and by adopting the approximation that $W_p \approx$ its largest term, which will be nearly the squared overlap integral at the m through which on average the transition takes place, namely, the $m = \langle m \rangle_p$ of Eq. (59).

In our example, explicit evaluation of this approximate form gives

$$\eta_{gu}|_{G_u} \approx \left[1 + 0.8 \cdot 10^9\, W_6 + 1.24 \cdot 10^8 \exp(-8.430\hbar\omega/kT)\right]^{-1}. \tag{71}$$

In the evaluation Eq. (71), the approximated activation energy is $8.430\hbar\omega$, of which $8\hbar\omega$ is the zero phonon energy and the activation energy to the crossover would be $10.40\hbar\omega$. Thus the Mott activation energy is only ~20% too large and could be used, with an adjusted preexponential term, to give a reasonable fit to the transition rate.

This case describes qualitatively the precipitous sequential quenchings of Eu^{3+} emissions in oxysulfides[1,2,12], the ^5D Sm^{2+} emissions in BaClF[30], and the $^5D_3$ Tb^{3+} emissions in oxysulfides.

e. Two-Step Quenching with a Slow Second Step. "Slow" here means slow compared to bottom crossover rates, i.e., $< 10^{10} s^{-1}$. If one deletes the t curve in Figure 6, then the $u \rightarrow g$ emissions would be competitive with such a $u \rightarrow v \rightarrow g$ transition. The particular example in this figure has as the second step the broad band $v \rightarrow g$ emission, but it is equally encountered as a $v \rightarrow g$ thermally activated outside crossover.

The analysis in Section III.A.3.d leads now to

$$\eta_{gu}|_{G_u} = \eta_{gu}|_{G_v} = \left[1 + \{(v \rightarrow g)/(u \rightarrow g)\} r^p\right]^{-1} = \frac{1}{1 + (R_{gv}/R_{gu}) r^p},$$

$$\eta_{gu} + \eta_{gv} = 1. \tag{72}$$

This quantum efficiency is shown in Figure 7 by the curves labeled $u \rightarrow v \rightarrow g$. Also shown is the gradual buildup with temperature of the $v \rightarrow g$ emission intensity, the curve labeled η_{gv}.

This case describes qualitatively the slow replacement of $^2E \rightarrow A_2$ Cr^{3+} narrow line R emission in Al_2O_3 with broad band $^4T_2 \rightarrow {}^4A_2$ emission with increasing temperature.[24,25,10]. It also explains the replacement of Sm^{2+} $^5D_2 \rightarrow {}^7F$ narrow-line emissions with broad band $4f^55d \rightarrow {}^7F$ emission[30] in SrF_2:Sm^{2+} In this latter case[31], the $4f^55d$ state is only 550 cm^{-1} above the $^5D_0$ state, and the $4f^55d \rightarrow {}^7F$ radiative rate is fast (10^7 s^{-1}) compared to the $^5D_0 \rightarrow {}^7F$ rate (10^2 s^{-1}). The replacement is then steeper than the $u \rightarrow v \rightarrow g$ curve shown in Figure 7 and occurs at low temperature, 80K.

f. Low-temperature Tunnelling Crossover. Figure 8 shows a low-temperature tunnelling transition, namely, the $u \rightarrow v$ transition which competes with $u \rightarrow g$ emission.

We describe this $u \rightarrow v$ transition in Table 4, with the parameters $\hbar\omega = 200\text{cm}^{-1}$, $S_0 = 30.25$, $p = 2$, $A_{vu} = 10^{13}\text{s}^{-1}$.

This $u \rightarrow v$ tunnelling transition is significantly underestimated by the Mott activation energy expression with the crossover energy at 1307 cm^{-1} The crossover is at $n = 6.5$ while the average n utilized is near unity when a center with radiative rate 10^3 s^{-1} is being quenched. That this is so is evident in Table 4, since the radiative rate and nonradiative rate are then equal when $10^{10} W_2 = 1$, i.e., near 57K.

The tunnelling case describes qualitatively the low-temperature precipitous quenching of the Tm^{3+} $^3P_0$ (more likely $^1I_6$) $\rightarrow {}^3F_4$ narrow line emission in La_2O_2S[5,20].

IV. HARMONIC OSCILLATOR WAVEFUNCTIONS

IV.A. HERMITE POLYNOMIALS

The study of vibrations naturally leads to the harmonic oscillator. We will later obtain the harmonic oscillator wavefunctions in dimensionless variables. Our starting point is familiarity with Hermite polynomials, though we will derive the relation of these polynomials to the study of vibrations by demonstrating the solution of the Schrödinger equation.

We start from the generating function for these Hermite polynomials,

$$G_H(t,z_v) = e^{-t^2+2z_vt} = \sum_{m=0}^{\infty} \frac{t^m}{m!} H_m(z_v). \tag{73}$$

For familiarity one can note that

$$H_m(z_v) = \frac{\partial^m G_H}{\partial t^m}\Big|_{t=0} \tag{74}$$

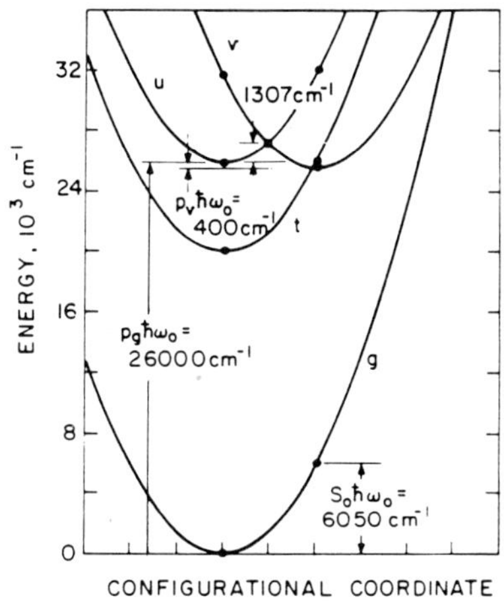

Figure 8. Low-Temperature Tunnelling Crossover

TABLE 4

Description of a Tunnelling Transition

n	$A^2_{n,2+n}$	$10^{10}(1-r)r^m A^2_{n,2+n}$				
		0K	57K	72K	86K	101K
0	3.334-11	0.333	0.331	0.327	0.32	0.32
1	8.253-9		0.552	1.484	2.84	4.47
2	6.520-7		0.294	2.147	8.00	20.27
3	2.286-5		0.070	1.379	10.01	40.82
4	4.040-4		0.008	0.446	6.31	41.43
5	3.775-3		0.000	0.076	2.10	22.23
6	1.845-2			0.007	0.37	6.24
7	4.298-2			0.000	0.03	0.84
8	3.482-2				0.00	0.04
9	0.061-2					0.00
	$10^{10} W_2$	0.333	1.255	5.856	29.98	136.7
	$\langle n \rangle_2$		1.101	2.066	2.903	3.570
	$\langle n \rangle$		0.007	0.019	0.036	0.061
	$kT/\hbar\omega$		0.20	0.25	0.30	0.35
	r		0.00674	0.01832	0.03567	0.05743

and, specifically,

$$H_0 = 1, \tag{75}$$

$$H_1 = 2z_v. \tag{76}$$

We get the recursion formula for the Hermite polynomials by differentiating Eq. (73) with respect to t:

$$\frac{\partial G}{\partial t} = G\cdot(-2t + 2z_v) = \sum_{m=1}^{\infty} \frac{t^{m-1}}{(m-1)!} H_m(z_v) = \sum_{m=0}^{\infty} \left[\frac{-2t^{m+1}}{m!} H_m + \frac{2z_v t^m}{m!} H_m \right] \tag{77}$$

The second and third expression in Eq. (77) can be equated using a common index:

$$0 = \sum_{k=0}^{\infty} t^k \left[\frac{-2H_{k-1}}{(k-1)!} + \frac{2z_v H_k}{k!} - \frac{H_{k+1}}{k!} \right]. \tag{78}$$

Equation (78) is an infinite series in t which is identically 0 and therefore each coefficient is itself 0. Therefore

$$-2kH_{k-1} + 2z_v H_k - H_{k+1} = 0. \tag{79}$$

Equation (79) is the desired recursion formula for the Hermite polynomials. With its use one can easily obtain

$$H_2 = 4z_v^2 - 2, \qquad H_3 = 8z_v^3 - 12z_v, \tag{80}$$

etc.

By differentiating Eq. (73) with respect to z_v, entering the sum in Eq. (73) for the exponential on the left hand side, and equating the coefficients of the powers of t, one obtains the expression

$$\frac{dH_n(z_v)}{dz_v} = 2nH_{n-1} \tag{81}$$

for the derivative of the Hermite functions.

IV.B. GENERATING FUNCTION FOR THE HARMONIC OSCILLATOR WAVEFUNCTIONS

The generating function for the harmonic oscillator wavefunctions can now be written down except for a normalization factor. We will include the normalization factor immediately and show later that the value chosen for it is correct.

$$G_v(t,z_v) = e^{-t^2 + 2z_v t + z_v^2/2} = \sum_{m=0}^{\infty} \frac{t^m}{m!} H_m(z_v)\, e^{-z_v^2/2} = \sum_{m=0}^{\infty} \frac{t^m}{m!} (m!2^m\pi^{1/2}x_v)^{1/2}\, v_m. \tag{82}$$

In Eq. (82), x_v has the dimension of length and its value will be given when v_m is shown below to satisfy the Schrödinger equation. This equation defines the normalized wavefunction as

$$v_m = H_m(z_v)\, e^{-z_v^2/2} (m!2^m \pi^{1/2} x_v)^{-1/2}. \tag{83}$$

We will first demonstrate the first three factors in the normalization constant of Eq. (83). We form the integral

$$\int_{-\infty}^{\infty} e^{-t^2+2tz_v-\frac{z_v^2}{2}} \cdot e^{-s^2+2sz_v-\frac{z_v^2}{2}} dz_v = \sum_{n,m=0}^{\infty} \frac{t^m s^n}{m!n!} \int_{-\infty}^{\infty} H_m(z_v)H_n(z_v)e^{-z^2} dz_v$$

$$= e^{2ts} \int_{-\infty}^{\infty} e^{-(z_v-t-s)^2} d(z_v - t - s) = \sqrt{\pi}\, e^{2ts} = \sqrt{\pi} \sum_k \frac{2^k t^k s^k}{k!}. \tag{84}$$

Equating the second and the fifth expression in Eq. (84), the ones involving the summations, gives the orthogonality expressions:

$$\int_{-\infty}^{\infty} H_m(z_v)H_n(z_v)e^{-z_v^2} dz_v = 0, \quad m \neq n,$$

$$= 2^k k! \sqrt{\pi}, \; m = n = k. \tag{85}$$

One might use the expression for H_3 in Eq. (80) and the known definite integrals of $z_v^k e^{-z_v^2}$ to confirm this normalization constant, namely, $2^3 \cdot 3! \sqrt{\pi}$. The first three factors in parentheses in Eq. (59) are now derived.

The x_v factor requires the demonstration of the solution to the Schrödinger equation which will be given below. To this end, we will obtain the expressions for the z operator and the d/dz operator, from them the raising and lowering operators, from them the differential equation satisfied by v_m. Finally, we show that this equation is indeed the Schrödinger equation for the harmonic oscillator in dimensionless units.

To obtain the z operator expression, one differentiates with respect to t the first and last expression for G_v in Eq. (82), replaces the exponential term in the derivative with this last expression,

$$(\pi^{1/2}x_v)^{1/2} \sum_{m=0}^{\infty} \left[\frac{-2t^{m+1}}{m!}(m!2^m)^{1/2} v_m + \frac{2t^m}{m!}((m!2^m)^{1/2} z_v v_m - \frac{mt^m - 1}{m!}(m!2^m)^{1/2} v_m \right] = 0, \tag{86}$$

eliminates the $(\pi^{1/2}x_v)^{1/2}$ prefactor, performs the indicated divisions,

$$\sum_{m=0}^{\infty} \left[-2t^{m+1}\left(\frac{2^m}{m!}\right)^{1/2} v_m + 2t^m \left(\frac{2^m}{m!}\right)^{1/2} z_v v_m - mt^{m-1}\left(\frac{2^m}{m!}\right)^{1/2} v_m \right] = 0, \tag{87}$$

sets the term in t^k equal to zero,

$$-\left(\frac{2^{k+1}}{(k-1)!}\right)^{1/2} v_{k-1} + \left(\frac{2^{k+2}}{k!}\right)^{1/2} z_v v_k - (k+1)\left(\frac{2^{k+1}}{(k+1)!}\right)^{1/2} v_{k+1} = 0, \qquad (88)$$

and, finally, divides by $(2^{k+1}/k!)^{1/2}$, obtaining

$$\sqrt{2}\, z_v v_k = \sqrt{k}\, v_{k-1} + \sqrt{k+1}\, v_{k+1}, \qquad (89)$$

the desired expression.

A related set of algebraic steps, viz., differentiating Eq. (82) with respect to z_v, adding twice this expression to Eq. (86) and thereby eliminating the z_v term, dividing by $(\pi^{1/2} x_v)^{1/2,}$ performing the indicated divisions, collecting terms in t^k and setting them equal to zero, and finally dividing by the factor $(2^{k+1} / k!)^{1/2}$, leads to the derivative expression

$$\sqrt{2} \frac{dv_k}{dz_v} = \sqrt{k}\, v_{k-1} - \sqrt{k+1}\, v_{k+1}. \qquad (90)$$

We now find the differential equation satisfied by the v_m. If one adds Eq. (89) to or subtracts it from Eq. (90) one gets the lowering or raising operators

$$O_- v_m = \left(\frac{d}{dz_v} + z_v\right) v_k = (2k)^{1/2} v_{k-1}, \qquad (91)$$

$$O_+ v_m = \left(\frac{d}{dz_v} - z_v\right) v_k = -(2k+2)^{1/2} v_{k+1}. \qquad (92)$$

The desired differential equation is obtained either by the operator O_+O_- or O_-O_+. For example,

$$O_-O_+ v_k = \left(\frac{d}{dz_v} + z_v\right)\left(\frac{d}{dz_v} - z_v\right) v_k$$

$$= \frac{d^2 v_k}{dz_v^2} - v_k - z_v^2 v_k = \left(\frac{d}{dz_v} + z_v\right)\left[-(2k+2)^{1/2} v_{k+1}\right] = -(2k+2) v_k \qquad (93)$$

or

$$\frac{d^2}{dz_v^2} v_k - z_v^2 v_k + (2k+1) v_k = 0. \qquad (94)$$

This is the Schrődinger equation for the harmonic oscillator, in dimensionless units, as we will now demonstrate. The dimensioned Schrődinger equation is

$$\left[\frac{\hbar^2}{2M}\frac{d^2}{dx^2} + E_{vm} - \tfrac{1}{2}k_v(x\text{-}a)^2\right]\Psi_m(x,k_v,a) = 0, \tag{95}$$

$$\text{with} \int_{-\infty}^{\infty} \Psi_m^2(x,k_v,a)dx = 1. \tag{96}$$

In Eqs. (95) and (96), Ψ has dimensions $l^{-1/2}$, x and a have dimensions l, k_v has dimensions mt^{-2}, and $\hbar$ is Planck's constant over 2π.

Let

$$E_{vm} = (m + \tfrac{1}{2})\hbar\omega_v, \tag{97}$$

$$\omega_v = \left(\frac{k_v}{M}\right)^{1/2}, \tag{98}$$

$$x_v^2 = \frac{\hbar}{\sqrt{Mk_v}}, \tag{99}$$

$$z_v = (x\text{-}a)/x_v, \tag{100}$$

$$\tfrac{1}{2}k_v x_v^2 = \tfrac{1}{2}\hbar\omega_v. \tag{101}$$

Then

$$dx = x_v dz_v, \tag{102}$$

$$\tfrac{1}{2}k_v(x\text{-}a)^2 = \tfrac{1}{2}k_v(z_v x_v)^2 = z_v^2\left(\frac{\hbar\omega_v}{2}\right), \tag{103}$$

$$\frac{\hbar^2}{2Mx_v^2} = \left(\frac{\hbar\omega_v}{2}\right)\left[\frac{\hbar}{Mx_v^2}\left(\frac{M}{k_v}\right)^{1/2}\right] = \left(\frac{\hbar\omega_v}{2}\right)\left[\frac{\hbar}{\sqrt{Mk_v}}\cdot\frac{1}{x_v^2}\right] = \frac{\hbar\omega_v}{2}. \tag{104}$$

Introducing Eqs. (96)-(104) into Eq. (95) gives Eq. (94) as asserted. The dimensionality of Ψ, as needed because of the dx term in Eqs. (96) and (102), forces the introduction of the $x_v^{1/2}$ term incorporated into the normalizing factor of v_m in Eq. (82).

When $a = 0$ equations analogous to Eqs. (97)-(104) with $v \rightarrow u$ are posed, e.g., for Eq. (100),

$$z_u = x/x_u. \tag{105}$$

Note especially that the z in Eq. (89) is z_v, namely, that z_v which relates Eqs. (94) and (95) correctly. We rewrite this equation both for the v and u functions:

$$\sqrt{2}\, z_v v_k = \sqrt{k}\, v_{k-1} + \sqrt{k+1}\, v_{k+1}, \tag{106}$$

$$\sqrt{2}\, z_u u_k = \sqrt{k}\, u_{k-1} + \sqrt{k+1}\, u_{k+1}, \tag{107}$$

This completes our derivation of the harmonic oscillator wavefunctions in the form used in this paper.

V. THE MANNEBACK RECURSION FORMULAS

V.A. INTRODUCTION

Manneback[32] first expounded the value of recursion formulas for studying harmonic oscillator wavefunctions. We give in this section his derivation, in our notation, for the recursion formulas for the overlap integrals involved in the optical transitions and, in our model, in the nonradiative transitions also.

We will consider unequal force constants and obtain the equal-force-constants results by particularizing the unequal-force-constants results. These recursion formulas for the overlap integrals will be shown to have importance beyond their use in computations. We will in later sections use them to obtain the recursion formula for the W_p and also many functions related to W_p.

V.B. THE OVERLAP INTEGRAL

The overlap integral in question is

$$A_{nm} = \int_{-\infty}^{\infty} u_n(z_u) v_m(z_v)\, dx = \langle u_n | v_m \rangle . \tag{108}$$

We recall that the dimensionality of u and v is $l^{-1/2}$ and of dx is l. We also relist the variables from Eqs. (7) and their u counterparts, namely,

$$z_u = x/x_u, \qquad z_v = (x-a)/x_v, \tag{109}$$

$$x_u^2 = \hbar/(Mk_u)^{1/2}, \qquad x_v^2 = \hbar/(Mk_v)^{1/2}, \tag{110}$$

$$\omega_u = (k_u/M)^{1/2}, \qquad \omega_v = (k_v/M)^{1/2}, \tag{111}$$

$$E_u = (n+\tfrac{1}{2})\hbar\omega_u, \qquad E_v = (m+\tfrac{1}{2})\hbar\omega_v. \tag{112}$$

V.C. THE GENERATING FUNCTION FOR THE OVERLAP INTEGRAL

The Manneback recursion formulas are derived from the generating function for the overlap integrals, which in turn are derived from the generating function of the harmonic oscillator wavefunctions, applied to each electronic state:

$$G_v(z_v) = e^{-t^2 + 2tz_v - z_v^2/2} = \sum_{m=0}^{\infty} \frac{t^m}{m!}\left[m!2^m x_v \pi^{1/2}\right]^{1/2} v_m(z_v). \qquad (113)$$

$$G_u(z_u) = e^{-s^2 + 2sz_u - z_u^2/2} = \sum_{n=0}^{\infty} \frac{t^n}{n!}\left[n!2^n x_u \pi^{1/2}\right]^{1/2} u_n(z_u). \qquad (114)$$

To derive the generating function for the overlap integrals, we will define

$$\left(1/x_{uv}^2\right) = \left(1/x_u^2\right) + \left(1/x_v^2\right), \qquad (115)$$

$$a_{uv} = a/x_{uv}, \qquad (116)$$

$$\tan\theta = \left(\frac{k_v}{k_u}\right)^{1/4} = \left(\frac{\omega_v}{\omega_u}\right)^{1/2} = \left(\frac{1/x_v}{1/x_u}\right) = \frac{x_u}{x_v}. \qquad (117)$$

From Eqs. (115) and (117) and the trigonometric identity, $\sec^2\theta = 1 + \tan^2\theta$, one sees

$$\frac{x_{uv}}{x_u} = \cos\theta, \qquad \frac{x_{uv}}{x_v} = \sin\theta. \qquad (118)$$

We now set down the generating function between the overlap integrals. Each step of this equation will be explained immediately after the equation itself is given.

$$G_A^{\cdot} = \int_{-\infty}^{\infty} G_v G_u dx = \sum_m \sum_n \frac{t^m}{m!}\frac{s^n}{n!}\left[m!n!2^{m+n} x_u x_v \pi\right]^{1/2} \int_{-\infty}^{\infty} u_n v_m dx$$

$$= \int_{-\infty}^{\infty} \exp\left[-t^2 + 2t\left(\frac{x-a}{x_v}\right) - ½\left(\frac{x-a}{x_v}\right)^2 - s^2 + 2s\left(\frac{x}{x_u}\right) - ½\left(\frac{x}{x_u}\right)^2\right] dx$$

$$= \int_{-\infty}^{\infty} \exp\left[-t^2 + 2t\,(y-a_{uv})\sin\theta - ½\sin^2\theta\,(y-a_{uv})^2\right] x_{uv}dy$$

$$- \int_{-\infty}^{\infty} \exp\left[s^2 + 2sy\cos\theta - ½\cos^2\theta\, y^2\right] x_{uv}dy$$

$$= \int_{-\infty}^{\infty} \exp\left[-½y^2 + y\,(2t\sin\theta + a_{uv}\sin^2\theta + 2s\cos\theta)\right]$$

$$\cdot \left[\,(-t^2 - ½a_{uv}^2\sin^2\theta - s^2 - 2ta_{uv}\sin\theta)\right] x_{uv}dy$$

$$= \exp\left[t^2 - ½\sin^2\theta\, a_{uv}^2 - s^2 - 2ta_{uv}\sin\theta + ½(2t\sin\theta + a_{uv}\sin^2\theta + 2s\cos\theta)^2\right]$$

$$\cdot \int_{-\infty}^{\infty} \exp\left[- ½ \left\{y - (2t\sin\theta + a_{uv}\sin^2\theta + 2s\cos\theta)\right\}^2\right] x_{uv}dy$$

$$= \sqrt{2\pi}\, x_{uv} \exp\left[½ a_{uv}^2(- \sin^2\theta + \sin^4\theta)\right] \cdot\exp\left[- t^2(1-2\sin^2\theta) - s^2(1-2\cos^2\theta)\right.$$

$$\left. + t(-2a_{uv}\sin\theta + 2a_{uv}\sin^3\theta) + s(2a_{uv}\sin^2\theta\cos\theta) + st(4\sin\theta\cos\theta)\right]$$

$$= x_{uv}\sqrt{2\pi}\exp(-\frac{1}{8}a_{uv}^2\sin^2 2\theta)e^{\delta_0}$$

$$= \sum_m\sum_n t^m s^n \left[\frac{2^{m+n}\pi}{m!n!}\right]^{\frac{1}{2}} \left[\frac{x_{uv}}{(\sin\theta\cos\theta)^{1/2}}\right] A_{nm}. \qquad (119)$$

The first line of Eq. (119) introduces the second expressions in Eqs. (113) and (114) for G_v and G_u. The second line introduces the first expressions in these equations; the third line uses the expressions for z_v and z_u in Eq. (109), the definition of a_{uv} in Eq. (116), the θ relations of Eq. (117), the definition

$$y = x/x_{uv} \qquad (120)$$

and its differential form, and the identities

$$(x-a)/x_v = (y-a_{uv})\cos\theta, \qquad (121)$$

$$x/x_u = y\sin\theta. \qquad (122)$$

The fourth line groups the terms in powers of y. The fifth line completes the square in the fourth expression and separates the terms independent of the integrating variable; the sixth line uses the integral

$$\int_{-\infty}^{\infty} \exp\left[- ½(y-b)^2\right]dy = 2\pi \qquad (123)$$

and collects terms in powers of t and s. The seventh line uses the definition

$$\delta_0 = -t^2\cos 2\theta + s^2\cos 2\theta - t(a_{uv}\sin 2\theta\cos\theta) + s(a_{uv}\sin 2\theta\sin\theta) + 2st\sin 2\theta \qquad (124)$$

and the trigonometric identities

$$\sin^4\theta - \sin^2\theta = -\frac{1}{4}\sin^2 2\theta, \tag{125}$$

$$\sin^3\theta - \sin\theta = -\tfrac{1}{2}\sin 2\theta\cos\theta, \tag{126}$$

$$\sin^2\theta\cos\theta = \tfrac{1}{2}\sin 2\theta\sin\theta. \tag{127}$$

The eighth line repeats the expression in the first line and uses the notation of Eq. (108), A_{nm}, for the overlap integral.

If now we equate the last two expressions of Eq. (119), cancel the $x_{uv}\pi^{1/2}$ term, and transfer the $(\sin\theta\cos\theta)^{1/2}$ term, we obtain

$$G_A = \frac{(\sin\theta\cos\theta)^{1/2}}{x_{uv}\pi^{1/2}} G_A^{\cdot} = \sqrt{\sin 2\theta}\exp\left[-\frac{1}{8}a_{uv}^2\sin^2 2\theta\right] e^{\delta_0}$$

$$= \sum_m \sum_n t^m s^n \left[\frac{2^{m+n}}{m!n!}\right]^{\frac{1}{2}} A_{nm}, \tag{128}$$

the desired generating function for the overlap integrals A_{nm}. In Eq. (128), δ_0 is given by Eq. (124).

V.D. THE RECURSION FORMULAS FOR THE OVERLAP INTEGRALS

We now derive two recursion formulas for the overlap integrals by taking derivatives of the generating function for them, Eq. (128), with respect to t and to s. After doing so, we replace the G_A with its Eq. (128) expression:

$$(-2t\cos 2\theta + 2s\sin 2\theta - a_{uv}\cos\theta\sin 2\theta)\sum_{m,n} t^m s^n \left[\frac{2^{m+n}}{m!n!}\right]^{1/2} A_{nm} = \sum_{m,n} m s^n t^{m-1}\left[\frac{2^{m+n}}{m!n!}\right]^{1/2} A_{nm}, \tag{129}$$

$$(2s\cos 2\theta + 2t\sin 2\theta + a_{uv}\sin\theta\sin 2\theta)\sum_{m,n} t^m s^n \left[\frac{2^{m+n}}{m!n!}\right]^{1/2} A_{nm} = \sum_{m,n} n s^{n-1} t^m \left[\frac{2^{m+n}}{m!n!}\right]^{1/2} A_{nm}. \tag{130}$$

We extract from each of these equations the term in $s^k t^l$, getting

$$-2\cos 2\theta\left[\frac{2^{k+l-1}}{k!(l-1)!}\right]^{1/2} A_{k,l} - 1 + 2\sin 2\theta\left[\frac{2^{k+l-1}}{(k-1)!l!}\right]^{1/2} A_k - 1,l\text{-}a_{uv}\cos\theta\sin 2\theta\left[\frac{2^{k+l}}{k!l!}\right]^{1/2} A_{k,l}$$

$$= (l+1)\left[\frac{2^{k+l+1}}{k!(l+1)!}\right]^{1/2} A_{k,l+1}$$

(131)

$$2\cos 2\theta\left[\frac{2^{k+l-1}}{l!(k-1)!}\right]^{1/2} A_{k}-1,l+2\sin 2\theta\left[\frac{2^{k+l-1}}{(l-1)!k!}\right]^{1/2} A_{k,l}-1\text{-}a_{uv}\sin\theta\sin 2\theta\left[\frac{2^{k+l}}{k!l!}\right]^{1/2} A_{k,l}$$

$$= (k+1)\left[\frac{2^{k+l+1}}{l!(k+1)!}\right]^{1/2} A_{k+1,l}$$

(132)

Dividing by the common factor, $(2^{l+k+1}/k!l!)^{1/2}$, we obtain the two Manneback recursion formulas

$$-\cos 2\theta\sqrt{l}\,A_{k,l-1} + \sin 2\theta\sqrt{k}\,A_{k-1,l} - \frac{a_{uv}}{\sqrt{2}}\cos\theta\sin 2\theta\,A_{kl} = \sqrt{l+1}\,A_{k,l+1}, \tag{133}$$

$$\cos 2\theta\sqrt{k}\,A_{k-1,l} + \sin 2\theta\sqrt{l}\,A_{k,l-1} + \frac{a_{uv}}{\sqrt{2}}\sin\theta\sin 2\theta A_{kl} = \sqrt{k+1}\,A_{k+1,l}. \tag{134}$$

A better notation comes from replacing $k,l \rightarrow m,n$ and introducing

$$k = \cos 2\theta, \tag{135}$$

$$k_{+} = \sin 2\theta, \tag{136}$$

$$S_u^{1/2} = \frac{a_{uv}\cos\theta}{\sqrt{2}}, \qquad S_v^{1/2} = \frac{a_{uv}\sin\theta}{\sqrt{2}}, \qquad \frac{S_v}{S_u} = \tan^2\theta = \frac{\hbar\omega_v}{\hbar\omega_u}, \tag{137}$$

$$a_{uv} = \{2(S_u+S_v)\}^{1/2}, \qquad \theta = \tan^{-1}\left[\left(\frac{S_v}{S_u}\right)^{1/2}\right]. \tag{138}$$

Both $S_u^{1/2}$ and $S_v^{1/2}$ are signed quantities, carrying the sign of a_{uv}. The two recursion formulas then are, in a rearranged order,

$$k_{+}\sqrt{n}\,A_{n-1,m} = \sqrt{m+1}\,A_{n,m+1} + k_{+}S_u^{1/2}A_{nm} + k\sqrt{m}\,A_{n,m-1}, \tag{139}$$

$$k_{+}\sqrt{m}\,A_{n,m-1} = \sqrt{n+1}\,A_{n+1,m} - k_{+}S_v^{1/2}A_{nm} - k\sqrt{n}\,A_{n-1,m}. \tag{140}$$

The equal force constants equations are obtained from Eqs. (139) and (140) by setting $\theta = 45°$, $k=0$, $k_{+}=1$, $S_u = S_v \equiv S = a_{uv}^2/4$,

$$\sqrt{n}\,A_{n-1,m} = \sqrt{m+1}\,A_{n,m+1} + S^{1/2}A_{nm}, \tag{141}$$

$$\sqrt{m}\,A_{n,m-1} = \sqrt{n+1}\,A_{n+1,m} - S^{1/2}A_{nm}. \tag{142}$$

For familiarity, from Eq.. (128), A_{00} for arbitrary θ is

$$A_{00} = (\sin 2\theta)^{1/2} \exp\left[-\frac{1}{8} a_{uv}^2 \sin^2 2\theta \right] \tag{143}$$

and for $\theta = 45°$ is

$$A_{00} = e^{-S/2}. \tag{144}$$

The A_{01} and A_{10} can be obtained directly from Eq. (129) and (130), respectively, evaluated at $t=s=0$. They are

$$A_{01} = -(a_{uv}\cos\theta \sin 2\theta/\sqrt{2})A_{00} = -k_{+}S_u^{1/2}A_{00} \tag{145}$$

and

$$A_{10} = (a_{uv}\sin\theta \sin 2\theta/\sqrt{2})A_{00} = k_{+}S_v^{1/2}A_{00}. \tag{146}$$

These equations can also be obtained somewhat heuristically from Eqs. (133) and (134) or (139) and (140) by setting all A terms with negative indices to zero. This procedure is indeed always correct, and all A_{0m} and A_{n0} are readily obtained by means of three-term recursion formulas, and also A_{11} by either of two such three-term recursion formulas. Then all other rows and columns of the A matrix can be obtained in turn. This use of the recursion formulas reduces by orders of magnitude the work of getting the A matrix with respect to evaluating them from some explicit equation, such as that of Hutchisson[33]

The u_n and v_m wavefunctions are symmetric and antisymmetric when their index is even or odd, respectively. Therefore, from visualizing the two displaced wavefunctions, one can see that A_{10} and A_{01} are equal but with opposite signs when the force constants are equal, and are unequal but still have the opposite signs when the force constants are unequal. In general, these properties are true for all A_{nm} when n-m is odd. When n-m is even, they are identical with equal force constants and their signs are the same with unequal force constants.

These equal-force-constants relationships are valuable enough to state explicitly:

$$\langle u_\alpha | v_\beta \rangle = (-1)^{\alpha-\beta} \langle u_\beta | v_\alpha \rangle = \langle v_\beta | u_\alpha \rangle = (-1)^{\alpha-\beta} \langle v_\alpha | u_\beta \rangle. \tag{147}$$

V.E. ADDITIONAL EQUAL-FORCE-CONSTANTS RECURSION RELATIONS

For equal force constants, one can obtain recursion formulas for other combinations of the A_{nm} from the Manneback recursion formulas, Eqs. (141) and (142). A simple diagram will elucidate the logic of this procedure.

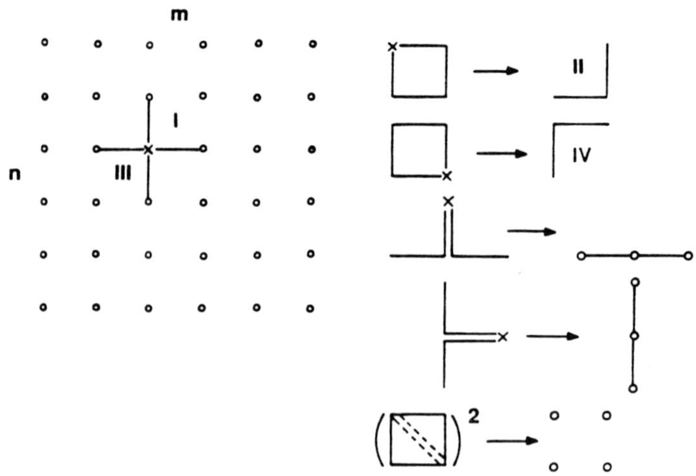

Figure 9. Recursion Relations

Figure 9 diagrams Eqs. (141) and (142) on a grid of the A_{nm} matrix. The two Manneback recursion formulas relate matrix elements most picturesquely labeled as by quadrants, I and III, respectively.

The several picture equations sketched beside the grid show combinations of these equations which share a specific matrix element in common. When this shared matrix element is eliminated, then the other orientations sketched are obtained.

To proceed according to the first picture equation, if we subtract $\sqrt{m}$ times Eq. (142) with $n \rightarrow n\text{-}1$ from $\sqrt{n}$ times Eq. (141) with $m \rightarrow m\text{-}1$, the term in $A_{n-1,m-1}$ is eliminated and we obtain

$$(m-n)A_{n,m} + \sqrt{S}\left(\sqrt{m}\,A_{n,m-1} + \sqrt{n}\,A_{n-1,m}\right) = 0 \tag{148}$$

or

$$\left(\frac{n-m}{\sqrt{S}}\right)A_{n,m} - \sqrt{m}\,A_{n,m-1} - \sqrt{n}\,A_{n-1,m} = 0. \tag{149}$$

Similarly, according to the second picture equation, subtracting $\sqrt{m+1}$ times Eq. (142) with $m \rightarrow m+1$ from $\sqrt{n+1}$ times Eq. (141) with $n \rightarrow n+1$ eliminates the $A_{n+1,m+1}$ term and leads to

$$\left(\frac{n-m}{\sqrt{S}}\right)A_{n,m} = \sqrt{n+1}A_{n+1,m} + \sqrt{m+1}\,A_{n,m+1} \tag{150}$$

The third picture equation combines Eq. (141) and Eq. (150) to eliminate the term in $A_{n-1,m}$ and gives

$$\left(\frac{n\text{-}m\text{-}S}{\sqrt{S}}\right)A_{nm} = \sqrt{m}\,A_{n,m-1} + \sqrt{m+1}\,A_{n,m+1} = \sqrt{2}<u_n|z_v|v_m>, \tag{151}$$

as is evident from Eq. (3).

The fourth picture equation combines Eq. (141) and Eq. (151) to eliminate the term in $A_{n,m+1}$ and gives

$$\left(\frac{m\text{-}n+S}{\sqrt{S}}\right)A_{nm} = \sqrt{n+1}\,A_{n+1,m} + \sqrt{n}\,A_{n-1,m} = \sqrt{2}\,<v_m|z_u|u_n>. \tag{152}$$

Expressions for the derivative operator analogous to Eqs. (151) and (152) can be obtained from Eq. (4), namely,

$$<u_n|\frac{d}{dz_v}|v_m> = A_{n,m,d/dz_v} = \sqrt{\frac{m}{2}}\,A_{n,m-1} - \sqrt{\frac{m+1}{2}}\,A_{n,m+1}, \tag{153}$$

$$<v_m|\frac{d}{dz_u}|u_n> = A'_{m,n,d/dz_u} = \sqrt{\frac{n}{2}}\,A_{n-1,m} - \sqrt{\frac{n+1}{2}}\,A_{n+1,m}. \tag{154}$$

V.F. THE W_p RECURSION FORMULA

We return now to the fifth picture equation of Figure 9. The open circles represent squares of the corresponding matrix element. By squaring the two recursion formulas sketched in such a way that the cross terms have common indices, one can eliminate this cross term and be left with a four-term relationship between squared A_{nm} terms. Thus, Eq. (141) and (142) are rewritten with the $S^{1/2}A_{nm}$ term isolated to force the correct cross terms. The relationship between squares which results will lead directly to the recursion formula for the W_p. The result is

$$(n\text{-}m)(A^2_{n-1,m-1} - A^2_{nm}) = S\,(A^2_{n,m-1} - A^2_{n-1,m}). \tag{155}$$

We then set $n\text{-}m = p$ and operate upon Eq. (155) with $\sum_m^{\infty}(1-r)r^m$ with the lower m limit specified either to

$$m_0 = max(0,-p), \qquad m_1 = max(1,-p). \tag{156}$$

dependent upon whether the $m=0$ term does not or does, respectively, lead to negative indices. The result is

$$S\sum_{m=m_1}^{\infty} r^m\,(1-r)\,A^2_{p+m,m-1} - S\sum_{m=m_0}^{\infty} r^m\,(1-r)\,A^2_{p+m-1,m}$$

$$= p\sum_{m=m_1}^{\infty} r^m\,(1-r)\,A^2_{p+m-1,m} - p\sum_{m=m_0}^{\infty} r^m\,(1-r)\,A^2_{p+m,m}. \tag{157}$$

Now, using the definition of W_p

$$W_p = \sum_{m=m_0}^{\infty} (1-r)\, r^m A^2_{p+m,m} \tag{158}$$

and redefining the summation index as required, we obtain from Eq. (157)

$$SrW_{p+1} - S\,W_{p-1} = r\,p\,W_p - p\,W_p. \tag{159}$$

For example, the first sum in Eq. (157) uses, in turn, $p+m = p+1+m\text{-}1$, $m\text{-}1 \rightarrow m'$, and $(m=m_1) \rightarrow (m'=m_0)$. We now use, recalling Eq. (56),

$$<m> = \frac{r}{1-r}, \qquad r = \frac{<m>}{<1+m>}, \qquad 1-r = \frac{1}{<1+m>}, \tag{160}$$

to get the desired recursion formula,

$$S <m> W_{p+1} + p\,W_p - S<1+m>\,W_{p-1} = 0. \tag{161}$$

The $<m>$ is Planck's measure of the temperature.

V.G. EXPLICIT SERIES EXPRESSION FOR THE W_p FUNCTION

The explicit expression for W_p is

$$W_p = \exp(-S<2m+1>) \sum_{j=\max(0,-p)}^{\infty} \frac{(S<m>)^j (S<1+m>)^{p+j}}{j!(p+j)!}. \tag{162}$$

This can be proved by substituting Eq. (162) into the recursion formula Eq. (161) and demonstrating that indeed the result is identically zero. Thus we are to prove that

$$\begin{aligned} &S <m> \sum_{j=\max(0,-p-1)}^{\infty} \frac{(S<m>)^j (S<1+m>)^{p+1+j}}{j!(p+1+j)!} \\ &+ p \sum_{j=\max(0,-p)}^{\infty} \frac{(S<m>)^j (S<1+m>)^{p+j}}{j!(p+j)!} \\ &- S <1+m> \sum_{j=\max(0,-p+1)}^{\infty} \frac{(S<m>)^j (S<1+m>)^{p-1+j}}{j!(p-1+j)!} = 0. \end{aligned} \tag{163}$$

We must consider three cases separately. The first is for $p>0$, when we extract the term from Eq. (153) in $(S<m>)^k (S<1+m>)^{p+k}$. Its coefficient is

$$\left[\frac{1}{(k-1)!(p+k)!}+\frac{p}{k!(p+k)!}-\frac{1}{k!(p-1+k)!}\right]=\frac{k+p-(p+k)}{k!(p+k)!}=0, \quad k>0,$$

$$\left[\frac{p}{p!}-\frac{1}{(p-1)!}\right]=0, \qquad k=0. \tag{164}$$

The second case is for $p=0$. Then the coefficient of $(S<m>)^k(S<1+m>)^k$ is

$$\left[\frac{1}{(k-1)!(k)!}-\frac{1}{k!(k-1)!}\right]=0. \tag{165}$$

The third case is for $p<0$. Then the coefficient of $(S<m>)^k(S<1+m>)^{p+k}$. is

$$\left[\frac{1}{(k-1)!0!}+\frac{-k}{k!0!}\right]=0, \qquad k=-p,$$

$$\left[\frac{1}{(k-1)!(p+k)!}+\frac{p}{k!(p+k)!}-\frac{1}{k!(p-1+k)!}\right]=\frac{k+p-(p+k)}{k!(p+k)!}=0, \quad k>-p. \tag{166}$$

Thus, in all possible cases, Eq. (153) is shown identically zero. It must next be proved that the expression Eq. (152) is the normalized solution of the recursion formula Eq. (151). To this end, we note that

$$\sum_{p=-\infty}^{\infty}\sum_{j=\max(0,-p)}^{\infty}\frac{(S<m>)^j(S<1+m>)^{p+j}}{j!(p+j)!}=\sum_{j=0}^{\infty}\sum_{p+j=k=0}^{\infty}\frac{(S<m>)^j(S<1+m>)^k}{j!(k)!}$$

$$=\exp(S<m> + S<1+m>) \tag{167}$$

and therefore Eq. (162) is the normalized solution of the recursion formula Eq. (161) and is a valid expression for W_p.

VI. ENERGY TRANSFER

The theories of energy transfer in the literature are concerned with the distance between the energy donor and acceptor, distributions for these distances as controlled by concentration, averaging over these distances, etc. These concerns enter the electronic factor of the transition rate. Our concern is with the nuclear factor, i.e., with the energy mismatch dependence of the rate and with the temperature dependence of these rates. Insofar as the previous authors treat specifically these two dependences they have come to approximate forms of the $W_p, W_{p,d/dz,}$ or $M_{p,d/dz}$ rates. In these treatments, and in ours also, the energy donor and the energy acceptor are treated as having vibrational modes totally independent of each other.

Our model is illustrated in Figure 10. Figure 10 shows one possible transition from donor to acceptor, namely, $d_m^*, a_l \rightarrow d_n, a_k^*$. This transition con-

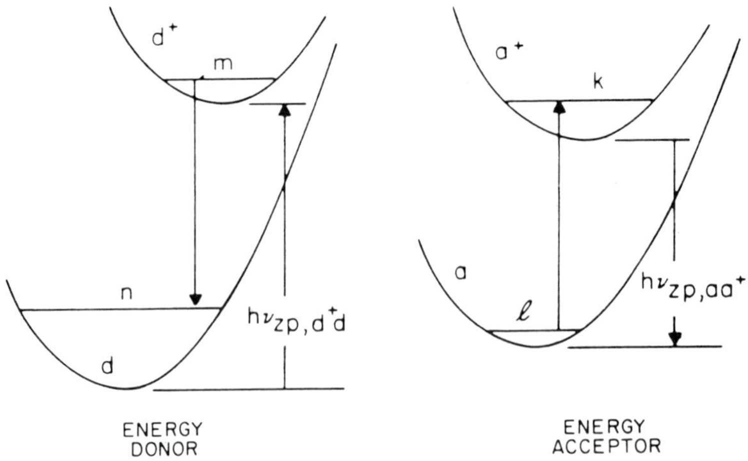

Figure 10. The Model of Energy Transfer from Donor to Acceptor

serves energy but generates p_{da} phonons in accordance to the energy balances,

$$p_{da}\hbar\omega = [(n-m)+(k-l)]\hbar\omega = h\nu_{zp,d^+d} - h\nu_{zp,a^+a} \tag{168}$$

The thermal Franck-Condon weight of all d→a transitions generating p_{da} phonons is

$$W_{p_{da}} = \sum_{m,l,n,k=0}^{\infty} (1-r_0)r_0^m(1-r_0)r_0^l <d_n|d_m^+>^2 <a_k|a_l^+>^2, \tag{169}$$

where the m,n,k,l indices are constrained to follow Eq. (168). Eq. (169) is recognized through Eq. (13) as

$$W_{p_{da}} = \sum_{p_d=-\infty}^{\infty} W_{p_d}(S_d, <m>)W_{p_a}(S_a, <m>), \tag{170}$$

where the indices are constrained by

$$p_d + p_a = p_{da}, \tag{171}$$

and is further recognized through Eq. (32) as itself a W_p function, namely,

$$W_{p_{da}}(S_d+S_a, <m>) = \sum_{p_d=-\infty}^{\infty} W_{p_d}(S_d, <m>)W_{p_a}(S_a, <m>) \tag{172}$$

We will have occasion also to consider transfer between centers with different θ, a_{uv} values. For example we are interested in energy transfer between a broad-band center whose emission is described with a U_{p_U}

function of Eq. (25) and a narrow-line center whose absorption is described with a small-S W_p function. In general, these distributions can be combined on any phonon energy metric by interpolating each distribution at the metric energies. We choose to use the vibrational levels of the acceptor and obtain the distribution

$$C_{p_{da}} = \sum_{p,p'} W_p U_{p'},$$

$$p\hbar\omega + p'\hbar\omega_u = p_{da}\hbar\omega, \tag{173}$$

to give a measure of the transfer rate.

This completes our formulas for energy transfer.

VII. RARE EARTHS IN OXYSULFIDES

VII.A. THE Eu ENERGY LEVEL DIAGRAM AND QUALITATIVE BEHAVIOR

Figure 11 shows the single-configurational-coordinate (SCC) energy level diagram appropriate for Eu^{3+} in La_2O_2S, Y_2O_2S, and LaOCl. The ^7F $4f^6$ states are not shown. They are split into seven J states spanning about 6000 cm^{-1} with J between 0 and 6. They would be placed with the zero of energy at the minimum of the $^7F_0$ parabola and with only a minor offset, with $S\hbar\omega < 100 cm^{-1}$. The first set of excited $4f^6$ states are the ^5D states, split into five J states between 0 and 4. The first four of these states have been identified; the $J = 4$ state and the other $4f^6$ states will be referenced only by their absorption energies in nanometers.

In each host there is an offset charge-transfer state (CTS), placed in the one diagram appropriately for the broad absorption band observed in each host.

The diagram ignores the small crystal field splittings of the various states and the degeneracies and splittings of the CTS.

Table 5 gives the energies of some of the Eu^{3+} states relative to $^7F_0$. These energies are approximate averages of the several crystal-field states for this ion.

From Figure 11 one can expect narrow line absorptions from $^7F_0$ at 0K and also from the thermally populated ^7F_j states at higher temperatures into the ^5D_j states. There will also be a broad band absorption into the CTS, only from $^7F_0$ at 0K but also from the thermally populated ^7F_j states at higher temperatures. In the absence of experimental knowledge of these electric dipole matrix elements, and not having had the opportunity to develop a theoretical treatment of them in the vein of the Judd-Ofeld analysis, we take them here to be all equal. One can expect also narrow line emissions from the various ^5D states to the lower ^7F states. As might be anticipated from the diagram, the broad band emission from the CTS to the lower ^7F states are not seen because of the much faster CTS $\rightarrow$ ^5D_j feedings and the thermal inaccessability of the CTS from the lowest ^5D states.

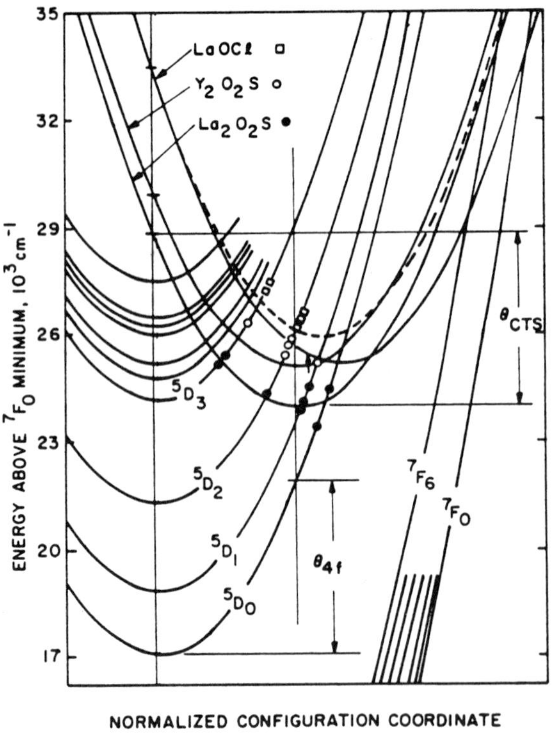

Figure 11. SCC Diagram for Eu^{3+} in Several Hosts

Figure 12 shows the excitation spectrum of $(Y_{0.999},Eu_{0.001})O_2S$ $^5D_0$ and $^5D_3$ emissions at 77K. These spectra have not been corrected for the source intensity which peaked near 380 nm. These spectra show the narrow 4f→4f absorptions into states above $^5D_3$ and the broad-band ^7F→CTS at higher energies.

TABLE 5

$^7F_0 \rightarrow {}^7F_i, {}^5D_i$ Energies, wavenumbers

	i	Y_2O_2S	La_2O_2S
^5D	3	24210	24270
	2	21360	21410
	1	18940	18900
	0	17140	17180
^7F	4	2890	2890
	3	1700	1700
	2	1190	1190
	1	340	340

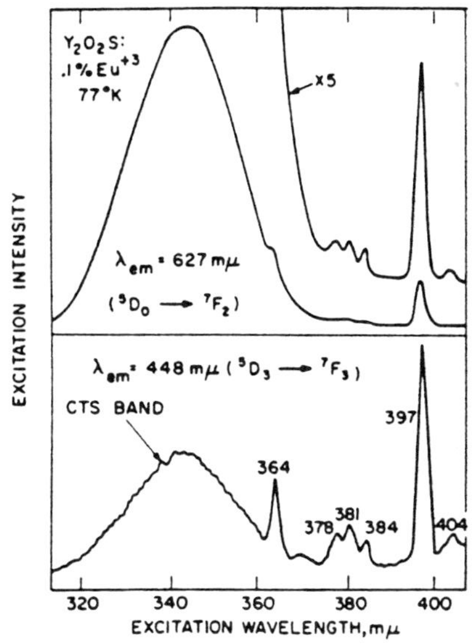

Figure 12. Excitation Spectra for $(Y_{0.999},Eu_{0.001})_2O_2S$

Figure 13 shows the emission spectra generated in this phosphor at 77K under two different excitation wavelengths, noted in the figure. These wavelengths are into the CTS (top curve) and into a state above $^5D_3$ which cascades directly and totally into $^5D_3$ (bottom curve). These are again spectra uncorrected for the detector spectral response, which decreased by a factor of 3 to the shorter wavelengths. The notation (*i,j*) is used for the transition $^5D_i \rightarrow {}^7F_j$. One important characteristic of CTS excitations is visible in Figure 13, namely, that the $^5D_3$ emissions are excited at 397 nm. but not with the CTS, although the CTS is above the 397 nm. state. Populating the CTS leads to bypassing $^5D_3$ and the $4f^6$ levels above $^5D_3$ and to direct feeding of the lower ^5D states via CTS → $^5D_2$, $^5D_1$, and $^5D_0$ transitions. Besides confirming that indeed the lower ^5D states are fed

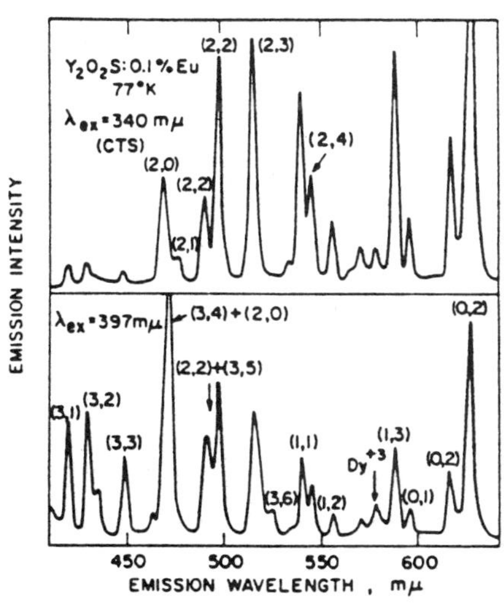

Figure 13. Emission Spectra for $(Y_{0.999},Eu_{0.001})_2O_2S$

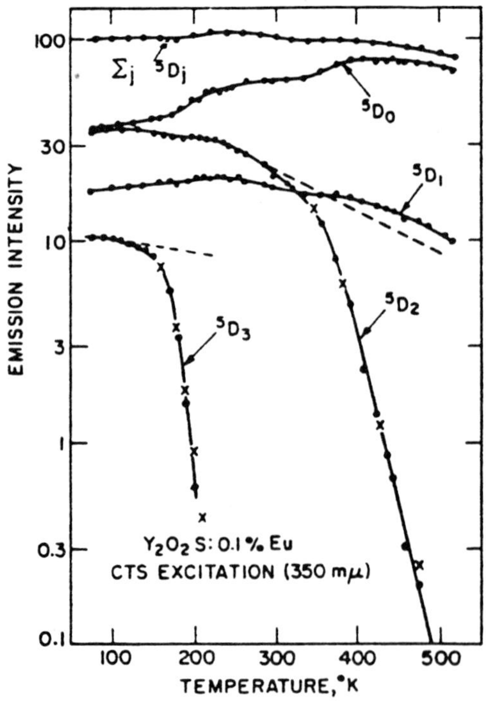

Figure 14. Thermal Quenching of $(Y_{0.999},Eu_{0.001})_2O_2S$ Emissions

through the CTS-^{5}D crossovers near the CTS minimum, this performance suggests the model for studying quantitatively the CTS feeding fractions into the ^{5}D manifold, namely, by assessing the relative rates of these bottom crossovers through the equal-force-constants SCC-model W_p functions, or the unequal-force-constants U_{pu} functions, or even the derivative-operator expressions in one or in many coordinates and in one or in many frequencies. Here, we restrict our attention to the Condon-operator W_p and U_{pu} functions.

In addition, as was first reported by Wickersheim, Buchanan, and Yates [34], the ^5D_j emissions quench precipitously with increasing temperature in the then unexpected order: $^5D_3$ at the lowest temperature, then $^5D_2$, then $^5D_1$ as the temperature rises. Figure 14 shows this sequential quenching of the $(Y_{0.999},Eu_{0.001})O_2S$ emissions. Our work established these quenchings as two-step quenchings through the higher offset CTS, i.e., as the case presented in Section III.A.3.d. and pictured in Figure 6.

VII.B. FEEDING FRACTIONS

Excitation into a particular ^5D_j state leads to ^5D_j emissions and also, via at low concentrations a sequence of small-offset multiphonon emission transitions, to emissions originating in all lower ^5D_k states. At higher concentrations there is the added known two-center energy transfer process for populating these lower ^5D states, namely, $(^5D_i, ^7F_0) \rightarrow (^5D_k, ^7F_l)$, where the indices are chosen so as to conserve energy. In either case or even in any combination of them at a given temperature, there is a unique pattern of emission intensities for each ^5D_j excitation and a different unique pattern for CTS excitation. Let q_{ij} and q_{iCTS} be the quantum efficiencies for emission from ^5D_i for ^5D_j and CTS excitations, respectively. Let α_j be the quantum efficiency for transferring CTS excitation into the ^5D_j level. Then

$$q_{iCTS} = \sum_j q_{ij}\alpha_j, \qquad i,j = 0,1,2,3. \tag{174}$$

Eq. (174) can be inverted to give the α_j in terms of the q_{iCTS} and q_{ij}.

For low temperatures, as at 77K here, $\sum_i q_{ij}$ was close to unity. Thus q_{ij} was equal to the fraction of photons emitted from ^5D_i for ^5D_j excitation, and q_{iCTS} is the fraction of photons emitted from ^5D_i for CTS excitation. The inversion then gives α_j which sum to unity over j, and these are the feeding fractions.

In our model, we take

$$\alpha_j = \frac{(2j+1)A_{5D_j,CTS}U_{p_{U,j}}}{\sum_{i\ \mathrm{unquenched}}(2i+1)A_{5D_i,CTS}U_{p_{U,i}}} \approx \frac{(2j+1)U_{p_{U,j}}}{\sum_{i\ \mathrm{unquenched}}(2i+1)U_{p_{U,i}}} \tag{175}$$

Eq. (175) uses $U_{p_{Uj}}$, not $U_{p_{Uj,d/dz}}$, nor the multicoordinate-model $M_{p_{Uj,d/dz}}$, nor the multiple-frequency-model $D_{p_{Uj,d/dz}}$. Furthermore we take all $A_{5D_i,CTS}$ equal. In this weighting by the simple degeneracy factor and equal electronic integrals, we ignore the detailed selection rules and variable transition integrals among component transitions. This latter approximation, as also for the absorption band fits of Section VII.E below and for the quenching fits of Section Vii.F below, introduces the more serious error and probably accounts for most of the discrepancies between measured and observed feeding fractions, quenching temperatures, and temperature-dependent bandshapes. Moreover, it will be seen that the discrepancies encountered tend to point to crossovers occuring before thermal equilibrium is established: there is a consistent pattern of a few percent direct feeding at low temperature through the higher crossovers not expected from a thermal population of the CTS vibrational levels.

These discrepancies do not negate the important general conclusions reached here.

In Table 6 are listed the observed feeding fractions in these three hosts from the CTS of Eu^{3+} into the ^5D_i states.

TABLE 6

Observed CTS → ^5D_i Feeding Fractions in Oxysulfides:Eu

Host	%Eu	T,K	α_3	α_2	α_1	α_0
La_2O_2S	0.5	77	0.02	.30	0.65	0.04
Y_2O_2S	0.1	77	0.20	.65	0.15	0.02
	0.1	295		.59	0.37	0.03
	2.0	295		.60	0.40	0.00
	2.0	495		.46	0.53	0.02

VII.C. EFFICIENCIES UNDER QUENCHING CONDITIONS

In analogy with Section III.A.3.d we list the efficiency expressions $(\eta_i)_j$, $(\eta_i)_{CTS}$, of ^5D_i emissions for excitation into ^5D_j or into the CTS, respectively. These are solutions of the steady state relationships among the concentrations of the various states according to

$$\begin{bmatrix} G_{CTS} \\ G_3 \\ G_2 \\ G_1 \\ G_0 \end{bmatrix} = \begin{bmatrix} (CTS\to 3,2,1,0) & -(3\to CTS) & -(2\to CTS) & -(1\to CTS) & -(0\to CTS) \\ -(CTS\to 3) & (3\to CTS,2,F) & 0 & 0 & 0 \\ -(CTS\to 2) & -(3\to 2) & (2\to CTS,1,F) & 0 & 0 \\ -(CTS\to 1) & 0 & -(2\to 1) & (1\to CTS,0,F) & 0 \\ -(CTS\to 0) & 0 & 0 & -(1\to 0) & (0\to CTS,F) \end{bmatrix} \begin{bmatrix} n_{CTS} \\ n_3 \\ n_2 \\ n_1 \\ n_0 \end{bmatrix} \tag{176}$$

The solutions are obtained in the temperature region of $^5D_3$ quenching by setting to zero the $(2\to CTS)$, $(1\to CTS)$, $(0\to CTS)$, and all but either the G_3 or G_{CTS} rates. In the region of $^5D_2$ quenching, $(3\to CTS) >> (3\to 2,F)$ and n_3 can be eliminated from the set of unknowns using the second equation. Similar considerations simplify Eq. (176) for η_2 and η_1. The results are

$$(\eta_3)_3 = \frac{(3\to F)n_3}{G_3} = \left[1 + \frac{(3\to 2)}{(3\to F)} + \frac{(CTS\to 2,10)}{(CTS\to 3,2,1,0)}\cdot\frac{(3\to CTS)}{(3\to F)}\right]^{-1}$$

$$\equiv \left[1 + \frac{(3\to 2)}{(3\to F)} + \beta_3(3\to CTS)\right]^{-1},$$

$$(\eta_3)_{CTS} = \frac{(3\to F)n_3}{G_{CTS}} = \frac{(CTS\to 3)}{(CTS\to 3,2,1,0)}(\eta_3)_3,$$

$$(\eta_2)_2 = \frac{(2\to F)n_2}{G_2} = \left[1 + \frac{(2\to 1)}{(2\to F)} + \frac{(CTS\to 10)}{(CTS\to 2,1,0)}\cdot\frac{(2\to CTS)}{(2\to F)}\right]^{-1}$$

$$\equiv \left[1 + \frac{(2\to 1)}{(2\to F)} + \beta_2(2\to CTS)\right]^{-1},$$

$$(\eta_2)_{CTS} = \frac{(2\to F)n_2}{G_{CTS}} = \frac{(CTS\to 2)}{(CTS\to 2,1,0)}(\eta_2)_2,$$

$$(\eta_1)_1 = \frac{(1\to F)n_1}{G_1} = \left[1 + \frac{(1\to 0)}{(1\to F)} + \frac{(CTS\to 0)}{(CTS\to 1,0)}\cdot\frac{(1\to CTS)}{(1\to F)}\right]^{-1}$$

$$\equiv \left[1 + \frac{(1\to 0)}{(1\to F)} + \beta_1(1\to CTS)\right]^{-1},$$

$$(\eta_1)_{CTS} = \frac{(1\to F)n_1}{G_{CTS}} = \frac{(CTS\to 1)}{(CTS\to 1,0)}(\eta_1)_1, \qquad (177)$$

When our interest is in the precipitous quenching itself, we discount the weak temperature dependences due to other loss processes, e.g., the f → f transitions, and define "quenching efficiencies" $(\eta'_i)_i$ as the ratio of the efficiency to the efficiency if no ^5D_i → CTS transitions were occurring. Thus

$$(\eta'_i)_i = (\eta_i)_i\left[1 + \frac{(i\to i-1)}{(i\to f)}\right] \equiv \frac{1}{1 + \beta'_i(i\to CTS)} \equiv \frac{1}{1 + \beta'_i A_{vu} V_{p_{v_i}}},$$

$$\beta'_i = \left[\frac{(CTS\to i-1,\ldots,0)}{(CTS\to i,i-1,\ldots,0)\,(i\to F)}\right] \Big/ \left[1 + \frac{(i\to i-1)}{(i\to f)}\right] \qquad (178)$$

VII.D. ABSORPTION SPECTRA AT T > 0K

Our attention will be limited to fitting the long-wavelenth portion of the ^7F → CTS absorption band. We then can ignore the unknown complex electronic structure of this CTS state and be content with assessing the energy only of the lowest such state.

Even so, we must take into account the absorptions from the higher ^7F_j levels. We do so by assuming them at 340, 1190, 1700, and 2890 cm^{-1} above $^7F_0$ for $j = 1,2,3,4$, respectively, as shown in Table 5, and then giving them thermal (Boltzmann) and also $(2j+1)$ degeneracy weights. The energy placements were approximate, insofar as they correspond to 2, 7, 10, and 17 $\hbar\omega$ for the $\hbar\omega = 170 cm^{-1}$ value chosen for the equal-force-constants fit below.

The absorption strength calculated for every wavelength then includes the contributions from each of the ^7F_j states, each contribution being a V_{p_v} with its p_v appropriate for its initial state energy.

Once again, the electronic dipole moment transition integral is assumed identical for all ^7F_j → CTS absorptions.

VII.E. THE FIT OF THE ABSORPTION DATA

1. Gaussian Approximations. In our earliest work, we have described the Eu absorptions as fitted by a superposition of Gaussians from

the thermally weighted ^7F_j states, with each Gaussian sharing common parameters A, $h\nu_0$, σ according to

$$P_0(h\nu) = A\exp[-(h\nu - h\nu_0)^2/\sigma^2], \qquad \sigma = 2(kT_{eff}\theta_e^2/\theta_g)^{1/2},$$

$$kT_{eff} = E_{zp}\coth(E_{zp}/kT), \tag{179}$$

where kT_{eff} is an effective thermal energy which approaches the zero-point vibrational energy E_{zp} of the ground state at low temperatures and kT at high temperatures. The θ_e and θ_g are identical with our $S_v\hbar\omega_v$ and $S_u\hbar\omega_u$ of Figure 1, respectively. In our notation, $h\nu_0 = ½(\hbar\omega_u - \hbar\omega_v) + h\nu_{zp,vu} + S_v\hbar\omega_v$, i.e., the energy between the parabolas at the ground-state equilibrium configuration, and $h\nu = ½(\hbar\omega_u - \hbar\omega_v) + h\nu_{zp,vu} + p_v\hbar\omega_v$.

These expressions stem from Klick and Schulman[35] in their semiclassical treatment of the luminescent center. We used this treatment in early work before we appreciated the quantum mechanical expressions set forth in this book. In the better notation of this book, these equations become, using Eq. (18),(30), and (9) and giving the correct normalization factor,

$$P_0(h\nu) = \left(\frac{S_u}{2\pi S_v^2 < 2n+1 >_v}\right)^{1/2} \exp\left[-\left(\frac{(p_v - S_v)^2 S_u}{2S_v^2 < 2n+1 >_v}\right)\right]. \tag{180}$$

$$G(p) = \left(\frac{1}{2\pi\sigma^2}\right)^{1/2} \exp\left[-\left(\frac{(p-S)^2}{2\sigma^2}\right)\right]. \tag{181}$$

Because these Gaussians do fit the observed spectrum and because, using Eq. (18), Gaussians are approximate forms of W_p functions for large S, the W_p distributions equivalent to these Gaussians will also fit. Indeed, an adequate fit will be provided by every V_{p_v} distribution for parabola placements which maintain the $h\nu_0$ and the slope of the excited-state parabola at the ground state minimum configuration. There is freedom to choose the vibrational frequency as well for each such parabola placement. Explicitly, the allowed fits satsify, using Eqs. (1),

$$E_u(h\nu) = S_u\hbar\omega_u\left(\frac{x}{a_0}\right)^2 = E_0 = S_0\hbar\omega_0\left(\frac{x}{a_{45}}\right)^2,$$

$$E_{v,\theta} = E_{v,min,\theta} + S_v\hbar\omega_v\left(\frac{x}{a_\theta} - 1\right)^2,$$

$$\left.\frac{dE_{v,0}}{dx}\right|_{x=0} = -\left(\frac{2S_v\hbar\omega_v}{a_\theta}\right) = \left.\frac{dE_{v,45}}{dx}\right|_{x=0} = -\left(\frac{2S_0\hbar\omega_0}{a_{45}}\right). \tag{182}$$

Having then specified a value for $S_0\hbar\omega_0$, we have then the relationships

$$\left(\frac{a_0}{a_{45}}\right) = \left(\frac{S_u\hbar\omega_u}{S_0\hbar\omega_0}\right)^{1/2},$$

$$S_v\hbar\omega_v = \left(\frac{a_0}{a_{45}}\right)S_0\hbar\omega_0 = \left(S_u\hbar\omega_u S_0\hbar\omega_0\right)^{1/2} = \frac{S_0\hbar\omega_0}{\tan^4\theta},$$

$$S_u\hbar\omega_u = \frac{S_v\hbar\omega_v}{\tan^4\theta},$$

$$E_{v\,min,0} = E_{v,min,45} + S_0\hbar\omega_0\left[1 - \frac{1}{\tan^4\theta}\right]. \tag{183}$$

VII.F. BANDSHAPES

As one searches for the most perceptive fit of these model functions to all experimental data, one often wishes to describe the same absorption and emission spectra both with equal and with unequal force constants. Such descriptions might keep the sum of the first moments and the product of the second moments constant, i.e., using the first two of Eqs. (30) and (31) truncated to only their first terms and evaluated at 0K, would have

$$S_v\hbar\omega_v + S_u\hbar\omega_u = 2S\hbar\omega,$$

$$\left[\frac{\omega_v}{\omega_u}S_v(\hbar\omega_v)^2\right]\left[\frac{\omega_u}{\omega_v}S_u(\hbar\omega_u)^2\right] = \left[S(\hbar\omega)^2\right]^2. \tag{184}$$

We use the $(\theta, a_{uv}) \longleftrightarrow (S_u, S_v, \omega_u, \omega_v)$ relationships in Eqs. (117), (137), and (138) to get

$$\hbar\omega_u = \hbar\omega\left[\frac{1+\tan^4\theta}{2\tan^3\theta}\right],$$

$$\hbar\omega_v = \hbar\omega_u\tan^2\theta,$$

$$S_u = S\left[\frac{4\tan^3\theta}{(1+\tan^4\theta)^2}\right],$$

$$S_v = S\left[\frac{4\tan^5\theta}{(1+\tan^4\theta)^2}\right],$$

$$a_{uv}(\theta) = a_{uv}(\theta = 45°)\left[\frac{2\tan^3\theta(1+\tan^2\theta)}{(1+\tan^4\theta)^2}\right]^{1/2}.$$

$$h\nu_{zp,vu}(\theta) - h\nu_{zp,vu}(\theta = 45°) = S_u\hbar\omega_u - S\hbar\omega + \tfrac{1}{2}(\hbar\omega_v - \hbar\omega_u) \tag{185}$$

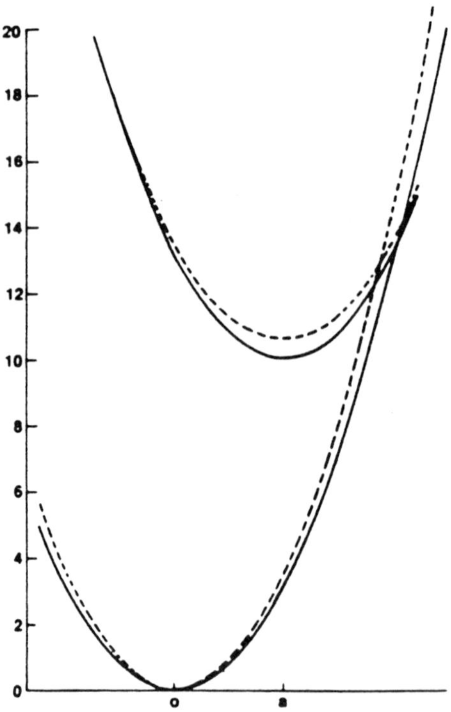

Figure 15. Single-Configurational Models with Similar Optical Bands

Figure 15 shows a typical example of these similar centers. Both equal-force constant curves and $\theta = 42°$ curves are shown accurately drawn to follow Eqs. (184) and (185). The offset parameters are $a_{uv} = 6$ for $\theta = 45°$ and $a_{uv} = 5.886$ for $\theta = 42°$. The phonon energies are $\hbar\omega = 400\text{cm}^{-1}$ for $\theta = 45°$ and $\hbar\omega_u = 1.135\hbar\omega = 454.1\text{cm}^{-1}$, $\hbar\omega_v = 0.920\hbar\omega = 368.1\text{cm}^{-1}$ for $\theta = 42°$. Thus $S = 9$ for $\theta = 45°$ and $S_u = 9.568$, $S_v = 7.757$ for $\theta = 42°$. Through these choices of the parameters, the sum of the first moments is thus maintained at $18\hbar\omega$ and the product of the second moments is maintained at $81(\hbar\omega)^4$. The last of Eq. (185) keeps the energy difference between the parabolas constant at the upper-state minimum. The zero-phonon line position is shown in the figure at 10000 for $\theta = 45°$ and at

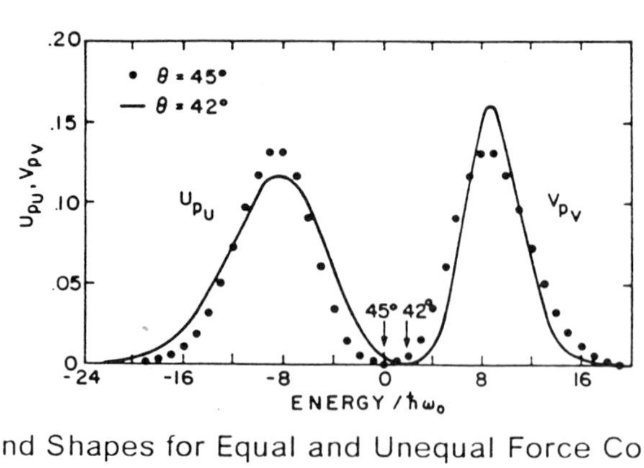

Figure 16. Band Shapes for Equal and Unequal Force Constants Cases

$h\nu_{zp,vu}(\theta=42°) - h\nu_{zp,vu}(\theta=45°) = 652.12cm^{-1}$ for $\theta = 42°$, or $h\nu_{zp,vu}(\theta=42°) = 10652.1cm^{-1} = 23.457\hbar\omega_u$. Figure 16 gives the U_{pu} and V_{pv} band shapes at 0K for $\theta = 45°$ (points) and for $\theta = 42°$ (curves). When, as here, the sum of the first moments and the product of the second moments are maintained equal, then for $\theta=42°$ the absorption is narrower and the emission is broader, and the zero-phonon energy is shifted slightly to higher energy. The small differences between these lineshapes are often of secondary importance compared to the differences predicted for other properties such as the thermal quenching behavior or the feeding fractions from an offset state into small-offset states.

1. Oxysulfides. For La_2O_2S and Y_2O_2S, we have fit the absorption bands originally using the Klick-Schulman semiclassical model equation, Eq. (179) with equal force constants, with $S_u\hbar\omega_u = S_v\hbar\omega_v = 4800cm^{-1}$ for both hosts and with $h\nu_0 = 29900cm^{-1}$ and $= 28800cm^{-1}$ for Y_2O_2S and La_2O_2S, respectively. The corresponding W_p distributions, namely those listed in Table 7 under the columns headed by $u \rightarrow v = {}^7F_0 \rightarrow$ CTS and $\theta=45°$, will also provide a fit. The fit is shown in Figure 17.

Table 7 gives values for the quantities listed in the first column. The starred ones are fundamental; those not starred can be derived from the fundamental parameters. For the moment, we concentrate on the starred quantities θ, a_{uv}, $\hbar\omega_u$, $h\nu_{zpvu}$ in the third and fifth columns, for $\theta=45°$. From these, using Eq. (9), we obtain $\hbar\omega_v$. For $\theta=45°$, $S_0=S_u=S_v$ and then Eq. (9) yields S_0. At the three temperatures shown in Figure 17 we know $<m>$ from Eq. (14) and we then know the $W_p(S_0, <m>)$ distribution from, e.g., the first of Eq. (14). The proper p_j for ${}^7F_j \rightarrow$ CTS absorption at some energy $h\nu_{abs}$ is known from the energy balance Eq. (53); the $h\nu_{zpvu}$ and the chosen $\hbar\omega_u$ are listed in Table 7; and the ${}^7F_j$ energy are listed in Table 5. The computed W_{p_j} of Eq. (14) is weighted by the partition function, namely, $(2j+1)exp\{-E({}^7F_j)/kT\} / [\sum_i (2j+1)exp\{-E({}^7F_i)/kT\}]$. The thus weighted contributions from all initial ${}^7F_j$ states are summed and plotted as the curves in Figure 17. The points in this figure are the experimental measurements taken from Reference 1.

We are led especially from the observed feeding fractions to an alternative fit involving unequal force constants. The alternative fit, shown in Figure 18, is that in the second and fourth columns of Table 7, involving $\theta=44°$.

Here it has been found necessary also to mimic anharmonic effects, i.e., expansion of the lattice with increasing temperature, by allowing the position of the CTS to decrease slightly with temperature, by $5kT/2$ units, as shown in Table 7. We find striking evidence forcing us to adopt this variation in the energy position of the offset states in ruby.

Once again for unequal force constants, the starred quantities in Table 7 are the fundamental ones. From them, using Eqs. (9) and (14), the numerically computed V_{pv} distributions of Eq. (25), the p_j selection and the

TABLE 7

Parameters for Fitting Oxysulfide:Eu $^5D \rightarrow$ CTS Data

	Y_2O_2S		La_2O_2S	
u	$^7F_0$	$^7F_0$	$^7F_0$	$^7F_0$
v	CTS	CTS	CTS	CTS
$\cdot\theta$	44°	45°	44°	45°
$\cdot a_{uv}$	7.58	10.62	8.32	10.62
S_u	14.87	28.2	17.91	28.2
S_v	13.86	28.2	16.70	28.2
$\cdot \hbar\omega_u$	295	170	295	170
$\hbar\omega_v$	275	170	275	170
$S_v\hbar\omega_v$	3810	4800	4590	4800
$\cdot h\nu_{zp\,vu}$	26300-5kT/2	24900	24700-5kT/2	23600
$\cdot\beta' A_{uv}$	2.5 + 8	2.3 + 7	2.5 + 8	2.3 + 7

thermal and degeneracy weighting discussed in the $\theta = 45°$ fit above, we obtain the curves of Figure 18. The points in this figure are those of Figure 17, originally from Reference 1.

Both $\theta = 45°$ and $\theta = 44°$ fits are deemed adequate: the selection will be made from fitting the quenching temperatures and feeding fractions in these phosphors.

VII.G. FITTING THE QUENCHING DATA

1. Oxysulfides. Figures 19 and 20 show the fits to the La_2O_2S and Y_2O_2S quenching data for the equal-force-constants case of the third and

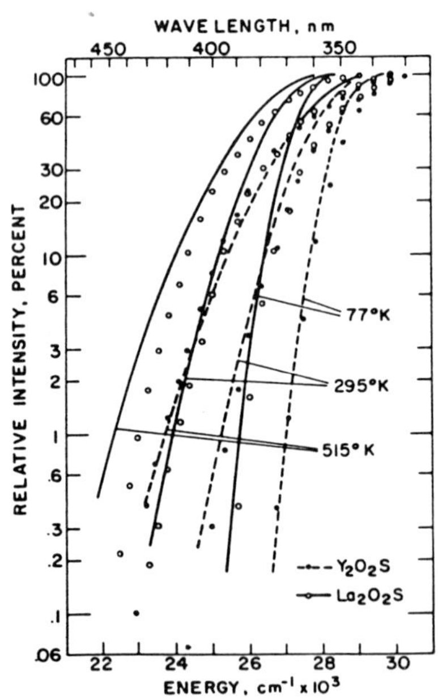

Figure 17. Equal Force Constants Fit of Oxysulfide Absorption Bands

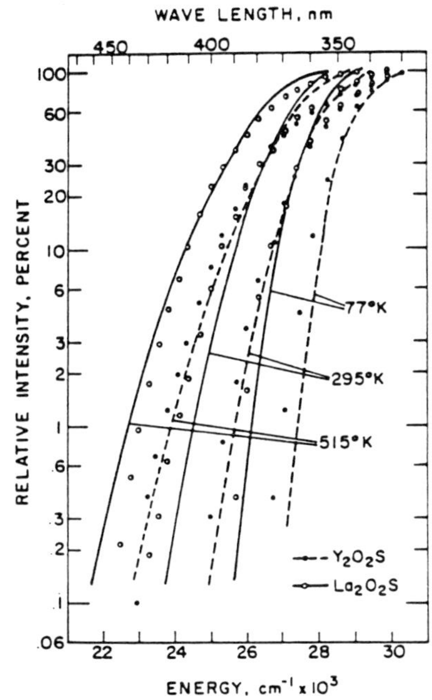

Figure 18. 44° Fit of Oxysulfide Absorption Bands

fifth columns of Table 7 and for the unequal-force-constants case, involving a downward shift in the position of the CTS with temperature, of the second and fourth columns of this Table.

The curves are the Eq. (178) "quenching efficiencies" with the V_{p_V} values calculated through Eq. (25) and the parameters of Table 7. The points are ratios of data points taken from Figs. 4,5, and 8 of Reference 2. Both fits are deemed adequate, although the unequal-force-constants fit seems slightly better.

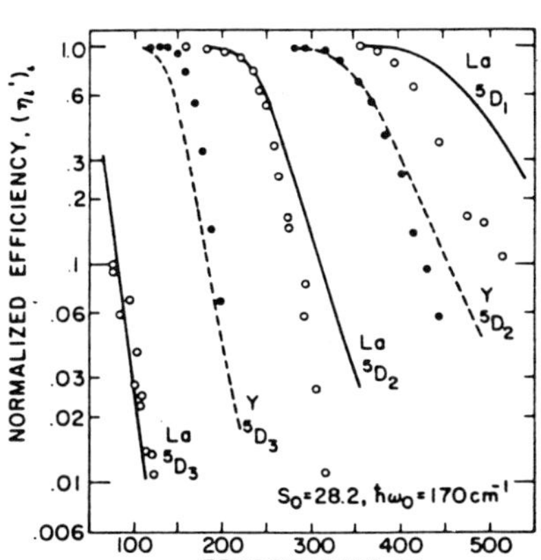

Figure 19. Equal Force Constants Fit of Oxysulfide Quenching Data

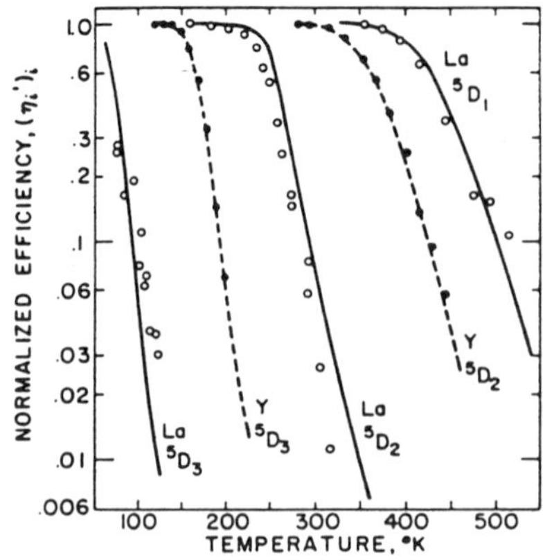

Figure 20. 44° Fit of Oxysulfide Quenching Data

2. Fitting the $(\eta_i)_i$ Finally, we return to Eqs. (177). In Figure 21 $(\eta_i)_i$, the measured efficiencies of the $^5D_i \rightarrow {}^7F$ emissions in $(Y_{0.999},Eu_{0.001})O_2S$ when excited into the emitting states, are shown as a function of temperature. The data points are from Figs. 4, 5, and 6 of reference 2. The plots are the calculated curves using Eqs. (177) with the parameters in Table 8 to describe the 4f → 4f transitions and with the parameters in Table 7 to describe the $^5D \rightarrow$ CTS quenchings.

The first three columns of Table 8 all give reasonable fits to $(\eta_1)_1$. We choose the second column because we believe that $A_{uv} \approx 10^{12}$ and that $R_{uv} \approx 10^4$. The $(\eta_2)_2$ and $(\eta_3)_3$ are then calculated after a search for the S_0 which best fits each's low-temperature 4f → 4f nonradiative rates.

Several points are worth making. First, Moos[27] has discussed the general forbiddenness of $^5D_1 \rightarrow {}^5D_0$ transitions relative to $^5D_2 \rightarrow {}^5D_1$ and $^5D_3 \rightarrow {}^5D_2$ transitions. This forbiddenness is seen to be a factor of ~3 in the case studied here.

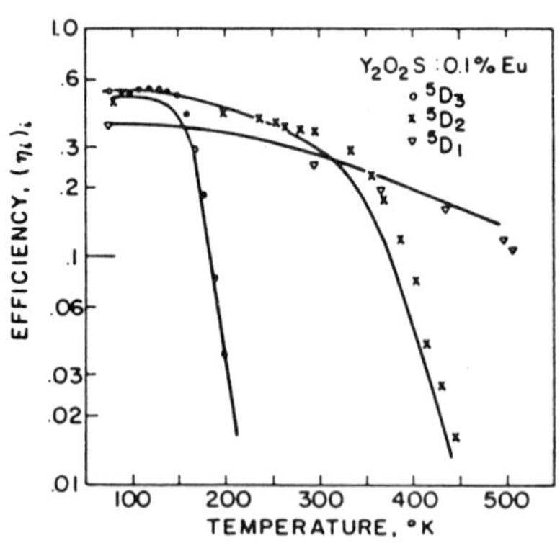

Figure 21. Measured and Calculated Efficiencies in $(Y_{0.999}Eu_{0.001})_2O_2S$

TABLE 8

Parameters for Fitting $Y_2O_2S:Eu^{3+}$ Data 4f → 4f Transitions

u	$^5D_0$	$^5D_0$	$^5D_0$	$^5D_1$	$^5D_2$
v	$^5D_1$	$^5D_1$	$^5D_1$	$^5D_2$	$^5D_3$
$\cdot\theta$	45°	45°	45°	45°	45°
$\cdot a_{uv}$	0.155	0.346	0.632	0.632	0.894
S_0	0.006	0.03	0.10	0.101	0.20
$\cdot\hbar\omega_0$	438	435	429	435	435
$S_0\hbar\omega_0$	2.62	13.0	42.9	43.5	67.0
$\cdot h\nu_{zp\,vu}$	1800	1800	1800	2420	2860
$\cdot A_{uv}/R_{gv}$	6.6+10	1.+8	1.+6	1.+8	1.+8

The second point is that in visually discounting the slow 4f → 4f transitions and thus obtaining experimental "quenching efficiencies" to be compared with Eq. (178), no gross violence has been done, for the same description of the ^5D - CTS interactions has been used in both.

Third, an additional advantage of the unequal-force-constants fitting of the quenching performance in oxysulfide hosts, shown in Figure 20, over the equal-force-constants counterpart in Figure 19, is the absolute value of the $(\eta_3)_3$ data points. Because we have not followed these emission intensities below 77K, we do not know the experimental normalization factor. Therefore we have normalized these emission intensities so as to bring them close to the fitted theoretical $(\eta_3)_3$ curves in both cases. Our belief is that the Figure 20 efficiencies are closer to the correct values as a function of temperature. This constitutes the additional advantage tending toward establishing the unequal-force-constants description.

Fourth, we have chosen to describe the CTS - $4f^6$ interactions with phonon energies near 300 cm^{-1} and the $4f^6$ - $4f^6$ interactions with above 400 cm^{-1} phonon energies. This is found to be a general trend, that the fast large-offset transitions are dominated by the lower-energy phonons while the slow small-offset transitions are dominated by the larger-energy phonons. It is to be expected that some multi-frequency model functions, such as those in Section II.F, will encompass both performances simultaneously. However, it is very likely that the description thus formed will not prove particularly enlightening, in that the disjoint performance seen will be interpreted in disjoint descriptions of the corresponding single-mode *A* and *S* parameters for these transitions.

VII.H. FITTING THE FEEDING FRACTIONS

For the oxysulfides, the feeding fractions allow a final selection of a most reasonable description of the Eu^{3+} center. Table 9 shows the calculated α_j values at several temperatures for the two descriptions of this center in these oxysulfide hosts in Table 7.

TABLE 9

CTS → ^5D_i Feeding Fractions in Oxysulfides:Eu

Host	%Eu	T,K	α_3	α_2	α_1	α_0	
La_2O_2S	0.5	77	0.02	.30	0.65	0.04	obs.
		77	0.00	0.31	0.63	0.06	calc. for $\theta = 44°$
		295		0.32	0.57	0.11	calc. for $\theta = 44°$
		77	0.00	0.02	0.91	0.07	calc. for $\theta = 45°$
		295		0.21	0.67	0.12	calc. for $\theta = 45°$
Y_2O_2S	0.1	77	0.20	.65	0.15	0.02	obs.
	0.1	295		.59	0.37	0.03	obs.
	2.0	295		.60	0.40	0.00	obs.
	2.0	495		.46	0.53	0.02	obs.
		77	0.13	0.83	0.04	0.00	calc. for $\theta = 44°$
		295		0.85	0.14	0.01	calc. for $\theta = 44°$
		77	1.-6	0.61	0.38	0.003	calc. for $\theta = 45°$
		295		0.61	0.37	0.02	calc. for $\theta = 45°$

Once again, the crucial data are the large observed α_3 for $(Y_{0.999},Eu_{0.001})O_2S$ at 77K and the large observed α_2 for $(La_{0.995},Eu_{0.005})O_2S$. Here, the $\theta = 44°$ descriptions are clearly preferable to the $\theta = 45°$ descriptions. The remaining discrepancies are, for $(La_{0.995},Eu_{0.005})O_2S$, indicative of pre-thermal-equilibrium crossovers into $^5D_3$. For $(Y_{0.999},Eu_{0.001})O_2S$, a slightly larger $\theta \approx 44.2°$ might afford a slightly better fit but there would still be some 10% feeding into $^5D_3$ to ascribe to pre-thermal-equilibrium crossovers. Of the two descriptions considered, however, the $\theta = 44°$ parameters clearly is the better choice.

TABLE 10

Optical Transition Energies, cm^{-1}, in $La_2O_2S:Tm^{3+}$

Transition[a]	Excitation	Emission
$^1I_6 \rightarrow {}^3H_6$	34000 34500	...[b]
$^1I_6 \rightarrow {}^3H_4$		28325
$^1I_6 \rightarrow {}^3H_5$		25650 25750
$^1I_6 \rightarrow {}^3F_4$		21325 21475
$^1D_2 \rightarrow {}^3H_6$	27500 27625 27850	27350 27500
$^1D_2 \rightarrow {}^3H_4$		21850 21925 22050
$^1D_2 \rightarrow {}^3H_5$		18975 19175 19300
$^1D_2 \rightarrow {}^3F_4$		14875 14975 15025
$^1G_4 \rightarrow {}^3H_6$	20800 20950 21025	20600 20800
$^1G_4 \rightarrow {}^3H_4$		15175 15225 15350

[a] Observed crystal field states have not been included in this assignment.
[b] $^1I_6$ emissions have been observed only for direct excitation into $^1I_6$ itself; after excitation into the host lattice absorption band, the $^1I_6$ level is poorly fed. The resolution used here was too low to allow recording the $^1I_6 \rightarrow {}^3H_6$ emission.

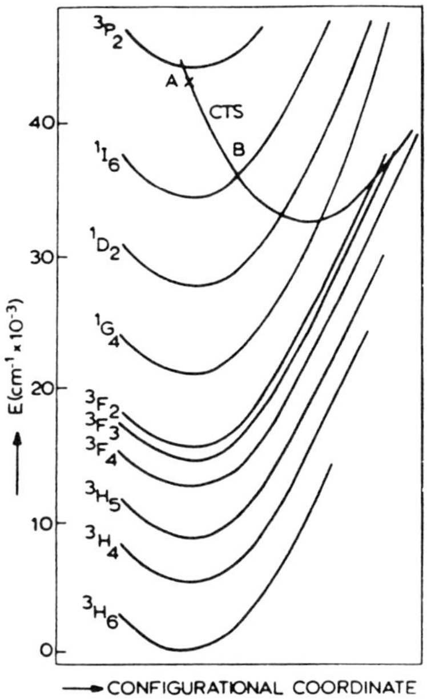

Figure 22. Energy Level Diagram for $La_2O_2S:Tm^{3+}$

VII.I. $La_2O_2S:Tm^{3+}$

1. The Experimental Behavior. The energy level diagram for Tm^{3+} in a La_2O_2S host lattice[20] is shown in Figure 20. The $4f^{12}$ states are placed as reported in Reference 35, ignoring crystal field splittings.

Strong host-lattice absorption prevents direct observation via typical absorption or excitation spectra of any level above 35500 cm^{-1}. Below 35500, however, excitation and emission transitions are seen at energies, listed in Table 10, which are consistent with the diagram. The spectra supporting the energy values listed in Table 10 are in Figures 23-25.

Figure 25 shows the nonlinear two-photon excitation (TPE) spectrum of $^1D_2 \rightarrow {}^3H_6$ emission at 27400cm^{-1}. Two spectra are shown, the weaker one with a 50% transparency filter, the stronger one without such a filter in the excitation beam. The excitation was the output of a tunable dye laser (Molectron DL 200) pumped with a nitrogen laser (Molectron UV 14). The dyes used for output between 18500 and 24000 cm^{-1} were Coumarin 120 (Exciton), 7D4MC (Molectron), and Coumarin 485 (Exciton). The intensities of the output from the various dyes was measured using Lumogen T rot GG (BASF) as a standard.

The ratio of these two curves is a factor of 4 and the excitation is therefore two-photon nonlinear absorption directly into states at twice the energy of the scale.

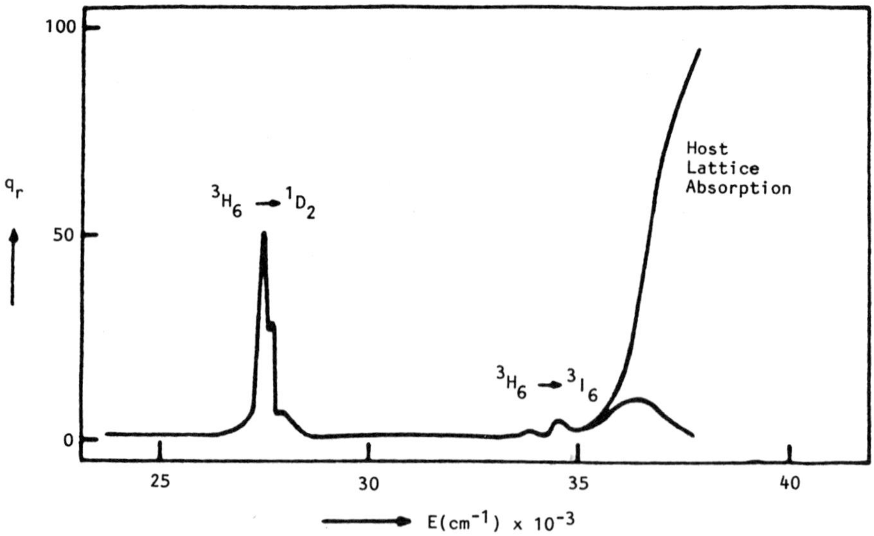

Figure 23. Excitation Spectrum for $La_2O_2S{:}Tm^{3+}$ $^1D_2 \rightarrow {}^3H_4$ Emission at 21930 cm^{-1} at 5K.

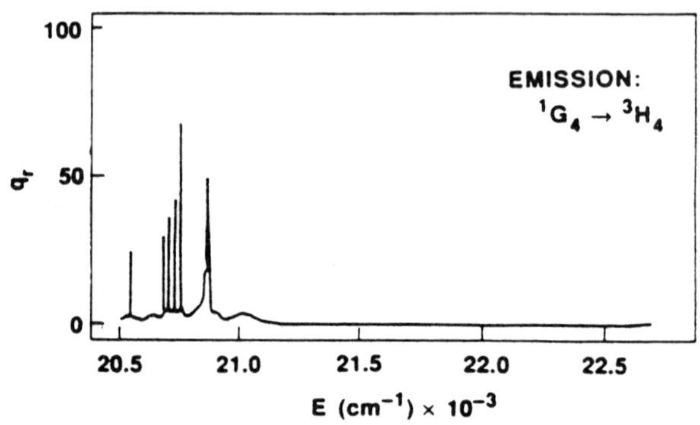

Figure 24. Excitation Spectrum for $La_2O_2S{:}Tm^{3+}$ $^1G_4 \rightarrow {}^3H_4$ Emission at 15380 cm^{-1} at 5K.

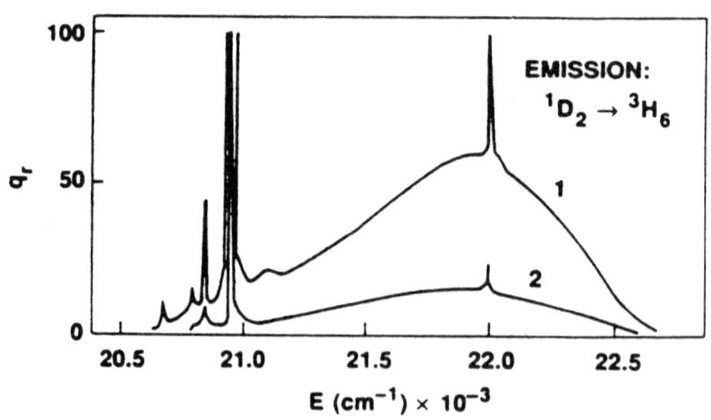

Figure 25. Two-Photon Excitation of $La_2O_2S{:}Tm^{3+}$ $^1D_2 \rightarrow {}^3H_6$ Emission at 27400 cm^{-1} at 5K.

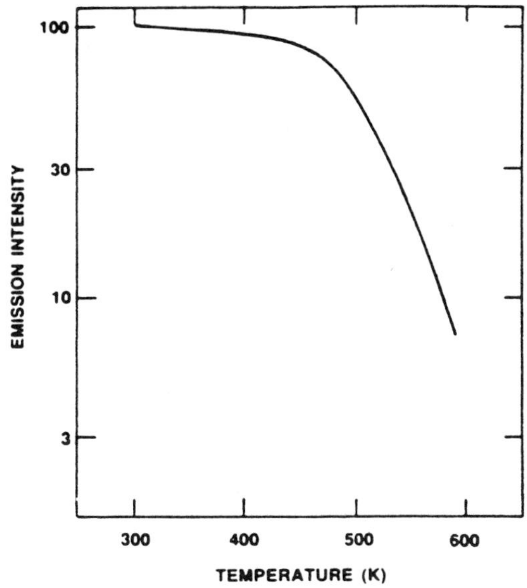

Figure 26. Thermal Quenching of $La_2O_2S:Tm^{3+}$ $^1D_2 \rightarrow {}^3H_4$ Emission at 457 nm.

In comparing Figures 24 and 25, we see three features of importance. First, the sharp lines between 20500 and 21000 cm^{-1} have too much overlap not to ascribe the Figure 24 sharp lines to excited state absorption from $^1G_4$ into the broad-band state. Second, the broad band itself, here ascribed to the CTS, is reached by the nonlinear two-photon absorption of the exciting laser beam without any intermediate absorption. Third, the sharp line near 22000 cm^{-1} is also a nonlinear two-photon absorption into a small-offset a Tm^{3+} $4f^{12}$ state, here ascribed to $^3P_1$. The nonlinear absorption in both cases is made evident by the factor of four in the intensity of the two curves of Figure 24, which are themselves related by a factor of two in the excitation intensity.

Figure 26 shows the quenching of the Tm^{3+} $^1D_2 \rightarrow {}^3H_4$ emissions at 457 nm. under $^3H_6 \rightarrow {}^1D_2$ excitation at 363 nm. The quenching is precipitous near 500K. In Ref. 5 it was stated that what were described as $^3P_0$ emissions and are better described as $^1I_6$ emissions quench precipitously at ~40K. There, indeed, these two quenchings were stated to follow the quenching behavior,

$$I(T)/I_0 = [1 + A\exp(-E/kT)]^{-1} \qquad (186)$$

with A ~ 10^7 and $E = 6000cm^{-1}$ and $400cm^{-1}$, for $^1D_2$ and $^1I_6$ quenchings, respectively.

2. Fitting with W_p Functions. The first point to determine is that the broad band absorption shown in Figure 25 cannot explain the observed sequential quenching of $^1I_6$ at 40K and of $^1D_2$ at 500K.

The best chance, which fails, is to explain this absorption curve as indeed two bands, the first at 42500 cm^{-1} with a width at half height equal to the twice the two-photon width at quarter height, namely 2200 cm^{-1}, and the second at 43700 cm^{-1} with a width at half height 2400 cm^{-1}. Using the relationships

$$\sigma^2 = \frac{(\Delta_{1/2})^2}{8\, ln2} = S < 2m+1 > (\hbar\omega)^2 \approx S(\hbar\omega)^2, \tag{187}$$

noting that $S\hbar\omega$ will be larger the smaller $\hbar\omega$ is, and finally adopting the smallest plausible $\hbar\omega$, namely 100 cm^{-1}, we are led to $S = 87.28$, $S\hbar\omega = 8728cm^{-1}$. Since $h\nu_{zp,vu}$ for $^1I_6$ is 34000 cm^{-1}, and even allowing for anharmonicity by introducing a downward shift by (5/2)kT in the CTS placement, we find, through the relations

$$E_{zpvu,CTS}(T) = 42500cm^{-1} - S\hbar\omega + \frac{1}{2}\hbar\omega - \frac{5}{2}kT,$$

$$p = (E_{zpvu,CTS}(T) - E_{zpvu,l})/\hbar\omega \tag{188}$$

that $p = 2.65$ at 50K and that at this temperature $W_p = 1.30 \cdot 10^{-15}$. No reasonable value for the electronic integral will give a nonradiative rate that will compete against a millisecond time constant radiative process. That is, no quenching of this level is predicted near 40K. Furthermore, no variant of this cited parameter set, aimed at describing a state with its absorption maximum near 42500 cm^{-1}, both gave nonradiative rates which with reasonable electronic integrals compete against emission and simultaneously gave a sharp increase of this nonradiative rate near 40K.

There is a level placement which will indeed explain the observed sequential quenchings. We take $S = 16$, $\hbar\omega = 200cm^{-1}$, $E_{zpvu,CTS} = 34200cm^{-1}$, which, with $E_{zpvu,l} = 34000cm^{-1}$, gives from Eqs. (188) $p_l = -1, \Delta_{1/2} = 1884$. We ascribe values of radiative lifetimes of the order 10^{-4} and $10^{-3.5}$ to $^1I_6$ and to $^1D_2$, respectively. We allow a downward shift in the CTS placement according to

$$E_{zpvu,}CTS(T) = E_{zpvu,CTS}(0K) - 2.8kT \tag{189}$$

and we take $A_{vu} \doteq 10^{12.5}$ for both $^1I_6 \rightarrow$ CTS and $^1D_2 \rightarrow$ CTS transitions. Table 11 gives the quenching efficiency of $^1I_6$ emissions predicted and Table 12 gives those for $^1D_2$ emissions.

In Tables 11 and 12 we do indeed find $^1I_6$ quenching at ~40K and $^1D_2$ quenching at ~500K.

TABLE 11

Quenching of $^1I_6$

T,K	10	35	40	45	50
W	.572-18	.496-9	.145-8	.351-8	.758-8
$N = 10^{12.5}W$	1.81-6	1.57 + 3	4.59 + 3	1.11 + 4	2.40 + 4
$\eta = R/(R+N)$	.98	.86	.68	.47	.29

TABLE 12

Quenching of $^1D_2$ Tm^{3+} Emissions in La_2O_2S through the CTS

T,K	300	350	400	450	500	550	600
$E_{zpvu,CTS}(T)$	33616	33519	33422	33324	33227	33130	33032
$E_{zpvu,D}$	27500	27500	27500	27500	27500	27500	27500
$p_{CTS,D}$	-30.558	-30.094	-29.608	-29.121	-28.635	-28.148	-27.662
$W_{\|p_{CTS,D}\|}$	.00570	.00575	.00804	.01050	.01295	.01536	.01773
r^p	1.862-13	1.796-11	5.615-10	8.174-9	6.961-8	4.018-7	1.730-6
W_{p_D}	1.062-15	1.032-13	4.513-12	8.579-11	9.014-10	6.171-9	3.069-8
$R = 10^{12.5}W_{p_{CTS,D}}$	.00335	.3263	14.271	271.3	2850.	19514.	97050.
η	1.0	1.0	.996	.92	.53	.14	.03

The question now arises whether anything in the absorption spectra gives evidence of such a state. From Eq. (10.12) we estimate the width at half height for this band to be near 1900 cm^{-1}. A band shaped as the one hand-sketched into Fig. 36 would have its peak absorption at ~37000 cm^{-1}, near 34200 + 3200 = 37400 cm^{-1}, would have its width at half height at 1900 cm^{-1}, and would account for some of the gradual rise of the host lattice absorption in this spectral region, i.e., the very weak shoulder on the onset of the host-lattice absorption curve in Fig. 36 near this energy.

The demand, because of the $^1I_6$ and $^1D_2$ quenchings, for placing an offset state at this energy with this $S\hbar\omega$ offset is in our opinion strong. The circumstantial evidence would be strengthened if the amplitude of this shoulder should be found to depend on the Tm^{3+} concentration. It would also be convincing to demonstrate that when $^1I_6$ emissions quench, they are replaced with $^1D_2$ emissions through two-step $^1I_6 \rightarrow$ CTS $\rightarrow$ $^1D_2$ processes.

It is to be expected that two-photon absorptions might have their transition selection rules so that some states are not coupled well to the ground state. The assignment of the states uncovered by the two photon experiments is an aim of research in itself. Their lifetime against decay into the host-lattice electronic states would be of great interest.

VII.J. Y_2O_2S:Yb^{3+}

1. The Experimental Behavior. The experimental behavior[37,21,38] of Yb^{3+} in Y_2O_2S:Yb^{3+} is understandable in terms of the energy level diagram shown in the insert in Figure 27. In this figure, the two $4f^{13}$ $^2F_{5/2}$ and $^2F_{7/2}$ states are labeled u and u'. The state labeled v is the offset CTS of Yb^{3+}

Also shown in this diagram are computed optical spectra: the narrow line seen in both emission and absorption, two broad band emissions labeled $v \rightarrow u'$ and $v \rightarrow u$, and the broad band absorption labeled $u \rightarrow v$. These computed spectra are in agreement with the positions and shapes at 300K, as observed both by Nakazawa and by us. Higher CTS are seen in the absorption spectrum but are ignored in Figure 27. The zero-phonon energies appropriate to the transitions labeled A,B, and C in the energy diagram are shown as corresponding vectors along the energy scale .

When excitation is into the CTS at ~30000-35000 cm^{-1} at low temperatures, both broad bands $v \rightarrow u'$ and $v \rightarrow u$ as well as the narrow line $u' \rightarrow u$ are seen in emission. At low Yb^{3+} concentrations, the ratio of the $v \rightarrow u'$ and $u' \rightarrow u$ integrated intensities is unity. One is seeing the sequence of two emissions $v \rightarrow u' \rightarrow u$ competing with the direct $v \rightarrow u$ emission.

As the temperature is raised, nonradiative $v \rightarrow u'$ transitions take over from both broad band transitions. The $u' \rightarrow u$ intensity grows with temperature and the $v \rightarrow u'$ intensity decreases with temperature.

At higher Yb^{3+} concentrations, an additional depopulation of the CTS is observed at low temperatures, due to the energy transfer $v,u \rightarrow u',u'$. Then

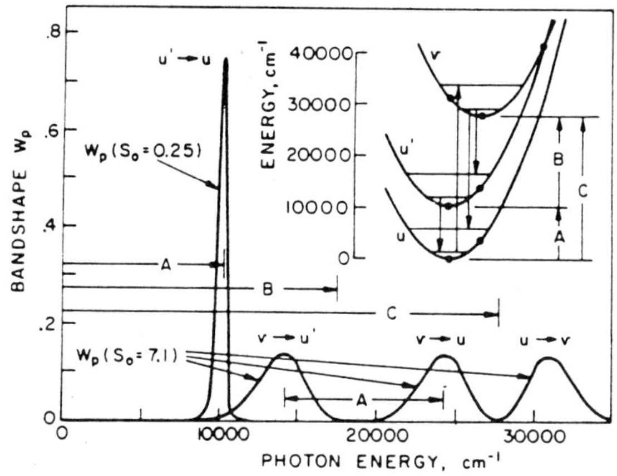

Figure 27. Optical Transitions of $Y_2O_2S:Yb^{3+}$ and the SCC Energy Diagram

both broad bands are decreased in intensity and the line emission is increased, even at low temperatures.

Figure 28 shows schematically the nonradiative $v \rightarrow u'$ transitions which become dominant at high temperature and the energy transfer $v,u \rightarrow u',u'$ transitions which become dominant at high Yb^{3+} concentrations. Note especially that the energy transfer occurs between a large-offset center and a small-offset center, and has a large energy mismatch, that is, creates many phonons. Nevertheless this transfer can dominate over emission.

The CTS emissions are found to quench near 250K through $v \rightarrow u'$ nonradiative transitions. Also, at 77K, the quantum efficiencies of the $v \rightarrow u$ and $v \rightarrow u'$ emissions are near unity for $(Y_{.9995}Yb_{.0005})_2O_2S$ and $\sim$ 0.75 for $(Y_{.99}Yb_{.01})_2O_2S$.

The decay curves of these two compositions were given by Nakazawa[60] as shown in Figure 29. Note that these decay curves are distinctly nonexponential. We shall return to this later in this Chapter. At any rate, the radiative decay time for these emissions is of the order of 0.4 μs.

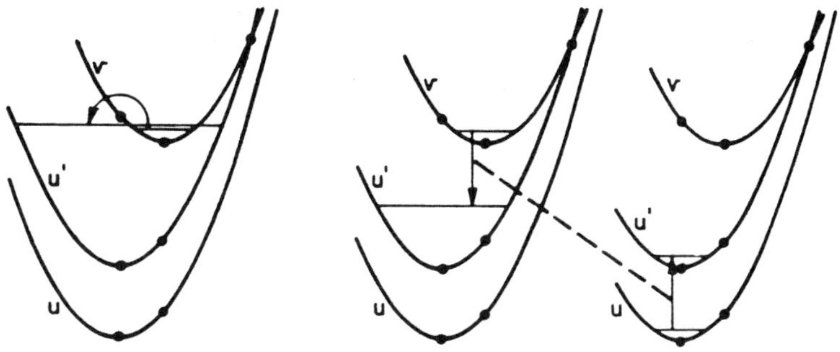

Figure 28. Nonradiative and Energy Transfer Transitions in $Y_2O_2S:Yb^{3+}$ and the SCC Energy Diagram

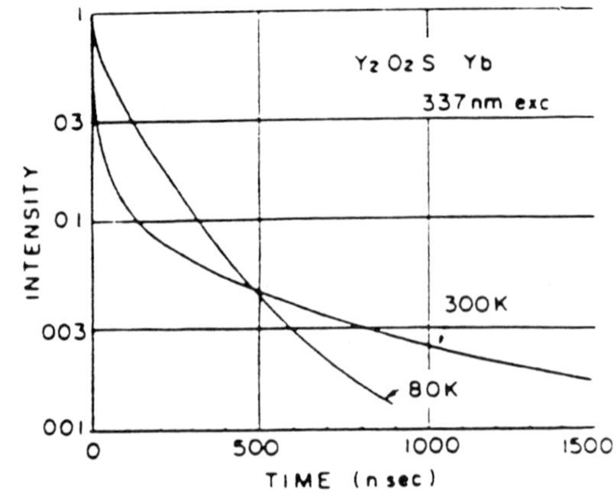

Figure 29. Decay Curves for Yb^{3+} CTS Emissions in Y_2O_2S

2. Fitting the Optical Spectra with SCC Model Functions. For the narrow-line transition $u \rightarrow u'$ and its inverse, we use equal force constants with $S = 0.25$ and $\hbar\omega = 500\text{cm}^{-1}$.

We give in Table 13 both Nakazawa's and our fitting parameters for the broad absorption and emission bands. Nakazawa used the semiclassical model of Eq. (179) and we have translated his parameters into the notation of this book as in the corresponding Eqs. (179)-(182). The required parameters are those starred. From them, using Eq. (9) we determined θ, then S_u, S_v and $\hbar\omega_u$, and from them a_{uv}. The zero-point energy was chosen to fit the absorption and emission energy placements. The second moments in absorption were calculated using the second of Eq. (31) truncated to the first term.

The $\theta = 45°$, $S_0 = 7.1$, $\hbar\omega = 500\text{cm}^{-1}$ fit of the second column was our best equal-force-constants fit of the data. The three other columns are our unequal-force-constants fits with the same sum of first moments and products of second moments over absorption and emission, through Eqs. (184) and (185). Our finding was that the $\theta = 44°$ fit was closest among these coarse-grid selections for the 300K lineshapes. All five would be acceptable to us, but we would have some reluctance to admit θ below 43° since such phonon inequalities have not been needed in any other case.

The agreement in the second moments between Nakazawa's and our 44° fit at 300K demonstrates the reproduceability of the properties of Yb^{3+} in this host. One can see in his spectra the downward shift in the position of the CTS with increasing temperature, and this should be introduced in his placement of the $h\nu_{zpvu.CTS}$. We see some preference for his 300 cm^{-1} phonon energy in the observed temperature dependence of the linewidth; but the phonon energy which explains the optical band shapes can be found different from and sometimes smaller than that which explains quenchings.

TABLE 13

Fitting the $4f^{13}$ - CTS Optical Bands of $Y_2O_2S:Yb^{3+}$

	Nakazawa	Ours			
θ	43.584	45°	44°	43°	42°
a_{uv}	7.054	5.329	5.318	5.283	5.228
$\hbar\omega_u$. cm^{-1}	300.	500.	519.02	541.42	567.58
S_u	13.	7.1	7.316	7.467	7.548
$\cdot S_u\hbar\omega_u$	3900	3550	3797	4043	4284
$\cdot\hbar\omega_v$	271.7	500.	484.02	470.81	460.15
S_v	11.778	7.1	6.823	6.493	6.119
$\cdot S_v\hbar\omega_v$	3200	3550	3302	3057	2816
$h\nu_{zpvu,CTS}$	29800	29000-6.5kT	29266-6.5kT	30021-6.5kT	30522-6.5kT
$<2n+1>_{80K}$	1.048	1.000248	1.000177	1.000118	1.000074
$<(p-<p>)^2>_v(80K)$	11.179	7.102	6.363	5.646	4.961
$\sigma^2.cm^{-1}(80K)$	825240	1775440	1490720	1251503	1050471
$<2n+1>_{300K}$	1.6219	1.2000	1.1810	1.1610	1.1407
$<(p-<p>)^2>_v(300K)$	17.301	8.520	7.514	6.555	5.659
$\sigma^2.cm^{-1}(300K)$	1557100	2129940	1760412	1453097	1198197

3. Fitting the Quenchings with SCC Model Functions. Our interest is in simultaneously understanding the optical bands, the $v \rightarrow u'$ quenching at ~250K, and the low-temperature $v,u \rightarrow u',u'$ energy-transfer quenching at high Yb^{3+} concentrations. To this end, we have explored the descriptions of the $v \rightarrow u'$ quenching transitions in Table 13. We describe the energy transfer rate as in Eq. (173), namely, the combination of the large-offset $v \rightarrow u'$ and the small-offset $u \rightarrow u'$ distributions for generating the phonons needed for the energy offset in the $v,u \rightarrow u',u'$ transition. For equal force constants, this combined distribution is, through Eq. (172), itself a W_p function with its S the sum of the two component S's. For unequal force constants, this sum must be computed numerically.

We take anharmonicity and subsequent lattice expansion into account by lowering the CTS proportional to kT as shown. We adopt those values of $A_{vu'}$ and A_{xfer} which in our opinion are at the upper limit of plausibility so that the nonradiative rates are as large as possible. The rates are then described, for equal force constants, by $p = 37.6 - .013kT$, $S = 7.1$, $A_{u'v} = 10^{14}$ and, using Eq. (172), $p = 17.2 - .013kT$, $S = 7.35$, $A_{xfer} = 10^8$, respectively, for the $v \rightarrow u'$ nonradiative rate and the $v,u \rightarrow u',u'$ energy transfer rate.

At other θ values, all parameters are determined by Eqs. (184) and (185) from the equal-force-constants set in Table 13. In Table 14 we give some of the output of this Eqs. (184),(185) arithmetic for selecting the p values in the calculation of the $v \rightarrow u'$ rate at some representative temperatures.

In Figure 30 we give the nonradiative rates thus calculated as a function of temperature. The dotted curves for each θ are the calculated $v \rightarrow u'$ rates. The full curves give the sum of the nonradiative $v \rightarrow u'$ and energy-transfer $v,u \rightarrow u',u'$ rates. These latter rates must compete near 250K with the observed radiative rate of $\approx 2.5 \cdot 10^6$. It is seen that the $\theta = 44°$ fit, however good it is for the optical spectra, cannot afford a fit to this quenching. One needs some value near $\theta = 42°$. Secondly, the energy-transfer rate at 77K must be a quarter of the radiative rate, or near $0.6 \cdot 10^6$ in order to have the observed $\eta = 0.75$ for the 1% concentration samples at this temperature. This requirement is also met by the $\theta = 42°$ choice. We are thus in the position that we can understand the optical bandshapes by using $\theta = 43.6°$ and $\hbar\omega_u = 300\text{cm}^{-1}$ but are forced, against our better judgment as to plausible values for the Manneback angle, to $\theta = 42°$ and $\hbar\omega_u = 576.6\text{cm}^{-1}$ for the energy transfer behavior and the thermal quenching behavior.

The defeat is, however, a lead into an effect beyond the SCC model. The nonexponential decays shown by Nakazawa were interpreted correctly by him as evidence of CTS breakup into a free hole and a deeply trapped electron as Yb^{2+} If the dissociated CTS are dominantly nonradiatively recombined, then this rate must be added to the two nonradiative rates considered thus far to assess the competition between radiation and loss processes. AT 300K, as much as 90% of the emission betrays the influence of this process, and thus the calculated total nonradiative rates

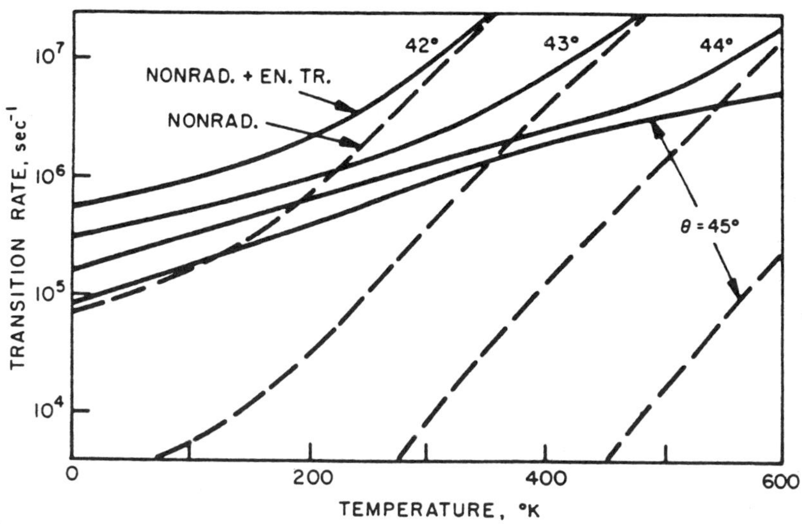

Figure 30. Nonradiative Rates vs. Temperature in $Y_2O_2S{:}Yb^{3+}$

TABLE 14

Selecting the Parameters for Fitting the $Y_2O_2S{:}Yb^{3+}$ Quenchings

θ	45°	44°	43°	42°
a_{uv}	5.329	5.318	5.283	5.228
$\hbar\omega_u$, cm^{-1}	500.	519.02	541.42	567.58
$\hbar\omega_v$	500.	484.02	470.81	460.15
$h\nu_{zpvu}(^2F_{7/2})$	10000	10000	10000	10000
$h\nu_{zpvu\ CTS}$	29000-6.5kT	29266-6.5kT	30021-6.5kT	30522-6.5kT
$\Delta E(80K)$	18838	18904	19660	20160
$p_U(80K)$	37.277	36.423	36.311	32.106
$kT/\hbar\omega_v(80K)$	.11120	.11487	.11810	.12083
$\Delta E(200K)$	18097	18362	18892	19619
$p_U(200K)$	36.193	35.378	34.893	34.565
$kT/\hbar\omega_v(200K)$	.27800	.27818	.29524	.30208
$\Delta E(250K)$	17871	18137	18893	19393
$p_U(250K)$	35.741	34.944	34.895	34.167
$kT/\hbar\omega_v(250K)$	.34750	.35898	.36905	.37760
$\Delta E(300K)$	17645	17910	18666	19167
$p_U(300K)$	35.289	34.507	34.476	33.769
$kT/\hbar\omega_v(300K)$	.41700	.43077	.44286	.45312

shown in Figure 30 must be raised by even an order of magnitude. Then the calculated nonradiative rates for all θ values become large enough to explain the observed quenchings. One may well find that $\theta = 43.6°$, with $A_{u'v}$ adjusted downwards by somewhat less than one order towards more plausible values, and even perhaps with $\hbar\omega_u$ smaller than the $500 cm^{-1}$ value used here, will afford a fit to the observed quenchings.

Such dissociation phenomena have been discussed[3,4] in detail for $Y_2O_2S:Eu^{3+}$. We do not have the detailed experimental information required to assess this dissociation with finality in the Yb^{3+} phosphors here.

REFERENCES

1. Struck, C. W. and Fonger, W. H., J. Lum. 1/2:456(1970).

2. Fonger, W. H. and Struck, C. W., J. Chem. Phys. 52:6364(1970).

3. Fonger, W. H. and Struck, C. W., J. Elect. Soc. 118:273(1971).

4. Struck, C. W. and Fonger, W. H., Phys. Rev. B4:4(1971).

5. Struck, C. W. and Fonger, W. H., J. Appl. Phys. 42:4515(1971).

6. Fonger, W. H. and Struck, C. W., J. Elect. Soc. 118:273(1971).

7. Fonger, W. H. and Struck, C. W., J. Lum. 8:452(1974).

8. Struck, C. W. and Fonger, W. H., J. Chem. Phys. 60:1988(1974).

9. Fonger, W. H. and Struck, C. W., J. Chem. Phys. 60:1994(1974).

10. Fonger, W. H. and Struck, C. W., Phys. Rev. B11:3251(1975).

11. Struck, C. W. and Fonger, W. H., J. Lum. 14:253(1976).

12. Struck, C. W. and Fonger, W. H., J. Chem. Phys. 64:1784(1976).

13. Struck, C. W. and Fonger, W. H., J. Lum. 10:1(1975).

14. Struck, C. W. and Fonger, W. H., J. Chem. Phys. 60:1988(1974).

15. Struck, C. W. and Fonger, W. H., J. Chem. Phys. 63:1533(1975).

16. Fonger, W. H. and Struck, C. W., J. Lum. 17:241(1978).

17. Fonger, W. H. and Struck, C. W., J. Chem. Phys. 69:4171(1978).

18. Struck, C. W. and Fonger, W. H., J. Lum. 18/19:101(1979).

19. Struck, C. W. and Fonger, W. H., Phys. Rev. B19:4400(1979).

20. Bleijenberg, K. C., Kellendonk, F. A., and Struck, C. W., J. Chem. Phys. 73:3586(1980).

21. Fonger, W. H. and Struck, C. W., in *Radiationless Processes* ed. B. Di-Bartolo,(Plenum, New York,1980), p.471.

22. Condon, E. U., Phys. Rev.32:858(1928).

23. Mott, N. F., Proc. Roy. Soc. (London) A167:384(1938).
Gurney, R. W. and Mott, N. F., Trans. Faraday Soc. 35:69(1939).

24. Kisliuk, P. and Moore, C. A., Phys. Rev. 160:307(1967).

25. Misu, A., J. Phys. Soc. Japan 19:2260(1964).

26. Partlow, W. D. and Moos, H. W., Phys. Rev. 157:252(1967).
Riseberg, L. A., Gandrud, W. B., and Moos, H.W., Phys.Rev.159:262(1967).
Riseberg, L. A. and Moos, H.W., Phys.Rev.174:429(1968).

27. Moos, H. W. , J. Luminescence 1/2:106(1970).

28. Chamberlain, J. R., Paxman, D. H. , and Page, J. L., Proc. Phys. Soc. 89:143(1969).
Weber, M. J., Phys.Rev.171:283(1968).
Fong, F. K., Naberhuis, S. L., and Miller, M. M., J. Chem. Phys. 56:4020(1972).

29. Kiel, A., in *Proc. Third Intern. Conf. Quantum Electronics, Paris, 1963* ed. P. Grivet and N. Bloembergen,(Columbia University Press, New York, 1966), p. 765.

30. Mahbub'ul Alam, A. S. M. and Di Bartolo, B., J. Chem. Phys. 47:379 0(1967).

31. Sorokin, P. P., Stevenson, M. J., Lankard, J. R., and Petit, G. D. , Phys. Rev. 127:503(1962).

32. Manneback, C., Physica 17:1001(1951).

33. Hutchisson, E., Phys. Rev. 36:410(1930), 37:45(1931).

34. Wickersheim, K. A. , Buchanan, R. A., and Yates, E. C., Proc. Conf. Rare Earth Res. 7th, Coronado, California, 1968:835(1969).

35. Klick, C. C. and Schulman, J. H., in *Solid State Physics* Eds. Seitz, F. and Turnbull, D., Vol.5 (Academic Press, New York, 1967) p. 97.

36. Wortman, D. E., Leavitt, R. P., and Morrison, C. A., J. Phys. Chem. Solids 35:591(1974).

37. Buchanan, R. A., private communication at Gordon Research Conference on Luminescence, 1972. Dr. Buchanan first informed us of the two broad emission bands in $Y_2O_2S:Yb^{3+}$ separated by the 10000 cm^{-1} $^2F_{5/2}$ to $^2F_{7/2}$ energy difference. Our investigations were conducted the few months after this Gordon Conference. One of us (WF) discussed this case with Dr. Nakazawa at the Paris, 1978 International Luminescence Conference.

38. Nakazawa, E., J. Lum. 18/19:272(1979).

ADVANCES IN NON-RADIATIVE PROCESSES IN SOLID STATE LASER MATERIALS

F. Auzel

C.N.E.T. - Laboratoire de Bagneux
196 Av. H. Ravera 92220 BAGNEUX - FRANCE

ABSTRACT

Non radiative transitions are generally considered as well recognized drawbacks in laser materials because they are known to reduce quantum efficiency. However, some positive aspects of non-radiative processes are always intricately involved with the negative ones in laser materials. The design of a good laser material results from a careful compromise between both aspects.

Starting from a classification of the different non-radiative processes of the ion-phonon and of the ion-ion types, the lectures shall discuss their negative and positive roles at different points in the pumping and laser cycle.

Different cases pertaining to fix and to tunable frequency laser material systems shall be presented, bringing problems to points where they reach the present frontier of research in their fundamental as well as their applied aspects.

I. INTRODUCTION

When non-radiative transitions are considered in relation to laser materials, the first step is of course to look at them as the negative process in competition with the radiative transition used in the laser action. However, dwelling a little more on the problem, it can be soon realized that though being an important effect it is not so important as to prevent laser action. More and more frequently recent advances in laser materials are considering ions which are different from the two archetypes : Cr^{3+} and Nd^{3+}.

These new lasers require more careful attention during the laser material design because they are further away from ideal model systems. For instance, pulsed laser action has been obtained in system with very low quantum efficiency at the laser transition : in the case of YAG:Er at 2.9 μm, it is as low as .3% [1], yet output at the Joule level has been obtained [2].

Advances in Nonradiative Processes in Solids
Edited by B. Di Bartolo, Plenum Press, New York, 1991

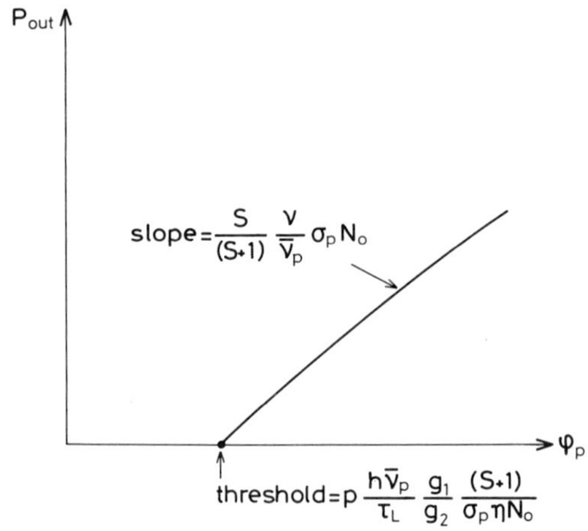

Fig. 1 . Theoretical output versus pumping flux for an ideal 4-level laser.

It may be useful to remember that in the 4-level laser case, the laser behaviour with respect to basic material parameters can be simply modelized [3]. A summary of this is presented on Fig. 1 giving the output power versus pumping flux. Interestingly the differential efficiency ("slope efficiency") of the laser is independant of the quantum efficiency at the laser transition :

$$\eta_{diff} = \frac{S}{S+1} \; \frac{\upsilon}{\bar{\upsilon}_p} \; \sigma_p N_0 \tag{1}$$

where S is the output coupling parameter defined as τ_L/τ_R where τ_L describes the material losses and τ_R the necessary loss due to the output coupling of the laser mirrors : υ is the laser frequency, $\bar{\upsilon}_p$ the average pumping frequency ; σ_p is the pumping cross-section and N_o the chemical concentration of active ions in the host medium.

Yet the output power as well as threshold are inversely proportional to quantum efficiency (η) :

$$P_{out} = \frac{ph\upsilon}{\tau_L \eta} S \left[\frac{\phi_p}{\phi_{th(Rmax)}(S+1)} \; -1 \right] \tag{2}$$

where $\phi_{th(Rmax)}$ is the pumping flux ϕ_p at threshold for two R_{max} mirrors (S=0) ; for given output coupling S($\neq$ 0), threshold pumping flux is given by :

$$\phi_{th} = \phi_{th(Rmax)} \;\; (S+1) = p \frac{g_1}{g_2} \; \frac{h\nu_p}{\tau_p \sigma_p q N_o} \tag{3}$$

where p is the volumic density of modes in the cavity for the given spectral width of the gain medium.

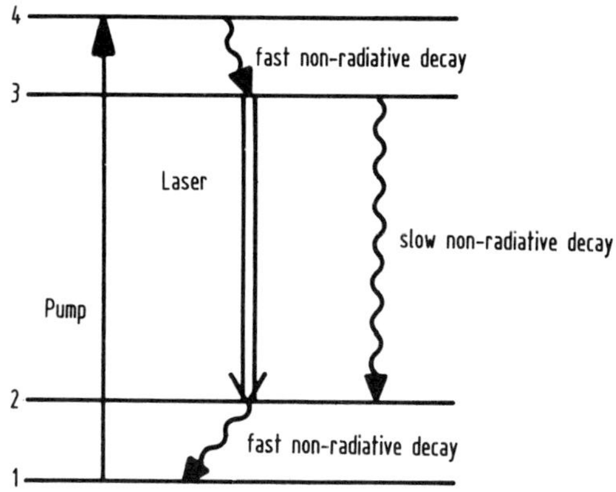

Fig. 2 . Ideal 4-level laser scheme with respect to non-radiative processes.

Such equations have been established under the assumption that (see Fig. 2) non-radiative processes used to populate the laser emitting level 3 from the excited level 4, is much more probable than radiative emission between 3 and 2 below threshold and that such is the case for non-radiative transition between final laser level 2 and ground state 1. Here we see in a simple way the necessity for efficient non-radiative processes at certain point in the pumping cycle.

In this example, the question arises : are such requirements not in contradiction with the existence of a high quantum efficiency at the laser transition ? In order to answer such question and similar ones for tunable laser materials where conditions are still more entangled we need first to recall briefly the basic non-radiative processes.

In chapter II.A we treat processes based on ion-phonon interactions ; then in II.B, non-radiative processes based on ion-ion interactions and in II.C multiphonon transfers which involve simultaneously ion-ion and ion-phonon processes.

In chapter III we discuss in more details the non-radiative processes in lasers and in chapter IV we shall present recent advances which have been evolved since the previous 1977 school on the subject [4] and the publication of the excellent review of R. Engleman [5] ; such advances constitute we believe the present frontier of research.

II. DIFFERENT TYPES OF NON-RADIATIVE PROCESSES

II.A. Processes based on ion-phonon interactions

1. Non-radiative multiphonon decays

The most basic process for energy loss is the multiphonon non-radiative decay. Following an absorption of photons heat only comes out of the sample. It can be viewed theoreticaly in a simple way : it takes its roots in the approximation left when one uses the Born-Oppenheimer description of energy levels in solid state.

This description is based on the adiabatic approximation in which it is assumed that the electrons constitutes a fast subsystem with respect to the slower lattice motion. Born-Oppenheimer (B.O) states are "stationary states" of the system only within the approximation : real states are non stationary which means that excitation cannot be distinguished for part of it between "excited" and "ground" state. This is the essence of the multiphonon non-radiative process.

Taking the usually neglected term in the adiabatic approximation as the "Non-adiabatic promoting hamiltonian" leads for non-radiative transitions between B.O states of general form :

$$\psi_{iv}(r,Q)=\phi_i(r,Q)\theta_{iv}(Q) \tag{4}$$

to the well known result for non-radiative probability between electronic states $i \rightarrow f$:

$$W_{i\rightarrow f}=\frac{2\pi}{\hbar}\sum_{v'v"}P_{v'}\Sigma_s R_s^2(f,i)/<X_{fv"s}/\frac{\partial}{\partial Q_s}/X_{iv's}>/^2$$

$$\times\prod_{j\neq s}/<X_{fv"j}/X_{iv'j}>/^2\delta(E_{fv"}-E_{iv'}) \tag{5}$$

where $P_{v'}$ is the Boltzman factor for the distribution among initial vibrational levels v' ; $R_s(f, i)$ is the electronic part of the non-adiabatic hamiltonien :

$$R_s(f,i)=-\hbar^2<\phi_f/\frac{\partial}{\partial Q_s}/\phi_i> \tag{6}$$

and $X_{iv's}$ are harmonic oscillator wave function approximation of the vibrational wave functions :

$$\theta_{iv}=\prod_N^s X_{iv_s}(Q_s)$$

Now we can distinguish between "promoting" modes s responsible for inducing the electronic transition, and the accepting modes j which are sinks for the electronic energy non-radiatively dissipated.

The "promoting modes" (s) are characterized by the non vanishing integrals :

$$<X_{fv"s}/\frac{\partial}{\partial Q_s}/X_{iv's}> \tag{7}$$

The "accepting" modes (j) are characterized by the non vanishing Franck-Condon Factors :

$$<X_{fv"j}/X_{iv'j}> \tag{8}$$

Such factors corresponds to the number of vibrational quanta into which the electronic energy is dissipated.

The summation over v' and v" in (5) has been calculated by Huang and Rhys [6] by series expansion of the harmonic oscillator wave functions with displaced coordinates, use of Condon approximation and of a recurring formula between integrals followed by careful combinatorial analysis of terms pertaining to N-phonon process :

$$W_N(i \to f) = \frac{\hbar^2}{\omega_m}\{\frac{1}{2}Z^2(\bar{n}R_{N+1} + (\bar{n}+1)R_{N-1}$$

$$+|Y|^2((\bar{n}+\frac{1}{2})^2 + \frac{1}{2}\bar{n}(\bar{n}+1))R_N - |Y|^2(\bar{n}+\frac{1}{2})$$

$$\times(\bar{n}R_{N+1} + (\bar{n}+1)R_{N-1} + \frac{1}{4}|Y|^2(\bar{n}^{2}R_{N+2}$$

$$+(\bar{n}+1)^2R_{N-2})\}\delta(E_j - E_i) \tag{9}$$

where

$$R_N = \exp(-(2\bar{n}+1)S_o)(\frac{\bar{n}+1}{\bar{n}})^{N/2} I_N(2S_o\sqrt{\bar{n}(\bar{n}+1)}) \tag{10}$$

$$Z^2 = (\frac{\omega_m}{\hbar})\sum_s R_s^2(f,i) \tag{11}$$

$$Y = (\frac{1}{\hbar\omega_m})\sum_s R_s(f,i)(A"_s - A'_s) \tag{12}$$

$$S_o = \frac{1}{\omega_m^2\hbar^2}\quad \Sigma(A"_s - A'_s)^2 \tag{13}$$

in which ω_m is the frequency of the mode s and has been assumed to be the same for each s and to be constant whatever the electronic state is. I_N is the Bessel function with imaginary argument of order N. A", A' are the linear coupling constants to the lattice. S_o is the so-called Huang-Rhys electron-phonon coupling parameter.

Neglecting the differences in phonon order (N-1), (N-2), N, because they were dealing with the very strong coupling case of F-centers, Huang and Rhys approximated (9) by :

$$W_{NR}\quad (i \to f) \simeq \frac{\hbar}{\omega_m}\quad (\bar{n}+\frac{1}{2})\quad Z^2R_N\delta(E_f - E_i) \tag{14}$$

assuming N ≃ (N-1) ≃ (N-2).

Z^2 contains the electronic matrix element for the promoting operator $\partial/\partial Q_s$.

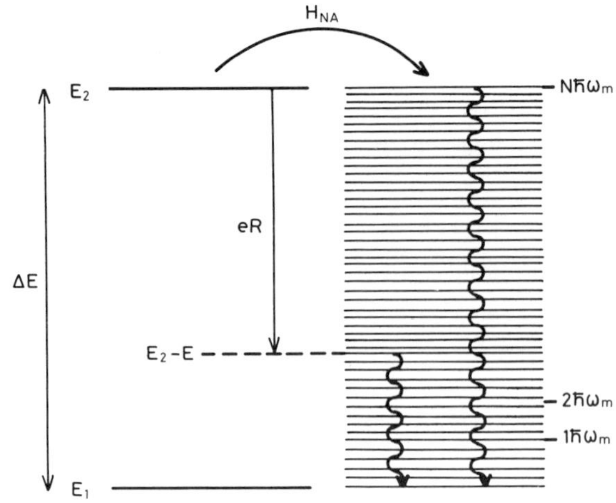

Fig. 3 . Energetical comparison between radiative (eR) and non-radiative (H_{NA}) decays from accepting terms point of view.

So in eq. (14), the multiphonon non-radiative decay probability is the product of a scaling factor Z^2 by an energy shape given by R_N containing the well known Bessel Function I_N generally used [4] [5] [7] ; see for instance the lecture by C. Struck in this school.

We have shown [8] that this R_N shape function has essentially a statistical meannig rather than a physical one : it indicates the way it is possible to accomodate the energy loss into vibrational modes for a given N-phonon process. Then it is very general and this accounts for the generality of the energetical shape behaviour of other multiphonon processes such as multiphonon emission or absorption side bands which we discuss in the next subchapter.

2. Non-radiative processes in radiative multiphonon side-bands

Because energy states of electron-phonon coupled systems are described by B-O states such as (4), the electronic energy depends parametrically on Q the vibrational coordinates. Because of the non-zero value of the Huang-Rhys coupling parameter (13) when linear couplings (A', A") are different for electronic ground and excited states, the radiative transition between such displaced state is given by :

$$P_{rad} = \frac{2\pi}{\hbar} < i/eR/j >^2 \quad R_N \delta(E_j - E_i - h_\nu) \tag{15}$$

eR is the electric dipole operator which by analogy with (14) can be said to promote the radiative transition ; R_N plays the same role as the non-radiative "accepting" part in (14);

$h\nu_e$ is the emitted photon energy. Clearly there is a strong analogy between eq. 14 and eq. 15 which can be pictured as on Fig. 3 : in both case the energy shape is governed by R_N. A difference between non-radiative and radiative transition is that the promoting part is of spontaneous emission or of induced absorption origin in radiative transition whereas it is from vibrational origin in non-radiative one. We shall come back again on that point in Chapter IV.

Also the energies involved in the δ functions are different : in non-radiative transition the whole energy difference $(E_f - E_i)$ is involved in the non-radiative part, whereas in (15) only $(E_f - E_i - h\upsilon)$ is taken up by the non-radiative loss. In absorption the δ function has argument $(h\upsilon - E_f + E_i)$ representing the non radiative part. The sum of these two energies losses represent the "Stockes shift" often given by : $h\upsilon_a - h\upsilon_e \simeq 2S_o\hbar\,\omega$.

3. <u>The R_N function and its approximations</u>

We have seen that in order to obtain the simplification of eq.(9) into (14), Huang and Rhys have assumed to have N >> 2 loosing some features due to promoting modes and to the significance of the (N-1), (N-2) terms. At low temperature their physical meaning can be viewed in the following way : in this case energy cannot be absorbed from the lattice and in order to breakdown the adiabaticity, the promoting modes take their energy from the only available energy, the energy gap. This reduces the number of phonons involved by at most two ; the action of the promoting and accepting modes being simultaneous. Such features as we shall see in IV.A are being rediscovered but have been known for a long time by many authors [4].

Now considering the R_N function, it can be simplified in some cases.

The Bessel function in R_N for the N-phonon term can be approximated for two separate cases :

i) Small coupling, low temperature. ($x \ll N$ and $N > O$). Writing for simplification :

$$x = 2S_o\sqrt{\bar{n}\,(\bar{n}+1)},$$

one has

$$I_N(x) \sim \frac{(n/2)^N}{N!} \tag{16}$$

ii) Strong coupling, high temperature. ($x \gg |N|$ and $N^2 \gg 1$) :

$$I_N(x) = (\frac{2\pi}{N})^{1/2} \exp(x - N^2/2x). \tag{17}$$

In solid state laser materials we are interested either in rare earth ions or in transition metals ones or in colored centers as active centers. They have quite different

TABLE I

Typical Huang-Rhys parameter (S_o) values according to transition types.

Transition type	f - f	f - d	d - d	s - p
active center type	rare-earth (trivalent)	rare-earth (bivalent)	transition metals	F-centers
S_o	10^{-2} - 10^{-1}	1 - 6	1.5 - 10	10 - 120

electron-phonon coupling parameters (S_o) and approximations have of course to be adapted to such coupling strength. In order to give an idea of this the typical value of coupling strength are given in table I.

If we are interested in trivalent Rare-earth ion we shall consider the first approximation.

From it, one has a simple recurrence between the N-phonon (N-1) phonon and (N-2) phonon terms in eq. (9) at low temperature :

$$R_N = \exp(-S_o)\bar{n}^{-N/2}(S_o\bar{n}^{-1/2})^N/N! = \exp(-S_o)S_o^N/N! \tag{18}$$

This form of R_N is the so-called "Pekarian" function [9] and :

$$R_{N-1} = R_N N/S_o$$
$$R_{N-2} = R_N N(N-1)/S_o^2 \tag{19}$$

from which eq. (9) is written at low temperature :

$$W_{i\to f} \approx \frac{N^2 S_o^N}{S_o^2 N!} \tag{20}$$

In this approximation (small coupling, multiphonon, low temperature) the results of Huang and Rhys are equivalent to the one of Miyakawa and Dexter [10] for non-radiative transitions :

$$W_{NR} i \to f = (\frac{2\pi}{\hbar})R^2(1-\frac{N}{So})^2(\frac{So^N}{N!})\delta(\Delta E - N\hbar\omega) \tag{21}$$

Because :

$$(1-\frac{N}{S_o})^2 = (\frac{N}{S_o})^2 \text{ for } N > S_o \tag{22}$$

Now developping N! by Stirling Formula :

$$N! = \sqrt{2\pi N}(N/e)^N \tag{23}$$

one obtains the well known exponential gap law for non-radiative transitions :

$$W_{NR} i \to f = \frac{2\pi}{h}R^2 \exp -\alpha_{NR}\Delta E \tag{24}$$

with :

$$\alpha_{NR} = (\hbar\omega_M)^{-1}(1-\frac{2}{N})[\ln(\frac{N}{So})-1] \qquad (25)$$

in which Δ E the energy gap below a given state is spanned by N phonons of energy $\hbar\omega_m$.

Usually the promotting terms having been either considered as a constant or neglected because as in Huang Rhys N >> 2, the non-radiative gap law parameter is written at T = 0K :

$$\alpha_{NR} = (\hbar\omega_m)^{-1}[\ln\frac{N}{So}-1] \qquad (26)$$

Same gap law has been found to apply with α given by eq. (26) for side bands excitation [11] in the trivalent Rare-earth case, just because of the existence of the R_N function in eq. (15).

Now if we are interested in transition metal ions, or bivalent rare-earth, the approximation of the exponential gap law is no longer valid and the full R_N expression has to be considered. However a convenient approach is to take the R_N series form as given by the series expension of the Bessel function I_N:

$$R_N = e^{-(2\bar{n}+1)S_o}\frac{(\bar{n}+1)^{N/2}}{\bar{n}^{N/2}}\left(\sum_{S=0}^{\infty}\frac{(S_o^2\bar{n}(\bar{n}+1))^{s+N/2}}{s!(N+s)!)}\right) \qquad (27)$$

$$= e^{-(2\bar{n})S_o}S_o^N\frac{e^{-S_o}}{N!}(\bar{n}+1)^N\sum_{S=o}^{\infty}\frac{N!S_o^{2s}\bar{n}^s(\bar{n}+1)^s}{s!(N+s)!}$$

$$= e^{-S_o}\frac{S_o^N}{N!}(\bar{n}+1)^N e^{-2\bar{n}S_o}[1+\frac{S_o^2\bar{n}(\bar{n}+1)}{(N+1)}+\ldots] \qquad (28)$$

where $\bar{n} = [\exp(\hbar\omega_m/kT)-1]$ the phonon occupation number at temperature T.

For $S_o < 1$, and not too high temperature, we can retain only the first term : that is for a "Pekarian function" at 0°K, we associate a temperature dependence of the form :

$$(\bar{n}+1)^N e^{-2\bar{n}S_o}$$

Then for small S_o the approximation at temperature T can be written :

$$R_N \sim \frac{e^{-So}S_o^N}{N!}(\bar{n}+1)^N e^{-2\bar{n}S_o} \qquad (29)$$

Valid for trivalent rare-earth ions [4].

In the case of bivalent rare-earth or transition metal ions, we use eq. 28 keeping in the braket a number of terms which as a rule of thumb is of the order of the magnitude of S_o [12].

For large values of S_o as the one encountered in s → p transition and high T, the approximation of an Arrhemius law with an activation energy ($\Delta E^2/2S_o\hbar\omega$) as given by eq. 17 is obtained where ΔE is the electronic energy to be dissipated.

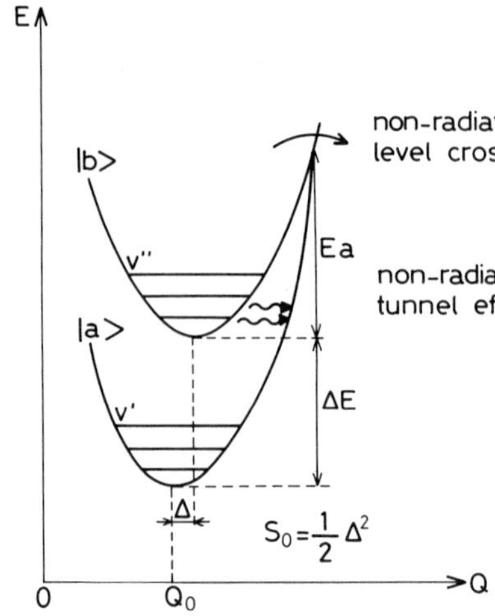

Fig. 4 . Non-radiative process limits in a configuration diagram picture ; $E_R = \Delta E^2/2 S_o \hbar\omega$.

This is just the Von Hippel-Seitz-Mott theory for non radiative transition [13]. This theory is based upon the representation by a configuration diagram where the non-radiative transition is considered to happen at the crossing point of the energy curves (see Fig. 4). Reaching this point by thermal activation of the vibrating modes, corresponds to the necessary activation energy for the strong coupling high temperature case. Then the complete R_N function describes both the small coupling case non-radiative process operating by a tunneling effect occuring before the crossing point and the crossing point behaviour for stronger coupling and temperatures as depicted on Fig. 4.

II.B. Processes based on ion-ion interactions

Besides interaction of one ion with lattice vibrations, one ion may also interact with another ion either because the concentration is increased which reduces the average distance between them or because they tend to form clusters. Then two types of non-radiative cross-relaxation may happen : down-conversion or self-quenching and up-conversion or APTE effect towards undesired states.

1. Non-radiative energy transfers and diffusion to non-radiative centers ; down-conversion cross-relaxation and self-quenching

Energy transfer occurs in a systems when absorption and emission do not take place within the same center. It may occur with the successive emission and absorption of a real photon, or without the exchange of a real photon in which case it is said to be a non-radiative transfer. Having discussed at length such effect in previous Erice schools [14] [15], we shall recall briefly the basic points.

Let us consider the simple case of two ions, each with one excited electronic state separated from its electronic ground state by nearly equal energy. With suitable interaction between the two electronic systems, the excitation will jump from one ion to the other before one is able to emit a quantum of fluorescence. The mutual interactions are the Coulomb interactions of the Van der Waals type between the two ions.

However, Dexter pointed out [16] that the theory should include higher multipole and exchange interactions. In fact, for an isolated atom, one can consider the transition probability as decreasing as $(a_o / \lambda)^{2n}$ where a_o is the Bohr radius, λ the wavelength, and n an integer representing the multipole order.

Whereas, in an energy transfer process with a dependence on near-zone interactions, the transition probabilities drop off as $(a_o /R)^{2n}$ where R is the separation of the interacting ions. R can be as much a three orders of magnitude smaller than λ, so that the energy transfer effect tends to be more pronounced in systems with forbidden transitions. This holds true to for ions with transitions forbidden to first order, such as transition metal and lanthanide ions.

The energy transfer probability for electric multipolar interactions can be more generally written as [16] :

$$P_{SA} = \frac{(R_o/R)^S}{\tau_s} \qquad (30)$$

where τ_s is the actual lifetime of the sensitizer excited state, including multiphonon radiative decay, and R_o is the critical transfer distance for which excitation transfer and spontaneous deactivation of the sensitizer have equal probability.

s is a positive integer taking the following values :

s = 6 for dipole-dipole interactions

s = 8 for dipole-quadruple interactions

s = 10 for quadrupole-quadrupole interactions.

It should be noted that, for dipole-dipole interactions, differences between radiative and non-radiative transfer lay essentially in that for radiative transfer there is no critical R_o depending only upon concentration. The variation goes as R^{-2} instead of R^{-6}, and the sensitizer lifetime does not depend upon the distance R.

Now, to be able to calculate effectively $P_{SA}(R)$, eq. (30) is not very useful, because R_o cannot be easily obtained theoretically. Applying Racah's tensorial methods straightaway at the beginning of the calculation of Dexter, eq. (31), permits one to develop calculations analoguous to Judd's theory for radiative transitions. The case of the multipolar interactions was treated in this way by Kushida [17] and calculations for magnetostatic, and exchange interactions were performed in our group [18].

The general form obtained is :

$$P_{SA}(R) = \frac{2\pi}{\hbar} \frac{I}{g_S^* g_{\AA}} \sum_{l_1 l_2} | C_{l_1 l_2} < S^* \| U^{(l_1)} \| S^\circ >< \AA \| U^{(l_2)} \| A^* > |^2 \quad (31)$$

with $I = \int \gamma_S(E)\Gamma_A(E)\,dE$, the so called spectral overlap integral.

where $g_S^*(g_{\AA}^*)$ is the degeneracy of the S^* (Å) level, $\gamma_S(E)$ ($\Gamma_A(E)$) is the normalized shape function of emission (absorption) spectrum. $U^{(l)}$ are the tensorial operators already seen for Judd's theory [19]. $|C_{l_1 l_2}|^2$ can be considered as a parameter analoguous to T_λ for oscillator strength [19].

This expression for transfer probabilities has the following advantages :

- Radial and orbital parts have been separated
- Only a few reduced matrix elements need be calculated. They are the same ones for the three interactions we consider (for any interaction leaving spins unchanged).
- Comparison between two different interactions can be made through comparison of $C_{l_1 l_2}$ coefficients. They are independent of the states involved in the transfer and we call them $E_{l_1 l_2}, M_{l_1 l_2}$ and $X_{l_1 l_2}$ for electrostatic, magnetostactic and exchange interactions, respectively.
- Forced electric dipole transitions as calculated by Judd's method, can be induded in (31).
- This expression gives also a single mathematical form regardless of the interactions, which is a convenient result.

The somewhat complicated expressions for the different $C_{l_1 l_2}$ of 4f electrons are given in [18]. But we can note that :

- For electrostatic interaction ($E_{l_1 l_2}$) the $l_1 = 1$, $l_2 = 1$ term, corresponding to dipolar-dipolar interaction, being zero in first order, necessitates the introduction of Judd's T_λ parameters. The $E_{l_1 l_2}$ values are between $E_{22} = 30$ cm^{-1} for quadrupole-quadrupole intensities to $E_{66} = 3\times10^{-1}$ cm^{-1}, but all contain some dipole-dipole part due to the T_λ.
- For magnetostatic interactions ($M_{l_1 l_2}$) only terms with $l_i = 1,3,5$ are nonzero and have the following order of magnitude :

  $M_{11} = 1$ cm^{-1} , $M_{55} = 2\times10^{-7}$ cm^{-1}.
- For exchange interactions ($X_{l_1 l_2}$) we have $1 < l_1 < 6$, giving estimates of 1 to 10^{-1} cm^{-1} for the coefficients.

These results show that exchange or magnetostatic interactions can be found in cases of small dipole-dipole and quadrupole electrostatic interactions if the matrix elements allow them.

Up to this point we have been dealing with the microscopic case of two ions interacting with one another. To discuss the case of real macroscopic samples with many ions and to obtain a link with experimental facts, a statistical analysis of the energy transfer is necessary.

We have then to think about the overlap integrals that, we have seen, arise in all transfer between two ions. In the microscopic case we are sure that the lineshapes involved can be only due to some homogeneous broadening even for transfer between two identical ions in different lattices sites.

In the macroscopic case, we can measure absorption and emission spectra taking into account all broadening processes averaged over the whole sample ; for instance the inhomogeneous broadening process due to emission and absorption at centers in different lattice sites. Then the overlap integral measured experimentally from the usual spectra is a measure in excess of the real overlap since we take into account emission and absorption of centers at any distances, even those which cannot interact. The error is the largest for the processes at shortest interacting distance (exchange), and so is certainly negligible for radiative transfer, since photons can travel a much larger distance than the spread of the spatial disorder. The error is also smaller for systems with small inhomogeneous broadening having centers in only one type of lattice site, that is, without disorder (crystals versus glasses).

Fluorescence line-narrowing technique could give some idea about the homogeneous part of an emission line, but the statistical analysis for the whole sample should still be performed. Supposing only a sensitizer-activator interaction, an average transfer efficiency can be calculated [20] his was stutied in some detail by Inokuti and Hirayama [21]. They considered the number of activators situated at random in a sphere around a sensitizer such that the activator concentration is constant when the volume of the sphere and the number of activator ions considered goes to infinity. Then the average probability for transfer from one sensitizer to any acceptor is :

$$W_{SA} = N_A \int_{R_{\min}}^{\infty} P_{SA}(R) 4\pi R^2 dR \tag{32}$$

Introducing eq. (32) into the expression for the intensity emitted by all sensitizers, each with different activator neighborhood, they obtained the following relation for the intensity decay of the emission of the sensitizer surrounded by many activators :

$$I(t) = \exp - \frac{t}{\tau_{so}} - \Gamma\left(1 - \frac{3}{S}\right)\frac{C}{C_o}\left(\frac{t}{\tau_{so}}\right)^{3/s} \tag{33}$$

where τ_{so} is the decay constant of the sensitizer in the absence of activators, C is the activator concentration : C_o is the critical activator concentration and s the parameter of the multipolar interaction.

The comparison between experimental decay and this theoretical expression has been widely used to determine the index of the multipolar interaction involved. This theory is valid only when there is no sensitizer-to-sensitizer transfer or activator-to-activator transfer. This formulation, therefore, has to be modified for high concentrations of sensitizers and activators. Then rapid energy migration between sensitizers or between activation is possible, because of the perfect resonance conditions. The general result is complicated [22] but, for large t, I(t) decays exponentially [23] :

$$I(t) = \exp\left(-\frac{t}{\tau_s} - \frac{t}{\tau_D}\right) \tag{34}$$

Then, two cases can be distinguished :

i) In one case, spontaneous decay of excited sensitizers, diffusion among sensitizers and energy transfer between sensitizers and activators are of about the same order of magnitude.

For sufficiently long times and dipole-dipole interactions one has [23] [24] :

$$\tau_D^{-1} = VN_S N_A \tag{35}$$

with $V = 8\pi C^{1/4} C_{SS}^{3/4}$

N_S is the sensitizer concentration ; N_A the activator concentration : C is the sensitizer-activator energy transfer constant, such as $C = R_o^s/\tau$; C_{SS} is the sensitizer-sensitizer transfer constant.

ii) For high sensitizer concentration, the diffusion rate can be faster than spontaneous sensitizer decay or sensitizer-activator energy transfer. The limiting step is no longer diffusion and D appears to saturate with increasing donor concentration ; each activator experiences the same excited sensitizer neighbourhood. We take R as the minimum distance between sensitizers as permitted by the lattice ($R = R_{min}$). One has :

$$\tau_D^{-1} = UN_A \tag{36}$$

with U a constant depending on the type of interaction as discussed earlier in this section through eq.(31).

Another approach to the macroscopic case is the use of the well-known rate equations which deal with the population of ions in a given state. This was used as a phenomenological approach in studies of lasers. The applicability of those equations in relation to the Inokuti and Hirayama statistics has been discussed by Grant [25].

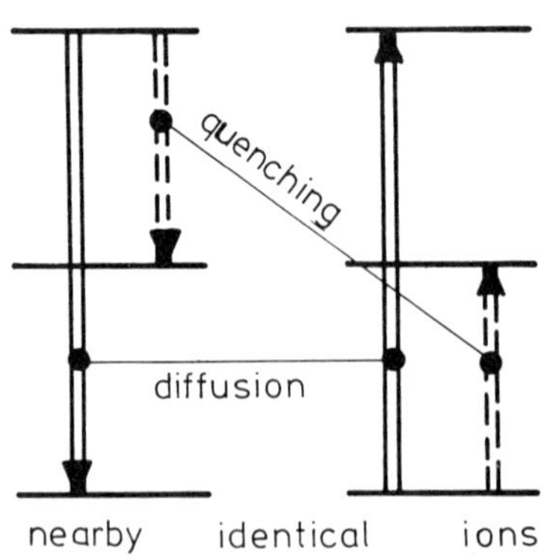

Fig. 5 . Two cross-relaxation processes : diffusion and quenching by down conversion.

The basic result of Grant is that the energy transfer probability is proportional to the activator concentration :

$$W_t = UN_A \tag{37}$$

this result is the same as obtained by fast-diffusion studies (eq. (36)).

An interesting role of diffusion is that the decays are again exponential, as when ions are not interacting. This validates the use of rate equations.

Cross-relaxation term usually refers to all types of energy transfer occuring between identical ions. In such a case the same kind of ion is sensitizer and activator.

As shown in Fig. 5, cross relaxation may give rise to the diffusion process already considered between sensitizers when the involved levels are identical or to self-quenching when they are different. In the first case there is no loss of energy whereas in the second case there is a loss or a change in the energy of the emitted photons.

Also in case level b be coupled to level a by a strong non-radiative decay, the cross relaxation (c,a) $\rightarrow$ (b,b) is the energy transfer providing the "self-quenching" of level b which impeds the photon emission of level c. Besides this, the cross-relaxation (c,a) $\rightarrow$ (a,c), providing diffusion, may help dissipating the energy of level c in case one of the ion in the sample has its level c coupled by energy transfer to some unknown "poisonous centers".

2. Up-conversion and excited state absorption

Several up-conversion processes exist which may either deviate the pump energy from the useful pumping cycle of the laser or depopulate the laser emitting level. They all have in common that they can be considered as n-photon processes more or less in resonances with the ion levels of the laser material.

Energy schemes for 2-photon up-converting processes are presented on Fig. 6 by order of decreasing efficiency as shown in Table II.

Since we are dealing with non linear effects, the quantum efficiency as usually defined by the ratio of number of photon out over the number in, increases with the excitation level ; then it has to be normalized. In the case of Table II, it has been normalized to 1 Wcm^{-2} of incident power.

A simple inspection of Fig. 6 and Table II shows that for highest efficiencies, photons have to interact with the medium a longer time, which is practically realized by the existences of resonances.

As shown, the APTE effect (Addition de Photon par Transferts d'Energie) is the most efficient up-conversion processes because it is fully resonant and involves energy transfers with diffusion. Cooperative effects take place in closely coupled pairs and have been shown to happen even is systems with concentration of active ions as small as $5 10^{18}$ cm^{-3} [27][28] because of the existence of clusters.

However in laser materials the tendancy is to have much larger concentration (10^{19} - 10^{22} cm^{-3}) in which case diffusion may take place and practically only the first two processes, APTE and excited state absorption (ESA), have to be considered.

TABLE II

Comparison between different 2-photons up-converting processes [26]

Mechanisms	Efficiency (cm^{-2}/W)	Materials
A.P.T.E.	~ 10^{-3}	YF_3:Yb:Er
2-step absorption	~ 10^{-5}	SrF_2:Er
cooperative sensitization	~ 10^{-5}	YF_3:Yb:Tb
cooperative luminescence	~ 10^{-8}	Yb PO_4
S.H.G.	~ 10^{-11}	K D P
2-photon absorption + emission	~ 10^{-13}	CaF_2:Eu^{2+}

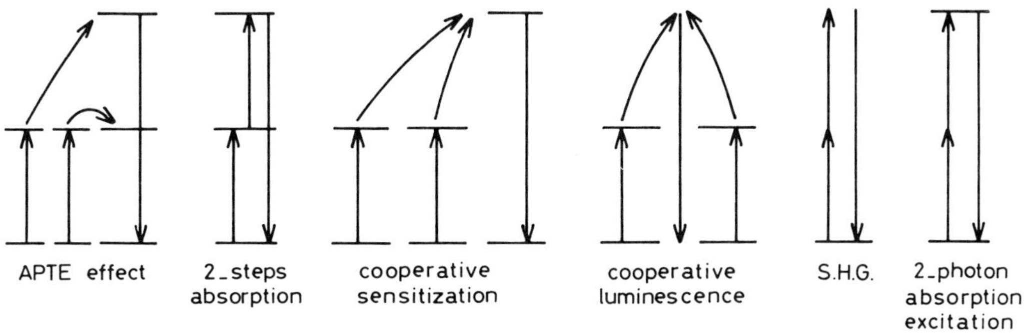

Fig. 6 . Energy schemes for different 2-photon up-conversion processes.

ESA is just ordinary absorption from a populated excited state. APTE has to be considered as a sequential energy transfer, the first step being of the type described in II.B.1.

Before 1966 all considered transfers were toward and activator in its ground state. I proposed then [29] to consider cases where activators were already in an excited state as shown on Fig. 7.

Such energy transfer are possible provided the two nearby ions are already in an excited state and that their interaction $W_{SA} > \tau_e^{-1}$; τ_e being the activator excited state lifetime. Diffusion between sensitizer ions taking always place at considered concentration, rate equations are justified to treat APTE processes and the energy transfer probability can be simply written for instance from eq. (37).

$$W_{SA} = U_{SA} N_{jA} \tag{38}$$

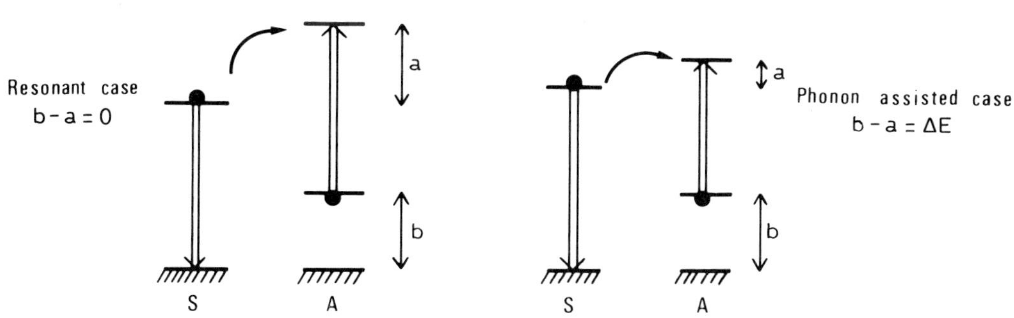

Fig. 7 . Essential step for APTE processes in resonant and non resonant cases.

II.C. Processes based on ion-ion interactions mediated by ion-phonon ones : multiphonon assisted transfers

If now we consider two ions with excited states of different energies, the probability for energy transfer should drop to zero where the overlap integral $\int g_S(\upsilon)g_A(\upsilon)d\upsilon$ vanishes. However it is found experimentally that energy transfer may take place without phonon-broadened electronic overlap provided that the overall energy conservation is maintained by production or annihilation of phonons with energies approaching $k\theta_d$, where θ_d is the Debye temperature of the host matrix [30]. Then for small energy mismatch (~ $100cm^{-1}$) energy transfer assisted by one or two phonons can take place [31][32]. However, in energy transfer between rare-earths, energy mismatches as high as several thousand reciprocal centimeters are encountered. This is much higher than the Debye cutoff frequency found in normally encountered hosts, so that multiphonon phenomena have to be considered here.

Miyakawa and Dexter [10] [33] showed that it is still legitimate to write the probability of energy transfer in a form analoguous to eq. 31, where ρ (E) is taken as S_{SA} the overlap of the lineshape functions for emission in ion S and absorption of ion A, including the multi-phonon sidebands in the lineshape. It is necessary to consider each partial overlap between the m-phonon emission line shape of ion S and the n-phonon absorption lineshape of ion A. This mathematical assumption has been given a physical meaning by our demonstration of the existence of multiphonon sidebands for trivalent rare-earth ions, even in such a case of very small electron-phonon coupling [11] [34] even when they cannot be seen by usual spectroscopy.

Along the same lines as for vibronic-sideband studies, S_{SA} can be expressed as follows :

$$S_{SA} = \sum_N e^{-(S_{oS}+S_{oA})}\frac{(S_{oS}+S_{oA})^N}{N!}\sigma_{SA}(0,0,E)\delta(N,\Delta E/\hbar\omega) \tag{39}$$

where S_{oS} and S_{oA} are the respective lattice coupling constants for the ions S and A, N the order to the multiphonon process with $N = \Delta E / \hbar\, \omega_m$, ΔE the energy mismatch between both ions, and $\hbar\, \omega_m$ the phonon cutoff frequency. σ_{SA} (0,0,E) is the zero-phonon overlap integral between S and A. A "Pekar" function is still found.

The expression for S_{SA} with an energy mismatch of ΔE for small S_o constant and for an occupation number n= $(\exp(\hbar\,\omega / kt) - 1)^{-1}$ not exceeding unity at the operating temperature, can be approximated with Stirling's formula by :

$$S_{SA}(\Delta E) = S_{SA}^{(o)}e^{-\beta\Delta E} \tag{40}$$

where S_{SA} (0) is the zero-phonon overlap between S and A in the case where there is no energy mismatch between the two ions. β is given by :

$$\beta = (\hbar\omega)^{-1}\,(logN/S_o(\bar{n}+1) - \log\left(1-\frac{S_{oA}}{S_{oS}}\right) \equiv \alpha_{NR} - \gamma \simeq \alpha_{NR} - ln2 \qquad (41)$$

involving α_{NR} the non-radiative decay parameters and assuming identical electron-phonon coupling for ions A and S. This exponential dependence on energy-mismatch agrees well with experiment [35].

Such multiphonon assisted transfers are operating both for activator in ground or in excited state (APTE case) as depicted on Fig. 7.

III. NON-RADIATIVE PROCESSES AND LASER FUNCTIONING

III.A. Positive aspects versus negative ones in fixed frequency lasers

The typical general level scheme of an ideal 4-level fixed frequency laser is given on Fig. 2.

In such scheme pumping transition is connected to laser emitting state by fast non-radiative decay ; the laser final state itself is also connected to ground state by a fast non-radiative decay in order to obtain a 4-level functioning.

1. Pumping simultaneously at several wavelengths

The first conditions can be generalized to several excited state above the laser emitting state. Then several wavelengths may be used to pump the laser state and this is particularly necessary when incoherent broadband pumping sources are used such as black-body types (quartz iode, flashlamps). Of course since non-radiative processes are used, the wavelengths further remote from the useful excitation create heat dissipation at the expense of the overall efficiency. However the laser effect is obtained even at such price (e.g. YAG:Nd^{3+} pumped by lamp.) and if one consider for instance YAG:Nd^{3+} [36], Fig. 8 shows that quantum efficiency for pumping is undependant of pumping wavelength.

2. Practical conditions for a 4-level laser material

The ideal situations of Nd^{3+} is due to the fact that the energy scheme of Nd^{3+} presents one single rather large gap (~ 6000 cm^{-1}) between $^4F_{3/2}$ and $^4F_{15/2}$ on each side of which there are closely packed levels separated by less than 2000 cm^{-1} ; together with maximum phonon frequency of the order $\hbar\,\omega_m$ ~ 700 cm^{-1} for YAG.

Making use of the exponential gap law described in II.A, such ideal compromize can be reflected by the following inequalities :

In order to insure fast decay between excited states one needs :

$$\Delta E_e/2 \lesssim \hbar\omega_m \qquad (42)$$

where ΔE_e is a general energy gap between excited states above laser state.

In same maner the fast decay from the final laser state implies :

$$\Delta E_g/2 \lesssim \hbar\omega_m \qquad (43)$$

In order to avoid thermal reabsorption through such energy gaps at given temperature T :

$$kT < \Delta E_e, \Delta E_g \qquad (44)$$

Now to keep a good quantum efficiency at the laser transition (h υ) one needs as a rule of thumb derived from the exponential gap law :

$$4\hbar\omega_m < h\upsilon \qquad (45)$$

such inequalities are summarized as :

$$\frac{kT}{2} < \frac{\Delta E_g}{2}, \frac{\Delta E_e}{2} \leq \hbar\omega_m < \frac{h\upsilon}{4} \qquad (46)$$

which is nearly verified for YAG:Nd^{3+} for instance as :

$$\frac{207}{2} < \frac{2000}{2}, \frac{1000}{2} \leq 700 < 2500 \quad (\text{in}\, cm^{-1})$$

Now if a fluoride host is considered there could be a problem due to the smaller phonon frequency of such hosts. It is typically the case for the $LiYF_4$:Er^{3+} laser at 2.7 μm ($^4I_{11/2} \rightarrow {}^4I_{13/2}$ transition). Because the laser frequency (~ 3300 cm^{-1}) is at smaller energy inequality (46) is not fulfilled :

$\Delta E_g/2 = 3000 cm^{-1}$ is not smaller than h υ/4.

This is reflected in the longer lifetime of the final state $^4I_{13/2}$ as shown on Fig. 9. The 4-level behaviour is likely to stop soon after begining because of the bottleneck of population in $^4I_{13/2}$ created by the laser effect itself.

3. Lifetime engineering

As shown on Fig. 10. a careful choice of the Er^{3+} concentration allows some correction to the lifetime inequality since $^4I_{13/2}$ is more sensitive to self-quenching than to $^4I_{11/2}$. It is also known [37] that adding Ho^{3+} reduces $^4I_{13/2}$ lifetime more efficiently than $^4I_{11/2}$. Also the dynamics of excitation may give some correction to the limitation in $^4I_{13/2}$ deactivation either by reabsorption of the pump (by ESA) from $^4I_{13/2}$ or by up-conversion (APTE effect) emptying of this state [38] [39].

The selective reduction of the $^4I_{13/2}$ lifetime is the object of current researches.

III.B. The vibronic tunable laser case ; the compromize between tunning range and laser threshold

Since multiphonon side-bands provide the gain shape in vibronic tunable laser materials, when searching for a laser material with a large tuning range, a large energy effective phonon mode ($\hbar\omega$) with a large S_o should be looked for. However in cases where the same mode is also the accepting mode of the multiphonon non-radiative process induced by H_{NA}. eq. 29 shows also that large S_o and large h ω imply that non-radiative transitions may take place even for large separation between electronic levels (ΔE), "short circuiting" in some way radiative transitions even at T=OK, (Figure 3).

1. Experimental determination of S_o and effective hω for non radiative transitions

Non-radiative probabilities are obtained by fluorescence lifetime measurement versus temperature :

$$W_{NR}(T) = \frac{1}{\tau(T)} - \frac{1}{\tau_o}$$

where τ_o in the radiative lifetime. Then a fit with eq. 28 gives S_o and $\hbar\omega_{eff}$. Results are presented on Figures. 11 and 13. respectively for MgF : 2 % Ni^{2+} and 10 % Ni^{2+} and for $MgAl_2O_4$:2 % Ni^{2+}. τ_o is determined by absorption oscillator strength measurement with the assumption of being constant with temperature. This means that we assume that there are no dynamically induced transitions.

The validity of this assumption can be estimated from the following absorption integrated intensity ratios for different temperatures (Table III), as compared with coth ($\hbar\omega$/2kT) which would be given for the dynamically radiative induced part [40].

Though some variation is found, it is systematically smaller than what should be for dynamically induced transitions and justify us to neglect it in a first approximation.

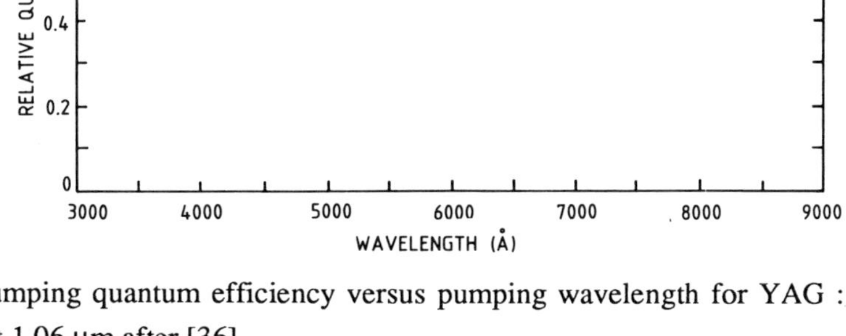

Fig. 8 . Pumping quantum efficiency versus pumping wavelength for YAG : Nd^{3+} (1 %) emission at 1.06 μm after [36].

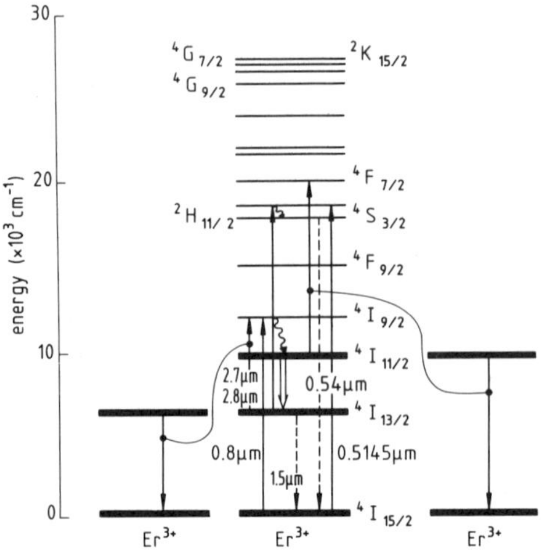

Fig. 9 . Energy levels of Er^{3+} involved in 2.7 - 2.8 μm laser emission [38].

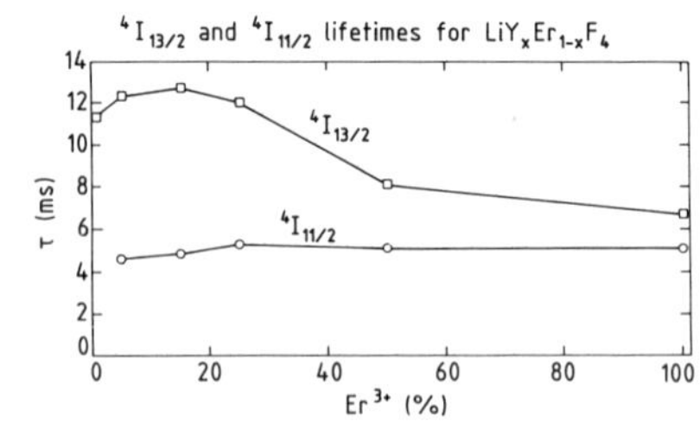

Fig. 10 . Lifetime evolution with concentration for $LiY_xEr_{1-x}F_4$ as laser material for $^4I_{11/2}$ -> $^4I_{13/2}$ transition under weak excitation.

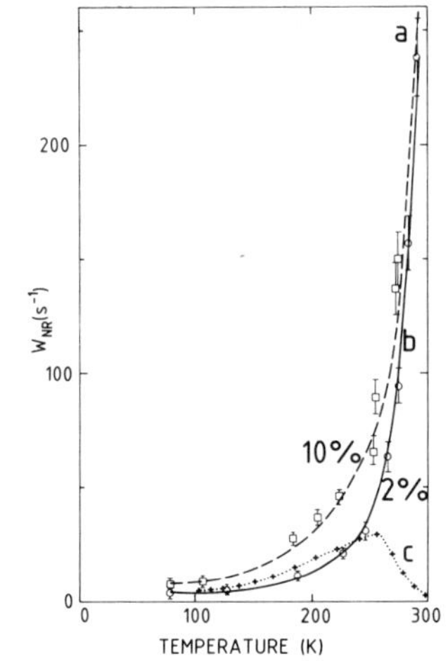

Figure 11 . Multiphonon non-radiative probability W_{NR} versus temperature and fit with eq. 28 for a) MgF_2:10% Ni^{2+} and b) MgF_2:2% Ni^{2+} ; curve c is a-b.

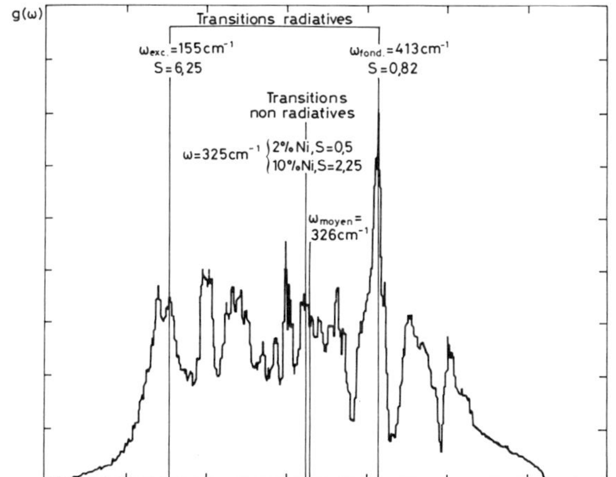

Figure 12 . Effective phonon energies and coupling strength for non-radiative and radiative multiphonon transition in MgF_2:Ni^{2+} in relation with phonon density of modes [41].

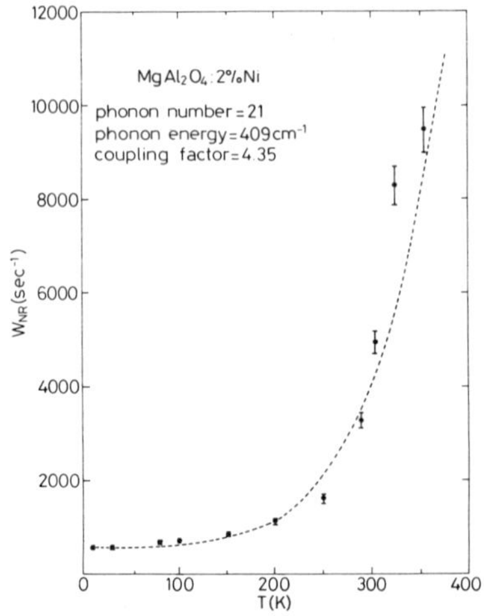

Figure 13 . Multiphonon non-radiative probability fit to eq. 28 for $MgAl_2O_4$:2% Ni^{2+}.

TABLE III

$^3A_2 \rightarrow {}^3T_2$ integrated intensity ratios normalized to 10K

T / Sample	10 K	80 K	290 K
MgF_2:2%Ni	1	1.08	1.24
coth ($\hbar\omega/2kT$)	1	1.006	1.5
$MgAl_2O_4$:2%Ni	1	1.01	1.13
coth ($\hbar\omega/2kT$)	1	1.01	1.3

2. <u>Determination of S_o and $\hbar\omega$ for radiative transitions</u>

Since we used a one-dimensional configuration diagramm though we are in a multidimensional one, we are justified to see wether or not mediating modes are different for radiative and non-radiative transitions. Huang has pointed out a few years ago [42] that since energy spanned by the multiphonon process for non-radiative and radiative transition are much different, (figure 3) the coupling and energy of mediating mode should also be different.

Assuming electron-phonon coupling values larger than 1 one can fit the widths of the multiphonon absorption and emission bands versus T with the following relations :

$$W_e(T) = W_e(0)\,[\tanh(\hbar\omega_{ex}/2kT)\,]^{-1/2}$$
$$W_a(T) = W_a(0)\,[\tanh(\hbar\omega_f/2kT)\,]^{-1/2} \tag{47}$$

respectively for emission and absorption band where $\hbar\omega_{ex}$ is for the electronic excited state and h ω_f for the ground one. Now it is possible to obtained the corresponding coupling value by looking at energy differences between band maxima and 0-phonon line :

$$\Delta E_e = S_e\hbar\omega_f$$
$$\Delta E_a = S_a\hbar\omega_{ex} \tag{48}$$

Results are presented in Figures 12 and 14 respectively for $MgF_2:Ni^{2+}$ and $KZnF_3:Co^{2+}$ on known phonon density of modes [41] [43].

The common feature is that mediating modes for non-radiative transitions correspond in each case to the average phonon frequency, though the coupled effective mode for radiative transitions may be at higher or lower energy. There seems to be a tendancy for coupling value to be the stronger the lower the phonon energy.

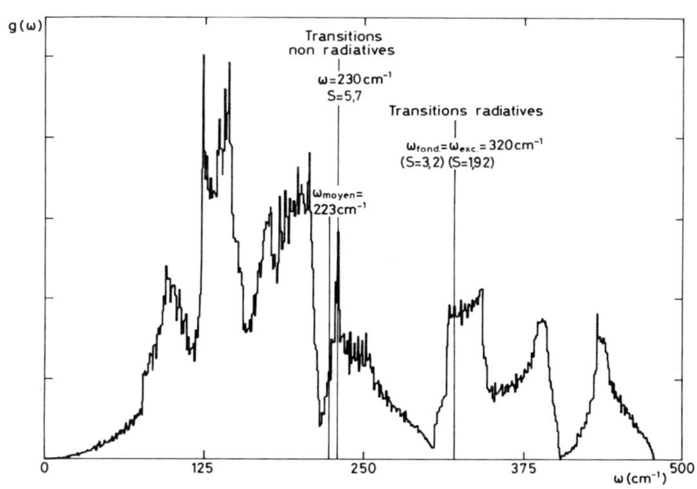

Figure 14 . Effective phonon energies and coupling strength for non-radiative and radiative multiphonon transitions in $KZnF_3:Co^{2+}$ in relation with known phonon density of modes [43].

3. Merit factor for multiphonon processes in tunable laser materials

Hopefully, phonon modes have been seen to be different for non-radiative and radiative transitions and consequently conditions for large tuning range (radiative multiphonon process) and good efficiency (non-radiative multiphonon process) may be decoupled. The compromize is to look for high ratios :

$$F = \frac{S_e \omega_f}{S_o \omega_{nr}} \tag{49}$$

which maximize simultaneously tuning range ($S_e \hbar\omega_g$) and quantum efficiency ($1/S_o \hbar\omega_{nr}$).

Results are presented in Table IV :

TABLE IV

Multiphonon processes merit factor (F) for different tunable laser materials

Materials	$MgF_2 : Ni^{2+}$	$KZnF_3 : Ni^{2+}$	$MgAl_2O_4 : Ni^{2+}$	$MgF_2 : Co^{2+}$
doping level	2 - 10 %	0.55 %	2 %	1.5 - 5.3 %
F	2.1 0.46	0.14	0.56	2.13 0.86

Of course F characterizes only part of the useful properties for a laser material : it does not contain cross-sections.

IV. RECENT ADVANCES AND EXAMPLES TAKEN FROM THE FRONTIER OF RESEARCH

IV.A. PROMOTING MODES AND EFFECTIVE ENERGY GAP REDUCTION

As already pointed out in II.A.3., Huang and Rhys had assumed equality of (N-1), (N-2) with N in their final calculation. Taking into account such difference is equivalent to the reduction of the energy gap by the energy of at most 2 phonons as we had pointed out in the previous school [4]. Such reduction comes from the necessity to provide energy to the promoting modes which initiate the non-radiative transition. It can be viewed as a phonon enhancement of the non-radiative transition when comparing with radiative ones where there is no need to promote the radiative transition by phonon energy when of course it is allowed. In case the transition be forbidden in first order and allowed by the existence of vibrations, we predicted by analogy that the same enhancement could be observed for Franck Teller radiative transitions [4].

However, such role of promoting modes for non-radiative transitions have been "rediscovered" empirically [44] pointing to the interesting fact that the preexponential factor for the exponential gap law is nearly constant for various materials when the reduced effective energy gap ($\Delta E \sim 2\hbar\omega$) is used instead of the plain gap (ΔE). As shown in Fig. 15 there is a crossing point for the various materials with preexponential factor of value about $10^7\,s^{-1}$ for 4f-4f transitions ; it corresponds to the energy differences $\Delta E \sim 2\hbar\omega$. So the used of the "effective" gap reduces the spreading of preexponential factors by 5 orders of magnitude. This is a new result of practical importance. The physics behind such fact can be understood by the relationship which seems to exist between radiative and non radiative electronic matrix elements in eq.11 and 14 as developped below.

IV.B. <u>Relationship between radiative and non-radiative probabilities</u>

The possibility to connect the radiative matrix element to the electronic matrix elements appearing through the non-radiative hamiltonian was first pointed out by Huang and Rhys [6]. They could express Z^2 in eq.11 in term of the electric dipole matrix element $<|eR|>$.

However such property has been looked for in more details [44] by Van Dijk and Schuurman since a seminar at the preceeding school [4]. The connexion is obtained by noting that both electronic matrix elements in (5) and (14) may be given in term of the interaction potential V(r, R) between electrons and nuclei.

$$<\phi_e|\frac{\partial}{\partial R_i}|\phi_e'>=<\phi_e|\frac{1}{\Delta E(R)}\frac{\partial}{\partial R_i}V(r,R)|\phi_e'> \tag{50}$$

and

$$<\phi_e|\sum_j r_j|\phi_e'>=\frac{h^2}{2m\Delta E(R)}<\phi_e|\frac{\Sigma_j\frac{\partial\, V(r,R)}{\partial rj}}{\Delta E(R)}|\phi_e'> \tag{51}$$

Similarity between both matrix elements comes from the symetry in r_{el} and R_a in :

$$V(r,R)=-\sum_{a,el}\frac{Z_a e^2}{|r_{el}-R_a|} \tag{52}$$

where Z_a is the nuclear charge of atom a.

For the trivalent rare-earth case, taking the exponential gap law approximation for the Franck-Condon Factors, the relationship between non-radiative and radiative probabilities though eq. (50) and (51) can be obtained practically as below [44] :

$$P_{NR}=\frac{40\pi c^3}{M_i\Delta E_o}CP_R\exp-(\Delta E_o-2\hbar\omega_{\max})\alpha_{NR} \tag{53}$$

where C as a value between 1 and 10, M_i is the mass of the enghbouring atoms of the rare-earth center. That is the order of magnitude of the preexponential factor can now be estimated for P_R the radiative probability. Here the promoting mode energy is taken into account by $2\hbar\omega_{max}$. Of course when the transition is of magnetic dipole type the relationship does not hold. An experimental verification of relationship (53) can be found in the case of transition $^5D_1 \rightarrow {}^5D_0$ (Eu^{3+}) which is forbidden even through odd crystal field mixing : the corresponding non-radiative transition has been found 2 order of magnitude below the exponential gap law [45].

Assuming that radiative probabilities in 4f -4f systems are given by oscillator strength with a constant order of magnitude ($f \sim 10^{-6}$) one can now understand why the preexponential factor for non-radiative transition in 4f-4f system is also a constant.

IV.C. Up-conversion and Auger like effect in Lanthanides doped semi-conducting materials

Frequently now, active ions are embedded into a semi-conducting host. Such is the case of the well known electroluminescent system ZnS:Mn [46] or CdF_2:Gd^{3+} or Tb^{3+} [47] ; also more recently even III-V compounds doped with Yb^{3+}, Er^{3+} are considered [48].

Then it has been demonstrated [46] [47] that Auger effect may be a more efficient non-radiative process than multiphonon non-radiative decay studied up to this point. A simple experimental proof of the role of conduction electrons in the non-radiative process is that high resistivity ZnS:Mn is an efficient phosphor whereas low-resistivity one is not [46].

Such effect is found to have a propabilities proportional to the carrier concentration n [47] [49] :

$$P_A = C_A \, n \tag{54}$$

which is valid as well for free carrier or when the Auger process results in the ionization of the shallow donors.

A critical carrier concentration (n_o) is defined by equality between radiative emission and Auger quenching :

$$C_A n_o = \frac{1}{\tau} \tag{55}$$

As shown on Fig. 16 such Auger process can be view as an up-conversion process of APTE type from the localized donor to the delocalized carrier or to the localized shallows donor. Calculations of the corresponding energy transfers have been performed by Langer [47] assuming for free carrier the coulomb interaction. This leads to a critical carrier concentration of the form :

$$n_o = 4\pi^{5/2} \left(\frac{a_o m_o}{\alpha m^*}\right)^{1/2} \quad n_r^5 \lambda_o^{-7/2} \tag{56}$$

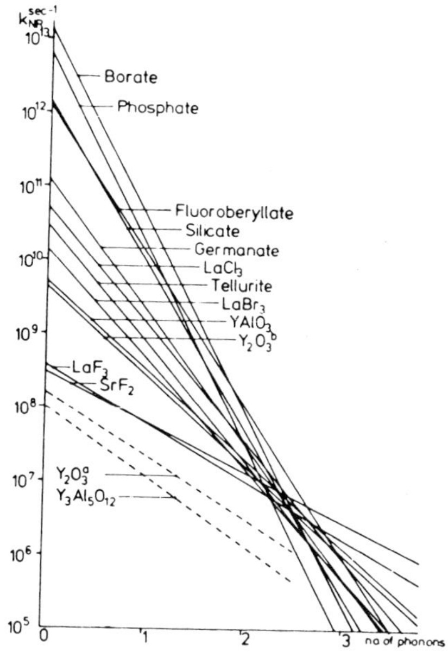

Figure 15 . Multiphonon non-radiative decay versus energy gap given in number of phonon after [44].

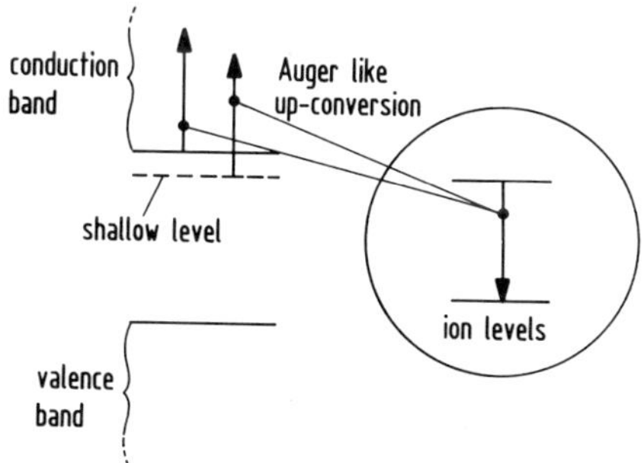

Fig. 16 . Auger-like non-radiative process in semi-conducting materials.

where λ_o is the quenched transition wavelength, a_o is the Bohr radius (= 0.529 Å) ; α=1/137 ; n_r is the refractive index. The integration over semi-conductor band states is performed in the parabolic approximation : $E(k) = \hbar^2 k^2/2\, m^*$.

For most semi-conductor $2 < n_r < 4$ and with λ_o=1μm one obtains : $2\ 10^{14} < n_o < 6\ 10^{15} cm^{-3}$, showing that such non-radiative process is effective at quite low carrier concentration .

Recently we gave estimate for quantum efficiency of the $^4F_{5/2} \rightarrow {}^4F_{7/2}$ transition of Yb^{3+} in InP [50] ; we show that it is much lower (~ 10^{-4} - 10^{-3}) than for GaAs:Er^{3+} for which estimates is about a few 10 %. For crystal field strength reasons it can be shown that Yb^{3+} enters readily the InP host whereas Er^{3+} stays in an insulating compound protecting it from the free carrier influence. The Auger process above explains quite well the difference in quantum efficiencies estimates which are obtained.

Such Auger process can also be applied to the case of quenching by a metallic surface near an emitter ; again the description can be of the energy transfer type [51].

IV.D. Concentration enhanced electron-phonon coupling ; role of spin-orbit coupling in diffusion for system with strong electron-phonon interaction ; phonon correlation.

1. Effect of concentration on the measure of S_o (Example of MgF_2:Ni^{2+}).

Usually electron-phonon coupling in transition-metals ions in considered as a one-center process. However by fitting the thermal behaviour of measured multiphonon non radiative transition probabilities to a one effective phonon model, we have observed an increase in coupling strength (Huang-Rhys parameters S_o) with concentrations (Figure 12). Physically it could be view as an increase in the spatial region probed by the electrons which are exchanged during the ion-ion interaction, as would be the case for electrons with larger extension wave functions. Yet, even though thermal behaviour is correctly described (Figure 11), low temperature non-radiative probability should vary more than found when S_o varies from 0.5 to 2.25.

Taking the probability difference at same temperature for two concentrations (2 % and 10 %), Figure 11, a bump is found at 250K which corresponds to the shape also found for self-quenching rate versus temperature (Figure 17). Consequently an increase in coupling can be view, as an increase in diffusion with concentration. This points to the special role plays at different stage by mode at ~ 160-165 cm^{-1} -~ k 250 and to the intricacy of the different non-radiative processes with radiative ones when diffusion is involved.

Such effect of concentration on electron-phonon coupling had already been notice in Eu^{3+} doped systems for excitation spectra [52] and for luminescence ones [53], explanation was through diffusion between nearby centers increased by the presence of a charge transfer state. Though self-quenching by electron-phonon coupling enhancement could have been suspected [53], in the case of MgF_2:Ni^{2+} such effect is directly observed.

2. Self-quenching and role of spin-orbit coupling in diffusion for systems with medium to high S_o

Self-quenching properties are important limiting processes in laser materials. Though rather widely studied in 4f ions systems [3], not much is known of self quenching of 3d

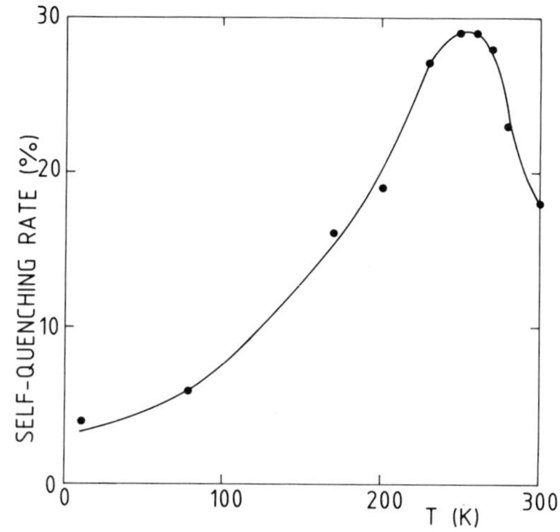

Figure 17 . Role of temperature on self-quenching rate MgF_2:Ni^{2+} ; rate $(\tau_{0.2\%} - \tau_{10\%}/\tau_{10\%}))$

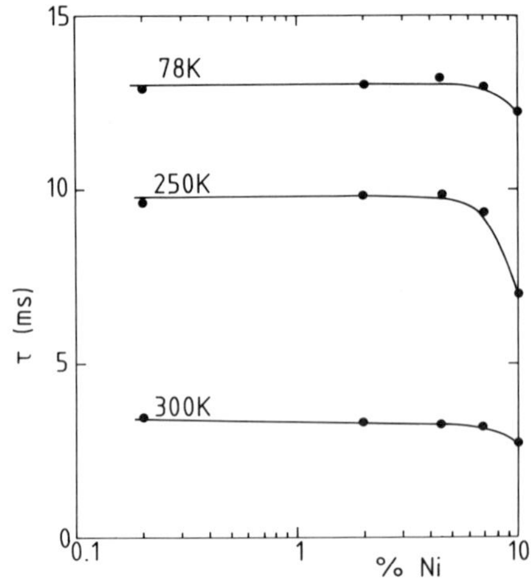

Figure 18 . Effect of concentration on the $^4T_2$ ($MgF_2:Ni^{2+}$) lifetime.

ions involved in vibronic laser materials [54] such as Ni^{2+} and Co^{2+}. We shall report here on a study of the role of active ion concentration on quantum efficiency [54]. Examples are on $MgF_2:Ni^{2+}$ (figure 18) with some results on $MgF_2:Co^{2+}$. Since the laser emitting state considered in such tunable laser systems is the first electronic excited state, the usual cross-relaxation considered for 4f laser systems cannot be advocated here ; other effects such as diffusion to sinks, and concentration enhanced electron-phonon coupling are investigated. All those processes are mediated by larger values (S_o=0.5 to 9) than in 4f (S_o=0.05 to 0.1).

When cross relaxation cannot be considered, resonant diffusion to sinks whatever they could be is the most likely source of self-quenching. Due to Stokes-shift the spectral overlap takes place for identical ions at low temperature only through the O-phonon lines : then the stronger the coupling, the smaller the diffusion.

Comparing concentration for onset of self-quenching in ruby and in $MgF_2:Ni^{2+}$ or Co^{2+} demonstrates clearly such general behaviour (Table V).

However, in MgF_2 there exist some important spin-orbit splittings (ΔE) which may interfer with this simple view : also the coupled phonon modes ($\hbar\omega$) may be in resonance with ΔE. To demonstrate this effect, we use simple one coupled phonon line shapes. Without spin-orbit splittings and supposing the two identical ions to interact independently with the phonon mode, one may write for the spectral overlap :

$$I = \int G_a(E) G_e(E)\, dE \tag{57}$$

with

$$G_a(E) = e^{-s} \sum_{p=-\infty}^{\infty} e^{P\theta} Ip\left(\frac{S_o}{\mathrm{Sinh}\,\theta}\right) \delta(E - \upsilon_o - p\hbar\omega) \tag{58}$$

Table V

Self-quenching (S.Q.) onset concentrations for various electron-phonon coupling (S_o)

Materials	$Al_2O_3 : Cr^{+3}$	$MgF_2 : Ni^{3+}$	$MgF_2 : Co^{3+}$
S.Q. onset concentration (cm^{-3})	6.10^{19}	3.10^{21}	5.10^{20}
So $\hbar\omega$ (cm^{-1}) (Radiativ. or Non Rad.)	≤ 0.02 375 (R)	6.2;05-2.2.;0.8 160 ; 325 ; 413 (R) (NR) (R)	4.9;2,7-;4-7 139 ; 270 ; 320 (R) (NR) (R)
ΔE (cm^{-1})	29	~ 165	130 - 1398

as usual for the absorption line shape, where $S = S_o(2n + 1)$, $\theta = \hbar\omega/2kT$, Ip is the modified Bessel function of order p ; n, $\hbar\omega$, S_o are respectively the occupation number for the phonon mode of energy hω with Huang-Rhys coupling S_o at 0K ; υ_o is O-phonon transition energy. G_e (E) is obtained by changing p → -p ; n → (n+1) ; (n+1) → n. From this, the overlap integral, is found to be :

$$I = e^{-2s}(2I_1^2 + I_o^2) \tag{59}$$

where the approximation $I_{p=0} = 0$ for $S_o/\sinh\theta < p$ has been assument to hold when $S_o = 0.5$; $\hbar\omega = 160$ cm^{-1} which is the mode found to be coupled in radiative transitions [54] : and for T<400K the range to of our studies.

Assuming a spin-orbit splitting ΔE for excited state (59) becomes :

$$I_{SO} = e^{-2s}(2I_1^2 + I_o^2)(1 + e^{-\Delta E/kT}) \tag{60}$$

Now when the order of magnitude of ΔE falls in the range of phonon modes coupled to radiative transitions, fortuitous resonances may happen ($\Delta E = \hbar\omega$) in which case more terms enter I_{SO} corresponding to $\delta^2(E - \upsilon_o - p\hbar\omega)$ with p=0,1,2 and (60) becomes :

$$I_{SOR} = e^{-2s}[(2I_1^2 + I_o^2)(1 + 2e^{-2\theta}) + 4I_1I_oe^{-\theta} \tag{61}$$

Such increases in overlaping are shown on Figure 19 demonstrating this effect which explain the general behaviour found in Table V : decrease in overlaping induced by electron-phonon coupling may be compensated by spin-orbit splittings in resonance with phonons ($T \neq 0$).

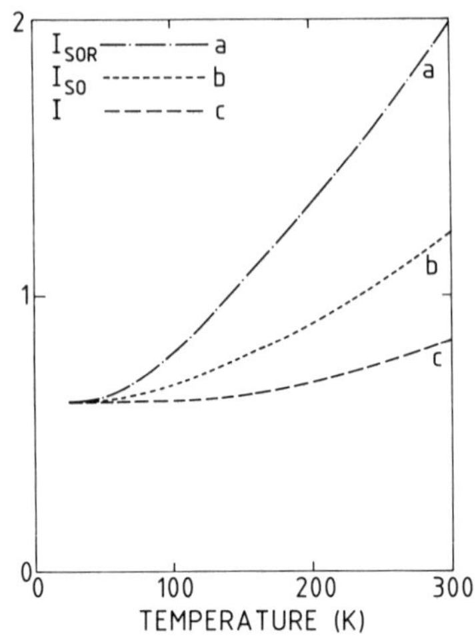

Figure 19 . Overlap integral functions I, I_{SO}, I_{SOR} versus temperature.

To summarize this chapter we can say that both approaches : diffusion to sinks and coupling enhancement are physically equivalent except for the final dissipative step. In the first case it is some non-radiative sink, for the second one it is the phonon field ; the first leaves a possibility of improvement, the second one is of a more fundamental limiting nature.

Now role of concentration in diffusion of energy may be found at two different stages : the usual one corresponds to the macroscopic statistical effect of concentration for which diffusion probabilities are found proportional to sensitizer and/or activator concentration [25] ; the other stage corresponds to the phase relationship between two sites for a given vibrations [55] which we have neglected in eq. 39 and (57) : an increase of the overlap integral itself should be found. Taking average elastic constants for fluorides, an increase overlap by a factor 1.5 could be predicted when increasing Ni^{2+} from 2 to 10 % in MgF_2.

This last effect is the phonon correlation taken into account by Kohli and Liu in their study of energy transfer between magnetic impurity [55]. Though known for a long time it has been generally ignore in other works. We think that in heavily doped systems it could play a non negligeable role.

IV.E. Stimulated non-radiative transitions and energy transfers ; cooperative radiationless transitions and superfluorescence

1. Stimulated non-radiative processes

It has been recently proposed [56] that under a high intensity laser pumping field stimulated non-radiative transitions or/and stimulated energy transfers between two ions could take place. Calculations are similar to resonance Raman scattering however with a very small energy difference ~0.01 - 0.1 cm^{-1} between incident and scattered laser beam. Because considered interactions are electron-radiation and electron-phonon or ion-ion types, there appears a multiplication factor to the usual non-radiative decay or transfer probability such that :

$$W_{Stimul.NR} = R W_{N.R.} \qquad (62)$$

and

$$W_{Stim.a \to b} = R\, W_{a \to b} \qquad (63)$$

with

$$R \sim \varepsilon^4 \frac{(\mu)^4}{\Delta^4} \qquad (64)$$

where ε is the pumping electric field ; μ the electric dipole of the transition between the two considered level ; Δ is their with. With MW laser, for $<\mu> = 1$ Debye and $\Delta = 1 - 5$ cm^{-1}, R can reach 10^5 to 10^8. This indicates that such effect could be more important than spontaneous energy transfer or non-radiative decay. It has been proposed [56] that a trace of the existence of this effect could be the line narrowing of excitation spectra of stimulated emission of $Na_5\ La(MoO_4)_4 : Nd^{3+}$ powders [57]. However because stimulated emission is non-linear at threshold such excitation spectra narrowing can be expected as was demonstrated for other non-linear processes such as up-conversion [26].

2. Cooperative non-radiative processes or "supernonradiance"

This proposal [58] is the superradiance equivalent for phonons of the photon superradiance of Dieke in the same way as stimulated nonradiative decay was the equivalent of stimulated emission for photons.

The basic result of "supernonradiance" is the same as for superradiance :

The non-radiative decay should be :

$$W_{COOP\,NR} \sim \frac{1}{4} N^2 W_{NR} \qquad (65)$$

where W_{NR} is given by eq.21 and N is the number of cooperating atoms which have been prepared in an excited state. In photon superradiance the sample can be larger than λ^3 ; it is believed, that supernonradiance could survive in large sample too, provided the phonon mean free path (l_c) be larger than the distance between two centers (δ). In this connection it is interesting to observe that if the efficiency of the non-radiative channel for the single atom decay is much greater than the radiative one so that fluorescence is not normally observed, and if the smallness of l_c ($<\delta$) precludes supernonradiance, then the enhancement of the radiative channel by superradiance of photon from same prepared N cooperating atoms may cause fluoresence from levels that normally decay non-radiadively. Though there exists up to now no experimental proof of this type of process in Lanthanide ions, the possibility is may be not so far remote since we have obtained quite recently indication of superfluoresence (incoherently excited superradiance) for the $^4I_{11/2} \rightarrow {}^4I_{13/2}$ transition of Er^{3+} [60] in a case where we have certainly the condition $l_c < \delta$ and a low quantum efficiency of about 6 %.

It means that superfluoresence could be the indication of a bottleneck in non-radiative decay, an opposite effect to stimulated non-radiative decay.

V. CONCLUSION

Though the physics of basic non-radiative processes are well understood, due to the complexity of real situations, their actual description is still rather rough because the number of considered effective phonon modes is always very limited (usually even to one).

The full implication of such processes in laser materials is only now being realized and the simple descriptions of this lecture are useful in this context.

However in a number of cases the approximate descriptions used up to now are clearly unsuffisient : such is the case for the apparent increase of the electron-phonon coupling parameter following an increase in active ion concentration.

Also the implications of the existence of the connexion between radiative and non-radiative matrix elements have not yet been pushed to their full consequences.

In the developping interest for doped semi-conductors the Auger-like up-conversion begins only to be recognized as a very efficient non-radiative process for ions internal transitions.

Finally, proofs of the existence of stimulated non-radiative and "supernonradiant" transitions have yet to be found by experimentalists and consequences of their existence have still to be imagined.

REFERENCES

[1] F. Auzel, D. Meichenin and H. Poignand, Electron. Letters 34 909 (1988)

[2] Laser Focus World Buyer's guide ed. M.R. Levitt, Penwell Publishing Co. Westford MA USA (1989), p.126

[3] F. Auzel, in Spectroscopy of Solid-State laser-Type Materials ed. B. Di Bartolo Plenum Publishing Co, (1987), p.293

[4] F. Auzel, in Luminescence of Inorganic Solids ed. B. Di Bartolo (Plenum Publishing Co (1978) p. 67

[5] R. Engleman, Non-Radiative Decay of Ions and Molecules in Solids North-Holland, Amsterdam (1979)

[6] K. Huang and A. Rhys. Proc. Roy. Soc. A204 406 (1950)

[7] Y. Toyozawa, in Dynamical Processes in Solid State OPtics ed. R. Kubo, Benjamin (1967) p.90

[8] F. Auzel, in Radiationless Processes ed. B. Di Bartolo and Velda Goldberg, Plenum publishing Co New-York (1980) p. 213

[9] Y.E. Perlin, Sov. Phys. Uspecki 6 542 (1963)

[10] T. Miyakawa and D.L. Dexter, Phys. Rev. B1 2961 (1970)

[11] F. Auzel, Phys. Rev. B13 2809 (1976)

[12] F. Auzel and J.M. Breteau, Journal de Physique 48 C7-451 (1987)

[13] N.F. Mott and R.W. Gurney, Electronic Processes in Ionic Crystals, the Clarendon Press, Oxford (1948)

[14] F. Auzel, in Eneergy Transfer Processes in Condensed Matter ed. B. Di Bartolo, Plenum publishing co NYC (1984) p.497

[15] F. Auzel, ibidem. p. 511

[16] D.L. Dexter, J. Chem. Phys. 21 836 (1953)

[17] T. Kushida, J. Phys. Soc. Jpn. 34 1318 (1973)

[18] J.F. Pouradier and F. Auzel J. de Phys. 39 825 (1978)

[19] B.R. Judd, Phys. Rev. 127 750 (1962)

[20] F. Auzel, Ann. Telecom. 24 363 (1969)

[21] M. Inokuki and F. Hirayama, J. Chem. Phys. 43 1978 (1965)

[22] M. Yokota and O. Tanimoto, J. Phys. Soc. Jpn, 22 779 (1967)

[23] M.J. Weber, Phys. Rev. B4 2932 (1971)

[24] N.V. Artomonova, C.M. Briskim, A.L. Burshtein, L.D. Zusman and A.G. Skleznen, Sov. Phys. JETP 35 457 (1972)

[25] W.J.C. Grant, Phys. Rev. 109 648 (1958)

[26] F. Auzel, in Pare Earth Spectroscopy, eds. B. Trzebiatowska, J. Legendziewicz and W. Strek, World Scientific, Singapore (1985), P. 502

[27] J.C. Vial, R. Buisson, F. Madeore and M. Poirier, J. de Phys. 40 913 (1979)

[28] M.P. Miller and J.C. Wright, J. Chem. Phys. 71 324 (1979)

[29] F. Auzel C.R. Acad. Sci. (Paris) 262 1016 (1966)

[30] J.D. Axe and P.F. Weller, J. Chem. Phys. 40 3066 (1964)

[31] R. Orbach, in Optical Properties of ions in Crystals eds. H.M. Crosswhite and H.W. Moos Intersciences Publ. (1967) p.445

[32] R. Orbach, in Optical Properties of ions in Solids ed. B. Di Bartolo, Plenum NYC (1975) p.355

[33] T. Miyakawa and D.L. Dexter, Phys. Rev. B1 70 (1970)

[34] F. Auzel, in Elements de Transition, Elements Lourds dans les solides. ed. F. Gaume, Editions du CNRS, Paris (1977) p.143

[35] M.J. Weber, Phys. Rev. 138 54 (1973)

[36] R.A. Brandewie and C.L. Telk, J. Optic. Soc. Am. 57 1221 (1967)

[37] J. Rubin, A. Brenier, R. Moncorgé and C. Pedrini, J. de Phys. 48 161 (1987)

[38] F. Auzel, S. Hubert and D. Meichenin, Appl. Phys. Lett. 54 681 (1989)

[39] G.J. Kintz, R. Allen and L. Esterowitz, Appl. Phys. Lett. 50 1553 (1987)

[40] J.F. Donegan, T.J. Glynn, G.F. Imbusch and J.P. Remeika, J. of Lum. 36 93 (1986)

[41] R. Almairac, Thèse, U. de Montpellier (1975)

[42] K. Huang, J. of Lum. 31/32 738 (1984)

[43] M. Rousseau, J.Y. Gesland, R. Russi and G.A. Barbosa, J. Phys. 45 1773 (1984)

[44] J.M.F. Van Dijk and M.F.H. Schuurmans, J. Chem. Phys. 78 5317 (1984)

[45] L.A. Riseberg and H.W. Moos, Phys. Rev. 174 429 (1968)

[46] N.T. Gordon and J.W. Allen, Solid State Com. 37 441 (1981)

[47] J.M. Langer and Le Van Hong, J. Phys. c : Solid State Phys. 17 L923 (1984)

[48] H. Hennen, J. Schneider, G. Pomrenke and A. Axman, Appl. Phys. Lett. 43 943 (1983)

[49] J.M. Langer, J. of Lum. 40/41 589 (1988)

[50] F. Auzel, A.M. Jean-Louis and Y. Toudic, J. of Appl. Phys. (to be published) (1989)

[51] D.H. Waldeck, A.P. Alivisatos and C.B. Harris, Surf. Science 158 103 (1985)

[52] T. Hoshina, S. Imanaga and S. Yokono, J. of Lum. 18/19 88 (1979)

[53] F. Auzel, G.F. de Sa and W.M. de Azevedo, J. of Lum. 21 2 (1980)

[54] F. Auzel, J.M. Breteau and D. Meichenin, J. of Lum. 40/41 595 (1988)

[55] M. Kohli and N.L. Huang Lin, Phys. Rev. B9 1008 (1974)

[56] Luo Zundu, in Excited states of transition Elements, eds. B. Trzebiatoroska, J. Legendziewicz and W. Strek, World Scientific, Singapore (1988), P. 279

[57] V.M. Markuskev, V.F. Zolin and C.M. Briskina, Sov. J. Quantum Electron 16 281 (1986)

[58] M.M. Miller and F.K. Fong, J. Chem. Phys. 59 1528 (1973)

[59] R.H. Dicke, Phys. Rev. 93 99 (1954)

[60] F. Auzel, S. Hubert and D. Meichenin, Europhys. Lett. 7 459 (1988)

MODEL CALCULATIONS OF NONRADIATIVE MULTIPHONON PROCESSES IN INORGANIC INSULATING SOLIDS

C.J. Donnelly and G.F. Imbusch

Department of Physics
University College
Galway, Ireland

ABSTRACT

We examine the predictions of models employed to describe nonradiative multiphonon decay processes between electronic states of optically-active centres in inorganic insulating solids. We initially consider the single configurational coordinate model in the linear harmonic approximation, next the case of anharmonic vibrational potentials is taken into account, and finally we consider the case of different force constants in the electron-lattice coupling for the two electronic states. Numerical values are compared for the different cases.

I. INTRODUCTION

The presence of nonradiative multiphonon processes in electronic centres in solids can be advantageous for the rapid efficient optical pumping of metastable levels, but it can also be deleterious to practical application when it competes with the desired radiative transition. Consequently, an understanding of the nonradiative process is important for the proper design and fabrication of efficient luminescent materials. A variety of theoretical treatments and models have been employed to analyze the nonradiative process. We shall look at the theoretical framework and numerical predictions of some of these models, in particular their abilities to explain the observed dependencies on temperature and energy gap. We attempt to take account of anharmonic effects by obtaining numerical solutions of the relevant Schrödinger equation and employing these wavefunctions to evaluate the nonradiative decay process. The effect of different electron-lattice force constants for the two electronic states is examined using the recursion method of Fonger and Struck [1].

Advances in Nonradiative Processes in Solids
Edited by B. Di Bartolo, Plenum Press, New York, 1991

II. THE SINGLE CONFIGURATIONAL COORDINATE MODEL IN THE LINEAR HARMONIC APPROXIMATION

II.A. Weak Electron-Lattice Coupling

The nonradiative multiphonon process causes the transfer of energy from an electronic centre (ion, molecular ion, colour centre, etc.) to vibrational energy of the lattice in which the electronic centre resides. To describe this process quantum mechanically it is helpful if we can decouple the electronic and vibrational subsystems. If the coupling between the two subsystems is weak, we can write the Hamiltonian as the sum of distinct electronic and vibrational terms (H_{el} and H_{vib}, respectively) plus a weak electron-lattice interaction (H_{e-l}). This latter term can be written as an increasing power series in the strain ε :

$$H_{e-l} = V_1\varepsilon + V_2\varepsilon^2 \quad .... \qquad (1)$$

Such a description is valid for describing nonradiative transitions between $4f^n$ states of triply-charged rare earth ions in solids.

Because of the weak electron-lattice coupling the nonradiative process itself is weak; a process in which electronic energy is converted into six or more phonons can rarely compete with the radiative process on the same rare earth ion. For such weak coupling the nonradiative process can be treated by perturbation theory, in which the strain operator ε causes the creation of one phonon. If p phonons are needed to bridge the electronic energy gap E ($E = p\hbar\omega$), the ε^p operator is required. The larger the value of p the higher the order of the perturbation, and the smaller the nonradiative decay rate. Thus it is found that in general the nonradiative process between the $4f^n$ levels of triply-charged rare earth ions proceeds by the lowest order process and hence utilizes the highest energy lattice phonons. The decay rates at low temperatures ($W_{nr}(0)$) on different rare earth ions in a specific host lattice are found to follow an exponential gap law, $W_{nr}(0) \propto \exp(-\alpha E)$ or $W_{nr}(0) \propto \exp(-\beta p)$, and the dependence of $W_{nr}(0)$ on p is found to be remarkably similar for different host materials, as Fig. 1 shows. For crystalline materials the <u>average</u> behaviour can be described by the formula $W_{nr}(0) \simeq 10^7.\exp(-2(p-2.4))$, which shows that the decay rate falls by about a decade when p increases by one.

II.B. Strong Electron-Lattice Coupling

If the coupling between the electronic and vibrational subsystems is not weak, the mathematical decoupling of the two subsystems is a difficult task and involves making a number of approximations in the quantum description of the system. When the coupling is strong the conversion of electronic to vibrational energy is a much more efficient process; a nonradiative multiphonon process involving the generation of 30 or more phonons may compete with the radiative process.

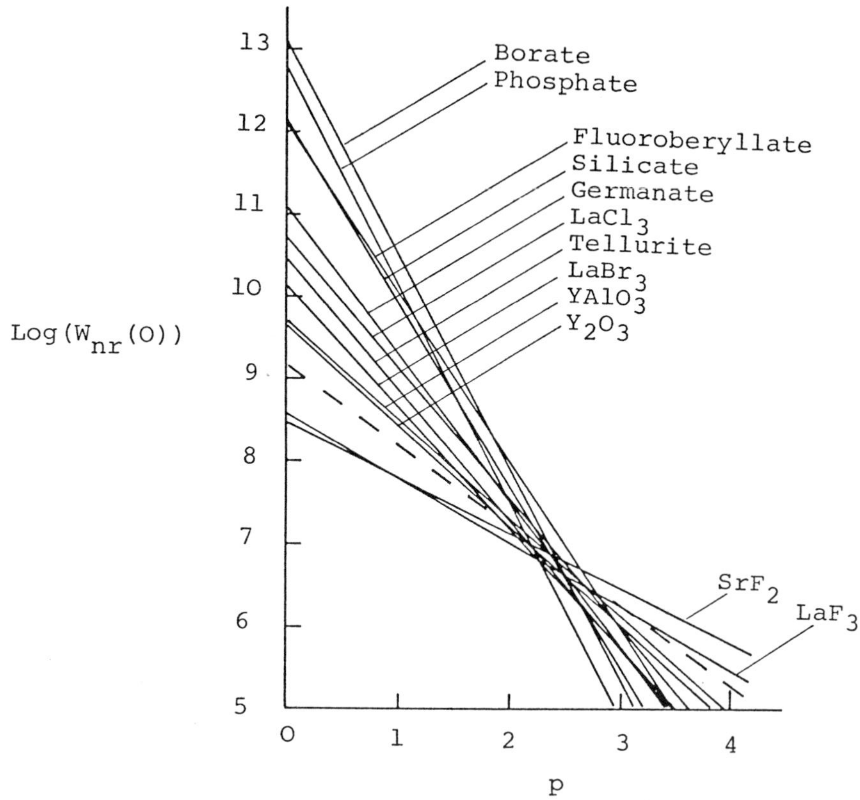

Fig. 1. Measured values of $W_{nr}(0)$ for transitions on rare earth ions in various hosts are made in the region from $p \simeq 3$ to $p \simeq 5$, and for each host the measured values show an exponential dependence on p. The lines shown are linear extrapolations to lower p values of these experimentally measured slopes, and they show the conjunction at $p \simeq 2.4$.

Using the adiabatic approximation [2] we can separate the two subsystems, ending with a configurational coordinate picture, such as that in Fig. 2, which describes an electronic system interacting with a single vibrational mode of the lattice. Electronic state b is above a. Q is the normal coordinate of the particular mode of vibration. If one is dealing with a totally-symmetric breathing mode vibration, then Q is proportional to the separation of the electronic centre from the neighbouring ions of the crystalline environment. If we assume that we can represent the full spectrum of lattice vibrations by a single representational mode, then we have the single configurational coordinate (SCC) model. The potential energy curves in Fig. 2 have an anharmonic character. A major mathematical simplification results if we can replace these by harmonic curves (parabolae), and an additional simplification ensues if it can be assumed that the frequencies of the harmonic vibrations are the same for the two electronic states, a and b. This is the case of equal force constants. We then have a SCC model in the harmonic approximation with linear coupling [3,4].

The situation is now described by Fig. 3; the parameters are the phonon energy ($\hbar\omega$), the size of the energy gap (E) or the number of phonons p ($p = E/\hbar\omega$), and the displacement energy E_d, which is related to the relative configurational coordinate shift (horizontal shift) of the two parabolae (Fig. 3). We write $E_d = S\hbar\omega$, where S is the Huang-Rhys parameter. The electronic-vibrational quantum state is written $|e,n\rangle = |e\rangle|n\rangle$, where $|e\rangle$ represents the electronic state and $|n\rangle$ the lattice state, n being the number of phonons present in the vibrational mode.

We consider the nonradiative process from $|b,m\rangle$ to $|a,n\rangle$. If the interaction causing the transition is purely electronic, the process is represented on Fig. 2 by the horizontal arrow. At T = O K, where only the $|b,0\rangle$ state is occupied, the transition is $|b,0\rangle \rightarrow |a,p\rangle$, where $p = E/\hbar\omega$, and since the transition is caused by some purely electronic operation, H', the relevant matrix element is

$$\langle a,p|H'|b,o\rangle = \langle a|H'|b\rangle \langle p|o\rangle \quad (2)$$

which contains the overlap integral $\langle p|0\rangle$. The nonradiative decay rate can then be written as

$$W_{nr}(0) = R|\langle p|0\rangle|^2 = R \exp(-S)S^p/p! \quad (3)$$

where R is an electronic factor. $|\langle p|0\rangle|^2$ is referred to as the zero-temperature Franck-Condon factor, F_p, and the properties of the harmonic oscillator wavefunctions are used to evaluate $|\langle p|0\rangle|^2$ in equation (3).

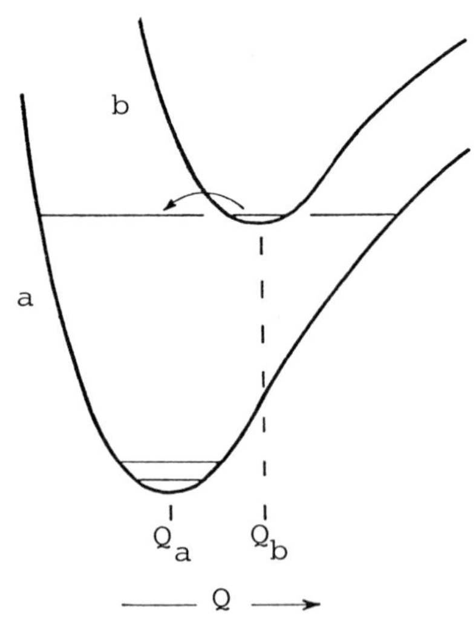

Fig. 2. The potential energy curves for a specific vibrational mode can be displaced horizontally when the electronic state changes. Q is the configurational coordinate of the mode. We define $\Delta Q = Q_b - Q_a$

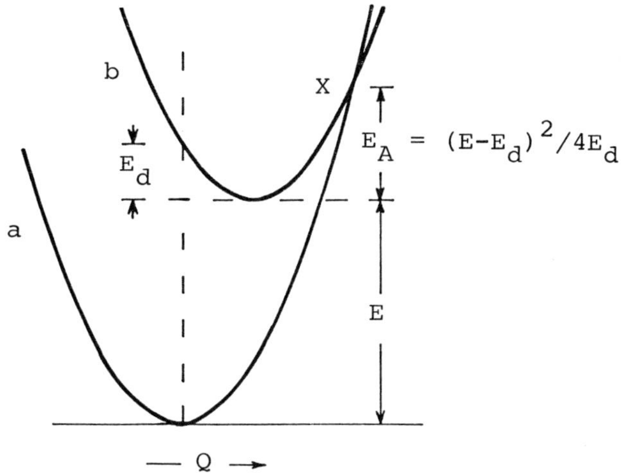

Fig. 3. SCC model in the harmonic approximation with linear coupling. $E_d = S\hbar\omega$, where ω is the phonon frequency for the a and b electronic states.

When the temperature is raised so that the higher vibrational states $|b,m\rangle$ are occupied, the decay rate involves a thermal average over the squares of the overlap integrals $\langle p+m|m\rangle$. Thus the evaluation of overlap integrals is central to our analysis. We can evaluate the overlap integrals in two ways: (i) using the analytical formula [3]

$$\langle n|m\rangle \;=\; \exp(-S/2)(-\sqrt{S})^{m-n}(m!/n!)^{\frac{1}{2}}\; L_n^{m-n}(S) \qquad (4)$$

where $L^m(x)$ are Laguerre polynomials, or (ii) using the recursion method of Struck and Fonger [1,4], who show how to calculate $\langle n|m\rangle$ in terms of $\langle 0|0\rangle$ (which has the value $\exp(-S/2)$).

The nonradiative decay rate at temperature T can be written

$$W_{nr}(T) = R.F_p \qquad (5)$$

where F_p is now the thermal Franck-Condon factor. F_p, being a thermal average over the squares of the overlap integrals, is given by

$$F_p = \frac{\sum\limits_m |\langle p+m|m\rangle|^2 \exp(-m\hbar\omega/kT)}{\sum\limits_m \exp(-m\hbar\omega/kT)} \qquad (6)$$

This can be evaluated using the recursion formulae of Struck and Fonger or from the analytical expression [3]:

$$F_p = \exp(-S(1+2\bar{m}))((1+\bar{m})/\bar{m})^{p/2}\ I_p(2S\sqrt{\bar{m}(\bar{m}+1)}) \qquad (7)$$

where $\bar{m} = (\exp(\hbar\omega/kT)-1)^{-1}$ is the thermal occupancy of phonons, and $I_p(x)$ is the modified Bessel function. The formulation of Struck and Fonger has the advantage that it includes the case of different vibrational frequencies for the two electronic states (i.e. the case of unequal force constants); analytical solutions for the case of unequal force constants are quite complicated [3].

If the interaction causing the nonradiative transition is assumed to be due to the terms in the Hamiltonian which were ignored on the adiabatic approximation, and if one adopts the Condon approximation (i.e. weak dependence of electronic states on the configurational coordinate), then the interaction matrix element involves vibrational terms such as $\langle p|\partial/\partial Q|0\rangle$, and these lead to a transition rate which can be written approximately as

$$W_{nr}(T) \simeq R'[(1+\bar{m})F_{p-1} + \bar{m}\ F_{p+1}] \qquad (8)$$

where the F_p's are the thermal Franck-Condon factors, and the electronic factor R' differs from that (R) in equation (5).

Recently [5-9] it has been shown that if one goes to higher order in the calculation, beyond the Condon approximation, the calculated decay rate can be much larger than that calculated using the Condon approximation. Using this calculation for the SCC model in the harmonic approximation with equal force constants, the formula for W_{nr} is found to contain factors F_p, F_{p-1}, F_{p+1}, F_{p-2}, F_{p+2} [5], and taking only the largest components we obtain

$$W_{nr}(T) \simeq R''S(1+\bar{m})^2\ F_{p-2} \qquad (9)$$

where R" is now a different electronic factor.

The temperature dependence of $W_{nr}(T)$ in these formulae (5, 6, 9) is contained in the F_p factors and in the $\bar{m}$ and $(1+\bar{m})$ factors. If p >> 1 the temperature dependence is similar for all cases, and one can take equation (5) as a general formula with which to analyze the temperature dependence of the nonradiative decay rate. Despite the many approximations inherent in it, equation (5) is used by many workers to describe nonradiative processes when the coupling between electronic and vibrational subsystems is not small [10-12].

To see how closely equation (5) approximates to the values measured experimentally, we consider the transition between the $^4T_1$ and $^4T_2$ states of Co^{2+} in $KMgF_3$, which appears to be a very suitable example. This has been examined by Sturge [13], who showed that at low temperatures the radiative and nonradiative rates are approximately of the same strength, while the

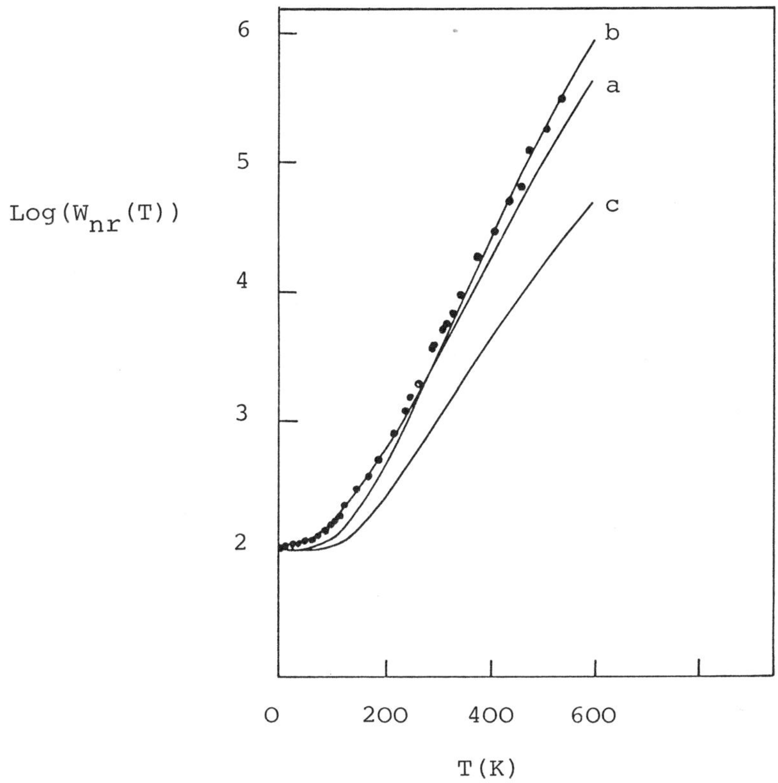

Fig. 4. Nonradiative decay rates for the $^4T_2 \rightarrow {}^4T_1$ transition of Co^{2+} in $KMgF_3$. The points are the measured values of Sturge [13]. Curve (a) gives the calculated values of Sturge using the full spectrum of phonon modes. Curve (b) gives the values calculated on the SCC model. Curve (c) gives the values calculated on the SCC model if the variation of the gap with temperature is ignored. (The measured values and curve (a) are taken from Sturge [13].)

nonradiative rate becomes dominant as the temperature is raised. Sturge worked in the harmonic approximation, but he generalized the theory to take the continuous nature of the phonon spectrum into account. The points in Fig. 4 show the measured values of $W_{nr}(T)$, solid curve (a) shows the temperature dependence of $W_{nr}(T)$ calculated by Sturge and fitted to agree with the measured value at $T = 0$, and there is excellent agreement. Solid curve b shows the prediction of the SCC model (equation (5)), using $S = 2.3$ and $\hbar\omega = 370\ cm^{-1}$, which is approximately the centre frequency of the phonon spectrum. The calculations for solid curves (a) and (b) take account of the variation of the energy gap with temperature. Curve (c) shows the predictions of the SCC model if the variation of the energy gap with temperature is ignored. Although the agreement with the SCC model (curve (b)) is not nearly as good at low temperatures as that obtained by taking the full spectrum of phonons into account, the overall agreement with the predictions of the simple SCC model is reasonable and would support the use of the

SCC model to explain the temperature dependence of the W_{nr} values. As was pointed out by Sturge [13], the phonons involved in the SCC model of the nonradiative decay are those in the centre of the phonon spectrum; they are not the highest frequency phonons, as is generally the case for nonradiative transitions on rare earth ions. Reasonable agreement is similarly found between the variation with temperature of W_{nr}, as calculated by the SCC model, and that measured experimentally for many other transition metal ion systems [10-12]. Returning to the case of $KMgF_3:Co^{2+}$, although Sturge found that the calculated temperature dependence of W_{nr} was in good agreement with the measurements, the <u>absolute</u> values of the calculated decay rates were too large by two orders of magnitude, in comparison with the measured values.

II.C. The Activation Model

Some workers have employed an activation formula [14,15], $W_{nr} \propto \exp(-E_A/kT)$, to account for their experimental results. The activation energy, E_A, is given by $E_A = (E - E_d)^2/4E_d$, and this value of E_A, as Fig. 3 shows, is the energy from the bottom of the excited states parabola to the crossing point of the two parabolae, X. The physical picture is that the system must be raised thermally to the X level so as to cross from the b to the a parabola.

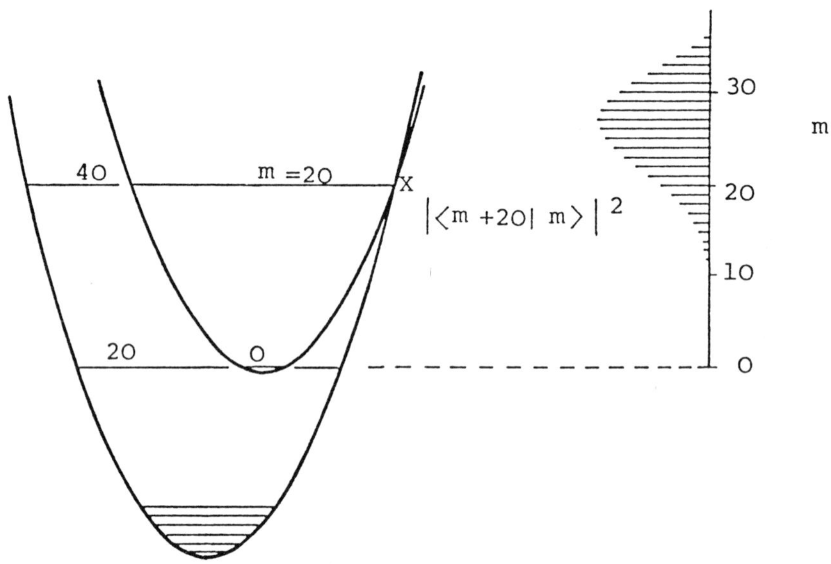

Fig. 5. The horizontal lines on the right gives the relative values of $<m+20\ m>^2$ for m values between 0 and 36. The X point corresponds to m = 20.

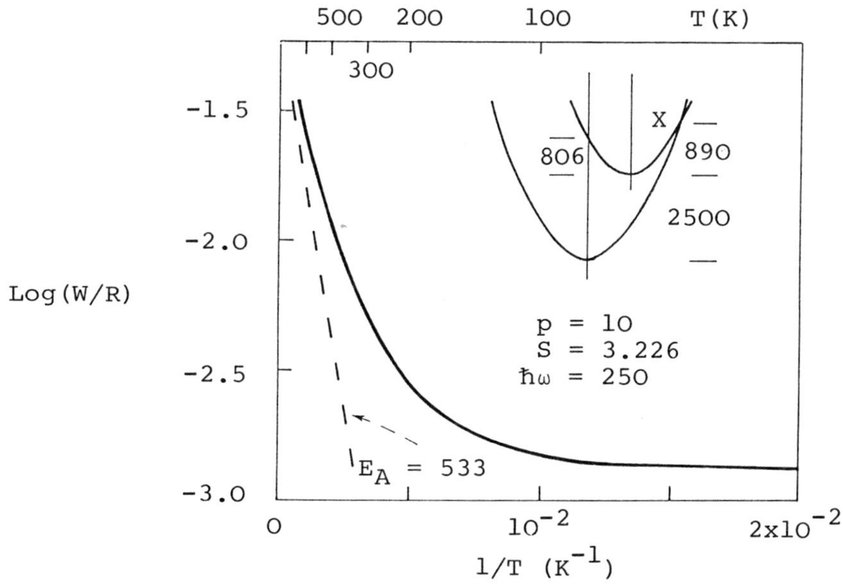

Fig. 6. Log-lin plot of W_{nr}/R against $1/T$ where W_{nr} is calculated on the SCC model with $p = 10$, $S = 3.226$, $\hbar\omega = 250$ cm^{-1}. The energies are given in cm^{-1}, and the W_{nr} values are given in s^{-1}. The broken line corresponds to $E_A = 533$ cm^{-1}.

We can look into the validity of this activation formula by checking whether the overlap model shows such an activation behaviour over some range of temperatures. The increase in W_{nr} with increasing temperature on the overlap model comes about because $|\langle m+p|m\rangle|^2$ increases as m increases from zero. This is shown in Fig. 5 for the case of $p = 20$, $S = 3.372$ and $m = 20$ at the crossing point. The values of $|\langle 20+m|m\rangle|^2$ increase substantially as m increases from zero, $|\langle 30|10\rangle|^2$ being about six orders of magnitude larger than $|\langle 20|0\rangle|^2$. We see that relatively large values of the square of the overlap integral are found below the crossing point, and we observe that the maximum value occurs at $m = 27$, rather than at $m = 20$, the crossing point.

To see whether the overlap model gives values of $W_{nr}(T)$ which follow an activation behaviour over some limited temperature range and for some effective activation energy, we calculate $W_{nr}(T)/R = F_p$ (equation 5), taking $p = 10$, $\hbar\omega = 250$ cm^{-1}, and $S = 3.226$, for which $E_A = 889$ cm^{-1} and the crossing occurs at $m=3$. We plot $\log(W_{nr}(T)/R)$ against $1/T$, and the results are shown in Fig. 6. We see that only at high temperatures is there any semblance of an activation behaviour, and here the effective activation energy is 533 cm^{-1}.

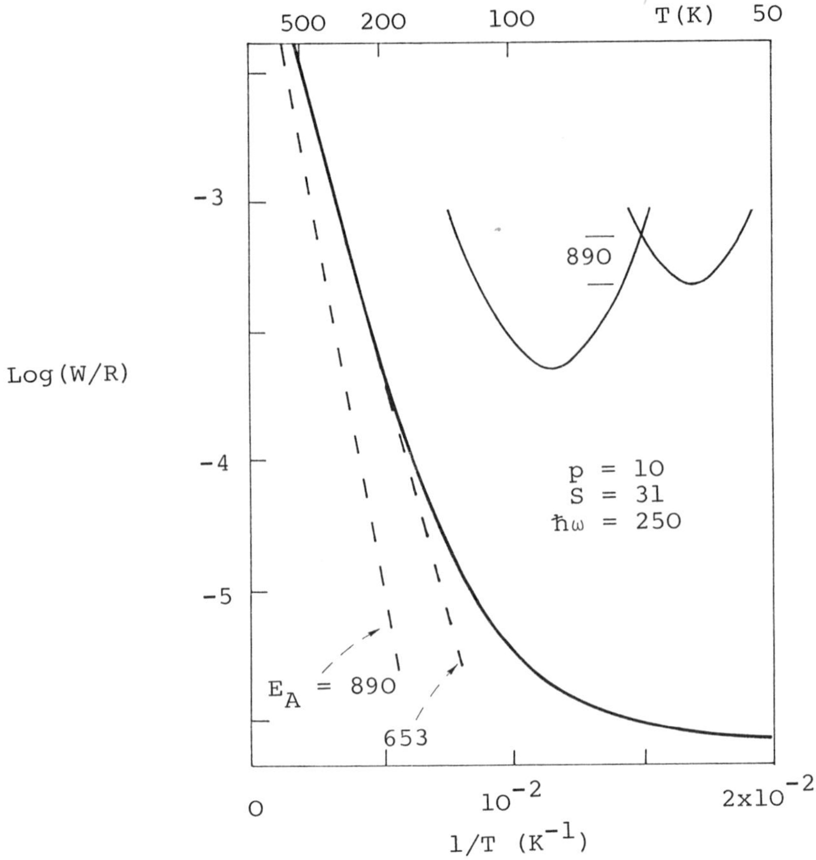

Fig. 7. Log-lin plot of W_{nr}/R against $1/T$ for the case where the higher energy parabola is outside the lower. Energy values are given in cm^{-1}.

For the same value of p we obtain the same value for the theoretical activation energy (E_A = 889 cm^{-1}) when the upper parabola is outside the lower (Fig. 7) if we take S = 31. One might expect closer similarity between the predictions of the overlap and activation models in this case. We calculate $W_{nr}(T)$ using the overlap model and we plot $\log(W_{nr}/R)$ against $1/T$ in Fig. 7, but we observe that, whereas an activation behaviour is followed over a larger range of temperature, the activation formula is not a good description at low temperatures. This poor agreement has been pointed out by Fonger and Struck [16]. We note that, particularly at small values of S, the activation formula is quite poor. That it is completely invalid at low temperatures is clear from its prediction that $W_{nr}(0) = 0$ in all cases, a prediction which is clearly at variance with experiments.

A quantum mechanical analysis by Englman and Jortner [17] has shown that in the strong coupling limit, W_{nr} is described by an activation formula of the form

$$W_{nr}(T^*) = c(T^*)^{-\frac{1}{2}}\exp(-E_A/kT^*) \qquad (10)$$

where

$$T^* = (\hbar\omega/2k)\coth(\hbar\omega/2kT) \qquad (11)$$

It is of interest to see how closely the overlap model adheres to this modified activation model (equations 10 and 11) for the typical S values encountered experimentally. On Figs. 8 and 9 we have plotted $\log(W_{nr}\sqrt{T^*}/R)$ against $1/T^*$, where W_{nr} is calculated from the ovelap model, for the two cases considered previously : $p = 10$, $\hbar\omega = 250$ cm^{-1}, $S = 3.226$ (Fig. 8), and $p = 10$, $\hbar\omega = 250$ cm^{-1}, $S = 31$ (Fig. 9). As expected, the modified activation model shows better agreement with the overlap model, but agreement is still poor at low temperatures.

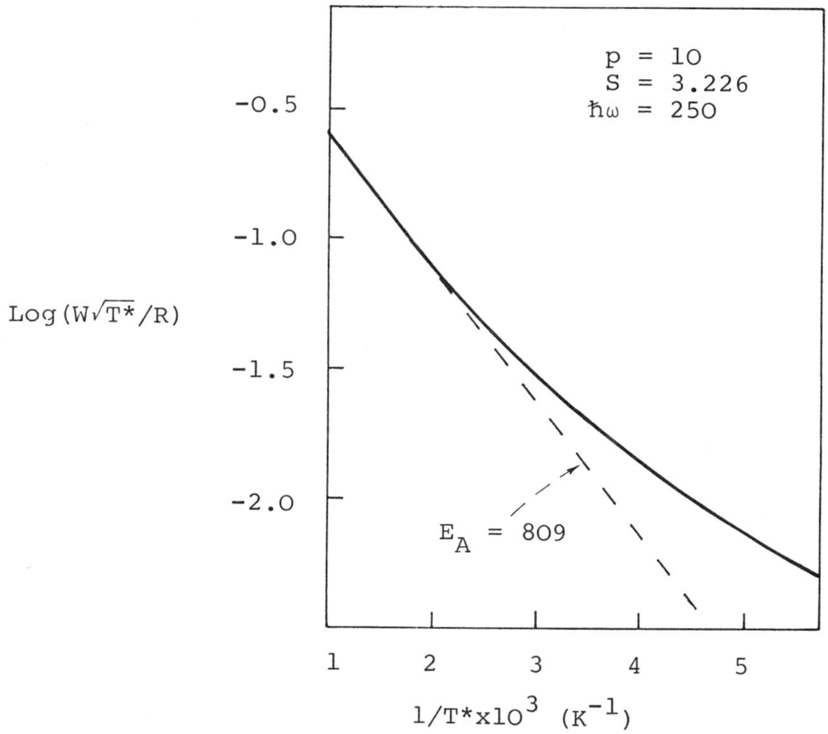

Fig. 8. Log-lin plot of $W_{nr}\sqrt{T^*}/R$ against $1/T^*$ where W_{nr} is calculated on the SCC model. Energy values are given in cm^{-1}.

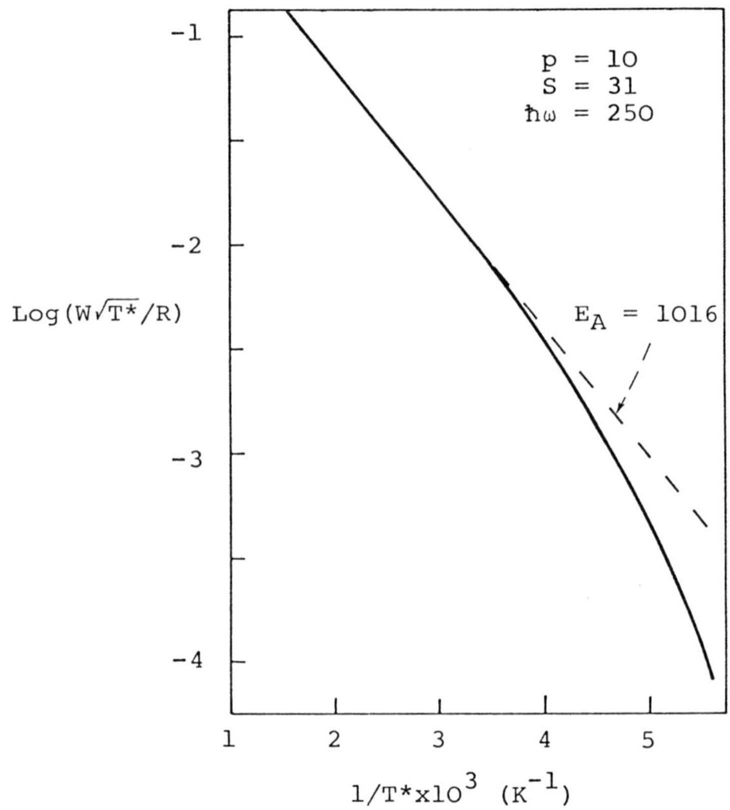

Fig. 9. Log-lin plot of W_{nr} T*/R against 1/T*, where W_{nr} is calculated on the SCC model. Energy values are given in cm^{-1}.

II.D. Application to Rare Earth Ions

We now enquire if the overlap model can be used for the case of weak coupling ($S \simeq 0.1$) appropriate for transitions between $4f^n$ electronic levels of triply-charged rare earth ions.

When the argument of the Bessel function in equation (7) is much less than unity we can make the approximation

$$I_p(x) \simeq (x/2)^p (p!)^{-1} \tag{12}$$

Thus for S << 1 we obtain from equation (7)

$$W_{nr}(T) \simeq R \exp(-S(1+2\bar{m})(1+\bar{m})^p S^p/p! \tag{13}$$

The temperature dependence is contained mainly in the term $(1+\bar{m})^p$, and such a term was first proposed by Kiel [18] and found experimentally [19]. We note that to the same degree of approximation the W_{nr} formulae (8) and (9) also contain a $(1+\bar{m})^p$ term.

For T = 0 we have $W_{nr}(0) = R \exp(-S)\, S^p/p!$ which is valid for all values of S. Making use of the Stirling approximation $[\ln(p!) \simeq \ln\sqrt{2\pi p} + p\ln(p) - p]$ we find

$$\ln(W_{nr}(0)) \simeq \ln R - S - \ln\sqrt{2\pi p} - p(\ln(p/S)-1) \qquad (14)$$

For a fixed value of S we can use this formula to examine how $W_{nr}(0)$ varies with p, that is, with the size of the energy gap. To the extent that one can ignore the variation of the $\ln\sqrt{2\pi p}$ factor, we obtain from equation (14)

$$W_{nr}(0) \simeq C \exp(-\beta p) \qquad (15)$$

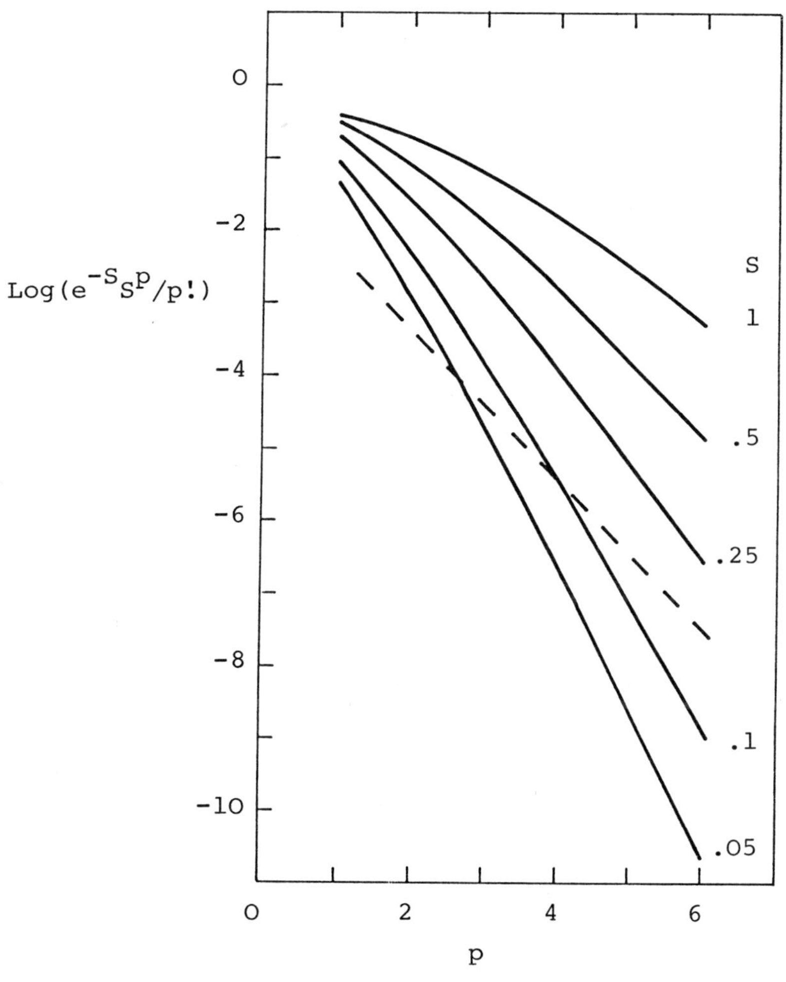

Fig. 10. Log-lin plots of $W_{nr}(0)/R$ against p for low values of S. The broken line shows the average slope obtained from the experimental measurements on rare earth ions in crystals, shown in Fig. 1.

where $\beta = (\ln(p/S)-1)$. This has the appearance of an exponential gap law, but it is only an approximation since β is not constant. To check how closely the overlap model approximates to an energy gap law <u>for a fixed small value of S</u> we plot $\log(W_{nr}(0)/R)$, which is $\log[\exp(-S)S^p/p!]$, against p, for a number of values of S (Fig.10). We recall that for rare earth transitions the measured nonradiative processes involve 2-6 phonons and $S \simeq 0.1$. We observe from Fig. 10 that for $S \simeq 0.2$ the dependence of $\log W_{nr}$ on p is linear to good approximation over this range of p. This must not, however, be regarded as a theoretical justification for the experimentally-observed exponential gap law, as Struck and Fonger have emphasized [20]. The broken line in Fig. 10 gives the slope of $W_{nr}(0)$ (<u>measured</u>) against p; this is the average over a large number of crystalline materials taken from Fig. 1. The measured slope does not agree with the theoretical curves for $S \simeq 0.1$. However, we ought not to expect that the different energy gaps for the transitions on different ions in a given host material will all have the same value of S; it is more likely that as the energy gap (and hence p) increases, the parabola offset (and hence the value of S) increases. A change in S from 0.1 to 0.2 between a small gap ($p \simeq 3$) amd a large gap ($p \simeq 6$) would give agreement with the experimentally-measured slope.

III. THE EFFECT OF ANHARMONIC POTENTIALS

It is convenient to define a dimensionless parameter q to replace the configurational coordinate Q, so that the harmonic potential can be written

$$V_H(q) = KQ^2/2 = \hbar\omega q^2/2 \tag{16}$$

where

$$q = Q\sqrt{K/\hbar\omega} \tag{17}$$

and K is the harmonic force constant.

We consider two types of anharmonic potential, a perturbed harmonic oscillator potential, V_{AH}, and a Morse potential, V_M [13]. These potentials, which reduce to V_H (equation 16) in the limit of small q, are given by

$$V_{AH}(q) = \begin{cases} \hbar\omega q^2/2 - \hbar\omega c q^3 & q < 1/3c \\ \hbar\omega/54c^2 & q \geq 1/3c \end{cases} \tag{18}$$

$$V_M(q) = \hbar\omega/2a^2 \, (1 - \exp(-aq))^2 \tag{19}$$

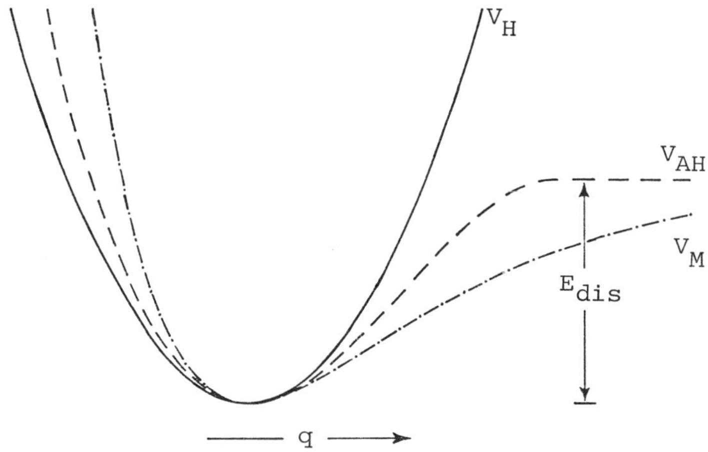

Fig. 11. A comparison of V_H, V_{AH} and V_M for $E_{dis} = 50\ \hbar\omega$.

where ω is the vibrational frequency for small amplitude oscillations, and $\hbar\omega/54c^2$ and $\hbar\omega/2a^2$ are the dissociation energies, E_{dis}, for V_{AH} and V_M, respectively. These are shown, together with V_H, in Fig. 11 for $\hbar\omega = 500\ cm^{-1}$ and $E_{dis} = 50 \cdot \hbar\omega = 25{,}000\ cm^{-1}$. Since V_M asymptotically approaches E_{dis} with increasing q, it may not be as suitable a potential as V_{AH} for describing the situation in a solid.

Straightforward and accurate numerical techniques are now available for solving the Schrödinger equation for any one-dimensional potential [21]. Such a technique was employed to obtain the energy levels and wavefunctions for V_{AH} and V_M.

We consider the nonradiative decay process from $|b,m\rangle$ to $|a,n\rangle$ for the case where $q_b < q_a$, as shown in Fig. 12; the same type of potential is assumed for the ground (a) and excited (b) electronic states. The value of $q_a - q_b$ was taken as $2\sqrt{2}$, which corresponds to a Huang-Rhys parameter of S = 4 in the linear harmonic approximation. The electronic energy gap, E, was taken as $10{,}000\ cm^{-1}$. As in the case of the linear harmonic approximation, the nonradiative decay rate at temperature T was calculated using the formula

$$W_{nr}(T) = R \frac{\sum_m |\langle n|m\rangle|^2 \exp(-E_m/kT)}{\sum_m \exp(-E_m/kT)} \qquad (20)$$

where E_m is the energy of the mth vibrational level (relative to the m = 0 level) for the excited electronic state b. For the anharmonic potentials, the overlap integrals $\langle n|m\rangle$ were calculated using the numerically-generated wavefunctions. For a given value of m the nth vibrational level for the ground electronic state was selected as that level whose energy E_n

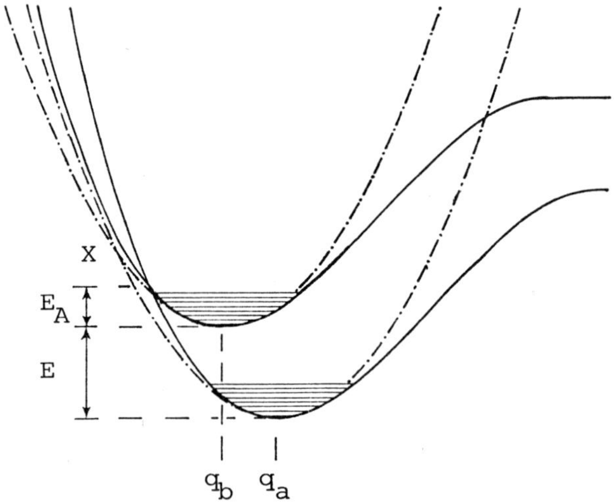

Fig. 12. Configurational coordinate diagram for harmonic potential, V_H (broken curves) and for anharmonic potential, V_{AH} (solid curves) for $E = 10000\ cm^{-1}$, $\hbar\omega = 500\ cm^{-1}$, $E_{dis} = 25000\ cm^{-1}$ and $q_a - q_b = 2\sqrt{2}$. The crossing of the V_{AH} potentials occurs at lower energy than for the V_H potentials.

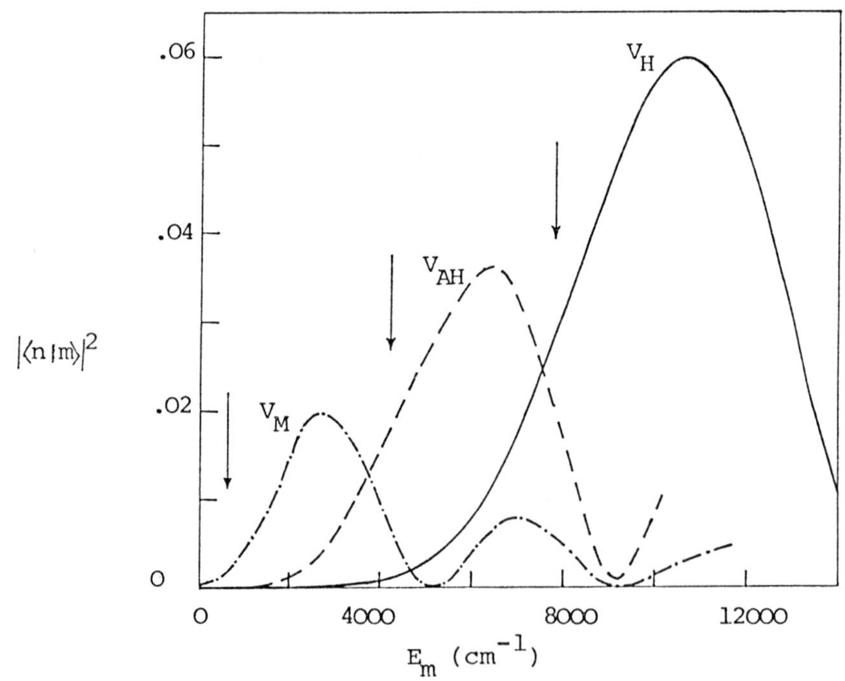

Fig. 13. The curves are drawn through the $|\langle n|m\rangle|^2$ values for each of the three potentials. The arrows indicate the E_m values at the crossing of the potentials in each case.

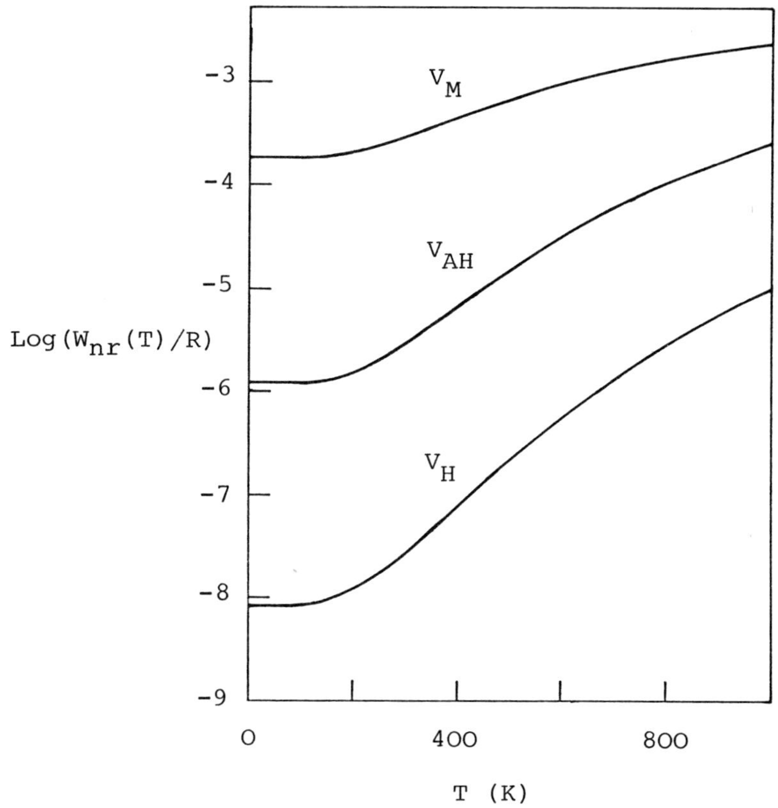

Fig. 14. Relative values of $W_{nr}(T)$ for the three potentials, for the same parameters as in Fig. 12.

(relative to the n = 0 level) was closest to $E_m + E$. The values of the squares of the calculated overlap integrals are plotted against E_m in Fig. 13 for both of the anharmonic potentials and for the harmonic potential. The arrows indicate the values of E_m at which the ground state and excited state potentials cross. We observe that the crossing occurs at lower energy for the anharmonic potentials than for the harmonic potential, and consequently the values of $|\langle n|m\rangle|^2$ are larger at small values of E_m for the anharmonic potentials. The calculated decay rates, $W_{nr}(T)$, are plotted against T in Fig. 14. As expected from Fig. 13, W_{nr} is largest for the Morse potential and smallest for the harmonic potential. The presence of anharmonicities is seen to greatly affect the nonradiative decay rates; when $q_a > q_b$ the absolute values of W_{nr} can be greater, by orders of magnitude.

In the converse case (Fig. 15), where $q_b > q_a$, the anharmonic potentials do not cross except for very large values of $q_b - q_a$. For $\hbar\omega$ = 500 cm^{-1} and $E_{dis} = 50\hbar\omega$, $q_b - q_a$ would need

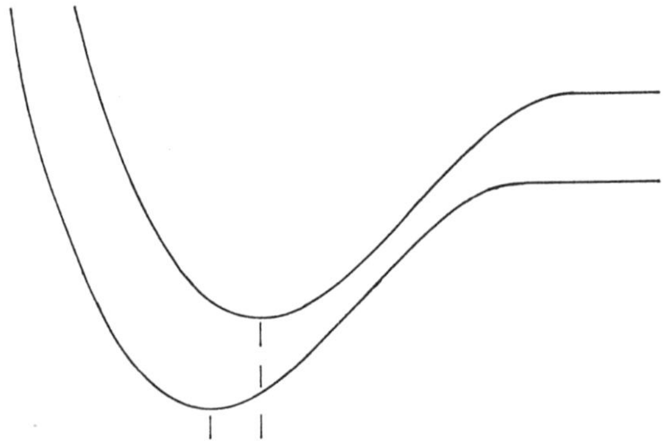

Fig. 15. The anharmonic potentials, V_{AH}, do not cross for parameter values as before but with $q_b - q_a = +2\sqrt{2}$.

to be greater than 5 for the ground and excited state V_{AH} potentials to cross; this corresponds to S = 12.5 for the linear harmonic model For V_M an even larger value of $q_b - q_a$ is required before the ground and excited state potentials cross. Consequently, the values of the overlap integrals for small $q_b - q_a$ are considerably smaller than the values for the harmonic case. For the parameters chosen previously, S = 4, $\hbar\omega = 500\ cm^{-1}$, $E_{dis} = 50\ \hbar\omega$, but now with $q_b - q_a = +2\sqrt{2}$, the W_{nr} values are orders of magnitude smaller for the anharmonic potentials than in the harmonic case. The significant reduction in W_{nr} values due to anharmonicities when $q_b > q_a$ was pointed out by Sturge [13].

IV. HARMONIC MODEL WITH DIFFERENT FORCE CONSTANTS

We finally consider the case of different force constants, in which the vibrational frequencies for the upper and lower electronic states are different. This is most easily handled using the recursion method of Fonger and Struck [1]. We define $\gamma = \omega_b/\omega_a$, the ratio of the vibrational frequencies for the b and a electronic states. The parameters θ in the treatment of Fonger and Struck is given by $\tan\theta = \sqrt{\gamma}$. We define S by writing for the ground state potential $K(Q_b - Q_a)^2/2 = S\hbar\omega_a$; this reduces to the standard Huang-Rhys parameter S when $\gamma = 1$. The value of $\langle 0|0\rangle$ is now given by $\langle 0|0\rangle = \sqrt{\sin 2\theta}.\exp(-S\gamma/2)$, and the other overlap integrals are obtained from the recursion relations [1]. Writing $W_{nr}(T) = R\ F_p$ (equation 5), we can evaluate F_p by using these overlap integrals in equation (20). This evaluation has been made over the temperature range 0 - 1000K for a number of values of γ for the case where S = 4, $\hbar\omega_a = 500\ cm^{-1}$, and the zero-vibrational levels of b and a differ by 15,000 cm^{-1}. The results are plotted in Fig. 16. We see that there is a significant change in the values of W_{nr} when γ changes from unity.

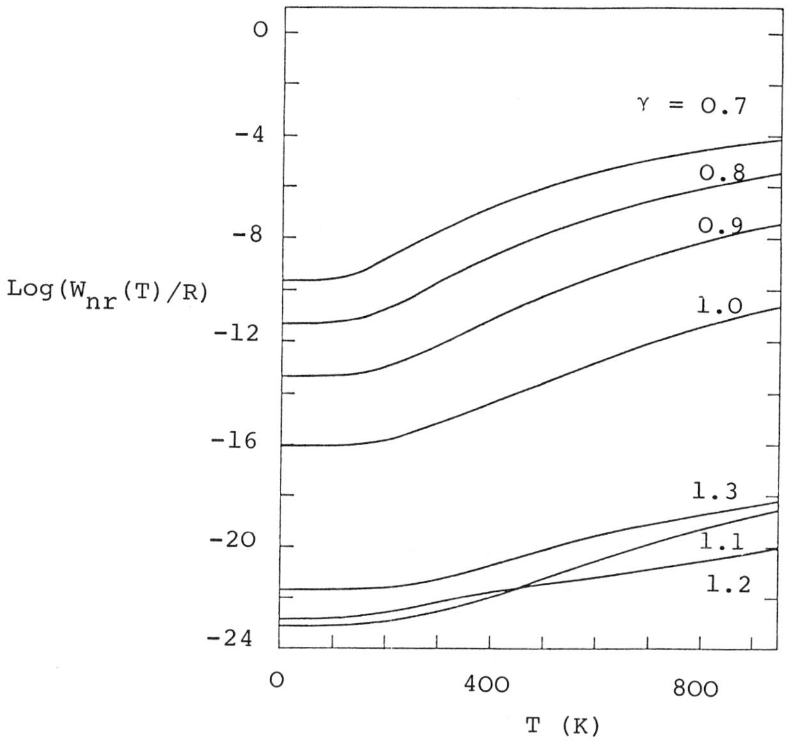

Fig. 16. $W_{nr}(T)$ for γ values between 0.7 and 1.3, $\hbar\omega = 500\ cm^{-1}$ and $E = 15000\ cm^{-1}$.

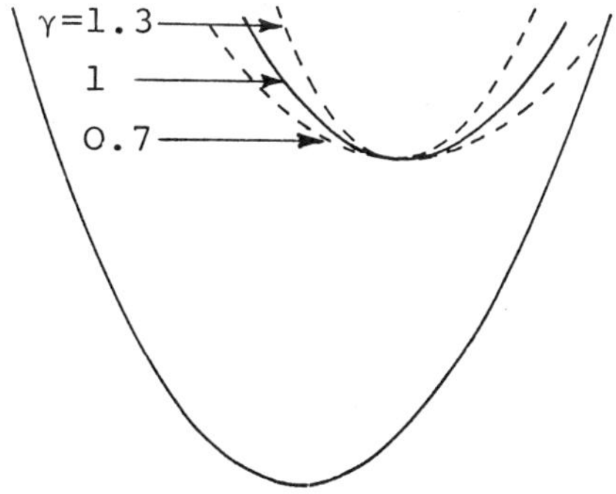

Fig. 17. Shape of the upper harmonic oscillator for γ = 0.7, 1.0, 1.3.

The relevant harmonic oscillator parabalae are drawn for $\gamma = 0.7$, 1.0, and 1.3 in Fig. 17. For $\gamma = 0.7$ the crossing of the b and a parabolae occurs at a lower energy than for $\gamma = 1$, and the W_{nr} values are considerably larger. Conversely, for $\gamma = 1.3$ the crossing of the parabolae occurs at a higher energy than for $\gamma = 1$, and the W_{nr} values are considerably smaller. A reduction in the vibrational frequencies of the $t_{2g}(\zeta)$ mode for the excited $^4T_2$ states of Cr^{3+} in ordered perovskites relative to the $t_{2g}(\zeta)$ vibrational frequency for the $^4T_2$ ground state was invoked by Bartram et al. [22] to explain the observed thermal quenching of chromium luminescence in these materials. The measured nonradiative decay rate in these materials is found to be 6 - 8 orders of magnitude larger than that calculated on the harmonic model with equal force constants.

V. CONCLUSION

For rare earth ions in solids the optically-active 4f electrons are shielded from the ionic environment by the outer closed electronic shells of the rare earth ion, and the electronic-vibrational coupling is weak. The nonradiative transition across the energy gap E occurs by the release of the smallest number (p) of the highest frequency lattice phonons ($\hbar\omega$), $E = p\hbar\omega$, and overall the decay rates can be described by an empirical formula $W_{nr}(T) \simeq 10^7(1+m)^p \exp(-\beta(p-2.4))$, where $\beta \simeq 2$, and m is the thermal occupancy of phonons of frequency ω.

When the optically-active electrons are on the outside of the ion the electronic state is strongly affected by the vibrations of the ionic environment, and, conversely, the environmental configurations and the frequency of vibration can be affected by changes in the electronic state of the ion. Although the temperature dependence of $W_{nr}(T)$ is often in good agreement with the predictions of a harmonic model based on a linear coupling to a single vibrational mode, the absolute values of W_{nr} calculated on this model can be off by order of magnitude from the measured values. Anharmonic effects as well as differences in the vibrational frequencies for the two electronic states are shown to significantly affect the strength of the nonradiative multiphonon process.

REFERENCES

1 W.H. Fonger and C.W. Struck, J. Chem. Phys. 60, 1994 (1974).
2 F. Auzel in :Luminescence of Inorganic Solids (Ed. B. di Bartolo) Plenum Press, New York, 1978; B. Di Bartolo and R.C. Powell, Phonons and Resonances in Solids, Wiley, New York, 1976.
3 T.H. Keil, Phys. Rev. A140, 601 (1965).
4 C.W. Struck and W.H. Fonger, J. Chem. Phys. 60, 1988 (1974).
5 E. Gutsche, J. Lumin 24/25,689 (1981); Phys. Stat. Sol. (b) 109, 583 (1982).
6 K. Peukar, R. Enderlein, A. Schenk, and E. Gutsche, Phys. Stat. Sol. (b) 109, 599 (1982).
7 K. Huang, Sci. Sin. 24, 27 (1981).
8 G. Helmis, Ann. Phys. (Leipzig) 19, 41 (1956).
9 R. Passler, Czech. J. Phys. B24, 322 (1974).

10 J.F. Donegan, F.J. Bergin, T.J. Glynn, G.F. Imbusch, and J.P. Remeika, J. Lumin. 35, 57(1986).
11 J.F. Donegan, T.J. Glynn, and G.F. Imbusch, J. Lumin. 36, 93 (1986).
12 G.F. Imbusch and R. Kopelman in: Laser Spectroscopy of Solids (Eds. W.M. Yen and P.M. Selzer). Springer-Verlag, Berlin, 1981.
13 M.D. Sturge, Phys. Rev. B8, 6 (1973).
14 N.F. Mott, Proc. Roy. Soc. (London) A167, 384 (1938).
15 C.C. Klick and J.H. Schulman, in :Advances in Solid State Physics (Eds. F. Seitz and D. Turnbull), Academic Press, New York, 5, 97 (1957).
16 C.W. Struck and W.H. Fonger, J. Lumin. 10, 1 (1975).
17 R. Englman and J. Jortner, Mol. Phys. 18, 145 (1970).
18 A. Kiel in: Third Intern. Conf.on Quantum Electronics, Paris, 1963 (Eds. P. Grivet and N. Bloembergen), Columbia University Press, New York, 765 (1964).
19 H.W. Moos, J. Lumin. 1/2, 106 (1968).
20 W.H. Fonger and C.W. Struck, J. Lumin. 17, 24 (1978).
21 J.S. Bokman, Amer. J. Phys. 40, 1511 (1972).
22 R.H. Bartram, J.C. Charpie, L.J. Andrews, and A. Lempicki, Phys. Rev. B34, 2741 (1986).

ADVANCES IN NONRADIATIVE PROCESSES IN SEMICONDUCTORS

R. G. Ulbrich

4. Physikalisches Institut
der Universität Göttingen
Göttingen, Fed. Rep. Germany

ABSTRACT

The concepts of elementary excitations, electronic band-structure, and scattering mechanisms of electrons and holes in semiconductors are reviewed at an elementary level. Several aspects of the relaxation process of nonequilibrium charge carriers in the direct gap semiconductor gallium arsenide are discussed: electron-electron and electron-phonon coupling rates, and their competition with radiative interband transitions are summarized. Two specific experiments are discussed in detail: "hot electron" luminescence under conditions of cw excitation above the band gap and nonlinear, transient "pump-probe" optical spectra excited with ultrashort (70 fs) laser pulses.

I. INTRODUCTION

The usual classification of solids -- insulators, metals and semiconductors -- is based on an electronic property: the existence and magnitude of a gap in the excitation spectrum. The fundamental "band gap" determines all other properties of a solid, its electrical and thermal conductivity, the optical spectra, etc.. At first sight the structural aspects are less decisive: in fact all three classes exist in ordered and disordered structures, i.e. in crystalline and amorphous form. The mean free path, the degree of localization and other dynamical properties of charge carriers in semiconductors, however, depend crucially on structural details, notably the deviations from perfect symmetry in the presence of impurities and defects.

Because of their technological potential for electronic and opto-electronic devices [1], crystalline semiconductors have been thoroughly studied, in experiment and theory, for more than four decades. Si, Ge and some of the III-V compound semiconductors are by now the best understood solids, and this is true as well for the impurities and defects in these materials. Most recent developments aim at the "engineering" of gap energies E_{GAP} with controlled layer growth and the fabrication of lateral structures down to the atomic scale [2] in group IV (Si, Ge), III-V- and II-VI-semiconductors. Optical transitions, i.e. the absorption or emission of photons with $\hbar\Omega \geq E_{GAP}$ play a key role in these investigations, and are closely connected with the electronic excitation spectrum [3]. Many other important dynamical processes in solids involve no real photons, and may be summarized, as in this course, with the term "nonradiative". Such processes originate from (i) electron-phonon coupling and (ii) the electron-electron interaction. In semiconductors, both mechanisms control the effective relaxation of momentum and kinetic energy of mobile charge carriers. They provide, on one hand, the damping and dissipation which is necessary for stationary transport of electrons and holes, and on the other hand determine the actual lifetimes and spectral lineshapes of electron-hole (e-h) pair transitions [4].

In this article we will briefly discuss at an elementary level the concepts of elementary excitations, electronic band structure, total energy, and scattering mechanisms. We will then focus on two different aspects of the relaxation process of highly non-equilibrium charge carriers in the III-V semiconductor gallium arsenide:

(i) The injection of "hot" carriers into conduction and valence bands via optical absorption and the subsequent cooling through emission of optical and acoustical phonons. A fairly consistent description of the temporal and spatial evolution of the energy relaxation process has emerged in the last years, at least for intrinsic (i.e. undoped) semiconductors and in the case of low excited carrier densities.

(ii) Ultrafast carrier relaxation phenomena on the 100 fs time scale, that is well below the shortest lattice vibration period. Here the concepts are less well developed and most experiments are done under conditions of high excitation with ultrashort laser light pulses. The predominant role of Coulomb scattering between photoexcited carriers, its dynamic screening by highly nonequilibrium carrier distributions, and intervalley scattering of electrons between different conduction band extrema will be discussed in detail. Closely related nonlinear optical effects, the spectral "hole burning" and the "bleaching" of edge absorption are briefly discussed.

II. ELEMENTARY EXCITATIONS, THEIR INTERACTIONS, AND ELECTRON-HOLE PAIR TRANSITIONS IN SEMICONDUCTORS

The concept of elementary excitations in solids is remarkably successful: despite the very large number of strongly interacting electrons and nuclei involved in the process of "excitation", the terms "electron", "phonon", "exciton" etc. keep a well-defined and plausible meaning and are taken as the natural basis for description of charge transport, heat propagation and optical transitions in semiconductors. The elementary excitations may be visualized as (small amplitude) superpositions of the normal modes of the system. The compact notation suppresses all details due to the microscopic origin of the interactions [5].

The simplest basis to describe the excitation "electron" inside a crystalline semiconductor is constructed from one-electron states classified by their wave vector $\vec{k}$, their energy E and their spin $\vec{s}$. In the beginning we assume translational symmetry within the crystal, so that all electronic wave functions can be represented as the product of a phase factor $e^{i\vec{k}\cdot\vec{r}}$ and the atomic orbital-like term $u_{\vec{k}}(\vec{r})$, which is cell-periodic. Bloch's theorem states that all eigenstates with respect to energy can be written in such a form. This approach assumes that a fictitious rigid, and translationally invariant crystal potential $V_o(\vec{r})$ is either given a priori or found in a self-consistent procedure. Obviously this can only be an approximate description: the motion of the nuclei and the strong interaction of one electron with all others, which leads to correlated electron motions, are not contained in this so-called "one-electron" or "single-particle" scheme. Its lasting success rests on its simplicity and the wealth of practical results since its invention at the beginning of solid state physics [6].

Electron band structures $E(\vec{k})$ and wave functions $\Phi(\vec{r})$, which are based on the single-electron approximation, are available for practically all relevant semiconductor materials [7]. A variety of methods has been developed in three decades to find eigenvalues and eigenfunctions of the periodic crystal potential problem [8, 9, 10]. Fig. 1 shows as an example the one-electron band structure of GaAs, obtained from pseudo-potential calculations [11]. Semiempirical methods and interpolation schemes, where band gap and critical point energies, and other data are taken from experiment, have led to a remarkable precision in certain parameters of $E(\vec{k})$. Very well known are the curvatures of $E(\vec{k})$ at the band extrema ("effective masses"), their deviations from spherical shape ("warping") and parabolic dependence ("non-parabolicity") [12].

Electron band structure $E(\vec{k})$ in this context is the approximate description of the many-particle system "solid" with one extra electron added to the empty, available conduction band states, or for one missing electron ("hole") in the otherwise occupied valence band states (see Fig.2). The valence electron charge density $\rho(\vec{r})$ is given by the sum over all occupied single-electron levels

$$\rho(\vec{r}) = e \sum_i | \phi_i |^2 \tag{1}$$

One conceptual difficulty of semi-empirical band structure calculations is their fit to experimental optical spectra: the latter contain inevitably just those pair correlations, which are a priori not included in the one-electron scheme [13].

To understand the cohesive properties of semiconductors, i.e. their lattice constants, compressibility, etc., we need a more refined description of the ground state. The central quantity is the total energy E_{tot} of the ground state as a function of electronic and vibronic motion. If E_{tot} is differentiated with respect to displacements in the positions r_i and r_j of any two nuclei, one obtains the (static) force constants and anharmonic couplings for the chosen types of vibronic motion. From here it is a lengthy, but conceptually clear way to understand phonon spectra, their elastic anisotropy, and anharmonicity effects, in short: the full microscopic description of the semiconductor [14].

For the sake of simplicity, dynamical properties of crystal lattices are usually discussed without explicit consideration of the electronic degrees of freedom. The adiabatic approximation of Born and Oppenheimer [15,16] takes the large mass ratio between electrons and nuclei to separate their equations of motion. This procedure leads, in parametrized form, to the electron-phonon coupling, i.e. the dependence of electronic energies on the configuration of nuclei, resp. phonon amplitudes.

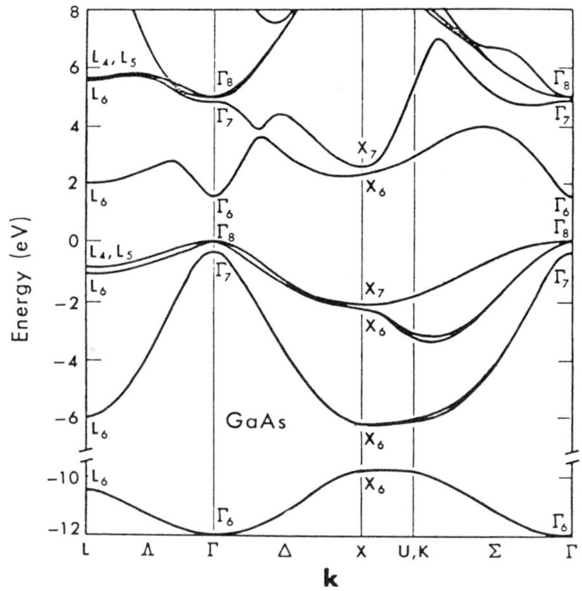

Fig.1. One-electron band structure $E(\vec{k})$ of the direct gap semiconductor GaAs, calculated with the semi-empirical pseudopotential method (from Ref.11).

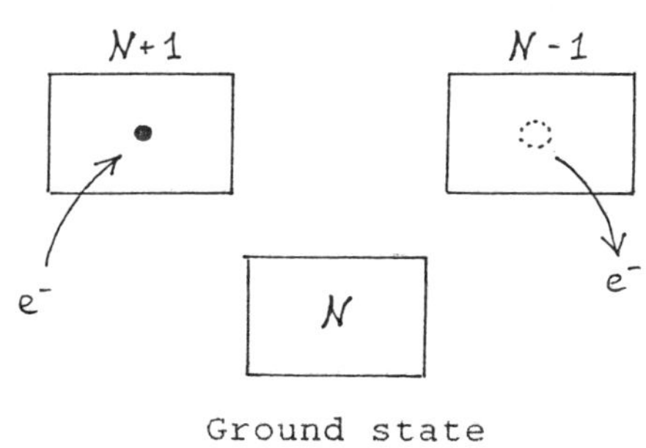

Fig.2. The concept of single-particle energies in a semiconductor: "electron" added to conduction band (upper left), "hole" created in valence band (upper right). The ground state is constructed from N occupied one-electron levels.

All interactions between bare electrons and nuclei which contributed to the static $V_o(\vec{r})$ only in averaged form, will appear in a more advanced description as effective interactions between the elementary excitations: (i) the electron-electron-interaction, in its simplest form given by the statically screened Hückel-Debye potential [17] and (ii) the different types of electron-lattice coupling (see below).

In other words, any deviation from perfect periodicity, induced by the presence of other excited electrons, or the lattice deformation due to a phonon, or an impurity atom, leads, in principle, to a damping of the elementary excitations. They are no longer stationary states. The corresponding damping mechanisms can be characterized by a scattering rate Γ (or a cross section) and we may proceed in analyzing the different possible ways of energy and momentum transfer between the "one-electron" states of $V_o(r)$ and other electronic and vibronic motions of the system. One may, in analogy to particle scattering, classify into elastic or inelastic processes, small angle and large angle scattering. If the damping $h\Gamma$ turns out to be much smaller than the characteristic (kinetic) energy of the one-electron state, it is natural to identify this state with the elementary excitation (or quasi-particle) "electron" or "hole" [5].

II.A. Interactions: Electron-Electron

Let us first examine the simplest case: carrier-carrier collisions, the scaled analog of Rutherford scattering. In a typical semiconductor with effective electron mass $m_e^* = 0.1\ m_o$ and mass ratio $m_h^*/m_e^* = 10$, and with N extra electrons of density n in the conduction band, which have been introduced via doping or by optical excitation, the screened Coulomb interaction between two such electrons is approximately

$$V_{e-e} = (e^2/r)\ \exp(-r/L), \quad (2)$$

where the characteristic length

$$L = 6.9\ (T_e/n)^{1/2} \quad (3)$$

is the Debye-Hückel-screening radius. This result is rigorously valid for a single charge (at rest) embedded in a non-degenerate equilibrium carrier distribution, which is characterized by a carrier temperature T_e [18]. It leads to a scattering rate

$$\Gamma = C\ n\ m^{-1/2}\ T_e^{-3/2} \quad (4)$$

where C depends weakly on T_e and n in the interesting ranges 10...100 K and $10^{12}...10^{17}$ cm^{-3}. Taking parameters $n = 10^{16}$ cm^{-3}, T_e = 300 K, we find L = 40 nm, and $\Gamma = 10^{-11}$ s^{-1}. It is clear that for higher densities of free carriers the electron-electron scattering becomes the dominating inelastic scattering effect.

A certain class of collisions between three particles (e-e-h, e-h-h) are called Auger transitions. They annihilate one e-h pair with energy E_{gap} and transfer this energy to the third particle, electron or hole. For a three-particle process the rate is proportional to n^2, with a coefficient $C = 10^{-30}$ $cm^6 s^{-1}$ in Si [19]. Nonradiative Auger transitions may occur in the continuum of delocalized band states (free-to-free) or in free-to-bound or bound-to-bound configuration. The transition rates are very sensitive to details in the wavefunctions (overlap integrals) and accurate calculations are difficult to obtain [20].

The long-range character of the Coulomb interaction supports also collective motions of electrons, so-called plasmon modes. Their frequency is

$$\omega_p = (4\pi e^2 n / e_o m^*)^{1/2}. \qquad (5)$$

In an excited semiconductor there are different types of such plasma oscillations. In addition to the valence charge plasmon, which describes the motion of all valence charges relative to the positively charged ions, one observes also collective motions of excited conduction band electrons, or valence band holes or both (bipolar e-h plasmons) [21]. For doping and/or excitation levels in the 10^{17} cm^{-1} range, the plasma frequency is in the infrared part of the spectrum (10 ... 100 meV) [18]. The valence charge density of 10^{23} cm^{-1} gives for the valence charge plasmon ω_p = 10 ... 20 eV, comparable to the plasma frequency in a typical metal.

Plasmon emission by highly energetic charge carriers is an efficient nonradiative relaxation channel [22]. Especially in small band-gap semiconductors it is the dominant interband recombination mechanism.

II.B. Interactions: Electron-Phonon

Electron-phonon coupling is characterized by the relatively small phonon energies in solids. The vibronic spectrum has an upper cut-off (Debye frequency), and the effective coupling strength is weighted with the phonon density-of-states $D(E) \sim E^3$. Since electron wave vectors are typically small compared with π/a, the maximum wave vector at the Brillouin zone edge, whereas phonon momenta have to be large to reach high energy and high density-of-states, one obtains a specific energy dependence [23] of the damping functions. For carrier kinetic energies E < $\hbar\omega_{LO}$ (optical phonon energy) the weak coupling to acoustic phonons gives relaxation times of the order of nanoseconds. At the above $\hbar\omega_{LO}$ this time drops by four orders of magnitude! The optical phonon energy marks a distinguished threshold. Electrons above this energy loose energy rapidly by emission of optical phonons. The mismatch in group velocities leads to a process similar to Cerenkov radiation (polaron-"stripping") [24]. The marked energy dependence has important consequences for nonlinear "hot electron" transport, especially in polar semiconductors [25].

All parameters of the one-electron band structure depend on the lattice constant a_o. Longitudinal acoustic phonons are the quanta of a compression wave, which changes a_o periodically in space and time. The corresponding change in the energy of a band electron defines the so-called deformation potential constant E_{Dp} [26].

$$a_o \, dE_o/da = E_{DP} \tag{6}$$

E_{DP} has the simple quantum mechanical interpretation: the decrease of the lattice constant increases the kinetic energy of all electrons with orbitals localized around the ionic cores (the reverse is true for the non-bonding orbitals). An inspection of the square-well potential model of a solid (Kronig-Penney) with parameters a_o = 0.5 nm and $E_c - E_v$ = 2 eV (two lowest states) gives indeed E_{DP} = 5 eV, which is just in the range of empirical values found for the deformation potential of the fundamental gap in direct gap semiconductors [27].

The procedure characterizes each electron state, occupied or empty, with a deformation potential constant. Piezospectroscopy and photoemission-spectroscopy have provided many of these constants (and differences between them) in relevant semiconductors. But even in well-known materials there are open questions with respect to the proper interpretation of experimental results in terms of deformation potential constants [28].

A second general class of electron-phonon coupling occurs in polar crystals, which show the piezo-electric effect. The long-range Coulomb interaction between the dielectric polarization field of a longitudinal optical lattice vibration and an electron (or hole) leads to a relatively strong coupling [29]. The same mechanism couples also to acoustic phonons (transverse and longitudinal, depending on crystal direction) and is the dominant intrinsic dissipation mechanism of electrons in III-V and II-VI compound semiconductors at very low temperatures [30].

Compared with ionic crystals, intrinsic semiconductors have relatively weak electron-phonon-couplings. There is no tendency for self-trapping of carriers [31], and the scattering times are long. Together with the high group velocities due to small effective masses

$$v_{gr} = (2\, E_{kin}/m^*)^{1/2}, \qquad (7)$$

this leads to very large mean free path lengths (v_{gr} is 2.10^7 cm/s for an electron of E_{kin} = 10 meV and $m^* = 0.07\ m_o$).

Let us examine as an example the model semiconductor GaAs: it has a Fröhlich coupling constant of 0.07. This implies relatively small polaron corrections (of the order of percent) to the effective mass below the threshold for optical phonon emission. The best available bulk material has (unintentional) impurity contents of less than 10^{14} cm^{-3}, that is more than 2 μm inter-impurity distance. At helium temperatures, with carrier kinetic energies well below the polaron threshold, one expects from the known electron-acoustic phonon coupling a mean free path of 5 μm for electrons [32]. The enormous potential for "ballistic" transport over large distances, which can now be controlled by lithography and man-made crystal structures, is obvious. A purely quantum-mechanical electronic device, based on injection and resonant tunneling and/or reflection in suitable double or multiple quantum well structures may be feasible [33].

II.C. Excitation and Interband Transitions

The non-thermal excitation of electrons in a semiconductor can be accomplished in two complementary ways: the first is direct injection through contact barriers (or injection of an e-beam), the second is optical excitation via absorption of photons which induce the transition of a valence band electron into a conduction-band state. The former method has been recently refined in a series of elegant experiments [34] by Heiblum and Levi. Their work aimed at controlled tunneling through narrow barriers to inject hot carriers and study their ballistic transport in suitably layered structures. The latter method is standard in optical spectroscopy; its quantum mechanical description is based on either the momentum or the electric dipole operator. It usually connects the p-like valence band orbitals with the s-like conduction band wave functions, and creates an electron-hole pair per incident photon. The final state interaction between electron and hole is responsible for so-called excitonic fine structure at the absorption edge of all semiconductors [35,36]. Excitons are bound e-h pair states with binding energies in the 1 ... 100 meV range.

An important difference between optical transitions in a semiconductor and a transition in a single atom is the quantitative difference in phase coherence time. An atom, usually an isolated object, undergoes relatively few phase-destroying collisions with other scatterers; it has a relatively long dephasing time T_2. In a semiconductor there is inevitably and simultaneously the motion of electron and hole, in opposite direction and with different phase velocities, and thus a rapid loss of (local) coherence induced by the motion. In other words, the pronounced delocalization of both e and h wavefunctions leads to a strong "cross" relaxation among different atomic sites. The concept of coherent e-h pair excitation and propagation in semiconductors has been worked out recently by A. Stahl and co-workers [37].

Optical excitation is a reversible process; emission of light after an absorption process is always probable. In most cases, however, (we exclude here stimulated emission) the reverse process is relatively weak compared with the competing relaxation processes. The interplay between the (weak) electron-photon and the relatively strong electron-electron- and electron-lattice interactions (see above) determines the radiative efficiency of an excited semiconductor [38].

II.D. Radiative versus Nonradiative Transitions

Close to the band extrema the energy dispersion $E(\vec{k})$ is proportional to k^2, so that radiative intraband transitions, which have to be "vertical" in k-space are not allowed in first order. Another excitation with finite k-vector is needed to make the transition allowed. Free-carrier infrared absorption is an example, where acoustic phonons and the deformation potential coupling provide the necessary momentum conservation for intraband radiative transitions in (doped) semiconductors.

The localization of the electron wavefunction at an impurity site may contribute to momentum conservation in free-to-bound radiative processes: the corresponding electric dipole moment is connected with the envelope wave function (in general s-type in the ground state) of the impurity bound electron and its continuum wave functions (of p-character) in the final state. It is convenient to scale the parameters of the well documented case of the photoionization of the hydrogen atom. The time constant for a radiative transition from the conduction band into the lowest bound state of an effective-mass-like donor is of the order of 10^{-6} s, long compared with the time constants of e-phonon coupling (see above) and cascade capture. With increasing spatial extent of the envelope wave function (i.e. less binding to the center) the transition becomes more effective [23].

Radiative transitions where both initial and final states are bound to impurities (so-called pair transitions) depend critically on the symmetry of the corresponding band states, the two envelope wave functions, and on their mutual spatial overlap. Donor-acceptor pair transitions (e.g. with neutral initial and ionized final states) cause a net charge transfer in real space. The Coulomb interaction between the ionized donor and acceptor in the final state leads to a characteristic 1/r dependence of emission photon energies ("pair spectra").

III. ENERGY AND MOMENTUM RELAXATION OF "HOT" CARRIERS

III.A. Intrinsic Crystals

In general electron-phonon coupling dominates the dissipation of energy from (or into) a carrier in a given band. Photon-mediated intraband- processes (like IR-absorption through free carriers) are kinematically forbidden and usually weak. So the term "nonradiative" is appropriate for the dominant intra-band relaxation mechanisms.

As outlined above, there are two distinct physical mechanisms for the electron-lattice coupling:

(i) The omnipresent deformation potential mechanism. Its short-range nature is caused by the confinement effect of the electrons in the crystal potential $V_o(\vec{r})$. The coupling constant has a typical value 10 eV, and the transition rate depends linearly on electron kinetic energy [23];

(ii) The long-range coupling to longitudinal optical phonons and also piezoelectric active acoustic phonons in polar (i.e. zincblende) semiconductors. This electrostatic coupling leads to an inverse dependence on phonon wavevector [23].

A comparison with the transition rate for interband recombination in a direct gap semiconductor (e.g. GaAs) shows the strong dominance of nonradiative processes: the radiative rate R_{rad} (in a degenerate e-h pair distribution, i.e. with each phase space cell $d^3\vec{k}$ filled with one electron resp. one hole) is $3\times10^9\ s^{-1}$ for an interband matrix element P = 29 eV [39]. It is not incidental that the interband transition rate in GaAs is almost identical to the corresponding rate of a 2p to 1s transition in atomic hydrogen (0.3 ns). In fact the radiative rates for all direct gap semiconductors differ by not more than a factor of 3.

As a consequence, electron (and hole) energy relaxation proceeds on the 100 fs to 100 ps time scale within the bands, before any appreciable amount of radiative interband recombination occurs, almost independent of injection energy. Only a small fraction of both electron and hole distributions, characterized by the branching ratio

$$b = R_{rad} / R_{nonrad} \tag{8}$$

recombines by emission of photons with energy

$$\hbar\omega = E_e + E_{gap} + E_h \tag{9}$$

where E_e and E_h are electron and hole kinetic energies, respectively.

The emission spectrum of optically excited GaAs (a direct gap semiconductor) is shown in Fig.3 on a logarithmic scale as a function of excess photon energy over E_{gap}. The spectrum is dominated by e-h pairs with low kinetic energies, thermalized in distributions of approximate width 3/2 kT at the band edges. On the high energy side the exponential tail of the nondegenerate carrier distributions changes into a plateau. This qualitative change marks the onset of "hot" luminescence: in the plateau region the e-h pairs have undergone only a few scattering events (with phonons), and they are not in thermal equilibrium with the majority of carriers at the band edges when they recombine. The remaining spectral structure is a direct measure of the branching ratio as a function of kinetic energy and has been analyzed to provide information on band structure [40] and intervalley scattering [41]. Pronounced peaks mark the cascade of LO phonon inelastic scattering, and depend on the

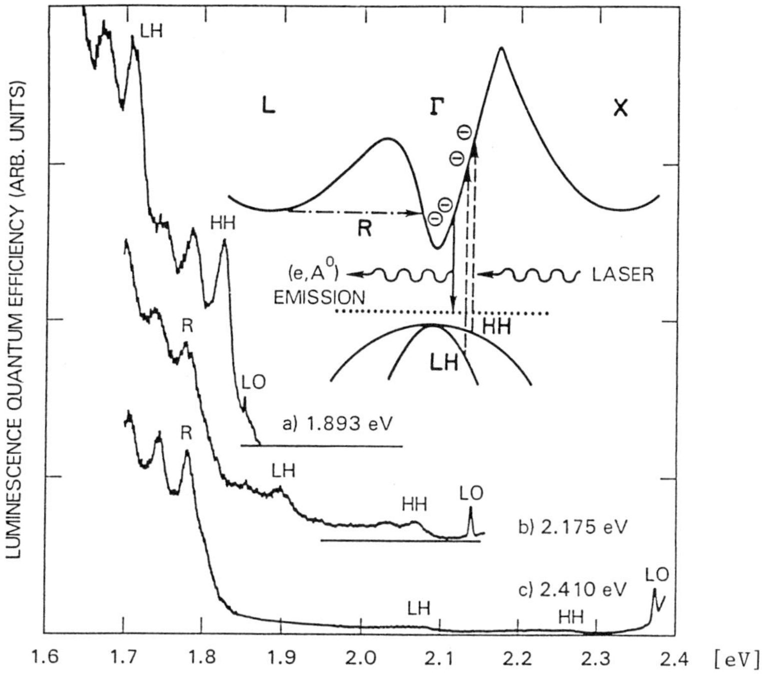

Fig.3. Hot electron emission spectrum of Mg-doped GaAs at lattice temperature T = 2K. the spectra were excited as indicated. The inset explains the origins of the three sets of periodic peaks (see text). From Ref. 41.

excitation photon energy. Their energetic "memory" and degree of linear and circular polarization can be used to determine accurately electron-phonon scattering times [41].

Indirect semiconductors, like Si (or Ge), have no efficient interband recombination process. The intrinsic lifetime for e-h pairs is 35 ms (20 μs), and is controlled by the coupling of e-h pairs to zone-edge TA and TO phonons [42]. The large (and qualitative) difference among direct and indirect gap semiconductors is due to the symmetry of the lowest conduction band minimum (the valence bands are practically identical). Fig. 4 shows schematically the relevant cell-periodic wavefunctions $u_{\vec{k}}(\vec{r})$ and phase factors $e^{i\vec{k}\cdot\vec{r}}$ for Si and GaAs in ordinary and reciprocal space. In the dipole approximation (i.e. neglecting the small wavevector of the emitted light) the transition is forbidden in first order of e-photon coupling. Only second order processes, like emission of a phonon and a photon, give a finite rate. It has been shown

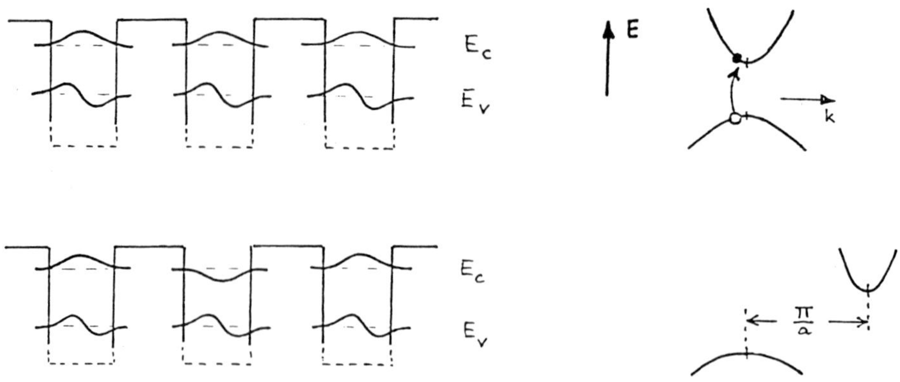

Fig.4. Direct and indirect energy gap in semiconductors: for $k_e = k_h$ (top) each cell contributes to the dipole transition from E_V to E_C. An indirect gap (bottom) causes alternating phases of U_C (k) and there is no net effect (Schematic).

that both variants, electron-phonon- and hole-phonon-coupling play a role in the intrinsic recombination process of Si and Ge.

Any perturbation of the symmetry of the crystal lattice e.g. by the presence of a surface, an interface, or an impurity may change the lifetime of a pair appreciably. The same is true for the lack of intensity of inelastic light scattering in certain "forbidden" geometric configurations [43].

III.B. The Role of Impurities

The incorporation of defects or substitutional impurities in a crystal lattice brings, in general, an essential modification of all wave functions in the vicinity of the center. The perturbation of crystal translational symmetry causes localization of extended states with a more or less deep binding to the center. The amount and extension of lattice distortion around the impurity and the related change of valence charge density (compared with the unperturbed crystal) depends on the electronic configuration of the impurity atom and its hybridization with the host wavelengths.

Ab initio calculations of binding energies, wave functions and matrix elements between "free" carrier states and bound impurity states are difficult to obtain. Only a few systems have been tackled up to now [44]. Much more is known on the limiting case of pure Coulomb centers with extended wave functions, especially donors, which can be treated in the so-called "Effective-Mass-Approximation" [45]. The common degenerate valence band structure composed of directed orbitals with total angular momentum F = 3/2 (electron spin included) leads to complicated acceptor spectra and a more elaborate description [46].

Not much is known on impurity-phonon coupling from theoretical calculations. Centers with a large amount of lattice relaxation were for a long time ignored in semiconductor physics and are still poorly understood [47]. A prominent and ubiqitious prototype of these so-called "deep" centers is the EL2 center in GaAs. It controls effectively the compensation (i.e. capture of mobile carriers) of practically all known crystalline material. In spite of many experimental and theoretical attempts to understand its electronic and structural configuration, this deep center is still not understood.

Transition probabilities ("capture rates") of mobile carriers into deeply bound impurity states vary over many orders of magnitude -- depending on temperature and the charge state of the center. There are only phenomenological descriptions of the capture process in terms of electron-lattice coupling [47].

IV. ULTRAFAST CARRIER RELAXATION: PHASE-SENSITIVE EFFECTS

Very recent progress in the field of passively mode-locked dye lasers with large spectral bandwidth has opened a qualitatively new field of time-resolved optical spectroscopy: pulse durations down to 8 fs, corresponding to only four cycle times of visible light, with electric fields in the range 10^{10} V/cm have been achieved [48]. Typical lattice vibrational times T_o of semiconductors are in the 100 to 200 fs range. This implies that electronic transitions can be investigated in "frozen lattice" configurations: the inertia of the lattice causes an effective freeze-out of ionic motion on such time scales.

With appropriate experimental techniques, which are all derivatives (variants) of the pump-probe-method in transmission-, reflectance- or emission configuration, the sequence of momentum and energy relaxation of optically excited e-h pairs has been investigated on the ultrashort time scale [49]. Four-wave mixing methods incorporate more than two incident excite beams and exploit the dispersive and absorptive properties of optically induced gratings of excitation density within the semiconductor sample [50]. The essential new feature in the 100 fs time domain is the qualitative importance of phase coherence between pump field, induced dielectric polarization, and probe electric field. In full analogy with an optically pumped 2 level system one expects saturation effects (induced transparency) and photon echo (free induction decay). Both phenomena have been reported recently in GaAs [51,52].

The continuum of e-h pair states above the gap energy has a relatively high density of states and thus very efficient relaxation channels. Many of the above-mentioned "two-level" properties are therefore masked by strong damping and could not be identified as well resolved as in free atoms.

The strong interest in ultrashort light pulses has to do with their potential to probe high-field carrier transport: nonlinear optical properties of e-h distributions (at high enough densities in the bands or close to the band edges) can be analyzed in the time and energy domain to extract details of carrier dynamics and spatial transport in microstructured devices. High field electrical transport and optical excitation on the 100 fs time scale are closely related phenomena: the description of charge transport beyond the Boltzmann equation contains all elements of the optical response mediated through inter and intraband electronic processes.

Key element in the description of optical and charge transport properties are the single-particle distribution functions f_e and f_h (in energy or momentum representation). A proper account for the phases of both wave-like excitation has to be made [53]. Here it is sufficient to think of f as the time-averaged probability to find the corresponding phase space volume $d^3\vec{r}\ d^3\vec{k}$ filled with one electron (or hole). All excited electrons and holes interact via a screened Coulomb interaction; the Pauli principle restricts all possible scattering processes such that f takes instantaneously the value 0 or 1. Static screening with the Debye-Hueckel expression (see above) assumes only equilibrium distributions f_o. More realistic treatments take the nonequilibrium distributions into account (which in turn modify the polarizability of the total system) and try to treat the whole problem in a self-consistent manner [54].

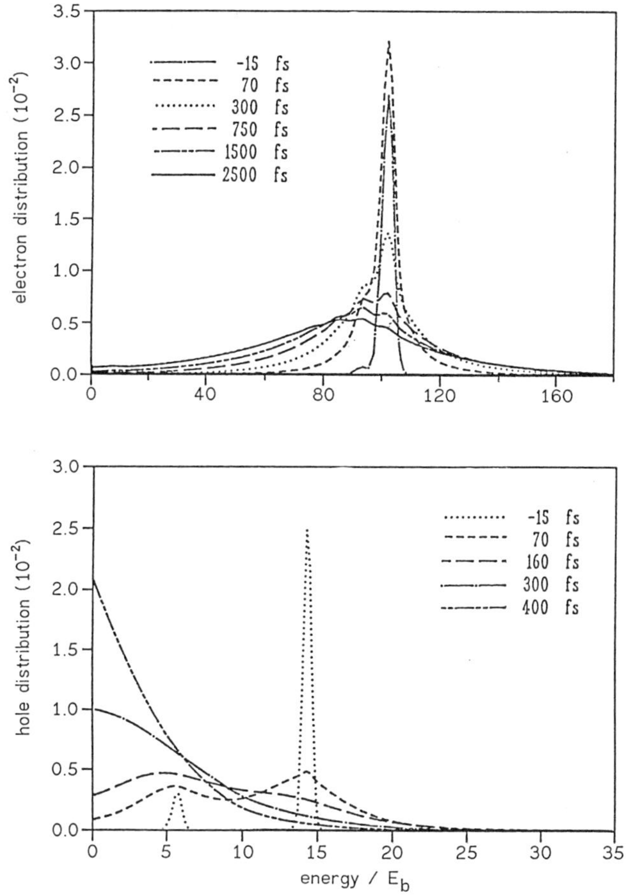

Fig.5. Energy relaxation of optically excited electron-hole pairs. The energy scale is in units of the exciton binding energy E_B = 4 meV in GaAs. From Ref. 55.

Fig. 5 shows the result of such a calculation for the case of photoexcited e-h pairs in GaAs. Carrier-carrier scattering between electron and holes, the emission and absorption of LO phonons and the coupling of e-h pairs to the driving light field have been included. The parameters are excitation energy 2 eV, n = $10^{17} cm^{-3}$, effective masses m_e = 0.07 m_o and m_b = 0.3 m_o [55]. The calculations show explicitly the net cooling effect with increasing time (i.e. the drop of average carrier kinetic energy) and the gradual randomization of initial nonequilibrium carrier distributions (i.e. the increase in energetic width) caused by the intercarrier collisions. The interplay between purely dissipative

LO-phonon emission, which brings the center of the distributions to lower energy, and the more or less symmetric spreading due to e-e collisions is clearly visible. There is a slight asymmetry because of the density-of-states factor in the scattering probability.

The occupation factors f_e and f_h can be determined quantitatively from optical measurements: the connection between an occupied valence band state and an empty conduction band state (we discuss the absorption process) is mediated in first order perturbation theory through the interband matrix element $\langle c|p|v\rangle$. The driving light field is regarded as a small perturbation. For a given light frequency one has to sum over all possible resonances, i.e. electron hole pair states with the same momentum (vertical transitions). The occupation of initial and final states is taken into account through the factors f_e and f_h in the denominator, so that

$$\varepsilon(\omega) = 1 + \frac{4\pi e^2}{h\, m^2} \sum_v \frac{\langle v|p|c\rangle\langle c|p|v\rangle \cdot (1-f_h) \cdot (1-f_e)}{(\omega_d - \omega_v)^2 - \omega^2} \qquad (10)$$

describes the dielectric response function of the excited solid at the probe frequency [56]. The expression eq.5 is a simple quantum mechanical analog of the Drude-Lorentz formula for the response of a (linear) system to an external harmonic perturbation. Any spectral structure in the occupation f_e and f_h will show up in $\varepsilon(\omega)$, as a dispersive and/or absorptive feature in all optical spectra. It turns out that reflectivity spectra are dominated by the real part of $\varepsilon(\omega)$, whereas transmission spectra give information on the imaginary part [55]. Nonlinear changes in the primary optical spectra dR/R_o or dT/T_o, even if they are small, are sensitive probes of the overall energy distributions f_e and f_h.

Fig. 6 shows the measured and calculated change in reflectivity for the parameters discussed above. The abscissa is the relative time delay between exciting and probe light pulses, which had both 70 fs duration and could be tuned separately within a 3 nm wavelength interval. The e-h relaxation processes cause a rapid shift of the spectral "hole" burnt into the continuum pair spectrum towards lower energies. The combined effects of energy dissipation through LO phonon emission and broadening of the distributions leads to the characteristic fast spike in the nonlinear reflectivity. The slow decay at longer delay times (in the

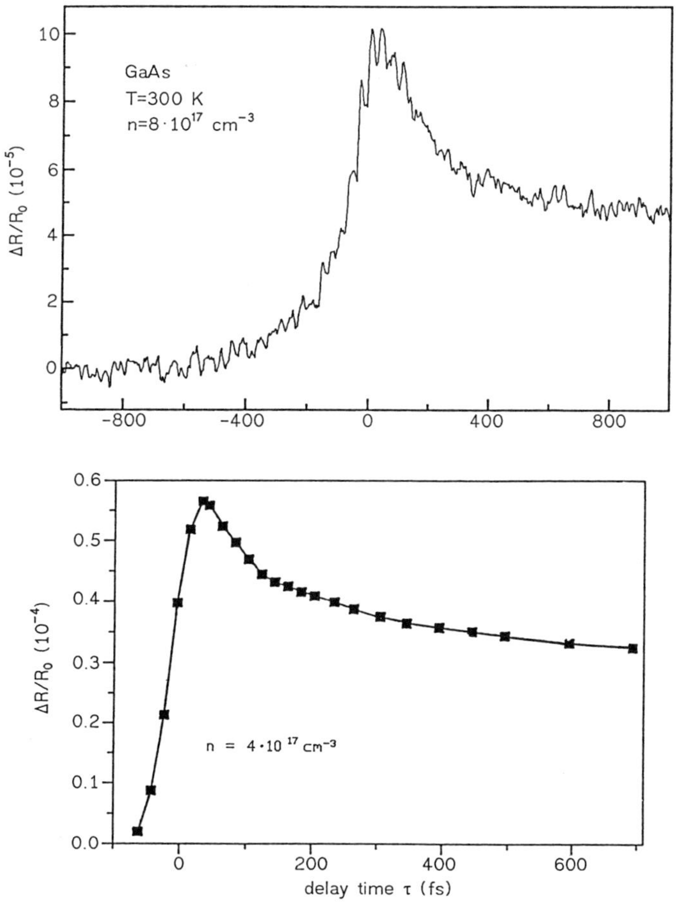

Fig.6. Measured (top) and calculated (bottom) transient nonlinear reflectivity spectra of GaAs. The abscissa is the delay time between pump and probe pulses of 70 fs duration. From Ref. 55.

range 0.5 to 5 ps) is caused by the energy relaxation of pairs towards the gap energy. There is a range of kinetic energies, where the efficient coupling to LO phonons (with a threshold of 36 meV, see above) is only indirectly mediated through the high energy tail of the carrier distributions. Another effect, which has not been included in the calculations, the volume expansion of the excited plasma, causes a drift velocity of the order of 10^7 cm/s and gives a typical time constant of 10 ps at the chosen carrier density. Within 100 fs, carriers travel ballistically approximately 200 nm at the initial injection energy, that is less than the optical absorption depth at 2 eV photon energy.

REFERENCES

1. S.M. Sze, Physics of Semiconductor Devices, Second Edition, (Wiley, New York, 1981) Ch.1.
2. M.J. Kelly and C. Weisbuch, The Physics and Fabrication of Microstructures and Microdevices (Springer Proceedings in Physics Vol. 13, Berlin, 1986) Part II.
3. F. Wooten, Optical Properties of Solids (Academic Press, New York, 1972) Ch.5.
4. F. Bassani and G. Pastori Parravicini, Electronic States and Optical Transitions in Solids (Pergamon Press, Oxford, 1975).
5. S. Nakajima, Y. Toyozawa and R. Abe, The Physics of Elementary Excitations (Springer Series in Solid-State Sciences, Vol.12, Berlin, 1980) Ch.2.
6. F. Seitz, The Modern Theory of Solids (McGraw Hill, New York, 1940).
7. G.B. Bachelet, D.R. Hamann, and M. Schlüter, Phys. Rev. B26, 4199 (1982).
8. J.C. Phillips, Bonds and Bands in Semiconductors (Academic Press, New York, 1973) Ch.5.
9. M.L. Cohen and J.R. Chelikowski, Electronic Structure and Optical Properties of Semiconductors (Springer, Berlin, 1988) Ch.1.
10. Local Density Approximations in Quantum Chemistry and Solid State Physics (eds. J.P. Dahl and J. Avery, Plenum Press, 1985).
11. R.R.L. Zucca, J.P. Walter, Y.R. Shen and M.L. Cohen, Solid State Commun. 8, 627 (1970).
12. U. Roessler, Solid State Commun. 49, 943 (1984).
13. W. Hanke, in: Advances in Solid State Physics Vol.19, (ed. J. Treusch, Vieweg, Braunschweig, 1978) p.43.
14. D.J. Chadi, Proc. 19th Int. Conf. Semiconductors, Warsaw, 1988 (ed. W. Zawadzki, Inst. of Physics, Polish Academy of Sciences, 1988) p.3.
15. M. Born and R. Oppenheimer, Ann. d. Physik 34, 457 (1927).
16. See the article by B. DiBartolo in this book.
17. P. Debye and E. Hueckel, Phys. Z. 24, 185 (1923); ibid. 305.
18. P.M. Platzman and P.A. Wolff, Waves and Interactions in Solid State Plasmas (Academic Press, New York, 1973) Ch.1.
19. G. Benz and R. Conradt, Phys. Rev. B16, 843 (1977).
20. D.B. Laks, G.F. Newmark, A. Hangleiter and S.T. Pantelides, Phys. Rev. Lett. 61, 1229 (1988).
21. J. Appel and A.W. Overhauser, Phys. Rev. B26, 507 (1982).
22. R.P. Huebener, Rep. Prog. Phys. 47, 175 (1984).
23. B.K. Ridley, Quantum Processes in Semiconductors (Clarendon Press, Oxford, 1982) ch.3.
24. R.P. Feynman, Statistical Mechanics (W.A.Benjamin, Reading, 1972) Ch.8.
25. See e.g. Proc. 5th Int. Conf. on Hot Carriers in Semiconductors, Boston, 1987 (Pergamon Press, New York, 1988), Ch.4.
26. J. Bardeen and W. Shockley, Phys. Rev. 80, 72 (1950).
27. Sadao Adachi, J. Appl. Phys. 58, R1 (1985).
28. G.S. Cargill, III, J. Angilello, and K.L. Kavanagh, Phys. Rev. Lett. 61, 1748 (1988).
29. H. Fröhlich, in: Polarons and Excitons (ed. C.G. Kuper and G.D. Whitfield, Oliver and Boyd, 1963) p.1.
30. J.D. Wiley, in: Semiconductors and Semimetals Vol.10, (ed. R.K. Willardson abd A.C. Beer, Academic Press, New York, 1975) Ch.2; D.Rode, ibid., Ch.1.
31. Ref.5, Ch.7.5.
32. C.M.Wolfe, G.E. Stillman, and W.T. Lindley, J. Appl. Phys. 41, 3088 (1970).
33. Physics of Quantum Electron Devices (ed. F. Capasso, Springer Series in Electronics and Photonics Vol.28, 1990).

34. See M. Heiblum, Ref.25, p.617; ibid. J. Hayes, p.619.
35. R.J. Elliot, in: Polarons and Excitons (ed. C.G.Kuper and G.D. Whitfield, Oliver and Boyd, 1963) p.269.
36. J.Dow, in: New Developments in the Optical Properties of Solids (ed. B. Seraphin, North Holland, 1988) Ch.2.
37. A. Stahl and I. Balslev, Electrodynamics of the Semiconductor Band Edge (Springer, 1986) Ch.4.
38. J.I. Pankove, Optical Processes in Semiconductors (Dover Publications, New York, 1971) Ch.7.
39. P. Lawaetz, Phys. Rev. B4, 3760 (1971).
40. G. Fasol and H.P. Hughes, Phys. Rev. B33, 2953 (1986).
41. R.G. Ulbrich, J. Kash and J. Tsang, Phys. Rev. Lett. 62, 949 (1989)
42. G.A. Thomas, E.I. Blount, and M. Capizzi, Phys. Rev. B19, 702 (1979).
43. G. Abstreiter, M. Cardona, A. Pinczuk, in: Topics in Applied Physics Vol. 54 (Springer, Berlin, 1984) p.5.
44. M. Jaros, Phys. Rev. B16, 3694 (1977).
45. W. Kohn, in: Solid State Physics Vol.5 (Academic Press, New York, 1957) p.257.
46. A. Baldereschi and N.O. Lipari, Phys. Rev. B8, 2697 (1973); S.T. Pantelides, Rev. Mod. Phys. 50 (1978).
47. C.H. Henry and D.V. Lang, Phys. Rev. B15, 989 (1977).
48. R.L. Fork, C.H. Brito Cruz, P.C. Becker, C.V. Shank, Opt. Lett. 12, 483 (1987).
49. D.R. Grischkowski, J. Opt. Soc. Am. B2, 582 (1985) and references therein.
50. H.J. Eichler, P. Günter, D.W. Pohl. Laser-induced Dynamic Gratings (Springer, Berlin, 1986) Ch.2.
51. W.H. Knox, D.S. Chemla, G. Livescu, J.E. Cunningham, and J.E. Henry, Phys. Rev. Lett. 61, 1290 (1988).
52. P.C. Becker, H.L. Fragnito, C.H. Brito Cruz, R.L. Fork, J.E. Cunningham, J.E. Henry, and C.V. Shank, Phys. Rev. Lett. 61, 1647 (1988).
53. Ref. 24, Ch.2.7.
54. W. Schaefer, in: Advances in Solid State Physics Vol.28 (ed. U. Roessler, Vieweg, Braunschweig, 1988) p.63.
55. G. Boehne, T. Sure, R.G. Ulbrich and W. Schaefer, Phys. Rev. B41, 1990.
56. See Ref.3, Ch.7.2.
57. T. Kuhn and G. Mahler, Phys. Rev. B39, 1194 (1989).

RADIATIVE AND NONRADIATIVE PROCESSES IN COLOR CENTERS

G. Baldacchini

ENEA, Centro Ricerche Energia Frascati
C.P.65 - 00044 Frascati, Rome, Italy

ABSTRACT

The optical properties of color centers in alkali halides have been used in the past and are still used today in some noteworthy applications, e.g., dosimetry, information storage, and tunable near infrared lasers. However, a satisfactory description of the emission properties after excitation is still lacking. Thus, in order to reach a better understanding of the luminescent phenomena, the proper experimental results in F centers are reviewed and compared with the prevailing theory of the crossover model. This comparison shows that a complete agreement between theory and experiments is far from being reached.

I. INTRODUCTION

The optical cycle of color centers performed following excitation in their typical absorption band is one of the most interesting properties of this kind of point defect in alkali halides. After excitation and subsequent lattice relaxation, a luminescence process can occur, followed by another relaxation toward the ground state. The radiative transitions of the optical cycle and the reorientational properties of axial centers have been used in the past and are still used today in some noteworthy applications, such as dosimetry, optical information storage, and tunable near infrared lasers.

However, the phenomenon of luminescence has not yet been fully clarified, especially as far as the nature of the emitting state and its de-excitation processes are concerned. The emission of the simplest among all color centers, the F center (an electron trapped in an anion vacancy), can be quenched by high temperatures, high electric fields, high F center concentration, F_2 centers, ion impurities, and other factors which have not been clarified satisfactorily up to now.

This paper is intended to be primarily didactical and, at the same time, to deliver a general picture of the emission and related properties as it stands today. As a consequence, after an initial account of the history of color centers and their main applications in Sec.II, which helps to correctly place the color centers in the broader field of solid state physics, it has been deemed necessary to give a general description of the F and the more simple F aggregated centers in Sec.III. However, people who work in this field or know about it do not need to read this material in detail, but can rather pass directly to Sec.IV where the radiative and nonradiative processes in F centers are described in some detail and classified according to their properties and to a somewhat personal judgement. Finally, in Sec.V a comparison of the experimental results with the crossover model, devised long ago to describe the radiation and radiationless transitions in solids, is presented and the apparent shortcomings of the model and the latest results are also critically discussed.

II. HISTORICAL BACKGROUND AND APPLICATIONS

The birth date of color center activities goes back to the past century when a crystal of sodium chloride was colored at high temperatures in sodium vapors [1]. However, that result and others obtained subsequently, also in connection with the discovery of natural radioactivity and X rays which both produced coloration in several crystals [2], were considered mere scientific curiosities up to the twenties, when a systematic research was started at Gottingen by R.W. Pohl and coworkers.

Before the forties, color centers in alkali halides were well enough characterized as far as the light absorption, which produces the coloration and the photoconductivity properties, was concerned [3,4]. However, there were also expectations for fluorescence in the near infrared after illuminating the absorption band in the visible region [5].

After World War II, the research on this subject was resumed and a lively debate started about the existence of emission [6] after optical excitation of the F center, which is the prototype of all color centers. Eventually, luminescence was found [7,8], and the optical cycle of the F center and other kinds of centers was later on almost completely clarified [9].

In the meantime, the applicative side of this field of research was on the move. An earlier attempt at Los Alamos from 1943 to 1945 to use the coloration of an alkali halide crystal to measure the artificial radioactivity failed partially because the coloration faded at room temperature [10]. At the same time, at the Carnegie Institute of Technology the process of coloration was also studied for the realization of cathode ray screens for radar, but never really employed in favor of other more stable materials [11].

Later on, the first stimulated emission was observed in an F_A center (an F center associated to an alkali ion impurity) in KCl:Li [12]. However, this important result was largely disregarded at the time mostly because laser technology was not yet well developed. Only later was the work resumed in a more systematic way with the realization of a c.w. laser tunable in the range 2.5 - 2.9 μm by using the same kind of centers [13]. Contemporaneously, the possibility of storing images in a three-dimensional holography was demonstrated in a crystal of KCl:Na containing F_A centers [14]. Moreover, the existence of zero phonon lines inhomogeneously broadened allowed *hole burnings* to be made by means of narrow band lasers. Such an effect, demonstrated for the first time on F_3^+ centers (an ionized triangular array of three F centers set in the (111) plane) in NaF, opened the way to a new and powerful system for optical information storage [15].

Nowadays, the coloration of some alkali halide crystals is currently used to measure the dose of radiation absorbed by the human body, the holographic information storage is receiving little attention, the hole burning technique is still under study [16], and the color center laser field has reached its maturity with still signs of further developments.

Extensive research has been carried out during the last decade on investigating the potential usefulness of several types of defects as active laser materials. As a result of this effort, it is possible today to cover the spectral range between 0.8 and 4 μm entirely with c.w. tunable lasers [17,18,19].

Figure 1 gives an idea of this situation. The maximum power of color center lasers averaged within each family of centers is displayed as a function of the emission wavelength. The data refer to crystals cooled mostly at liquid nitrogen temperature. The working range of color center lasers spans a region where there are no other similar tunable lasers. Indeed, dye lasers from the visible range arrive with difficulty at ~1 μm, while Pb tunable diode lasers start at ~3 μm towards longer wavelengths [20].

Color center lasers also have a very narrow linewidth [21], which makes them most suitable for high resolution molecular spectroscopy. Their emission around 1.5 μm has also been used to study light propagation in optical fibers, which have a minimum absorption and a zero group velocity dispersion at that wavelength. Recently, by using the nonlinear properties of optical fibers, a soliton laser has been realized with an output pulse of 19 fs duration [17].

In conclusion, it is evident from the previous digression that the optical properties and, in particular, the luminescent ones, of color centers in alkali halides have found wide applications in various fields besides stimulating a great interest in their basic knowledge. Hence, these two factors together provide sound justifications for a detailed study of the radiation and radiationless processes of color centers in alkali halides.

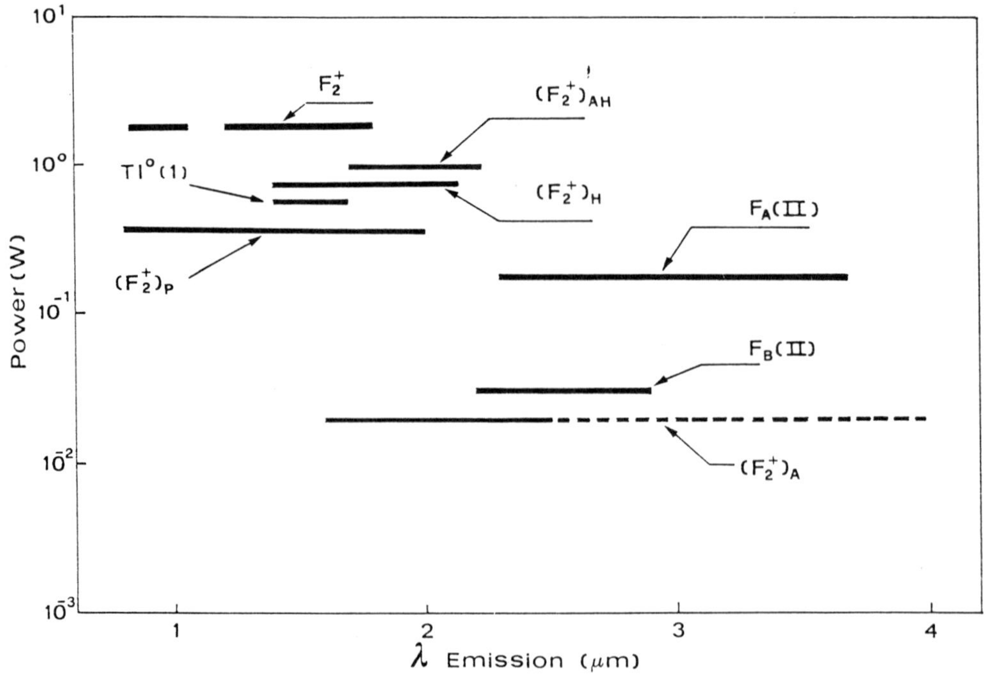

Fig.1 Continuous wave (c.w.) laser emission at LNT of various families of color centers. The dashed curve represents pulsed systems.

III. COLOR CENTERS IN ALKALI HALIDES

III.A. Introduction to Color Centers

Color centers in alkali halides have been studied for a long time, and as a lot of review papers and books have been written on the argument, we refer to only a few of them for a detailed description of their production and mechanical, electrical, and optical behaviour [5,6,22-26]. Only a general account of the basic properties is given in this work in order to make a continuous and consequential lecture possible.

Alkali halide crystals have a cubic symmetry and a face-centered (Li, Na, K, Rb halides and CsF) and body-centered (Cs halides) structure. They are very good electric insulators with an energy gap of several eV and are transparent from the far infrared to the ultraviolet region of the electromagnetic spectrum. The last property makes them very useful for the study of point defects in the lattice, whose electronic transitions are easily observed as absorption or emission bands in the energy gap. The absorption of the defects gives rise to a typical coloration of the crystal, which, for the F center, varies from yellow in NaCl to violet in KCl. This coloration is the reason why the names *color center* and *F center* (F is the initial of the German word Farbe=color) have been given to such defects.

III.B. F Centers

The simplest point defect is the F center (an electron trapped in an anion vacancy), which is also a prototype case study for all the other more complex centers, somewhat like to the hydrogen atom in the atomic and molecular spectroscopy fields. The electron of the F center moves around the positive charge of the anion vacancy, Fig.2a, and so it forms hydrogen-like bound states 1s, 2p, etc., in analogy with the atomic case. The absorption from the ground state, 1s, to the first excited state, 2p, corresponds to the main absorption band of the F center, Fig.2b. However, the analogy with the hydrogen atom ends here. Indeed, the new distribution of the electronic charge is not in electrostatic equilibrium with the local crystal fields, so the first positive neighboring ions move outward in order to compensate for the charge variation. This process is usually called relaxation and at its completion, the electron reaches the relaxed excited state (RES). From this peculiar state, the electron decays with the emission of light, the typical F center luminescence, to the unrelaxed ground state and, eventually, after a further relaxation process, it reaches the ground state (GS) again.

The complete excitation/de-excitation process, which is usually known as the optical cycle, can be illustrated much more effectively and realistically with a configuration coordinate diagram. The two electronic states 1s and 2p are plotted in Fig.2c as a function of the normal coordinate Q, which is a measurement of the ionic displacement of the dominant lattice mode. In the approximation of the harmonic oscillator, the two states can be described by two parabolas with the two minima displaced with respect to each other because of the different lattice equilibrium around the defect. The displacement is also a measurement of the electron-lattice coupling which is a very important parameter for the optical cycle, as we will see shortly. At low temperatures, the electron of the F center occupies the bottom of the 1s parabola, the GS. Optical excitation produces a vertical transition (Frank-Condon principle) to the 2p state in a highly excited vibrational level. The vibronic system, electron + phonons, relaxes very quickly, $\sim 10^{-12}$ s, to the equilibrium position corresponding to the minimum of the 2p parabola, the RES. The electron can remain here a relatively long time, ~ 1 μs, before returning to the unrelaxed GS with a vertical transition accompanied by light emission. In the end, a lattice relaxation completes the optical cycle to the GS.

The diagram of Fig.2c shows that both relaxation processes release energy to the lattice, contrary to what is implied schematically and incorrectly by Fig.2b for the first relaxation. Moreover, this simple model explains not only the Stokes-shift of the absorption and emission bands, but also their broad bands whose linewidth is a function of the temperature. Indeed, more than one vibrational state is involved in the quasi-vertical

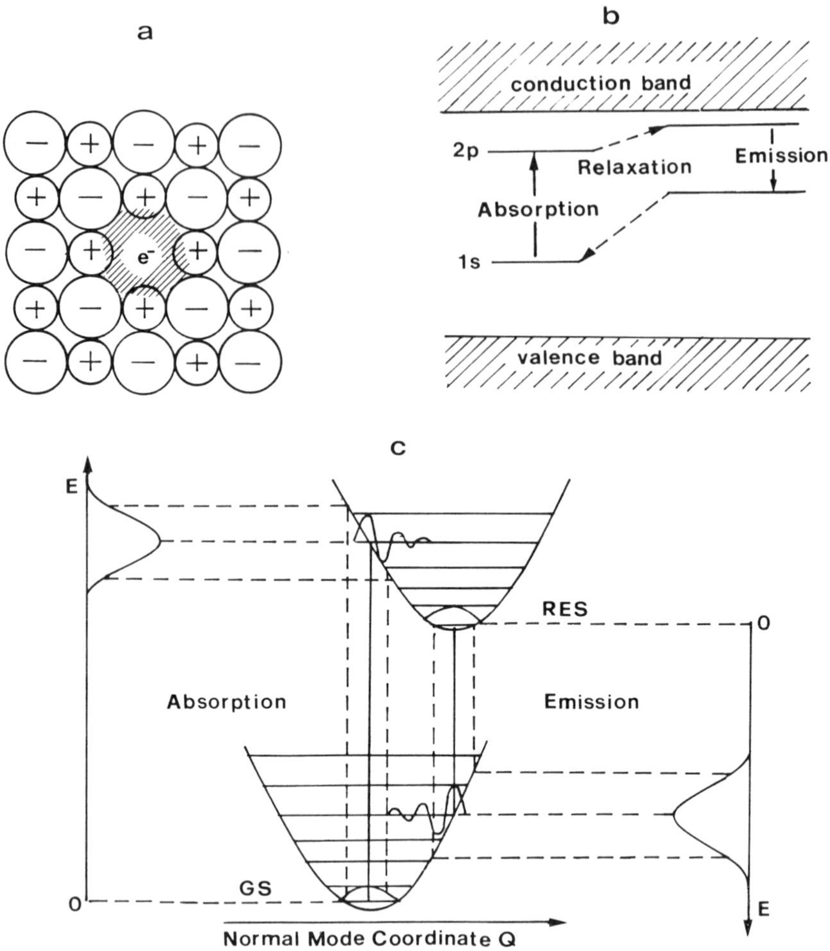

Fig.2 Description of the F center in alkali halides: a) lattice model; b) main electronic states and optical cycle; c) configuration coordinate diagram.

transition, and this number increases with temperature, following Boltzmann's law. The width of the absorption and emission bands is given by the following equation:

$$W = W_0 \left[\coth \frac{h\nu}{2kT} \right]^{1/2} , \tag{1}$$

where W_0 is the width (FWHM) at low temperatures, h is Planck's constant, ν is the vibrational frequency of the lattice mode coupled with the electron, k is Boltzmann's constant, and T(K) is the absolute temperature of the crystal. Figure 3 gives the absorption and emission bands of the F center in KCl for various temperatures [27]; not only the broadening, but also the peak positions of the bands show a temperature dependence which is not accounted for by Eq.(1).

The simple descriptive model used so far should not deceive the reader about the complexity of the F center, although extended experimental and theoretical work has led to a high degree of understanding. The complexity

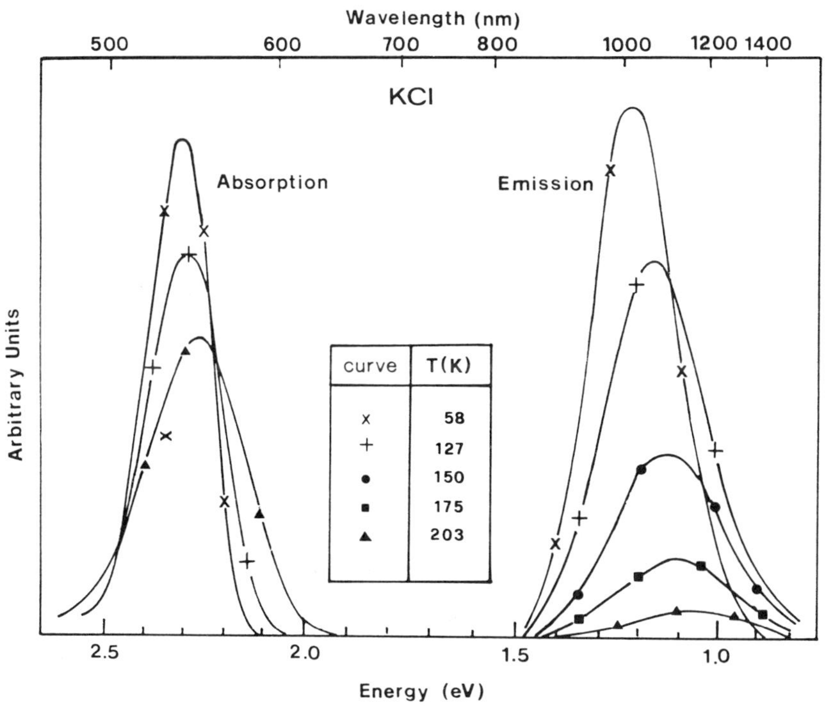

Fig.3 F center absorption and emission bands in KCl as a function of temperature [27].

of the F center is made evident by the mere fact that there is no unique model able to explain at the same time different parameters of the optical cycle.

In the ground state, the electron density is so strongly localized within the vacancy that the F center can be treated like an electron in a potential box, a well-known problem in quantum mechanics. The separation of the first two quantum states, corresponding to the 1s and 2p levels, is given by

$$E_{2p} - E_{1s} = \frac{3\pi^2\hbar^2}{8ma^2} \sim a^{-2} \tag{2}$$

where a is the nearest-neighbor separation. This expression, which is known as the Mollwo-Ivey law, fits the experimental data of the absorption peak energy, E_A, of the known F centers surprisingly well (see Fig.4), with only a slightly different exponent, ~1.8 instead of 2.0 [9,28].

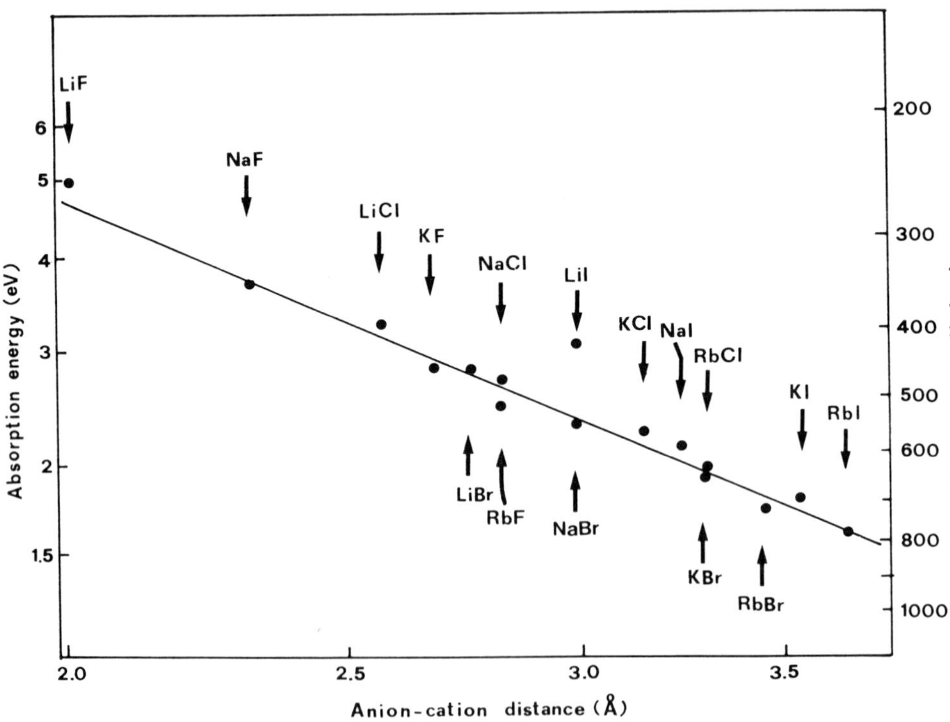

Fig.4 Peak position of the F center absorption as a function of nearest-neighbor distance. The figure has been taken from Ref.[28] except for a small correction in the case of NaI and the insertion of LiI [29].

However, Eq.(2) fails completely if applied to the emission process [9]. Indeed, the electron in the RES is not confined within the vacancy, so it should be treated as moving in a Coulomb potential field with effective mass m* and dielectric constant at high frequency ϵ_∞. The separation of the first two quantum states can be derived by the Rydberg law and is given by

$$E_{2p} - E_{1s} = \frac{3e^4 m^*}{128\pi^2 \hbar^2 \epsilon_\infty^2} \sim \epsilon_\infty^{-2}. \tag{3}$$

The experimental data of the emission peak energy, E_E, reported in Fig.5, show that the functional dependence of Eq.(3) is verified well enough [30].

As a conclusion, it should be kept in mind that the F center is and remains a *difficult* center. However, it posesses a simple structure and is one of the best known centers among any other point defect systems from the experimental and theoretical point of view, so it is often used as a model case to understand other, more complex centers.

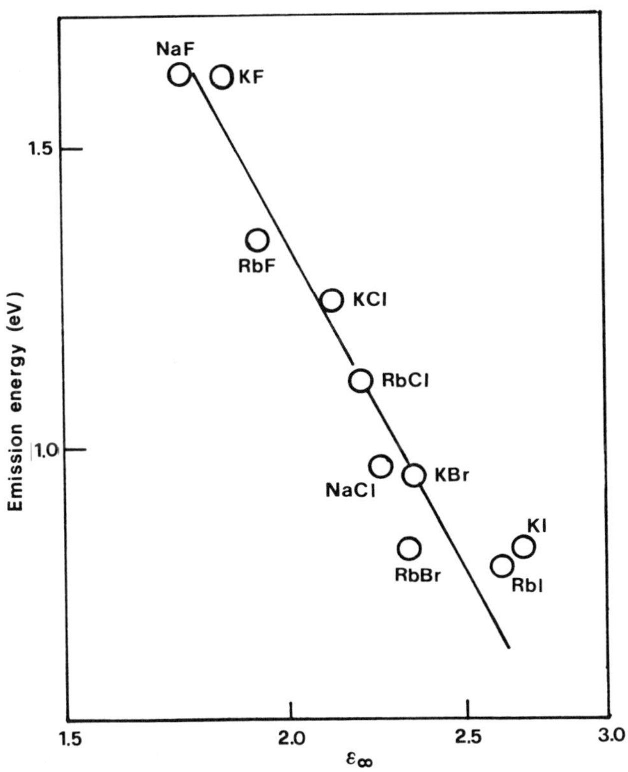

Fig.5 Low temperature values of emission peak energy vs ϵ_∞ [30].

III.C. F Aggregated Centers

An F center can be attached to another point defect giving origin to a new and more complex center. Figure 6 gives the lattice structure and the main optical properties for a few of them that can be considered the most simple color centers after the F center.

Two F centers can be aggregated along the (110) direction forming an F_2 center, formerly M center. The absorption band is split in two, the first almost unchanged in respect to that of the F center due to dipole transitions normal to the center axis, the second moved toward the infrared due to the dipole transition parallel to the center axis. The emission remains similar to that of the F center, slightly moved toward the infrared.

An F center can be associated to an alkali-ion impurity in a (100) lattice position, forming the F_A center which can be of type I or type II, depending on the impurity. Both types, which have an axial symmetry, display an absorption split into two components due to the transitions normal and parallel to the center axis. However, type I (with an Na^+ ion, for example)

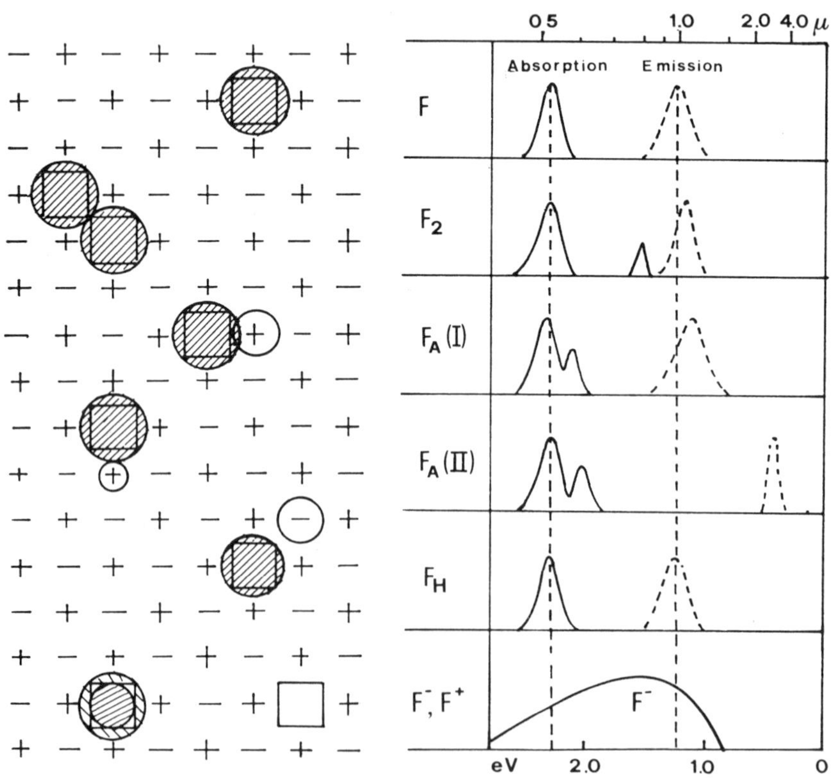

Fig.6 Structural models, absorption and emission behavior of the F center and of simple aggregated centers at low temperature in KCl.

luminescence is similar to that of the F center, while type II (with a Li^+ ion for example) luminescence is moved towards the infrared and much narrower with respect to the F center luminescence. Moreover, both centers can be reoriented under polarized light excitation with rates which are temperature dependent for type I and independent for type II centers [31].

An F center can be associated to an anionic impurity, atomic or molecular ion, in a (110) lattice position forming an F_H center. Apparently, neither the absorption nor the emission change appreciably if the substitutional ion is H^- [32]. However, it has recently been found that with practically no variation in absorption there are profound changes in the emission properties in the case where the negative impurity is a molecular ion [33]. The absorption is also split in a few cases [34] and sometimes in a very peculiar manner [35].

An F center can be joined by the electron of another F center giving rise to the F^- center, formerly F'. In the formation process, a vacancy is left behind, the so-called F^+ center, formerly α center. The F^- center has a very broad absorption band and no luminescence at all [36]. It is not usually a very stable center and is formed during photostimulated processes as a transient species with a very important role in the optical cycle.

From these primary F aggregates, it is possible to build more complex centers which are not mere scientific curiosities. Indeed, by looking at Fig.1, we can easily realize that most of the defects used as laser active centers are described in Fig.6, or they derive from them by further aggregation. For instance, the F_2^+ is an ionized F_2 center, while the $(F_2^+)_{A/H}$ is an F_2^+ center associated with and stabilized by a positive/negative ion. $(F_2^+)_p$ is an F_2^+ center *perturbed* by intrinsic and extrinsic defects. F_B is an F_A center associated with another alkali-ion impurity. Tl°(1) is a neutral Tl atom associated to an anion vacancy.

The optical properties of these color centers, such as the peak position of bands, their temperature broadening, lifetime, etc., are peculiar to each kind of center, but their behavior can be described fairly well by an F center-like model. For this reason, only the prototype F center is considered in some detail in the remaining sections of the paper.

IV. QUENCHING OF F CENTER LUMINESCENCE

IV.A. Classification of Quenching Factors

The optical cycle of the F center has been described in the previous section by assuming a radiative electronic transition from the RES to the unrelaxed ground state. However, this is not always the case because it has been extensively observed that a certain number of factors can greatly influence the emission up to a complete absence of luminescence, which is often referred to as the *quenching effect*. The emission is strongly

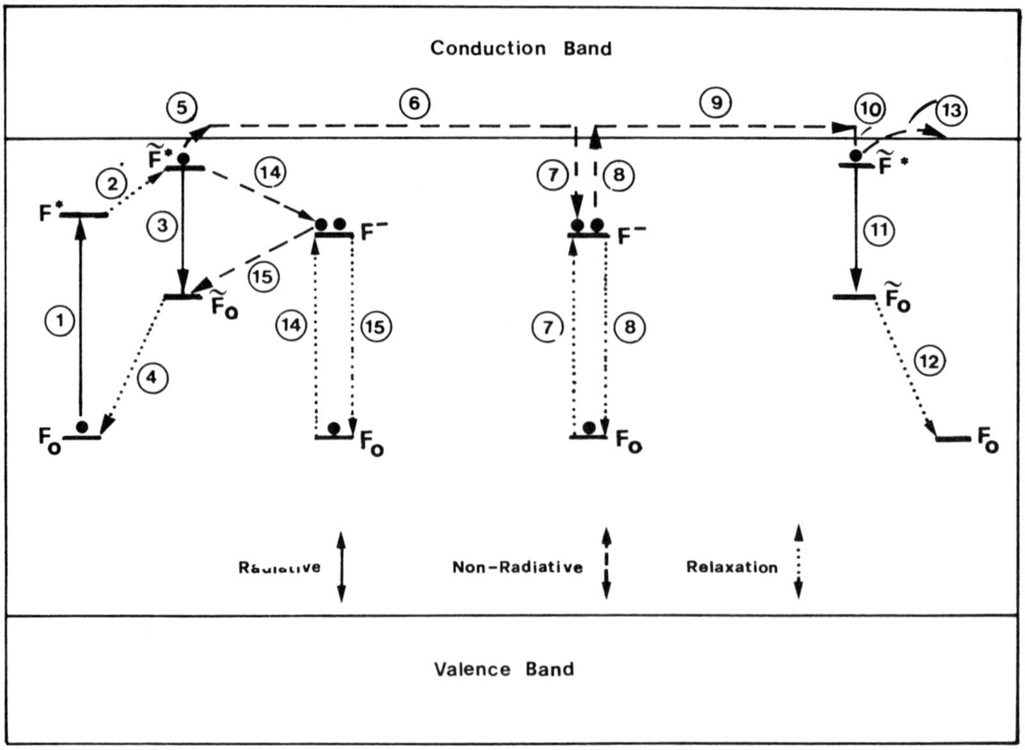

Fig.7 Electronic energy level diagram with illustration of the radiative, nonradiative and relaxation transitions for F center systems.

influenced by the temperature, electric field, F center concentration, F_2 centers, ionic impurities, all of which are *extrinsic factors*, and by nonradiative de-excitation, which is an *intrinsic factor*. The previous classification, in some respects quite arbitrary, stems from the fact that the extrinsic factors can mostly be avoided, while the intrinsic ones belong to the color center system itself in such a way that they are always present.

The interest of this paper is mainly focussed on the nonradiative de-excitation processes, but a plain description of the extrinsic processes not only has been deemed necessary for a general understanding of the argument, but it also shows peculiar processes which are of paramount importance for the nonradiative transitions. In the following, the matter is restricted only to F centers in pure and intentionally doped crystals, the latter only with impurity ions that are relevant to the matter treated here. However, most of the phenomena described can also apply to other more complex centers.

IV.B. <u>Extrinsic Processes</u>

In Figs. 2b and 2c, the radiative and nonradiative transitions of the F center have been described in a very simple way in order to smoothly introduce the ideal optical cycle. However, things are much more complicated and Fig.7 is an attempt to sketch the real situation as far as we know it to date.

In a crystal with a low concentration of F centers after light excitation "1" and relaxation "2", the electron reaches the RES, $\tilde{F}^*$, from where, besides luminescence "3" and relaxation "4", which closes the optical cycle, electron emission into the conduction band "5" is also possible. Indeed, the RES is usually a very shallow state, ~0.1 eV from the conduction band, so thermal excitation [37] or electric field ionization [38] can be very efficient mechanisms of activation. The free electron in the conduction band "6" can join an F center in the GS, F_0, by forming an F^- center, from which it can be emitted again into the conduction band "8" by thermal, electrical, and optical excitation. The free electron "9" can also be captured by an ionized F center through the RES "10", and is either ionized again "13" or reaches the GS after emission "11" and relaxation "12". However, whatever the final fate of the electron in the conduction band is, the luminescent process "3" is greatly affected by electric and thermal ionization.

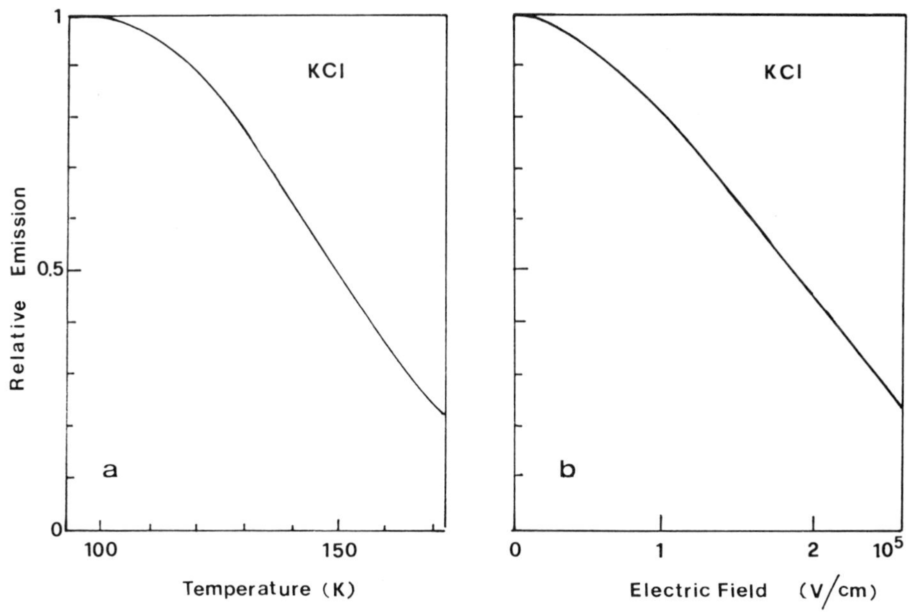

Fig.8 Temperature (a) and electric field (b) dependence of the F center luminescence in KCl. In both samples there are ~2×10^{17} F/cm^3, and in (b) the temperature is kept at T = 93 K [38].

Figure 8 displays the quantum efficiency of the luminescence of F centers in KCl as a function of temperature (a) and electric field (b) and shows that the two processes behave in the same way, as expected.

In the case of a high concentration of F centers, the electron in the RES, $\tilde{F}^*$, can also tunnel to a nearby F center "14" with a contemporary formation of an F^- center. Subsequently, the electron can tunnel back to the unrelaxed GS, $\tilde{F}_0$, "15" by destroying the F^- center, and it eventually reaches the relaxed GS, F_0. In doing so, the electron bypasses the radiative transition "3" thus producing the concentration quenching of the F luminescence [39].

Figure 9 displays the behavior of the luminescence vs the F center concentration in KCl at low temperature. It can be easily derived from similar measurements that the quantum efficiency decreases proportionally to 1/r, where r is the average distance between the F centers [40]. This concentration quenching model has also been verified by magnetic field experiments which are sensitive to the spin polarization of the electron. Due to the Pauli exclusion principle, the ground state of the F^- should be a singlet state, i.e., two electrons with spins antiparallel. Hence the nonradiative transition "14" should be forbidden at low temperatures and high magnetic fields. This effect has been really observed in a highly quenched KCl crystal [41].

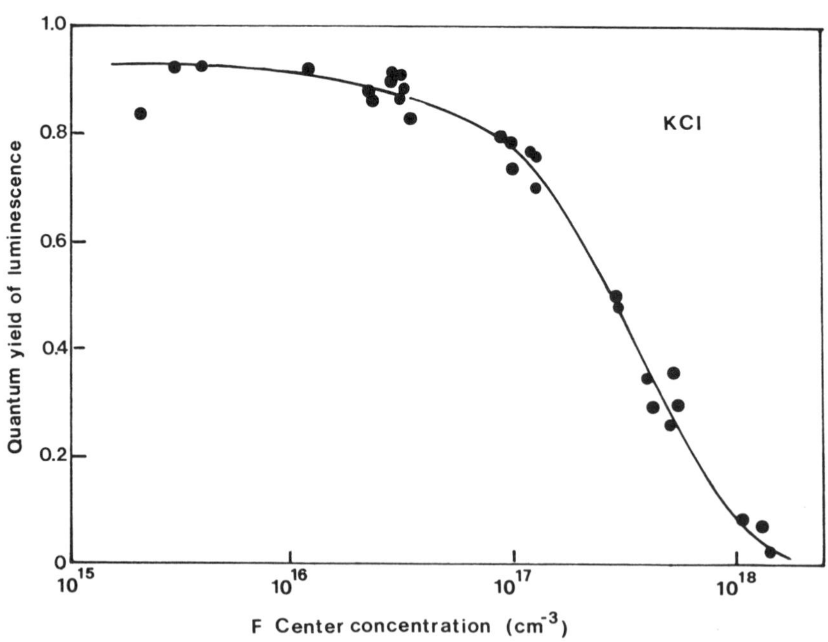

Fig.9 Quantum efficiency of F center luminescence as a function of F center concentration in KCl at T = 90 K [39].

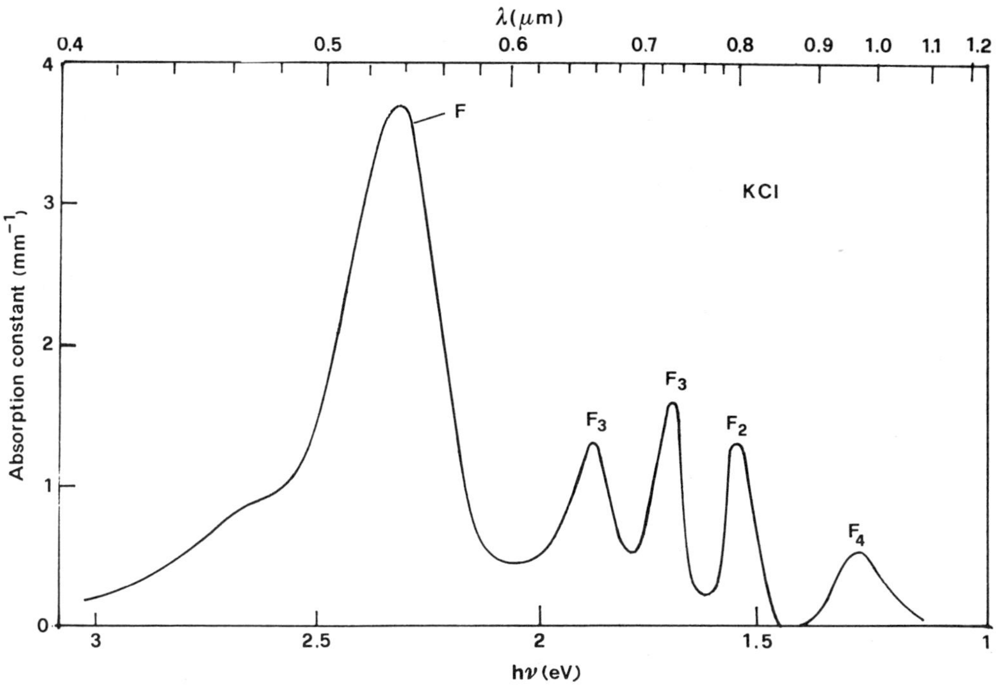

Fig.10 Optical absorption due to various electron-excess color centers [42].

When an alkali halide crystal containing F centers is illuminated in the F band at room temperature, a procedure known as optical aggregation, the F band absorption decreases and other bands appear, especially at its low energy side [42], see Fig.10. They are due to the aggregation of two F centers, F_2, three F centers, F_3, four F centers, F_4, etc. In particular, the F_2 center, which is the first aggregate to be formed, has optical transitions underlying the F band, besides the main one in the red region of the spectrum. When the F, F_2, F_3,... centers are present in the same crystal, the F center emission disappears, while the F band excitation gives rise to new emission bands characteristic of F_2, F_3,... centers [43].

Figure 11 shows the emissions of such a crystal pumped always in the F band. The typical luminescence of F centers at 1 μm decreases by increasing the F_2 and F_3 concentrations up to a complete quenching for the most converted $F+F \rightarrow F_2+F \rightarrow F_3$, sample. This very effective quenching effect can either be due to a direct absorption of the F light by the numerous F_2 bands which lie under the F band, or to a nonradiative energy transfer from F to F_2 centers [9]. The F_2 center triplet configuration [44,45] seems to play a decisive role in the latter process, although the whole phenomenon has not yet been clarified. Anyway, the F_2 centers are very effective in killing the

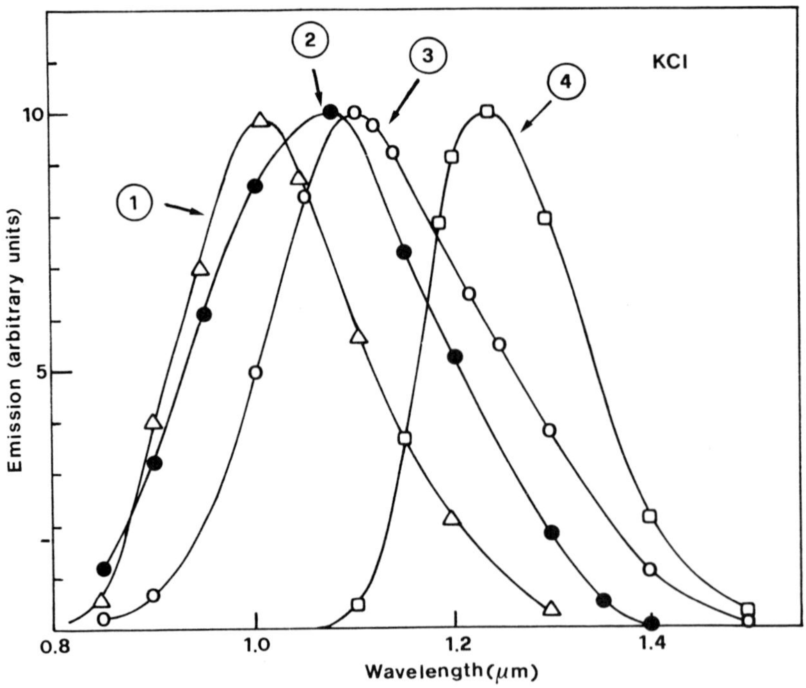

Fig.11 Emission spectra excited in the F band at T = 77 K. The crystal of curve 1 contains only F centers. The crystal of curve 2 contains F and F_2(M) centers. The crystal of curve 3 contains more F_2(M) centers and some F_3(R) centers. The crystal of curve 4 contains a much higher concentration of F_3(R) centers [43].

F luminescence and at the same time in opening new radiative pathways. Thus, this quenching effect should be addressed more properly as an energy exchange process among different color centers.

Similar processes, but with vastly different results, happen when molecular ion impurities show up in the crystal. Ions like OH^- and CN^-, the former often found as an unwanted impurity in all alkali halides, can occupy a lattice position replacing a negative ion (Fig.12). They can also attach themselves to F centers, leading to the formation of $F_H(OH^-)$ and $F_H(CN^-)$ centers, which have been introduced summarily in the previous section (Fig.6). These new centers can be formed by optical aggregation at a temperature slightly below 0 °C in order to avoid the formation of F_2 centers. Detailed experimental and theoretical studies have shown that in both systems there is an efficient transfer of the electronic energy of the excited F center to the vibrational stretching mode of the OH^- or CN^- ion. In the case of OH^-, as far as we know to date, the radiative energy is completely lost in relaxation channels, while for CN^- a vibrational fluorescence has been found, in a few cases with a full quantum efficiency transfer.

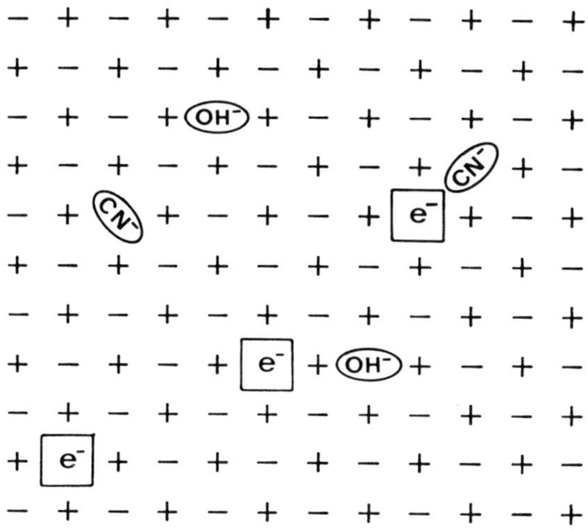

Fig.12 Molecular ions in a KCl crystal isolated and aggregated to F centers.

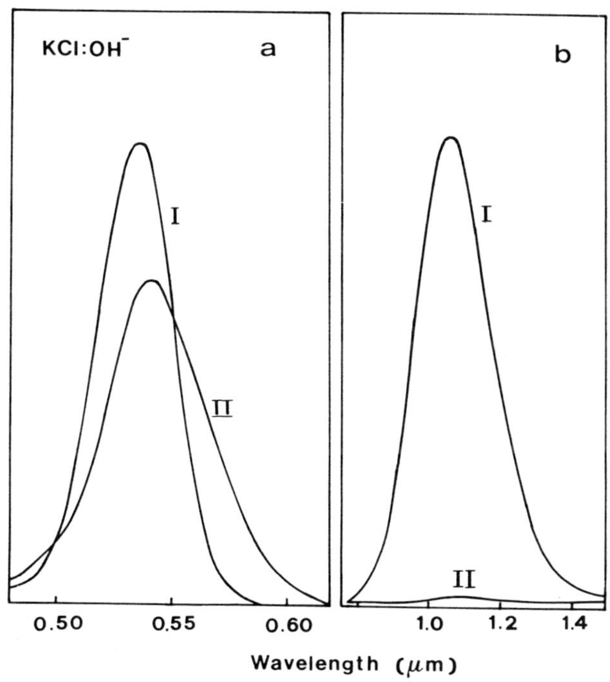

Fig.13 Electronic absorption (a) and emission (b) of additively colored $KCl:OH^-$ crystal (2.8×10^{-3} OH^-) at T = 10 K. I, quenched crystal; II, after optical aggregation of F centers to ions [33].

The absorption and emission properties of $F_H(OH^-)$ in KCl at low temperature are reported in Figs.13a and 13b respectively. Configuration I corresponds to the situation without aggregation, i.e., F centers and impurities statistically distributed. In this case the absorption and emission lineshapes look very similar to the ones displayed by a pure crystal (see Fig.3). At a closer examination, the quantum efficiency of the emission is smaller than in a pure crystal, showing a kind of interaction between F centers and the statistically dispersed OH^- ions. Configuration II corresponds to the aggregated centers and the absorption decreases with a marked broadening at the low energy side, and at the same time, the emission decreases strongly. A comprehensive picture of this kind of result is reported in Fig.14. The continuous line gives the luminescence of F centers as a function of OH^- concentration, while the dashed line refers to the $F_H(OH^-)$ centers in the same crystal. These results show that the distance between the F centers and the OH^- impurities is a critical parameter, as has been established in a simple **step model** theory [33]. The nonradiative relaxation following excitation takes place in a very short time, typically 1 ns or less [46].

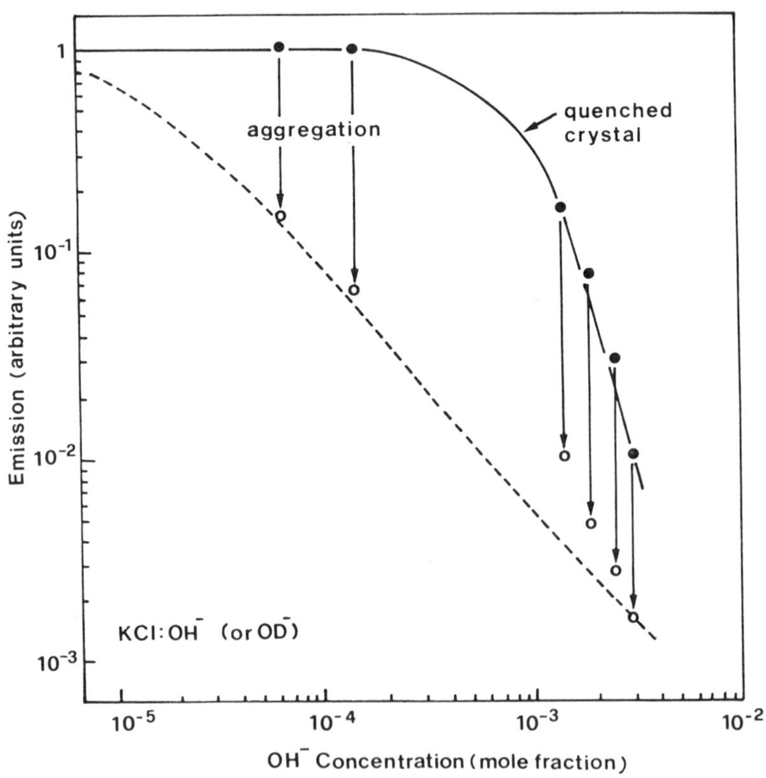

Fig.14 Plot of F center emission as a function of OH^- concentration for the quenched and optically aggregated KCl crystals at T = 77 K [33].

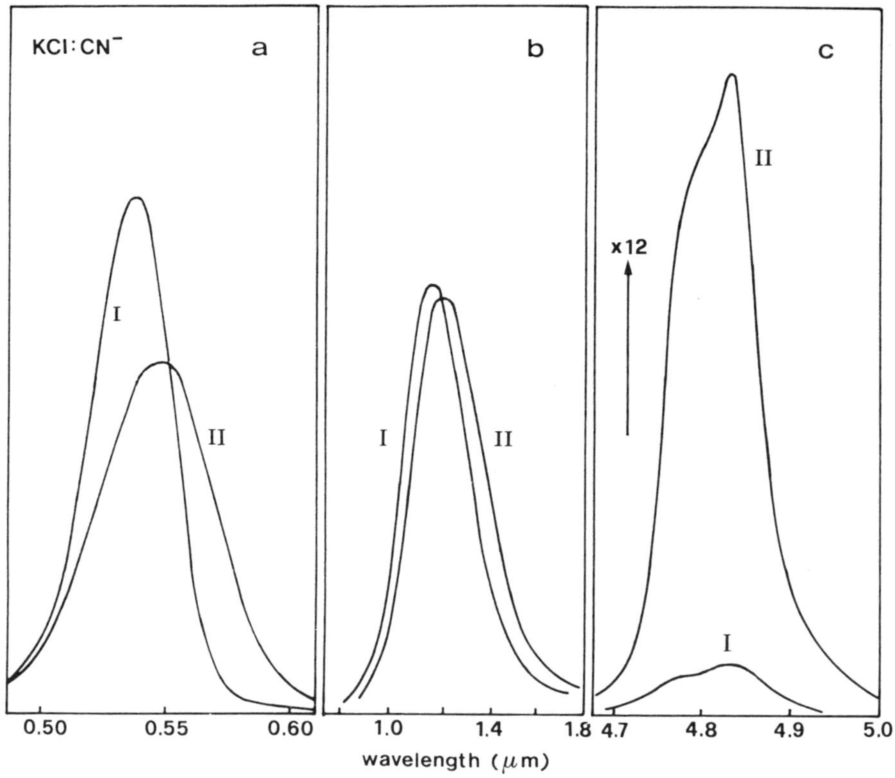

Fig.15 Electronic absorption at T = 10 K (a), emission at T = 80 K (b), and vibrational emisison at T = 80 K (c) of additively colored $KCl:CN^-$ crystals (8×10^{-3} CN^-). I, quenched crystal: II, after optical aggregation of F centers to CN^- defects [47].

Figure 15 shows the absorption and the emission properties of $F_H(CN^-)$ in KCl at cryogenic temperatures [47]. As far as the electronic absorption is concerned (a), it behaves similarly to the OH^- case. However, electronic emission is not quenched at all and only a small shift is observed after aggregation (b). Moreover, a sizeable vibrational emission is measured after aggregation. This fluorescence, with a lifetime of the order of ms, has a quantum yield of a few percent. These results are typical of all hosts with the NaCl structure, while in the case of CsCl, after aggregation the absorption is split into two bands, the normal F emission is totally quenched, and a strong vibrational emission appears at ~4.8 μm with quantum efficiency about five times higher than that of the original F luminescence [34]. All this evidence indicates a complete transformation of electronic excitation into CN^- vibrational excitation followed by emission. This luminescence process is so efficient that a laser has been realized with $F_H(CN^-)$ centers in CsCl [48] and recently in CsBr [49], as well.

TABLE 1

Relevant parameters, dielectric constant ϵ_∞, peak energy E_E, halfwidth W_E, excitation energy ΔE, and quantum efficiency η, for the luminescence of the F center in several alkali halides. The various parameters refer to LHeT and have been taken from Ref.[23] unless otherwise specified. Values of η for RbI and RbBr are not reported, but should be of the order of unity.

HOST	ϵ_∞	E_E (eV)	W_E (eV)	ΔE (meV)	η
LiI	3.80				
LiBr	3.17				
NaI[a]	2.93	0.57	0.12	30	0.005
LiCl[b]	2.78	1.56	0.43	40	0.6
KI	2.62	0.83	0.19	110	1[e]
RbI	2.59	0.81	0.15	82	
NaBr[a]	2.59	0.63	0.21	61	0.002
RbBr	2.34	0.87	0.19	110	
KBr	2.34	0.92	0.22	135	1[c]
NaCl	2.34	0.97	0.34	74	0.3[c]
RbCl	2.19	1.09	0.24	130	1[c]
KCl	2.19	1.21	0.26	150	1[c]
RbF	1.96	1.33	0.33	80	0.9[d]
LiF	1.96			160	0[d,f]
KF	1.85	1.66	0.38	138	0.3[d]
NaF	1.74	1.66	0.39	55	0.4[c]

[a]51; [b]52; [c]59; [d]30; [e]60; [f]61

IV.C Intrinsic Processes

By taking special care during the growth of the crystals from ultrapure materials [50], during the coloration methods, and by avoiding aggregation processes which lead to the formation of F_2, F_A, F_H, etc..., centers, it is possible to obtain pure F centers which can then be studied at low temperatures. In this way, only the radiative process "3" or nonradiative processes from the RES, $\widetilde{F}* \rightarrow \widetilde{F}_0$, and eventually during the relaxation, "2"→"4", can take place after excitation "1" (Fig.7). Let us therefore see what the present experimental situation is regarding the emission properties. The peak energy of the radiative transition from the RES in a hydrogen-like model is expected to be inversely proportional to the square of the dielectric constant at high frequency, see Eq.(3). It therefore appears logical to list in Table 1 all the alkali halide crystals, apart from the Cs halides, in order of decreasing ϵ_∞.

The optical properties of the F centers in the crystals NaI, NaBr [51], and LiCl [52] have been thoroughly investigated in the last decade after a new review and analysis of the occurrence or quenching of the emission process [53], so they deserve some further description.

Luminescence has been found in a LiCl crystal grown from extra pure material and zone refined, colored with electrons, 25 MeV and 15 μA, at LNT. Attempts to color the crystal with additive and electrolytic methods have failed. Under excitation of the F band centered at 3.3 eV, two prominent emissions are observed at 1.56 eV and 2.31 eV, the latter probably associated to Na impurities. The parameters of the luminescence at 1.56eV, half width, peak energy, intensity, and lifetime, 1.53 μs at LHeT, are characteristic of F center luminescence and it was attributed accordingly [52].

NaBr and NaI crystals were grown from ultrapure material, zone refined, and always handled in a dry atmosphere because of their high hygroscopicity. Electron or X irradiation and normal additive methods did not produce satisfactory coloration. In the end, they were colored by heating them at ~600°C in Na vapors and ~6 atmospheres of H_2. In this way U centers (substitutional H^- centers) were produced and subsequently converted into F centers by UV irradiation at RT. Under F irradiation, a very low quantum efficiency emission with a double spectral structure was measured [51], see Figs. 16 and 17 for NaI and NaBr respectively. New and interesting phenomena have also been observed, such as the formation of a stable F^- band on the high energy side of the F band, which is a relevant property and is considered in detail in the last section. The optical properties of the luminescence prompted us to identify the peak at lower energy as due to the normal RES emission, while the luminescence at higher energy is due most probably to a radiative recapture process from a higher RES. This attribution has been confirmed recently in NaI by magnetic circular dichroic measurements in the luminescence [54], and in NaBr by radiationless electronic relaxation studies with picosecond optical pulses [55].

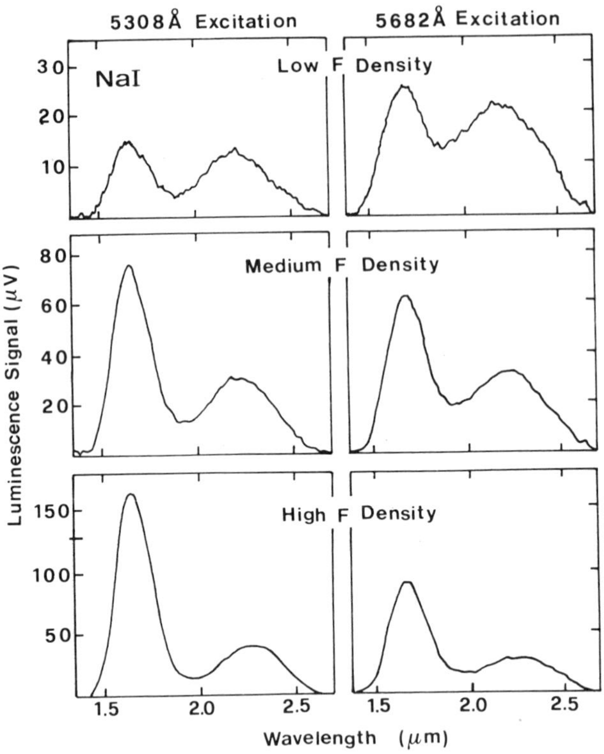

Fig.16 Emission spectra observed at T = 10 K in pure F centers of three different concentrations in zone-refined NaI under laser excitation at 5308 and 5682 Å.

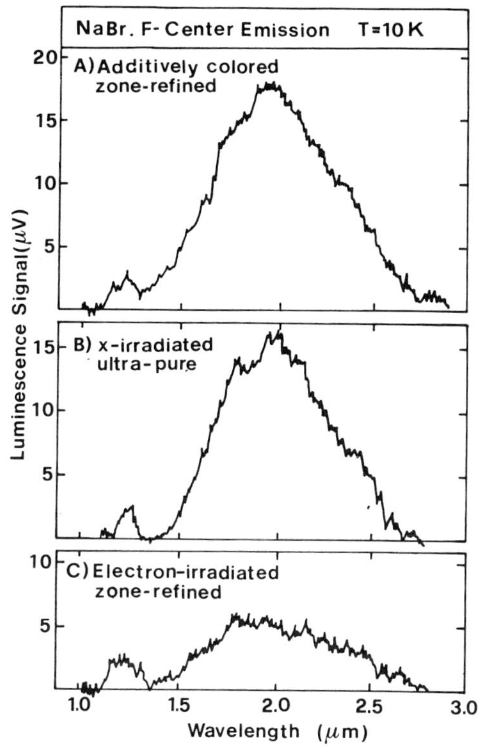

Fig.17 Emission spectra obtained under 5309 Å excitation at T = 10 K in NaBr crystals with F centers produced by three different techniques.

Figure 18 reports the values of the peak energy, E_E, of the F luminescence in NaI, NaBr, LiCl, and other crystals for a comparison with the linear slope, expected from Eq.(3) and already reported in Fig.5, as a function of the dielectric constant at high frequency ϵ_∞. Moreover, the ionization energy, ΔE, has been reported in Fig.19 as a function of the inverse square of the static dielectric constant ϵ_s, whose values refer to low temperatures [56]. It has been shown that a linear slope can roughly fit the experimental data of the known F centers [57], with the possible exception of LiF.

The two plots show that the data of NaI and NaBr fit the expected behavior, while the data of LiCl remain away from the linear slope in both cases. These results cast some doubts on the attribution to F centers of the radiative band of LiCl reported in Table 1. It should also be noted that the luminescence decay time in LiCl shows an unexpected dependence on the coloration method, e^- or γ irradiation [58]. Thus the LiCl data should be taken with extreme caution, although at the moment we cannot completely exclude

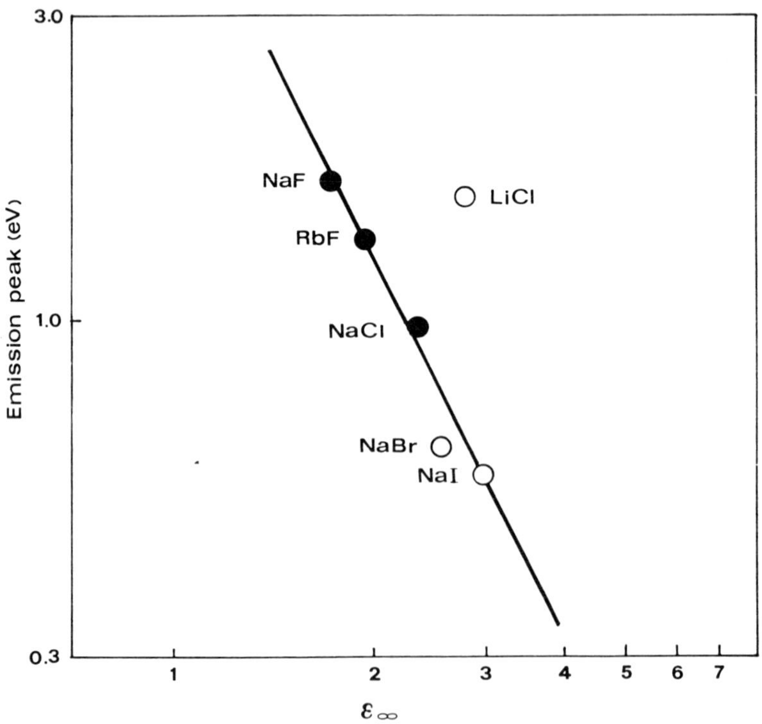

Fig.18 Double logarithmic plot of the F center emission energy at low temperatures vs the high frequency dielectric constant ϵ_∞. See also Fig.5.

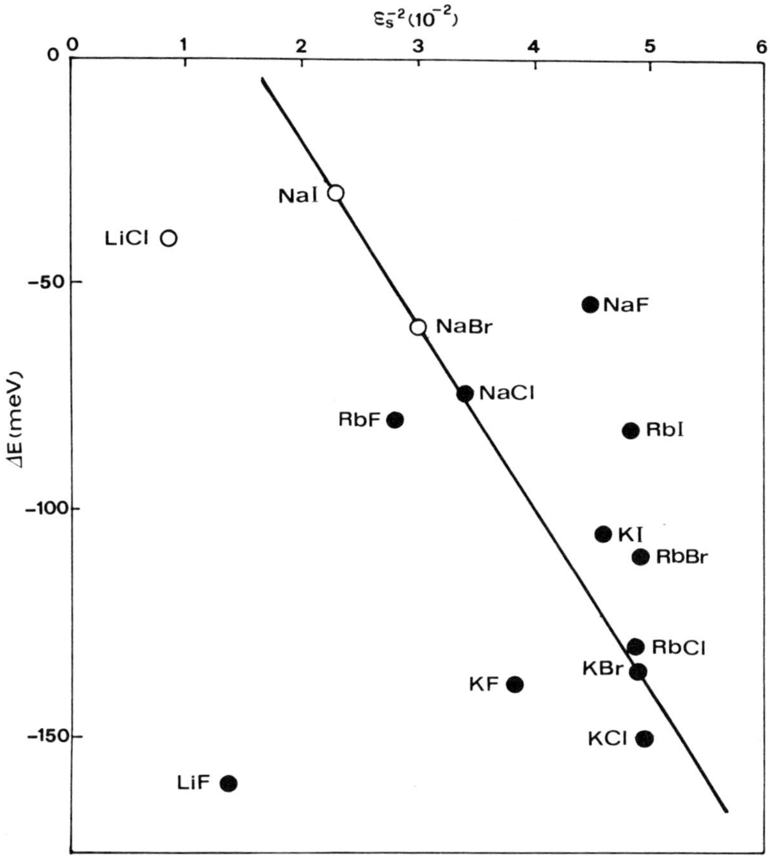

Fig.19 Plot of the gap energy ΔE as a function of the inverse square of the static dielectric constant ϵ_s. The straight line is the best fit to ΔE values as discussed in Ref.[57].

for the Li halide crystals a different behavior from the rest of the alkali halides (see also Fig.4 for the absorption band position), or that the parameters used for the plots are not reliable enough.

Anyway, Table 1 shows that no emission has been detected so far in the Li halide crystals, with the possible exception of LiCl, and that the other F centers have an almost full quantum efficiency, η, except in the case of NaI and NaBr where η is about two orders of magnitude lower. In conclusion, these data show very clearly that a radiationless process, competing with the radiative transition "3" in Fig.7, is at work in several host crystals. In the next section a critical discussion on such processes is made in connection with the known experimental facts.

V. DISCUSSIONS AND CONCLUSIONS

V.A. Crossover Model and Luminescence

In Sec.2 the lattice model of the F center and its optical cycle have been introduced with simple schemas which were justified at that point by the necessity of having to describe clearly the basic phenomena in a preparatory way. For instance, the configuration coordinate diagram in Fig.2c, which refers to a general case, does not take into consideration a possible interaction between the GS and the RES parabola. A much more realistic diagram, which also includes the conduction band, is plotted in Fig.20, where $E_R = S \cdot h \cdot \nu_{phon}$, S is the Huang-Rhys factor and ν_{phon} is the frequency of the lattice mode coupled with the electron. The case depicted in Fig.20, the crossing point X lies energetically slightly below point B reached by the F center electron after optical excitation, belongs to the strong coupling limit, S>6 [26]. In KCl, for instance, S = 36, i.e., thirty-six phonons are produced during the relaxation B → C. The case of weak coupling, S < 1, the point X is well above B, applies only for a few F aggregated centers, but not for F centers.

Using this configuration coordinate model, which takes into account only one phonon mode and one excited state, Dexter, Klick, and Russel (DKR) [62] argued that the position of the point X is responsible for the occurrence or quenching of the emission process. They postulated that emission will always occur for $E_X > E_R$, while for $E_X < E_R$ the system can *cross over* between the two states during the relaxation process, so that the RES in C is never reached and the luminescent transition C → D is killed. Indeed, the electron from X relaxes directly to A, following the GS parabola.

Subsequently, the DKR criterion for luminescence was revisited by Bartram and Stoneham [53] after F centers had been studied extensively in terms of their absorption and emission properties. They introduced the parameter $\Lambda = E_R/E_A$ and concluded that whether there was luminescence or not depended on the value of Λ; more exactly

$$\begin{array}{ll} \Lambda \le \frac{1}{4} & \text{Presence of Luminescence} \\ \Lambda > \frac{1}{4} & \text{Absence of Luminescence or Quenching} \end{array} \qquad (4)$$

Values of Λ were calculated from available experimental data and a trend was also found between Λ and the ratio r_-/r_+, where $r_+(r_-)$ is the radius of the positive (negative) ion.

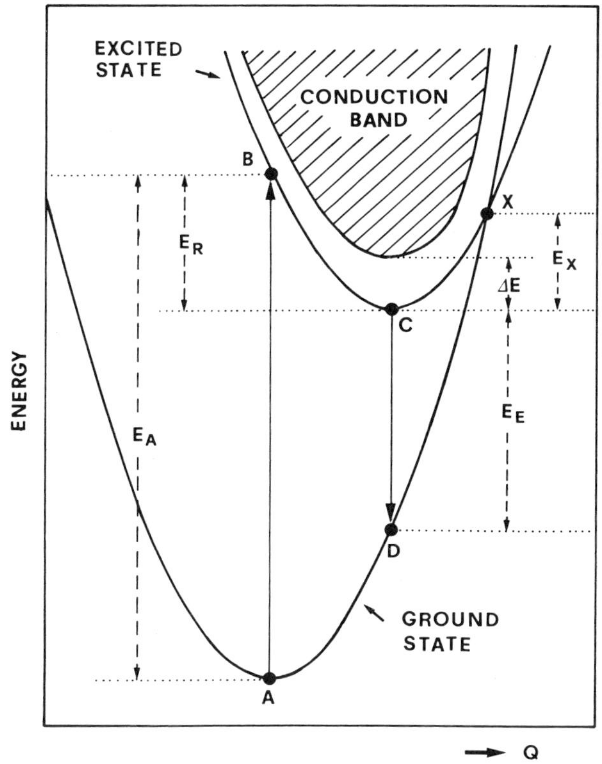

Fig.20 Energy diagram for the ground and excited states of an F center as a function of the configuration coordinate Q.

TABLE 2

Values of Λ, r_+/r_-, and emission quantum efficiency η for alkali halides with rock salt structure. No F luminescence has been observed in LiI, LiBr, LiF, and almost certain none in LiCl. Experimental data on RbI and RbBr have not been reported in the literature, but they should not be vastly different from unity.

HOST	Λ	r_+/r_-	η Theory BS[a]	η Theory LS[b]	η Experiment Table 1
LiI	0.831	3.22		0.348	-
NaI	0.384	2.24	0.018	0.006	0.005
NaBr	0.375	2.00	0.020	0.024	0.002
LiCl	0.371	2.66	0.022	0.065	0.6 ?
LiBr	0.358	2.88	0.026	0.051	-
LiF	0.323	1.96	0.040	0.071	-
NaCl	0.260	1.85		0.608	0.3
KI	0.231	1.65		~1	1
KBr	0.223	1.47		~1	1
RbI	0.211	1.48		~1	~1
KF	0.189	1.00		~1	0.3
KCl	0.188	1.36		~1	1
RbCl	0.182	1.22		~1	1
NaF	0.175	1.36		~1	0.4
RbF	0.173	0.90		~1	0.9
RbBr	0.162	1.32		~1	~1

[a]Ref. [66]; [b]Ref. [67]

Later on, following the observation of weak luminescences in NaI, NaBr [63, 51] and LiCl [64,52] systems, the condition (4) was somewhat softened and quantum efficiency was calculated for each host [65]. Indeed, when the vibronic system (electron+phonons) crosses point X during cooling, there is a probability W_x that the system will continue to follow the excited vibronic ladder, and a probability W_g that the system changes for the ground vibronic ladder. The ratio between the two probabilities, R, is called the branching ratio and is given by [66]

$$R = \frac{W_x}{W_g} = \frac{(1-2\Lambda)^2}{3}, \tag{5}$$

while the quantum efficiency of the luminescence can be written as

$$\eta = \frac{R}{1+R}, \tag{6}$$

which clearly reduces to unity when $R \gg 1$, i.e., $W_x \gg W_g$, meaning an almost total absence of crossover effects.

Moreover, the crossover of the excited electron could be not only effective at the crossing point X, but also over a large interval of energies above and below X. As a consequence, by taking into account this possibility as well, Leung and Song [67] estimated a slightly different quantum efficiency.

The values of Λ, r_-/r_+, and η from theories and experiments are given in Table 2, where the crystal hosts are listed in order of decreasing Λ. It is evident that the criterion (4) is valid in dividing the crystals into two groups of luminescent and nonluminescent hosts. Theoretical refinements also agree with the observed weak luminescence in NaI and NaBr but not with the strong one in LiCl, and this is a further source of doubt about the nature of the measured luminescence [52].

V.B. Experimental Evidence and Radiationless Pathways

Although theory and experiments seem to agree with each other more than satisfactorily, if theoretical approximations and experimental errors are taken in due account, there is a fundamental aspect not yet clarified which throws some doubt on the correctness of the DKR model. Following the crossover mechanism, only a fraction of the F center electrons, or none at all in the case of total quenching, should reach the RES, point C in Fig.20. On the contrary, there is experimental evidence in NaI and NaBr, where the F luminescence is strongly quenched, that all the excited electrons reach the RES. As clear, this is not a minor point in the quest to clarify the emission phenomena; therefore, it deserves a detailed description. For the sake of clarity and conciseness, only the case of NaI is treated here, but most of the facts also apply to NaBr [51].

Upon irradiation of the F band in NaI, a new band appears at its high energy side, which is also remarkably stable, see left side of Fig.21. The optical conversion can be completely reversed by light irradiation into the new band at a wavelength which is outside the F absorption. The new band has been identified as an F^- band following detailed experiments [51,68-70], so this denomination will be used from now on. Optical conversion rates $F \leftrightarrow F^-$ are reported in Fig.22 as a function of temperature and in crystals of different F center concentration and coloration methods.

In order to justify the energy position of the F^- band on the blue side of the F center band, the $F^- \rightarrow F$ temperature independent conversion, the stability of the F^- band, and the quenching of the luminescence, a new energy level diagram, which differs from that used for normal F centers in Fig.7, has been sketched in Fig.23. The F^- level has been placed deeper in the

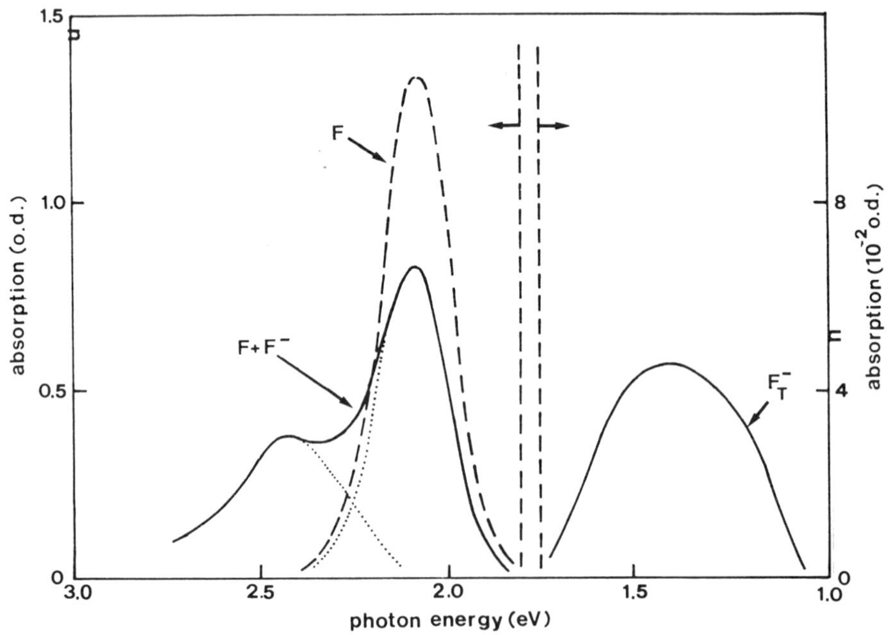

Fig.21 Absorption spectrum in colored NaI at T = 1.9 K. The crystal initially contains only F centers whose absorption (dashed curve) is centered around 2.08 eV. Upon irradiation at 6328 Å, the F band absorption decreases, while two new bands appear around 2.4 eV and 1.4 eV (continuous curves). Note the different o.d. scale in the two parts.

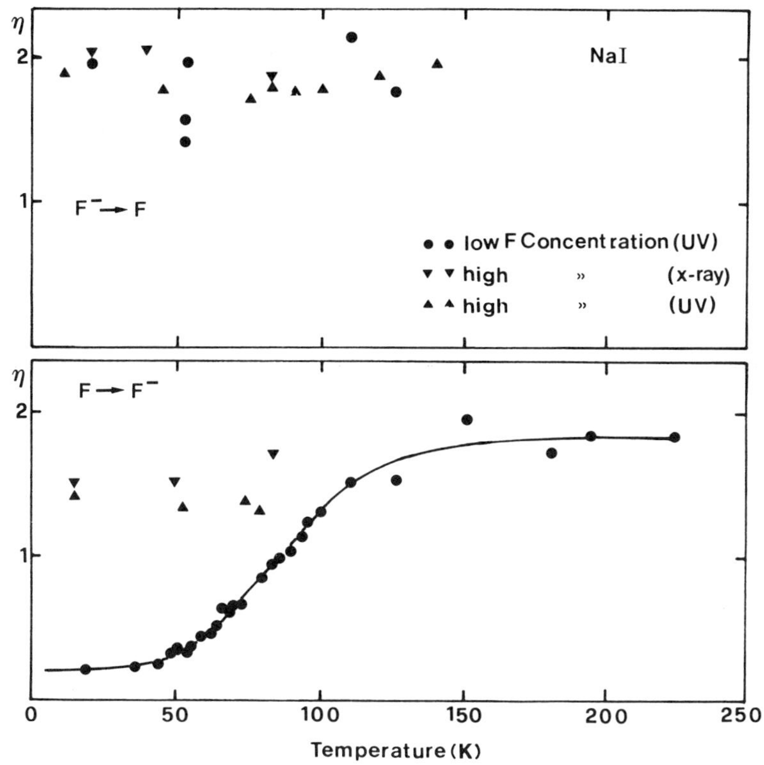

Fig.22 Summary of optical conversion rates as a function of temperature, measured in crystals of different F center concentration and coloration methods in NaI.

energy gap, eliminating the tunneling "15" in Fig.7, and a radiative "16" and two nonradiative transitions "3a" and "3b" have been added.

At low concentration of F centers, F^- can be formed only through thermal excitation "5"→"6"→"7", while the back conversion "8"→"9"→"10"→"11"→"12" is temperature independent. At high concentration of F centers, the tunnelling process "14" takes over and both conversions are temperature independent. The fact that the conversion $F \rightarrow F^-$ reaches almost full quantum efficiency, $2F \rightarrow F^-$, at high temperatures assures that *all the electrons reach the RES after optical excitation "1" and relaxation "2"*.

F center emission "11" can also be excited through F^- excitation "8", and the same luminescence efficiency as observed under optical F excitation has been found. The ionization energy, ΔE = 30 MeV, derived by the $F \rightarrow F^-$ conversion rate vs the temperature, Fig.22, is lower than E_X = 105 MeV [71], so *all the electrons from the minimum of the conduction band*, Fig.20, *reach the RES* directly without passing in the vicinity of point X.

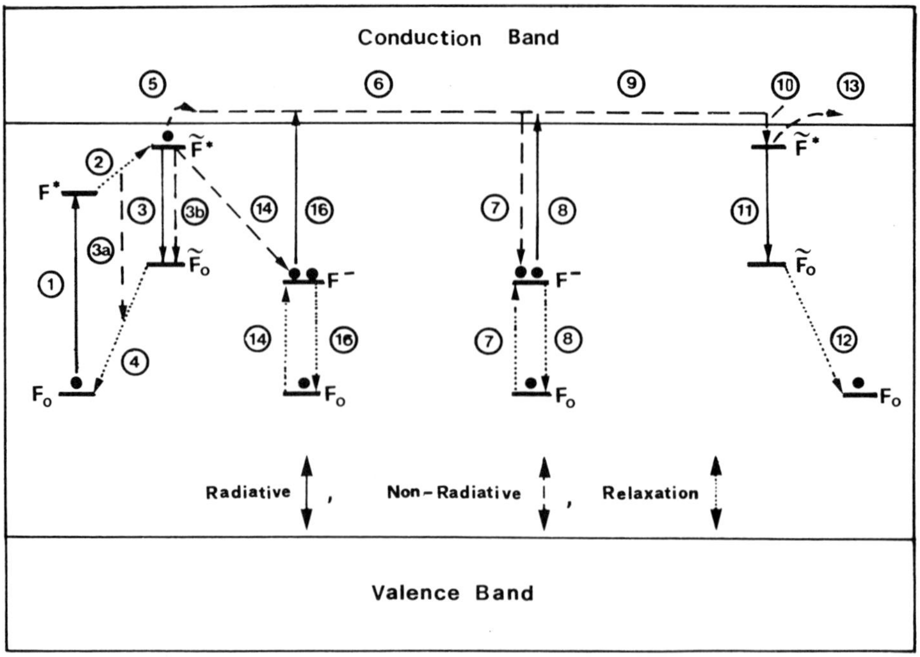

Fig.23 Electronic energy level diagram with illustration of the radiative, nonradiative, and relaxation transitions for F centers in NaBr and NaI.

The preceding results indicate that the *excited electrons always reach the RES* from where a nonradiative process takes place "3b" in competition with thermal ionization "5", radiative de-excitation "3", and tunneling process "14". Nonradiative de-excitation during relaxation "3a", compatible with the crossover model, has not been observed in these measurements.

However, the very important conclusions reached so far rely on the fact that the actual experimental results have been ascribed to an F center system. More exactly, the following questions arise. Is the absorption band at 2.4 eV really an F^- band? Is the emission at 0.55 eV due to a pure F center system?

In order to answer the first question, we made accurate magneto-optical measurements [68-70]. Figure 24 shows the rate of the supposed $F \rightarrow F^-$ conversion as a function of the magnetic field. Because of the singlet nature of the F^-, spins antiparallel, such conversion should be suppressed at high magnetic fields and low temperatures, where all the spins are parallel. The experimental points show only a decrease to about one half the maximum signal. However, if ϵ is the spin mixing parameter and δ the probability to form a triplet F^- center, F_T^- (spins antiparallel), the rate of the $F \rightarrow F^-$

conversion in a magnetic field can be written as follows [41]:

$$\frac{W(H)}{W(0)} = 1-\alpha P^2 = 1-\left[(1-2\epsilon)\left(\frac{1-\delta}{1+\delta}\right)\right]P^2,$$

where P is the static polarization of the electron spins.

In a separate set of experiments, aimed at revealing magnetic circular dichroism, MCD, in absorption we obtained $\epsilon = 0.12$ [68]. At the same time, accurate absorption measurements on the low energy side of the F band led to the discovery of a weak band (see right side of Fig.21), which because of its spectral shape, lifetime, optical kinetics, MCD, etc., has been assigned to the triplet state [69,70]. Figure 25 shows the magnetic circular dichroic signal of F and F_T^- centers and, as expected, there is no MCD signal around 0.5 μm for the F_s^- (singlet) center. For the sake of completeness, the variation of the first moment for the F_T^- signal is $\Delta E \cong 3$ meV, to be compared with $\Delta \cong 15$ meV of the F center [72]. The relative probability of formation of the two F^- bands centered at 1.40 and 2.45 eV can be calculated directly from their area if the same oscillation strength is assumed for both. The result is $\delta = 0.25$. Thus the value of the reducing factor in (7) is $\alpha = 0.46$, which compares fairly well with the experimental value of Fig.24, $\alpha = 0.48$.

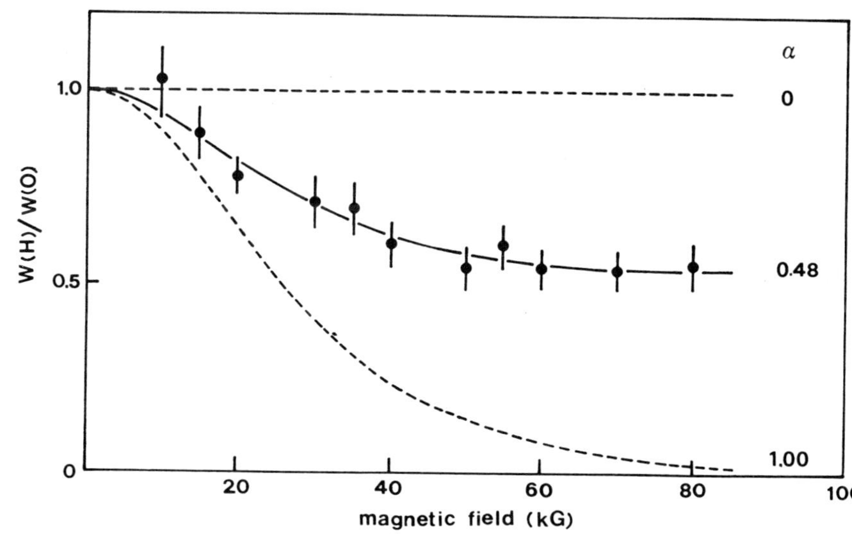

Fig.24 Rate of the initial conversion between F and F^- centers W(H) normalized to W(0) for NaI at T = 1.0 K. The experimental results (black dots with error bar) are compared with the expected theoretical behavior, $1 - \alpha P^2$ (see Eq.(7)), plotted for different values of the parameter α (dotted and continuous curves).

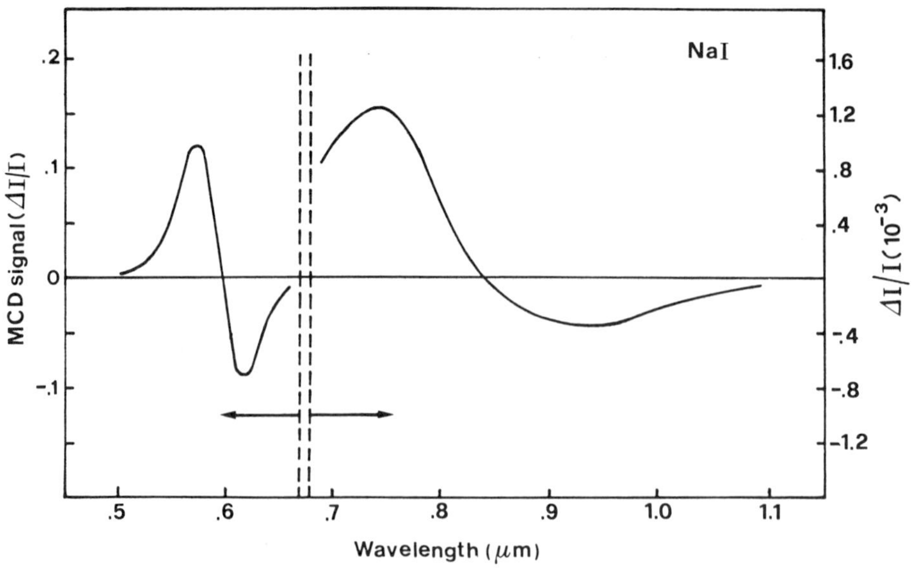

Fig.25 The MCD absorption spectrum of F and F_T^- centers at H = 50 KG and T = 1.8 K. Note the variation of the vertical scale on the right side.

With the discovery of a triplet state, the electronic energy levels were modified as in Fig.26, where the spins are also indicated. The triplet state F_T^- can be reached directly from the RES $\widetilde{F}^*$, and the transition to the conduction band produces the absorption band observed on the right side of Fig.21, where the absorptions from the F ground state and the F_s^- singlet state are also displayed. The triplet state, however, is not stable and it decays spontaneously with a lifetime of 540 sec at LHeT.

As far as the luminescence is concerned, we made accurate studies of the magnetic circular dichroic effects in emission [54,73]. If $I^{+(-)}$ is the intensity of the circularly right (left) polarized luminescence, the MCD signal is given by

$$S_{MCD} = \frac{I^+ - I^-}{I^+ + I^-} = C_d H + F_p P \ , \tag{8}$$

where C_d is the diamagnetic constant, H the magnetic field, F_p the paramagnetic constant, and P is the electron spin polarization in the RES. Usually the diamagnetic and paramagnetic signals can be measured separately. Table 3 reports the diamagnetic results in NaI for the two luminescences, and in other hosts and different color centers [54,73-76]. It can be clearly observed that the value of the diamagnetic constant $|C_d|$ for the luminescence at 2.25 μm is similar to that of the other known F centers, confirming the

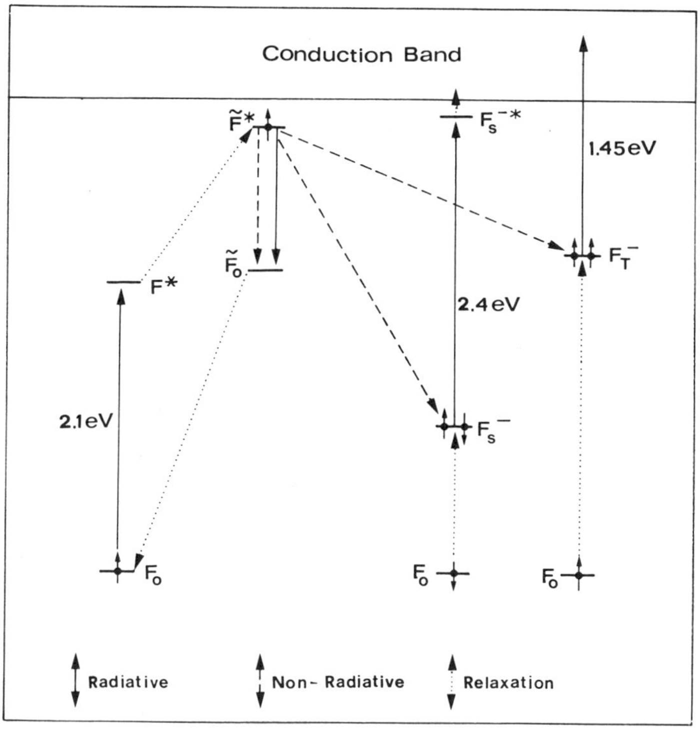

Fig.26 Electronic energy level diagram with illustration of radiative, nonradiative and relaxation transitions for F centers in NaI. The electron of the optically excited F center can tunnel into a neighboring F center of opposite or equal spin-orientation, forming the singlet, F_s^-, or triplet, F_T^-, state.

TABLE 3

Experimental data of the diagmagnetic constant $|C_d|$ (10^{-8} G^{-1}) relative to the emissions of known F, F_A centers, and to the two luminescences observed in colored NaI.

CRYSTAL	CENTER			
	F	$F_A(I)$	$F_A(II)$	?
KF	6[a]			
KCl	9[b]			
	11[c]			
	13[d]			
KCl:Na		12[d]		
KCl:Li			1[d]	
KBr	17[c]			
KI	19[c]			
NaI (2.25 μm)				15[d]
NaI (1.65 μm)				0.5[d]

[a]Ref.[74]; [b]Ref.[75]; [c]Ref.[76]; [d]Ref.[54,73]

attribution of this luminescence to the radiative de-excitation "3" of the normal RES. For comparison, I have also reported the $|C_d|$ values of $F_A(I)$ and $F_A(II)$ centers, which show how sensitive the MCD technique is to the change in the nature of the RES. Indeed, $F_A(I)$ is a slightly perturbed F center, while $F_A(II)$ is a strongly perturbed F center [31]. It is also indicative that the second luminescence in NaI at 1.65 μm has a strongly reduced diamagnetic effect, as in the $F_A(II)$ center case.

Using these arguments, the two questions are answered positively and, therefore, *as far as the experimental results in NaI and NaBr are concerned, they support a nonradiative de-excitation from the RES and not at the crossover point X during relaxation.*

Recently, the emission of F centers by pumping on the high energy side of the F band was observed to decrease at the onset of the crossover process [77,60], thus favoring the crossover model. However, as these experiments introduce new pathways for the de-excitation in Fig.20, adding a new complication to an already complex picture, these results cannot be interpreted without ambiguities [71].

V.C. New Results and Conclusions

The relaxation process "2" and the radiative "3" and nonradiative "3a" de-excitation have recently been the subject of detailed theoretical and experimental investigation. As far as the theory is concerned, the cascade of phonons during relaxation seems to conserve a high degree of coherence [78] with consequences which are not at all clear, and which do not favor a particular model at the moment. Further results from this *ab initio* approach are expected with great interest [79]. On the experimental side, it was possible to measure the recovery time of the GS population after a sudden F band excitation in NaBr [55]. The results indicate the existence of a higher relaxed excited state, which most probably is the one delivering the luminescence at 1.2 μm, while the RES lifetime has been established as being ~4 ns at T = 10 K. This value should be compared with our predicted lifetime of ~0.3 ns, calculated from the $F \rightarrow F^-$ conversion rate and by choosing the attempt frequency of the process equal to that of KCl [51]. The agreement is sufficiently good, taking into consideration the previous approximation, while for the crossover model we would have expected a much shorter value.

In conclusion, there is strong experimental evidence that after excitation all the electrons reach the RES from where both radiative and nonradiative de-excitation take place.

This result is contrary to the expectations of the crossover model which, however, remains a useful empirical criterion to classify properly radiative or nonradiative processes, as evident from Table 2. This contradiction is not astonishing because another radiationless process which could be taken into consideration is the tunnelling from C horizontally to the GS parabola [80], see Fig.20. This process depends on the height and width of the barrier much the same as the crossover model depends on E_x, thus there is a parallelism between the two processes [71].

However, a challenge still remains on the theoretical and experimental side. With regard to the latter, measurements of fluorescent lifetimes would be critical for the theory, and the determination of the quantum efficiency of the missing emissions in Table 2 and the measurement of absorption from the RES after a sudden excitation are also at the frontier of this old but well-alive field of research.

REFERENCES

1. H. Rose, Pogg. Ann. 120, 1 (1863).
2. E. Goldstein, Zi. Instrumentenk. 16, 211 (1896).
3. R.W. Pohl, Proc. Phys. Soc. London 49, extrapart 3 (1937).
4. R. Hilsch, Proc. Phys. Soc. London 49, extrapart 40 (1937).

5. N.F. Mott and R. W. Gurney, Electronic Processes in Ionic Crystals, Dover Publ. Inc., New York, 1964, Chap. IV. This book was first published by Oxford University Press in 1940.

6. F. Seitz, Rev. Mod. Phys. 26, 7 (1954).

7. Th. P.J. Botden, C.Z. van Doorn, and Y. Haven, Phylips Res. Repts. 9, 469 (1954).

8. Ch. Becher and H. Pick, Nachr. Akad. Wiss. Goettingen IIa, 167 (1956).

9. W.B. Fowler in Physics of Color Centers, W.B. Fowler, ed., Academic Press, New York, 1968, Chap.2.

10. E. Segré, private communication (1980).

11. F. Seitz, A.P.S. March Meeting St. Louis (1989).

12. E. Fritz and E. Menke, Solid State Commun. 3, 61 (1965).

13. L.F. Mollenauer and U.H. Olson, Appl. Phys. Lett. 24, 386 (1974).

14. H. Blume, T. Baden, and F. Luty, Opt. Commun. 12, 147 (1974).

15. R.M. Macfarlane and R.M. Shelby, Phys. Rev. Lett. 42, 788 (1979).

16. W.E. Moerner, F.M. Schellenberg, G.C. Biorklund, Prasad Kaipa, and F. Luty, Phys. Rev. B32, 1270 (1985).

17. L.F. Mollenauer in Tunable Lasers, L.F. Mollenauer and J.C. White, eds., Springer-Verlag, Berlin, 1987, Chap.5.

18. R. German in Handbook of Solid-State Lasers, Peter K. Cheo, ed., Marcel Dekker, Inc., New York, 1989, Chap.5.

19. G. Baldacchini, Cryst. Lattice Defects Amorph. Mat. 18, 43 (1989).

20. W. Demtröder, Laser Spectroscopy, Springer-Verlag, Berlin, 1982, Chap.7.

21. G. Litfin and R. Beigang, J. Phys. E: Sci. Instrum. 11, 984 (1978).

22. J.H. Shulman and W.D. Compton, Color Centers in Solids, Pergamon Press, Oxford, 1963.

23. Physics of Color Centers, W.B. Fowler, ed., Academic Press, New York, 1968.

24. A.M. Stoneham, Theory of Defects in Solids, Clarendon Press, Oxford, 1975.

25. Y. Farge and M.P. Fontana, Electronic and Vibrational Properties of Point Defects in Ionic Crystals, North-Holland Publ. Company, Amsterdam, 1979.

26. W. Hayes and A.M. Stoneham, Defects and Defect Processes in Nonmetallic Solids, Wiley Interscience Publ., New York, 1985.

27. F. Luty, Halbleiterprobleme Band VI, Verlag Friedr. Vieweg & Sohn, Braunschweig, 1961.

28. F. Agullo-Lopez, C.R.A. Catlow, and P.D. Townsend, Point Defects in Materials, Academic Press, London, 1988.

29. R.H. Dawson and D. Pooley, Phys. Status Solidi 35, 95 (1969).

30. L. Bosi, P. Podini, and G. Spinolo, Phys. Rev. 175, 1133 (1968).

31. F. Luty in Physics of Color Centers, W.B. Fowler, ed., Academic Press, New York, 1968, Chap.3.

32. Y. Kondo and F. Luty, Solid State Commun. 40, 325 (1981).

33. L. Gomes and F. Luty, Phys. Rev. B30, 7194 (1984).

34. Y. Yang, W. von der Osten, and F. Luty, Phys. Rev. B32, 2724 (1985).

35. G. Baldacchini, S. Botti, U.M. Grassano, L. Gomes, and F. Luty, Europhys. Lett. 9, 735 (1989).

36. M. Georgiev, F' Centers in Alkali Halides, Springer-Verlag, Berlin, 1988.

37. H. Feder, M. Hunger, and F. Luty, J. Phys. Chem. Solids 22, 299 (1961).

38. F. Luty, Z. Physik 153, 247 (1958).

39. A. Miehlich, Z. Physik 176, 168 (1963).

40. H. Pick, Nuovo Cimento (Suppl.) 7, 498 (1958).

41. F. Porret and F. Luty, Phys. Rev. Lett. 26, 843 (1971).

42. C.Z. van Doorn, Phylips Res. Repts. Suppl. 4 (1962).

43. J. Lambe and W.D. Compton, Phys. Rev. 106, 684 (1957).

44. H. Engstrom, Phys. Rev. B11, 1689 (1975).

45. J.M. Ortega, Phys. Rev. B19, 2369 (1979).

46. Du-Jeon Jang, T. Corcoran, M.A. El-Sayed, L. Gomes, and F. Luty in Ultrafast Phenomena V, Springer-Verlag, Berlin, 1986, p.286.

47. Y. Yang and F. Luty, Phys. Rev. Lett. 51, 419 (1983).

48. W. Gellermann, Y. Yang, and F. Luty, Opt. Commun. 57, 196 (1986).

49. W. Gellermann and F. Luty, Opt. Commun. 72, 214 (1989).

50. A. Scacco, contribution in this volume.

51. G. Baldacchini, D.S. Pan, and F. Luty, Phys. Rev. B24, 2174 (1981).

52. K. Takiyama, J. Phys. Soc. Japan 44, 1627 (1978).

53. R.H. Bartram and A.M. Stoneham, Solid State Commun. 17, 1593 (1975).

54. G. Baldacchini, U.M. Grassano, A. Scacco, and K. Somaiah, accepted for publication in Il Nuovo Cimento (1989).

55. M. Leblans, C. Sierens, W. Joosen, and D. Schoemaker, Abstracts, Int. Conf. on Defects in Insulating Crystals, August 29 - September 2, 1988, Parma (Italy), p.41.

56. J.R. Hardy, A.M. Karo, Lattice Dynamics and Statics of Alkali Halide Crystals, Plenum Press, New York, 1979.

57. G. Spinolo, Phys. Rev. 137, A1495 (1965).

58. T. Fujita, K. Takiyama, F. Nishiyama, and M. Nishi, Phys. Status Solidi b99, K97 (1980).

59. S. Honda and M. Tomura, J. Phys. Soc. Japan 33, 1003 (1972).

60. M. Hirai and S. Wakita, Semiconductors and Insulators 5, 231 (1983).

61. L. Bosi and M. Nimis, Phys. Status Solidi b149, K89 (1988).

62. D.L. Dexter, C.C. Klick, and G.A. Russel, Phys. Rev. 100, 603 (1955).

63. G. Baldacchini and F. Luty, Abstracts, Int. Conf. on Defects in Insulating Crystals, 9 - 14 October 1977, Gatlinburg, USA, p.25.

64. K. Takiyama, T. Fujita, A. Fujii, and M. Nishi, Abstracts, Int. Conf. on Defects in Insulating Crystals, 9 - 14 October 1977, Gatlinburg, USA, p. 418.

65. A.M. Stoneham and R.H. Bartram, Solid State Electronics 21, 1325 (1978).

66. R.H. Bartram and A.M. Stoneham, Semiconductors and Insulators 5, 297 (1983).

67. C.H. Leung and K.S. Song, Solid State Commun. 33, 907 (1980).

68. G. Baldacchini, G.P. Gallerano, U.M. Grassano, and F. Luty, Phys. Rev. B27, 5039 (1983).

69. G. Baldacchini, G.P. Gallerano, U.M. Grassano, and F. Luty, Lett. Nuovo Cimento 36, 495 (1983).

70. G. Baldacchini, G.P. Gallerano, U.M. Grassano, and F. Luty, Radiation Effects 72, 153 (1983).

71. F. Luty, Semiconductors and Insulators 5, 249 (1983).

72. C.H. Henny and P. Slichter in Physics of Color Centers, W.B. Fowler, ed., Academic Press, New York, 1968, Chap.6.

73. G. Baldacchini, U.M. Grassano, A. Scacco, and K. Somaiah, Abstracts, 8th General Conf. of the Condensed Matter Div. of EPS, 6 - 9 April 1988, Budapest, p.139.

74. M.P. Fontana and D.B. Fitchen, Phys. Rev. Lett. 23, 1497 (1969).

75. M.P. Fontana, Phys. Rev. B2, 4304 (1970).

76. G. Baldacchini, U.M. Grassano, and A. Tanga, Phys. Rev. B16, 5570 (1977).

77. S. Wakita, Y. Suzuki, and M. Hirai, J. Phys. Soc. Japan 50,. 2781 (1981).

78. G. Consolati and N. Terzi, Abstracts, Int. Conf. on Defects in Insulating Crystals, August 29 - September 2, 1988, Parma (Italy), p.43.

79. M. Dominoni and N. Terzi, Contribution in this volume.

80. P.F. Moulton in Laser Handbook, M.Bass and M.L. Stitch, eds., North-Holland Publ. Company, Amsterdam, 1985, Chap.2.

NONRADIATIVE TRANSITIONS BETWEEN ELECTRONIC LEVELS OF OPTICALLY-ACTIVE IONS IN GLASS

G.F. Imbusch

Department of Physics
University College
Galway, Ireland

ABSTRACT

We examine the nonradiative multiphonon relaxation mechanism between electronic levels of optically-active ions in glasses, and we compare the values of the relaxation rate with that found in crystals.

I. INTRODUCTION

When an optically-active ion in a solid is raised to an excited state, it may lose its excitation by radiative emission (Fig. 1(a)), or by a number of nonradiative processes : by multiphonon emission to the next lower state (1(b)), by cross-relaxation (1(c)), or by energy transfer resulting in loss of excitation to some nonradiative centre in the solid (1(d)). A number of these deexcitation processes may be operating simultaneously. Processes (c) and (d) can be eliminated by

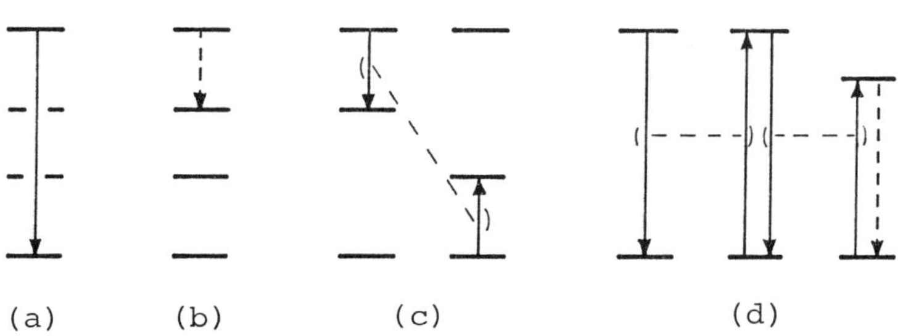

Fig. 1. Deexcitation processes out of the excited state. (a) is the radiative process, (b) is the nonradiative multiphonon process, (c) shows cross-relaxation with an adjacent similar ion followed by transfer to a centre which then decays nonradiatively.

Advances in Nonradiative Processes in Solids
Edited by B. Di Bartolo, Plenum Press, New York, 1991

going to small concentrations of active ions, while processes (a) and (b) are intrinsic to the particular ion-host combination. We will confine our attention to the multiphonon decay mechanism (b). This process has been studied intensively in optically-active ions in crystals for the past 25 years, extensive experimental data are available, and these data have been the subject of various theoretical analyses. In this paper we will examine the corresponding transitions between electronic levels of ions in glass, enquiring, in particular, whether the models which find application for crystalline solids can also be employed for glasses.

II. GLASS STRUCTURES

Glass is an amorphous material; it is a continuous random network of polyhedra (such as SiO_4 or BF_4 tetrahedra, BO_3 triangles, etc.). These are the glass formers. In simple glasses these polyhedra are connected at corners to form a random three-dimensional network. The more interesting glasses from a spectroscopic viewpoint contain additional modifier compounds (such as Na_2O, Al_2O_3, AlF_3) which modify the network of formers by breaking the bonding between some of the polyhedra, thereby opening up the glass structure. Minor additives (such as Sb_2O_3 and As_2O_3) help to improve the physical properties of the glass. The large variety of compositions allows the material scientist to attempt to design glass for specific application [1]. In addition, glass can be cast in large pieces or drawn into thin fibers, as appropriate to a specific application.

Fig. 2 is a two-dimensional representation of a sodium silicate glass [2]. The larger circles represent the sodium modifier ions. When such glasses are grown with small amounts of optically-active ions, these ions enter the glass structure mainly as network modifier ions. Glass has a variety of sites which these ions can enter, and these sites differ from each other in the arrangement of the neighbouring ions. Consequently, the energy levels of the active ions in these sites are not identical. Under broadband optical pumping, whereby ions in all the different sites are excited, the luminescence is inhomogeneously broadened, being a composite of emissions from ions in the different sites.

Glasses in general spectroscopic use are based on oxide (silicate, borate, phosphate, germanate, tellurite), fluoroberyllate, fluorophosphate, and fluoride formers. More recently, heavy metal fluoride [3] and chalcogenide glasses have been used as hosts. The spectroscopic properties of optically-active ions in these various matrices have been summarized by Reisfeld [4]; extensive references are contained in this review.

III. RARE EARTH IONS IN GLASS

Radiative transitions between $4f^n$ electronic states of rare earth ions in crystals are characterized by very sharp optical transitions - a consequence of (i) the weak coupling between the 4f electrons and the ionic lattice environment of the rare earth ion, and (ii) the fact that the rare earth ions

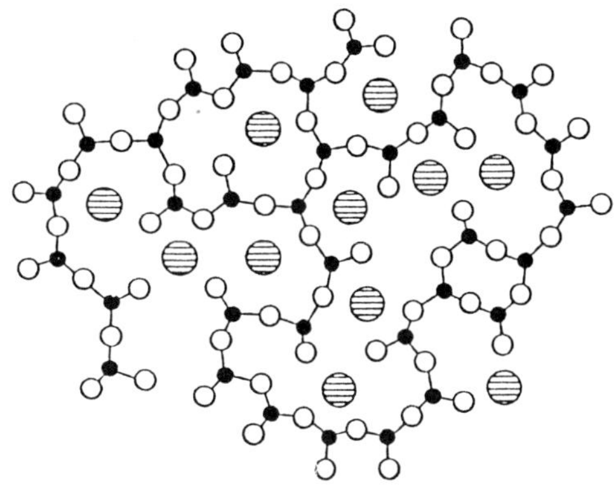

Fig. 2. Two-dimensional representation of a sodium silicate glass. The triangles of small filled and open circles show the silicate tetrahedra, the large hatched circles represent Na^+ ions [2].

generally enter identical sites in the crystal. Nonradiative transitions on these ions have been studied intensively following the pioneering experimental work of Kiel [5], Riseberg and Moos [6], and Weber [7]. This study is facilitated by the detailed knowledge of the electronic states in question and by the weakness of the electron-lattice coupling. The nonradiative decay rate at low temperatures is given to good approximation by the empirical formula:

$$W_{nr} \simeq 10^7 \exp[-\beta(p-2.4)] \tag{1}$$

where $\beta \simeq 2$. p is the number of phonons involved in the relaxation process, and $p = E/\hbar\omega$, where E is the energy gap to the next lower level, and $\hbar\omega$ is the phonon energy. A number of theoretical models have been advanced to account for the exponential dependence on E [5,8,9,10].

In contrast to the situation in crystals, rare earth ions in glass find a range of site types available to them, and this leads to relatively large inhomogeneous broadening. The way in which this complicates the analyses of the spectra and of the radiative and nonradiative processes on the rare earth ions is discussed in the next section.

III.A. Optical Spectroscopy of Rare Earth Ions in Glass

The inhomogeneous broadening of rare earth ion transitions in glass is illustrated by the low temperature $^4I_{9/2} \rightarrow {}^4F_{3/2}$ absorption on Nd^{3+} ions in different glasses shown in Fig. 3 [11]. For comparison, the same absorption transition in Nd^{3+}-doped YAG is also shown. The splitting of the $^4F_{3/2}$ level as well as the average frequency of the transition vary between the different glass hosts. Even though the transitions on any one ion are sharp, the transition frequencies vary from site to site in the same glass host, and the overall spectrum shows a large degree of inhomogeneous broadening.

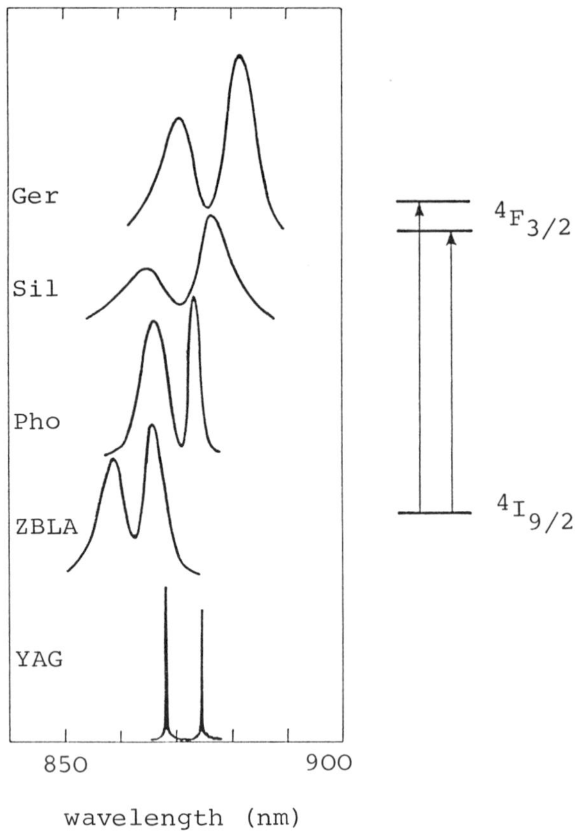

Fig. 3. Low temperature absorption of Nd^{3+} in various glasses and in YAG (adapted from [11]). ZBLA is a zirconium barium fluoride glass.

Site-to-site variations can be studied by the technique of narrow band laser excitation [2]. This technique selectively excites a subset of the rare earth ions - those whose absorption frequencies are within the laser bandwidth - and the resulting sharp luminescence can be studied spectroscopically. Using a tunable narrowband laser Brecher and Riseberg [12] excited subsets of Eu^{3+} ions in a silicate glass throughout the $^7F_0$ -> $^5D_0$ absorption transition. The resulting $^5D_0$ -> $^7F_{0,1,2}$ luminescence transitions are relatively sharp and show the position of the split levels of the $^7F_1$ and $^7F_2$ states. Brecher and Riseberg sought a structural model of the environment of the Eu^{3+} ion which would account for the observed spectral variations. The distribution of levels was found to be consistent with a gradual and continuous change from largely eightfold to largely ninefold coordination. This model is also found to be applicable to glasses consisting of a wide range of tetrahedral oxidic network formers (silicates, phosphates, borates, aluminates, and combinations), but large differences are found when the network forming anion is changed [2]. Fig. 4 shows the site-to-site variations in the Stark levels in the $^7F_{0,1,2}$ states of Eu^{3+} in a fluoroberyllate glass [13] (these are drawn relative to the emitting $^5D_0$ level) for laser

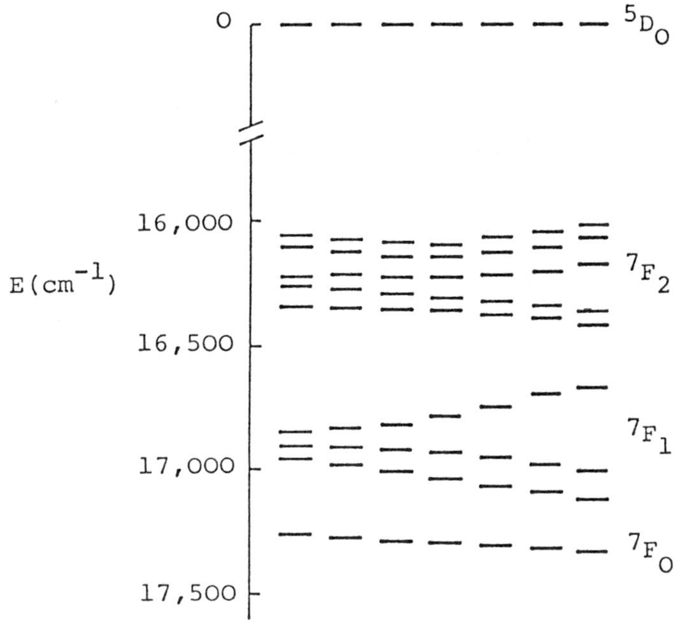

Fig.4. Site-to-site variation in crystal field levels of Eu^{3+} in fluoroberyllate glass, as discussed in the text.

excitation at seven distinct frequencies spanning the inhomogeneously-broadened $^7F_0 \rightarrow {}^5D_0$ absorption transition.

Fig. 5(a) shows the inhomogeneously-broadened $^4I_{9/2} \rightarrow {}^2P_{1/2}$ absorption transition of Nd^{3+} in a silicate glass [14] with a width of around 100 cm^{-1}. Laser excitation into ten different parts of this absorption transition excites ions in distinct sites, and the resultant $^4F_{3/2} \rightarrow {}^4I_{9/2}$ luminescence transitions show variations in frequency, in radiative decay rate, and in nonradiative decay rate. Fig. 5(b) shows the measured low temperature nonradiative decay rates ($W_{nr}(0)$) for ions in different sites in the glass. In addition, the radiative decay rate varies from 1.6×10^3 s^{-1}, after excitation at the short wavelength, to 3.6×10^3 s^{-1}, after excitation at the long wavelength. Thus, there can be large variations in the radiative and nonradiative decay rates and in the quantum efficiency within the same transition in the glass. Studies of transition rates and efficiencies in rare earth ions in glass generally do not seek the detail shown in Fig. 5; transition rates and efficiencies are generally measured using broadband excitation sources, and the measured values are averages over the distribution of sites in the glass.

III.B. Experimental Measurements of Nonradiative Decay Rates

If a specific rare earth level is populated by pulsed excitation, the population of excited ions (N) decays by radiative (W_r) and nonradiative (W_{nr}) processes:

$$N = N_o \exp(-Wt) \quad : \quad W = W_r + W_{nr} \qquad (2)$$

One measures W experimentally. The radiative rate, W_r, can usually be calculated if the Judd-Ofelt parameters have been determined for this rare earth-host system [15, 16]. (These parameters are determined by measurements of the absorption strengths). Thus the value of $W_{nr} = W - W_r$ is obtained.

Figure 6 illustrates the technique. Ions are raised by pulsed excitation to level 2, from which they decay at a rate

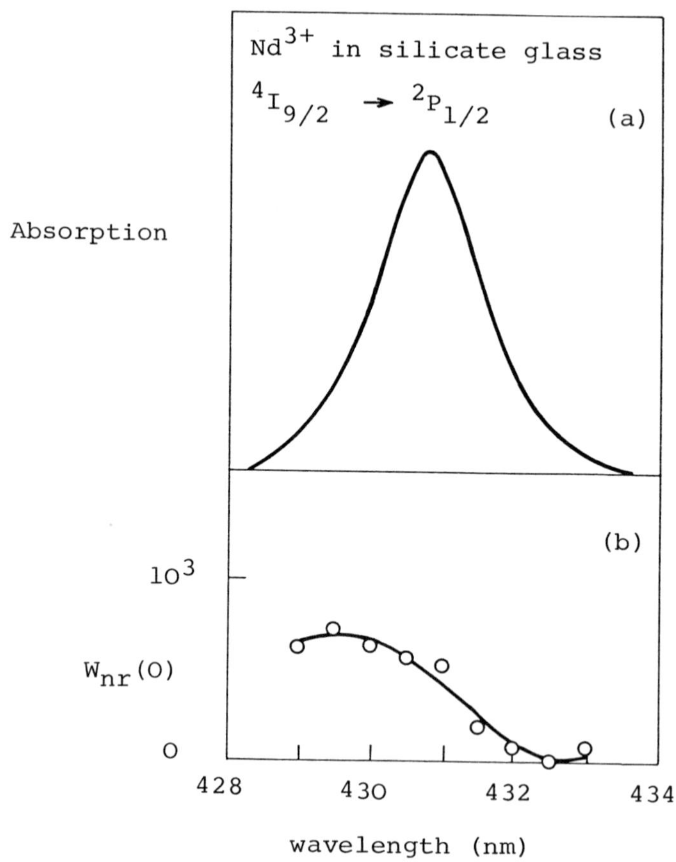

Fig. 5. (a) The inhomogeneously-broadened $^4I_{9/2} \rightarrow {}^2P_{1/2}$ absorption transition of Nd^{3+} in a silicate glass [14], showing a width of 100 cm^{-1}.
(b) Measured nonradiative decay rate for narrowband excitation at ten different frequencies within the inhomogeneously-broadened absorption transition.

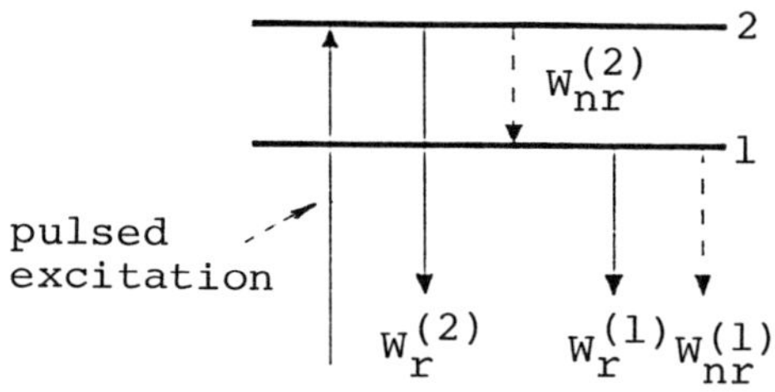

$$W_1 = W_r^{(1)} + W_{nr}^{(1)}$$

$$W_2 = W_r^{(2)} + W_{nr}^{(2)}$$

$$N_1(t) = C\left[\exp(-W_1 t) - \exp(-W_2 t)\right]$$

$$t_{max} = \ln(W_2/W_1)/(W_2 - W_1)$$

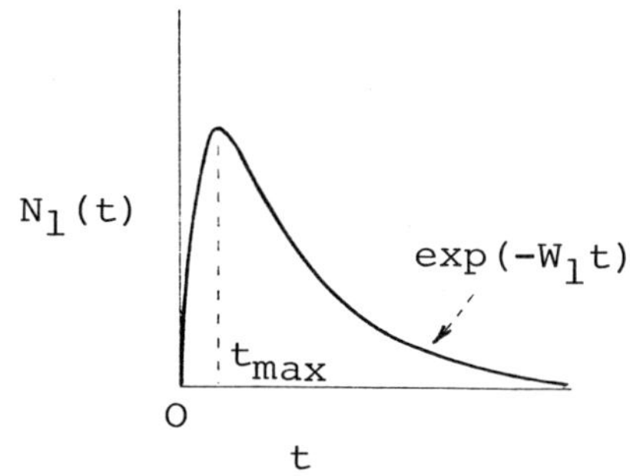

Fig. 6. Decay pattern of the population N_1 in excited level 1 after pulsed excitation into level 2. In this hypothetical case $W_2 > W_1$.

$W_2 = W_r^{(2)} + W_{nr}^{(2)}$, the nonradiative rate being to the nearby level 1. Particularly if the gap between levels 2 and 1 is small, $W_{nr}^{(2)}$ can be much larger than $W_r^{(2)}$, and the luminescence from level 2 can be exceedingly weak and difficult to observe. Nonradiative decay from level 2 to level 1 results in a build-up of population, N_1, in the latter level.

Monitoring N_1, by monitoring the luminescence from level 1, gives the pattern shown in Fig. 6: a fast build-up, at rate W_2, followed by a slower decay at rate $W_1 = W_r^{(1)} + W_{nr}^{(1)}$. Thus W_1 and W_2 can be measured experimentally, and $W_{nr}^{(1)}$, $W_{nr}^{(2)}$ found by subtracting the respective radiative decay rates.

As an alternative technique, the quantum efficiency of a given level can be measured, and the W_{nr} value determined once the radiative rate is known.

III.C W_{nr} Values for Rare Earth Ions in Glasses

The first quantitative studies of nonradiative processes on rare earth ions in glasses were carried out by Reisfeld and co-workers [16]. Extensive studies were later carried out by Layne, Lowdermilk, and Weber [17] and by Layne and Weber [18]. From the earliest experiments on oxide glasses, it was clear that a smaller number of luminescent states are observed for rare earth ions in these glasses than is the case for rare earth ions in oxide crystals. Since radiative decay rates on rare earth ions in crystals and glasses are of the same order of magnitude [17], this implies an enhanced nonradiative decay rate in glasses. As an illustration, we consider the $^5D_J \rightarrow {}^7F_{J'}$ transitions on Eu^{3+} ions in solids, the approximate separations of the J terms are shown in Fig. 7. In many crystalline hosts the low temperature luminescence from the $^5D_{3,2,1}$, levels is strong, indicating that the $^5D_J \rightarrow {}^7F_{J'}$ radiative transitions are able to compete successfully with the nonradiative transitions between the ^5D_J and $^5D_{J-1}$ levels. In the oxide glasses the corresponding luminescence transitions are exceedingly weak, the intensity decreasing in the order borate < phosphate < germanate < tellurite [16]. The enhanced nonradiative decay in the oxide glasses is attributable to the presence of high frequency vibrations in these glasses, in the range of 700 - 1400 cm^{-1}, larger than that found in common crystalline host materials.

We shall now examine the experiments of Layne, Lowdermilk, and Weber [17] and of Layne and Weber [18]. Their measurements were carried out on Nd^{3+}, Er^{3+}, and Tm^{3+} in a number of glasses, and with their apparatus they were able to measure decay rates as high as 10^8 s^{-1}. For each glass W_{nr} values were determined for different energy gaps (E), the rates varying over about five orders of magnitude. For each glass a plot of log W_{nr} against energy gap (to the next lower level) is linear to good approximation, showing an exponential dependence of W_{nr} on energy gap, just as is found for crystalline host materials. The exponential dependence of W_{nr} on the energy gap in the case of glass hosts was first pointed out by Reisfeld [16]. Fig. 8 shows the linear dependences for each of the glasses investigated by Layne, Lowdermilk, and Weber. We also show in

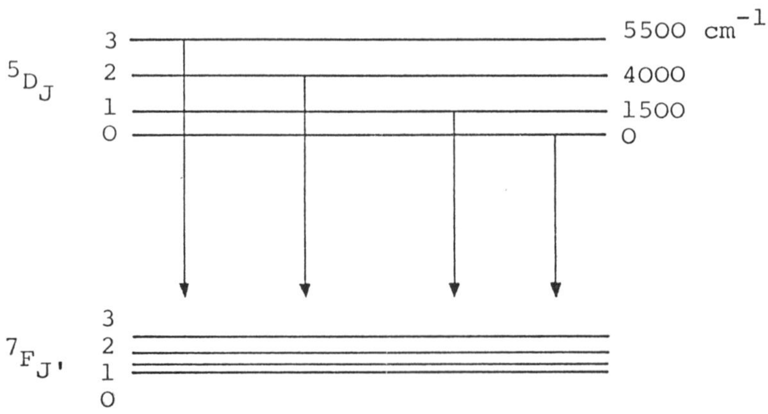

Fig. 7. Approximate energy level diagram for the $5D_J$ and $^7F_{J'}$ levels of Eu^{3+} in a solid.

Fig. 8 the straight line fit to the measured W_{nr} values obtained by Reisfeld [20] for rare earth ions in ZBLA, a heavy metal fluoride glass in which the effective phonon energy is 500 cm^{-1}, much smaller than the effective phonon energies in the oxide glasses. In each case the straight lines show the best fit to the data. For comparison, we also show the comparable straight line fits for the crystalline hosts $LaBr_3$, LaF_3, and Y_2O_3 [19]. For a given energy gap we see that the W_{nr} values are generally much larger for the glasses than for the crystalline materials.

The more meaningful parameter for classifying the decay process is the number of phonons (p) required to bridge the energy gap. $p = E/\hbar\omega$, where $\hbar\omega$ is the energy of the effective phonons involved in the relaxation. In order to determine p we need to determine the frequencies of these phonons. A theory of the vibrational spectra of glasses is given by Brawer [21], who showed that the complete vibrational spectrum of the glass can be found from the first order Raman spectrum. The Raman spectra of the glasses in question are shown in Fig. 9 [17,18]. For each glass the highest frequency vibration is associated with the symmetric stretching mode of the glass network formers.

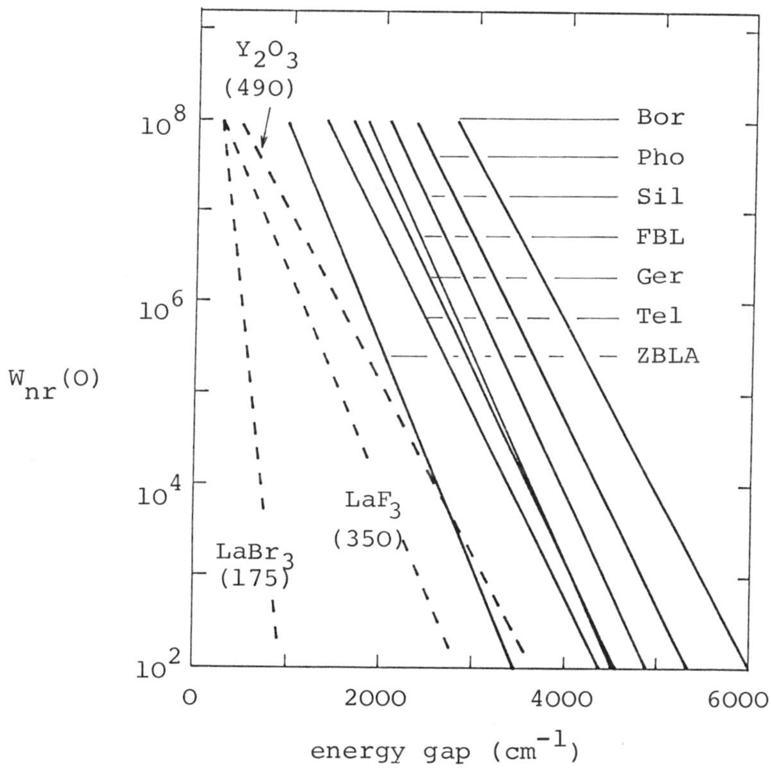

Fig. 8. Dependence of $W_{nr}(0)$ on energy gap for a number of glass hosts [17,18,20]. Values for some crystalline hosts are given for comparison [15]. The highest frequency vibrations for the crystalline hosts are given in cm^{-1}.

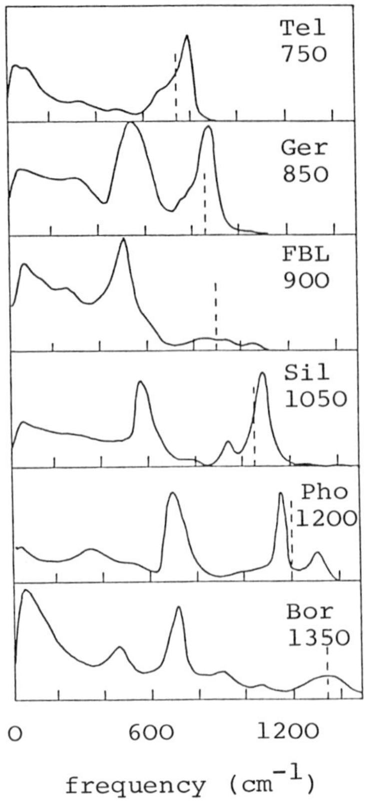

Fig. 9. Raman spectra of the glass hosts used in the experiments of Layne, Lowdermilk, and Weber [17,18]. The broken vertical lines indicate the approximate highest frequency vibrations, and these values are given in cm^{-1}. (Adapted from [17] and [18]).

The broken vertical line in each case gives the approximate highest frequency vibration, and this value (in cm^{-1}) is listed in the figure. We will assume that it is the phonons of highest frequency which are involved in the multiphonon nonradiative decay process. Some of the frequency values chosen here differ slightly from the values chosen by Layne et al. and by Reisfeld. In the case of the fluoroberyllate glass (FBL), the highest frequency is taken as 500 cm^{-1} by Reisfeld [4], while a value of 900 cm^{-1} is chosen in this analysis. The highest frequency vibrations (in cm^{-1}) for the crystalline materials are given in Fig. 8.

In Fig. 10 we have replotted the straight lines of Fig. 8 to show the exponential dependence of $W_{nr}(0)$ on p. The glass data are shown as solid lines; data for some crystalline materials are drawn as broken lines. The lines, for both the glass and the crystalline materials, show the interesting tendency to cross at $W_{nr} \simeq 10^7$ and $p \simeq 2.4$ [6]. The slope for the glasses, however, seems to be larger than for the crystalline materials. This is also seen in the similar but more extensive plot given by van Dijk and Schuurmans [10]. We can say that <u>on average</u> for the oxide and fluoride glasses the relaxation rate is given by $W_{nr}(0)\quad 10^7 \exp(-4(p-2.4))$, while for crystalline hosts $W_{nr}(0)\quad 10^7 \exp(-2(p-2.4))$.

Although the oxide and fluoride glasses show a strikingly similar dependence of W_{nr} on p, the situation is quite different for chalcogenide glasses. For Nd^{3+} transitions in ALS ($3Al_2S_3$, 0.872 La_2S_3, 0.128 Nd_2S_3) and in GLS ($3Ga_2S_3$, 0.85 La_2S_3, 0.15 Nd_2S_3), Reisfeld [20] finds that the multiphonon relaxation rate has an exponential dependence on p, but it differs from that found for the oxide and fluoride glasses, the rate varying approximately as $W_{nr}(0)\quad 10^5 \exp(-(p-2.4))$. Reisfeld [20] suggests that the different behaviour for the chalcogenide glasses may be due to the strong covalent bonding between the rare earth ion and the surrounding ligands, which contrasts with the strongly ionic coupling with oxide and fluoride glasses.

In the analysis of the oxide and fluoride glasses above it has been assumed that generally the highest frequency phonons are effective in the relaxation process. We can obtain a check on the frequency of the effective modes by examining the

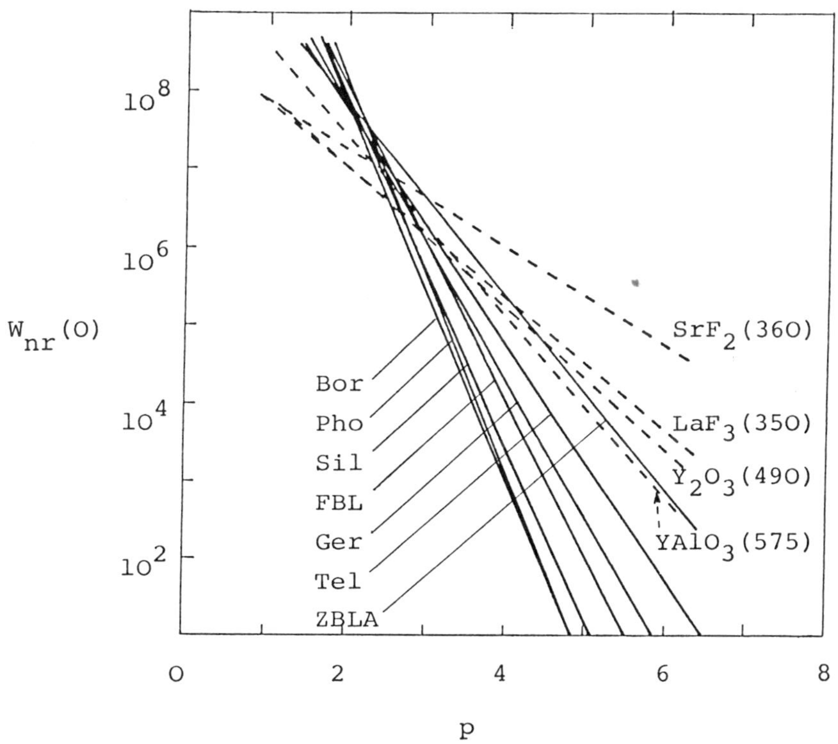

Fig. 10. Dependance of $W_{nr}(0)$ on p. The data for $YAlO_3$, Y_2O_3, and LaF_3 are taken from Riseberg and Weber [15], those for SrF_2 are taken from Reisfeld [20].

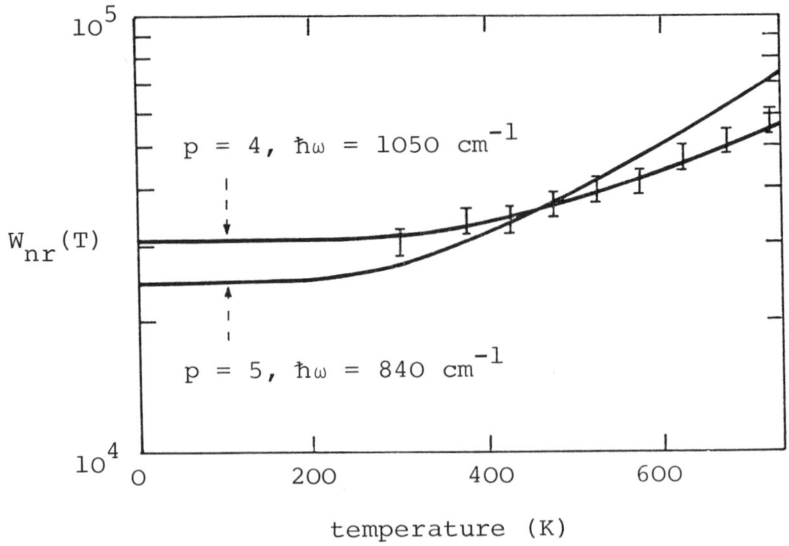

Fig. 11. Temperature dependence of the $^3F_4 \rightarrow {}^3H_5$ multiphonon relaxation rate for Tm^{3+} in silicate glass. Two fits to formula (3) are shown. (data from Layne et al. [17])

temperature dependence of W_{nr}. In the small coupling limit, applicable for rare earth ions, the temperature dependence of the relaxation rate across an energy gap E in which p phonons of frequency ω are involved is given by

$$W_{nr}(T) = W_{nr}(0)(1 + n)^p \qquad (3)$$

where $n = (\exp(\hbar\omega/kT)-1)^{-1}$ and $p = E/\hbar\omega$. The experimental values of $W_{nr}(T)$ across the approximately 4000 cm^{-1} energy gap between the $^3F_4$ and $^3H_5$ levels of Tm^{3+} in a silicate glass [17] are shown in Fig. 11. The solid curves represent eqn.(3) for two cases : $p = 4$, $\hbar\omega = 1050\ cm^{-1}$, and $p = 5$, $\hbar\omega = 840\ cm^{-1}$. The former is the better fit, and this is consistent with the value chosen for the phonons assumed to be effective in the multiphonon relaxation process ($\hbar\omega = 1050\ cm^{-1}$, Fig. 9). A similar consistency is found for the other oxide glasses investigated by Layne, Lowdermilk, and Weber [17].

It can be argued that since the Raman spectrum shows the vibrational modes of the host material, the modes which show up strongest in the Raman spectrum are not necessarily those which interact strongly with the rare earth ion. The vibrational modes which interact strongly with the rare earth ion show up in the vibrational sidebands accompanying the sharp zero-phonon transitions. Vibrational sidebands accompanying sharp zero-phonon transitions between $4f^n$ states are usually quite weak, and their observation can be complicated by overlapping zero-phonon lines to different Stark levels of the final J-state manifold.

Gd^{3+}, with its ground S state (with negligible splitting) and its intense $^6P_{7/2} \rightarrow {}^8S_{7/2}$ luminescence transition, is an excellent ion to probe the vibrational spectrum [22]. Vibronic sidebands accompanying this transition on Gd^{3+} ions in metaphosphate glasses were examined by Hall et al. [22]. In order to overcome the large inhomogeneous broadening and obtain sharp zero-phonon transitions they employed narrowband laser excitation which excited a subset of the Gd^{3+} ions into the $^6P_{7/2}$ level. The resultant narrowband luminescence zero-phonon line was accompanied by a clear vibrational sideband. Fig. 12(a) shows this sideband spectrum for Gd^{3+} in $Al(PO_3)_3$ glass, while the H-V Raman spectrum is shown in Fig. 12(b). The high frequency symmetric stretching mode of the PO_2 unit ($\simeq$ 1200 cm^{-1}) is strongly represented in both spectra, the sideband spectrum verifying that this mode interacts strongly with the Gd^{3+} ion.

Fig. 13 shows the vibrational sideband accompanying the $^6P_{7/2} \rightarrow {}^8S_{7/2}$ zero-phonon transition on Gd^{3+} in LaB_3O_6 glass [23], showing again a strong coupling to the 1350 cm^{-1} stretching vibration of the BO_4 tetrahedra in this glass. In these two examples we find strong coupling with the highest frequency vibrational modes of the glass formers, which supports the viewpoint that multiphonon relaxation of rare earth ions in

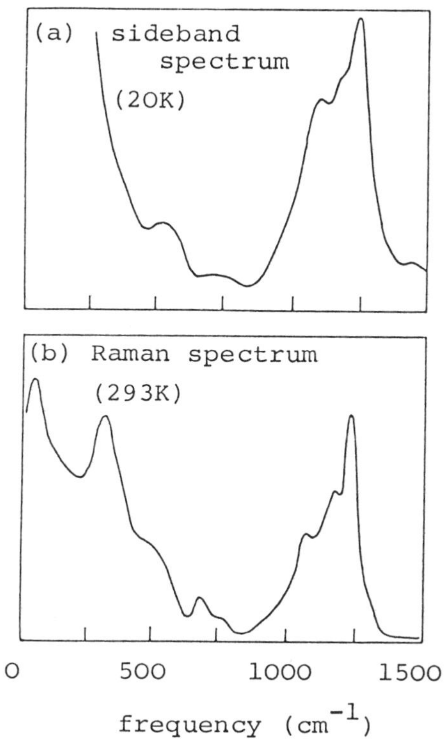

Fig. 12. (a) Vibrational sideband accompanying the $^6P_{7/2} \rightarrow {}^8S_{7/2}$ zero-phonon transition of Gd^{3+} in $Al(PO_3)_3$ glass.
(b) Raman spectrum of $Al(PO_3)_3$ glass [22].

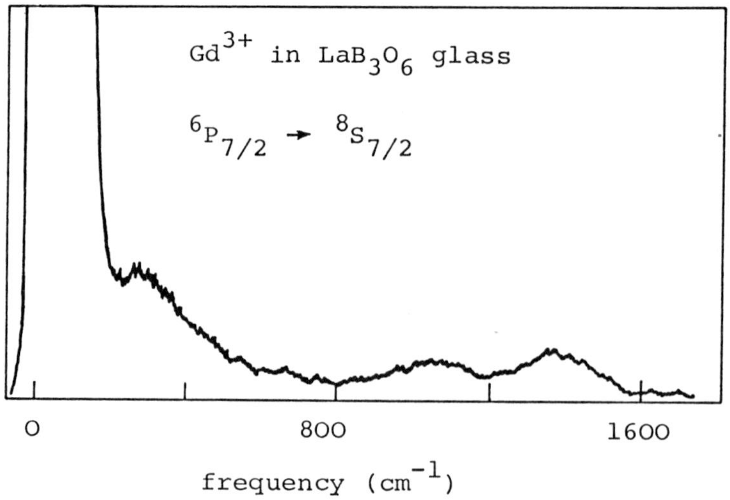

Fig. 13. Vibrational sideband accompanying the low temperature $^6P_{7/2} \rightarrow {}^8S_{7/2}$ transition on Gd^{3+} in LaB_3O_6 glass. Narrowband laser excitation was employed to obtain a sharp zero-phonon line [23].

glasses operates through the generation of the smallest number of highest frequency phonons required to bridge the energy gap to the next lower electronic level.

III.D. Summary for Rare Earth Ions in Glass

The multiphonon decay rate between $4f^n$ levels of rare earth ions in oxide and fluoride glasses behave in a very similar way to that found for crystalline materials. There is an exponential dependence on the energy gap, the highest frequency modes seem to be effective, and the dependence of $W_{nr}(T)$ on temperature and number of phonons (p) is given by

$$W_{nr}(T) \simeq A(1 + n)^p \exp[-\beta(p - 2.4)] \qquad (4)$$

where $n = (\exp(\hbar\omega/kT) - 1)^{-1}$, $A \simeq 10^7 \ s^{-1}$, and $\beta \simeq 4.0$. A is approximately the same for glasses and crystals, while β is about a factor of 2 smaller for crystals.

For oxide glasses the large energy of the phonons engaged in the nonradiative process means that the probability of nonradiative decay across the same energy gap is much larger for these glasses than for crystalline materials. This creates a limitation to the efficient long wavelength luminescence from oxide glasses. The more recently discovered [3] heavy metal fluoride glasses (e.g. ZBLA) with their smaller phonon energies and low attenuation properties in the near-infrared (at around 2.5 μm) present the possibility of extending efficient rare earth luminescence of glasses further into the infrared.

IV. TRANSITION METAL IONS IN GLASS

In contrast to the case of rare earth ions, whose active 4f electrons are partially shielded from interaction with the environment, for transition metal ions the active 3d electrons are on the outside of the ion, so that the $3d^n$ electronic states are much more sensitive to variations in the ionic environment. As a consequence, the electron-lattice interaction is much larger for transition metal than for rare earth ions. There are two consequences of this. One is a stronger nonradiative process for transition metal ions in crystals and in glasses than is the case for rare earth ions. The second is that, due to the wide range of sites in glass, the inhomogeneous broadening of optical transitions in glass is much more severe for transition metals than for rare earth ions. This large inhomogeneous broadening complicates the analysis of the absorption and luminescence spectra.

We observe also, as a general rule, that the quantum efficiency of luminescence of transition metal ions in glass is less than for crystalline host materials. As evidence for this we point to the number of active media for solid state lasers employing Ti^{3+}, Cr^{3+}, V^{2+}, Co^{2+} in crystalline materials, noting that laser action has not so far been achieved in any transition metal doped glass.

By far the largest amount of spectroscopic work on transition metal doped glasses has been carried out on the Cr^{3+} ion, and we shall concentrate on an analysis of the spectroscopic properties of chromium-doped glasses and on trying to understand the reasons for their low quantum efficiency. Afterwards we shall review the situation with other ions in glass.

IV.A. Spectroscopy of Cr^{3+} Ions in Glass

The essential features of the luminescence of Cr^{3+} ions in glass have been established by a number of workers [24-29]. Chromium luminescence in most crystalline hosts is characterized by a high quantum efficiency ($\simeq 1$), and such materials serve as active media for fixed frequency laser action (e.g. the $^2E \rightarrow {}^4A_2$ transition in ruby) and for tunable laser action (e.g. the $^4T_2 \rightarrow {}^4A_2$ transition in alexandrite). In contrast, the room temperature quantum efficiency of chromium in glasses has never been found to exceed 25% [30,31] and is more often found to be very small.

The energies of the low-lying levels of Cr^{3+} in octahedral sites are given in the Tanabe-Sugano diagram in Fig. 14. Dq represents the strength of the octahedral ligand field, and B is a Racah parameter. The luminescence transition occurs between the ^2E and/or $^4T_2$ state and the ground state. The vertical broken line represents a typical average ligand field configuration, and this equilibrium configuration is essentially the same for the ground $^4A_2$ and for the excited ^2E states. When an ion in this configuration absorbs a photon of light and is raised to the $^4T_2$ state, its coupling with the ionic environment is different from that in the $^4A_2$ and ^2E states, and as a result the ion in the $^4T_2$ state along with the ions in the immediate environment change to a new equilibrium configuration of lower

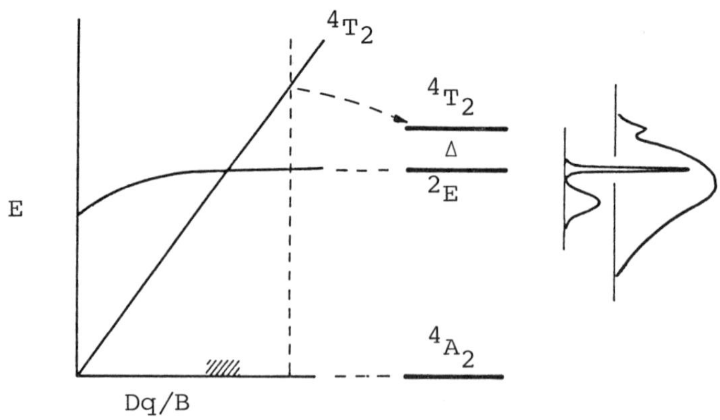

Fig. 14. Variation of the low-lying energy levels of Cr^{3+} with octahedral crystal field strength. The vertical broken line represents a typical average value of Dq/B for $^4A_2$ and ^2E states. Ions in the $^4T_2$ state adopt a different average value. The luminescence spectra of the $^2E \rightarrow {}^4A_2$ and $^4T_2 \rightarrow {}^4A_2$ transitions are represented on the right of the figure.

energy. Δ is the energy separation between the zero-vibrational level of this relaxed $^4T_2$ state and the zero-vibrational ^2E level (Fig. 14). If Δ is positive the relaxed $^4T_2$ state is above ^2E, the ion is said to be in a "high field site", and at low temperatures the luminescence is due to the $^2E \rightarrow {}^4A_2$ transition, which is characterized by sharp intense zero-phonon lines (R-lines) and one-phonon sidebands. For negative values of Δ luminescence is due to the $^4T_2 \rightarrow {}^4A_2$ transition, and the ion is said to be in a "low field" site. The site occupied by the Cr^{3+} ion is usually one of distorted octahedral symmetry, and as a result each of the ^2E and $^4T_2$ levels is split; the splitting of the ^2E level gives rise to the R_1 and R_2 zero-phonon lines.

Whereas Cr^{3+} ions in a crystalline host can enter one or two specific types of site, in a glass the Cr^{3+} ions may find a variety of types of site. The hatched area in Fig. 14 shows the range of Dq/B values in the different sites in a silicate glass. Most ions enter low field sites, each of which emits in a broad transition, while a small fraction enter high field sites, and each of these ions emit sharp R-lines and one-phonon sidebands. Fig. 15 contrasts the luminescence of Cr^{3+} in crystalline and glass hosts. Cr^{3+} ions in MgO mainly occupy high field sites of perfect octahedral symmetry, and at low temperatures the luminescence of these ions appears as a sharp $^2E \rightarrow {}^4A_2$ zero-phonon R-line and a one-phonon sideband (Fig. 15(a)). Some ions enter low field sites of distorted octahedral symmetry (rhombic sites) and the resulting low-temperature broadband $^4T_2 \rightarrow {}^4A_2$ luminescence is shown in Fig. 15(b). Fig. 15(c) shows the luminescence from Cr^{3+} in lithium lime silicate glass (ED-2).

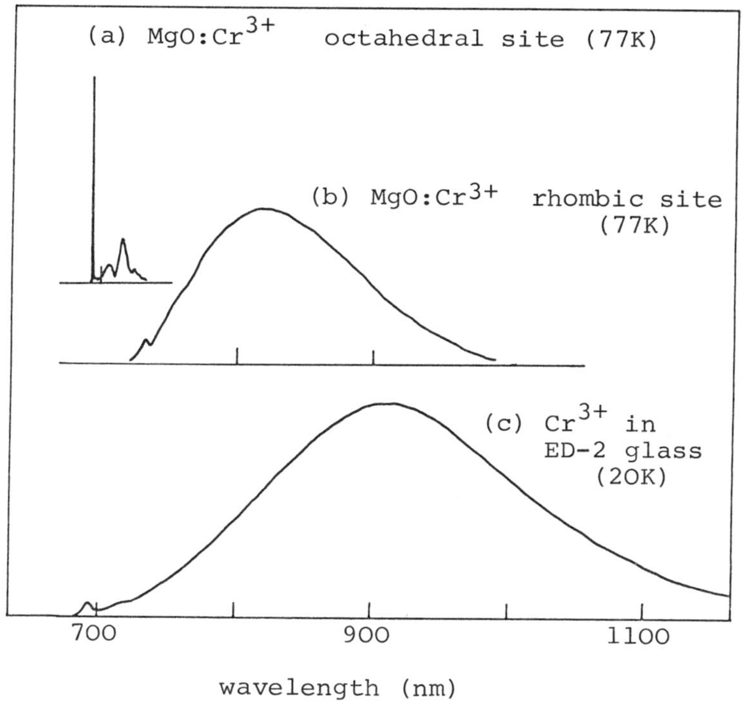

Fig. 15. Luminescence from Cr^{3+} ions in (a) octahedral sites (high field) in MgO, (b) rhombic sites (low field) in MgO, (c) high field and low field sites in ED-2 glass.

The weak feature at around 690 nm is a composite made up of sharp $^2E \rightarrow {}^4A_2$ transitions from Cr^{3+} ions in different high field sites. The width of this feature ($\simeq 200\ cm^{-1}$) is to be contrasted with the sharp (width $< 1\ cm^{-1}$) R-line in the octahedral site in MgO, and this attests to the inhomogeneous broadening in the glass. The predominant glass luminescence is a composite of $^4T_2 \rightarrow {}^4A_2$ bands from Cr^{3+} ions in various low field sites in the glass, and the large spread of the luminescence seen in Fig. 15(c) again attests to the inhomogeneous broadening in the glass. The radiative decay pattern of the R-line of MgO is a simple exponential with a decay time of 11.6 ms, the decay of the broadband luminescence from the rhombic sites is also single exponential, with a decay time of 35 μs. In contrast, the decay patterns of the composite R transitions at 690 nm and of the broadband luminescence of the glass are not single exponentials. This is clear from Fig. 16 which shows the decay patterns at 20K. The multiexponential nature of the decay of the broadband luminescence points to a range of decay rates from ions in different sites, the very fast rates occurring on ions in which there is a significant nonradiative decay process, in addition to the radiative decay process, depopulating the excited state.

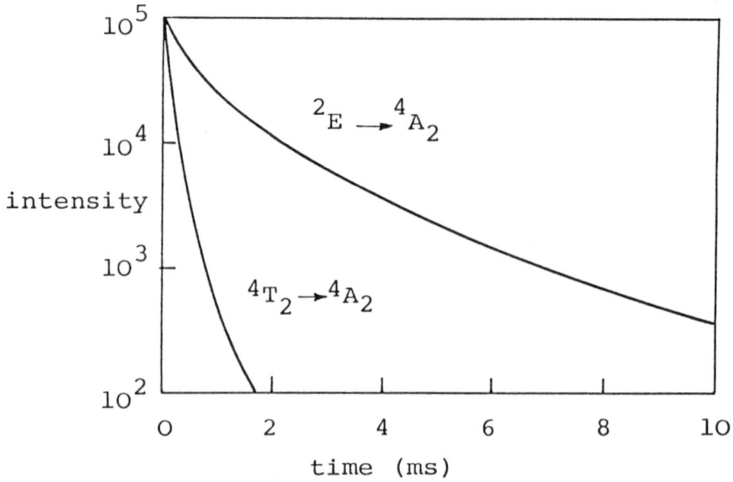

Fig. 16. Decay patterns of the luminescence intensity from Cr^{3+} ions in high field sites ($^2E \rightarrow {}^4A_2$) and low field sites ($^4T_2 \rightarrow {}^4A_2$) in ED-2 glass at 20K.

As illustration of this nonradiative decay we refer to the interesting study by van Die [32] on chromium-doped borate glass. Fig. 17(a) shows the luminescence at 20K after excitation into the peak of the absorption band. This luminescence consists of a weak composite $^2E \rightarrow {}^4A_2$ transition at 690 nm from ions in high field sites and a predominant broadband emission from ions in low field sites, only the initial part of which is seen in the figure. Almost all of the ions enter low field sites. Fig. 17(b) shows the absorption spectrum, the $^4A_2 \rightarrow {}^4T_2$ absorption transition peaking at 610 nm. Two excitation spectra are shown (Fig. 17(c)), one for luminescence monitored at 690 nm (from high field sites), the other for luminescence monitored at 800 nm (from low field sites). We compare this $^4A_2 \rightarrow {}^4T_2$ transition on low field sites measured in excitation with the analogous absorption transition (Fig. 17(b)). The peak of the excitation band occurs at higher energy (570 nm) than the peak of the absorption band (610 nm). This is taken as indicating that the ions in low field sites of lowest energy have lowest quantum efficiency and hence their contribution to the excitation spectrum is less than that of the higher energy ions [32]. We illustrate this schematically in Fig. 18, which shows the relative positions of the $^4T_2$ and ^2E emitting levels in four representative sites: a - d. Ion (a) is in a high field site, and emits relatively sharp luminescence from the ^2E level. Ions (b), (c), and (d) are in low field sites and emit broadband luminescence, but ion (d), whose $^4T_2$ level is lowest of all, has the lowest quantum efficiency and only contributes weakly to the luminescence. The decay rate for these lowest energy ions is larger than for the other low field ions - because of an enhanced nonradiative process - hence the luminescence from these ions decays more rapidly, and a pronounced blue shift with increasing time after excitation can be observed in the time-resolved $^4T_2 \rightarrow {}^4A_2$ luminescence [30].

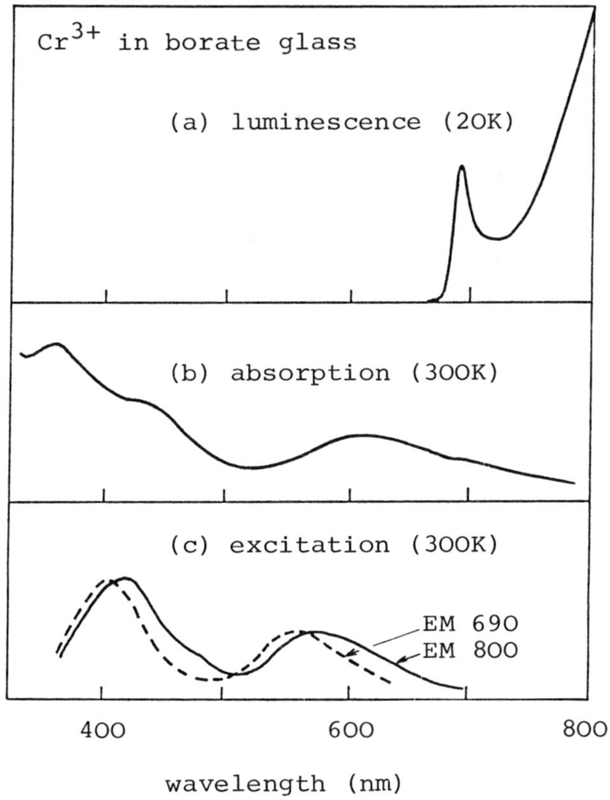

Fig. 17. Luminescence, absorption, and excitation spectra of Cr^{3+} in borate glass (after van Die [32]).

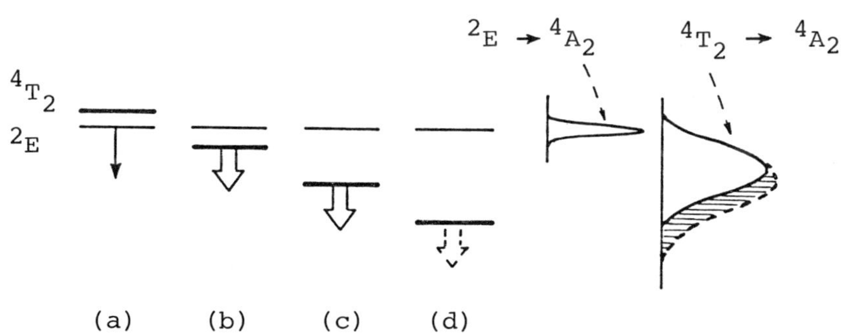

Fig. 18. Schematic representation of the emissions from four representative sites. The energy of the $^4T_2$ level falls significantly as Dq decreases. Because of their very low quantum efficiency, the ions of lowest energy contribute little to the overall luminescence, the shaded area represents the loss of this emission.

As temperature is raised from 20K to 300K the overall luminescence intensity decreases markedly, and the decay times shorten, indicative of an enhanced nonradiative decay rate at the higher temperatures.

IV.B. Quantum Efficiency Measurements for Chromium-doped Glasses

The quantum efficiency of a large number of chromium-doped glasses has been measured by Andrews et al. [30], by Reisfeld and her co-workers [4], by van Die and Blasse [31], and by Seelert [33]. Typical measured values of the room temperature quantum efficiency are gathered in Table 1. We observe no correlations between quantum efficiency and Dq values - the energy gap between $^4T_2$ and $^4A_2$ should scale approximately with Dq. Nor do we find any correlation with the maximum frequency of the glass vibrations. This is in contrast to the case of rare earth ions in glass, where there was a clear exponential gap dependence, and where the nonradiative process proceeds by emission of the smallest number of highest energy phonons. As a result, $W_{nr}(0)$ values for rare earth ions in tellurite glass are over an order of magnitude smaller than for silicate glass (Fig. 8). In contrast, the nonradiative decay rate of Cr^{3+} in tellurite glass is an order of magnitude larger than for silicate glass.

By carrying out fluorescence line narrowing studies on Cr^{3+} in high field sites on ED-2 silicate glass, Bergin et al. [36] obtained a spectrum of the vibrational sideband accompanying the sharp $^2E \rightarrow {}^4A_2$ transition on ions in a small subset of high field sites. The sideband (Fig. 19) shows a strong coupling of the Cr^{3+} ion to low frequency modes, and rather weak coupling to the high frequency vibrations of the silicate units ($\simeq$ 1050 cm^{-1}). If the chromium ions are coupled mainly to low frequency modes, then we should expect no correlation between W_{nr} and the highest frequency phonons of the glass.

Table 1. Room temperature quantum efficiency of some chromium-doped glasses.

Glass	QE(%)	Maximum phonon energy (cm^{-1})	Dq(cm^{-1})*	Ref.
Borate	1	1350	1640	30
Phosphate	10 22	1200	1470	30 34
Silicate	17	1050	1520	30
Germanate	25	850	1600	31
Tellurite	1	750	1544	30
ZBLA	0	500	1500	30,35

*taking energy of $^4A_2 \rightarrow {}^4T_2$ absorption peak as 10Dq.

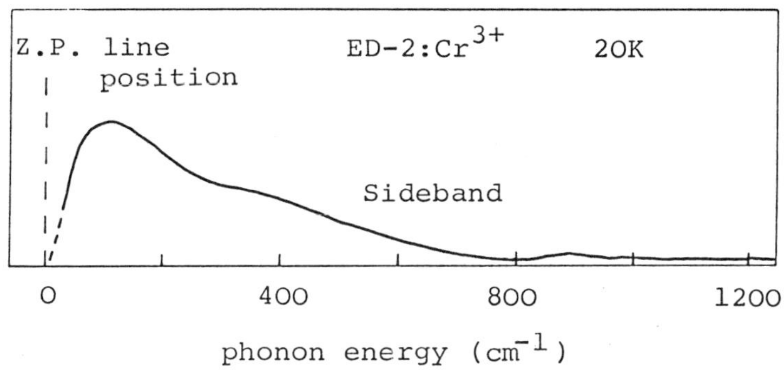

Fig. 19. Vibrational sideband accompanying the sharp $^2E \rightarrow {}^4A_2$ zero-phonon line of Cr^{3+} in a high field site in ED-2 glass. Fluorescence line narrowing techniques were employed to overcome the large inhomogeneous broadening [36].

IV.C. The Nature of the Sites occupied by Cr^{3+} Ions in Glass

From considerations of energy we expect that Cr^{3+} ions strongly favour sites of six-fold octahedral coordination. The observed absorption and emission spectra of chromium-doped glasses are consistent with the viewpoint that the Cr^{3+} ions are in sites of near-octahedral symmetry. We might expect, then, that in glasses there would be a large range of near-octahedral sites available to the Cr^{3+} ions; the ions in these various sites would differ in average energy as well as in the splitting of the ^2E level. For Cr^{3+} in different crystalline materials the measured $R_1 - R_2$ separations vary from 0 to 278 cm^{-1}. Thus we might expect a similar range of $R_1 - R_2$ separations in each glass. The fluorescence line narrowing experiments of Bergin et al. [36] on Cr^{3+} in ED-2 glass showed, on the contrary, that all ions in high field sites appear to have approximately the same R_1-R_2 separation, ($\simeq 55 \pm 25$ cm^{-1}), suggesting that ions in high field sites are in sites whose symmetries are quite similar to each other. Experiments by Seelert and Strauss [33,37] reveal that ions in low field sites in ED-2 glass - and the vast majority of ions enter low field sites - also appear to occupy a small number of specific sites. This raises the question as to what determines the site occupied by Cr^{3+} ions in glass.

One viewpoint [38] is that, in contrast to the case of crystalline materials where the sites available to the dopant ion are determined by overall crystallographic requirements, in glasses local bonding considerations determine the environment of the Cr^{3+} ion. In addition, we must bear in mind that a glass is a looser structure than a crystal. The existence of low frequency tunneling modes in glass [39,40] show that, even at low temperatures, the ions of the glass can relax from one quasi-equilibrium local configuration to another. We might expect that Cr^{3+} ions in the $^4A_2$ state, with their strong preference for sites of six-fold octahedral coordination, will arrange to be in such an ionic environment during the formation of the glass. When, under optical pumping, the Cr^{3+} ion is raised to the $^4T_2$ state, the electronic orbital structure changes ($t^3 \rightarrow t^2e$ orbitals), hence the Cr^{3+} ion's coupling with the surrounding ions changes. Given the loose nature of the glass structure and the strong influence of local bonding requirements, it is possible that in glasses, when the Cr^{3+} ion is raised to the $^4T_2$ state, the decrease in energy in going to the relaxed excited state is larger than that which occurs when the ion is in a crystal. Such a large configurational coordinate change in glass can significantly affect the nonradiative relaxation from the relaxed excited state.

There are two reasons for this. First, the larger the configurational coordinate change the smaller the $^4T_2 \rightarrow {}^4A_2$ energy separation, and we have observed that Cr^{3+} ions with longest wavelength have tended to have the smallest quantum efficiency (Figs. 17 and 18). Second, the larger the configurational coordinate shift the larger the value of $S\hbar\omega$, where S is the Huang-Rhys parameter. If we assume that the vibrational modes of the glass which couple to the Cr^{3+} ion have approximately the same frequencies as for crystalline materials, then the $^4A_2 \leftrightarrow {}^4T_2$ transitions of Cr^{3+} ions in glass are characterized by a larger value of S than for crystals. All theories of nonradiative multiphonon relaxation indicate that W_{nr} increases significantly as S is increased. A model based on the harmonic approximation with linear coupling shows that if S changes from 4 to 5, the value of $W_{nr}(0)$ increases by over two orders of magnitude. That a large configurational coordinate change leads to an enhanced nonradiative process has been pointed out by Blasse [41].

If such a proposed large configurational coordinate shift occurs for Cr^{3+} in glass, it should show up as a large Stokes shift between the $^4A_2 \leftrightarrow {}^4T_2$ transitions in absorption and emission. Measurements of the Stokes shift for Cr^{3+} in six representative oxide glasses give an average value of 4950 cm^{-1} [38], whereas the measured Stokes shift in a large number of crystalline oxides averages 2900 cm^{-1} [38]. On this viewpoint, then, we would attribute the low quantum efficiency of chromium-doped glasses to enhanced nonradiative multiphonon processes brought about by the large configurational coordinate relaxation when the Cr^{3+} ion is in the $^4T_2$ state. This large configurational coordinate relaxation is a consequence of the loose nature of the glass structure and of the relative importance of local bonding requirements in determining the arrangements of ions in a glass.

IV.D. Luminescence Efficiency of Other Ions in Glass

Divalent manganese has the largest energy gap, around 16,000 cm^{-1}, between ground and first excited state of any transition metal ion. In crystalline materials, where Mn^{2+} enters octahedral or tetrahedral sites, the $^4T_1 \rightarrow {}^6A_1$ emission transition across this gap has high efficiency. The drawback with Mn^{2+} is that it possesses no broad spin-allowed absorption transitions for effective optical pumping.

Mn^{2+} in octahedral sites in phosphate [42], silicate [32], borate [32], and ZBLA glasses [43,35] appears to have high quantum efficiency. Luminescence is also observed from Mn^{2+} ions in tetrahedral sites in silicate glasses [32]. Rather suprisingly, Mn^{3+} luminescence is found for manganese in silicate and borate glasses, and some Mn^{4+} luminescence is found in borate glass [32,44].

Fe^{3+} is isoelectronic with Mn^{2+}, and luminescence attributable to the $^4T_1 \rightarrow {}^6A_1$ transition on two Fe^{3+} centres in a fluorozirconate glass has been reported by Snow et al. [45,46]. A multi-phonon nonradiative process appears to cause a quenching of the luminescence with increasing temperature on one of these centres.

Although the energy gaps between electronic levels in Ni^{2+} are small ($\simeq$ 5000-6000 cm^{-1}), this ion can be an efficient emitting centre in crystalline materials. In glasses its quantum efficiency is at best around 1% at low tempeatures, and it decreases further with increasing temperature [47].

Co^{2+}, which has the longest wavelength emission of any transition metal ion when it enters octahedral sites in crystalline hosts, does not seem to emit any luminescence in glasses [47].

The luminescence of $Mo^{3+}(4d^3)$ in phosphate glass, where it enters near-octahedral sites, has been studied by Weber et al. [48]. Since Dq/B is much larger for Mo^{3+} than for Cr^{3+}, the $^4T_2$ level is well above ^2E, and the luminescence is due to the $^2E \rightarrow {}^4A_2$ transition across a gap of around 10,000 cm^{-1}. This luminescence appears as a broad band, approximately 500 cm^{-1} in width, with a weak composite $^2E \rightarrow {}^4A_2$ zero-phonon transition in evidence only at very low temperatures. From absorption measurements the radiative decay rate is estimated as $\simeq$ 250 s^{-1} [48], but the observed decay rate is much larger, indicative of strong nonradiative multiphonon processes. The decay pattern is highly nonexponential, and the decay rate increases by almost one order of magnitude in going from the high energy side to the low energy side of the luminescence. This is as expected, the nonradiative multiphonon rate increases as the energy gap decreases.

The uranyl ion (UO_2^{2+}). Although not a transition metal ion, the uranyl ion, a strongly-emitting centre, has an electron-lattice coupling similar to the transition metal ions. Luminescence across the 20,000 cm^{-1} gap between the first excited state and the ground state of the uranyl ion in crystalline materials is a very efficient process at room temperature. In glasses this luminescence is highly efficient at low temperatures, but falls to 30-40% at room temperature, the value varying with the type of glass [49,32,50]. An explanation is advanced [32] in terms of a large lateral displacement of the second excited configurational coordinate parabola in the glasses, which permits a cross-over of excitation to the ground state parabola at an energy lower than in the case of UO_2^{2+} in crystalline hosts.

IV.E. Summary for Transition Metal Ions in Glass

For transition metal ions, with their strong electron-lattice coupling, nonradiative multiphonon processes are effective over large energy gaps; processes involving the generation of 30-50 phonons can compete successfully with radiative processes. With such large energy gaps the role of anharmonicities in the vibrational potential can be important. Further, the strong electron-lattice coupling admits of changes in force constants when the electronic state changes. These two effects, which are difficult to identify, can greatly affect the relaxation rate. Thus it is not surprising that for transition metal ions in glasses, as well as in crystals, it is not possible to make the same broad generalizations as for rare earth ions.

Although the nonradiative multiphonon relaxation process for a given transition metal ion across the same energy gap is stronger when the ion is in glass than when in a crystal, it is not clear that this is due to the presence of high frequency vibrational modes of the glass. Rather it would seem to be related to the loose structure of glass and the ability of the local ionic complex about the transition metal ion in the glass to adjust to changes in the ion-lattice coupling when the electronic state changes.

V. LUMINESCENCE FROM CHROMIUM IONS IN TRANSPARENT GLASS CERAMICS

In comparison with crystalline hosts, in which its quantum efficiency is generally high, room temperature luminescence of Cr^{3+} in glasses is characterized by a low quantum efficiency. On the other hand, glass can be much more easily cast into large pieces, which is particularly useful when preparing active media for large laser systems or developing luminescence solar concentrators. Materials which potentially combine the good properties of glasses and crystals are the transparent glass ceramics. These are fabricated by thermally inducing devitrification of appropriately formulated glasses to produce submicron-size crystallites embedded in a residual glass matrix. These microcrystallites can be made so much smaller than the wavelength of visible light that light scattering is greatly reduced. To the eye these ceramics appear as clear and as transparent as glass. In one such material the mineral mullite ($2Al_2O_3.SiO_2$) is precipitated from the original mother glass [51], and chromium ions which are present in small amounts in the mother glass diffuse to the crystal phase during ceramic formation. Thus the Cr^{3+} ions in these ceramics have the high quantum efficiency characteristic of crystalline hosts. The development of these transparent glass ceramics involves major programmes in materials science and spectroscopy. These are discussed by Andrews, Lempicki, and coworkers [51,52,53], Beall et al. [54], Reisfeld and Jorgensen [55], and Durville et al. [56].

ACKNOWLEDGEMENTS

I am grateful to Profs. G. Blasse and E. Strauss for permitting me to make use of unpublished spectra and measurements from their laboratories.

REFERENCES

1. S.E. Stokowski in : Lasers, Spectroscopy and New Ideas (Eds. W.M. Yen and M.D. Levenson), Springer-Verlag, Berlin, 1987, p. 47.

2. M.J. Weber in : Laser Spectroscopy of Solids (eds. W.M. Yen and P.M. Selzer), Springer-Verlag, Berlin, 1981, p. 189.
3. A. Lecoq and M. Poulain, Verres Refract. 34, 333 (1980).
4. R. Reisfeld in : Spectroscopy of Solid State Laser-type Materials (Ed. B. di Bartolo), Plenum Press, New York, 1987, p. 343.
5. A. Kiel in : Quantum Electronics (Eds. P. Grivet and N. Bloembergen), Columbia University Press, New York, 1964, p. 765.
6. L.A. Riseberg and H.W. Moos, Phys. Rev. Letters 19, 1429 (1967); Phys. Rev. 174, 429 (1968).
7. M.J. Weber, Phys. Rev. 171, 283 (1968).
8. H.W. Moos, J. Luminescence 1,2, 106 (1970).
9. W.E. Hagston and J.E. Lowther, Physica 70, 40 (1973).
10. J.M.F. van Dijk and M.F.H. Schuurmans, J. Chem. Phys. 78, 5317 (1983).
11. J.M. Pellegrino, W.M. Yen, and M.J. Weber, J. Appl. Phys. 51, 6332 (1980).
12. C. Brecher and L.A. Riseberg, Phys. Rev. B13, 81 (1976).
13. C. Brecher and L.A. Riseberg, Phys. Rev. B21, 2607 (1980).
14. C. Brecher, L.A. Riseberg, and M.J. Weber, Phys. Rev. B18, 5799 (1978).
15. L.A. Riseberg and M.J. Weber in : Progress in Optics, vol. XIV (ed. E. Wolf), North-Holland, Amsterdam, 1976, p. 89.
16. R. Reisfeld, Structure and Bonding, 22, 123 (1975).
17. C.B. Layne, W.H. Lowdermilk, and M.J. Weber, Phys. Rev. B16, 10 (1977).
18. C.B. Layne and M.J. Weber, Phys. Rev. B16, 3259 (1977).
19. M.J. Weber, Proc. Tenth Rare Earth Conference 2, 932 (1973).
20. R. Reisfeld in : Radiationless Processes (Ed. B. Di Bartolo), Plenum Press, New York, 1980, p. 489.
21. S. Brawer, Phys. Rev. B11, 3173 (1975).
22. D.W. Hall, S.A. Brawer, and M.J. Weber, Phys. Rev. B25, 2828 (1982).
23. Hans Verwey (University of Utrecht), unpublished.
24. T. Bates and R.W. Douglas, J. Soc. Glass Tech. 48, 289 (1959).
25. R.J. Landry, J.T. Fournier, and C.G. Young, J. Chem. Phys. 46, 1285 (1967).
26. R.E. Tischer, J. Chem. Phys. 48, 4291 (1968).
27. M.J. Weber, E.J. Sharp, and J.E. Miller, J. Chem. Phys. Solids 32, 2275 (1971).
28. S.A. Brawer and W.B. White, J. Chem. Phys. 67, 2043 (1977).
29. A. Lempicki, L. Andrews, S.J. Nettel, B.C. McCollum, and E.I. Solomon. Phys. Rev. Letters 44, 12134 (1980).
30. L.J. Andrews, A. Lempicki, and B.C. McCollum, J. Chem. Phys. 74, 5526 (1981).
31. A. van Die, A.C.H.I. Leenaers, G. Blasse and W.F. van der Weg, J. Non-Cryst. Solids 99, 32 (1988).
32. A. van Die, thesis, University of Utrecht (1987).
33. W.R. Seelert, thesis, University of Oldenburg, (1987).
34. A. Kisilev and R. Reisfeld, Solar Energy 33, 163 (1984).
35. R. Reisfeld, M. Eyal, C.K. Jorgensen, A.H. Guenther, and B. Bendow, Chimia 40, 403 (1986).
36. F.J. Bergin, J.F. Donegan, T.J. Glynn, and G.F. Imbusch, J. Lumin. 34, 307 (1986).
37. E. Strauss (personal communication).
38. G.F. Imbusch, T.J. Glynn, and G.P. Morgan, to be published.
39. P. W. Anderson, B.I. Halperin, and C.M. Varma, Phil. Mag. 25, 1 (1972).
40. W.A. Phillips, J. Low Temp. Phys. 7, 351 (1972).
41. G. Blasse, Prog. in Solid State Chem. 18, 79 (1988).

42. R. Reisfeld, A. Kisilev, and C.K. Jorgensen, Chem. Phys. Letters 111, 19 (1984).
43. C.K. Jorgensen, Rev. Chim. Miner. 23, 614 (1986).
44. A.J. Faber, A. van Die, G. Blasse, and W.F. van der Weg, Phys. Chem. Glasses 28, 150 (1987).
45. E. Snow, J.A. Freitas, Jr., and U. Strom, Phys. Rev. B37, 10332 (1988).
46. G. Blasse, Phys. Rev. B39, 3408 (1989).
47. Y. Suzuki, W. A. Sibley, O.H. El Bayoumi, T.M. Roberts, and B. Bendow, Phys. Rev. B35, 4472 (1987).
48. M.J. Weber, S.A. Brawer, and A.J. DeGroot, Phys. Rev. B23, 11 (1981).
49. R. Reisfeld and C.K. Jorgensen, Structure and Bonding 49, 1 (1982).
50. R. Reisfeld and S. Neuman, Nature 274, 144 (1978).
51. L.J. Andrews, B.C. McCullum, S. Stone, D.E. Guenther, G.J. Murphy, and A. Lempicki, Final Report: Development of Materials for a Luminescent Solar Collector DOE/ER/04996-4, GTE Laboratories Incorporated, Waltham, Mass. (1983).
52. L.J. Andrews, B.C. McCollum, and A. Lempicki, J. Lumin. 24/25, 877 (1981).
53. L.J. Andrews, G.E. Beall, A. Lempicki, J. Lumin. 36, 65 (1986).
54. G.H. Beall, D.R. Cole, D.W. Hall, H.J. Holland, J.W.H. Schreuras, M. Edwards, R. Gillies, and A. Lempicki. Final Report: Physical and Optical Properties of Chromium-doped Glass Ceramics, US Dept. of Energy, DE-FG02-84ER45103 (1987).
55. R. Reisfeld and C.K. Jorgensen, Structure and Bonding, 69, 63 (1988).
56. F. Durville, B. Champagnon, E. Duval, G. Boulon, F. Gaume, A.F. Wright, and A.N. Fitch, Phys. Chem. Glasses 25, 126 (1984).

NONRADIATIVE PROCESSES IN LUMINESCENT MATERIALS: A MATERIALS SCIENTISTS VIEW

G. Blasse

Physics Laboratory
University of Utrecht
P.O.Box 80.000
3508 TA Utrecht
The Netherlands

ABSTRACT

Nonradiative processes in luminescent materials are of large importance for materials properties. Sometimes they are required to be fast, sometimes to be slow. In this contribution we will review an approach to nonradiative processes which is tractable for application to a wide range of materials, i.e. the materials scientist view. Physical models like the configurational coordinate diagram and the Förster-Dexter energy transfer approach will be used in a simplified form. Our approach will be illustrated by examples. The main stress will be on nonradiative transitions between two energy levels within one and the same luminescent species, and on energy transfer and especially energy migration. The compositions to be dealt with are mainly nonmolecular insulating compounds, but excursions to molecular species will be included.

I. INTRODUCTION

Luminescent materials cover a wide field of applications; luminescent lamps, cathode-ray tubes, x-ray photography, lasers, solar concentrators, electroluminescent display panels, luminescence immuno-assay, and so on [1,2]. One of the most important requirements for such a material is a high value of its light output. All energy absorbed by the material which is not emitted as radiation (luminescence) is dissipated to the crystal lattice (radiationless processes). It is, therefore, imperative to suppress those radiationless processes which compete with the emission process. There are, however, also nonradiative processes which favour a high light output, viz. those which take care of a more effective feeding of the luminescent activator and/or population of the emitting level.

Table I gives the quantum efficiencies of some important photoluminescent materials.

Advances in Nonradiative Processes in Solids
Edited by B. Di Bartolo, Plenum Press, New York, 1991

Table I. Quantum efficiencies q of some photoluminescent materials under ultraviolet excitation at room temperature.

Material	q(%)	ref.
Zn_2SiO_4–Mn^{2+}	68	3
YVO_4–Eu^{3+}	70	3
$CaWO_4$	70	3
$Ca_5(PO_4)_3(F,Cl)$–Sb^{3+},Mn^{2+}	70	3
$NaGdF_4$–Ce^{3+},Tb^{3+}	95	4

Note that the maximum obtainable value of 100% is by far not obtained. The complicated composition $NaGdF_4$–Ce,Tb [4] approaches this value best. As we will see below this is due to the fact that the radiationless feeding processes of the emitting Tb^{3+} ion are, although complicated, very effective, and the radiationless transition competing with the emission very ineffective. In $CaWO_4$, on the other hand, the emitting tungstate group is excited directly, and the 70% quantum efficiency is due to a ratio of the radiative and nonradiative decays of the emitting level of about 2:1.

These observations illustrate that it is important to have a good understanding of radiationless processes in order to understand, to improve and to predict luminescent materials. There are several approaches to this difficult problem. In our opinion all of these are useful. One of these is a completely theoretical approach. This may nowadays be of great use in understanding specific, simple situations. However, it cannot be used with much success in order to account for the properties of a large number of different materials.

There exist also general, approximative methods to study nonradiative processes. A good example is the chapter by Dr. C.W. Struck in this book. Our approach is probably the most simple one. The main characteristic is that the physical parameters are kept as simple as possible, whereas the chemical parameters are widely varied, i.e. by studying a large number of different materials we try to find out in howfar the simple physical model is still able to account for the results. The great advantages of this method are its simplicity and predictive ability.

This chapter reviews in section II the physical models which will be used. Sections III and IV describe nonradiative, intracentre transitions for the weak and the strong coupling case, respectively. Since the former is reasonably understood nowadays, we pay more attention to the latter which will be illustrated by many examples. Only isolated ions will be considered in these sections. Section V discusses the feeding processes of luminescent ions, viz. energy transfer and energy migration. This section takes into account the interaction which may exist between luminescent species. Finally section VI deals with some borderline cases.

II. MODELS

II.A. Configurational coordinate model

Let us consider a dopant ion in a host lattice and assume that it shows luminescence upon illumination. What we will have to discuss is the interaction of the dopant ion with the vibrations of the lattice. The environment of the dopant ion is not static: the surrounding ions vibrate about some average positions, so that the crystalline field varies. The simplest model to account for the interaction between the dopant ion and the vibrating lattice is the single-configurational coordinate model [5,6].

In this model we consider only one vibrational mode, viz. the so-called breathing mode in which the surrounding lattice pulsates in and out around the dopant ion (symmetrical stretching mode). This mode is assumed to be described by the harmonic oscillator model. The configurational coordinate (Q) describes the vibration. In our approximation it presents the distance between the dopant ion and the surrounding ions.

If we plot energy vs Q we obtain for the electronic states parabolae (harmonic approximation). This is presented in fig. 1 for the electronic ground state g and one electronic excited state e. Further Q_0 presents the equilibrium distance in the ground state, Q_0' that in the excited state. Note that in general these will be different! The g parabola is given by

$$V = \tfrac{1}{2}k\ (Q - Q_0)^2, \qquad (1)$$

where k is the force constant. Within the parabolae the (equidistant) vibrational energy levels have been drawn. They are numbered by n = 0, 1,2... The excited state parabola is drawn in such a way that the force constant is weaker than in the ground state. Since the excited state is usually weaker bound than the ground state, this is a representative situation.

Optical absorption corresponds to a transition from the g to the e state under absorption of electromagnetic radiation. Emission is the reverse transition. Let us now consider how these transitions have to be described in the configurational coordinate model. It is essential to remember that the wave function of the lowest vibrational state (i.e. n=0) is Gaussian, i.e. the most likely value of Q is Q_0 (or Q_0' in the excited state). For the higher vibrational states, however, the most likely value is at the edges of the parabola, i.e. at the turning points (like in the classic pendulum).

The most likely transition in absorption at low temperatures is from the n = 0 level in g, starting at the value Q_0. Optical absorption corresponds to a vertical transition, because the transition g $\longrightarrow$ e on the dopant ion occurs so rapidly that the surrounding lattice does not change during the transition (Born-Oppenheimer approximation). Our transition will end on the edge of parabola e, since it is there that the vibrational states have their highest amplitude. This transition, drawn as a solid line in fig. 1, corresponds to the maximum in the absorption band. However, we may also start at Q values different from Q_0, although the probability is lower. This leads to the width of the absorption band, indicated in fig. 1 by broken lines. It can be shown that the probability of the optical transition between the n = 0 vibrational level of the ground state and the n = n' vibrational level of the excited state is proportional to

$$\langle e(Q)|r|g(Q)\rangle\langle\chi_{n'}/\chi_0\rangle \quad (2)$$

where r presents the electric-dipole operator and χ the vibrational wave functions. The first term, the electronic matrix element, is independent of the vibrational levels; the second term gives the vibrational overlap. The transition from n = 0 to n' = 0 does not involve the vibrations. It is called the zero-vibrational transition (or no-phonon transition). Equation 1 shows that the effect of the vibrations is mainly to change the shape of the absorption line (or band), but not the strength of the transition (which is given by the electron matrix element).

What happens after the absorption transition? First we return to the lowest vibrational level of the excited state, i.e. the excited state relaxes to its equilibrium position, giving up the excess energy as heat to the lattice. The system of dopant ion and surroundings is then in the relaxed excited state. The emission transition can be described in exactly the same way as the absorption transition. This is indicated in fig. 1 in the same way as for the absorption transition. Finally the system relaxes within the g parabola to the lowest vibrational level.

If the temperature is not low, higher vibrational levels may be occupied thermally, so that we start the process not only from n = 0, but also from n = 1, and possibly from even higher levels. This leads to a further broadening of the absorption and emission bands, but does not change our arguments essentially.

The emission transition will usually be situated at lower energy than the absorption transition. This phenomenon is known as the Stokes

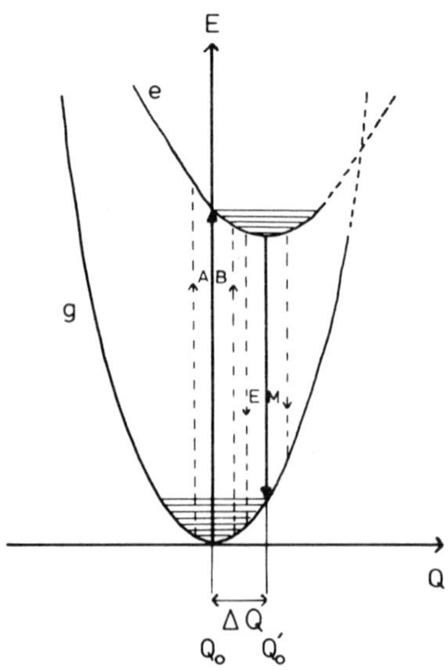

Fig. 1 The configurational coordinate diagram. The energy E is plotted vs the coordinate Q. Parabolae g and e refer to the ground state and excited state, respectively. Horizontal lines indicate the vibrational levels. The vertical transitions are discussed in the text. ΔQ gives the parabolae offset. The absorption transition is indicated by AB, the emission transition by EM.

shift. Only the zero-vibrational transition is expected to occur at the same energy in the absorption and emission spectra. The Stokes shift is a direct consequence of the relaxation processes which occur after the optical transitions. It is obvious that the larger $Q_0' - Q_0$ is, the larger the Stokes shift will be. If the two parabolae have the same shape and vibrational frequency, it is possible to define a parameter S (the so-called Huang-Rhys parameter) as follows

$$\tfrac{1}{2}k(Q_0' - Q_0)^2 = S\hbar\omega \qquad (3)$$

where $\hbar\omega$ is the energy difference between the vibrational levels. The Stokes shift is then given by

$$\Delta E_S = k(Q_0' - Q)^2 - \hbar\omega = 2S\hbar\omega \qquad (4)$$

The parameter S measures the interaction between the dopant ion and the vibrating lattice. Equation 4 shows that if S is large, the Stokes shift is also large. Equation 3 shows that S is immediately related to the offset of the parabolae in the configurational coordinate diagram (fig.1). This offset, $\Delta Q (= Q_0' - Q)$, may vary considerably as a function of the dopant ion and as a function of the vibrating lattice, as we will see below.

It can be shown that the relative intensity of the zero-vibrational transition ($n_g = 0 \rightarrow n_e = 0$) is exp(-S) [5,6]. We can now divide our luminescent centres in three classes, viz.

(i) those with weak coupling, i.e. $S < 1$, so that the zero-vibrational transition dominates the spectrum,

(ii) those with intermediate coupling, i.e. $1 < S < 5$, so that the zero-vibrational transition is observable, but not the strongest line in the absorption or emission band,

(iii) those with strong coupling, i.e. $S > 5$, so that the zero-vibrational transition is so weak that it is not observable in the spectra. This case is also characterized by large Stokes shifts.

Figure 2 shows three emission spectra which are representative of each case. Characteristic examples of case (i) are the trivalent rare earth ions. The value of S is so small for these ions, that the spectra consist in good approximation of the zero-vibrational transitions only. Figure 2a gives as an example the emission spectra of the Eu^{3+} ion in LaB_3O_6 and of the Gd^{3+} ion in LaB_3O_6. The former consists of many transitions which are all purely electronic; the latter consists of one strong electronic line at about 310 nm, whereas the weak repetition at about 325 nm is a vibronic transition. Actually the energy difference between these two lines corresponds to the vibrational stretching frequency of the borate group in LaB_3O_6.

A characteristic example of case (ii) is the uranyl ion (UO_2^{2+}). The $n_e = o \longrightarrow n_g = 2$ line dominates in the spectrum (fig. 2b). The tungstate ion (WO_4^{2-}) is a good example of case (iii). The very broad emission spectrum (see fig. 2c) does not show any vibrational structure at all, the Stokes shift is very large (~ 16.000 cm^{-1}) and the zero-vibrational transition is not observable, not even at the lowest possible temperatures nor for the highest possible resolving powers.

Finally we draw attention to the fact that the single configurational coordinate diagram is only an approximation. In practice there is more than one vibrational mode involved and the system is not harmonic. Therefore the value of S is not so easy to determine as suggested above. However, for a general understanding the simple model is extremely useful as we will see below.

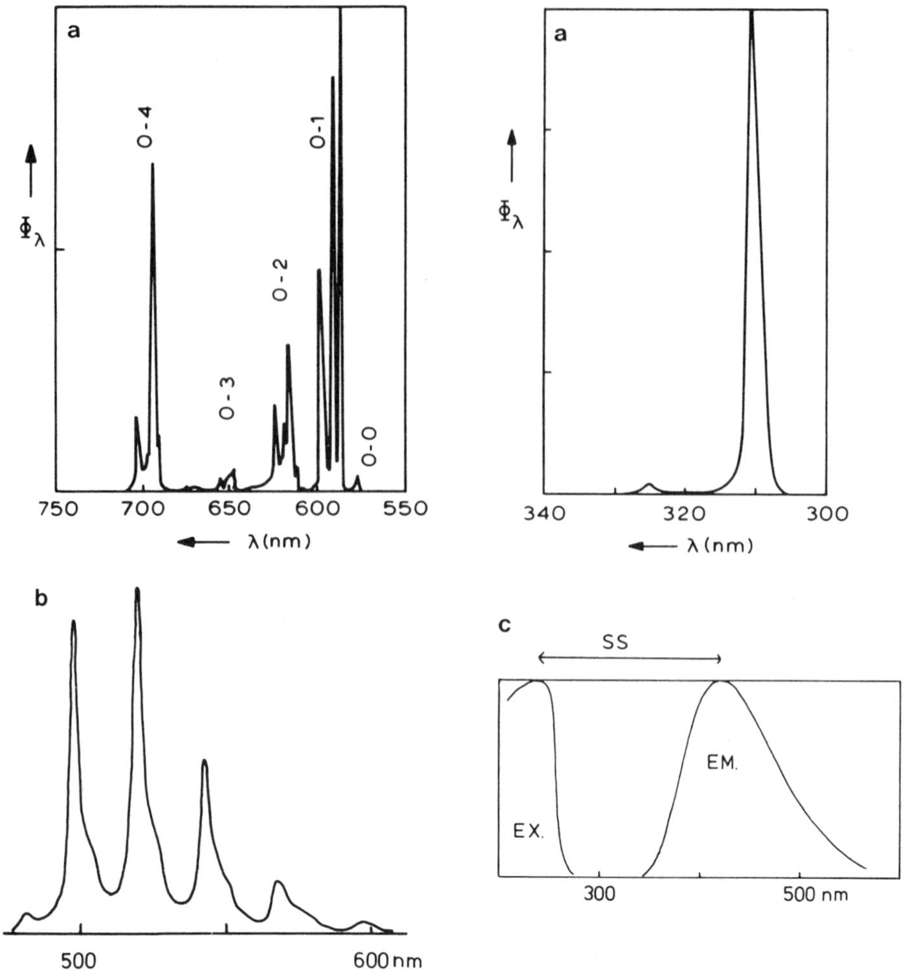

Fig. 2 Emission spectra for several coupling strengths.

a. Weak coupling strength (S<1). Emission spectrum of the Eu^{3+} ion in LaB_3O_6 (red spectral region; the notation J–J′ stands for the transition $^5D_J \rightarrow {}^7F_{J'}$) and of the Gd^{3+} ion in LaB_3O_6 (ultraviolet spectral region). In these, and all other figures of emission spectra below, Φ_λ denotes the spectral radiant power per constant wavelength interval in arbitrary units.

b. Intermediate coupling strength (1<S<5). Emission spectrum of the UO_2^{2+} ion in $SrZnP_2O_7$ at 300 K. Note the progression in the uranyl stretching vibration.

c. Strong coupling strength (S>5). Emission (EM) and excitation (EX) spectrum of the tungstate group in $CaWO_4$. SS indicated the Stokes shift.

Up to this point it was assumed that the return from the excited state to the ground state is radiative. This is usually not the case. Actually there are many centres which do not luminesce at all. Let us consider the configurational coordinate diagrams of fig. 3 in order to understand the relevant physical processes. Fig. 3a presents essentially the same as fig. 1. Absorption and emission transitions are well possible and Stokes shifted relative to each other. The relaxed-excited-state may, however, reach the crossing of the two parabolae if the temperature is high enough. Via the crossing it is possible to return to the ground state in a nonradiative manner. The excitation energy is then completely given up as heat to the lattice. This model accounts for the thermal quenching of luminescence (the latter being essentially a low-temperature phenomenon).

In fig. 3b the parabolae are parallel (S=0) and will never cross. It is impossible to reach the ground state in the way described for fig. 3a. However, also here nonradiative return to the ground state is possible if certain conditions are fulfilled, viz. the energy difference ΔE should be equal to or less than 4–5 times the higher vibrational frequency of the surroundings. In that case this amount of energy can excite simultaneously a few high-energy vibrations, and is then lost for the radiative process. Usually this nonradiative process is called multi-phonon emission.

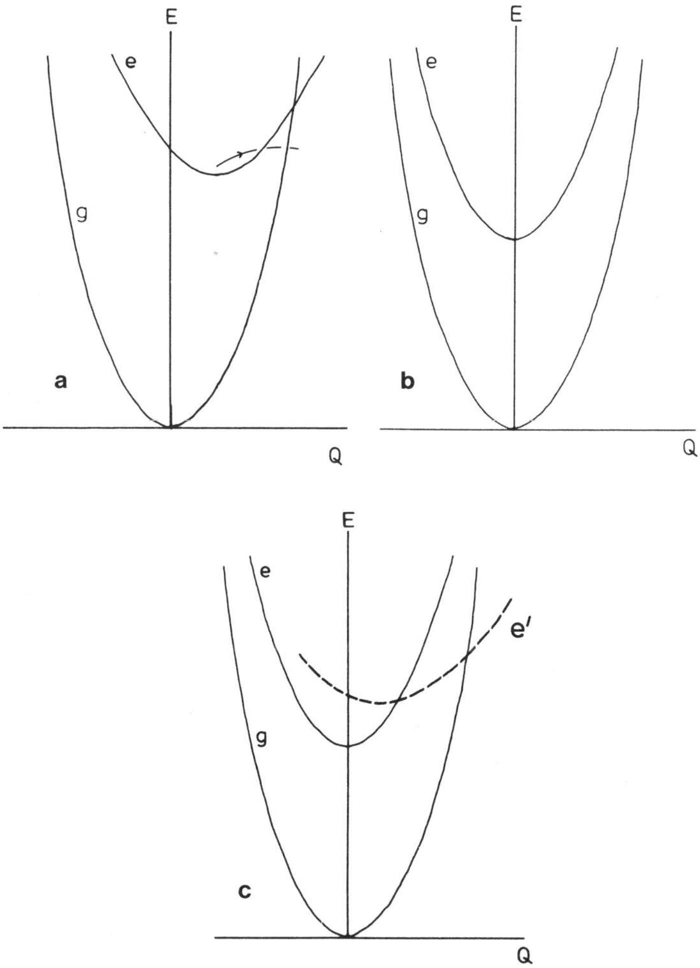

Fig. 3 Configurational coordinate diagram illustrating nonradiative transitions. g indicates the ground state parabolae, e and e′ the excited state parabolae. In fig. 3 the arrow indicates a nonradiative transition (thermally activated) from g to e. In fig. 3 excitation into e′ is followed by the emission transition e to g. See also text.

In fig. 3c both processes are possible in a three-parabolae-diagram. The parallel parabolae will belong to the same configuration, so that they are connected by forbidden optical transitions only. The third one originates from a different configuration and is probably connected to the ground state by an allowed transition. This situation occurs often. Excitation (absorption) occurs now from the ground state to the highest parabolae in the allowed transition. From here the system relaxes to the relaxed excited state of the second parabola. Fig. 3c shows that the nonradiative transition between the two upper parabolae is easy. Emission occurs now from the second parabola (line emission). This situation is found for $Al_2O_3:Cr^{3+}$ ($^4A_2 \longrightarrow {}^4T_2$ excitation, $^4T_2 \longrightarrow {}^2E$ relaxation, $^2E \longrightarrow {}^4A_2$ emission), Eu^{3+} ($^7F \longrightarrow$ charge-transfer-state excitation, charge-transfer-state to ^5D relaxation, $^5D \longrightarrow {}^7F$ emission), and Tb^{3+} ($^7F \longrightarrow 4f^75d$ excitation, $4f^75d \longrightarrow {}^5D$ relaxation, $^5D \longrightarrow {}^7F$ emission).

In general the temperature dependence of the nonradiative processes is reasonably well understood. However, the magnitude of the nonradiative rate is not, and cannot be calculated with any accuracy except in the weak-coupling case. The reason for this is that the temperature dependence stems from the phonon statistics which is known. However, the physical processes are not accurately known. Especially the deviation from parabolic behaviour in the configurational coordinate diagram (anharmonicity) may influence the nonradiative rate with many powers of ten [7]. However, it will be clear that the offset between the two parabolae (ΔQ) is a very important parameter for the nonradiative transition rate. This rate will increase dramatically if ΔQ becomes larger.

II.B. Energy transfer

If luminescent centres come together, they may show interaction with each other which results in new phenomena. Consider two centres, S and A, with a certain interaction. The relaxed-excited-state of S may transfer its energy to A. This energy transfer has been treated by Förster and Dexter and is nowadays well understood [8].

Dexter, following the classic work by Förster, considered energy transfer between a donor (or a sensitizer) S and an acceptor (or activator) A in a solid. This process occurs if the energy differences between the ground- and excited states of S and A are equal (resonance condition) and if a suitable interaction between both systems exists. The interaction may be either an exchange interaction (if we have wave function overlap) or an electric or magnetic multipolar interaction. In practice the resonance condition can be tested by considering the spectral overlap of the S emission and the A absorption spectra. The Dexter result looks as follows:

$$P_{12} = \frac{2\pi}{\hbar} \, |\langle 1,2^*|H_{SA}|1^*,2\rangle|^2 \int g_S(E) g_A(E) dE. \qquad (5)$$

Here the integral presents the spectral overlap, H_{SA} the interaction Hamiltonian and $|j\rangle$ and $|j^*\rangle$ are the electronic ground- and excited state functions, respectively, with $j = 1,2$. Here 1 refers to S and 2 to A. The distance dependence depends on the interaction mechanism.

A high transfer rate, i.e. a high value of P_{12}, requires a considerable amount of:
(i) resonance, i.e. the S emission band should overlap spectrally the A absorption band(s),
(ii) interaction, which may be of the multipole-multipole type or of the

exchange type. Only for some specific cases the interaction type is known nowadays.

Not always all of the excitation energy is transferred. If only part of it is transferred, this is called cross-relaxation [8].

This type of research is nowadays directed towards pairs of ions with a given distance. This can be done applying site-selective laser spectroscopy with high resolving power, so that ions with slightly different surroundings are excited selectively. An example is the study of LaF_3-Pr^{3+} by Vial and Buisson [9].

Up till here we considered energy transfer between two species S and A. If we consider transfer between two identical ions, for example between S and S, the same considerations can be used. If transfer between two S ions occurs with a high rate, what will happen in a lattice of S ions, for example in a compound of S? There is no reason why the transfer should be restricted to one step, so that we expect that the first transfer step is followed by many others. This can bring the excitation energy far from the site where the absorption took place: energy migration. If in this way the excitation energy reaches a site where it is lost nonradiatively (a killer or quenching site), the luminescence efficiency of that composition will be low. This phenomenon is called concentration quenching [1,8].

Energy migration in concentration systems has been an issue of research in the last decade. Especially since lasers became available, the progress has been impressive [10].

III. NONRADIATIVE PROCESSES IN ISOLATED LUMINESCENT SPECIES. THE WEAK-COUPLING CASE

Nonradiative transitions in the weak-coupling approximation are probably the best understood nonradiative processes. The experimental data relate mainly to the rare-earth ions, as far as their sharp-line transitions are considered (i.e. intra-$4f^n$-configuration transitions). The topic has been discussed in many books and review papers [1,3,10-13]. We summarize as follows:

For transitions between levels of a $4f^n$ configuration the temperature dependence of the nonradiative rate is given by

$$W(T) = W(0)(n + 1)^p, \qquad (6)$$

where W(T) is the rate at temperature T, $p = \Delta E/\hbar\omega$, ΔE the energy difference between the levels involved, and

$$n = \left[\exp(\hbar\omega/kT)-1\right]^{-1} \qquad (7)$$

W(0) is large for low p, i.e. for small ΔE or high vibrational frequencies. Further

$$W = \beta \exp\left[-(\Delta E-2\hbar\omega_{max})\alpha\right] , \qquad (8)$$

with α and β constants, and ω_{max} the highest available vibrational frequency of the surroundings of the rare earth ion. This is the energy-gap law in the revised form by Van Dijk and Schuurmans [13], which makes it possible to calculate W with an accuracy of one order of magnitude.

Let us illustrate this with some consequences. In aqueous solutions or in hydrates the rare earth ions do not emit efficiently with the exception of Gd^{3+} (ΔE=32.000 cm^{-1},$\omega_{max} \simeq$ 3.500 cm^{-1}). For Tb^{3+} ($\Delta E \simeq$ 15.000 cm^{-1}), and especially Eu^{3+} ($\Delta E \simeq$ 12.000 cm^{-1}), the quantum efficiencies are depressed, the other rare earth ions do practically not emit at all. For solids this can be nicely studied in the host lattice $NaLa(SO_4)_2H_2O$ where the rare earth site is coordinated to one H_2O molecule only. The q values are as follows:

Gd^{3+}	q = 100%
Tb^{3+}	70%
Eu^{3+}	10%
Sm^{3+}	~ 1%
Dy^{3+}	~ 1%

In the classic oxide glasses the rare-earth ions do not emit efficiently, since $\omega_{max} \simeq$ 1.000 – 1.200 cm^{-1} (silicates, borates, phosphates). Only Gd^{3+}, $Tb^{3+}$and Eu^{3+} show efficient luminescence. This situation changes drastically by using fluoride or chalcogenide glasses, where ω_{max} is considerably lower. A very interesting lattice in this connection is $Eu_2Mg_3(NO_3)_{12}\cdot 24H_2O$ [14]. At first sight the large amount of water molecules is expected to quench the Eu^{3+} emission completely. However, the Eu^{3+} ions are bidentately coordinated by six nitrate ions which shield them from the water molecules. The quantum efficiency is high.

Ions like Eu^{3+} and Tb^{3+} may emit from higher excited states: Eu^{3+} not only from $^5D_0$ (red), but also from $^5D_1$ (green) and $^5D_2$ (blue). However, this depends critically upon the host lattice. In Y_2O_3–Eu^{3+}, for example, all these emissions are observed, since $\omega_{max} \simeq$ 600 cm^{-1}. In borates and silicates, however, they are not.

This can be well studied by laser spectroscopy. An example is $NaGdTiO_4$:Eu^{3+} [15]. The time dependence of the Eu^{3+} emission in $NaGdTiO_4$ upon excitation into the $^5D_1$ level of Eu^{3+} is as follows: 10 μs after the excitation pulse the emission originates mainly from the $^5D_1$ level, but after longer time the $^5D_1$ intensity has decreased and that of $^5D_0$ increased. The decay curves of the $^5D_0$ emission show a build-up. From these data the rate for $^5D_1 \longrightarrow {}^5D_0$ decay is found to be 1.3×10^4 s^{-1} at 4.2 K. Its temperature dependence is given by $(n + 1)^p$ as argued above. The value of p turns out to be 5, the vibrational frequency involved being 347 cm^{-1}. This corresponds to the Eu–O stretching vibration. At 300 K this nonradiative rate is about 4×10^4 s^{-1}. The values of the nonradiative rate exceed that of the radiative $^5D_1$–^7F_J rate (~10^3 s^{-1}), so that the nonradiative process dominates and the emission occurs mainly from the $^5D_0$ level. In compounds with higher phonon frequencies available (e.g. borates, silicates), the $^5D_1$ emission of Eu^{3+} is usually hardly detectable, since the value of p is then much lower.

The Tb^{3+} ion may not only emit from $^5D_4$ (green), but also from $^5D_3$ (blue). ΔE is about 5000 cm^{-1}, much larger than in the case of Eu^{3+}. Diluted Tb^{3+} systems show, therefore, always some blue Tb^{3+} emission, unless ω_{max} is very high.

We will now pay some attention to an ion which is not studied frequently, viz. Gd^{3+} ($4f^7$). Its energy level scheme is given in fig. 4. The excited levels are in the ultraviolet and the corresponding transitions have low oscillator strengths. As a consequence accurate spectroscopy can only be performed with ultraviolet tunable lasers and/or x-ray excitation (for emission). The ground level is orbitally nondegenerate ($^8S_{7/2}$) which is an advantage for several reasons.

The emission transition $^6P_J \rightarrow {}^8S$ occurs over an energy gap which is about 32.000 cm^{-1}. Nonradiative transitions cannot compete with this radiative one because it occurs over such a large ΔE. Even water molecules ($\nu \sim 3500$ cm^{-1}) are not able to quench the Gd^{3+} emission [16,17,18]. The emission can only be quenched by transfer to other luminescent centres [17].

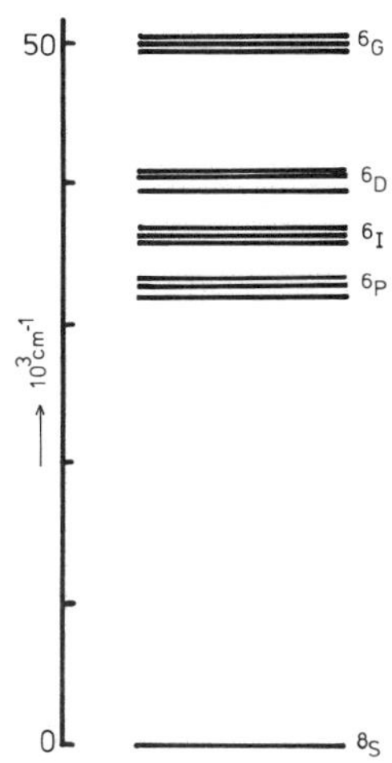

Fig. 4 Energy level scheme of the Gd^{3+} ($4f^7$) ion.

In some host lattices emission has also been observed from the higher excited levels ^6I_J, ^6D_J, and even ^6G_J. However, this depends strongly on the host lattice. If the maximum vibrational frequency is low, and the host lattice transparant, emission can be observed from all these levels, up to the ^6G_J level (at about 205 nm). There is even evidence for a higher level with 186 nm emission. Fig. 5 gives an example. This has been observed for GdF_3, $BaGdF_5$, and even for $GdPO_4$ [19]. In the case of borates and hydrates, however, all these emissions are quenched in favour of the ^6P_J emission. This is a clear demonstration that higher-frequency vibrations promote the radiationless transitions to the ^6P_J levels.

There is in the literature only one example of a quantitative treatment of these processes, viz. for $LiYF_4$–Gd^{3+} [20]. The temperature dependence of the $^6I_J \rightarrow {}^6P_J$ nonradiative transition was determined by measuring the decay time of the $^6I_j \rightarrow {}^8S$ emission transition, and could be fitted to the formulae given above. Imbusch et al. [21] were able to improve the fit by taking into account the occupation of higher ^6I_J levels at higher temperatures (fig. 6).

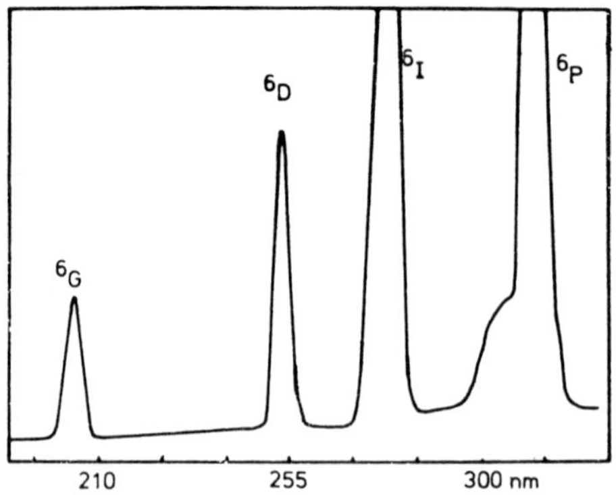

Fig. 5 The x-ray excited emission spectrum of GdF_3 at room temperature. The notation ^6G, ^6D, ^6I and ^6P stands for the transitions $^6G_{7/2} \rightarrow {}^8S$, $^6D_{9/2} \rightarrow {}^8S$, $^6I_{9/2} \rightarrow {}^8S$ and $^6P_{7/2} \rightarrow {}^8S$.

A very special type of a nonradiative process is the occurrence of vibronic emission lines on the lower energy side of the $^6P_{7/2} - {}^8S$ emission line. Fig. 7 gives several examples. Part of the emission energy is used to excite vibrations in the surroundings of the Gd^{3+} ion, i.e. it is lost nonradiatively. In $Cs_2NaGdCl_6$ [22] and MgF_2:Gd [23] the

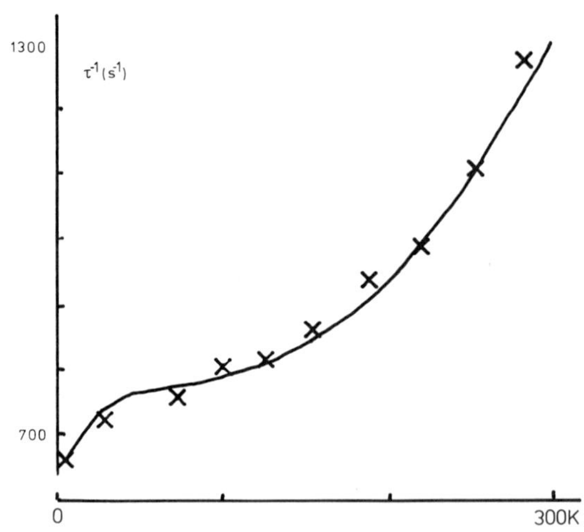

Fig. 6 Temperature dependence of the nonradiative $^6I_J \rightarrow {}^6P_J$ transition rate. Experimental points and best fit using the expressions in the text. The compound under study is $LiYF_4$–Gd^{3+}. The measurements are from A.J. de Vries, the fit from G.F. Imbusch.

vibronic intensity is about equal to that of the electronic line, in YOCl–Gd [24,25] and $Gd_2(SO_4)_3 \cdot 8H_2O$ [26] it is at least 10%.

In the former two the Gd^{3+} ion occupies a centre-of-symmetry site, i.e. the $^6P_{7/2} \rightarrow {}^8S$ electronic transition is only allowed as magnetic-dipole transition. In the latter two there is no inversion symmetry and the electronic line contains also a large electric-dipole contribution. In all cases the vibronic intenstity is about equal to that of the magnetic-dipole transition.

It is not the place here to review the vibronic spectra of the rare-earth ions which is a complicated topic anyhow [27,28]. However, we draw attention to a few facts, viz.

(i) in some cases the vibronic lines may be unexpectedly intense [29].

(ii) the strongest vibronic lines accompany always the hypersensitive electronic lines, i.e. those which are determined by the Judd-Ofelt parameter Ω_2 [30].

(iii) in some cases, however, vibronics are also observed on transitions determined by Ω_6 where they present Franck-Condon replicas [25]. This has been observed in the oxyhalides, for example LaOCl-Gd.

(iv) the vibronics contain information about the surroundings and the coupling strength to the surroundings. In $NaGd(SO_4)_2\ H_2O$ the water vibronic to the $^6P_{7/2} - {}^8S$ emission line is much stronger than in $Gd_2(SO_4)_3.8H_2O$, even after correction for the number of water molecules. This points to a stronger coupling in the former sulfate which can be related to the weaker hydrogen bond in that compound.

Another nonradiative process occurs in the process of anti-Stokes emission observed for Gd^{3+}. For example, if the $^6P_{7/2}$ level is excited, one may observe emission from the higher levels [31]. The emission intensity increases quadratically with laser excitation density, indicating a two-photon process. The excited $^6P_{7/2}$ level which lives relatively long absorbs another photon which brings the ion rather high up in energy. From that level (the nature of which is unknown) nonradiative decay to lower levels occurs resulting in higher-level emission.

A complete different spectroscopy is two-photon absorption spectroscopy of Gd^{3+} without an intervening level. This, however, does not contain a nonradiative step, so that we refer the reader to the literature [32,33,34].

IV. NONRADIATIVE PROCESSES IN ISOLATED LUMINESCENT SPECIES. THE STRONG-COUPLING CASE

This section starts by illustrating how important the value of the parameter offset is for the nonradiative transition rate. We will use several examples from different fields of chemistry for this purpose.

First we consider $CaWO_4$, a well-known x-ray phosphor for seventy years. The luminescent group is the tungstate group [1], an outspoken example of a centre for which the strong-coupling scheme holds (see also fig. 2c). $CaWO_4$ is a very efficient luminescent material at room temperature. The isomorphous $SrWO_4$, however, does not emit at 300 K, but had to be cooled down in order to reach a reasonable quantum efficiency. Also $BaWO_4$ has the same crystal structure, but even at 4.2 K it does not emit with high efficiency. Nevertheless the ground state properties of the tungstate group in these compounds (distances, vibrational frequencies) are practically equal. The strongly different radiationless

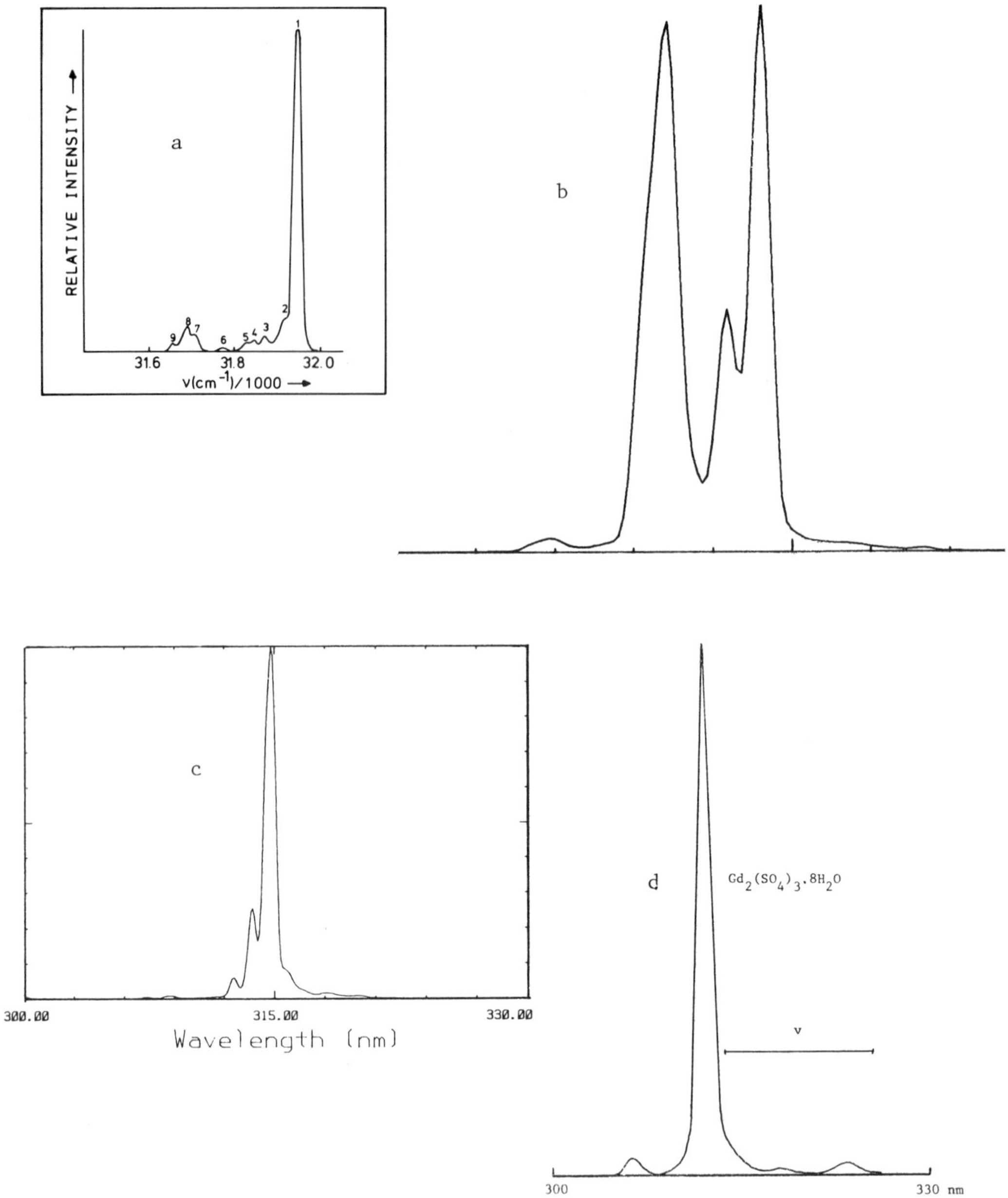

Fig. 7 Emission spectra of the Gd^{3+} ion in several host lattices showing the vibronic sidelines. Measurements at room temperature. The spectra refer to (a) $Cs_2NaGdCl_6$, (b) MgF_2:Gd, (c) LaOCl:Gd, and (d) $Gd_2(SO_4)_3.8H_2O$. Spectra (b) and (d) by courtesy of Dr. L.H. Brixner.

processes have to be ascribed to a difference in ΔQ, i.e. the parabolae offset. Since the ionic radii of Ca^{2+}, Sr^{2+}, Ba^{2+} decrease in that order, it seems obvious to assume that this is the reason why the offset increases, i.e. why the rate of the radiationless processes increases, as observed experimentally. The softer the surroundings, the larger is ΔQ.

Table II. Thermal quenching of the uranate luminescence of ordered perovskites A_2BWO_6–U^{6+}. Compare also fig. 8. Data from ref. 35.

A_2BWO_6–U		T_q(K)[1]	ΔQ(a.u.)[2]
A=Ba	B=Ba	180	10.9
Ba	Sr	240	10.6
Ba	Ca	310	10.2
Ba	Mg	350	10.0
Sr	Mg	350	10.0
Ca	Mg	350	10.0

1. Quenching temperature of the uranate luminescence.
2. ΔQ in arbitrary units, calculated by the Struck and Fonger method. ΔQ is put 10.0 for Ba_2MgWO_6.

There is a more impressive experiment to prove this simple model, viz. the luminescence in the ordered perovskites A_2BWO_6 where A and B are alkaline earth ions. Table II presents the quenching temperatures of the luminescence of the UO_6 group in these lattices [35]. Those for the WO_6 group run parallel [36]. These temperatures are used as a measure of the radiationless properties. The table shows that the radiationless rate does not depend on the nature of A, whereas that of B determines the value of this rate: the smaller the B ion, the higher the quenching temperature.

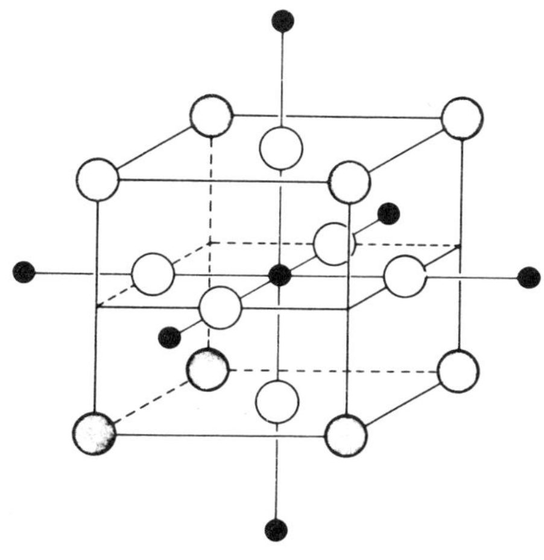

Fig. 8 The ordered perovskite structure of $A_2BW(U)O_6$.open circles: oxygen; black circles: U(W) (centre) and B; grey circles: A. Compare table II.

Table III. Stokes shift (10^3 cm^{-1}) of the band emission of some trivalent ions in the orthoborates MBO_3 (M=Sc, Y, La) After ref. 39.

	$ScBO_3$	YBO_3*	$LaBO_3$
$Ce^{3+}(4f^1)$	1.2	2.0	2.4
$Pr^{3+}(4f^2)$	1.5	1.8	3.0
$Sb^{3+}(5s^2)$	7.9	{ 14.5	19.5
		16.0	
$Bi^{3+}(6s^2)$	1.8	{ 5.1	9.3
		7.7	

*This lattice contains two sites for Y.

Fig. 8 shows that an expansion of the luminescent UO_6 (or WO_6) octahedron (i.e. the parabolae offset) is not directly counteracted by the A ion. However, the B ions are immediately involved, the angle U(W)–O–B being 180°. Table II shows also relative values of ΔQ calculated from the Struck and Fonger model [37]. It scales indeed according to predictions. Note that the total change in ΔQ is less than 10%. Since ΔQ is less than 0.1 Å for the uranate group [38], the variation in ΔQ in this series of compounds is less than 0.01 Å. This shows that small changes in ΔQ result in drastic changes in the nonradiative rates.

It is well known that luminescent materials with high quantum efficiencies and quenching temperatures have usually stiff lattices, so that expansion upon excitation is counteracted, i.e. ΔQ is as small as possible.

Table III shows for a series of borates how the Stokes shift, i.e. ΔQ, increases if the size of the host lattice cation increases [39]. In $ScBO_3$ the rare earth ions are strongly compressed and the surroundings are stiff. Small Stokes shifts result for Ce^{3+}, Pr^{3+} and Bi^{3+}, but not for the much smaller Sb^{3+}. Note, however, that the Stokes shift of the 4f–5d transitions is less sensitive to the surroundings than that of the 5s–5p transitions. If the data of table II are extrapolated to, for example, borate glasses, it can be excluded that we find efficient Sb^{3+} or Bi^{3+} emission, but for Ce^{3+} or Pr^{3+} this may still be the case. This is what has been observed experimentally [40].
Part of solid state chemistry is nowadays involved with what is called soft chemistry or soft materials. As a matter of fact these are not expected to luminesce, at least not when the luminescent centres are broad-band emitters. This has been shown to be the case, for example, for the isomorphous $Al_2(WO_4)_3$, $Sc_2(WO_4)_3$ and $Zr_2(PO_4)_2SO_4$. The Stokes shift of the tungstate and zirconate luminescence in these materials is enormous, viz. some 2 eV. The quantum efficiencies, even at 4.2 K are low [41]. The exact structural explanation has been discussed in the literature [41].

Another interesting phenomenon is the change in luminescence properties if one passes the fluid-liquid transition in a 4:1

ethanol-methanol solution of [(bpy)Re(CO)$_3$Cl] [42]. The absorption transition is not influenced, but the emission is at longer wavelengths in the liquid, i.e. the Stokes shift is larger (see fig. 9). This means that ΔQ increases if we go from the fluid to the less rigid liquid state. Simultaneously the quantum efficiency of the luminescence decreases.

This model in which the nonradiative transitions can be suppressed by a stiff surroundings, can be most elegantly tested by studying the luminescence of rare-earth cryptates [18,43-45].

The cryptand ligands are organic cages. They were synthesized for the first time by Lehn who obtained the Nobel prize for Chemistry in 1987 for this achievement (together with Cram and Pedersen [46]). Fig. 10 gives two examples. The 2.2.1 cryptand is just large enough to contain the Ce^{3+}ion, i.e. upon excitation the Ce^{3+} ion has not much space to expand. In fact the $[Ce \subset 2.2.1]^{3+}$ cryptate shows an efficient (broad-band) emission at room temperature with a small Stokes shift, in the solid state as well as in aqueous solution [11]. This is illustrated in fig. 11. The $[Ce \subset 2.2.2]^{3+}$ cryptate luminescence has a much larger Stokes shift. As a matter of fact the 2.2.2 cryptand offers a larger hole than the 2.2.1 cryptand.

Table IV shows the Stokes shift of the Ce^{3+} emission in several surroundings. In the 2.2.1 cryptand the Ce^{3+} Stokes shift is smaller than in some commercial Ce^{3+}–activated phosphors (Y_2SiO_5–Ce, Ca_2AlSiO_7–Ce). It becomes very small in $ScBO_3$ (see above) and in CaF_2 and $CaSO_4$ where it carries an effectively positive charge which will make the Ca-site smaller than it is on basis of the Ca^{2+} ionic radius.

In passing we note that the cryptands find nowadays an important application in the field of medical diagnostics which is also based upon their ability to suppress nonradiative transitions, viz. fluorescence immuno-assay. In order to detect biomolecules of a special nature, antibodies are used which are labelled with a fluorescent ion, for example Eu^{3+} [43], and which react specifically with the biomolecule of interest. The Eu^{3+} luminescence shows then where these biomolecules are residing. However, since investigations of this kind are performed in aqueous media, the Eu^{3+} luminescence is quenched as described in the

Table IV. The Stokes shift of the Ce^{3+} emission for several compositions.

Composition	Stokes shift (cm^{-1})
$[Ce^{3+} \subset 2.2.1]$	2100
$[Ce^{3+} \subset 2.2.2]$	4000
Ce^{3+} in aqueous solution	5000
$Y_3Al_5O_{12}$–Ce^{3+}	3800
Y_2SiO_5–Ce^{3+}	2500
$ScBO_3$–Ce^{3+}	1200

precious sector. To circumvent this problem, the Eu^{3+} ion is bound to an ultraviolet-absorbing cryptand which transfers its absorbed energy to the Eu^{3+} ion, and which shields the Eu^{3+} ion from the water molecules. By this shielding the radiationless processes in the weak-coupling scheme are depressed considerably. For this purpose a cryptand bpy.bpy.bpy is used [46]. The properties of $[Eu \subset bpy.bpy.bpy]^{3+}$ and $[Tb \subset bpy.bpy.bpy]^{3+}$ cryptate have been described in the literature [43,47].

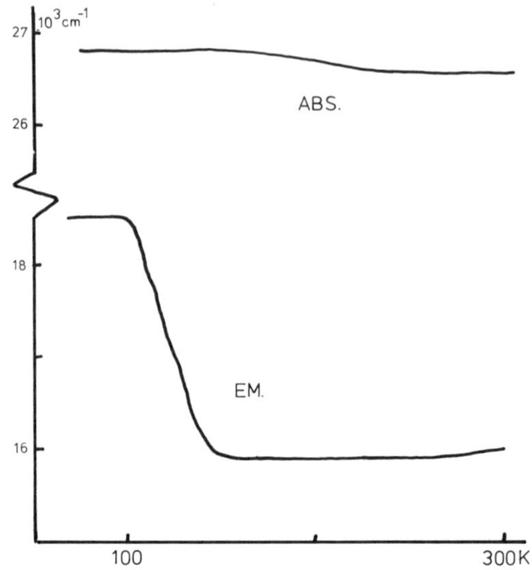

Fig. 9 The temperature dependence of the absorption and emission maxima of $[(bpy)Re(CO_3)Cl]$ in a 4:1 ethanol-methanol solution. After data from ref. 42.

Another clear example of a more complicated nature is the charge-transfer excitation of the Eu^{3+} luminescence, a process which is of large importance for applications. Consider the red phosphor in the three-clour fluorescent lamps (see also chapter by Dr. Smets). Its composition is Y_2O_3:Eu. Excitation at 254 nm in the charge-transfer state is followed by efficient red emission ($^5D_0$ – $^7F_2$) wither the $4f^6$ configuration. Fig. 12 shows the relevant configurational -coordinate diagram.

Note that the useful properties of Y_2O_3–Eu^{3+} are based on a fast radiationless process, viz. the transition from the charge-transfer state to the excited levels of the $4f^6$ configuration. For crystalline GdB_3O_6–Eu^{3+} the same model holds. This composition can also be obtained as a glass. Interestingly enough, charge-transfer excitation in the glass results in Eu^{3+} luminescence with a quantum efficiency which is an order of magnitude smaller than in the crystalline modification [48]. This is also the case at 4.2 K. This observation has been ascribed to a larger offset of the charge-transfer parabola in the glass than in the crystal, so that in the glass the charge-transfer state empties mainly into the ^7F ground-state manifold. Similar observations have been made by Oversluizen [45]. It is quite conceivable that the glass surroundings can counteract the expansion upon excitation less than the crystalline surroundings. The important consequence of this is that broad-band emission in glasses will have low efficiencies, unless the Stokes shift is small (see above).

A comparable situation, but now exclusively in crystalline lattices, occurs with the transition from the 4f5d to the $4f^2$ configuration of Pr^{3+} [50]. Also this phenomenon is of large technical importance. Fig. 13 gives the configurational coordinate diagram of Pr^{3+} in two different host lattices. If the offset of the 4f5d state is small, radiative return to the $4f^2$ configuration has a higher probability than the nonradiative transition to the $4f^2$ configuration. If the offset is large, 4f5d excitation leads to emission from the $4f^2$ configuration after a nonradiative 4f5d $\longrightarrow$ $4f^2$ transition. The former situation is encountered for YBO_3 [51], YOCl [52] and LaB_3O_6 [53], the latter for the apatite $Gd_{9.33}(SiO_4)_6O_2$ [50] and Gd_2O_2S [54].

The Pr^{3+} case has an advantage above that of Eu^{3+}, viz. the higher excited state can emit, so that the Stokes shift can be measured. This gives information on the relaxation and the parabolae offset. It was found that the nonradiative 4f5d $\longrightarrow$ $4f^2$ transition becomes of importance if the Stokes shift is larger than 3000 cm^{-1}.

The Pr^{3+} ion can be used successfully as a sensitizer in luminescent materials based on gadolinium compounds [50]. However, this is only possible if the 4f5d state of Pr^{3+} decays radiatively. Otherwise the nonradiative return to the $4f^2$ configuration occurs more rapidly than the Pb^{3+} $\longrightarrow$ Gd^{3+} transfer. Fig. 14 gives a schematical overview of the processes involved in such luminescent materials.

Our simple model predicts also that, if the parabolae offset is constant, the nonradiative transition will have a higher rate if the excited state is at lower energy. Or, in other words, broad excitation bands at low energies will usually imply low quantum efficiencies. Let us give a few clear examples. First we consider some organic Cu(I) complexes (see table V) [55]. Absorption and emission occurs here in broad metal-to-liqand charge-transfer transitions. The rule stated above is well substantiated. This implies that it is easier to find efficient blue broad-band emitters than red broad-band emitters, a well-known experience of many phosphor investigators.

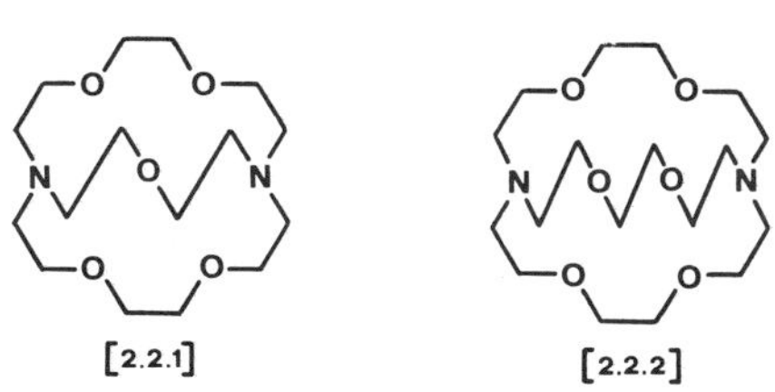

Fig. 10 The 2.2.1 and 2.2.2 cryptands

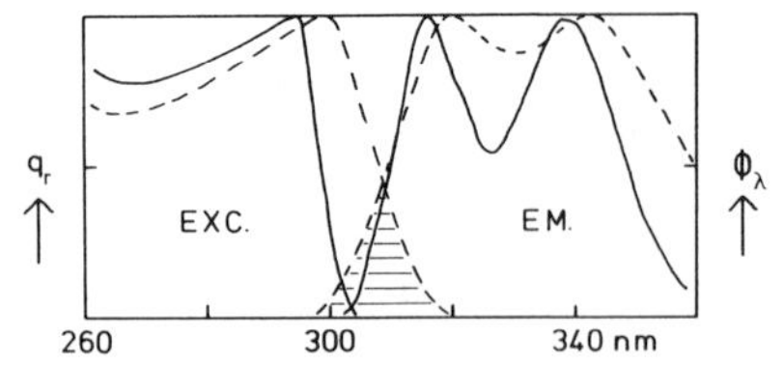

Fig. 11 Emission and excitation spectra of the luminescence of [Ce ⊂ 2.2.1]$^{3+}$. Drawn lines at 4.2 K, broken lines at 300 K. The spectral overlap (hatched area) is much larger at 300 K which makes energy migration possible.

Similarly, if the charge-transfer state of Eu^{3+} shifts to low energies, the quantum efficiency of the red emission upon broad-band excitation drops considerably (18). The composition LaOI—Eu^{3+} has this band at 28.000 cm^{-1} and a quantum efficiency of only 1% [56].

The more or less isoelectronic systems $CaWO_4$ and $CaMoO_4$ show broad band emission with maxima at 410 nm and 530 nm, respectively. Whereas $CaWO_4$ is very efficient at room temperature, $CaMoO_4$ is partly quenched [57].

We would now like to consider a few examples of medium or strong coupling which have been treated more quantitatively. Considerable progress has been made in understanding the excited state of transition-metal ion complexes with an empty d shell. Examples are

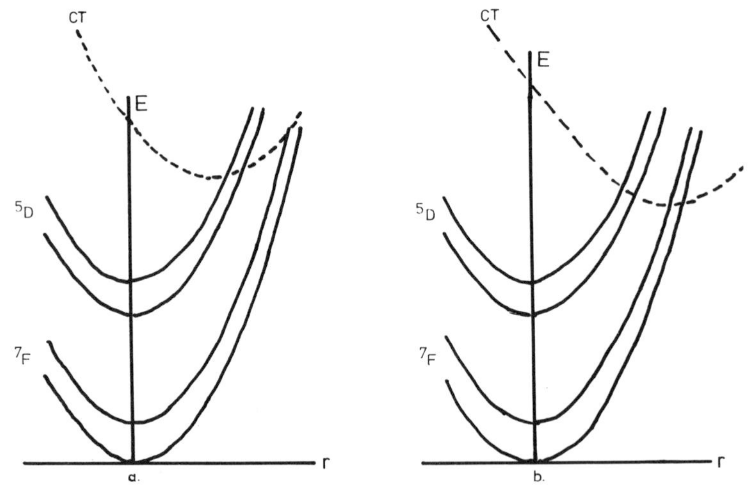

Fig. 12 Configurational coordinate diagram for Eu^{3+} in Y_2O_3(a) and in a borate glass (b). In the glass the charge-transfer state has such a large offset that excitation into this state results mainly in feeding of the ground-state ^7F manifold.

CrO_4^{2-}, VO_4^{3-}, and MoO_4^{2-}. Especially Van der Waals' group has contributed considerably [58-61]. The emission of these species is usually structureless with a large Stokes shift (strong-coupling case).

It has been realized since long that the transitions involved are of the charge-transfer type, and that the emitting state is a triplet state [62]. This follows from the long decay times observed. Further characterization is hardly possible due to the absence of any structure in the spectra. By performing EPR measurements in the excited state Van der Waals et al. were able to demonstrate directly the triplet character of this excited state.

Let us first consider $CaMoO_4$ [60]. In the ground state there is one MoO_4^{2-} group with S_4 site symmetry in the unit cell. However, in the excited state there are four magnetically equivalent MoO_4^{2-} groups in the unit cell. These show a strong trigonal distortion due to a static Jahn-Teller effect. This distortion can occur along any of the four Mo—O bonds. This means that the configurational coordinate diagram should take a Jahn-Teller active mode in account (probably the τ_2 mode). It is interesting to note that the interactions entering the problem decrease in the order: Jahn-Teller coupling > crystal field > spin-orbit coupling.

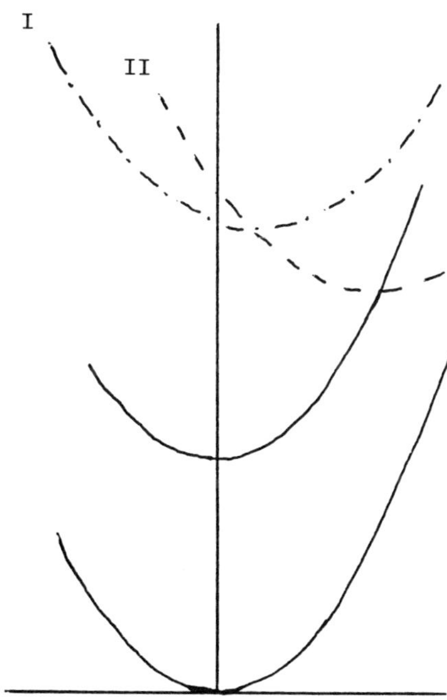

Fig. 13 Schematic configurational coordinate diagram for $Pr^{3+}(4f^2)$. Drawn lines relate to two levels of the $4f^2$ configuration; broken lines to two possible situations for the 4f5d configuration. Excitation into I yields d → f emission from I, excitation into II (with larger offset) results in relaxation to the $4f^2$ configuration and possible emission from that configuration.

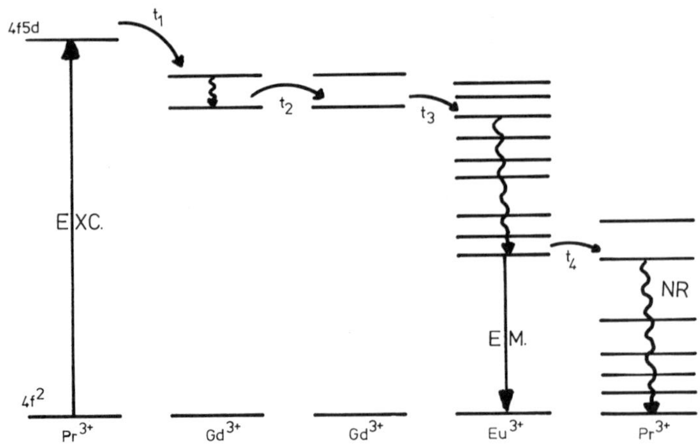

Fig. 14. The energy level scheme for $GdBO_3$–Pr^{3+}, Tb^{3+} with the relevant transfer processes.

Table V. Luminescence properties of Cu^+ complexes

Complex ion[a]	Absorption maximum (nm) in solution	Emission maximum (nm)	Quantum efficieny at 300 K	Decay time (μs)	
		300 K		4.2 K	300 K
$Cu(PPh_3)_2phen^+$	365	515	0.75	250	30
$Cu(PPh_3)_2dmp^+$	365	515	0.75	250	30
$Cu(PPh_2CH_3)_2dmp^+$	365	525	0.75	250	30
$Cu(PPh_3)_2bpy^+$	365	560	0.15	30	7
$Cu(PPh_3)_2biq^+$	415	605	0.17	30	7
$Cu(dmp)_2^+$	454	650	≈0.01	3	–
$Cu(dpp)_2^+$	439	≈710	<0.01	–	–
$Cu(biq)_2^+$	545	–	"0"	–	–

[a] PPh_3: triphenylphosphine; phen: 1,10-phenanthroline; dmp: 2,9-dimethyl-1,10-phenanthroline; $PPh_2^2CH_3$: methyldiphenyl-phosphine; bpy: 2,2'-bipyridine; biq: 2,2'-biquinoline; dpp: 2,9-diphenyl-1,10-phenanthroline

Very similar results have been observed for YVO_4 (with a VO_4^{3-} group). In $Ba_3(VO_4)_2$ the site symmetry is C_{3v}. Therefore only one or three excited species are expected. The latter possibility turned out to be the case. This confirms the model sketched.

The Stokes shift is determined by factors discussed above. There we had still ν_1 in mind as the more important mode. However, for the other modes the arguments run parallel. It is interesting that this way of reasoning predicts the three excited species for $Ba_3(VO_4)_2$, since the O_1^{2-} ion along the c axis is counteracted by a Ba^{2+} ion, but the three equivalent O_2^{2-} ions are not (fig. 15).

The importance of the Jahn-Teller effect in the excited state has also been demonstrated for other ions. First we mention the case of $Sb^{3+}(5s^2)$ en YPO_4 [63]. The relaxed excited state (^3P) shows two different minima (see fig. 16). This is due to the Jahn-Teller effect. At low temperatures the X minimum, which is populated by optical excitation, emits. At higher temperatures the barrier can be overcome and emission from the T minimum is observed too. At still higher temperatures thermal equilibrium occurs and the X emission reappears (see fig. 17). For the isoelectronic Te^{4+} ion ($5s^2$) the importance of the Jahn-Teller effect is demonstrated in another way. Fig. 18 shows the emission spectrum with an impressive vibrational progression in the Jahn-Teller active ν_2 vibration [64]. This shows also that the use of the ion- surroundings distance in the simple models sketched above is an oversimplification.

Another case where the influence of the Jahn-Teller effect has been demonstrated is the Cr^{3+} ion. Here we mention some aspects of impressive work by Güdel et al. [65,66]. They studied Cr^{3+} in the elpasolite structure where it has an undistorted octahdral coordination. The selected host lattices were $Cs_2NaInCl_6$, Cs_2NaYCl_6 and Cs_2NaYBr_6. Since the crystal field on the Cr^{3+} is weak, the emission is a broad-band $^4T_2 \longrightarrow {}^4A_2$ transition. This emission band gives at helium temperatures beautiful vibrational structure (fig. 19). The dominant intensity-giving vibrations are the bending modes t_{2u} and t_{1u}. On the false origins progressions are build in the e_g and a_{1g} mode. The occurrence of the e_g progression in a clear prove of a Jahn-Teller effect in the excited $^4T_2$ state. In several ways the authors derived values for the Huang-Rhys

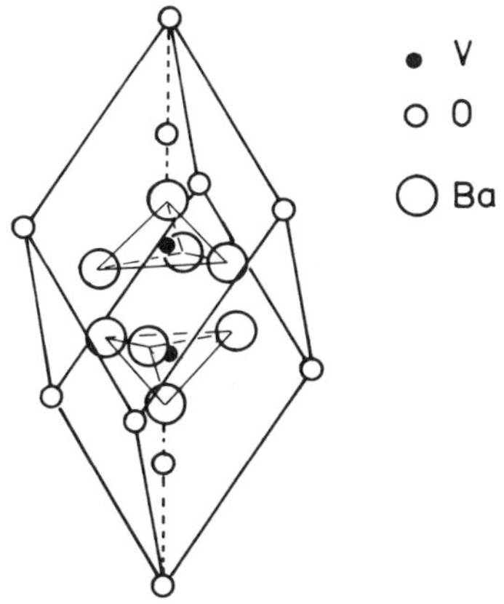

Fig. 15. The crystal structure of $Ba_3(VO_4)_2$. See discussion in the text. The oxygen ions along the vertical body diagonal are in between the vanadium ion and a barium ion.

factors and the parabolae offsets ΔQ (see table VI). In this way they were able to find the changes in metal-ligand bond lengths in the excited state relative to those in the ground state. Table VII gives these changes which are due to the combined effect of the a_{1g} and e_g displacements and which show that the excited CrX_6^{3-} octahedron is tetragonally distorted in the excited state. On the time scale of the experiments there is no indication of a static distortion, i.e. we are dealing with the dynamic Jahn-Teller effect.

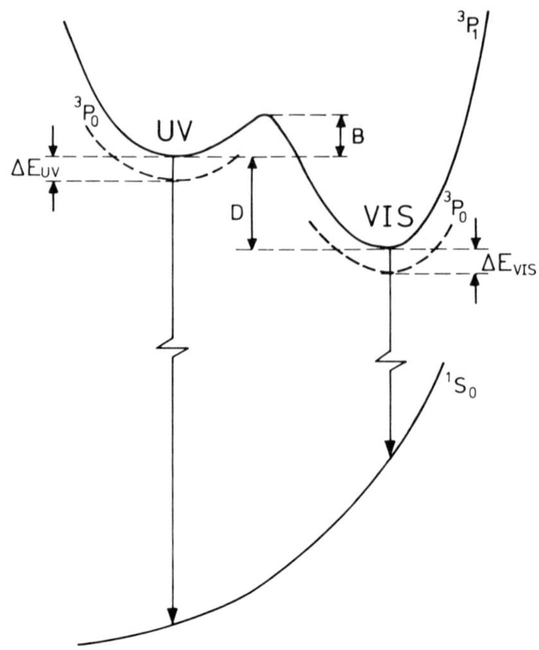

Fig. 16 Schematic configurational coordinate diagram for $LnPO_4:Sb^{3+}$ (Ln = Y,Sc,La). Note the two types of minima on the $^3P_1$ potential energy curve.

Table VI. Values of the Huang-Rhys factors (S) and the parabolae offset (ΔQ) for Cr^{3+} in several elpasolites (after ref. 65).

Host lattice	$S_{a_{1g}}$ (1)	S_{e_g} (2)	$S_{tot.}$ (3)	$\Delta Q_{a_{1g}}$ (Å)	ΔQ_{e_g} (Å)
$Cs_2NaInCl_6$	1.6	1.1	3.7	0.13	−0.13
Cs_2NaYCl_6	1.6	1.7	3.6	0.13	−0.15
Cs_2NaYBr_6			6.7	0.22	−0.22

(1) from intensity ratio in a_{1g} progression
(2) from intensity ratio in e_g progression
(3) from Stokes shift of the emission

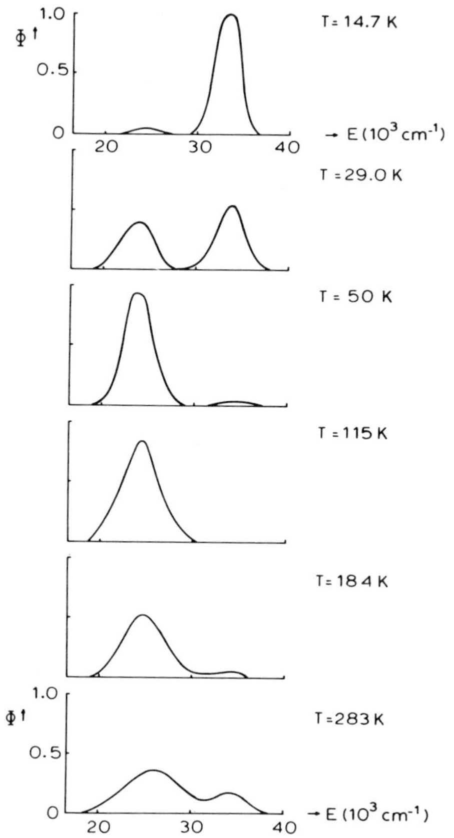

Fig. 17 Emission spectra of $YPO_4:Sb^{3+}$ as a function of temperature. After E.W.J.L. Oomen, thesis, Utrecht, 1987.

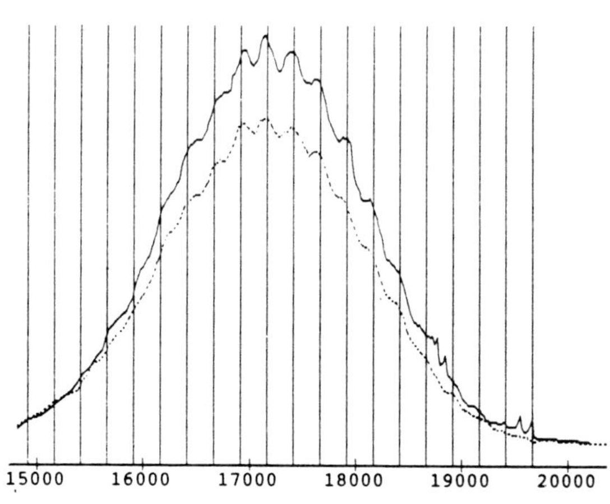

Fig. 18 Emission spectrum of Cs_2ZrCl_6–Te^{4+} showing an extended vibrational progression in the E_g mode. After H. Donker, thesis, Utrecht, 1989.

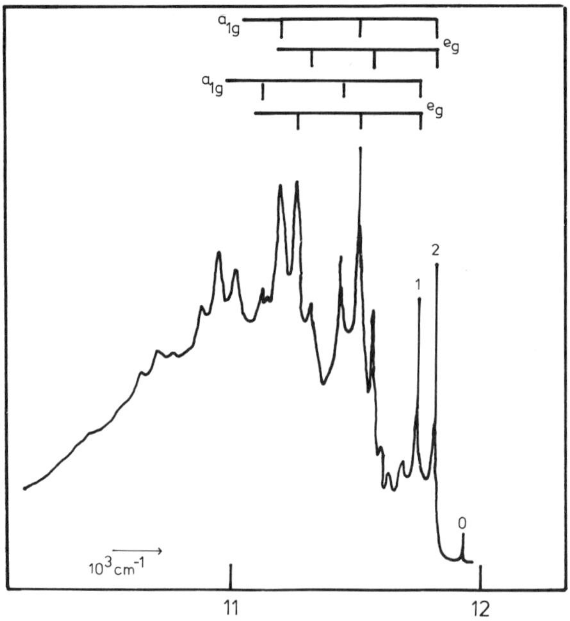

Fig. 19 The emission spectrum of $Cs_2NaInCl_6:Cr^{3+}$ showing the progression in the a_{1g} and e_g modes. Temperature 10 K. Line 0 is the electronic origin, lines 1 and 2 are the false origins due to coupling with t_{1u} and t_{2u}, respectively. After data from ref. 65.

In this way the authors had detailed information available in order to calculate the nonradiative $^4T_2 \longrightarrow {}^4A_2$ transition. They did this in several ways. Not successful were a single-coordinate Struck-Fonger model and an energy-gap law model. More successful were an activation-

Table VII. Distortions of the CrX_6^{3-} octahedron in the excited $^4T_2$ state with respect to the ground state (after ref. 65)

Host lattice	Δx,y (Å)	Δz (Å)
$Cs_2NaInCl_6$	0.09	−0.02
Cs_2NaYCl_6	0.10	−0.03
Cs_2NaYBr_6	0.13	−0.05

energy model and the Lauer-Fong model. It was assumed that the a_{1g} and e_g modes are the accepting modes because of the fact that only these two modes show a considerable parabolae offset.

In a similar way they studied the V^{3+} ion ($3d^2$) in the elpasolite structure. By performing several types of experiments they were able to derive a configurational coordinate diagram for V^{3+} in Cs_2NaYCl_6 and Cs_2NaYBr_6 [66]. Here it is of interest that they are able to fit the temperature dependence of the nonradiative transition rate to the formula [67]

$$P_{nr} = P_{nr}(0)\ \exp(-2nS)\ p!\ S^{-p}\left(\frac{n+1}{n}\right)^{p/2} . I_p\left(2S\left\{n(n+1)\right\}^{\frac{1}{2}}\right), \qquad (8)$$

where I_p is the modified Bessel function of order p, n the Bose-Einstein population for phonons of energy $\hbar\omega_{eff.}$, and p the number of phonons neccessary to bridge the energy gap involved. A good fit could only be obtained by assuming that the effective mode of distortion has contributions from the a_{1g} and the e_g mode.

Although Güdel et al. [65,66] present a satisfying complete picture of the d^3 ions in the elpasolite structure, it must be mentioned, as an illustration of the state-of-the-art, that Bartram et al. [68] came to different conclusions for Cr^{3+} in Cs_2NaYCl_6, K_2NaScF_6 and K_2NaGaF_6. They claim the best agreement for a two-configurational-coordinate model with linear coupling to an e_{1g} mode and quadratic coupling to a t_{2g} mode. For the latter there is no direct spectral evidence.

Imbusch et al. [69] have applied successfully the same formula for the temperature dependence of the nonradiative rate in $Y_3Al_5O_{12}$:Mn^{4+}($3d^3$), where the emission is due to the $^2E \longrightarrow {}^4A_2$ transition. The intensity of the vibronic sidebands is such that the weak-coupling model cannot be expected to work quantitatively. The Huang-Rhys factor S is found to be 0.35. With p=51 and $\hbar\omega$=300 cm^{-1} (which is far from the highest possible frequency) a good fit to the luminescence quenching was obtained. It should be stressed that these evaluations of the temperature dependence of the nonradiative rates do not even contain an estimation of the absolute rate which is still an impossible task. The temperature dependence contains the statistics, not the physics of the process.

Recently a new method to study the relaxation around photo-excited centres in solids was reported by Strauss and Walder [70]. The authors observed the effect of the spatial relaxation of the matrix around the excited centre in a condensed phase on the bulk optical index of refraction and use this information to derive the change in the equilibrium distance (i.e. ΔQ). For details the reader is referred to ref. 70. The experiments were performed on Mn^{2+} in water, i.e. on the aqua-complex $[Mn(H_2O)_6]^{2+}$. No luminescence was observed. Since the experimental life time was 4.6 μs, and the radiative life time of the order of milliseconds, the return to the ground state is completely dominated by nonradiative processes.

The Mn^{2+} centre was found to shrink upon excitation which is not unexpected, since the ground state is $^6A_{1g}$ ($t_{2g}{}^3e_g{}^2$) and the excited state $^4T_{1g}$ ($t_{2g}{}^4e_g$). On the other hand Cr^{3+} (in GdScGa-garnet) expands. The volume change in the case of $[Mn(H_2O)_6]^{2+}$ is $\Delta V = -7$ Å^3. This is solely due to the a_{1g} mode, because this is the mode which changes the volume of the complex. It is not a straight forward procedure to derive from this value of ΔV the corresponding value of ΔQ. The authors arrive at a value of -0.14 Å, which seems a very high value. The problem is which area belongs to the considered complex. If one take the average of the ionic radii of Mn^{2+} in the high-spin ($t_{2g}{}^3e_g{}^2$) and the low-spin state (t_2^5) we arrive at a $\Delta Q = -0.08$ Å. It is striking that these two completely different approaches yield values of the same sign and comparable value.

An appropriate way to close this section is the radiationless transition which may occur by electron transfer, because it certainly belongs to the strong-coupling scheme. A rare-earth ion may lose its excitation energy by energy transfer, but also by electron transfer to other centres [71]. The problem can be nicely illustrated on a molecular species, viz. the decatungstates $[RE.W_{10}O_{36}]^{9-}$ [72].

In the case of $EuW_{10}O_{36}^{9-}$ the Eu^{3+} ions shows luminescence with high quantum efficiency. In view of the arguments given above, $TbW_{10}O_{36}^{9-}$ and $GdW_{10}O_{36}^{9-}$ are expected to show even more efficient rare-earth ion emission. However, this is not the case. $GdW_{10}O_{36}^{9-}$ does not show any rare-earth emission at all , but only tungstate emission. The excited Gd^{3+} ion transfers its energy to the tungstate group from which emission occurs. More problematic is the fact that $TbW_{10}O_{36}^{9-}$ does not show efficient Tb^{3+} emission, nor efficient tungstate emission. Obviously the complex as a whole undergoes a nonradiative decay to the ground state which is not covered by the Förster-Dexter theory. This is quenching by <u>electron</u> transfer, in solution a well-known phenomenon [73] which in non-molecular solids is less general and has often been overlooked.

The quenching process occurs via an excited state which has a parabola in the configuration-coordinate diagram with a large offset relative to the ground-state parabola. In the case of the Eu^{3+} intraion decay this excited state was the (ligand-to-metal) CT state. However, also a metal-to-metal CT state can induce quenching if it is at low enough energy. It is, therefore, expected to be of importance for a complex species consisting of a metal ion which is easily oxidized and another metal ion which is easily reduced. The complex $TbW_{10}O_{36}^{9-}$ is an example, because the excited charge distribution Tb^{4+}–W^{5+} will not be at very high energy relative to the ground-state distribution Tb^{3+}–W^{6+}. This type of charge-transfer transitions has an analogy to the so-called outer-sphere charge-transfer transitions within ion pairs in solution [74]. Figure 20 shows schematically this quenching process. It has been observed for many ion pairs in solution, e.g. the quenching of the luminescence of [$Eu^{3+} \subset$ 2.2.1] by $M(CN)_6^{4-}$, where M = Fe, Os or Ru [75,76].

It has been little realized, however, that these processes occur also in non-molecular solids. The quenching of all luminescence in YVO_4–Tb^{3+} is a striking example. The excited state, responsible for the quenching, has the approximate charge distribution V^{4+}– Tb^{4+}. It is interesting to note that $YTaO_4$–Tb^{3+} is a luminescent material with a very high quantum efficiency, whereas $YNbO_4$–Tb^{3+} is in between the vanadate and the tantalate. This comparison shows that the rate of this quenching process decreases if the metal-to metal CT state shifts to higher energy (the 5th ionization potential of the host lattice metal increases in the sequence Ta, Nb, V).

The same arguments hold for Ce^{3+} and Pr^{3+}. The species $PrW_{10}O_{36}^{9-}$ only weakly luminesce due to this quenching process [72].

Sometimes solids offer special conditions to overcome this quenching process. The combinations Tb^{3+}–VO_4^{3-}, and even Ce^{3+}–VO_4^{3-} show efficient luminescence, and energy transfer according to the Förster-Dexter theory, in the host lattice $CaSO_4$ [77]. In this lattice there are pairs, e.g. Tb^{3+} on a Ca^{2+} site and V^{5+} on a neighbouring S^{6+} site. Since Tb^{3+} now carries an effectively positive charge and V^{5+} an effectively negative charge, the metal-to-metal charge transfer is relatively unfavourable and lies at a high enough energy to allow the radiative process.

Even couples of rare-earth ions can be quenched by this process, viz. Ce^{3+} and Eu^{3+} which quench each other's luminescence [70]. Here the metal-to metal CT state has Ce^{4+}–Eu^{2+} character.

Electron transfer does not necessarily lead to quenching of all possible luminescence. An interesting case is $Ba_6B_2W_3O_{18}$–Nd^{3+} (B=Gd, Lu, Y) reported recently [79]. The authors observed an absorption band at about 27.000 cm^{-1} which they ascribed to charge-transfer to oxygen. Comparison with the data for Eu^{3+}, Tb^{3+} and Pr^{3+} in same host lattice as given in ref. 37 excludes such an assignment. We ascribe this absorption band to electron transfer in the Nd^{3+}–W^{6+} complex. From this excited state the system returns to lower states. Since efficient Nd^{3+} emission was observed at ~ 10000 cm^{-1} from the $^4F_{5/2}$ level of Nd^{3+}, the electron-transfer state feeds obviously the $^4F_{5/2}$ level of Nd^{3+}. Since the absorption is strong, the resulting emission is also intense. In Fig. 20 the processes in $Ba_6B_2W_3O_{18}$–Nd^{3+} can be visualized as follows. Excitation: a $\longrightarrow$ d, relaxation: d $\longrightarrow$ b, emission: b $\longrightarrow$ a. The luminescence of the higher states is quenched, but that between lower levels is still possible.

V. ENERGY TRANSFER AND MIGRATION

In section II.B. we considered energy transfer between two species S and A. If we consider now transfer between two identical ions, for example between S and S, the same considerations can be used. If transfer between two S ions occurs with a high rate, what will happen in a lattice of S ions, for example in a compound of S? There is no reason why the transfer should be restricted to one step, so that we expect that the first transfer step is followed by many others. This can bring the excitation energy far from the site where the absorption took place: energy migration. If in this way the excitation energy reaches a site where it is lost nonradiatively (a killer or quenching site), the luminescence efficiency of that composition will be low. This phenomenon is called concentration quenching. This type of quenching will not occur at low concentrations, because then the average distance between the S ion is so large that the migration is hampered and the killers not reached.

Energy migration in concentrated systems has been an issue of research in the last decade. Especially since lasers became easily available, the progress has been great. Here we will first consider the case that S is an ion to which the weak-coupling scheme applies. In practice this case consists of the trivalent rare-earth ions. Subsequently we will deal with the case where S is an ion to which the intermediate- or strong-coupling scheme applies.

V.A. Weak-coupling scheme ions

At first sight transfer between identical rare-earth ions seems to be a process with a low rate, because their interaction will be weak in view of the well-shielded character of the 4f electrons. This is in fact true, and transfer is restricted to short distances, i.e. of the order of magnitude of 10 Å or less. However, not only are the radiative rates also small, but the spectral overlap can be expected to be large. This

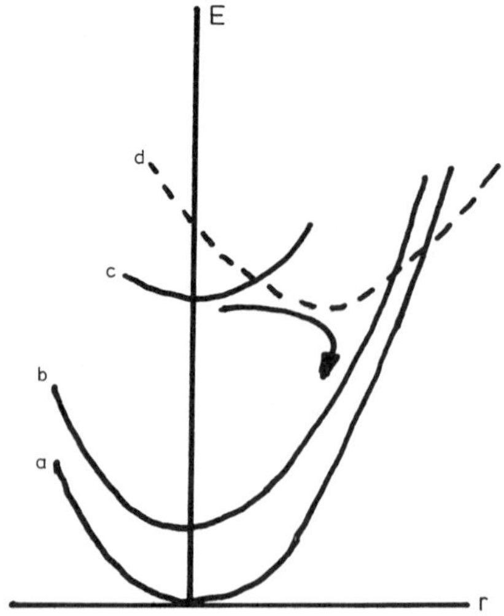

Fig. 20 Schematic representation of quenching by electron transfer. Curves a, b and c are states of the couple (A.B), state d is a charge-transfer state ($A^+.B^-$).

originates from the fact that $\Delta Q = 0$, so that the absorption and emission lines will coincide. In fact energy migration has been observed in many rare earth compounds, and concentration quenching becomes usually effective for concentrations of a few atomic percent of dopant ions. As an example we mention the transfer rate between Eu^{3+} ions and between Gd^{3+} ions which may be of the order of 10^7 s^{-1} if the distance is 4 Å or shorter. This has to be compared with a radiative rate of 10^2–10^3 s^{-1}. Consequently the excitation energy may be transferred more than 10^4 times during its life time.

This type of research uses pulsed and tunable lasers as an excitation source. The rare earth ion is excited selectively and its decay is analysed. The shape of the decay curve is characteristic of the physical processes in the compound under study [17]. For a detailed review the reader is referred to the literature [10]. Here we give some results for specific situations. We assume that the object of our study consists of a compound of a rare earth ion S which contains also some ions A which are able to trap the migrating S excitation energy by SA transfer.

First we consider some Eu^{3+} compounds. In $EuAl_3B_4O_{12}$ there is a three-dimensional Eu^{3+} sublattice with shortest Eu–Eu distance equal to 5.9 Å. At 4.2 K there is no energy migration, but at 300 K it occurs. The temperature dependence of the Eu–Eu energy transfer is exponential with $\Delta E \sim 240$ cm^{-1} [80]. This is due to the fact that the $^5D_0 - {}^7F_0$ transition is forbidden under the relevant site symmetry (D_3), so that the multipole interactions vanish. The distance of 5.9 Å is prohibitive for transfer by exchange interaction. At higher temperatures the $^7F_1$ level is thermally populated and multipolar interaction becomes effective. The experimental value of ΔE corresponds to the energy difference $^7F_0 - {}^7F_1$. The excitation energy makes 1400 jumps during its life time at 300 K with a diffusion length of 230 Å. In table VIII we have tabulated values of the transfer rate and the diffusion constant, together with those for other Eu^{3+}compounds [17].

Table VIII. Energy migration characteristics in some Eu^{3+} compounds at 300 K.

Compound	Shortest Eu–Eu distance	Diffusion constant ($cm^2\ s^{-1}$)	Hopping tie (s)	Dimensionality
$EuAl_3B_4O_{12}$	5.9 Å	8×10^{-10}	8×10^{-7}	3
$NaEuTiO_4$	3.7 Å	2×10^{-8}(a)	2×10^{-8}	2
$EuMgB_5O_{10}$	4.0 Å	$\sim10^{-8}$	$\sim10^{-7}$	1
$Eu_2Ti_2O_7$	3.7 Å	9×10^{-12}(b) 3×10^{-9} (c)	3×10^{-5}(b) 8×10^{-8}(c)	3
$Li_6Eu(BO_3)_3$	3.9 Å	2×10^{-9}	$\sim10^{-7}$	1
EuOCl	3.7 Å	5.8×10^{-10}	4×10^{-7}	2

(a) $D = 8\times10^{-11}\ cm^2\ s^{-1}$ at 1.2 K.
(b) values at 15 K.
(c) values at 43 K.

Samples of $EuAl_3B_4O_{12}$ which are so pure that the excited state does not reach a killer site during its life time, show efficient luminescence. Samples with a certain concentration of killer sites do not emit at 300 K. For example, crystals of $EuAl_3B_4O_{12}$, grown from a K_2SO_4/MoO_3 flux, do not luminescence, since the excitation energy is trapped by the Mo^{3+} impurities (concentration 25 ppm). The migration brings all of the excitation energy to a nonradiative sink. At 4.2 K the migration is hampered, and luminescence occurs.

Two-dimensional energy migration has been observed for $NaEuTiO_4$ [81] and $EuMgAl_{11}O_{19}$ [82]. One-dimensional energy migration was observed for $EuMgB_5O_{10}$ [83] and $Li_6Eu(BO_3)_3$ [84]. This fits the crystallographic data and was derived from the decay curve of the intrinsic Eu^{3+} emission.

In EuP_3O_9 a transition from one- to three-dimensional migration was observed, because this compound shows a phase transition [85].

The situation in Eu^{3+} compounds can be characterized as follows:
– for Eu–Eu distances larger than 5 Å exchange interaction becomes ineffective. Only multipolar ineractions can be of importance, and they are weak anyhow. If sufficiently pure, compounds in this class will luminesce efficiently. Examples are $EuAl_3B_4O_{12}$, $Eu(IO_3)_3$ and $CsEuW_2O_8$.
– for Eu–Eu distances smaller than 5 Å, exchange interaction is effective. Examples are the intrachain migration in $EuMgB_5O_{10}$ and $Li_6Eu(BO_3)_3$, and the migration in Eu_2O_3 [87].

Recently there has been a lot of interest in energy migration in Gd^{3+} compounds, because this opens interesting possibilities to obtain new, efficient luminescent materials [1,17]. The Gd^{3+} sublattice is sensitized and activated. The sensitizer absorbs efficiently ultraviolet radiation and transfers this to the Gd^{3+} sublattice. By energy migration in this sublattice the activator is fed and emission results. Absorption and quantum efficiencies of over 90% have been attained. The physical processes can be schematically presented as follows:

$$\xrightarrow{exc.} S \longrightarrow Gd^{3+} \xrightarrow{nx} Gd^{3+} \longrightarrow A \xrightarrow{emission}$$

A suitable choice of S is Ce^{3+}, Bi^{3+}, Pr^{3+} or Pb^{2+}. For A there are many possibilities: Sm^{3+}, Eu^{3+}, Tb^{3+}, Dy^{3+}, Mn^{2+}, UO_6^{6-}, and probably many more.

Here we consider one example more in detail, viz. $NaGdF_4$:Ce,Tb [4]. Figs 21 and 22 show the emission and excitation spectrum of interest. Fig. 21 shows that Ce^{3+} excitation results in predominantly Tb^{3+} emission. This is confirmed by the excitation spectrum of fig. 22. In Fig. 23 we give the emission spectrum of $NaYF_4$:Ce, Tb with the same dopant concentrations. There is practically no $Ce^{3+} \longrightarrow Tb^{3+}$ transfer. In this figure the Gd^{3+} levels are indicated by arrows, and it is seen that the spectral overlap is perfect. In fact the $Ce^{3+} \longrightarrow Gd^{3+}$ transfer has a high probability. Table IX gives some of the rates of importance in this system. Fig. 24 indicates the several transfer and radiative rates.

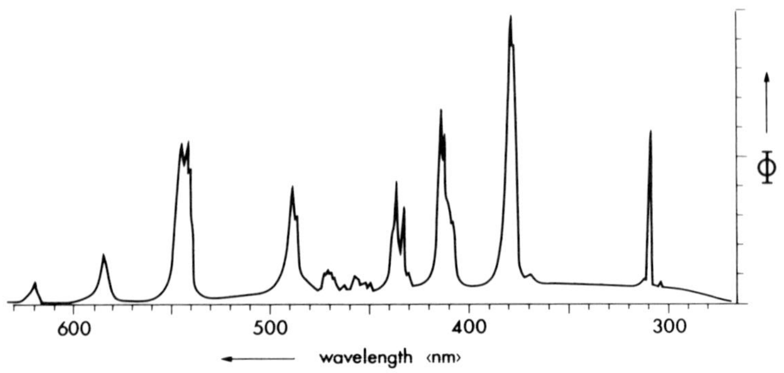

Fig. 21 The emission spectrum of $NaGdF_4$: Ce,Tb;300 K. Excitation of Ce^{3+} yields mainly Tb^{3+} energy migration.

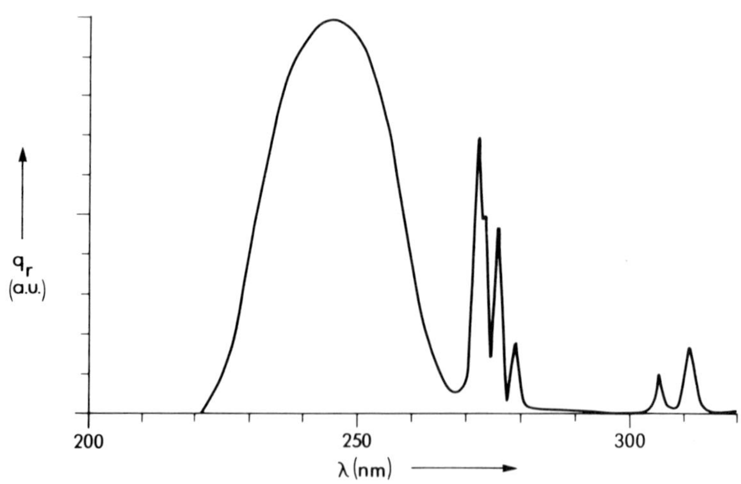

Fig. 22 The excitation spectrum of the Tb^{3+} emission of $NaGdF_4$:Ce,Tb at 300 K. Note the strong Ce^{3+} excitation band at 245 nm.

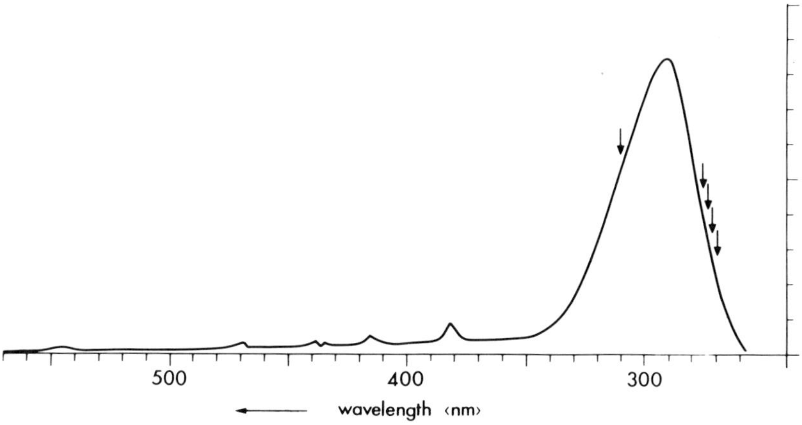

Fig. 23 The emission spectrum of $NaYF_4$:Ce,Tb at 300 K. Excitation of Ce^{3+} yields mainly Ce^{3+} emission. Arrows indicate the position of the Gd^{3+} levels.

An interesting aspect of energy migration in concentrated rare-earth compounds is the influence of the magnetic order which occurs in some cases on the migration process. This has been studied on compounds of Gd^{3+} and Tb^{3+} by Jacquier et al. [87–89]. As an example we mention $GdAlO_3$ and $TbAlO_3$ which become antiferromagnetic at 3.9 K and 3.8 K, respectively. In the paramagnetic phase the experimental decays are in agreement with fast diffusion energy transfer. Below the Néel temperature, however, the decays are no longer exponential and considerably slower. The migration has become diffusion limited. The diffusion constants reported are 1.6×10^{-9} cm^2 s^{-1} at 4.4 K for both compounds, but only 8×10^{-12} cm^2 s^{-1} and 8×10^{-14} cm^2 s^{-1} at 1.6 K for $GdAlO_3$ and $TbAlO_3$, respectively.

Table IX. Energy transfer and radiative probabilities in the system $NaY_{0.98-x}Gd_xCe_{0.01}Tb_{0.01}F_4$ at room temperature. Excitation is into the Ce^{3+} ion. The transfer rates are given for nearest neighbour distances (after ref. 4)

Probability*	Value in $NaGdF_4$
P_1	$(4\pm1)\ 10^7\ s^{-1}$
P_2	$10^8\ s^{-1}$
P_3	$130\ s^{-1}$
P_4	$10^{7\pm1}\ s^{-1}$
P_5	0
P_6	$4.\ 10^5\ s^{-1}$
P_7	$250\ s^{-1}$

*cf fig. 24

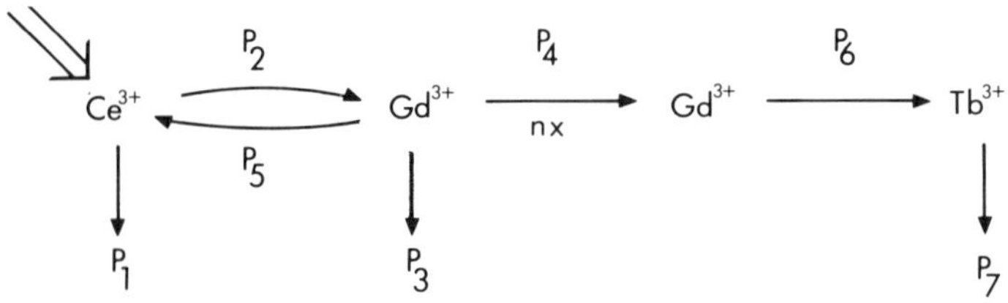

Fig. 24 Transfer and radiative processes in $NaGdF_4$:Ce,Tb. Figs. 21-24 are from H.S. Kiliaan, thesis, Utrecht, 1988.

In the antiferromagnetic phase the migration of excitation energy is slowed down, because nearest neighbour Gd^{3+}(or Tb^{3+}) ions are oriented antiparallel, which makes energy transfer by exchange interaction impossible.

In $EuAlO_3$ such an effect does not occur. The Eu^{3+} ions (ground state $^7F_0$) do not carry a magnetic moment. This compound is another example of a Eu^{3+} compound in which energy migration occurs down to the lowest temperatures [87] due to exchange interaction [90].

Let us now turn to energy migration in concentrated systems for which the weak-coupling scheme is no longer valid.

V.B. Strong-coupling scheme ions

In compounds of ions which belong to the strong-coupling scheme the situation is completely different from that in rare-earth compounds [1]. In the former, excitation of the luminescent centre results in strong relaxation of the excited state. This brings the excited state out of resonance with the neighbours in the ground state, and as a consequence energy transfer becomes impossible, especially at low temperatures. In the expression for the transfer rate the spectral overlap vanishes, so that the transfer rate becomes very low. In practice this can be measured by the Stokes shift: if the Stokes shift is large, the transfer rate will be low; if the Stokes shift is small, the transfer rate has a value which can compete with the radiative rate. It should be kept in mind that an increasing Stokes shift implies simultaneously an increase of the nonradiative rate for return to the ground state, as argued above.

Let us start this section with some examples of luminescent compounds of ions for which the strong-coupling scheme is clearly valid. These are compounds like vanadates and tungstates [62]. They contain oxidic anions with a central metal ion which has lost all its d electrons. Examples are VO_4^{3-}, MoO_4^{2-}, WO_4^{2-} and WO_6^{6-}. The Stokes shift of their emission is large (~15 .000 cm^{-1}). An example of the type of spectra was given in fig. 2c. Energy migration is hampered, often even up to room temperature. A well-known, more-or-less classic example is $CaWO_4$, which is an efficient x-ray phosphor since 70 years. Also the new x-ray phosphors $YTaO_4$ and $LuTaO_4$ belong to this class [91]. In these compounds the tungstate or tantalate group shows luminescence upon excitation. There is no migration of the excitation energy over the host lattice. However, as mentioned in the introduction, the quantum efficiency of the luminescence does not reach 100%, not even at 4.2 K, because the large parabolae offset (ΔQ) results in a certain amount of nonradiative loss which occurs without thermal activation.

In other compounds energy migration occurs at room temperature, for example YVO_4 and Ba_2MgWO_6. They have a smaller Stokes shift, viz. some 10.000 cm^{-1}. It is this property which makes YVO_4–Eu^{3+} an efficient phosphor at room temperature (but not at low temperatures): excitation into the vanadate group is followed by energy migration to Eu^{3+} centres.

There are other luminescent centres, like Ce^{3+} and Bi^{3+}, which belong either to the strong-coupling scheme or to the weak(er)-coupling scheme, depending on the host lattice. This can be seen immediately from the Stokes shift. The Bi^{3+} ion is a very impressive example, because the Stokes shift of its luminescence is more dependent on the surroundings than that of any other luminescent ion (see table X). Figure 25 shows the large difference between the spectral shape of the emission bands of Bi^{3+} in several compounds. The strong-coupling case (e.g. $Bi_2Al_4O_9$) shows broad band emission, the weak-coupling case (e.g. $NaYO_2$ –Bi^{3+}) narrow band emission with vibrational structure at low temperatures.

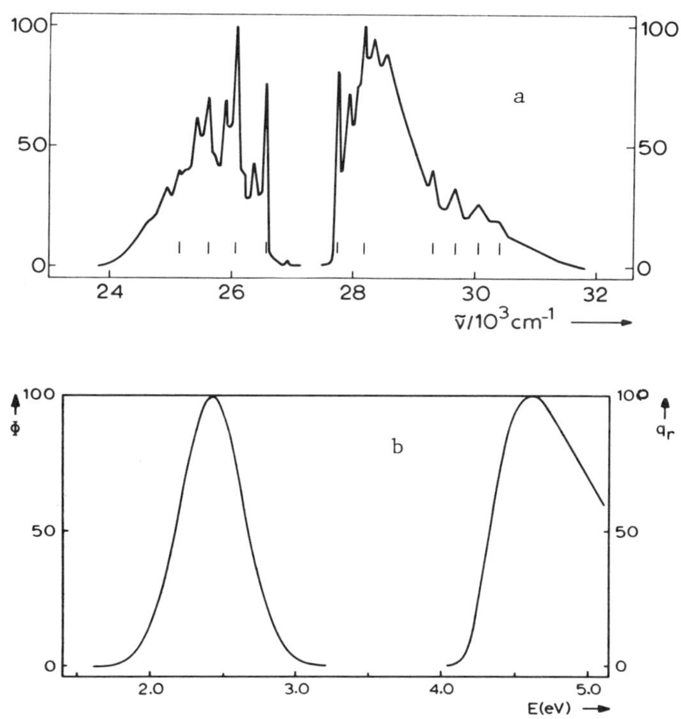

Fig. 25 Emission and excitation spectra of $NaYO_2$-Bi (a) and $Bi_2Al_4O_9$ (b) at low temperatures.

Table X. Stokes shift of the Bi^{3+} luminescence at 300 K in some compositions and compounds (after ref. 17). See also text.

Composition	Stokes shift (cm^{-1})	Concentrated compound	Luminescence of compound at 300 K
Cs_2NaYCl_6:Bi	800	$Cs_2NaBiCl_6$	–
$ScBO_3$:Bi	1.600	–	
$YAl_3B_4O_{12}$:Bi	2.700	–	
$LaBO_3$:Bi	8.500	–	
LaOCl:Bi	8.500	BiOCl	–
La_2O_3:Bi	10.800	Bi_2O_3	–
–	16.000	$Bi_2Al_4O_9$	+
–	17.600	$Bi_4Ge_3O_{12}$	+
–	20.000	$Bi_2Ge_3O_9$	–
$LaPO_4$:Bi	19.200	–	

*– (practically) no emission
+ efficient emission

The outspoken dependence of the Stokes shift on the host lattice is related to the electronic configuration of the ground state of the Bi^{3+} ion [1,17]. This will not be discussed here. For large Stokes shifts the nonradiative transition rates start to compete the radiative ones: $LaPO_4$–Bi is quenched at about 100K. It is illustrative to consider the luminescence of the bismuth compounds. If the Stokes shift is relatively small, the concentrated system is quenched. Concentration quenching in the systems $Cs_2NaY_{1-x}Bi_xCl_6$ and $Y_{1-x}Bi_xAl_3B_4O_{12}$, for example, occurs at 300K for $x < 0.01$. The Bi^{3+}–Bi^{3+} energy transfer occurs over large distances of several tens Å. If the Stokes shift is larger, efficiently emitting bismuth compounds result. An ouspoken example is $Bi_4Ge_3O_{12}$. Although chemically completely different from $CaWO_4$, the physics of their luminescence is very similar: in both cases an enormous reorganisation occurs after excitation. Actually, $Bi_4Ge_3O_{12}$ was first proposed as an x-ray phosphor, but nowadays it is applied in the L3 experiment at CERN in Geneva which needs 12.000 large single crystals of $Bi_4Ge_3O_{12}$.

Table X shows also that, if the Stokes shift becomes too large ($Bi_2Ge_3O_9$, 20.000 cm^{-1}), the compound is no longer luminescent at room temperature due to intra-centre nonradiative transitions. Note, therefore, that the reason why e.g. $Cs_2NaBiCl_6$ is quenched is completely different from that why $Bi_2Ge_3O_9$ is quenched (migration to killer centres vs intra-Bi^{3+} nonradiative loss).

In the case of Ce^{3+} the situation is less extreme, but nevertheless large variations occur as a function of the host lattice. The optical transitions are of the type 4f–5d, and are allowed as electric-dipole transitions. In $CeBO_3$ the Ce^{3+} emission at 300 K is quenched due to migration to quenching centres. However, isomorphous $LaBO_3$–Ce is an efficient luminescent material. The spectral overlap between the emission and absorption is large. On the other hand, CeF_3 is not quenched. Energy migration does not take place, at least not over large distances, because of a strong relaxation in the excited state. The same is true for the host lattice of the green emitting phosphor in the three-colour luminescent lamps, viz. $CeMgAl_{11}O_{19}$ [92].

More quantitative results are known for the solid cryptate [Ce ⊂ 2.2.1]$^{3+}$ $Cl_3.2H_2O$ [93] and $Y_{1-x}Ce_xAl_3B_4O_{12}$ [94]. In the cryptate (fig. 10) the Ce-Ce distance is large (~10 Å). At room temperature the spectral overlap is also large (see fig. 11). The Stokes shift is small. Therefore we meet here the exceptional case of energy migration in a metal coordination compound. The critical distance for transfer at 300 K is 18 Å. During its life time the excited state performs ~35 jumps, yielding a diffusion length of 60 Å. If the cryptate is pure, this does not lead to quenching and the migration remains unobserved in spectral measurements. However, small amounts of impurities lead to quenching.

In $(Y,Ce)Al_3B_4O_{12}$ the spectral characteristics are very similar to those of the cryptate, However, the Ce–Ce distance is much shorter, viz. 5.9 Å. This results in more efficient migration.

A well-known case of an ion in between the strong- and the weak-coupling scheme is the uranyl (UO_2^{2+}) group (see also fig. 2b). In uranyl compounds there is usually efficient migration of excitation energy. This has been beautifully studied by Denning and coworkers [95]. They were able to influence and control the trapping centres, a.o. by using ^{18}O. The reader is referred to the original literature for details.

Remarkably enough not much attention has been paid to energy migration in transition-metal-ion doped compositions in general. This problem is of interest for laser materials containing Ni^{2+} and Co^{2+}. Recently Auzel et al. [96] reported a study of this kind on MgF_2: Ni^{2+} and MgF_2:Co^{2+}. They show that there are two processes which are responsible for concentration quenching of the luminescence of the dopant ions. The first is energy migration to sinks, as discussed above. The second is an increase of the intracentre nonradiative rate with increasing dopant concentration. Due to exchange within pairs of ions, the electron-lattice coupling strength (i.e. S) increases. This is a new and interesting effect.

Finally we will consider some special, border line cases.

VI. BORDER LINE CASES

The discussion above was concerned with insulating solids which were for the greater part nonmolecular. However, there are many other types of compounds, and also of radiationless transitions. This is now illustrated on a few examples, viz. an organic compound in which a transition to the liquid-crystalline state occurs [97], the new scintillator BaF_2 [98], and finally the transition from insulator to semiconductor in the titanates [17,99].

In peripherally octa-n-dodecoxy substituted phthalocyanine the optically active phthalocyanine groups are stacked in linear columns (fig. 26). Due to the way of synthesis these compounds contain quenching centres. In the crystalline modification there is an angle between the flat, optically active groups. This angle diappears at the transition to the liquid-crystalline state. Fig. 27 gives the luminescence intensity as a function of temperature. At 4.2 K the quantum efficiency is higher than 50%. Obviously migration is of resticted importance. At higher temperatures quenching occurs since the migration becomes more important, so that the excitation energy reaches the killer sites. Above the phase transition the luminescence is even stronger quenched. The parallel orientation of the molecules favours faster migration, so that more excitation energy reaches the quenching sites. Note that the physical model is similar to the one used above for nonmolecular inorganic solids.

Recently there has been a lot of interest in the luminescence of BaF_2. Its crystals have a large potential as a scintillator material (detection of gamma rays). They show a luminescence at 220 nm with a very short decay time, viz. 600 ps. This short decay time offers the possibility of a good time resolution. This luminescence is of a new type (cross-over luminescence). Its nature has been unraveled by Russian investigators. Excitation with about 10 eV excites anion excitons. Upon annihilation these anion excitons show an emission at about 4.1 eV (300 nm). Excitation with about 18 eV excites cation excitons. These do not annihilate in a simple way, but by a so-called cross-over transition: an electron jumps from the F^- ion (2p orbital) to the hole in the 5p orbital of Ba^{2+} (see fig. 28). This is accompanied by emission at about 5.7 eV (220 nm) and weaker emissions at even higher energy. Since the energy difference between the 2p (F^-) and the 5p (Ba^{2+}) energy band is less than the band gap (~10 eV), the corresponding emission is observed as part of the intrinsic emission of BaF_2. The 220 nm emission shows practically no temperature quenching up to room temperature, whereas the 300 nm emission is for the greater past quenched under these conditions.

Finally we mention that in highly-concentrated luminescent materials there may occur a transition from localized to delocalized electrons. This is best illustrated on the titanates. Compounds like $MgTiO_3$, $BaTi(PO_4)_2$ and $La_2Ti_2O_7$ show the type of luminescence discussed under section V.B. The Stokes shift is large and energy migration does not occur. Table XI shows that if the Stokes shift becomes smaller and smaller, the quenching temperature of the titanates drops. TiO_2 is a clear extreme with line emission at the position of the band gap. In TiO_2 the configurational coordinate model has to be replaced by the energy band model. The line emission is due to free-exciton recombination. The band emission of the other titanates is a clear example of self-trapped exciton emission. For a further discussion, see ref. 17.

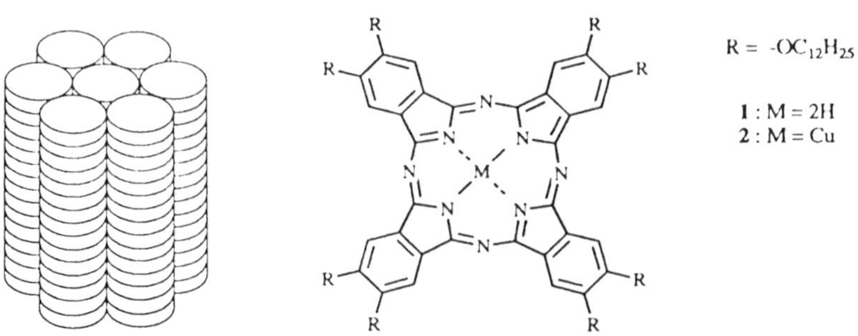

Fig. 26 Schematic representation of the columnar structure of the mesophase of compounds 1 and 2.

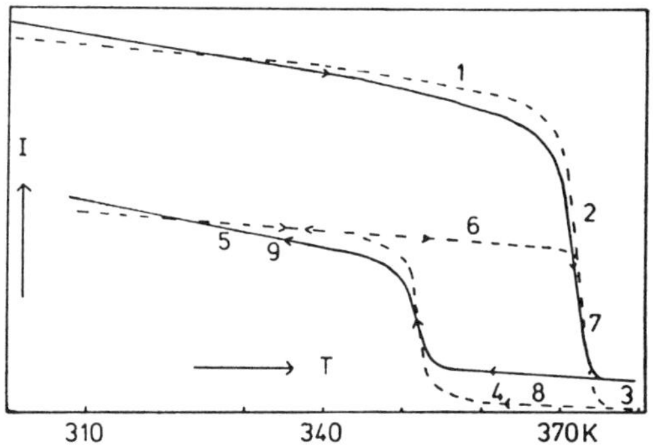

Fig. 27 Luminescence intensity of compound 1 (see fig. 26) around the solid ⟶ mesophase transition as a function of temperature. At the transition temperature the intensity drops strongly.

In conclusion, there are several types of nonradiative processes in luminescent materials. The two most important ones are energy transfer between two luminescent centres and intracentre nonradiative decay. The former are nowadays reasonably well understood and predictable. The latter are only understood and predictable in the case of ions for which the weak-coupling scheme is valid. For the others it is nevertheless possible to arrive at reasonable predictions by applying simple models to a series of luminescent compositions.

Table XI. Some data on the luminescence of titanates (after ref. 17)

Compound	E_g^a (eV)	Maximum of emission band (nm)^b	Stokes shift (eV)	T_q^c (K)
TiO_2	3.0	412 (line)	~0	12
$SrTiO_3$	3.2	520	0.9	35
$MgTi_2O_5$	3.6	490	1.7	100
$MgTiO_3$	3.7	410	1.7	100
$BaTi(PO_4)_2$	3.8	520	1.5	100
$La_2Ti_2O_7$	4.0	465	2.0	120

a. Value of the bandgap at room temperature
b. Values at 4.2 K
c. Quenching temperature of the luminescence

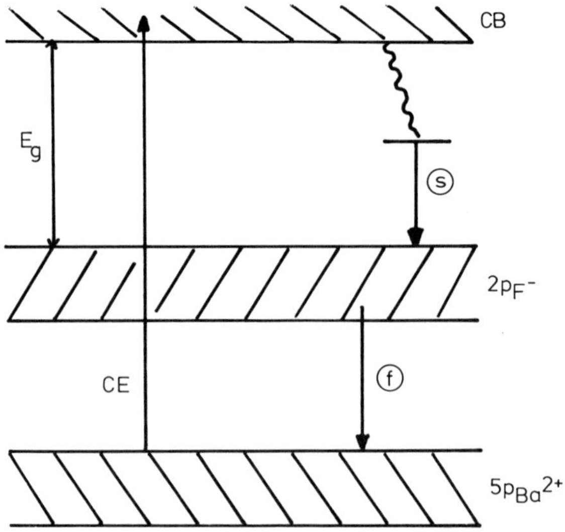

Fig. 28 Energy level scheme of BaF_2 (schematical). CB denotes conduction band, CE crossover excitation, E_g the band gap, s the slow emission and f the fast (crossover) emission.

ACKNOWLEDGEMENT

The author is indebted to Mrs. Biene Meijerman who made the manuscript camera ready.

REFERENCES

1. G. Blasse, editor, Mat. Res. Bull. 16, 3-4 (1987).

2. R.G. Papalardo, in Spectroscopy of Solid-State Laser-type Materials (Ed. B. Di Bartolo), Plenum Press, New York, 1987, p. 397.

3. See e.g. G. Blasse, in Radiationless Processes (Ed. B. Di Bartolo), Plenum Press, New York, 1980, p. 287.

4. H.S. Kiliaan, J.F.A.K. Kotte and G. Blasse, J. Electrochem. Soc. 134, 2359 (1987).

5. D. Curie, in Optical Properties of Ions in Solids (Ed. B. Di Bartolo), Plenum Press, New York, 1975, p. 71.

6. G.F. Imbusch, Chapter 1 in Luminescence Spectroscopy (M. Lumb, Ed.), Academic Press, New York, 1978.

7. M.D. Sturge, Phys. Rev. B8, 6 (1973).

8. See e.g. B. Di Bartolo (Ed.), Energy Transfer Processes in Condensed Matter, Plenum Press, New York, 1983.

9. J.C. Vial and R. Buisson, J. Physique Letters 43, L745 (1982).

10. See e.g. W.M. Yen and P.M. Selzer (Eds), Laser Spectroscopy of Solids, Springer Verlag, Berlin, 1981.

11. R. Reisfeld, Structure and Bonding 22, 123 (1975).

12. L.A. Riseberg and M.J. Weber, Progress in Optics (Ed. E. Wolf), Volume XIV, North-Holland Publ. Comp., Amsterdam 1976.

13. J.M.F. van Dijk and M.F.H. Schuurmans, J. Chem. Phys. 78, 5317 (1983); M.F.H. Schuurmans and J.M.F. van Dijk, Physica 123B, 131 (1984).

14. G. Blasse, G.J. Dirksen and J.P.M. Van Vliet, Rec. Trav. Chim. Pays-Bas 107, 138 (1988).

15. P.A.M. Berdowski and G. Blasse, Chem. Phys. Letters 107, 351 (1984).

16. W.T. Carnall, in Handbook of the Physics and Chemistry of the Rare Earths, Volume 3 (Eds. K.A. Gschneidner Jr. and Leroy Eyring), North Holland Publ. Comp., Amsterdam, 1979, p. 171.

17. G. Blasse, Progress in Solid-State Chemistry 18, 79 (1988).

18. G. Blasse and N. Sabbatini, Mat. Chem. Phys. 16, 237 (1987).

19. L.H. Brixner and G. Blasse, Chem. Phys. Letters, in press.

20. A.J. de Vries, M.F. Hazenkamp and G. Blasse, J. Luminescence 42, 275 (1988).

21. G.F. Imbusch, unplublished results obtained during stay at the University of Utrecht.

22. A.J. de Vries and G. Blasse, J. Chem. Phys. 88, 7312 (1988).

23. L.H. Brixner and G. Blasse, Mat. Res. Bull., in press.

24. J. Sytsma, G.F. Imbusch and G. Blasse, J. Chem. Phys., in press.

25. G. Blasse, J. Sytsma and L.H. Brixner, Chem. Phys. Letters 155, 64 (1989).

26. G. Blasse and L.H. Brixner, to be published.

27. B.R. Judd, Physica Scripta 21, 543 (1980).

28. J. Dexpert-Ghys and F. Auzel, J. Chem. Phys. 80, 4003 (1984).

29. G. Blasse, to be published.

30. B.R. Judd, Phys. Rev. 127, 750 (1962); G.S. Ofelt, J. Chem. Phys. 37, 511 (1962).

31. C. Linares, P. Jung, G. Boulon and F. Gaume, J. Less-Common Metals 93, 89 (1983).

32. M.C. Downer and A. Bivas, Phys. Rev. B 28, 3677 (1983).

33. B. Jacquier, Y. Salem, C. Linares, J.C. Gacon, R. Mahiou and R.L. Cone, J. Luminescence 38, 258 (1987).

34. J. Sytsma, J.W.M. Verweij, G.F. Imbusch and G. Blasse, J. Electrochem. Soc. 135, 537C (1988).

35. J.Th.W. de Hair and G. Blasse, J. Luminescence 14, 307 (1976); J. Solid-State Chem. 19, 263 (1976).

36. A.B. van Oosterhout, Phys. Stat. Sol. (a) 41, 607 (1977); thesis University of Utrecht, 1976.

37. K.C. Bleijenberg and P.A. Breddels, J. Chem. Phys. 72, 3539 (1980).

38. G. Blasse, J. Electrochem. Soc. 124, 1280 (1977).

39. G. Blasse, J.P.M. van Vliet, J.W.M. Verweij, R. Hoogendam and M. Wiegel, J. Phys. Chem. Solids, in press.

40. J.W.M. Verweij and G. Blasse, to be published.

41. G. Blasse, Y. Piffard and L. Struye, Chem. Phys. Letters 147, 514 (1988).

42. L.A. Worl and T. Meijer, Chem. Phys. Letters 143, 541 (1988).

43. N. Sabbatini, S. Perathoner, V. Balzani, B. Alpha and J.M. Lehn, in Supramolecular Photochemistry (ed. V. Balzani), Reidel, Dordrecht, 1987, p. 187.

44. N. Sabbatini and G. Blasse, J. Luminescence 40/41, 288 (1988).

45. G. Blasse, G.J. Dirksen, N. Sabbatini and S. Perathoner, Inorg. Chim. Acta 133, 167 (1987).

46. J.M. Lehn, p. 29 in Supramolecular Photochemistry (see ref. 43); Angew. Chemie 100, 91 (1988).

47. G. Blasse, G.J. Dirksen, D. van der Vorst, N. Sabbatini, S. Perathoner, J.M. Lehn and B. Alpha, Chem. Phys. Letters 146, 347 (1988); J. Phys. Chem. 92, 2419 (1988).

48. J.W.M. Verweij and G. Blasse, J. Non-Cryst. Solids 107, 49 (1988).

49. G. Oversluizen, Fall Meeting Electrochem. Soc., Abstract 711, 1047 (1986).

50. A.J. de Vries and G. Blasse, Mat. Res. Bull. 21, 683 (1986); 22, 1141 (1987).

51. A.M. Srivastava, M.T. Sobieraj, S.K. Ruan and E. Banks, Mat. Res. Bull. 21, 1455 (1986).

52. G. Blasse, G.J. Dirksen, W.E.J. van Kooten and C.A. van Walree, Chem. Phys. Letters 146, 343 (1988).

53. J.W.M. Verweij and G. Blasse, unpublished.

54. G. Blasse and G.J. Dirksen, J. Solid-State Chem. 73, 599 (1988).

55. G. Blasse, P.A. Breddels and D.R. McMillan, Chem. Phys. Letters 109, 24 (1984).

56. Li Youmo, Lu Hongde, Li Jiwen and Miao Xingin, J. Luminescence 35, 107 (1986).

57. R.C. Powell and G. Blasse, Structure and Bonding 42, 43 (1980).

58. W.A.J.A. van der Poel, M. Noort, J. Herbich, C.J.M. Coremans and J.H. van der Waals, Chem. Phys. Letters 103, 245 (1984).

59. W. Barendswaard and J.H. van der Waals, Chem. Phys. Letters 121, 361 (1985).

60. W. Barendswaard and J.H. van der Waals, Mol. Phys. 59, 337 (1986).

61. W. Barendswaard, thesis, University of Leiden, 1986.

62. G. Blasse, Structure and Bonding 42, 1 (1980).

63. E.W.J.L. Oomen, W.M.A. Smit and G. Blasse, Phys. Rev. B 37, 18 (1988).

64. P.J.H. Drummen, H. Donker, W.M.A. Smit and G. Blasse, Chem. Phys. Letters 144, 460 (1988).

65. R. Knochenmuss, C. Reber, M.V. Rajasekharan and H.U. Güdel, J. Chem. Phys. 85, 4280 (1986).

66. C. Reber and H.U. Güdel, J. Luminescence 42, 1 (1988).

67. See e.g. J.F. Donegan, F.J. Bergin, T.J. Glynn and G.F. Imbusch, J. Luminescence 35, 57 (1986).

68. R.H. Bartram, J.C. Charpie, L.J. Andrews and A. Lempicki, Phys. Rev. B 34, 2741 (1986).

69. J.F. Donegan, T.J. Glynn, G.F. Imbusch and J.P. Remeika, J. Luminescence 36, 93 (1986).

70. E. Strauss and S. Walder, Europhys. Lett. 6, 713 (1988).

71. See, for example, G. Blasse, p. 251 in ref. 8.

72. G. Blasse, G.J. Dirksen and F. Zonnevijlle, J. Inorg. Nucl. Chem. 43, 2847 (1981).

73. V. Balzani, F. Bolletta, M.T. Gandolfi and M. Maestri, Top. Corr. Chem. 75, 1 (1978).

74. See e.g. V. Balzani, N. Sabbatini and F. Scandola, Chem. Revs. 86, 319 (1986).

75. N. Sabbatini, A. Bonazzi, M. Ciano and V. Balzani, J. Am. Chem. Soc. 106, 4055 (1984).

76. N. Sabbatini and V. Balzani, J. Less-Common Metals 112, 381 (1985).

77. W.T. Draai and G. Blasse, Phys. Stat. Sol. (a) 21, 569 (1974); Chem. Phys. Lett. 25, 167 (1974).

78. G. Blasse, Phys. Stat. Sol. (a) 75, K41 (1983).

79. D. Wolf and S. Kemmler-Sack, Phys. Stat. Sol. (a) 93, 271 (1986).

80. F. Kellendonk and G. Blasse, J. Chem. Phys. 75, 561 (1981).

81. P.A.M. Berdowski and G. Blasse, J. Luminescence 29, 243 (1984).

82. M. Buijs and G. Blasse, J. Solid State Chem. 71, 296 (1987).

83. M. Buijs and G. Blasse, J. Luminescence 34, 263 (1986).

84. M. Buijs, J.I. Vree and G. Blasse, Chem. Phys. Letters 137, 381 (1987).

85. M. Buijs and G. Blasse, J. Luminescence 39, 323 (1988).

86. M. Buijs, A. Meijerink and G. Blasse, J. Luminescence 37, 9 (1987).

87. Y. Salen, M.F. Houbert, C. Linarès and B. Jacquier, J. Luminescence 40/41, 694 (1988).

88. M.F. Joubert, D. Jacquier and R.L. Cone, Phys. Rev. B 35, 8322 (1987).

89. R. Mahion, B. Jacquier and C. Madej, J. Chem. Phys. 89, 5931 (1988).

90. A.J. de Vries, J.P.M. van Vliet and G. Blasse, Phys. Stat. Sol. (b) 149, 391 (1988).

91. L.H. Brixner, Mat. Chem. Phys. 16, 253 (1987).

92. J.L. Sommerdijk and J.M.P.J. Verstegen, J. Luminescence 9, 415 (1974).

93. G. Blasse, G.J. Dirksen, N. Sabbatini and S. Perathoner, Inorg. Chim. Acta 133, 167 (1987).

94. F. Kellendonk, T. van den Belt and G. Blasse, J. Chem. Phys. 76, 1194 (1982).

95. J.R.G. Thorne, R.G. Denning, T.G. Barker and D.I. Grimley, J. Luminescence 34, 147 (1985); R.G. Denning and J.R.G. Thorne, Mol. Phys. 54, 701 (1985); R.G. Denning, J.O.W. Norris and P.J. Laing, Mol. Phys. 54, 713 (1985).

96. F. Auzel, J.M. Breteau and D. Meichenin, J. Luminescence 40/41, 595 (1988).

97. G. Blasse, G.J. Dirksen, A. Meijerink, J.F. van der Pol, E. Neeleman and W. Drenth, Chem. Phys. Letters 154, 420 (1989).

98. P. Schotanus, C.W.E. van Eijck, R.W. Hollander and J. Pijpelink, Nucl. Instr. Meth. A238, 564 (1984); Yu. M. Aleksandrov, V.N. Makhov, P.A. Rodnyl, T.I. Syreishchikova and M.N. Yakimenko, Sov. Phys. Solid State 26, 1734 (1984), and Ya. A. Valbis, Z.A. Rachko and Ya. L. Yansons, JETP Letters 42, 172 (1985).

99. G. Blasse, J. Solid State Chem. 72, 72 (1988).

TECHNIQUES FOR THE INVESTIGATION OF NONRADIATIVE PROCESSES

U.M. Grassano

Dipartimento di Fisica
Università di Roma — Tor Vergata
00173 Roma Italy

ABSTRACT

Nonradiative processes are generally studied by measuring the time delay of the emission or the change in the quantum yield that they introduce in the radiative processes. Nonradiative processes induced by external perturbations (concentration of defects, electric and magnetic fields, temperature...) are slower than the intrinsic processes. Among the latter are of particular interest, the intraconfigurational relaxations and the relaxations among the higher excited states of impurities and defects in ionic solids. The measurements of these processes has been made possible by the techniques of the ultrafast spectroscopy. Detailed description will be made of the techniques of pump and probe (with mode-locked lasers with picosecond and subpicosecond pulses), of streak camera and of phase correlation (with periodically modulated sources). Other nonradiative processes are excited by high energy photons. They produce the ejection from the sample of electrons (photoemission and Auger effect) or of atoms and ions related to the production of localized lattice defects.

I. INTRODUCTION

With the term "nonradiative phenomenon" or "radiationless process" one usually indicates the decay of an excited system toward its ground state without the emission of photons. The topic of nonradiative processes is very broad and many aspects of it will be treated in the lectures of this school. From the large literature existing on this field one can refer for example to the book of Englman [1] and to those edited by Di Bartolo [2] and by Grassano and Terzi [3]. In these lectures I will confine myself to the problem of excitation and deexcitation of localized impurities and defects in ionic solids.

As has been clarified by the previous lecturers the normal behavior of the optical cycle of such systems consists in an absorption AB, a phonon relaxation in the excited state BC, a radiative decay CD and another phonon relaxation in the ground state. A schematic diagram of the optical cycle of the F center (an electron bound to a negative ion vacancy in an alkali halide lattice) is shown in Fig. 1. The state labelled C is called Relaxed Excited State (RES). The luminescence emitted in the process in normally Stokes-shifted with respect to the absorption. The amount of the shift increases with the lattice relaxation in the excited state where the ions neighbouring the defect accommodate themselves

Advances in Nonradiative Processes in Solids
Edited by B. Di Bartolo, Plenum Press, New York, 1991

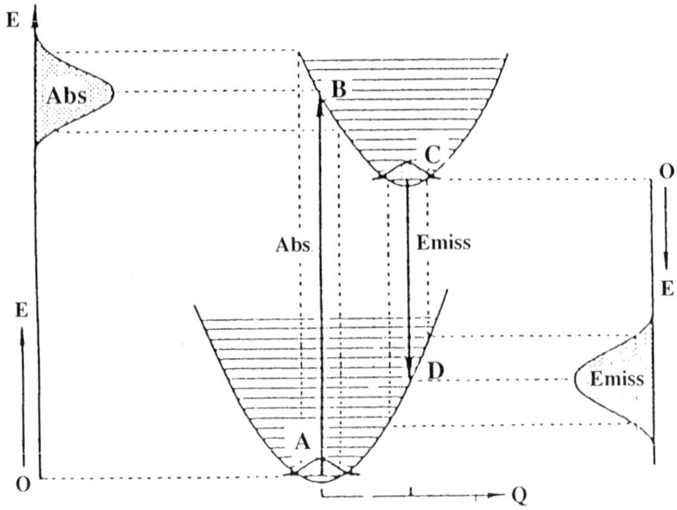

Fig. 1 Optical cycle of the F center.

into the new equilibrium distances, as determined by the excited state electronic wavefunction.

In presence of external perturbations (electric or magnetic fields, high temperature...) or of intrinsic peculiar configurations (crossing of the ground and of the excited state close enough to the RES...) the excited system return to the ground state without emitting photons: the optical excitation is followed by a nonradiative process. Also the processes BC and DA of Fig. 1 are radiationless processes in the sense that these relaxations are due to phonon emission. The phonons remain in the sample and no radiation comming from the crystal can be detected. The experimental techniques used in the investigation of radiationless processes cannot, in general, detect the direct process and are designed to measure some known property of the system, that is changed by the radiationless process. In most of the cases, one investigates the optical properties (emission spectra, lifetimes...) and measures the changes brought in by the nonradiative process.

II. NONRADIATIVE PROCESSES CAUSED BY EXTERNAL PERTURBATIONS

II.A. Lifetimes

As indicated above the presence of nonradiative phenomena can usually be deduced from the change in the radiative ones. For example the change in the lifetime of the F center with temperature is an indication of a thermally activated excitation of the F center to the conduction band, with successive radiationless return to the ground state. The total lifetime τ depends on the radiative lifetime τ_{rad} and on the probability of thermal ionization into the conduction band with activation energy ΔE:

$$\frac{1}{\tau} = \frac{1}{\tau_{rad}} + \frac{1}{\tau_0} e^{-\Delta E / kT} \qquad (1)$$

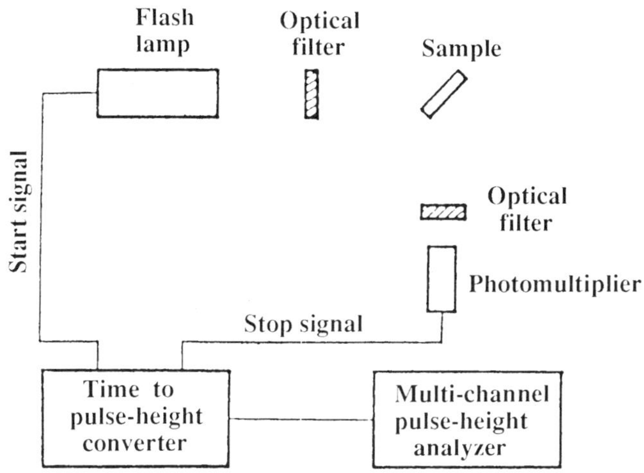

Fig. 2 Experimental set-up for lifetime measurements [4].

By defining the quantum efficiency for luminescence

$$\eta_R = \frac{\tau}{\tau_{rad}}, \tag{2}$$

one obtains

$$(1 - \eta_R)\frac{1}{\tau} = \frac{1}{\tau_0} e^{-\Delta E/kT}. \tag{3}$$

From the measure of τ and η_R as a function of T one can determine τ_{rad}, τ_0 and ΔE.

For the measurements of lifetimes the system is excited by a light pulse (usually a laser with pulses of $\simeq 1 - 10$ ns width, but also flash lamps that can be discharged with a repetition rate up to 20 kHz and give pulses of $\simeq$ 1 ns width). The lifetime can be deduced by the decrease with time of the luminescence measured by the detector. With low light signal and for an increased accuracy of measurement, one records with suitable electronics the delay between the exciting pulse (start) and the luminescence signal (stop). The statistical distribution of these time intervals is equivalent to the statistical distribution of the photons emitted at various times and from it, the lifetime can be deduced. A time to amplitude converter (TAC) with a multichannel analyzer allows the direct recording of the lifetimes from few nanoseconds to microseconds. For times shorter than nanoseconds different techniques are usually used as will be described below. An experimental apparatus for the measurement of τ and the results obtained for the F and F-aggregate centers in KCl are reported in Fig. 2 and 3 [4].

II.B. External perturbations

The effects of external perturbation inducing nonradiative phenomena are the easiest to observe. Let us describe briefly the technique of modulated absorption

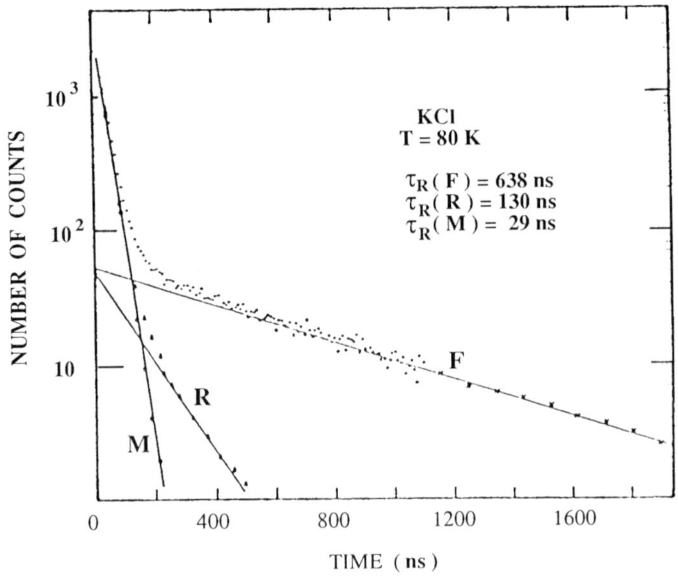

Fig. 3 Decay time of the F, M ($=F_2$) and R ($=F_3$) center luminescence [4].

used Chiarotti and Grassano [5] in the sixties and rediscovered more recently under the name of optical tagging [6].

The external perturbation, in this case an intensity modulated light beam, changes the population of the electronic levels, thus producing a change of the absorption and of the emission. In the original application an intense optical pumping with modulated intensity was used to excite F centers in KCl. The population of the F center ground state is periodically modulated at the same frequency of the pump and likewise that of the RES, but with opposite phase. From the RES, at high enough concentration of F centers, the electron can tunnel to a nearby F center forming an F′ center. (An F center with an additional electron). The formation of F′ centers quenches the F – center emission, because the F′ center reforms the F center by another nonradiative tunneling. (See Fig. 4 and 5). The modulated absorption of the F′ center has phase opposite to that of the transitions starting from the ground state of the F center (F,K, and bands). The modulated absorption is a means to detect transitions from the same level and to follow nonradiative reactions. To be able to measure the modulated absorption, one has to have strong pumping beams and frequencies of modulation slower that the characteristic times of the tunneling. Standard lock-in techniques are used in modulated experiments.

A different perturbation, an external magnetic field, can change the rate of the radiative processes. As observed by Porret and Luty [7] a strong magnetic field, polarizing the spins of the F– center electrons, prevents the F→ F′ tunneling because it is not possible to create F′ centers in the triplet state. In KCl indeed the F′ center has no bound triplet states. In the magnetic field, at low temperatures, the suppression of the F′ formation is detected by a new increase of the normal F– center luminescence. Effects analogous to the F→ F′

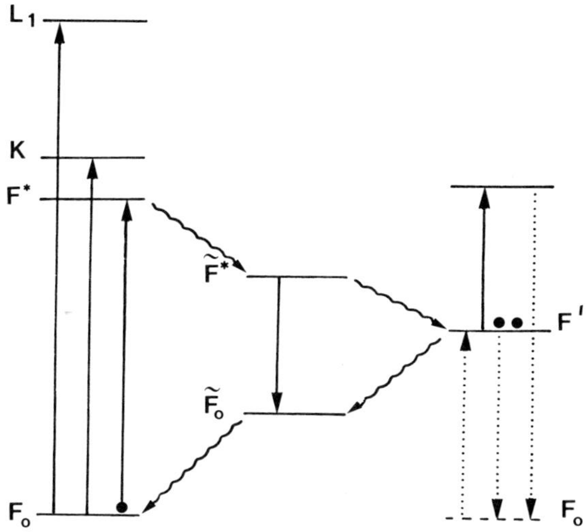

Fig. 4 Electronic energy-level diagram showing the F center absorption emission and the F→ F′ tunneling.

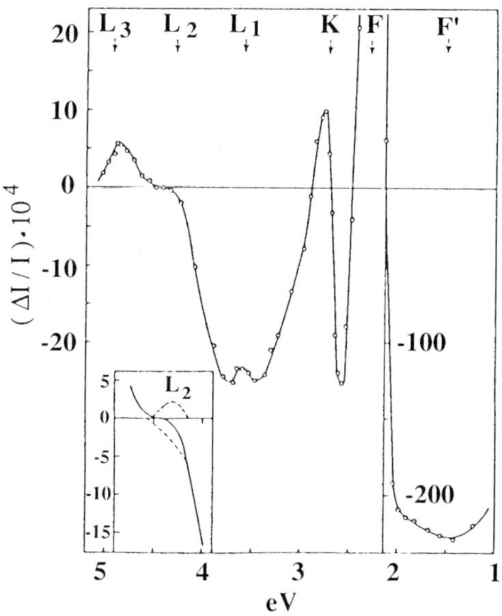

Fig. 5 Modulated absorption from the ground state of the F and F′ centers [5].

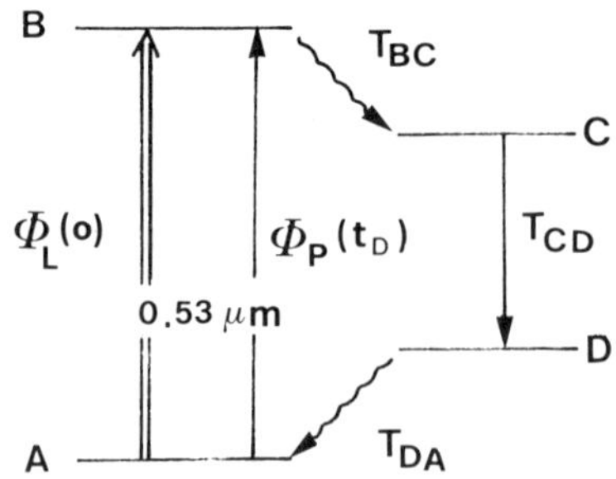

Fig. 6 Pulse and probe method [9]. The pulse $\phi_p(t_D)$ probes the ground state absorption with a delay t_D after the pumping pulse $\phi_L(0)$.

conversion can be observed in the energy transfer from the excited center toward an emitting center or an impurity. A large number of observations has been reported in laser materials with codoping with rare-earth impurities [8].

III. NONRADIATIVE INTRINSIC PROCESSES

The advent of mode-locked lasers with picosecond pulsewidth has opened new and exciting possibilities in the studies of the ultrafast processes in color centers and impurities such as the nonradiative decay in the optical cycle.

III.A. Pump and probe

The experimental technique most often used is called pump and probe and the idea behind is sketched in Fig. 6 [9]. A powerful pulse is used to pump the system to the excited level B. Successively the system (the F center in this case) after relaxation, emission, relaxation returns to the ground state. The probing pulse monitors the number of F centers present in the ground state at various times t_D following the excitation pulse. In a more complex version of the same technique [10] (see Fig. 7), the RES, populated at time t= 0 by the laser pulse $\phi_L(0)$, is depopulated by stimulated emission due to a second intense pulse $\phi_e(\theta)$ following the excitation after a delay θ. At the end the probe measures the ground state population at the time t_D, for example few picoseconds after the time θ.

An experimental set up for the pump and probe technique is shown in Fig. 8 [9]. The probing pulse, split from the main beam by the beam splitter S_1, in further split in two beams, one traversing the sample and one used for normalization. These pulses are used to measure the ground state absorption of the F centers, while the pump pulse changes its arrival time on the sample because of a variable delay line. The arrangement allows the pump to arrive on the sample at different times from 0 to 450 ps in advance of the probing pulses. The effect of the pump is an istantaneous increase of transmission lasting for times comparable to the radiative lifetime of the F center. However, in heavily additively colored samples, the transmission increases sharply after the pumping pulse, but decreases partially in a few tens of ps, to the plateau value. In these samples is therefore present a fast nonradiative channel for the repopulation of the ground state, much faster than the radiative lifetime. The

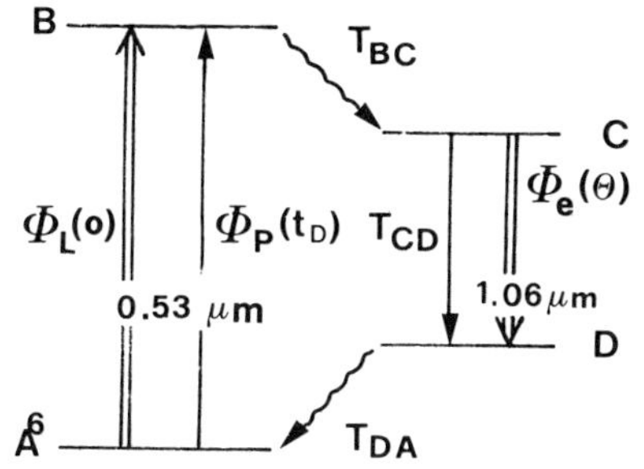

Fig. 7 Pulse and probe method [10]. The pulse $\phi_p(t_D)$ probes the ground state absorption after the pumping pulse $\phi_L(0)$ followed by the stimulating pulse $\phi_e(\theta)$ at the emission wavelength.

authors suggest that this behavior is due to a dipole-dipole interaction between spatially inhomogeneously distributed F- centers.

A further variation of the pump and probe technique was used by D. Schoemaker and coworkers [11]. A mode-locked Ar^+ ion laser producesin two different dye lasers, two pulses used respectively to pump and to probe the system. A double modulation of the two beams allows, with lock-in detection, to measure a change in transmission of 10^{-7}. The set up was used to probe the recovery of the ground state population after an optical cycle either in the $Ga^0(1)$ or $In^0(1)$ centers (formed by an atomic impurity close to

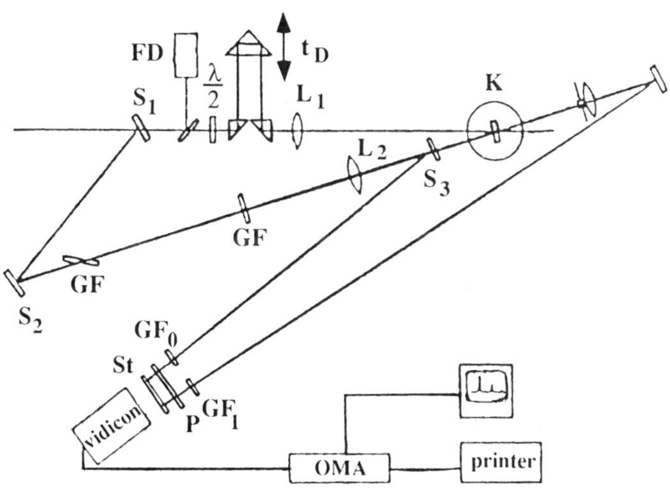

Fig. 8 Set-up for pulse and probe experiments [9]. FD reference photodiode; GF glass filter; K sample; L lens; P polarizer; S beam splitter; St Corning plate for registration.

a negative ion vacancy) [11] or in F center in NaBr [12]. In both cases the electronic deexcitation is nearly radiationless. For the metal impurities, the decay is assumed to be promoted by a quasilocalized impurity mode that connects the initial and final electronic state. In the case of F center in NaBr as observed by Baldacchini et al. [13], there are two emission bands. The decay from the lowest excited state is shorter than that of the normal F center (4 ns instead of 600 ns) and quenched by a fast phonon–assisted tunneling process between the excited and the ground state.

III.B. Up–conversion

A different and generally faster nonradiative transition in the optical cycle of a defect is the intraconfigurational relaxation such as those indicated by BC and DA in Fig. 1. The system excited by a photon to an excited state with a high vibronic quantum number, relaxes toward the RES, emitting phonons. At a later time, after the luminescent emission, the system in the ground state is again in a vibronically excited level and a new nonradiative intraconfigurational relaxation occurs.

Two different models describe this relaxation. In a configuration coordinate model the systems executes damped harmonic vibrations loosing one phonon at a time in the classical stationary points when the motion changes direction [14]. The second model describes the relaxation as an emission of a burst of coherent phonons in a time shorter than the period of the optical phonons [15]. It is therefore very interesting the experimental measure of the relaxation time. However the time characteristic of these processes are of the order of picoseconds and therefore very sophisticated techniques are needed.

The first experimental observation of the configurational relaxation was made in the F_A center in KCl:Li by Wiesenfeld, Mollenauer and Ippen [16] using the up–conversion light gate technique introduced by Mahr and Hirsch [17]. The idea is the following: a nonlinear crystal, such as ADP, KDP, $LiIO_3$ etc., is able to generate the sum of the frequencies of two light beams that enter the crystal with appropriate polarization. The two beams usually in the visible or infrared spectral region are thus up–converted into a violet or ultraviolet radiation that is filtered out from the two generating signals. In case of pulse beams the up–conversion occurs only when both pulses traverse the nonlinear crystal at the same time. The up–converting crystal in presence of a short pump pulse thus acts as an optical light gate that is "open" (for mixing) only when the pump pulse is on. Because the up converted power P_+ is proportional to the istantaneous signal power P_s (for fixed pump power P_p): $P_+ = k\, P_p\, P_s$, the temporal shape of the signal pulse can be mapped out by delaying the pump pulse (short) with respect to the arrival of the (longer) signal pulse and measuring the up–converted power as a function of delay. The up–conversion set–up is shown in Fig. 9 where the Ar^+ ion laser pulse is up–converted in presence of a dye laser.

This up–conversion technique was used in the study of the relaxation of the F_A center [16]. The F_A center (an F center perturbed by a nearest neighbour impurity ion) emits a near infrared luminescence and can be the active center of a color center laser. The time that the center needs after the excitation, to reach the RES is revealed by the increase of the probe beam intensity that is tuned at the F_A emission at 2.62 μm ($\hbar\omega_1 = 0,47\,\mathrm{eV}$). The experimental set up is shown in Fig. 10.

The crystal is pumped with a 0.7 ps long pulse from a mode–locked dye laser in the F_A absorption band at 0.61 μm ($\hbar\omega_2 = 2.03\,\mathrm{eV}$). Part of the same pulse is sent to a variable delay line and focused on a up–converting crystal. A phase–matched nonlinear crystal of $LiIO_3$ adds the intensity of the

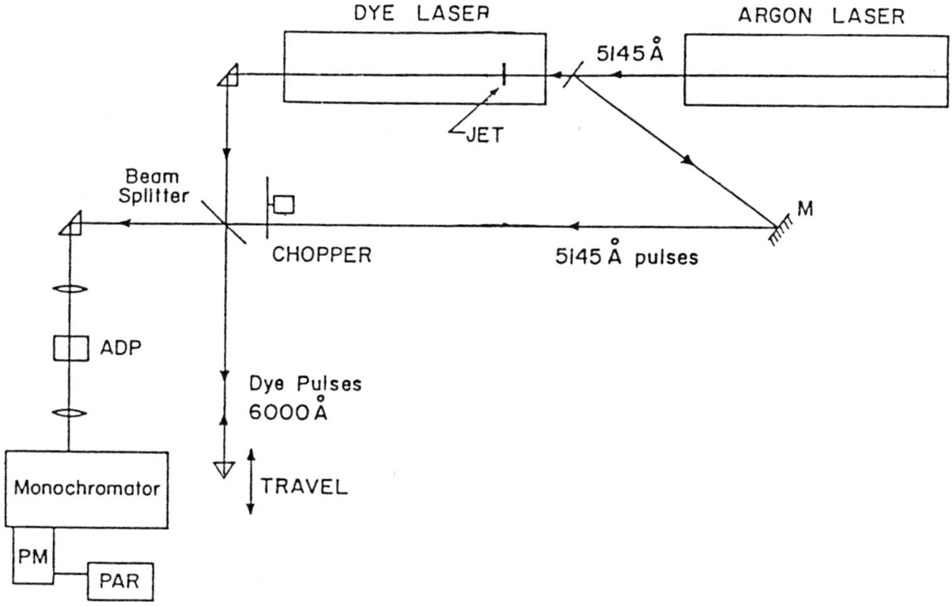

Fig. 9 Experimental set-up for up-conversion measurements [17].

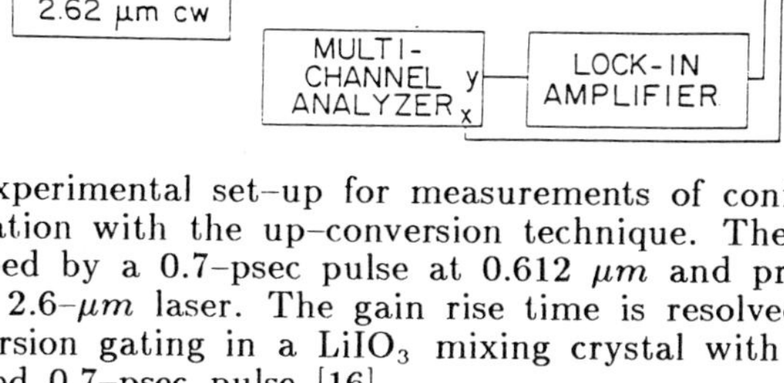

Fig. 10 Experimental set-up for measurements of configurational relaxation with the up-conversion technique. The sample is pumped by a 0.7-psec pulse at 0.612 μm and probed with a cw 2.6-μm laser. The gain rise time is resolved by up-conversion gating in a $LiIO_3$ mixing crystal with a time-delayed 0.7-psec pulse [16].

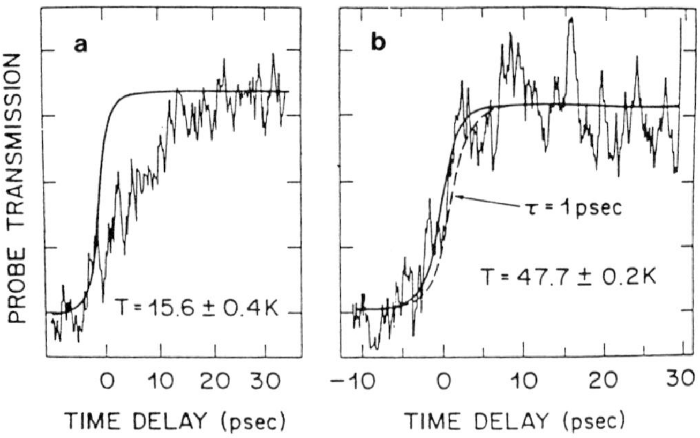

Fig. 11 Configuration relaxation in the F_A centers in KCl:Li. Time-resolved data. The rise of gain at 2.62 μm is shown for two temperatures: (a) 15.6±0.4 K, and (b) 47.7±0.2 K. The solid line is the instantaneous response of the system. The dashed line in (b) is the instantaneous response convolved with a 1.0-psec rise time. Rise time are 10±1 psec for (a) and <0.5 psec for (b) [16].

IR probe beam to that of the reference pulse. The up-converted signal at the frequency $\hbar\omega = \hbar\omega_1 + \hbar\omega_2 = 2.5\,\text{eV}$ (0.496 μm) can be filtered from the spurious components. Its intensity is proportional to the intensities of the summed signals and therefore to the istantaneous intensity of the I.R. beam at the time of arrival of the timing reference pulse. Changing the delay between the pumping pulse on the crystal and the reference "gating" pulse, from zero to few tens of picoseconds, allows the detection of the IR signal and reveals the time of its increase due to the population of the emitting RES after the relaxation. The time delay is controlled by a stepping motor which indexes a multichannel analyzer and the signal detected with phase sensitive electronics (note the low frequency chopper in the pump beam) is accumulated by repetitive scanning. In order to allow the F_A center to recover completely after each laser shot, it was necessary to decrease the repetition rate of the mode-locked laser with a cavity dumper that reduces the output frequency from $\simeq$ 100 MHz to 250 kHz.

Typical time resolved data are shown in Fig. 11. The full line shows the istantaneous response of the system which is obtained from the autocorrelation of the dye laser pulse. The signals measured at different temperatures gave lifetime for the rise of the gain of $\sim$ 13 ps at T$\rightarrow$ 0 K and shorter than 1 ps around 30 K. The temperature dependence of the lifetime is fitted to a model indicating the creation of 18 phonons, each of 47 cm^{-1} of energy. In this way however the experiment accounts only for 47 x 18 = 850 cm^{-1} of energy while the total relaxation in the excited state is of the order of 6000 cm^{-1}. Therefore other modes must be involved in the relaxation process.

The same authors indicate [18] that an exchange of the role of the lasers could be used to detect the relaxation in the ground state (see Fig. 12). The CW laser should be used this time to pump the centers and monitor the ground state absorption. A strong pulse at the emission wavelength would force a return of the population to the ground state through stimulated emission and, through the relaxation, the ground state is repopulated. The transmission of the CW laser will decrease, reflecting the return of the population to the ground state and the rate of this decrease would be a measure of the ground state relaxation time.

An experimental set-up analogous to that described in Fig. 10 was used by Knox, Mollenauer and Fork [19] for measuring the vibrational relaxation of the F_2^+ center (an ionized couple of two F- centers along the $< 110 >$ direction). This time the mode-locked laser generates 50 ns pulses at 625 fm in the absorption band of the F_2^+ center in LiF. The same laser focussed on a jet of a etylene glycol provides a continuum of pulses from which is selected the probe light around 865 nm in the emission band of the center. The intensity of the probe increases if the RES is populated and from the probe delay after the pumping, one can measure the relaxation time. The results obtained imply a relaxation shorter that 75 fs. The periods of the LO and TO phonons in LiF are 50 and 118 fs respectively and therefore, in this case at least, it appears that the relaxation takes place in times shorter than one phonon period.

Such a short lifetimes are to be expected in the systems strongly coupled with the host crystals. Relaxation times shorter than few ps have been measured with this or other techniques, in the F_2 center in NaF [20], in LiF [21] and in the Tl^+ center in KBr [20].

III.C. Streak camera

The situation is partially different for the negative ion impurities such as O_2^-, S_2^-, Se_2^- and NO_2^- in alkali halides host crystals because their own high frequency vibrational mode is weakly coupled to the host lattice. Relaxation times are in these cases of the order of tens of ps and can be revealed by a streak camera.

The heart of a streak camera is the image tube. In it the light striking the cathode produces a stream of electrons accelerated through a hollow anode towards the region of the deflecting plates. Two voltage ramps are applied to these plates and deflect the electrons toward different positions of the phosphor

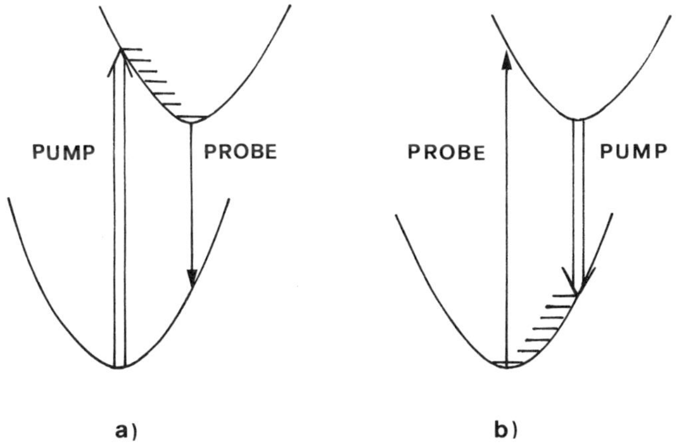

Fig. 12 Pump and probe pulses for measuring the configurational relaxation in the RES a) and in the ground state b).

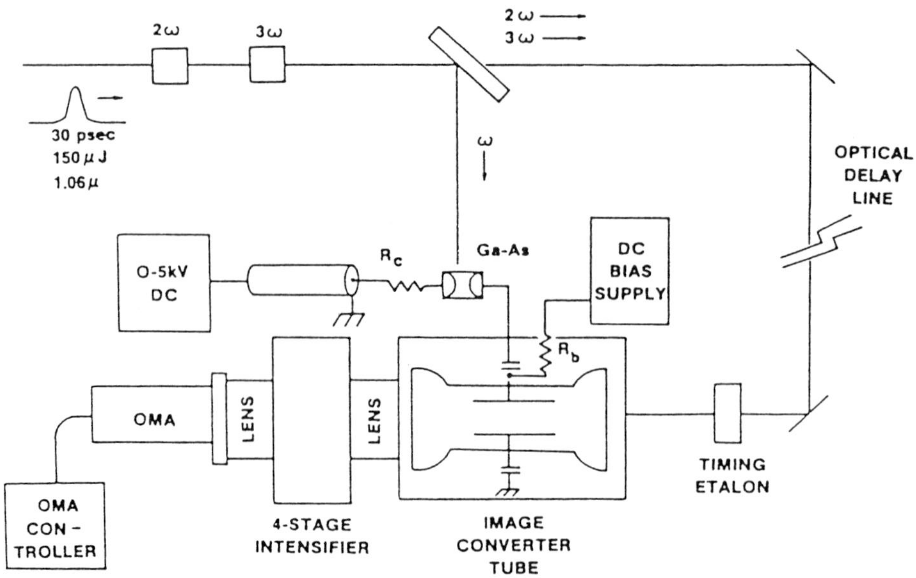

Fig. 13 Synchronized streak camera [22]. A frequency doubled Nd^{3+}; YAG single pulse is split and the infrared component ($\sim 100\mu$ J) is directed onto a 3 mm gap of GaAs:Cr, a high resistivity material. When excited by the 30 ps pulse, the GaAs becomes conducting and charges the deflection plates (through R_C) of an image–converter tube. The sweeping is thus synchronized to the green and UV components which provide excitation for samples placed in front of the image converter. The streaked signal is recorded with an intensified OMA and plotted. The sweepspeed is easily adjusted by varying the 0–5 kV supply or the resistor R_C.

screen. The temporal shape of the light pulse, reflected by the intensity of the photoelectrons, in thus spatially displayed on the screen. If the light intensity is very weak, it is difficult to detect the time resolved single shot. In case of repetitive signals one can trigger the camera applying the deflecting voltage in synchronism with the event repetition [22]. In this way one accumulates the readings and the integration improves the signal to noise ratio. If the synchronism has low jitter, on can obtain time resolution of $\simeq 2$ ps. An experimental configuration is reported in Fig. 13 [22]. If the signal is spectrally dispersed on the entrance slit of the streak camera, a two–dimensional detector can be coupled to the screen. Such a detector, for example a CCD, can record the temporally resolved spectra (and store on the x and y axes, the time and the wavelength dependence of the signal).

Among the results of radiationless decay measured with a synchronized streak camera I would like to mention the beautiful work of W.B. Knox [23] on the molecular ion O_2^-. The weak coupling of the O_2^- ion to the host lattice allows, the observation of the sharp vibronic transitions from the RES to the ground state in the different vibrational states ν. With obvious meaning of the symbols the transitions can be indicated by $|e, o> \rightarrow |g, \nu>$. Examples of fluorescence risetimes of O_2^- ion are plotted in Fig. 14. In the different host crystals the fluorescence risetime changes: from 15 ps in KCl to 41 ps in KBr, to 80 ps in RbI, while on the contrary the temperature dependence is very small. Before reaching the RES, the center can emit a "hot luminescence" i.e. a luminescence that originates not from the RES, but from a higher vibrational level of the excited state. The "hot luminescence" in KBr is clearly resolved only at short times (up to 30 ps) after the excitation and at times longer than 100 ps is totally disappeared while the normal emission reaches its full intensity [20].

An analogous set up has been used to observe the nonradiative relaxation of Ag^- centers in KI [24]. Risetime of the luminescence excited by a 2 ps laser pulse is analyzed shot-by-shot with a streak camera. As in the F_A center experiments, the phonon decay observed can account only for a fraction of the nonradiative energy loss and therefore a shorter (nonobservable) relaxation, with emission of high energy phonons, is postulated.

III.D. Phase fluorometry

A different technique in the study of emission lifetimes has been proposed by Spencer and Weber [25] and applied by our group to the study of lifetimes of color centers [26].

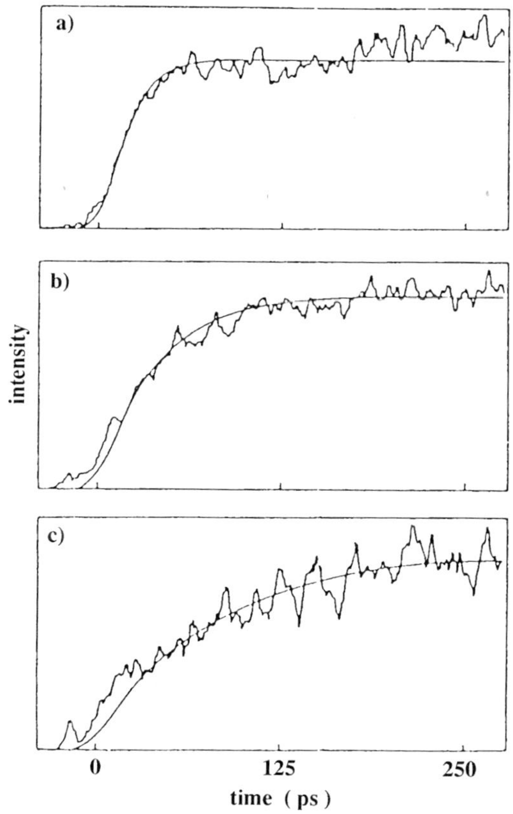

Fig. 14 Host dependence of fluorescence rise time of O_2^- ion at 20 K. Fluorescence for $\lambda > 645$ nm is collected: a) KCl, b) KBr, c) RbI [23].

The technique is based on the measurement of the phase shift and of the modulation of the emission with respect to the excitation. Let us suppose to excite a fluorescent sample with a light beam, the intensity of which is sinusoidally modulated at frequency ν. Also the intensity of the luminescence signal is modulated at the same frequency, but it shows a phase shift with respect to the excitation and a modulation μ (defined as the ratio between the ac. and the dc. part of the signal) smaller than that of the pumping source. The phase shift φ and the modulation ratio $M = \dfrac{\mu\,(\text{emission})}{\mu\,(\text{excitation})}$ contain all information about the emission response function I(t) of the system under investigation. In case of a single exponential decay of lifetime τ $[I(t) \propto \exp(-t/\tau)]$ one obtains the following expressions:

$$tg\,\varphi = \omega\tau, \quad M = \frac{1}{\sqrt{1+\omega^2\tau^2}}, \tag{4}$$

that relate φ and M directly with the decay time. The unknown parameter τ can be obtained independently from a measurement of either φ or M at a single frequency only. From the above relations one can see that, with a modulation frequency of 160 MHz and a sensitivity of 0.01 rad for the phase measurement, a lifetime of 10 ps can be readily measured. In case of multiple independent exponential decays with different lifetimes, the relations (4) have to be modified. The solution is not straightforward, but one can perform measurements of φ and M at different modulation frequencies and from their dependence on ω, by a fitting procedure, one can obtain the various lifetimes.

The most difficult requirement, on the experimental side, for a successful operation of a phase fluorometer is a stable modulated source. In recent year a mode-locked laser was used [27] or the synchrotron radiation from a storage ring [26] [28]. Both kind of sources emit a train of equally spaced pulses. It can be shown by the Fourier analysis, that the frequency content of this type of sources consists of a discrete set of frequencies, multiple of the fundamental frequency, that is the inverse of the time for a photon round trip inside the laser cavity or for the electron orbit inside the storage ring. Since the source is intrinsically modulated at all the harmonic of the fundamental frequency, the luminescence signal contains simultaneously the information about the response function at all the harmonics. Moreover, synchrotron radiation has a wide, continuous spectral distribution that allows to extend this method from the visible to the vacuum ultraviolet.

The experimental set-up is shown in Fig. 15. The photons emitted by the sample (signal) or split by a beam splitter from the pumping beam (reference) are detected by two identical photomultipliers. To avoid the measurement of the phase shift and of the modulation at the frequency ν (1-100 MHz), the gain of the photomultiplier is modulated at the frequency $\nu + \Delta\nu$ produced by a frequency synthetizer driven by the same oscillator that gives the clock to rf cavity driver of the storage ring. $\Delta\nu$ is usually chosen 24 Hz. Because of these intensity and gain modulations, the signals contain components at the sum and at the difference frequencies and these latter can be filtered with a low-pass filter. The phase shift between the two signals from the photomultipliers and the amplitude of their relative modulations are displayed by the detecting system. At each frequency, these measurements are alternatively performed for the sample and for a reference scatterer placed at the same position. In this way one takes into account the delay introduced by the different optical paths in the sample and reference channels.

An example of the results obtained for F and F_2 color centers in NaF is reported in Fig. 16. The preliminary results did not reach the ideal performances of the apparatus which should allow detection of delay times with the accuracy of 5 ps [28]. With small modifications of the apparatus, one can think of

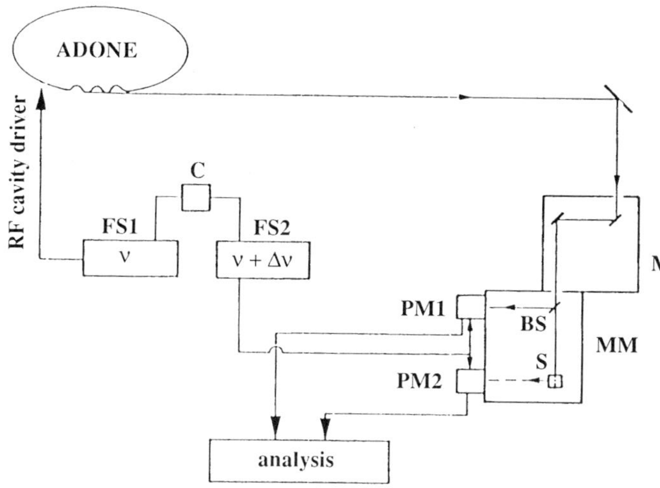

Fig. 15 Phase and modulation fluorometer using the synchrotron radiation pulses [26]. M monochromator, MM measuring module, BS beam splitter, S sample, PM1 and PM2 photomultipliers, FS1 and FS2 frequency synthetizers, C crystal oscillator [26].

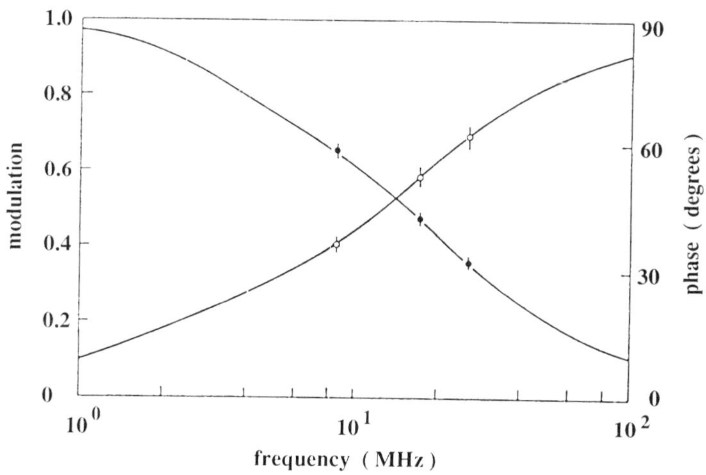

Fig. 16 Lifetimes of the F and F_2 centers in NaF. Frequency dependence of the phase (open circles) and modulation data (full circles) and best -fit curves using two decay times $\tau_1 = 10.2$ ns and $\tau_2 = 51$ ns. Excitation wavelength: 334 nm, emission wavelength: 673 nm [26].

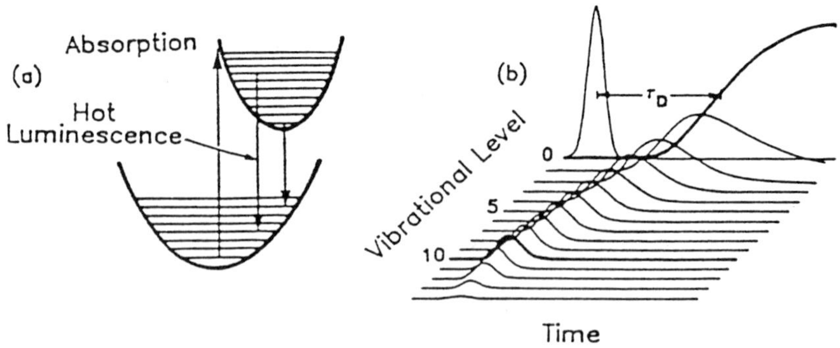

Fig. 17 Model for the "hot luminescence". This picture predicts hot luminescence on a fast time scale and a delay of the ordinary luminescence as a result of vibrational relaxation [20].

measurements of delay times between photons, emitted as hot luminescence and as ordinary luminescence revealing the time spent in radiationless decay. One could also measure the time lag of the same emission, but excited into different excited states such as the L, K and F bands of the F center or the A and C bands of the Tl^+-like impurities in alkali halides.

IV. HOT LUMINESCENCE

Another very interesting open question in relaxation processes of color centers is the emission of photons during the relaxation, before reaching the RES. These photons have energy higher than the ordinary luminescence and reflect two emission processes: the Resonant Raman Scattering close to the excitation and the hot luminescence (HL) covering the region between the excitation and the ordinary luminescence (OL) (see Fig. 17). radiative These are of cours processes and are investigated with the standard methods for the observation of the Raman scattering or of very weak signals. In particular one is interested in the observation of the polarization of the emitted light.

Nakamura et al. [29] have measured the HL of the F center in KCl with a time correlated single photon counter that allows to separate the fast (50 ps) HL from the slow (650 ns) OL. Contrary to earlier findings [30], they have demonstrated that the polarization of the HL is nearly constant (~ 35%) in an energy interval extending more than 5.000 wavenumbers from the excitation. This result rules out any drastic effect of the crossing of the 2p and 2s energy levels during the relaxation. To the mixing of these levels is assigned only the slight monotonous change of polarization from 37% at 500 cm^{-1} to 28% at 6.000 cm^{-1} from the excitation. The nonradiative processes are monitored also by a photocalorimetric technique that measures, beside the OL, the temperature change of the sample due to the phonon dissipation. The increase of temperature should depend linearly on the energy of excitation if the quantum yield for OL is constant. This dependence is observed for the F center in KCl but not in NaCl [31]. In this latter case the change of slope could be due to the onset of a nonradiative deexcitation channel during relaxation due to a crossing of the 2p and 1s state.

The spectral shape of the HL has been theoretically investigated. On descending the ladder of vibrational states, the F– center electron spend most of its time in the so-called classical turning point [14] [32]. The HL should contain structures separated by the phonon energy reflecting this mode of decay. Experimental evidences are difficult to obtain due to the smallnes of the signal and to the presence of spurious emissions such as that of the F_2^+ center. However they seem to have been observed not only for the F center [33] but also for the Tl^+ center in KI [34] and the F_2^+ center in LiF and NaF [35].

V. THERMAL TECHNIQUES

For completeness sake, I would like to mention the techniques aimed at the direct measure of the phonons produced during the radiationless processes.

They are essentially the measure of the acoustic waves generated by a periodic heating and cooling of the sample induced by photon absorption (photoacustic spectroscopy) and the measure of the temperature rise of the sample (photocalorimetry) [36]. These topics will be treated in other lectures at this school.

VI. NONRADIATIVE PROCESSES AND RADIATION DAMAGE

A completely different aspect of the nonradiative processes is the production of lattice defects. The energy absorbed by the crystals is not dissipated as phonons in the return to the ground state (as we have seen in the previous chapters), but is used to create lattice defects, the most common of which is a vacancy – interstitial pair.

The creation of point defects by ionizing radiations has been extensively studied in Alkali Halides using X-ray, electron or UV radiations [37], [38]. In a simplified description, the process is the following. The radiation creates excitons or, if the energy is high enough, electron-hole pairs. The hole is immediately self-trapped and the capture of the electron produces a self-trapped exciton (STE) (Fig. 18a). The STE can recombine radiatively with emission of σ and π luminescence (respectively from the singlet and the triplet recombination), but can also use its energy to form an F-H centers pair in a next nearest lattice sites. The H center is a neutral interstitial halogen atom. By a collisional sequence along the $< 110 >$ direction, a stable pair of defects is formed (Fig. 18b), while pairs that remain in a nearest neighbor configuration, are unstable against mutual annihilation. The F-H centers pair is readily formed if the STE is in an higher excited state (KCl and KBr for example); is formed only with a phonon assisted process if the STE is in its lowest excited levels (in NaCl and *KI*) [39], [40]. The measure of the efficiency of this nonradiative process is of course the rate of color center production.

The first measurements on the coloration efficiency by photon of few fixed energies gave a value of about 1–10 keV per F center produced, the larger value corresponding to less energetic photons [41]. A much more sensitive technique was later employed, monitoring the luminescence of the F center instead of their absorption. It was thus possible to measure the initial rate of coloration that showed a formation energy of about 1 keV per F center in KCl and KBr [42] at photon energies higher than 100 eV. A diagram of the experimental apparatus used by F.C. Brown et al. [42] is shown in Fig. 19.

Closely related to the mechanisms of radiation damage are the results obtained in the studies of atom desorption during the VUV irradiation [43]. The focused collisions long the $< 110 >$ direction leading to the formation of F and H center, causes the ejection from the crystal of halogen atoms or ions if the creation of the initial STE is close enough to the surface (see Fig. 18c). A second species emitted are neutral or ionized alkali atoms. The model for the alkali ions emission is not clear: either a localized excitation (e.g. a core hole) puts the ion in an highly anti-bonding excited state which leads to the ejection from the surface or else surface defects such as steps, kinks or vacancies are the topological points from which the emission is possible [44]. A much more

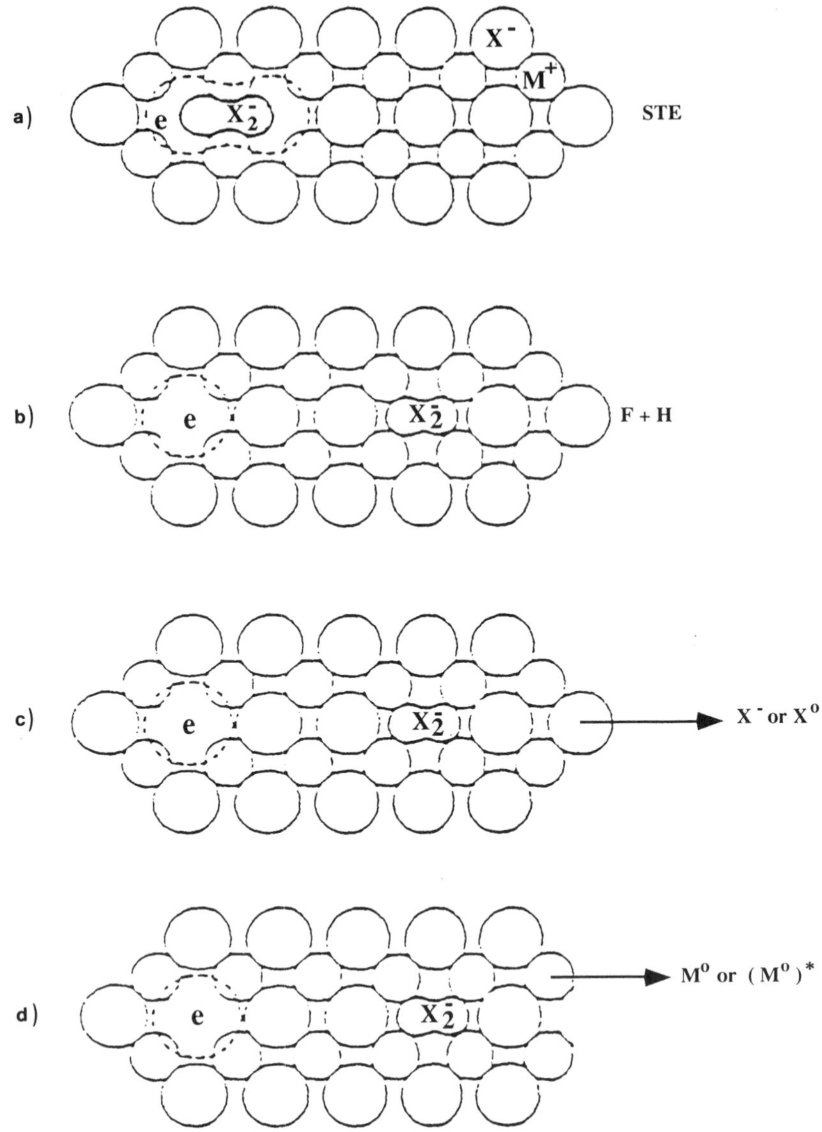

Fig. 18 Defect production and photostimulated desorption in alkali halides.
a) Creation of the precursor state: self trapped exciton;
b) Creation of a F–H pair;
c) Near the surface, halogen ion or atom ejection;
d) Neutralization of the alkali ion and thermal desorption in the ground or in the excited state.

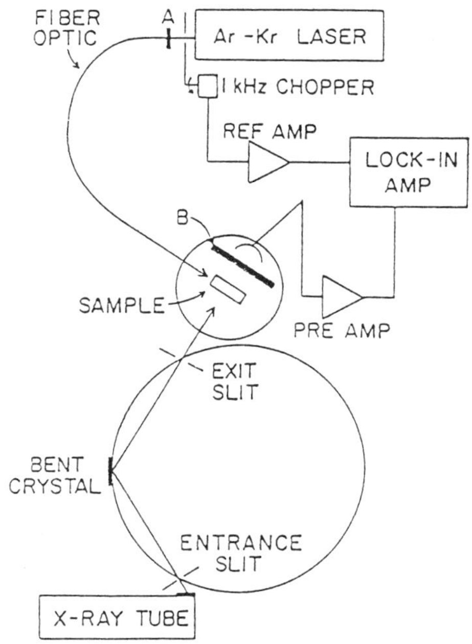

Fig. 19 The laser-induced luminescence set-up for measuring radiation defects. The instrumentation is shown in schematic form as installed on a Rowland circle bent crystal X-ray monochromator with rotating anode source. The apparatus was also used with a synchrotron radiation exciting source [42].

likely process is the desorption of neutral atoms either in the ground or in the excited state. The mechanism for the emission of neutral atoms in the ground state has been clarified [45] and is a thermal desorption from the crystal. The primary process of defect formation leads to the ejection of halogen atoms or ion and the alkali ion, undercoordinated, neutralized by the F center electron, can escape from the surface (Fig. 18d). The lack of spectral dependence of the desorption yield on the photon energy confirms the thermal nature of the process. A small fraction (10%) of the neutral atoms are desorbed in the excited state. The desorption mechanism is certainly nonthermal, but is not completely understood.

In the case of atom or ion desorption the emitted species are analyzed with a mass spectrometer or with time of light techniques, while their energy is measured with a cylindrical mirror analyzer (CMA). Neutral excited-state atoms where also detected by their fluorescence, while ground state neutral atoms were detected by laser induced fluorescence. In this way the products of photons-sample interaction are completely characterized.

Other aspects of the nonradiative deexcitation can be briefly mentioned: the photoemission of electrons excited above the ionization threshold, the Auger effect when a core excitation is not followed by the emission a X-ray or a VUV photon, but by the ejection of a valence - band electron. The techniques for these investigations are fully described in the books of electron spectroscopy [46], [47], and will not be further discussed here.

REFERENCES

1. R. Englman, Nonradiative decay of ions and molecules in solids, North -Holland, Amsterdam (1979).
2. Di Bartolo (ed.), Radiationless processes, Plenum Press, New York (1980).
3. U.M. Grassano and N. Terzi (eds), Excited-state spectroscopy in solids, North-Holland, Amsterdam (1987).
4. L. Bosi, C. Bussolati and G. Spinolo, Phys. Rev., B1, 890 (1970).
5. G. Chiarotti and U.M. Grassano, Phys. Rev. Lett., 16, 124 (1966) and Nuovo Cimento, 46, 78 (1966).
6. L.F. Mollenauer, Phys. Rev. Lett., 43, 1524 (1979).
7. F. Porret and F. Luty, Phys. Rev. Lett., 26, 843 (1971).
8. H.S. Kiliaan and G. Blasse in Proc. Symp. Luminescence Science and Technology, (C.W. Struck, B. Di Bartolo and W.M. Yen eds.), Electrochemical Soc. Inc. (1988), pag. 211.
9. M. Schubert and K. Vogler, Phys. Stat. Sol. (b), 101, 267 (1980).
10. S. Belke, M. Schubert and K. Vogler, Optics Commum., 38, 369 (1981).
11. C. Sierens, W. Joosen and D. Schoemaker, Phys. Rev., B37, 3075 (1988).
12. M. Leblans, W. Joosen and D. Schoemaker, Phys. Rev. (in press).
13. G. Baldacchini, D. Pan and F. Luty, Phys. Rev., B24, 2174 (1981).
14. A. Ranfagni, D. Mugnai and R. Englman, Phys. Rev., B23, 4140 (1981).
15. N. Terzi, J. Luminescence, 31/32, 194 (1984).
16. J.M. Wiesenfeld, L.F. Mollenauer and E.P. Ippen, Phys. Rev. Lett., 47, 1668 (1981).
17. H. Mahr and M. Hirsch, Optics Commun., 13, 96 (1975).
18. L.F. Mollenauer, J.M. Wiesenfeld and E.P. Ippen, Radiat. Effects, 72, 73 (1983).
19. W.H. Knox, L.F. Mollenauer and R.L. Fork in Ultrafast Phenomena V, (G.R. Fleming and A.E. Siegman eds.), Springer Verlag, Berlin (1986), pag. 277.
20. W.H. Knox and K.J. Teegarden, J. Luminescence, 31/32, 39 (1984).
21. V. Brückner, B. Schröder and K. Vogler, Optics Commun., 45, 52 (1983).
22. W. Knox and G. Mourou, Optics Commun., 37, 203 (1981).
23. W.H. Knox in Excited-state spectroscopy in solids, (U.M. Grassano and N. Terzi eds.), North-Holland, Amsterdam (1987), pag. 355.
24. B. Nikolaus and K. Schmitt, Appl. Phys., B36, 213 (1985).
25. R.D. Spencer and G. Weber, Ann. N.Y. Acad. Sci, 158, 361, (1969).
26. U.M. Grassano, M. Piacentini and N. Zema, Nuovo Cimento, 7D, 379 (1986).
27. J.R. Lakowicz, G. Laczko and I. Gryczynski, Rev. Sci Instrum., 57, 2499 (1986).
28. E. Gratton, D.M. Jameson, N. Rosato and G. Weber, Rev. Sci Instrum., 55, 486 (1984).
29. A. Nakamura, T. Sano, K. Kondo and M. Hirai, J. Phys. Soc. Japan, 56, 1603 (1987).
30. Y. Mori, R. Hattori and H. Ohkura, J. Phys. Soc. Japan, 51, 2713 (1982).
31. Y. Kondo, T. Noto, S. Sato, M. Hirai and A. Nakamura, J. Luminescence, 38, 164 (1987).
32. V.V. Hizhnyakov, Phys. Stat. Sol., (b)114, 721 (1982).
33. A. Freiberg, S. Savikhin and A. Anijalg, Optics Commun., 49, 34 (1984).
34. G.S. Zavt, V.G. Plekhanov, V.V. Khizhnyakov and V.V. Shepelev, JEPT Lett., 36, 288 (1982).
35. A. Mabud and F. Luty, Proc. Int. Conf. Defects in Insulating Crystals, Salt Lake City (1984), pag. 297.
36. P. Hess and J. Pelzl (eds.), Photoacustic and photothermal phenomena, Springer, Berlin (1988).
37. Ch. B. Lushchik, I.K. Vitol and M.A. Elango, Sov. Phys. Usp., 20, 489 (1977).
38. N. Itoh, Adv. Phys., 31, 491 (1982).
39. R.T. Williams, K.S. Song, W.L. Faust and C.H. Leung, Phys. Rev., B33, 7232 (1986).

40. M. Hirai, Y. Suzuki, H. Hattori, T. Ehara and E. Kitamura, J. Phys. Soc. Japan, 56, 2948 (1987).
41. M. Elango, C. Gahwiller and F.C. Brown, Solid State Commun., 8, 893 (1970).
42. F.C. Brown, B.R. Sever, N. Kristianpoller and J.P. Stott, Physica Scripta, 35, 582 (1987).
43. T.O. Woodruff, Semicond. and Insul., 5, 431 (1983).
44. T.R. Pian, M.M. Traum, J.S. Kraus, N.H. Tolk, N.G. Stoffel and G. Margaritondo, Surface Sci., 128, 13 (1983).
45. R.F. Haglund et al., Nucl. Instrum. Meth., B13, 525 (1986).
46. L. Ley and M. Cardona, Photoemission in Solids, vol. I and II, Springer, Berlin (1979).
47. G. Margaritondo and J.M. Rowe in Treatise on Analitical Chemistry, (I.M. Kolthoff and P.Y. Elving eds.), J. Wiley, New York (1986), part I, vol. 8, ch. 17.

NEW APPLICATIONS OF EFFICIENT PHOSPHORS

Bruno Smets

Philips Lighting
5600 JM Eindhoven
The Netherlands

ABSTRACT

Efficient phosphors are applied in X-ray intensifying screens, cathode-ray tubes and fluorescent lamps. The introduction of rare-earth phosphors has resulted in a drastic improvement of the performance of these devices. As far as X-ray imaging is concerned, all developments have focussed on an increase of the speed and resolution of the intensifying screens. For application in projection television tubes the phosphors should be able to stand high current densities. This condition is fulfilled by rare-earth based phosphors. In fluorescent lighting we see the introduction of new rare-earth based lamps with a high output in combination with an excellent colour rendition.

I. INTRODUCTION

Luminescent materials are widely used in the present day life. Their best known applications are in cathode-ray tubes and fluorescent lamps. X-ray phosphors also find daily use in X-ray intensifying screens. The large scale introduction of rare-earth based materials has revolutionized these three fields. In fluorescent lighting this has resulted in the introduction of the energy saving tricolour lamps as well as of the compact fluorescent lamps. Rare-earth based CRT phosphors can be operated at very high current densities, in this way the high luminous flux required for projection television can be obtained. Due to the application of rare-earth phosphors the exposure times in radiology can be drastically reduced, the health hazard involved with this diagnostic technique can be kept as small as possible in this way.

Disregarding the field of application two groups of luminescent materials can be distinguished: the phosphors with lattice excitation and the ones with direct excitation in the activator ions. The X-ray and CRT phosphors belong to the first category and as a matter of fact quite a lot of them are used in both applications. In virtually all of the phosphors applied in fluorescent lamps the exciting UV radiation is directly absorbed by the activator ions and no delocalised states are involved in the luminescence process.

In the first part of this paper we will deal with the phosphor exhibiting host lattice excitation. We will start with what these phosphors have in common: their excitation mechanism and ultimate efficiency. While in CRT phosphor most of the impinging energy is absorbed, this definitely does

Advances in Nonradiative Processes in Solids
Edited by B. Di Bartolo, Plenum Press, New York, 1991

not hold for X-ray phosphors. For X-ray phosphors the emission wavelength should lie between 300 and 550 nm, for CRT phosphors on the contrary the emission is confined to specific wavelength intervals in order to cover the entire colour range of the television tubes.

The UV phosphors will be dealt with in the second part of this paper. The evolution in this field has been hardly influenced by new developments in CRT or X-ray phosphors. In fluorescent lighting one aims at the creation of white light with a high efficiency and good colour rendition. A high efficiency and a good colour rendition to some extent are conflicting demands. This has resulted in the introduction of different lamp types. The use of a given type will depend on the exact application chosen.

II. LATTICE EXCITATION

II.A. Efficiency of phosphors under lattice excitation

The excitation mechanisms for X-ray phosphors and cathode ray phosphor are quite similar. The high energy ionizing radiation is absorbed by the lattice. In this process secondary charge carriers, i.e. electron hole pairs, are created. After thermalization these pairs recombine and this process results ultimately in the emission of a photon. For each absorbed electron or X-ray photon a large number of electron hole pairs will be formed and each thermalized electron hole pair can give rise to the emission of one photon by the activator ion.

The energy efficiency η of this overall process is given by Robbins [1] as

$$\eta = (1-r_b)\frac{h\upsilon_a}{E} S.\ Q, \qquad (1)$$

where r_b equals the fraction of the primary radiation scattered back, $h\upsilon_a$ the energy of the photon emitted by the activator ion, E the average energy required to generate one electron hole pair, S the transfer efficiency of the electron hole recombination energy to the activator, and Q the quantum efficiency of the activator center. The back scattering coefficient equals virtually zero for X-rays. For electrons on the contrary the coefficient will be in the order of ten to twenty percent. It is generally assumed that this factor does not vary greatly among practical CRT phosphors. The maximum efficiency obtainable η_{max} will be found in systems for which both the transfer and quantum efficiency equal one:

$$\eta_{max} \sim \frac{h\nu_a}{E} \qquad (2)$$

For a given activator ion the maximum efficiency therefore will be determined by E. The impinging primary particles loose their energy by the generation of optical phonons and by impact ionization. Therefore the average energy to create an electron hole pair will equal

$$E = E_i + E_{op}, \qquad (3)$$

where E_i is the ionization threshold, and E_{op} the average energy lost to optical phonons. It was found empirically that E is related to the band gap energy E_g by:

$$E = \beta . E_g \quad (4)$$

The factor β equals roughly 3 for semiconductors. The ratio of the energy lost to optical phonons and to ionization is given by Robbins [1] as

$$K \sim \left\{ \frac{1}{\varepsilon_\infty} - \frac{1}{\varepsilon_s} \right\} \frac{(h\upsilon_{op})^{3/2}}{1.5E_g} \quad (5)$$

where ε_∞ and ε_s are the high frequency and static dielectric constant respectively, and $h\upsilon_{op}$ the average energy of the optical phonons generated. The dependence of β on K is depicted in Fig. 1.

The coefficient β is only a shallow function of K for $K < 0.10$. Once both processes become of the same order of magnitude (K = 1) β steeply increases with K. Some data for typical phosphors are given in Table 1. From these data it follows that the highest efficiency obtainable is found for the sulfides, followed by the oxysulfides and the oxides. Both $h\upsilon_{op}$ and E_g increase in the same order, explaining why the maximum efficiency drops from 25 % for $ZnS:Ag^+$ to 14 % for $La_2O_2S:Eu^{3+}$, and even to 8 % for $Y_2O_3:Eu^{3+}$. Although the band gap for La_2O_2S and $CaWO_4$ are quite the same, a lower maximum obtainable efficiency is found for the latter compound. This is caused by the higher phonon frequency, resulting in almost a doubling of the coefficient β.

The experimentally determined efficiency η_{exp} corresponds fairly well with the maximum obtainable efficiency derived on the basis of this simple model. For most phosphors the agreement between the efficiencies measured under cathode ray and X-ray excitation is good. An exception to this rule is $CaWO_4$. X-ray efficiencies varying between 5 and 8 % have been reported [1-2]; the cathode ray efficiency equals only 3-4 % [1].

II.B X-ray phosphor

1. General principles

The high penetration depth of X-rays makes them highly suited as a diagnostic tool in the non-destructive analysis of large bodies. X-ray imaging has become one of the corner stones of today's medical practise. The technique however is also found in industrial applications such as the control of metal parts and nuclear materials.

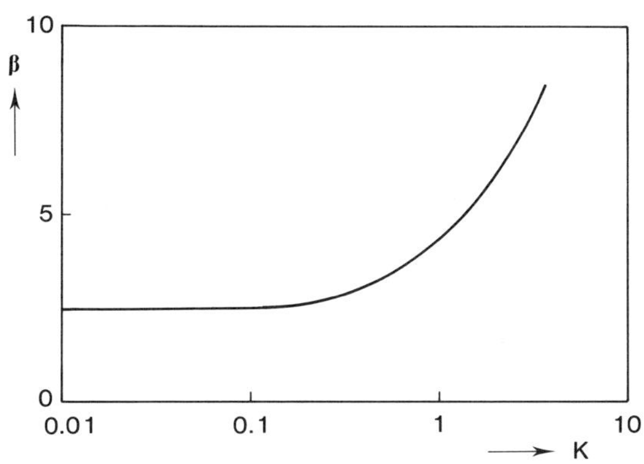

Fig. 1
The dependence of β on the loss ratio K [1]

Table 1: The energy efficiency of some phosphors [1]

phosphor	$h\upsilon_a$ (eV)	$h\upsilon_{op}$ (eV)	E_g (eV)	K	β	η_{max}	η_{exp}
$ZnS:Ag^+$	2.75	0.044	3.8	0.28	2.9	0.25	0.19-0.25
$La_2O_2S:Eu^{3+}$	2.0	0.057	4.4	0.80	3.9	0.12	0.11
$Y_2O_3:Eu^{3+}$	2.0	0.068	5.6	1.12	4.6	0.08	0.06-0.08
$CaWO_4$	2.9	0.112	4.6	2.61	7.0	0.09	0.03-0.08

X-rays can be generated by impinging electrons with a high energy of 70 keV on a tungsten anode. In this way a continuous X-ray spectrum is generated, extending from zero eV to the energy of the incident electrons. The maximum intensity will be found at energies which equal half the energy of the primary beam.

Almost immediately after the discovery of X-rays by Röntgen, he realized that this radiation is not very efficient in exposing photographic emulsions. A search for phosphors, which were able to convert the X-rays into visible radiation, was started. Only one year later Pupin found that the light output of X-rays could be enhanced with the aid of $CaWO_4$ [2]. In actual radiological applications a X-ray cassette is used, consisting of a double emulsion film in combination with two screens brought in close contact with the film (see Fig. 2).

All subsequent efforts in the field of X-ray phosphors have focussed on the gradual increase of the efficiency of the imaging system. Only with the advent of rare-earth based phosphors the predominance of $CaWO_4$ was seriously threatened. We will first consider in detail the old phosphor $CaWO_4$, which is still used in some applications. Then we will deal with $BaClF:Eu^{2+}$, $LaOBr:Tm^{3+}$, $YTaO_4:Nb^{5+}$ and $Gd_2O_2S:Tb^{3+}$.

2. $CaWO_4$

For the detection of an X-ray signal three processes are of major importance: the absorption of the X-rays by the phosphor, the emission of the visible photons and the detection of visible light by the photographic emulsion. The X-ray absorption and the luminescence efficiency for $CaWO_4$ are rather poor, its emission on the contrary matches very well the sensitivity of the photographic emulsion.

$CaWO_4$ crystallizes in the tetragonal $I4_1/a$ symmetry with a = 5.25 Å, c = 11.39 Å and Z = 4 [3]. An overview of the scheelite structure of $CaWO_4$ is given in Fig. 3. The calcium ions are eightfold coordinated by oxygen ions, all belonging to WO_4^{2-} tetrahedra.

The absorption spectrum for a 200 μm layer of $CaWO_4$ is given in Fig. 4. The decreasing signal from 30 to 70 keV is caused by excitation of

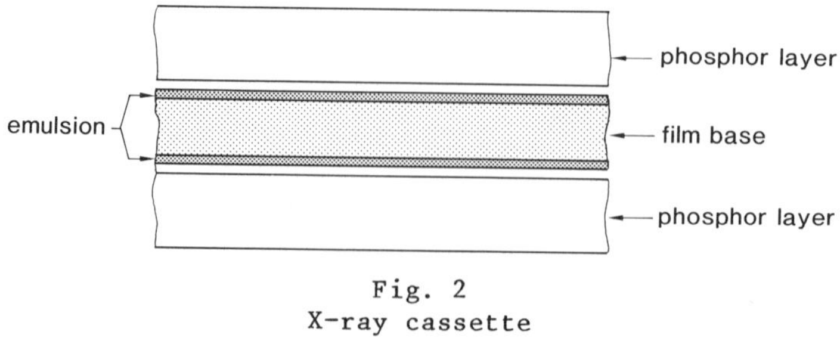

Fig. 2
X-ray cassette

tungsten L electrons. For energies of 70 eV or higher also tungsten K electrons can be excited. The mass absorption coefficient α obeys the following relation

$$\alpha = \sum_i c_i \frac{Z^4}{E^3} , \qquad (6)$$

where Z equals the atomic number, E the energy of the X-rays and c_i a proportionally constant for a given transition. The constant c_i is zero for energies smaller than the absorption edge of the transition. The absorption edge shifts towards higher energies with increasing atomic number. Because the absorption coefficient strongly depends on Z only the heavy atoms such as W (Z = 74) will largely contribute to the absorption. The overall absorption is fairly low. After Venema [4] the absorption of a cassette with a $CaWO_4$ layer of 100 mg/cm^2 would only equal 27 %.

$CaWO_4$ shows a broad band emission originating from the WO_4^{2-} groups, the top is located at 430 nm (Fig. 5). The overlap with the spectral response of a blue sensitive photographic emulsion D_λ is fairly high. The detection efficiency D, defined as $\int D_\lambda \phi \, d\lambda / \int \phi \, d\lambda$, equals 81 %. The energy efficiency of $CaWO_4$ is rather poor and only equals 5 % for commerial phosphors [2].

$CaWO_4$ also can suffer from a severe lag. After the excitation source has been switched of emission is still found. The intensity of this emission amounts to a few percent of the one found during excitation. This lag is caused by the trapping of charge carriers in defect centers. These traps are thermally emptied and the recombination energy is transferred to the luminescent center, i.e. the WO_4^{2-} groups. Because the lag spectrum differs from the normal spectrum, the charge carriers are trapped probably by or near distorted WO_4^{2-} groups. The delayed luminescence of the phosphor screen will cause the appearance of ghost images on subsequent exposures and therefore should be eliminated completely. This problem can be solved by adding $NaHSO_4$ during the phosphor synthesis. Probably the incorporation of SO_4^{2-} in the lattice creates shallow traps, which are completely emptied at room temperature [2].

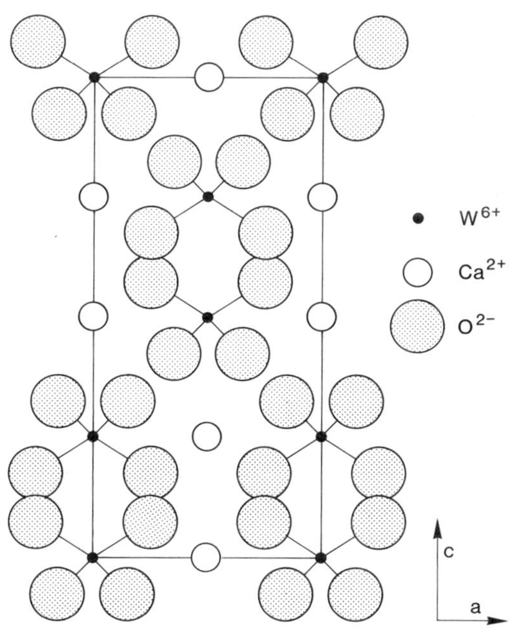

Fig. 3
Structure of $CaWO_4$

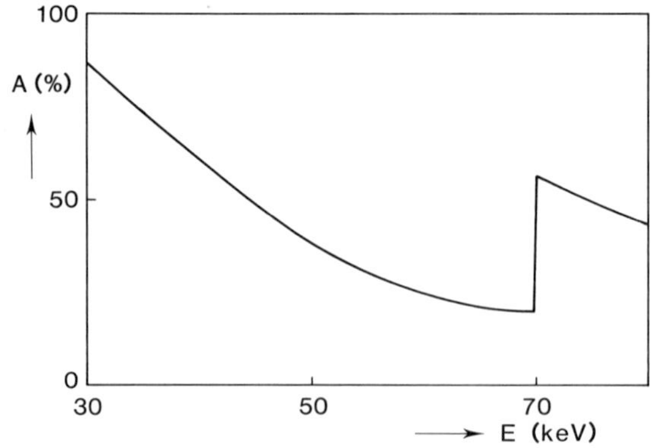

Fig. 4
Absorption spectrum of a 200 μm thick film of $CaWO_4$ [2]

An X-ray cassette can be characterized by its speed S. The higher the speed the lower the exposure time required to obtain the same darkening of the photographic emulsion. The speed will be directly proportional to the absorption strength A of the phosphor layer, the efficiency Q of the luminescent material and the detection efficiency D.

$$S \sim A.Q.D \qquad (7)$$

The absorption of course can be increased by increasing the thickness of the phosphor layer. This will, however, result in a decrease of the resolution (see Fig. 6). In a first approximation the resolution will be inversely proportional to the layer thickness. A phosphor with a higher intrinsic absorption strength and higher efficiency can be used to make a screen with a higher speed or with a higher resolution for a given speed.

3. $BaFCl:Eu^{2+}$ and $LaOBr:Tm^{3+}$

Both BaFCl and LaOBr crystallize in the P_4/nmm structure of PbFCl, with

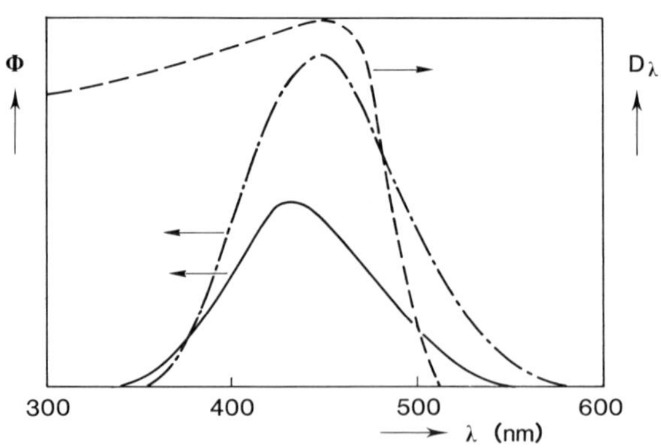

Fig. 5
Prompt (——) and delayed emission (x100) (-.-) of X-ray excited $CaWO_4$ [2] along with the spectral sensitivity (---) of the photographic emulsion. Φ equals the spectral power per unit wavelength interval in arbitrary units. D_λ is the spectral sensitivity of the photographic emulsion.

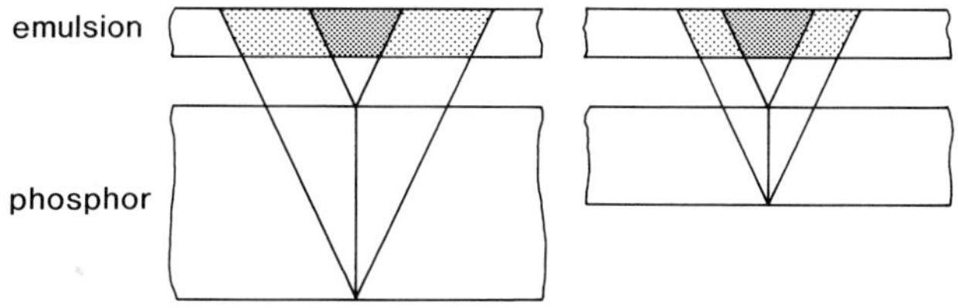

Fig. 6
The effect of the layer thickness on the resolution

a = 4.39 Å, c = 7.22 Å and Z = 2 for BaFCl and a = 4.15, c = 7.36 Å and Z = 2 for LaOBr respectively [6]. An overview of this structure is given in Fig. 7.

The absorption spectrum of BaFCl is given in Fig. 8. Because barium and lanthanum are neighbouring atoms in the periodic table, Z = 56 and 57 respectively, their absorption spectrum, which is almost completely determined by the heavy atoms, is nearly the same. The integral absorption of a screen with a phosphor layer of 100 mg/cm^2 equals 40 % for BaFCl and 41 % for LaOBr [4], which is much higher than the 27 % found for $CaWO_4$.

The emission spectrum under X-ray excitation of $BaFCl:Eu_{0.01}$ is given in Fig. 9. The energy efficiency of this process equals 13 %. This compound suffers from severe lag and, although this phosphorescence can be influenced by adapting the synthesis of the phosphor, it is very difficult to reduce it to an acceptable level. Also in the case of $BaFCl:Eu^{2+}$ the detection efficiency for a blue sensitive emulsion is very high and amounts to 96 % [4].

The occurence of defect centers makes this type of material, noteably with composition BaFBr:Eu, highly suited for use as a storage phosphor [7]. After illumination, image information is retained in the screen as trapped charge carriers. These traps can be emptied when stimulated by a 633 nm He/Ne laser beam scanning the phosphor screen. The photostimulated signal obtained in this way can be read out digitally. The major advantage of digital radiography is the enormous dynamic range of 5 decades, instead

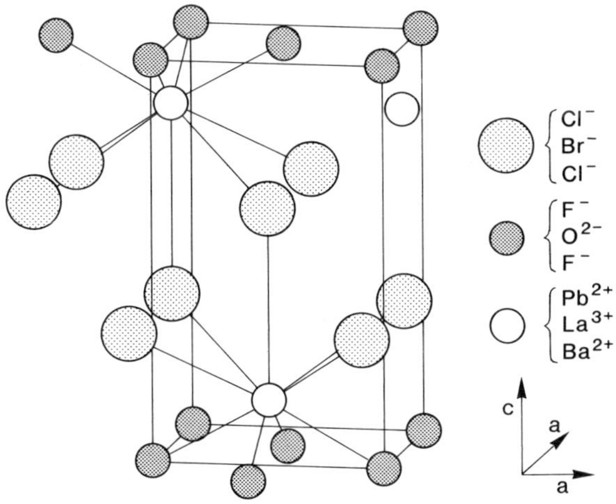

Fig. 7
The PbFCl structure of BaFCl and LaOBr

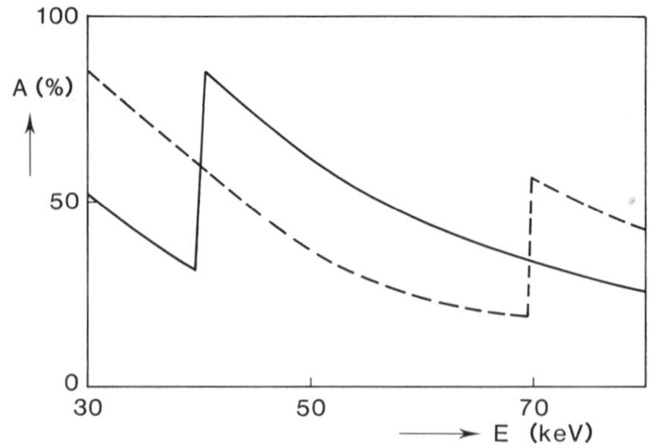

Fig. 8
X-ray absorption of 200 μm thick layer of BaFCl (——) and $CaWO_4$ (---) [2]

of only 1 à 2 decades for a photographic emulsion, and the ability to manipulate the image information [8].

A typical emission spectrum for LaOBr:Tm^{3+} is given in Fig. 10. The blue emission fits very well the sensitivity of the film, the detection efficiency being 85 % [4]. The energy efficiency of this phosphor is high and equals about 10 %. No lag is found with this phosphor.

The crystal growth of crystals with the PbFCl structure is highly anisotropic. They have a platy morphology, the c-axis being oriented normal to the plates. This property makes these crystals more difficult to pack and light guiding is observed toward the sides of the plates and consequently the sides of the screen, adversely affecting the image resolution (Fig. 11). By spray drying BaFCl:Eu^{2+} and subsequently fluxing it in an excess $BaCl_2$, a more polyhedral morphology can, however, be obtained [9]. Until now no method has been found to change the habitus of the LaOBr crystals.

Although in principle BaFCl:Eu^{2+} would be an excellent X-ray phosphor, it was only used for an interim period in intensifying screens in view of its lag. LaOBr:Tm, which also outperforms $CaWO_4$ as far as absorption and efficiency are concerned, shows no phosphorescence and is actually used as an X-ray phosphor [10]. In view of the platy habitus of LaOBr crystals the

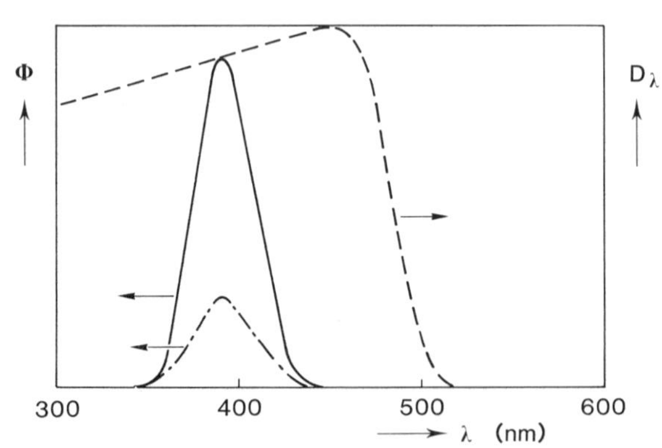

Fig. 9
X-ray excited emission (——) and lag (-.-.-) spectrum of BaFCl:Eu [2], along with the sensitivity (---) of a photographic emulsion

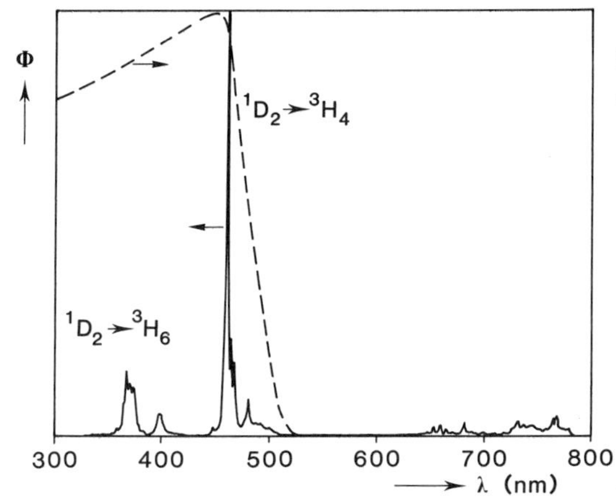

Fig. 10
Emission spectrum (——) of LaOBr:0.2 %Tm^{3+} along with the spectral sensitivity (---) of the photographic emulsion

resolution of the X-ray cassette will be rather poor and its major advantage as compared with $CaWO_4$ is the much higher speed obtainable.

4. M'-$YTaO_4$

The compound $YTaO_4$ exists in two crystallographic modifications. Both modifications are monoclinic. At temperatures above 1450°C the M-modification (C_2 or C_2/m) is found with a = 5.24 Å, b = 10.89 Å, c = 5.06 Å, β = 95.5° and Z = 4 [2]. This structure is closely related to the one of Scheelite and is also found for $YNbO_4$ [11]. At temperatures below 1450°C the M -modification (P_2/a) is stable with a = 5.30 Å, b = 5.45 Å, c = 5.11 Å, β = 96.5° and Z = 2 [12]. The structure of M'-$YTaO_4$ is depicted in Fig. 12. The yttrium atoms are eightfold coordinated by oxygen, the tantalum ions are in the center of a distorted octahedron. It is the octahedral coordination that distinguishes the M'-modification from the M-modification, in which Ta is tetrahedrally coordinated. No modification of $YNbO_4$ with the M -structure has been reported until now.

The absorption spectrum of M'$YTaO_4$ is given in Fig. 13. The K_α absorption edge of Ta (Z = 73) is located almost at the same energy as for W (Z = 74). The absorption on the average is, however, higher for $YTaO_4$. This is caused by the Y K_α absorption edge located at 17 keV. The integral absorption of an X-ray cassette with a 100 mg/cm^2 phosphor layer equals 40 % [4]. The luminescence efficiency for the M'-modification is almost a factor of two higher than for the M-modification [2].

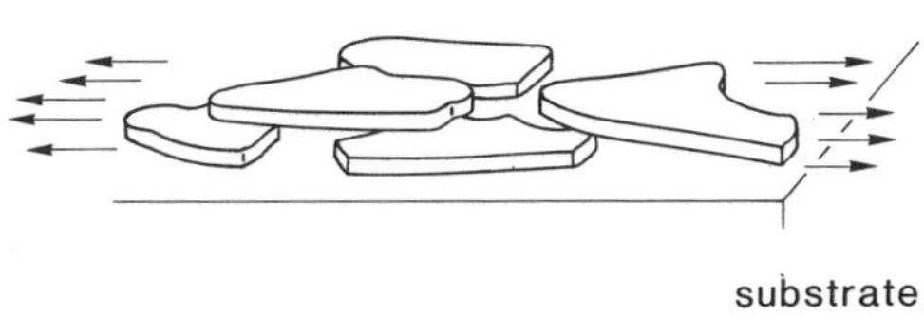

Fig. 11
Morphology of a LaOB-screen

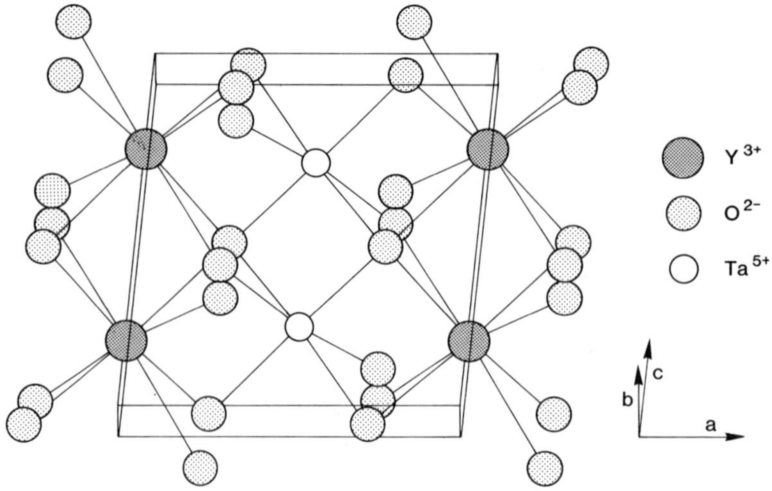

Fig. 12
M'$YTaO_4$ unit cell, showing 6-coordinated Ta and 8-coordinated Y ions

The tantalate emission in M'-$YTaO_4$ is located at a wavelength of 340 nm (Fig. 14). When small quantities of niobium are incorporated, all energy is transferred to the niobate groups emitting at 410 nm (Fig. 14). Niobium can be incorporated into the M'-modification up to a concentration of about 10 %. For higher Nb contents the M-modification will form. The substitution of 2 % niobium also results in an increase of the energy efficiency from 6 to 9 % [2]. The detection efficiency for M'-$YTaO_4$:Nb is about 95 %.

In view of the higher absorption and higher efficiency compared with $CaWO_4$,M'-$YTaO_4$:Nb is a much better phosphor than the tungstate. In contrast to LaOBr:Tm it is possible to obtain well formed polyhedral crystallites by firing M'-$YTaO_4$ in a Li_2SO_4 flux. Therefore it becomes feasible with this phosphor to reduce the layer thickness and simultaneously to enhance the image resolution.

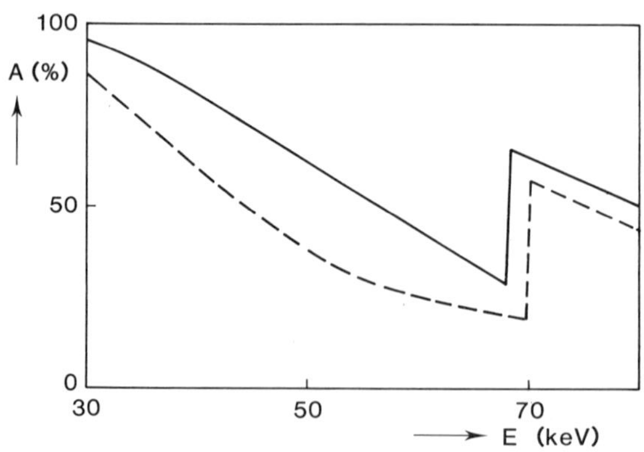

Fig. 13
X-ray absorption of a 200 μm thick layer of M'$YTaO_4$ (——) and $CaWO_4$ (---) [2]

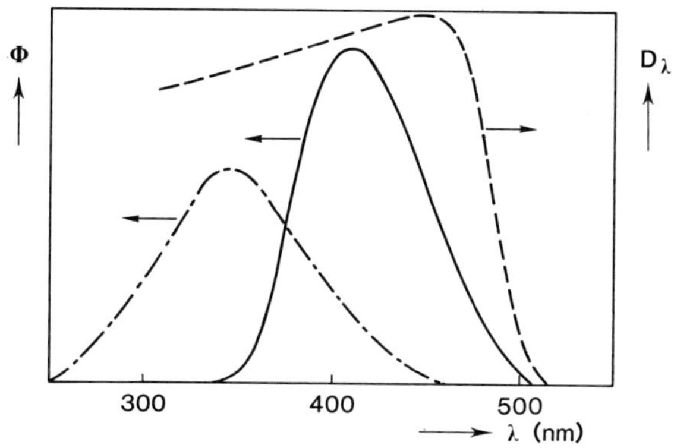

Fig. 14
Emission spectrum of $M'YTaO_4$ (-.-.-) and $M'YTaO_4:Nb_{0.02}$ (——) [2]
along with the spectral sensitivity of the film (---)

5. $Gd_2O_2S:Tb^{3+}$

All Ln_2O_2S compounds crystallize in the P_{3m1} hexagonal symmetry, for Gd_2O_2S a = 3.85 Å, c = 6.67 Å and Z = 1 [13]. Their structure is shown in Fig. 15. The compound is obtained by reacting Gd_2O_3, Tb_4O_7 and elemental S in a Na_2CO_3 flux at 1100°C; this results in the synthesis of nearly perfect polyhedra.

The X-ray absorption of Gd_2O_2S is much higher than for $CaWO_4$. This is caused by the higher density of Gd_2O_2S, where 2 out of 5 heavy ions are present instead of one out of six for $CaWO_4$. Because of the somewhat lower atomic number Z for Gd (Z = 64) the K_α edge is shifted to 50 keV (Fig. 16), enhancing the absorption in the region between 50 and 70 keV. The integral absorption of a 100 μm screen equals 38 % [4].

The emission spectrum for Gd_2O_2S:0.5 % Tb is given in Fig. 17. In view of the fairly low Tb^{3+} concentration apart from the green Tb^{3+} emission

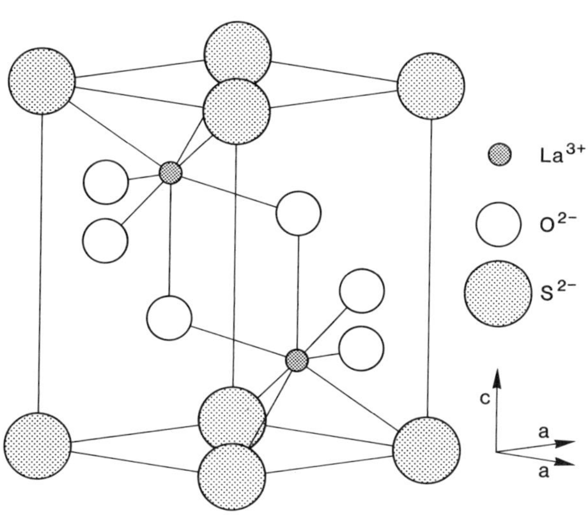

Fig. 15
The unit cell of Ln_2O_2S

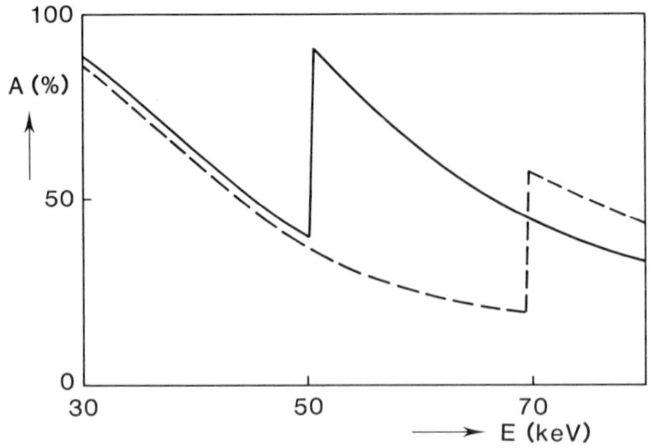

Fig. 16
Absorption spectrum of a 200 μm layer of Gd_2O_2S (——) and $CaWO_4$(---) [2]

($^5D_4$-^7F) also blue emission ($^5D_3$-^7F) is found. This phosphor is highly efficient. Its energy efficiency equals 15 % for the composition given. The detection efficiency for a blue sensitive photographic emulsion on the contrary is only 13 %.

The application of Gd_2O_2S:Tb^{3+} required the development of a green sensitive X-ray film. The spectral sensitivity of such an emulsion is also given in Fig. 17. Thanks to the use of a green sensitive film the detection efficiency for the Gd_2O_2S:Tb emission could be increased to 51 %. One of the problems found with green sensitive double emulsion films is the so-called print through. The green radiation is not absorbed by the polymer base and will cross over to the other side, thus adversely affecting the resolution of the image (Fig. 18). This print through is much smaller for blue or UV radiators, but can also be reduced for green radiation by the use of antihalation layers. In this way screens based on Gd_2O_2S:Tb are well suited for high resolution X-ray cassettes.

6. Present state of the art and future trends

An overview of the properties of the new rare-earth based phosphor

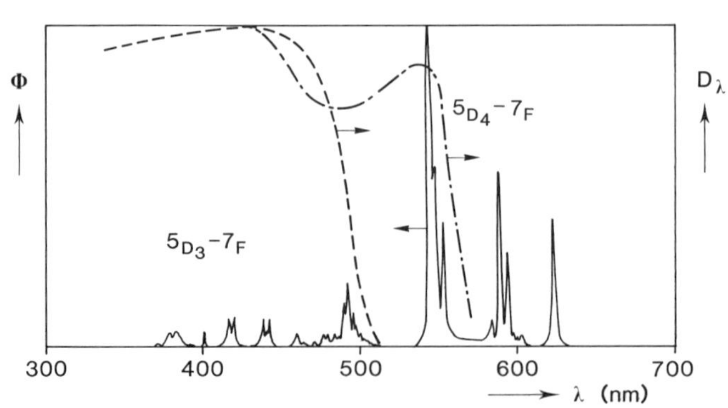

Fig. 17
Emission spectrum of Gd_2O_2S:Tb (——) along with the spectral sensitivity for blue (---) and green (-.-.-) X-ray films

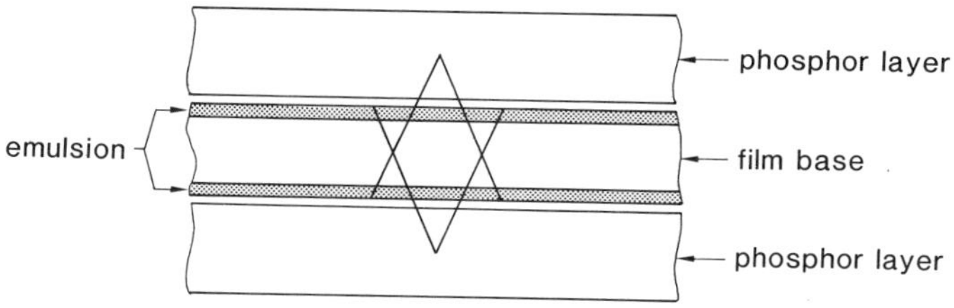

Fig. 18
Print-through in a double emulsion film

systems is given in Table 2. All show a much higher absorption and efficiency as compared with $CaWO_4$. The application of terbium-doped phosphors requires the use of a green sensitive film with a lower detection efficiency. The sensitivity increase with respect to $CaWO_4$ equals a factor of 3 for all rare-earth based phosphors. An increase in sensitivity can be used to increase either the speed or the resolution of the cassette. In view of the crystal habitus of LaOBr increasing the resolution has its limits and all developments have emphasized the high speeds obtainable. For M'-$YTaO_4$ and Gd_2O_2S:Tb on the contrary the attention has focussed on the improvement of the resolution.

As far as the future is concerned it is generally expected that the growth in the film market will decline in the coming years. New competing techniques have emerged such as magnetic imaging, ultra-sound and digital radiography, based on storage phosphors. None of them, however, is able to deliver high quality images for such a low price. Therefore over the coming years the demand for X-ray films will remain very substantial.

II.C. CRT phosphors

One can hardly imagine todays life without cathode-ray tubes (CRT). We find them at home in our TV set and in the office as computer terminals. Human vision and colour perception is restricted to a limited part of the electromagnetic spectrum, and even within the visible spectrum the eye sensitivity only peaks in certain definite regions. We shall deal first of all with colour vision. The principle of the colour television tube indeed is a natural consequence of the way people perceive colour. Then we will see how the application of rare-earth based phosphors has resulted in an enhancement of the luminous efficiency for a standard TV set and later on in the market introduction of large screen projection TV sets.

Table 2: Overview of the properties of rare-earth X-ray phosphors

	A (%)	Q (%)	D (%)	A.Q.D (%)	habitus
$CaWO_4$	27	5	81	1.1	polyhedra
LaOBr:Tm	41	10	85	3.5	plates
M'-$YTaO_4$	40	9	95	3.4	polyhedra
Gd_2O_2S:Tb	38	15	51	2.9	polyhedra

1. Colour triangle

The human eye contains three different receptors for colour vision with spectral sensitives $\bar{p}(\lambda)$, $\bar{a}(\lambda)$ and $\bar{t}(\lambda)$. Therefore, colour vision can be described using three independent functions such as p, a and r or linear combinations of these functions. In 1931 la Commission Internationale d Eclairage CIE introduced a system based on the functions $\bar{x}$, $\bar{y}$, $\bar{z}$ given as (Fig. 19):

$$\bar{x}(\lambda) = \left[1.359\ \bar{p}(\lambda) + 0.101\bar{t}(\lambda) - \bar{a}(\lambda)\right]/0.460 \qquad (7)$$

$$\bar{y}(\lambda) = \bar{p}(\lambda) \qquad (8)$$

$$\bar{z}(\lambda) = \bar{t}(\lambda) \qquad (9)$$

The observation of a light beam with spectral energy distribution $E(\lambda)$ can be described by the so-called tristimulus values defined as:

$$X = \int_\lambda \bar{x}\ E(\lambda)\ d\lambda \qquad (10)$$

$$Y = \int_\lambda \bar{y}\ E(\lambda)\ d\lambda \qquad (11)$$

$$Z = \int_\lambda \bar{z}\ E(\lambda)\ d\lambda \qquad (12)$$

Because colour perception is hardly influenced by the illumination level, it is not the absolute value of X, Y and Z, but rather their ratio which determines the colour impression. These ratios, the colour co-ordinates, are given as:

$$x = \frac{X}{X+Y+Z}, \quad y = \frac{Y}{X+Y+Z}, \quad z = \frac{Z}{X+Y+Z} \qquad (13)$$

In view of the fact that the sum of x, y and z adds up to one, a colour can be represented by the two colour coordinates x and y. Not all the values of x and y are possible, only the combinations lying within the colour triangle (Fig. 20) are found. The border line of the colour triangle is formed by the

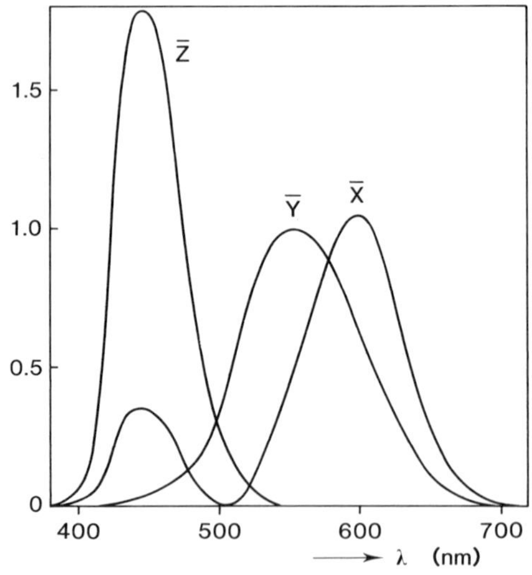

Fig. 19
The spectral sensitivity curves as defined by the CIE

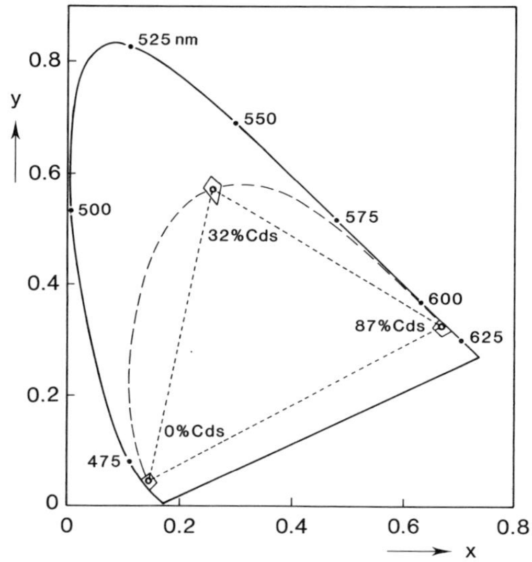

Fig. 20
CIE Chromaticity diagram: Monochromatic radiation (——), Locus of chromaticity coordinates for $(Zn,Cd)S:Ag^+$ phosphors with varying cadmium content (-----). The small quadrangles correspond with the EBU tolerance ranges.

colour points of monochromatic radiation. A colour, which can be obtained by mixing two colours with co-ordinates x_1, y_1 and x_2, y_2 will lie on the straight line connecting both points.

2. Colour television

As we have already seen in chapter II.A the ZnS based phosphors have a very high efficiency under cathode ray excitation. ZnS does exist in two crystallographic modifications [14], the cubic sphalerite structure ($F\bar{4}3m$, Z = 4, a = 5.41 Å) and the hexagonal wurtzite structure ($P6_3mc$, Z = 2, a = 3.82 Å, c = 6.26 Å). Both can be considered as a closed-packing of sulphur ions; the zinc ions are accomodated in the voids between the sulphur

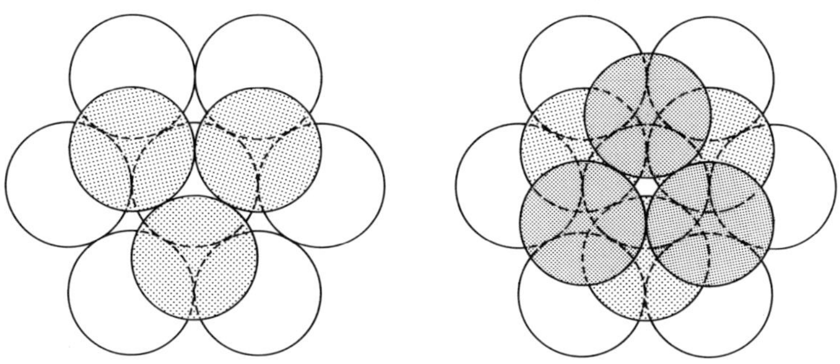

Fig. 21
The hexagonal and cubic structure of ZnS. The sulphur ions of the first layer are in white, of the second layer in light grey and of the third one in dark grey. In the hexagonal packing the third layer is the equivalent of the first one.

ions forming ZnS_4 tetrahedra. The sphalerite structure can be derived from a cubic closest packing (ccp), the wurtzite structure form an hexagonal closest packing (hcp) (Fig. 21). Because in both modifications the short rang structure is the same, their luminescence properties are quite similar.

When doped with Ag^+, ZnS shows a blue luminescence. By substituting part of the Zn by Cd, the emission can be shifted towards longer wavelengths (Fig. 22); for 32 % by weight CdS a green and for 87 % CdS a red luminescence is found. All colours lying inside the triangle interconnecting the colour points of these three phosphors (Fig. 20) can be obtained by mixing their emission spectra. In this way a very large part of the chromaticity diagram can be covered. In colour CRT's each pixel consists of a blue-, green- and red-emitting phosphor dot, each of them stimulated by a separate 20 keV electron gun. By adjusting the beam current of these guns the actual colour is generated. On the basis of $(Zn,Cd)S:Ag^+$ system the European Broadcasting Union EBU has defined three restricted areas within the colour triangle. The colour point of a CRT phosphor should lie within these regions, in order to comply with the broadcasting standards.

The major drawbacks of the (Zn,Cd)S system are the use of large amounts of toxic cadmium and the relatively low luminous efficiency of the red emitting phosphor. This low luminous efficiency is mainly caused by the width of the emission spectrum of the red phosphor. Indeed only the high energy tail of the emission spectrum overlaps with the eye sensitivity curve $\bar{y}$ (Fig. 23). A line emitting phosphor located at 615 nm would have the same colour point, but would match much better the $\bar{y}$ curve. Line emission in this region is only known for the Eu^{3+} and Sm^{3+} ion, the former being a far more efficient activator ion. The intensity of the $^5D_0$-$^7F_2$ emission lines, located at 610-620 nm, is very high in compounds such as $Y_2O_3:Eu^{3+}$, $YVO_4:Eu^{3+}$ and $Y_2O_2S:Eu^{3+}$. In view of the higher energy efficiency of the oxysulphides with respect to the oxides, the latter phosphor will yield the highest efficiency improvement in direct view CRT's.

The amount of Cd^{2+} in the green emitting compound can be reduced significantly by applying Cu^+ as a dopant. Indeed the Cu^+ acceptor levels are located at higher energies than the Ag^+ acceptor levels (Fig. 22). Therefore, $(Zn,Cd)S:Cu^+$ containing 7 % by weight CdS will show the same colour point as $(Zn,Cd)S:Ag^+$ with 32 % CdS.

3. Projection T.V.

Because we are dealing with a vacuum tube, the scaling up of the screen diameter will result in a very thick tube wall and consequently a highly massive device. The upper limit which can be reached with the present-day technology is a diameter of about 75 cm.

By using three small tubes, one for each primary colour, in combination with a lens system it is, however, possible to project the green, blue and red images on top of each other on a large screen of 2 m diameter (Fig. 24). In order to obtain high illumination levels on the large screen much

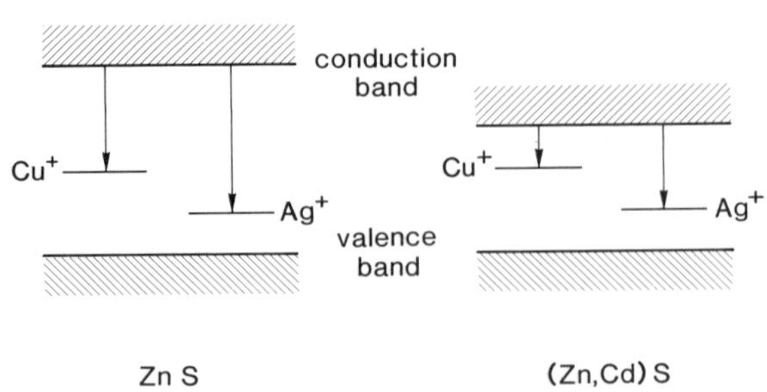

Fig. 22
Energy level diagram of $(Zn,Cd)S:Cu^+$ and $(Zn,Cd)S:Ag^+$

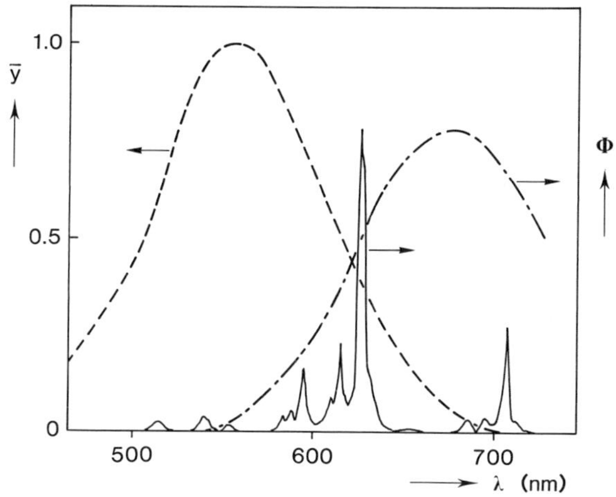

Fig. 23
Emission spectrum of (Zn,Cd)S:Ag^+ (80% CdS) (-.-) and $Y_2O_2S:Eu^{3+}$ (——). $\bar{y}$ gives the sensitivity curve of the human eye.

higher current densities have to be used in projection television (PTV) CRT's, than in direct view CRT's. Under these circumstances, the output no longer increases linearly with the excitation density and saturation of the luminescence process may occur.

Several processes can account for this non-lineair behaviour. The first one is ground state depletion of the activator ion. It was derived long ago by Bril [15] that saturation will occur under stationary excitation once:

$$I = \frac{N}{\tau\eta}, \qquad (14)$$

where I equals the generation rate of electron hole pairs per unit volume, N the activator ion density, τ the decay time of the excited state and η the transfer efficiency of the recombination energy of the electron hole pairs to the activator ion. A similar dependence on the activator concentration is also found under pulsed conditions. In PTV tubes the energy dissipation is very high. During a period as short as 0.1 μsec an energy of 120 Watt (30 kV

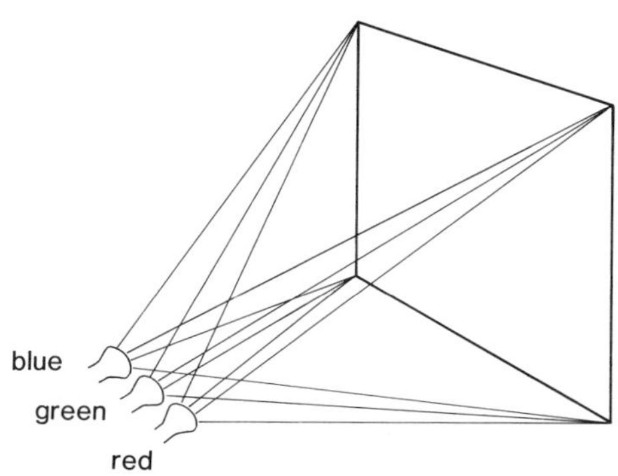

Fig. 24
Principle of PTV

x 4mA) is dissipated in a spot of only 200 μm in diameter. Even when the phosphor layer is brought in intimate contact with the glass substrate and water or air cooling is applied to the tube, the temperature of the phosphor layer will increase, which may result in thermal quenching. It was pointed out by Robertson et al. that the use of a substrate, with high thermal conductivity, such as $Y_3Al_5O_{12}$, instead of glass increases the linearity of the phosphor [16]. Excited state absorption and Auger type interactions between excited activator ions will also result in a lowering of the phosphor efficiency [17].

Because in ZnS phosphors the optimum activator concentration is very low, typical values are 100-1000 ppm, these phosphors will severely suffer from saturation. Although for oxide materials the efficiency is lower than for the highly efficient sulfides, it might be that the better linearity of these phosphors results in a higher efficiency at high current densities. This has resulted in the introduction of a number of new phosphor materials, which are solely used as PTV phosphors.

3.1. Green-emitting PTV phosphors

In Fig. 25 the luminance of a PTV tube as a function of the energy density is given for different green emitting phosphors. At low densities the energy efficiency for $ZnS:Cu^+$ equals 20 %. Typical figures for $LaOBr:Tb^{3+}$ and $Gd_2O_2S:Tb^{3+}$ are 14 and 12 % respectively. Different oxide phosphors are also used in PTV tubes, i.e. $Y_2SiO_5:Tb^{3+}$ (9 %) $InBO_3:Tb^{3+}$ (9 %) and $Y_3(Al,Ga)_5O_{12}:Tb^{3+}$ (9 %) [18]. Some of these compounds have been considered before, the structure of the last three will be discussed below.

Yttrium monosilicate crystallizes in the monoclinic space group $C_{2/c}$ with a = 10.52 Å, b = 6.82 Å, c = 14.59 Å, β = 122.26 and Z = 8 [19]. As can be seen from the structure in Fig. 25, four out of five oxygen ions are bound to silicon ions, forming SiO_4^{4-} tetrahedra. The fifth one occupies an isolated position in the structure. The compound therefore can best be considered as being an oxysilicate. It contains two yttrium site; the average Y-O distance equals 2.3 Å for both sites.

$Y_3Al_5O_{12}$ belongs to the garnet family. It forms cubic (Ia3d) crystals with a = 12.01 Å and Z = 8 [20]. Only one Yttrium site is present in this structure. The Y^{3+} ion occupies the center of a distorted cube (see Fig. 27). The Al^{3+} ions are accomodated in smaller tetrahedral and octahedral sites. Quite a lot of substitutions are feasible within this structure, such as the partial replacement of Al^{3+} by Ga^{3+} ions. This leads to an increase of the efficiency at low densities by about 30 %.

$InBO_3$ crystallizes in the calcite structure: space group R3c with a = 4.82 Å, b = 15.46 Å and Z = 6 [21]. The structure contains isolated planar BO_3^{3-} triangles. The In^{3+} ions are sandwiched between these BO_3 groups, the site symmetry being S_6 (Fig. 28).

In all these compounds except for $InBO_3$ the Tb^{3+} ions occupy a site without inversion symmetry. They will show more or less the same spectrum as far as the relative intensity of the $^5D_4$-^7F transitions is concerned, 50 % of the emission is located at 545 nm ($^5D_4$-$^7F_5$). Because in $InBO_3$ the Tb^{3+} ions occupy a site with inversion symmetry the intensity of the $^5D_4$-$^7F_5$ emission will be higher and amounts to about 75 % (see Fig. 29). In view of this property $InBO_3:Tb^{3+}$ will suffer to a smaller extent from chromatic abberation than the other green-emitting phosphors.

In all compounds the terbium concentration is in the one to ten percent range. Therefore, linearity will be much higher for these rare-earth phosphors as compared with ZnS:200 ppm Cu^+. For $Gd_2O_2S:Tb^{3+}$ the linearity is not as good as one would except on the basis of its terbium content. This unexpected behaviour is primarily caused by thermal quenching (Fig. 3C). Because LaOBr suffers somewhat more from thermal quenching than $Y_3Al_5O_{12}$, Y_2SiO_5 and $InBO_3$, its output will be smaller at densities of 10^{-1} J/cm^2. The linearity of the latter three phosphors is nearly the same [22] and is represented by one single line in Fig. 25.

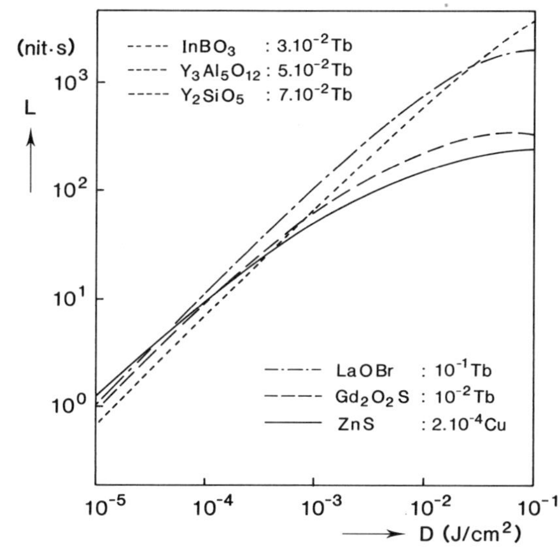

Fig. 25
Linearity of the green PTV phosphors

3.2. Red PTV phosphor

Above room temperature all Ln_2O_2S phosphors suffer severely from thermal quenching. Therefore, it is no surprise that the linearity of Y_2O_2S:5% Eu^{3+} is fairly low. A far better performance is found for Y_2O_3:5% Eu^{3+} (Fig. 31).

In the cubic Y_2O_3 structure (Ia3, a = 10.6 Å, Z = 16) [23] two clearly distinct yttrium sites are present. In both sites the surrounding oxygen ions occupy the corners of a (slightly distorted) cube. In view of the 3/2 ratio between oxygen and yttrium ions only six out of eight positions will be occupied. This results in a site with C_2 symmetry and a second one with S_6 symmetry (Fig. 30). Of all rare earth ions 25 % are in the S_6 position and 75 % in the C_2 position.

Because in a S_6 symmetry only magnetic dipole transitions are allowed,

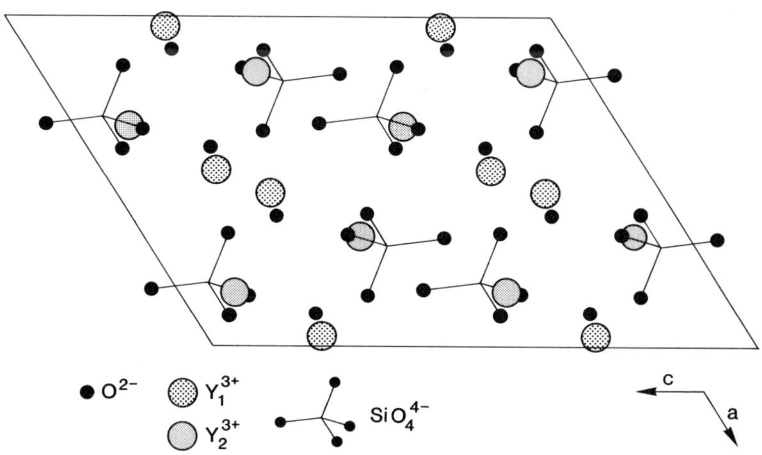

Fig. 26
The structure of Y_2SiO_5

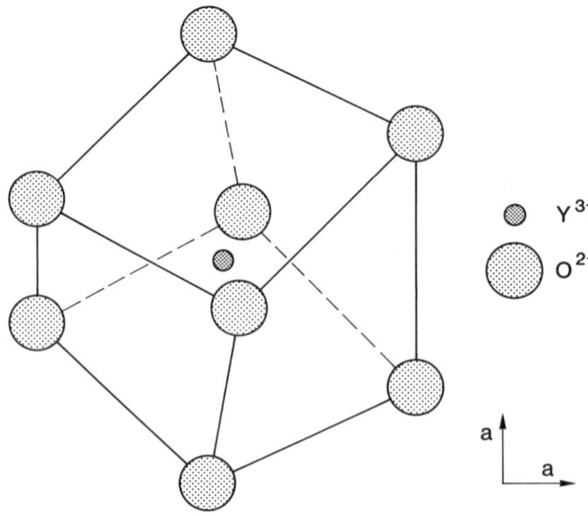

Fig. 27
Coordination of the Yttrium site in $Y_3Al_5O_{12}$

predominantly the $^5D_0-^7F_1$ transitions around 590 nm, will be found. In the C_2 site on the contrary, only the forced electrical dipole transition $^5D_0-^7F_2$ around 610 nm will be allowed. Instead of both emission lines, only one at 610 nm dominates the spectrum (Fig. 31). It appears that an efficient energy transfer occurs from the S_6 to the C_2 site in $Y_2O_3:Eu^{3+}$ [24]. Otherwise this efficient phosphor would not be suited for use as a red TV phosphor.

3.3. Blue PTV phosphor

In principle different activator ions could be used in a blue PTV phosphor. Apart from Ag^+ other candidates are Ce^{3+}, Eu^{2+} and Tm^{3+}. Doping with Ce^{3+} results in broad band emission [25]. Due to the occurrence of chromatic aberration in the lens system the use of a broad band phosphor has definitely its limitations.

Efficient Eu^{2+}-doped phosphors have been reported before [26]. The halfwidth of this emission is only 50 nm and can therefore be used in projection systems. Divalent europium phosphors however suffer from severe degradation even after only one minute of exposure to the electron beam. It seems that under the bombardment part of the Eu^{2+} is converted into Eu^{3+} [27].

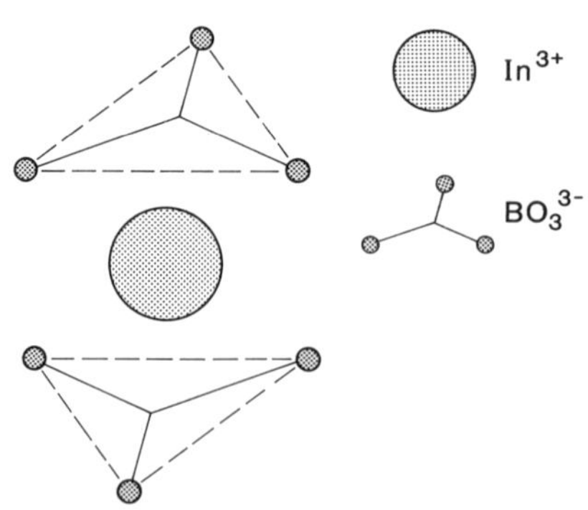

Fig. 28
The structure of $InBO_3$

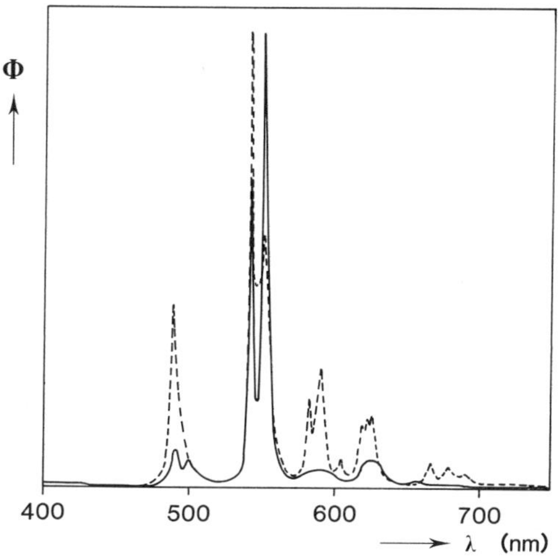

Fig. 29
Emission spectrum under CRT excitation of $Y_3(Al,Ga)_5O_{12}:Tb^{3+}$ (---) and $InBO_3:Tb^{3+}$ (——)

As we have seen before, Tm^{3+} gives rise to a blue line emission at 470 nm. Fairly efficient emission is found for example in $La_2O_2S:Tm^{3+}$ (10 %) and $LaOCl:Tm^{3+}$ (8 %). In both cases the optimum activator concentration is about 0.5 %. Due to cross-relaxation the efficiency will level off drastically at higher concentrations. In view of this fairly low activator concentration Tm^{3+}-doped phosphors will only show limited linearity.

None of the rare-earth phosphors can actually compete with $ZnS:Ag^+$; the ones that come closest are the Tm^{3+} phosphors. Although their linearity is considerably better than the one of $ZnS:Ag^+$, they can not surpass the efficiency of the latter phosphor at high densities.

4. Present state and future trends

Today in direct view TV sets only one rare-earth phosphor is applied, i.e. $Y_2O_2S:Eu^{3+}$. This is mainly caused by the much higher efficiency of ZnS phosphors as compared with rare-earth phosphors. Although the energy efficiency of $(ZnCd)S:Ag^+$ is almost a factor of two higher than the one of

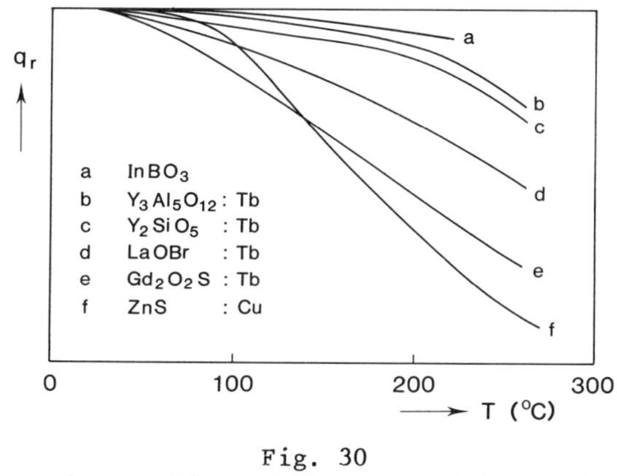

Fig. 30
Thermal quenching of the green CRT phosphors

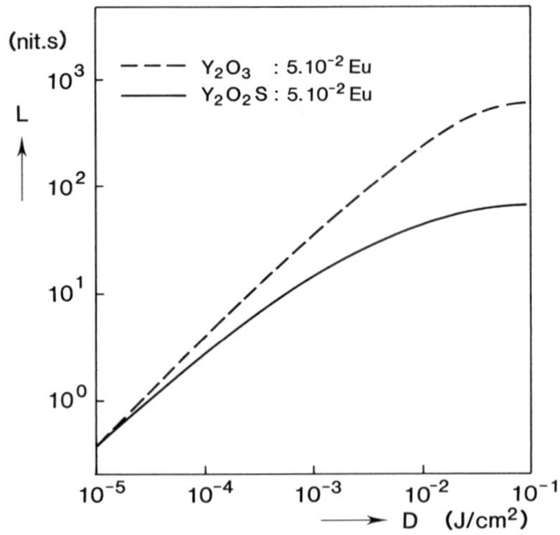

Fig. 31
Linearity of the red CRT phosphors

$Y_2O_2S:Eu^{3+}$, the luminous efficiency of the latter line emitter is superior. It was this higher luminous output that prompted the introduction of $Y_2O_2S:Eu^{3+}$ as the red CRT phosphor.

Rare-earth based phosphors regained interest with the introduction of projection television. In projection tubes the phosphors are subjected to much higher current densities than in standard CRT's. This results in saturation of the luminescence for the sulfide phosphors, a phenomenon which is far less pronounced for oxide materials with a higher activator content. The best choice for a red PTV phosphor is $Y_2O_3:Eu^{3+}$. For the green-emitting PTV phosphor a number of compounds might be used such as $Y_3(Al,Ga)_5O_{12}:Tb^{3+}$, $InBO_3:Tb$, $Y_2SiO_5:Tb^{3+}$ or $LaOBr:Tb^{3+}$. Until now, no blue phosphor with a better performance than $ZnS:Ag^+$ was found. Therefore the screen luminance will be limited by the blue-emitting component.

Over the past few years small television sets based on a liquid crystal display have appeared on the market. It is generally accepted that in the next decade the position of the CRT tube will not be threatened by this new development. Is there any future left for CRT tubes in the long run? The answer to this question will largely depend on whether or not one will be able to find an efficient blue emitting PTV phosphor.

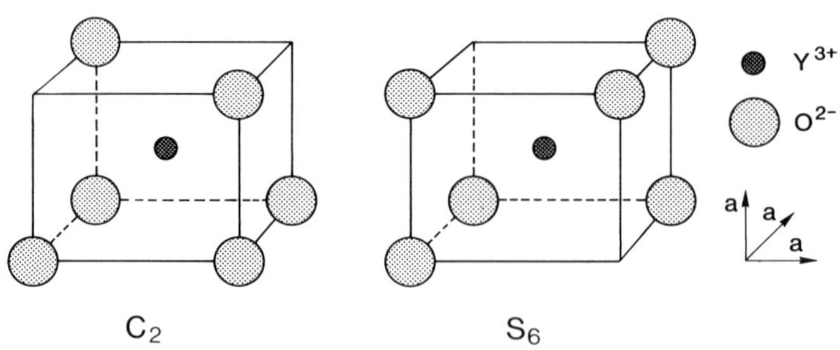

Fig. 32
The two rare-earth sites in $Y_2O_3:Eu^{3+}$

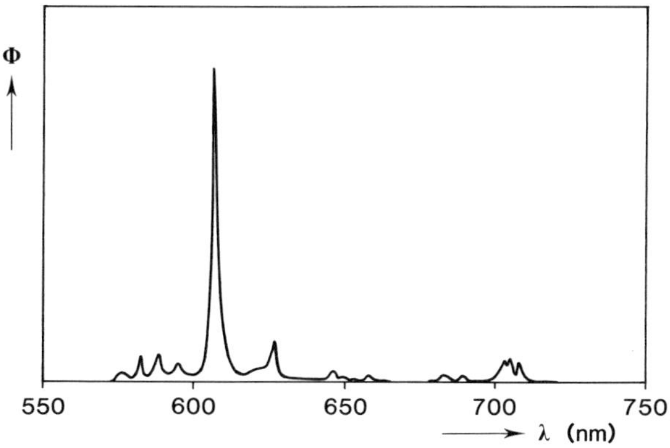

Fig. 33
Emission spectrum of $Y_2O_3:Eu^{3+}$

III. LAMP PHOSPHORS

Fluorescent lighting started already before the second world war. The UV radiation, generated by a low-pressure mercury discharge, was converted into white light by a phosphor layer sticking to the inner lamp wall. In this way a lamp was obtained four times more efficient than an incandescent lamp. The original phosphor blend was quickly replaced by a single phosphor, i.e. calcium halophosphate.

The advent of rare-earth based phosphors has altered drastically fluorescent lighting over the last decade. Higher efficiencies and a better to excellent colour rendition can be obtained in this way. This explains the large penetration of fluorescent lighting in stores and offices. Rare-earth phosphors can also be applied in lamps with a much smaller diameter, this has ultimately resulted in the introduction of compact fluorescent lamps. Thanks to their much smaller dimensions these lamps are highly suited for indoor applications.

III.A. Mercury discharge

A fluorescent lamp is filled with a noble gas at a pressure of 4.10^2 Pa, containing 8.10^{-1} Pa mercury. Due to collisions with electrons Hg-atoms are excited, and these atoms return to the ground state under the emission of UV radiation. Typically 85 % of the energy is emitted at 254 nm and 12 % at 185 nm. The remaining 3 % is found in the long wavelength UV and visible spectrum, notably at 365, 405, 436 and 546 nm.

In order to be of any practical use, a lamp phosphor should therefore be able to convert efficiently both the 185 nm and 254 nm radiation. The mercury lines emitted in the blue-green part of the spectrum will hardly contribute to the lamp output, but will affect the colour point of the emitted radiation. Due to the admixture of the visible radiation, the colour point of a lamp containing a single phosphor such as $Y_2O_3:Eu^{3+}$ will always lie inside the colour triangle.

III.B. White light

Lamps used for general lighting will emit white light, in this way our natural environment, i.e. the sun light, is imitated. The sun is a black body radiatior, its emission spectrum obeys Planck's equation:

$$E(\lambda) = \frac{A}{\lambda^5} \frac{1}{e^{-B/T_c} - 1} , \quad (15)$$

where A and B are constants λ equals the emission wavelength and T_c the temperature of the black body. With increasing T_c the colour of such a radiator will change from red, over faint orange to a hint of blue. The corresponding colour points are given by the black body locus (BBL). All the colours lying on the BBL are considered to be white. It is normal practice to characterize a lamp by its correlated colour temperature T_c, the temperature of a black body radiator with the same colour point (Fig. 34).

The simplest way to generate white light is by mixing blue and orange, this principle is used in halophosphate lamps. The mixing of the three primary colours: blue, green and red, results in highly efficient tricolour lamps. Blending a number of emission bands into a continuous spectrum will of course also result in white radiation, this principle is used in the Super de Luxe lamps.

Apart from its colour point colour rendition is a very important lamp characteristic, this property depends largely on the spectral energy distribution. It is determined by comparing the colour points of a set of test colours under illumination with the lamp to be tested and with a black body radiator. The colour rendering index (CRI) will equal 100 when the colour points are the same under illumination with both sources. The lower the CRI of the light source, the less natural the appearance of an object will be to the human eye.

III.C. Halophosphate lamps

The halophosphates with composition $Ca_5(PO_4)_3X$ (X = Cl,F) are closely related to hydroxy-apatite, the chief constituent of bones and teeth. It forms hexagonal crystals ($P6_3/m$, a = 9.37 Å, c = 6.88 Å, Z = 2) [28], its structure is given in Fig. 33. In this compound two different alkaline earth sites are present, the smaller Ca_I site and the larger Ca_{II} coordinated

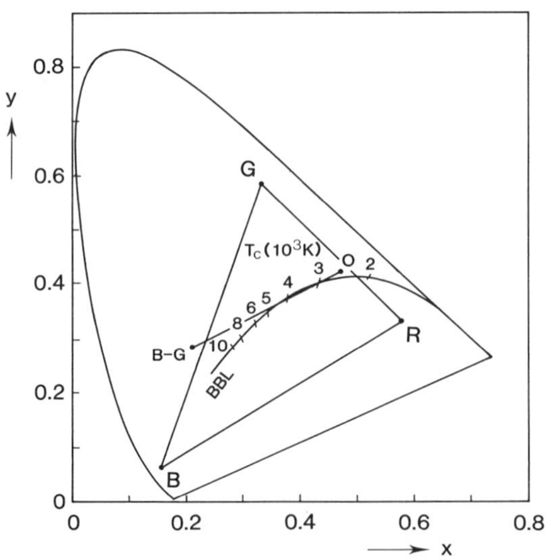

Fig. 34
CIE chromaticity diagram with black body locus (BBL). White light can be generated by blending blue-green and orange radiation, or by blending the three primary colours blue, red and green.

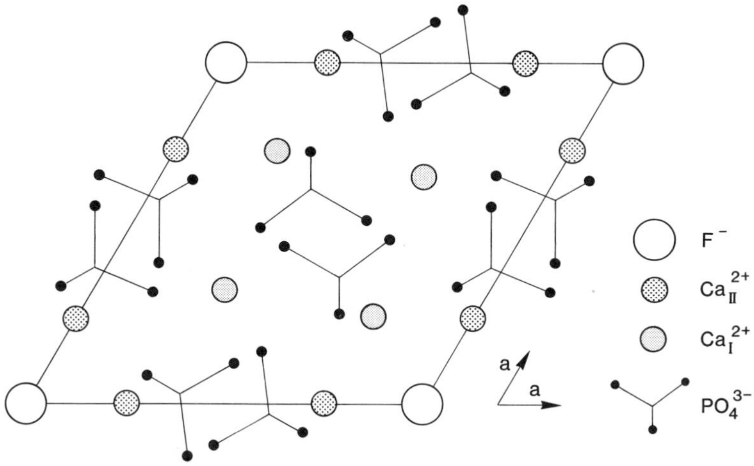

Fig. 35
Structure of $Ca_5(PO_4)_3X$. Due to the projection only three out of four P-O bonds are shown.

by both oxygen and halogen ions.

It was discovered by Jenkings et al. [29] that on doping with Sb^{3+} a highly efficient blue phosphor is formed. Oomen et al. recently pointed out that the high efficiency of this luminescence should be ascribed to the Jahn-Teller effect acting on the Sb^{3+} relaxed excited state [30]. It is also argued by these authors that the Sb^{3+} ions would preferentially occupy the larger Ca(II)-sites. When co-doped with Mn^{2+}, entering the smaller Ca(I)-sites [31], part of the absorbed energy is transferred to Mn^{2+} resulting in an orange luminescence (Fig. 36). By carefully adjusting the Sb^{3+} and Mn^{2+} concentrations a white emitting phosphor can be obtained with colour temperatures ranging between 6500 and 2700 K. The quantum efficiency of this phosphor equals 90 %, under 254 nm excitation. Under 185 nm excitation on the contrary the efficiency is only 50 %.

With the halophosphates a lamp output of 80 lm/W can be obtained. The CRI of these lamps, however, is rather low and equals only 50-60.

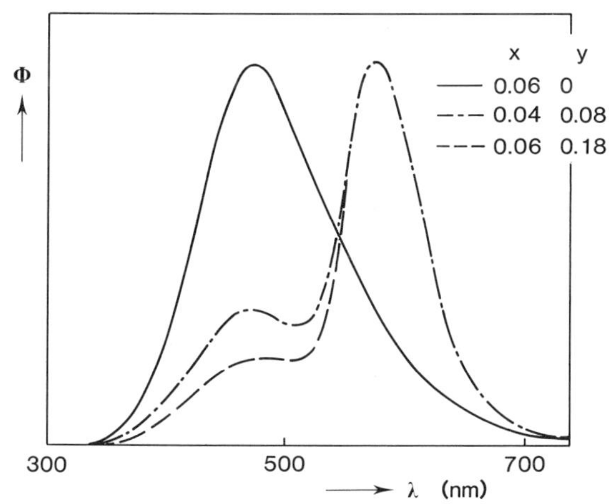

Fig. 36
Emission spectrum of $Ca_5(PO_4)_3(Cl,F):Sb_x^{3+},Mn_y^{2+}$

III.D. Tricolour lamps

About fifteen years ago it was predicted by Koedam and Opstelten [32], that by combining three phosphors, emitting in a narrow wavelength interval centered around 450, 550 and 610 nm respectively, a fluorescent lamp with an efficacy of 100 lm/W and a CRI of 80-85 could be obtained. A few years later such a lamp was actually realized using rare-earth based phosphors. This type of lamp is known as the tricolour lamp.

Narrow band emission or line emission is typically found with rare earth ions. Indeed, the half width of the emission band is appreciably smaller for Eu^{2+} than for Sb^{3+} phosphors, emitting in the blue part of the visible spectrum. Phosphors doped with trivalent rare earth ions typically show line emission, among them Eu^{3+} and Tb^{3+} are the best candidates for the red and green-emitting component respectively.

$Y_2O_3:Eu^{3+}$ fulfills all the requirements for a good red emitting phosphor. Its emission is located at 613 nm, it can easily be excited with a mercury discharge and its quantum efficiency comes very close to 100 %. The charge transfer band of Eu^{3+}, located at 240 nm [24], overlaps quite reasonably with the mercury excited at 254 nm. The 185 nm radiation is absorbed by the host lattice.

As far as the green- and blue-emitting tricolour phosphors are concerned the situation is somewhat more complicated. We will discuss them in separate sections.

1. The green-emitting Tb^{3+}-doped phosphors

In all green-emitting lamp phosphors the Tb^{3+} $4f^8 \rightarrow 4f^7d$ transition is located at fairly high energies, therefore the overlap with the main mercury radition at 254 nm is relatively small. In order to absorb efficiently the 254 nm mercury radiation a sensitizer such as Ce^{3+} has to be used. Originally $CeMgAl_{11}O_{19}:Tb^{3+}$ was applied but over the last few years two new components have emerged offering a somewhat superior output, i.e. $(La,Ce)PO_4:Tb^{3+}$ and $(Ce,Gd)MgB_5O_{10}:Tb^{3+}$.

$CeMgAl_{11}O_{19}$ has a magnetoplumbite-type hexagonal structure ($P6_3/mmc$, a = 5.57 Å, c = 22.00 Å, Z = 2) [33]. The crystal is built up out of spinel blocks with composition $MgAl_{10}O_{16}$. They are separated by an intermediate layer containing one Al^{3+} ion, one Ce^{3+} ion and three O^{2-} ions per spinel block (see Fig. 37).

The orthophosphates of the large rare-earth ions La^{3+} and Ce^{3+} are dimorphous. The monoclinic modification ($P2_1/n$, a = 6.76 Å, b = 7.00 Å, c = 6.44 Å, β = 103.63°, Z = 4 for Ln = Ce), with the monazite structure [34] yields highly efficient Tb^{3+} luminescence. Its structure is given in Fig.

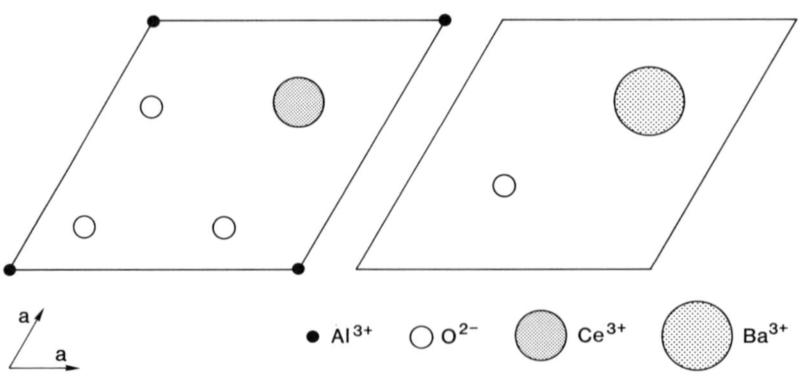

Fig. 37
The intermediate layer in the related structures of $CeMgAl_{11}O_{19}$ (a) and $BaMgAl_{10}O_{17}$ (b)

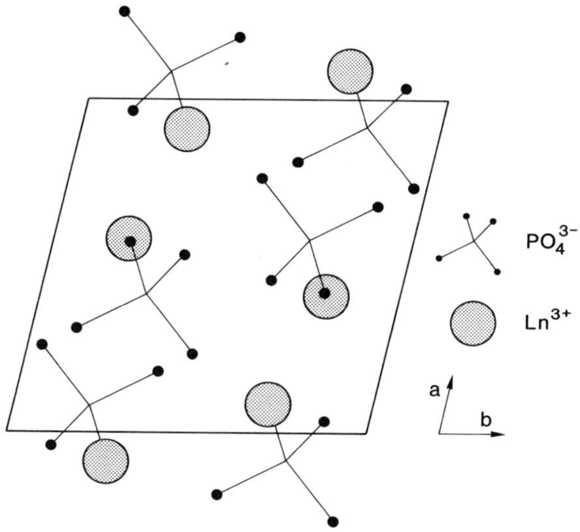

Fig. 38
The structure of monazite $LnPO_4$

38, it consists of isolated $PO_4{}^{3-}$ groups and Ln^{3+} ions occupying one crystallographic site.

Recently a new monoclinic borate compound with composition $LnMgB_5O_{10}$ was described by Saubat et al. ($P2_1/c$, a = 8.70 Å, b = 7.48 Å, c = 12.45 Å, β = 131.17°, Z = 4 for Ln = Gd) [35]. In this structure the rare earth ions form isolated zig-zag chains (Fig. 37). The shortest intrachain Gd-Gd distance equals 4 Å, the shortest interchain distance 6.4 Å. The boron network consists of BO_3 triangles and BO_4 tetrahedra. Especially the gadolinium compound yields an efficient phosphor when doped with Ce^{3+} and Tb^{3+}.

The emission spectra of the three compounds are given in fig. 40. In all of them some UV emission originating either from Ce^{3+} or Gd^{3+} is present. In view of the large Stokes-shift of the Ce^{3+} emission almost no

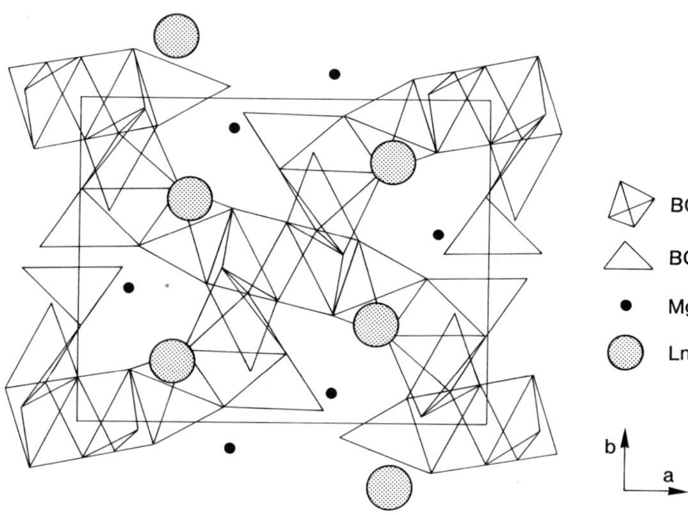

Fig. 39
The structure of $GdMgB_5O_{10}$

energy transfer will occur between Ce^{3+} ions in $CeMgAl_{11}O_{19}$ [36]. The cerium emission band at 330 nm overlaps the terbium $^7F_6 \rightarrow {}^5G_2, {}^5D_1, {}^5H_7$ absorption lines [37], making an efficient transfer between neighbouring Ce^{3+} and Tb^{3+} ions possible. In order to get an efficient overall transfer from Ce^{3+} to Tb^{3+} therefore fairly high Tb^{3+} concentrations should be used.

In $(Ce,La)PO_4$ the Stokes-shift of the Ce^{3+} emission is much smaller and energy migration over the Ce^{3+} sublattice will play an important role in the luminescence process [38]. Due to this migration even at relatively low terbium concentrations an efficient overall transfer is found between Ce^{3+} and Tb^{3+} (see Table 3).

In $LaMgB_5O_{10}:Ce^{3+}$ also a small Stokes-shift is found, resulting in energy migration over the Ce^{3+} sublattice [39]. The Ce^{3+} emission bands at 310 and 315 nm overlap quite well with the gadolinium $^8S-{}^6P$ absorption line. Therefore gadolinium will play an intermediate role in the energy migration from Ce^{3+} to Tb^{3+} [40]. Apart from one-dimensional intrachain energy transfer, also interchain transfer is found in $GdMgB_5O_{10}$ [41]. The more efficient energy transfer in Gd^{3+} containing phosphors is exemplified by the much lower UV emission found in $(Ce,Gd)MgB_5O_{10}:Tb^{3+}$ compared with $(Ce,La)PO_4:Tb^{3+}$ (see Table 3).

Table 3
Quantum efficiency (254 nm excitation) of Tb^{3+} phosphors

composition	U.V. (%)	Tb^{3+} (%)
$Ce_{0.67}Tb_{0.33}MgAl_{11}O_{19}$	5	85
$Ce_{0.45}La_{0.40}Tb_{0.15}PO_4$	7	86
$Ce_{0.3}Gd_{0.5}Tb_{0.2}MgB_5O_{10}$	2	88

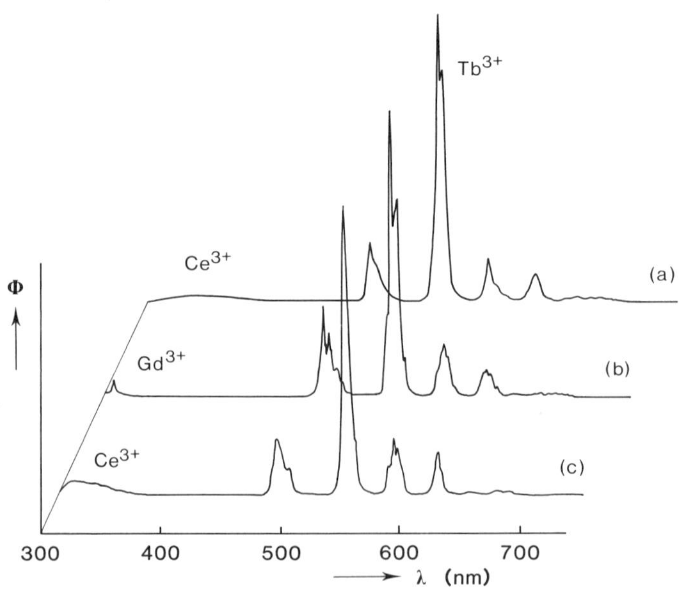

Fig. 40
Emission spectra of $CeMgAl_{11}O_{19}:Tb$ (a), $(Ce,Gd)MgB_5O_{10}:Tb$ (c) and $(La,Ce)PO_4:Tb$ (b)

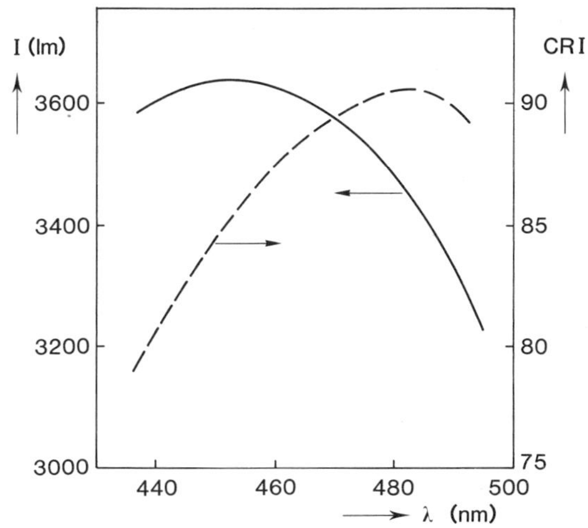

Fig. 41
Dependence of the lamp output of a TLD 36W bulb (T_c = 4000 K) on the emission wavelength of the blue component

2. Blue-emitting Eu^{2+} phosphors

The dependence of the lamp efficiency and CRI on the position of the blue emission band is given in Fig. 41. The highest efficiency is found for a blue phosphor emitting at 450 nm. The colour rendition increases drastically with increasing wavelength, this however occurs at the expense of the lamp efficiency. Because tricolour lamps are aiming at a high output in combination with good colour rendition only phosphors emitting between 440 and 460 nm are of practical interest. Three efficient phosphor with a quantum yield of 90 % are known in this wavelength interval, i.e.: $BaMgAl_{10}O_{17}:Eu^{2+}$, $Sr_3(PO_4)_5Cl:Eu^{2+}$ and $Sr_2Al_6O_{11}:Eu^{2+}$. Their emission spectra are given in Fig. 42.

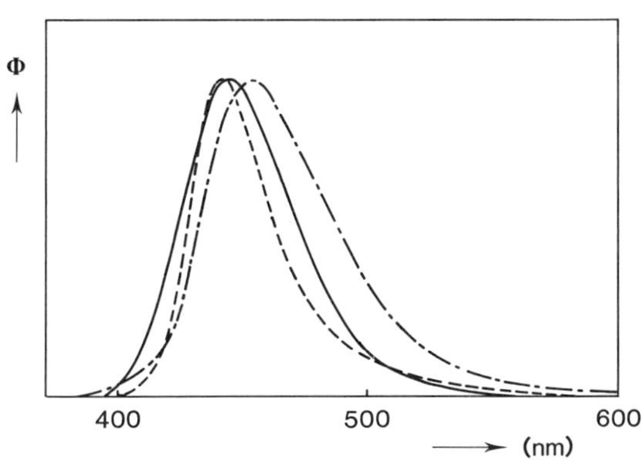

Fig. 42
Emission spectrum of $BaMgAl_{10}O_{17}:Eu_{0.1}$ (——), $Sr_5(PO_4)_3Cl:Eu_{0.05}$ (---) and $Sr_2Al_6O_{11}:Eu_{0.04}$ (-.-.)

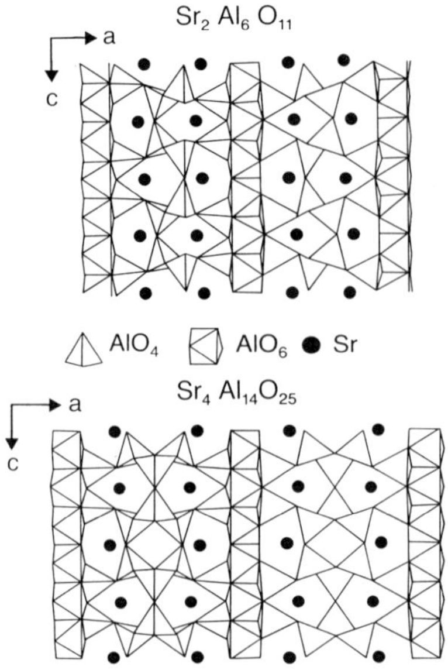

Fig. 43
Structure of $Sr_2Al_6O_{11}$:Eu and $Sr_4Al_{14}O_{25}$:Eu

$BaMgAl_{10}O_{17}$ forms β-alumina type crystals ($P6_3/mmc$, a = 5.64 Å, b = 22.65 Å, Z = 2) closely related to $CeMgAl_{11}O_{19}$. Both compounds contain identical spinel blocks, while Ce^{3+} is replaced by Ba^{2+} and AlO_3 by a single oxygen ion (Fig. 37). $Sr_5(PO_4)_3Cl:Eu^{2+}$ belongs to the halophosphates, mentioned before. The compound $Sr_2Al_6O_{11}:Eu^{2+}$ crystallizes in the orthorhombic P_{nmm} space group with a = 22.13 Å, b = 4.88 Å, c = 8.42 Å and Z = 4. Its structure is depicted in Fig. 43 and is built-up out of alternating layers containing either AlO_4 or AlO_6 groups. Two crystallographic Sr sites are present, in both cases the average oxygen strontium distance equals 2.7 Å.

III.E. Super de Luxe lamps

Tricolour lamps show only emission in restricted wavelength intervals. For objects with a reflection spectrum peaking outside these regions the colour appearance under illumination with a tricolour lamp will differ from the one under illumination with a black body radiator. Although a CRI of 85 guarantees a normal appearance for most objects, some typical colours will look unnatural under illumination with tricolour lamps. This has resulted in the introduction of the Super de Luxe lamps with a near perfect CRI of 95.

As was shown in Fig. 41 with a blue component emitting at 490 nm a colour rendition of 90 can be obtained in tricolour lamps. The colour rendition can be increased even further by using band instead of line phosphors for the red and green components. In this way a continuous spectrum extending from the blue to the deep red is obtained.

$Sr_4Al_{14}O_{25}$ doped with Eu^{2+} yields a blue emission located at 490 nm (Fig. 44), its quantum efficiency equals 90 %. The compound $Sr_4Al_{14}O_{25}$ crystallizes in the orthorhombic space group P_{mma}, with a = 24.78 Å, b = 8.49 Å, c = 4.89 Å and Z = 2. Actually its structure is closely related to the one of $Sr_2Al_6O_{11}$ (Fig. 41), except for the fact that instead of the tetrahedral plane a twinned tetrahedral plane is present, resulting in a longer a-axis. Due to the twinning the average Sr-O distance in site I becomes much smaller than in site II. Apart from the 490 nm band a second emission band is also found at 400 nm, the first one originates from Eu^{2+} in the Sr_I site the second from Eu^{2+} in the Sr_{II} site. Although the number

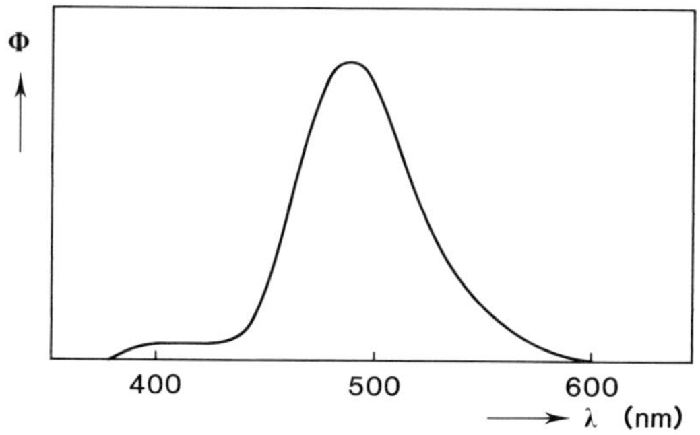

Fig. 44
Emission spectrum of $Sr_4Al_{14}O_{25}$:0.1Eu

of Sr_{II} sites equals the number of Sr_I sites the intensity of the 490 nm emission is much higher, this is caused by the efficient energy transfer from europium ions in site II to the ions in site I.

A broad-band red emitting phosphor can be realized in the pentaborate lattice. The $LnMgB_5O_{10}$ structure also contains small distorted octahedral Mg^{2+} sites, which can easily accommodate Mn^{2+} ions. In view of their coordination these Mn^{2+} ions will show a typical red luminescence around 630 nm. Direct sensitization of the Mn^{2+} emission by Ce^{3+} is not possible [42], but the emission can be sensitized via the Gd^{3+} lattice [43].

An efficient lamp phosphor with suitable broad-band green emission is not known until know. But such a band can be simulated by combining a Tb^{3+}-emission band such as the one of (Ce,Gd)MgB_5O_{10}:Tb^{3+} with the Mn^{2+} emission band of $Ca_5(PO_4)_3$(Cl,F):Sb^{3+},Mn^{2+}.

A mixture of $Sr_2Al_6O_{11}$:Eu^{2+}, (Ce,Gd)MgB_5O_{10}:Tb^{3+}, $Ca_5(PO_4)_3$(Cl,F):Sb^{3+},Mn^{2+} and (Ce,Gd)MgB_5O_{10}:Mn^{2+} will yield a lamp with an efficacy of 65 lm/W and a CRI of 95. Because Tb^{3+} and Mn^{2+} can be accommodated simultaneously in the $GdMgB_5O_{10}$ structure (Fig. 45) in reality three instead of four phosphors are used.

In order to obtain an excellent colour rendition a more or less continuous spectrum is a prerequisite. For low colour temperature lamps the major part of the blue emission will, however, originate from the sharp mercury lines adversely affecting the CRI. Only by lowering the intensity of the blue mercury lines a CRI of 95 can be obtained for lamps with a $T_c \lesssim$ 3000 K. This can be realized either by using a yellow pigment or, even more

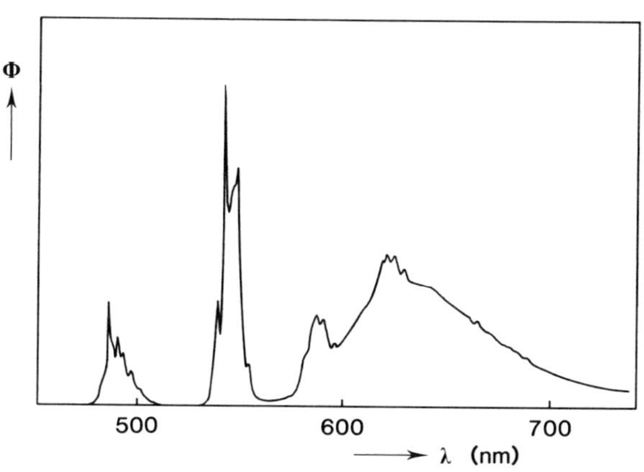

Fig. 45
Emission spectrum of $Ce_{0.2}Gd_{0.6}Tb_{0.2}Mg_{0.9}Mn_{0.1}B_5O_{10}$

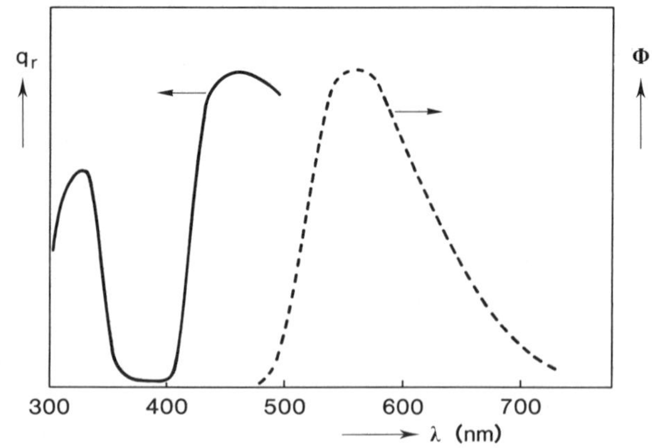

Fig. 46
Emission (---) and excitation (——) spectrum of $Y_3Al_5O_{12}:Ce^{3+}$

advantageously, by using a phosphor responding to the 435 nm mercury radiation.

Such a phosphor is known in the garnet system, i.e.: $Y_3Al_5O_{12}:Ce^{3+}$ [44]. The excitation and emission spectra of $Y_3Al_5O_{12}:Ce^{3+}$ are given in Fig. 44. The lowest lying Ce^{3+} 4f-5d absorption band in this compound is situated at 460 nm. On excitation in this band a yellow emission around 580 nm is found, the quantum efficiency of this process equals 80 %.

III.F. Compact fluorescent lamps

The application of rare-earth based phosphors has also resulted in a drastic improvement of the lamp maintenance as compared with the halophosphates. Lamp maintenance tends to decrease with increasing wall load (see Fig. 47). The wall load is more or less inversely proportional with the diameter of the discharge tube. Thanks to the higher stability of rare earth based phosphors the lamp diameter could be reduced, without affecting the lamp life.

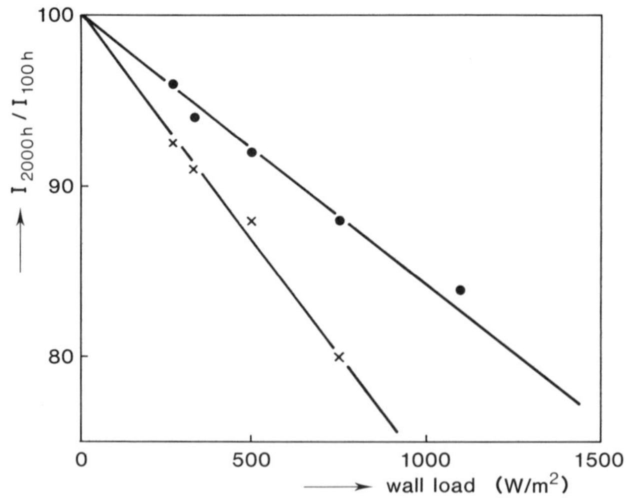

Fig. 47
Lamp maintenance, output after 2000 hrs relative to the output after 100 hrs of burning time, as a function of the wall load for a halophosphate (x) and a tricolour lamp (•)

Tricolour and Super de Luxe lamps are now available in a tube diameter of 25 mm (wall load of 330 W/m^2) instead of 35 mm (270 W/m^2). It was even possible to reduce the diameter of tricolour lamps to 10 mm (1100 W/m^2), a crucial step in the development of compact fluorescent lamps.

III.G. Present State and Future Trends

The introduction of rare-earth based phosphors has drastically altered fluorescent lighting over the last few years. Apart from the cheap halophosphate lamps we now have at our disposal the highly efficient tricolour lamps, with good colour rendition, and the Super de Luxe lamps, with lower output but excellent colour rendition. The large market penetration of the more expensive rare-earth phosphors in Europe and the Far East proves that people are willing to pay for a more pleasant illumination.

As we have seen these new phosphors almost perfectly meet all the requirements for application in fluorescent lamps: i.e. strong UV absorption, high efficiency and optimum emission wavelength. The performance of fluorescent lamps is not expected to improve drastically over the coming years. The major draw-back of the rare-earth based phosphors applied in fluorescent lighting is their relatively high price. Probably future efforts in lamp phosphor development will concentrate on a reduction of the phosphor costs.

REFERENCES

1. D.J. Robbins, J. Electrochem. Soc. 127, 2694 (1980)
2. L.H. Brixner, Mat. Chem. Phys. 16, 253 (1987)
3. R.W.G. Wyckoff, Crystal Structures, Vol. 3, Robert E. Krieger Publishing Company, p. 21 (1981)
4. H.. Venema, Radiology 130, 765 (1979)
5. A.L.N. Stevels and F. Pingault, Philips Res. Rep. 30, 277 (1975)
6. R.W.G. Wyckoff, Crystal Structures, Vol. 1, Robert E. Krieger Publishing Company, p. 295 (1982)
7. K. Takahashi, K. Kohda, J. Miyahara, J. Lumin. 31/32, 266 (1984)
8. K. Takahashi, Y. Shibahari and J. Migahara, Electrochem. Soc., 163rd meeting, Abstract 921 (1983)
9. A. Feretti, U.S. Patent 4524016 (1985)
10. J.G. Rabatin, ACS Symp. Ser. 164 , 203 (1981)
11. R.W.G. Wyckoff, Crystal Structures Vol. 3, Robert E. Krieger Publishing Company, p. 22 (1981)
12. L.H. Brixner and H.Y. Chen, J. Electrochem. Soc. 130, p. 2435 (1983)
13. R.W.G. Wyckoff, Crystal Structures Vol. 2, Robert E. Krieger, Publishing Company p. 3 (1986)
14. S.D. Scott and H.L. Barnes, Geochimica et Cosmochimica Acta 36, 1275 (1972)
15. A. Bril, Physica 15, 361 (1949)
16. J.M. Robertson, M.W. van Tol, W.H. Smits and J.P.H. Heynen, Philips J. Res. 36, 15 (1981)
17. W.F. v.d. Weg, J.M. Robertson, W.K. Zwicker and Th.J.A. Popma, J. Lum. 24/25, 633 (1981)
18. K. Ohno, Electrochem. Soc., 172 nd meeting, Abstract 1239 (1987)
19. C.M. Buisson and E.F. Bertaut, C.R. Acad. Sc. Paris Ser. B 264, 397 (1967)
20. M.L. Keith and R. Roy, Am. Mineral. 39, 1 (1954)
21. R. Wyckoff, Crystal Structures Vol. 2, Robert E. Krieger Publishing Company, p. 362 (1986)
22. B. Hintzen, Electrochem. Soc., 172nd meeting, Abstract 1240 (1987)
23. A. Fert, Bull. Soc. Franc. Mineral-Crist 85, 267 (1962)
24. M. Buijs, A. Meyerink and G. Blasse, J. Lum. 37, 9 (1987)

25. A.H. Gomes de Mesquita and A. Bril, Mat. Res. Bull. 4, 643 (1969)
26. A.L.N. Stevels, P. Vriens, A.W. Veenis and A.T. Vink, Electrochem. Soc., 155th meeting, Abstract 337 (1978)
27. D. Klaassen, D. de Leeuw and T. Welker, J. Lum. 37, 21 (1987)
28. O.R. Trautz, Ann. N.Y. Acad. Sci. 60, 696 (1955)
29. H.G. Jenkings, A.H. McKeag and P.N. Ranby, J. Electrochem. Soc. 96, 1 (1949)
30. E.W.J.L. Oomen, W.M.A. Smit and G. Blasse submitted for publication in Mat. Chem. Phys.
31. R.W. Warren, Phys. Rev. B2, 4383 (1970)
32. M. Koedam and J.J. Opstelten, Light. Res. Technol. 3, 205 (1971),
33. M. Gasparin, M.C. Saine, A. Kahn, F. Laville and A. Lejus, J. Solid State Chem. 54, 61 (1984)
34. R.G. Wyckhoff, Crystal Structures, Vol. 3 (2nd ed.), John Wiley, p. 32 (1965)
35. B. Saubat, M. Vlasse and C. Fouassier, J. Solid State Chem. 34, 271 (1980)
36. J.M.P.J. Verstegen, J.L. Sommerdijk and J.G. Verriet, J. Lumin. 6, 425 (1973)
37. R.C. Ropp, J. Electrochem. Soc. 116, 623 (1969)
38. J.C. Bourcet and F.K. Fong, J. Chem. Phys. 60, 4 (1974)
39. B. Saubat, C. Fouassier, P. Hagenmuller and J.-C. Bourcet, Mat .Res. Bull. 16, 193 (1981)
40. J. de Hair, J. Lumin. 18/19, 797 (1979)
41. G. Blasse, J. Solid State Chem. 62, 207 (1986)
42. M. Saakes, M. Leskelä and G. Blasse, Mat. Res. Bull. 19, 83 (1984)
43. M. Saakes, M. Leskelä and G. Blasse, Mat. Res. Bull. 19, 151 (1984)
44. M.V. Hoffman, J. IES 17, 89 (1977)

ENERGY TRANSFER IN Cr,Tm:YAG

A.M. Buoncristiani and G. Armagan
Christopher Newport College
Newport News, VA 23606 USA

B. Di Bartolo
Boston College
Chestnut Hill, MA 02167 USA

J.J. Swetits
Old Dominion University
Norfolk, VA 23508 USA

ABSTRACT

We describe a systematic analysis of energy transfer processes in crystals of YAG doped with varying concentrations of Cr and Tm. We use both spectral measurements and measurements of the temporal response to pulsed excitation to give independent determinations of the microscopic interaction parameter for Cr to Tm transfer. We discuss the different factors in influencing the temperature dependence of the Cr to Tm transfer. Finally, we determine the dependence of the Tm cross-relaxation rate on Tm concentration.

I. INTRODUCTION

Recently there has been an interest in the development of a solid state laser operating in the "eye-safe" region of the spectrum (wavelengths greater than 1.54 microns). This interest has led to renewed studies of the interionic processes among Cr, Tm and Ho ions in various crystalline hosts. The ${}^5I_7 \rightarrow {}^5I_8$ transition of the triply ionized Holmium ion is a good candidate for a 2-micron laser as it has a large cross-section and a long upper state lifetime. Thulium is expected to be a good sensitizer for Holmium because it has an efficient cross relaxation and energy transfer coupling to the upper laser level of Ho. Furthermore, Thulium absorbs around 780 nm so it could be pumped directly

Advances in Nonradiative Processes in Solids
Edited by B. Di Bartolo, Plenum Press, New York, 1991

by a semiconductor diode laser or indirectly by energy transfer from flashlamp pumped Chromium. In order to obtain pulses short enough for several of the proposed applications it is necessary to operate the laser in a Q-switched mode. Design of lasers to operate in this mode requires a thorough knowledge of the transfer rates for each of the various interionic processes coupling the energy of the pump to the laser transitions. We have initiated a systematic study of the energy transfer among Cr, Tm and Ho ions in YAG as a first step in the analysis of Q-switched operation of lasers based on the Ho ion sensitized by Cr and Tm. In this paper we shall focus on the Cr, Tm system, describing a general set of procedures aimed at obtaining detailed information on the energy transfer processes. We shall also mention the cross relaxation process in Tm. This work provides a clear example of the energy transfer processes discussed in references 1 and 2.

II. CHROMIUM TO THULIUM ENERGY TRANSFER

II.A. The Transfer Process

The energy levels of Cr and Tm are shown in Figure 1. We shall be interested in the following energy pathway. Energy absorbed in the broad Cr absorption bands ($^4T_1$, $^4T_2$) rapidly goes to the ^2E level. From Cr: ^2E the energy transfers to the Tm $^3F_2$ and $^3F_3$ levels. From there it nonradiatively populates the $^3H_4$ level. Tm may undergo a cross-relaxation $\left[^3H_4, {}^3H_6 \rightarrow {}^3F_4, {}^3F_4\right]$ to the $^3F_4$ level from which it feeds the upper laser level in Ho. In this article we shall focus attention on the Cr to Tm transfer and the Tm cross-relaxation.

We are considering a crystalline host in which two ionic species can carry an optical excitation; donor ions D are initially excited and may subsequently transfer this excitation to acceptor ions A. The rate of energy transfer, by a dipole-dipole transition, from a donor ion to an acceptor ion separated by a distance R is proportional to the inverse of the sixth power of R, [3]

$$W_{DA} = C^{(6)}/R^6. \tag{1}$$

The coefficient $C^{(6)}$ is called the microscopic interaction parameter. It is convenient to express it in terms of the effective decay time of the donor ion (in the absence of any acceptors) τ_0 and a characteristic interaction radius R_0

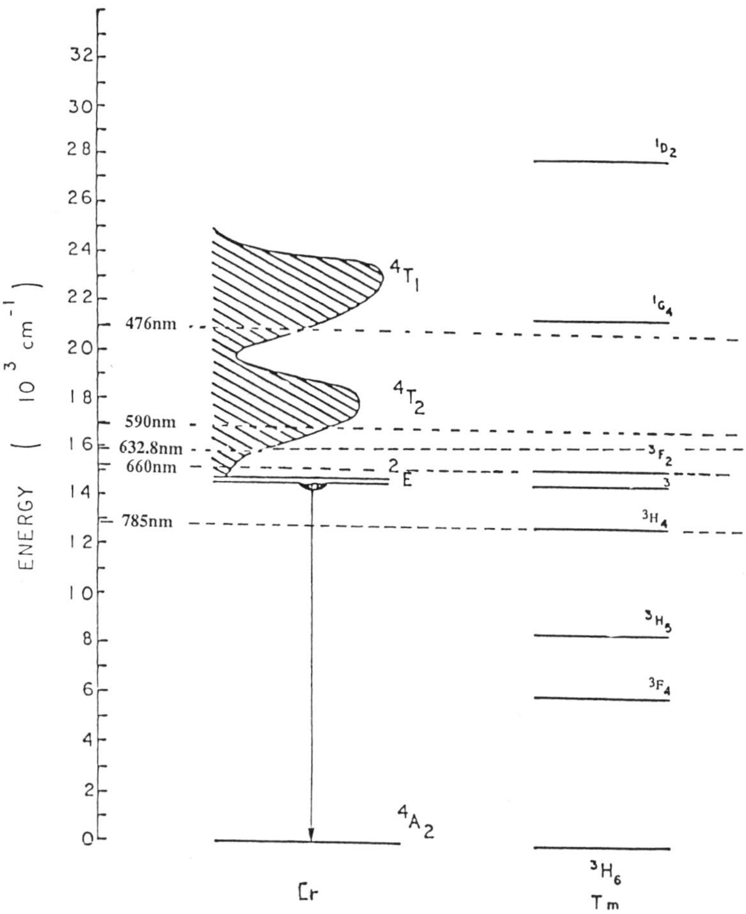

Figure 1. Energy levels of Cr, and Tm in YAG.

$$C^{(6)} = R_0{}^6/\tau_0. \qquad (2)$$

When the two ions are separated by R_0 the rate of energy transfer to an acceptor is equal to the decay rate for an isolated donor.

Generally speaking we are considering two types of measurements - steady state spectral measurements and measurements of the temporal response to pulsed excitation. We can use each type of measurement to get an independent determination of the inter-ionic interaction rate.

Experiments with pulsed excitation lead to effective decay times of the donor ion. This decay time in a sample without acceptors gives the lifetime of the donor level τ_0 and in a sample with acceptors it displays the lifetime shortening due to the energy transfer process. Experiments

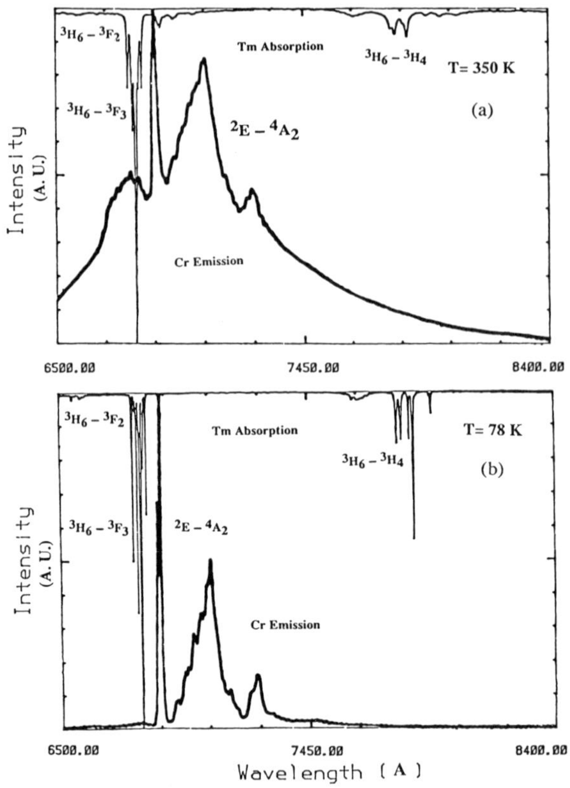

Figure 2. The Tm absorption spectrum and the Cr emission spectrum are shown for two temperatures.

with steady state excitation yield the emission spectra of the donor ions and the absorption spectra of the acceptor ion.

We should mention here that the donor ions may also transfer energy among themselves, diffusing the excitation. The more donor diffusion competes with donor to acceptor energy transfer the more difficult it is to extract information from the examination of the donor luminescence. This interference increases with increasing concentration of donors and leads to the different regimes discussed in reference 2.

II.B. Temperature Dependence of Energy Transfer

An important part of the experimental strategy demonstrated here involves examination of the variation of the various physical quantities with temperature. There are several ways in which the temperature dependence of these phenomena yield valuable information. First of all, there is some intrinsic interest in the temperature dependence of these factors. In addition, there is evidence that two phenomena may be related if they share the same temperature dependence. Finally, a statistical enhancement of the data accrues from the examination of the tendency of thermal variation.

II.C. Spectroscopic Measurements

As discussed in reference 1 there are three separate factors which contribute to the microscopic interaction parameter - the oscillator strengths for electronic dipole-dipole transitions in the donor ion emissions and in acceptor ion absorption and the spectral overlap between the donor emission and the acceptor absorption. The first of these, the donor oscillator strength, is proportional to the reciprocal of the decay time of the donor (in the absence of acceptors); this can be determined directly from the decay curves. The second, the acceptor oscillator strength is proportional to the integral of the acceptor absorption spectra over the wavelength range of the manifold. The third, the overlap integral can be calculated from the donor emission and the acceptor absorption spectra.

Figure 2 shows the Tm absorption spectrum and the Cr emission spectrum for two temperatures. While both spectra show broadening with temperatures the Cr broadening is much more extensive. There is a sideband developing on the high energy side of the Cr- Emission lines which overlaps with the ($^3H_6 \rightarrow {}^3H_4$) Tm absorption. The temperature dependence of the overlap integral is due primarily to the development of this sideband in resonance with Tm absorption. The Tm absorption does not change significantly with temperature and consequently the Tm oscillator strength remains constant with temperature. Using this spectral data we can determine the temperature dependence of the Cr-Tm overlap and the Tm absorption oscillator strength. The oscillator strength of the Cr emission is proportional to the reciprocal lifetime of Cr in the absence of Ho. The temperature dependence of each of these three factors is shown in Fig. 3. As can be seen there is little change of the Tm absorption integral with temperature over the range of temperature shown. There is a slight change in the overlap integral at lower temperatures reflecting the development of the anti-Stokes sideband. At the higher temperatures of interest in laser operation the dominant change in the microscopic interaction parameter is due to changes in the Cr emission oscillator strength.

II.D. Time Resolved Spectra

The time resolved spectra also gives evidence of the energy transfer. As can be seen in Figure 4, the temporal response to pulsed excitation for Cr alone is exponential. For crystals with higher Tm concentration this response becomes non-exponential. At early times it displays the rapid transfer of excitation from Cr to those Tm ions which are nearby. As time increases and the number of available nearby Tm ions decreases, the energy transfer rate diminishes and the decay curve settles into an exponential behavior. Evidence for energy migration among the Cr ions can be seen in the slopes of the decay curves at longer times. As the concentration of Cr increases the magnitude of the slopes increases.

As shown in (1) the time dependence of the decay curve is given by

$$I(t) = I_0\ \rho(t) \qquad (3)$$

where $\rho(t)$ is given explicitly in reference 1. Taking the natural logarithm of this we obtain an expression that is quadratic in $t^{1/2}$ and depends on three parameters.

$$\ln(I(t)) = \alpha - \beta\ t^{1/2} - \gamma\ t \qquad (4)$$

The measured data was fit to this expression to find the values of (α,β,γ) that minimize mean square error. As shown in reference 1 these parameters are

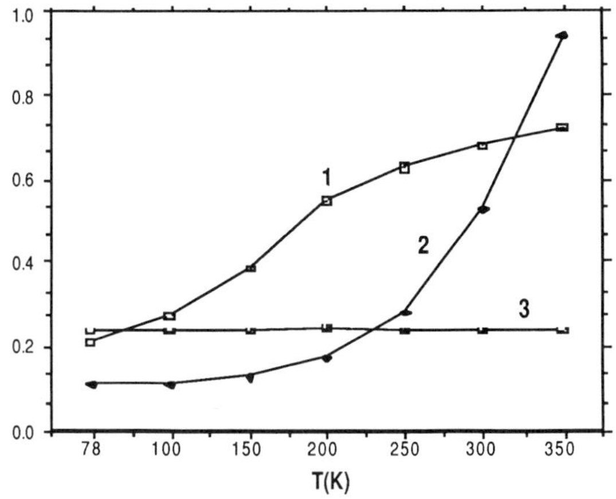

Figure 3. The time dependence of the three factors determining the microscopic interaction parameter. 1) Cr,Tm overlap integral $\times 10^{20}$ cm^{5}, 2) reciprocal lifetime of Cr (1/microsec), 3) Tm absorption coefficient integral $\times 10^{-3}$ (cm^{2}).

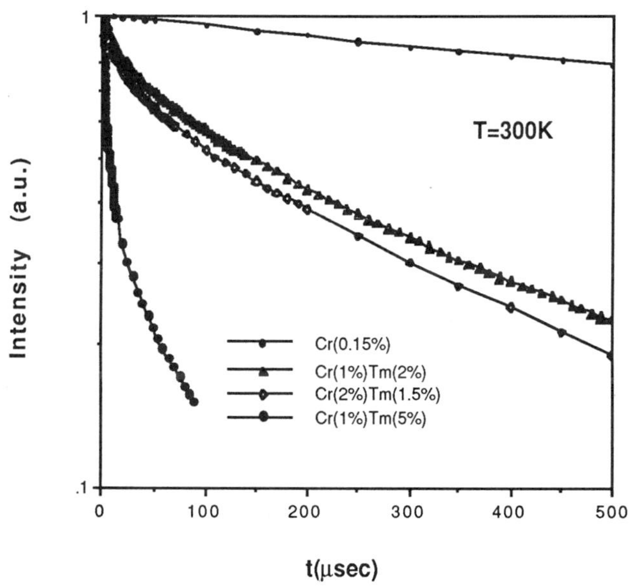

Figure 4. The temporal response to pulsed excitation for three samples of Cr and Tm doped YAG.

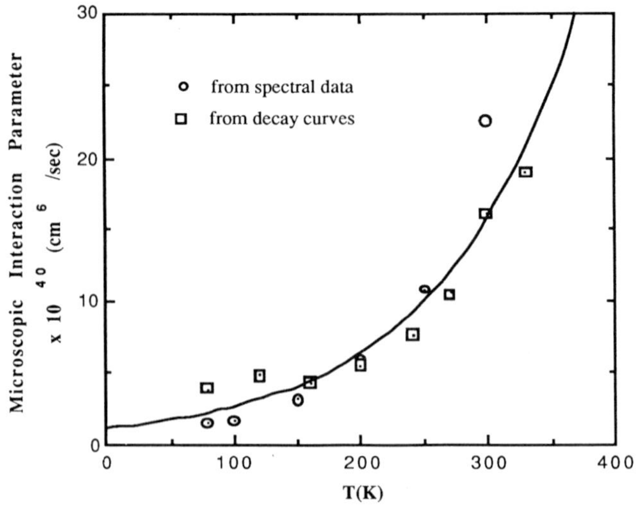

Figure 5. The microscopic interaction parameter determined by both spectroscopic measurements and by fitting the decay curves to temporal response predicted in Eq.(4).

$$\alpha = \ln (I_0)$$
$$\beta = (\pi/\tau_0)^{1/2} (C/C_0) \qquad (5)$$
$$\gamma = 1/\tau_0 + K_D$$

After obtaining the best fitting, we compare the measured life time with the coefficient of t.

II.E. Cr to Tm Microscopic Interaction Parameter

As we have discussed, it is possible to determine the microscopic interaction parameter by two independent techniques, one using spectroscopic data and the other from time resolved date. The microscopic interaction parameter calculated from both techniques is shown in Figure 5. As evident there is qualitative agreement between the results of the two methods. The curved line shown in the figure is a best fit to the data points; as discussed previously the temperature dependence of the microscopic interaction parameter is due to the three independent factors so that the shape of the fitted curve cannot be explained further.

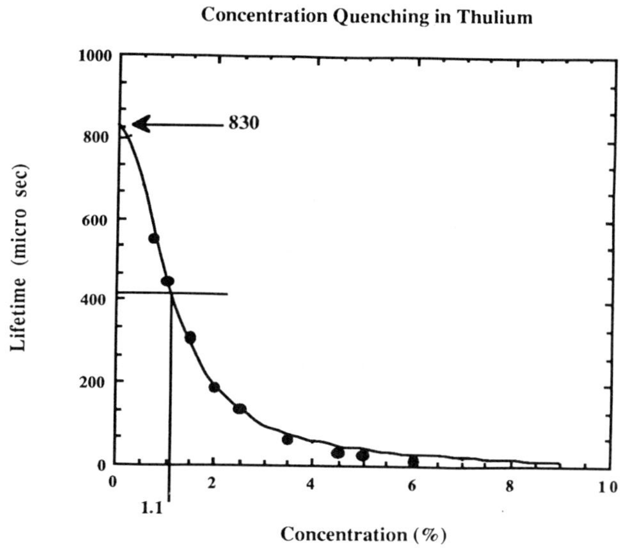

Figure 6. The lifetime of the $^3H_4$ level of Tm as a function of Tm concentration. The curve is then given by Eq.6.

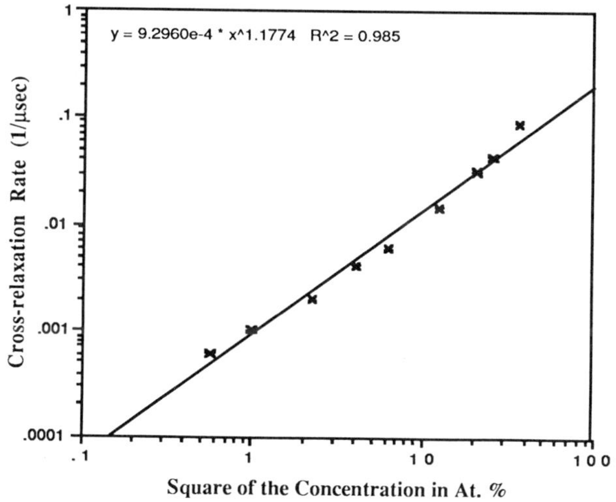

Figure 7. The rate of cross relaxation from the $^3H_4$ level of Tm as a function of the square of the Tm concentration.

III. THULIUM CROSS-RELAXATION

Finally, we wish to describe the Tm cross relaxation process. To determine the transfer rate we examine the concentration quenching of Tm. That is the decrease in lifetime of the Tm level with increasing concentration. This concentration quenching effect is shown in Figure 6. Fitting the concentration dependence of the lifetime to the quadratic formula

$$\tau = \tau_0 \Big/ \left[1 + (C/C_0)^2\right] \tag{6}$$

we can determine the intrinsic lifetime of Tm $^3H_4$ and the associated half concentration (the concentration of Tm at which the intrinsic lifetime is decreased by half). Using the intrinsic lifetime we can then separately determine the cross-relaxation rate. The concentration dependence of this rate is shown in Figure 7.

IV. SUMMARY

In this paper we have described an analysis of energy transfer processes occurring in Cr, Tm:YAG. We have used both spectral measurements and decay curves to provide separate determinations of the microscopic interaction parameter for the Cr to Tm transfer. By examining the temperature dependence of the various factors contributing to the microscopic interaction parameter we have been able to demonstrate that its temperature variation, above 200K, is due primarily to changes in the Cr lifetime. Concentration quenching in the Tm cross-relaxation led to a determination of the cross-relaxation rate.

This work was performed at the Solid State Laser Materials Laboratory, NASA Langley Research Center, Hampton, VA and supported by NASA grants NAG-1-796, NAG-1-955 and NAG-1-957.

REFERENCES

1. G. Armagan, B. Di Bartolo and A.M. Buoncristiani, J. of Luminescence, 44, 129 (1989).

2. G. Armagan, B. Di Bartolo and A.M. Buoncristiani, J. of Luminescence, 44, 141 (1989).

3. B. Di Bartolo, These Proceedings.

LUMINESCENCE AND NONRADIATIVE PROCESSES IN POROUS GLASSES*

Renata Reisfeld**

Department of Inorganic and Analytical Chemistry
The Hebrew University of Jerusalem
Jerusalem 91904 Israel

ABSTRACT

Porous glasses can be obtained by selective leaching of borosilicate precursors or by the new technique of the sol-gel process. The latter uses tetraalkoxysilanes as monomeric glass precursors. Inorganic ions or organic molecules exhibiting luminescence properties are incorporated into the porous glasses at room temperature. The radiative and nonradiative processes in the glasses and the connection between the luminescent properties and the interaction of the ions with the surrounding medium can be calculated from the luminescent behaviour of the emitting species. These properties will be discussed and the physics of the processes elaborated in detail.

I. INTRODUCTION

Recently there has been considerable interest in understanding the structure and dynamics in porous solids. Some of these solids can be obtained commercially. For example the most studied porous glass Vycor (R) 7930 [1] obtained from Corning is prepared by leaching out the boric oxide phase from borosilicate glass which has been thermally pretreated. It constitutes a porous open structure of effective area of 200 square meters per gram and densities of about 1.5 gm/cm^3. The diameters of the pores vary quite widely their average size being 40 Å. The pores form topologically connected, intermeshed random paths in the porous glass the stochastic geometry characteristics of which are amenable to description in terms of self similar structures.

Different types of porous glasses can be prepared by a sol-gel process [2] (and references therein). Depending on the preparation conditions the distribution of the pores and their size can be controlled.

The colorants, inorganic or organic, the behaviour of which we shall describe in this paper can be introduced into the glass either by soaking

* Partially supported by The Krupp and Volkswagen Foundations, the US Israel Binational Fund and the European Community
** Enrique Berman Professor of Solar Energy

or by adding them to the starting materials in the preparation by the sol-gel process.

II. PREPARATION OF AND CHARACTERIZATION OF POROUS GLASSES

The sol-gel process involves mixing metal organic reagents in solution with a solvent, water, and a catalyst at room or slightly higer temperatures. In this process a complex set of hydrolysis and polymerization reactions occur. The initially fluid solution becomes increasingly viscous as polymerization proceeds until a "wet solid" known as a gel is formed. This gel is composed of an interconnected oxide phase and pores which contain an alcohol/water/catalyst liquid. Drying the gel results in the expulsion of this liquid with substantial volume shrinkage of the gel, leaving a "dry" porous solid.

The overall reaction for tetraalkoxyorthosilane as exemplified here by tetramethylorthosilicate (TMOS) is [3]

$$nSi(OCH_3)_4 + mH_2O \rightarrow (-O\text{-}Si\text{-}O-)_n + 4nCH_3OH + (m - 2n)H_2O$$

in which water is consumed by

$$H_2O + Si(OCH_3)_4 \rightarrow \equiv Si - OH + CH_2OH$$

and released by

$$2\equiv Si\text{-}OH \rightarrow \equiv Si\text{-}O\text{-}Si\equiv + H_2O.$$

It is seen that the oxygen atoms which bridge the silicon atoms in the resulting SiO_2 matrix originate from the water molecules. It is also seen from the general reaction scheme that the theoretical minimal amount of required water is 2 equimolars (actually, a somewhat higher amount is required because of the surface silicons). In practice, high ratios of water/silane (w/s) are usually used, and it has been shown in several reports that the reaction route, the sol-gel transition and the structure of the final gel, all depend on the w/s ratio [4-20].

Table I. Properties of xerogels dried at 25°C for different catalysts [2].

Catalyst	25°C			600°C			
	Volume shrinkage (%)	Bulk density (g/cm^3)	percent porosity	Volume shrinkage (%)	Bulk density (g/cm^3)	percent porosity	Vickers hardness
HOAc	84.0	1.32	0.7	-	2.08	1.9	666.5
HCl	81.3	-	-	85.2	2.06	2.8	429
HNO_3	79.9	1.14	1.7	85.2	1.82	10.0	470
H_2SO_4	71.6	-	-	80.0	1.46	31.0	224
HF	78.4	0.54	56.7	82.7	0.71	67.0	75
NH_4OH	67.8	0.49	57.0	71.7	0.70	68.0	28
no catalyst	87.5	0.95	54.6	-	1.25	43.4	-

A number of homologous alkyl groups has been tried with corresponding alcohols. Also the silicate groups may be exchanged by borate groups. For example, a sodium borosilicate gel of composition $80SiO_2.15B_2O_3.5Na_2O$ wt% was prepared form tetraethylorthosilicate ($Si(OEt)_4$:TEOS) trimethyl borate ($B(OMe)_3$:TMB), and sodium methoxide (NaOMe), where Et and Me denote ethyl and methyl groups, respectively [21]. The compositions of the gel and glass are those for the starting material of the "Vycor" process which experiences phase separation under heat treatment.

The gelation process is influenced by many variables such as the raw materials used, the solvents, the catalyst, relative concentrations of different materials in the solution including H_2O, the temperature and even the design of the reactor [22,23]. Decreasing the pH of the solution also decreases the time of hydrolysis and gelation. Table I shows the influence of catalysts on the porosity of gels and the hardness after the gels prepared at 25°C have been heated to 600°C [2].

The main difference in pore size arises from the velocity of hydrolysis and gelation which is very fast with basic catalysis, resulting in large pores, and slow with acid catalysis [23] resulting in small pores. Analogous to the hydrolysis of esters [24] the mechanism of the basic hydrolysis is as follows:

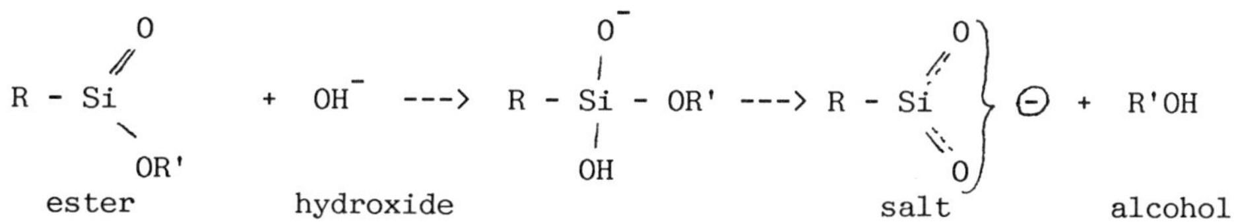

and the mechanism for acid catalytic hydrolysis is represented in the following eqilibrium:

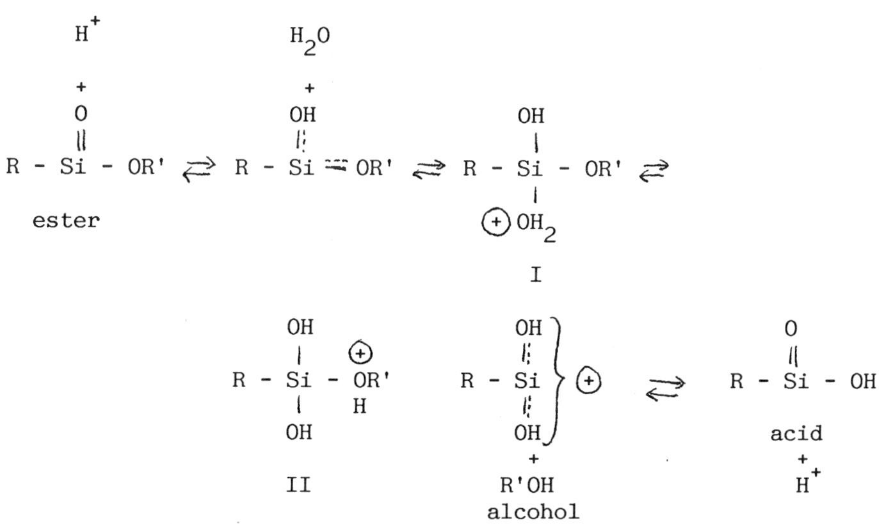

The dependence of pore size and pore distribution under experimental conditions is presented in Figs.1 and 2. Figure 1 presents the percentage porosity of the silicate glass prepared from TEOS as a function of pH [23] and Fig.2 presents the dependence of percentage porosity on temperature of gelation in borosilicate glass [21].

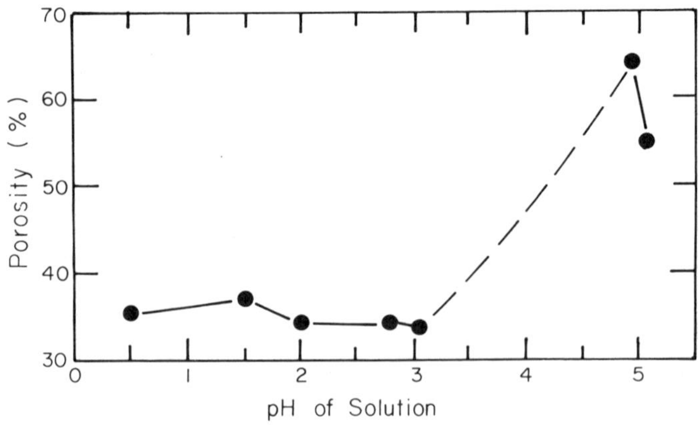

Fig. 1. Dependence of percentage of porosity on pH of solution

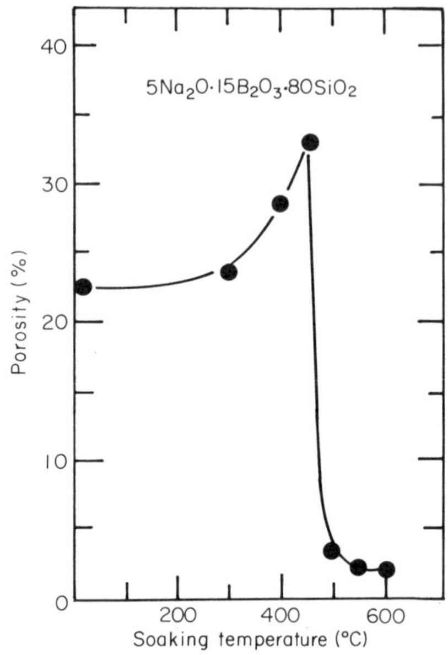

Fig. 2. Percentage porosity of silicate glass prepared from TEOS as a function of soaking temperature

A number of methods have been applied in order to characterize the morphology of porous materials [1,25]. These include physical adsorption small angle X-ray and neutron scattering (SAXS and SANS) and direct energy transfer (DET). Each of these techniques probes a different length scale of the pore morphology. Physical adsorption is sensitive to the local morphology and roughness of the pore surface, especially when adsorption on a solid is studied as a function of the size of the adsorbates [26,27]. However, it is frequently difficult to separate clearly the pure geometrical characteristics of the adsorption process thus making the adsorption results difficult to interpret [28]. Small angle X-ray and neutron scattering methods [29-31] cover a large range of length scales, typically between 5 and 5000 Å. A specific model is necessary, however, to obtain structural information at the local scale of the pore.

III. RADIATIVE TRANSITIONS OF INORGANIC IONS AND THEIR CHANGE WITH GELATION PROCEDURE

III.A. Cobalt

In this section we present the behaviour of the radiative and nonradiative transition probabilities of inorganic ions introduced into the glasses prepared by the sol-gel method and adsorbed into Vycor glass. The ions we are dealing with here are Co(II), Eu(III) and other rare earths and ruthenium(II) bipyridyl complexes.

Cobalt is known to exist in aqueous solution in an octahedral symmetry surrounded by 6 water molecules. Because of the existence of a center of inversion in octahedral complexes the intra d shell transitions are parity forbidden and the radiative transition probability of cobalt in water solution is extremely low. When a salt of cobalt, for instance, hydrated cobalt acetate, $Co(CH_3COO^-)_2.6H_2O$, or cobalt nitrate, $Co(NO_3)_2.6H_2O$, is introduced into a solution from which a sol-gel glass is formed a gradual change in the shape of the spectrum of cobalt and of transition probabilities is shown when cobalt exchanges the adjacent water molecules for surrounding silicate groups. As an example of such behaviour we may look on sol-gel glass prepared by hydrolysis and polycondensation of TMOS under acid catalysis to which 0.2 mole% of cobalt acetate was added, see Fig.3. The absorption spectrum of cobalt in the starting solution (of the following composition: 2.5 ml TMOS, 21.25 ml methyl alcohol, 1.25 ml 0.03 M.HCl and 0.2 mole cobalt acetate or nitrate per mole of TMOS dissolved in water or alcohol) exhibits a very small absorption peak around 510 nm due to octahedral cobalt. When heated to 60°C then some of the water molecules are expelled and the absorption band shifts to c 540 nm. The geometry is still very close to octahedral, however the transition probabilities are increased as a result of lowering in the symmetry. As the glass is heated gradually from 200°C to 800°C an

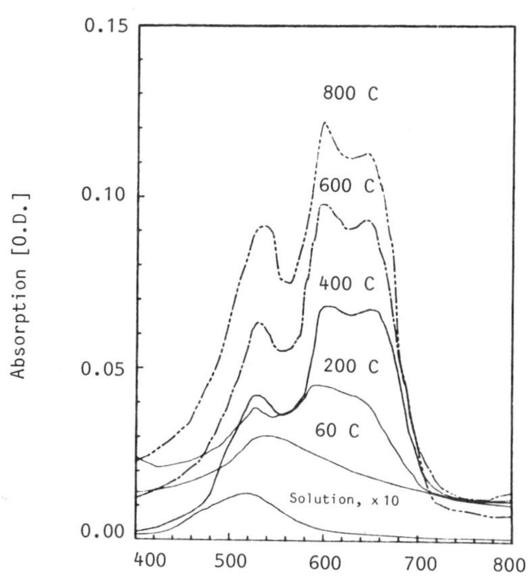

Fig. 3. Absorption spectra of cobalt in sol-gels after various heat treatments

increase in the absorption intensities (transition probabilities) is observed and the spectrum becomes structured revealing symmetries which are much lower than tetrahedral (which is also typical for cobalt in inorganic complexes), as the silicate groups gradually replace the water molecules and form first a porous glass around the cobalt and finally at 800°C a regular glass [32]. Similar behaviour is obtained when cobalt is put into the precursor solution as its nitrate salt.

III.B. Radiative and nonradiative transitions of rare earths in porous glasses

The spectra of rare earths in glasses are very well known [33,34]. The main difference between optical spectra of inorganic ions in amorphous materials and crystalline media of comparable composition is the inhomogeneous broadening due to the variety of sites in the former [35,36]. This is due to the fact that the dopant ions in glasses reside in a variety of environments and experience different perturbing local fields. Thus, instaed of identical electronic energy levels, there is a distribution of spectroscopic parameters which depend on the host glass composition and result in different radiative and nonradiative transition probabilities. These variations are evident in absorption and fluorescence spectra as inhomogeneously broadened lines when conventional light sources are used. The decay times of fluorescence from excited states of different sites normally exhibit a nonexponential behaviour. However, if fast diffusion of excitation energy occurs among the different sites, then the decay will become exponential with the fastest time constant available among the sites.

Laser-induced fluorescence line narrowing (FLN) enables, in some cases, the excitation of a subset of ions having the same energies. In the absence of accidental coincidence of electronic levels, a tunable pulsed laser can be used for determining the electronic levels and transition probabilities of a selected subset of the total ensemble of sites. The intraconfigurational 4f-4f transitions of rare earth ions are especially suitable in this type of investigation because of small electron phonon coupling. The degree of selectivity of the FLN technique depends on the ratio of homogeneous to inhomogeneous widths [37], which is of the order of 10^{-3} for rare earths of low temperatures.

III.C. Radiative transitions

For most practical purposes, the oscillator strengths and the connected radiative transition probabilities obtained from the absorption spectra correspond to some average values due to the total number of sites. This is also true for the nonradiative transition probabilities which are obtained from the measured average lifetimes and average quantum efficiencies of fluorescence.

The positions of the spectra of the rare earth ions which are due to intraconfigurational f-f transitions in amorphous materials are only slightly dependent on the host. However, a small shift of the free-ion levels to lower energies, the nephelauxetic effect, can be observed as a result of covalency between the rare earths and the matrix. The intensity of the transitions can be calculated by the Judd-Ofelt theory, in which the calculated matrix elements of the transition are combined with the experimentally obtained intensity parameters [33].

The optical transitions of rare earths in solids are predominantly of electric dipole character and their spectral intensities can be described using the Judd-Ofelt treatment. In this approach the line strength S of a transition between to J states is given by the sum of products of the empirical intensity parameters Ω_t and matrix elements of tensor operators $U^{(t)}$ of the form

$$S(J,J') = \sum_{t=2,4,6} \Omega_t \ |\langle aJ||U^{(t)}||bJ'\rangle|^2$$

The values of Ω_t are obtained from a least squares fit of measured and calculated absorption line strengths and typically have an experimental uncertainty of ~ ± 10%. The integrated intensities of the absorption bands yield Ω's which are an effective average over the different rare earth environments in the glass.

The most significant factor determining the Ω values is the strength of the odd-order terms in the expansion of the local field at the rare earth site. These in turn are affected by the nearest neighbour anion(s) and cations. For a given glass, former systematic changes of Ω_t have been observed with changes in the size and charge of network modifier ions [38]. Intensity parameters of Pr(III), Nd(III), Sm(III), Eu(III), Dy(III), Ho(III), Er(III) and Tm(III) in a variety of glasses in comparison to crystals are extensively tabulated in Ref. [34].

III.D. Nonradiative transitions

Excited electronic levels of rare earths in solids decay nonradiatively by exciting lattice vibrations (phonons) [40] and in water by OH vibrations. When the energy gap between the excited level and the next electronic level is larger than the phonon energy, several lattice phonons are emitted in order to bridge the energy gap. It was recognized that the most energetic vibrations are responsible for the nonradiative decay, since such a process can conserve energy in the lowest order. In glasses, the most energetic vibrations are the stretching vibrations of the glass-network polyhedra, and it was shown than these distinct vibrations are active in the multiphonon process [41] rather than the less energetic vibrations of the bond between the rare earth and its surrounding ligands. Later [42] it was demonstrated that these less energetic vibrations may participate in cases when the energy gap is not brigded totally by the high energy vibrations. The experimental results reveal that the logarithm of the multiphonon-decay rate decreases linearly with the energy gap, or the number of phonons bridging the gap. The theory of nonradiative decay by mutiphonon mechanism was first proposed by Kubo and Toyozawa [43] who proposed that the basic mechanism allowing such transitions is the correction in the Born-Oppenheimer (BO) approximation due to vibrational motion of ions which admixes the electronic wave function and causes transitions that represent stationary states in the zero-order BO approximation. For multiphonon processes, one has to proceed to higher order being $E/\hbar$ to get real transitions between the electronic states [44]. It should be noted that in contrast to the smallness of the radiative processes of high order, the nonradiative high order processes are quite high [39].

Since the OH vibrations are much more energetic around 3000 cm^{-1} as compared to the SiO vibrations which are around 1200 cm^{-1} in silicate glass, the interchange of water molecules by silicate groups during the gelation procedure or by adsorbing the rare earths on a Vycor glass, decreases significantly the nonradiative transitions and increases the radiative transitions, for instance the fluorescence of rare earth ions arising from f-f transitions in Pr(III), Eu(III), Tb(III), Dy(III), Ho(III), Er(III) and Tm(III) was observed from the ions adsorbed on porous Vycor glass or on pressed discs of fumed silica. The nonradiative relaxations of the excited states were depressed as a result of a strong chemical bond between the ions and silica acting as a ligand [45]. The same was also observed when europium tris-benzoyl acetonate was adsorbed on Vycor porous glass [46].

III.E. Europium

The fluorescence of Eu(III) is a sensitive probe for the gel-glass transformation. Silica-gel glass containing Eu(III) was prepared by hydrolysis of TMOS. The total fluorescence of Eu(III) from the $^5D_0$ electronic state exhibits a gradual increase and a relative increase of intensity of the hypersensitive transition $^5D_0 \rightarrow {}^7F_2$ as a function of time and temperature of dehydration of the gel prepared by hydrolysis of TMOS. This behaviour is explained by formation of a chemical bond between Eu(III) and the oxygens of the silicate group whereby Eu(III) acquires a symmetry similar to that in a regular glass obtained by melting the appropriate oxides. The degeneracy of the $^7F_1$ and $^7F_2$ states is removed in low-symmetry sites in the glass as reflected by the half-width of the emission bands [68].

III.F. Ruthenium bipyridyl in porous glass

Ruthenium bipyridyl ($Ru(bpy)_3^{2+}$) has been extensively studied in connection with its most interesting photochemistry and its possible application in utilization of solar energy by photochemical processes. The nature of the first excited luminescent state of $Ru(bpy)_3^{2+}$ is attributed to the metal-to-ligand charge transfer (mlct) of an electron residing on a single bipyridyl ligand. A small residual interaction between the ligands allows a migration of excitation from one ligand to another on a nanosecond to microsecond time scale. The lifetime of the excited complex reveals the quantum efficiency and nonradiative relaxation of the excited state due to triplet-triplet quenching by the atmospheric oxygen. Compared to the short lifetimes of 0.32 µsec of 10^{-4} mole/l water solution of $Ru(bpy)_3^{2+}$ the lifetimes of the complex of 10^{-3} mole/l concentration adsorbed on Vycor glass increase up to 0.49 µsec [46] and to 0.8 µsec on silica gel films [47] and up to 1.34 µsec in silica gel bulks [48] at the same concentration. The lower nonradiative relaxations from the excited

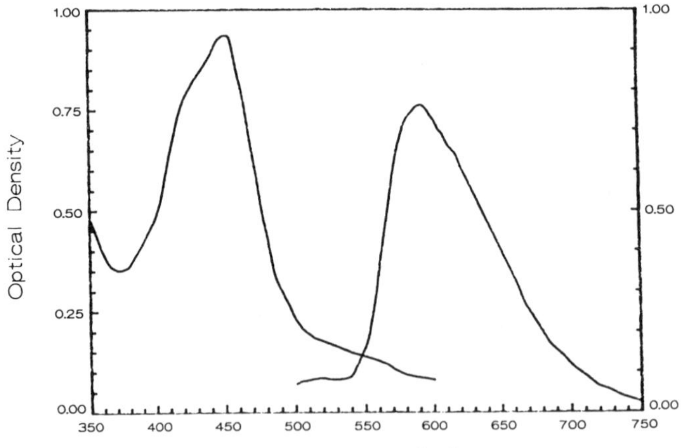

Fig. 4. Absorption and emission spectra of $Ru(bpy)_3^{2+}$ in sol-gel bulk.

Table II. Lifetimes of $Ru(bpy)_3^{2+}$ in various environments as function of concentration

Matrix		Concentration (M/l)	Lifetime (μsec)
Sol-gel film		$4.33 \cdot 10^{-3}$	1.40 ± 0.07
Sol-gel bulk		$7.00 \cdot 10^{-4}$	1.34 ± 0.05
Adsorbed on Vycor	(a)	$3.71 \cdot 10^{-7}$	0.70 ± 0.03
"	(b)	$2.54 \cdot 10^{-3}$	0.49 ± 0.02
"	(b)	$1.13 \cdot 10^{-3}$	0.45 ± 0.02
Water solution	(b)	$1.10 \cdot 10^{-3}$	0.41 ± 0.02
"	(a)	$0.2\text{-}1.0 \cdot 10^{-5}$	0.60 ± 0.03
"	(b)	$1.00 \cdot 10^{-5}$	0.32 ± 0.02
Fluoride glass		$1.00 \cdot 10^{-3}$	1.98 ± 0.02
		$1.00 \cdot 10^{-5}$	1.96 ± 0.02
		$3.00 \cdot 10^{-3}$	1.89 ± 0.02
PVA		$1.00 \cdot 10^{-3}$	1.74 ± 0.03

(a) Ref. [49]
(b) Ref. [46]
Fluoride glass - 52.2 SnF_2; 10.5 SnO; 5.1 PbF_2; 32.2 of 0.5 P_2O_5 as $NH_4H_2PO_4$ mole%
PVA - polyvinyl alcohol

states in the porous glasses are again a result of exchange of water molecules by silicate groups eliminating the presence of atmospheric oxygen. The lifetimes of $Ru(bpy)_3^{2+}$ in various media are presented in Table II. These lifetimes were calculated from the exponential part of the decay curves, starting at 200 nsec after the excitation pulse, in solids. The absorption and emission spectra are shown in Fig.4.

IV. RADIATIVE TRANSITIONS OF ORGANIC DYES IN GLASS MEDIA

Sol-gel technology which is a low temperature technique for preparation of single or multicomponent oxide glasses allows the introduction of organic molecules into the glass. The incorporation of organic dyes into glasses was first achieved in 1983 [47,50] and a large number of papers was published since on the subject. During the gelation procedure the organic dye, in most cases a laser dye, can be introduced into a glass whereby one can obtain a bulk glass [50] embedding the dye or a thin glass film including the dye deposited on a glass or plastic plate [47,51-53].

It has been shown experimentally that when organic dyes in general are introduced into a glass film their photostability is increased considerably in comparison with those dyes in solution or organic polymers. Also the amount of dimers decreases in glasses as the molecules are isolated in glass cages in a monomeric form.

The magnitude of the radiative transition probabilities of the dyes in glasses is sometimes higher and sometimes lower than in solution. The relative intensities depend on a variety of factors: whether the organic dye is of anionic, cationic or molecular form, the polarity of the excited versus the ground state of the dye, hydrogen bonds between the ground and excited states of the dye with the hydrogens of the solvent, polarity of the solvent, etc.

Table III. Molar extinction coefficients of band maxima of oxazine-170 in various solvents

Solvent	Molar extinction coeff.	Dielectric constant
Bulk glass	60,000	-
Ethanol	92,000	24.3
Methanol	85,000	32.6
Water	47,000	78.5

Table IV. Absorption and emission maxima, quantum efficiencies and lifetimes of rhodamine 6G in glasses as compared to solution

Solvent	Conc. (mole/l)	Abs. max. (nm)	Em. max. (nm)	Quantum efficiency	Lifetime (nsec)	Mole ext. coeff.	Ref.
sol-gel(tf)	5×10^{-3}	534	566	0.82	2.7±0.05	82000	*
sol-gel(b)	1×10^{-3}	534	566	0.82	2.9±0.01	82000	*
Fluoride(g)	~10^{-2}	538	560		3.0±0.1		*
H_2O	10^{-6}	526	553	0.88	3.95±0.1	78300	52
CH_3OH	10^{-6}	527	556	0.98	3.9	110000	52
$CHCl_3$	10^{-6}	530	556	0.75	3.6(1 day)	82000	52
PMMA(tf)	1×10^{-3}	534	563	0.75	2.7±0.05		*

tf - thin film
b - bulk
g - glass
* - M. Eyal, private communication

We shall present here an example of an anionic dye, fluorescein [51], cationic dyes, rhodamine 6G [52] and oxazine-170 [54,55], and molecular dyes BASF-241 and BASF-339 (the latter dyes dissolve only slightly in protonic solvents).

As an example of such a system we may examine rhodamine 6G (Rh6G) incorporated in a sol-gel film. The absorption and emission spectra of Rh6G in sol-gel glass is presented in Fig.5 [52]. This material in sol-gel glasses has been suggested for luminescent solar concentrators (LSC) [52]. Based on the experimental data of absorption and emission of Rh6G in sol-gel glass, its quantum efficiency of 0.95, molar extinction coefficient of 82,000 and the overlap of the absorption of Rh6G with the solar spectrum using a 50 micron thick film deposited on a plate having refractive index of 1.5, a Monte Carlo computer method was performed for an optical efficiency of a plate having dimensions of $1m^2$. Details of the computation procedures may be found in reference 53. The optical efficiencies of such plates are approximately 15%.

While Rh6G is not stable enough to be used in a practical device for LSCs it can serve as a model for computation in sol-gel glasses. Much better results can be obtained with the photostable dye BASF-241 the spectrum of which is presented in Fig.6, in combination with additional dyes, provided it can be introduced into the sol-gel system. Preparation and laser operation of a new tunable solid state laser based on BASF-241 in the spectral range of 530-630 nm has been reported recently [56]. Also a combination of dyes increases the overlap of absorption with the solar spectrum followed by an increase in optical efficiency.

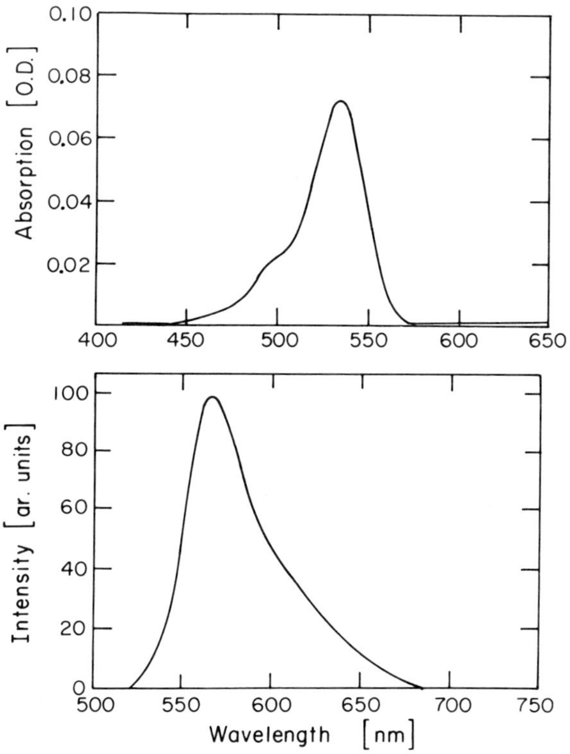

Fig. 5. Absorption and emission spectra of Rh6G in sol-gel film.

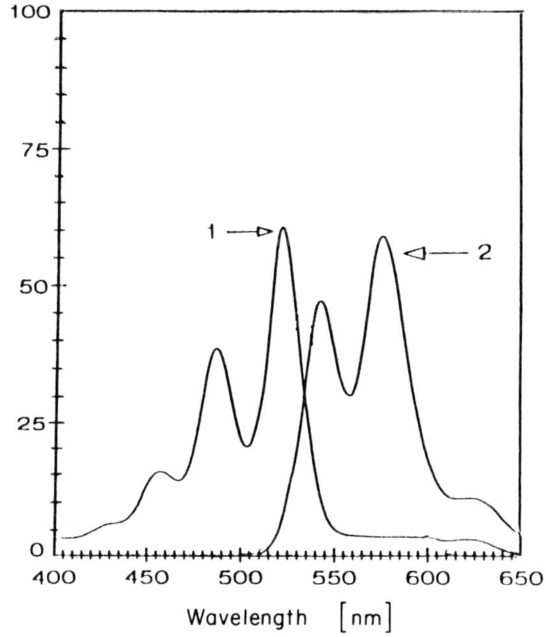

Fig. 6. Absorption and emission spectra of BASF-241 in a composite glass.

The composite glasses were prepared by the method described in ref [57]. The dye BASF-241 was dissolved in methylmethacrylate (MMA) at a concentration of about 10^{-3} M/L. The sol-gel bulks were then impregnated by immersion in the dye doped solution of the monomer catalyzed by 2% benzoyl peroxide. After the impregnation, the bulks were re-immersed in the solution of MMA catalyzed by 0.5% of benzoyl peroxide in a sealed container for 4 days at 60°C. The samples were then withdrawn, cleaned and polished manually to a reasonable surface quality to obtain slabs of typical dimensions of 10x2x2 mm.

The absorption spectra of the samples were measured on a Perkin-Elmer Lambda 5 double beam spectrophotometer and corrected emission spectra on a Jasco FP 550 spectrofluorimeter. Absorption and emission spectra of a sample are presented on Fig. 6.

Lifetime of the dye in the laser slab was determined by measurement of the decay curve of the dye emission after the excitation by a pulse from a PRA Nitrogen Laser (0.6 nsec.FWHM, 1.4 mJ/pulse, 20 Hz repetition rate). The decay signals were filtered on a McPherson Monochromator, amplified on a Hammamatsu microchannel plate and digitized by a fast Tektronix digitizer. Overall response of the system was less than the laser width. The signals were then stored in a computer memory and deconvolved from the laser pulse. Precision of the method is ±0.1 nsec. [58]. Decay of the luminescence is exponential, with lifetime of 4.30 ± 0.1 nsec. In parallel the radiative (natural) lifetime of the dye using the Strickler and Berg formula [59] was determined, resulting in 4.4 nsec., showing that the quantum yield of the dye in the composite material is near unity.

The laser action was tested as follows; A frequency doubled Nd:YAG laser radiation was focused transversely on the slab by spherical and cylindrical lenses to a line of 5x0.2 mm. The cavity was formed by using a 100% reflecting metallic mirror on one side and a 30% reflecting output coupler on the other side of the slab. The laser threshold was less than 60 µJ and the slope efficiency at 580 nm was 3.5 %.

A variety of perylene dyes emitting in a wide spectral range are being prepared and will be tested in the near future.

The influence of the hydrogen ion concentration on radiative and nonradiative probabilities can be seen in oxazine-170 incorporated in a sol-gel film. Fig.7 presents the absorption spectra of oxazine-170 in a sol-gel glass film prepared under different catalyzing conditions. The full curve corresponds to the absorption spectrum of the cationic species. In this experiment the film was prepared with HCl as catalyst and oxazine exists totally in its cationic form (II). The broken curve presents the same concentration of the dye in its neutral form (I). Here the film was prepared without any addition of acid, the only catalyst being the H ions lost by the cationic form [54]. Fig. 8 shows the corresponding emission spectra of the two forms.

As can be seen from Fig.7 the radiative transition probability of the neutral form is much lower than that of the cationic form due to the decrease of the electric dipole and the transition probability in the molecular form. The same also can be seen in Fig.8. Contrary to this behaviour the nonradiative transition of the molecular form is much higher than in the cationic form as can be seen by comparing the quantum efficiencies of the corresponding luminescence which is 0.12% in the molecular form and 45% in the cationic form [60].

The various forms of oxazine-170 are:

Neutral I

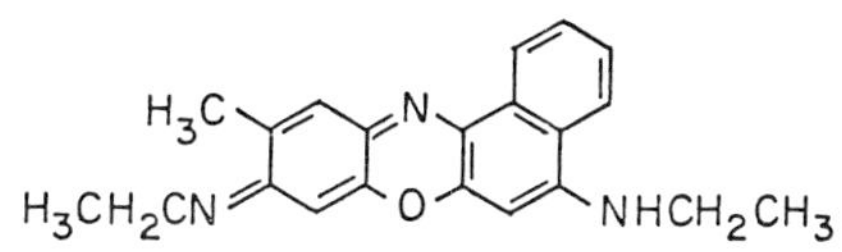

Cationic II

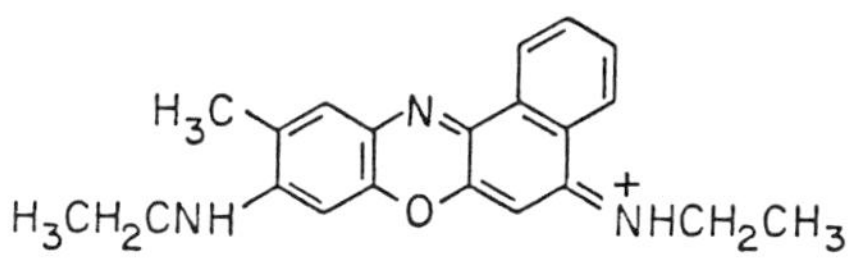

Di-cationic III

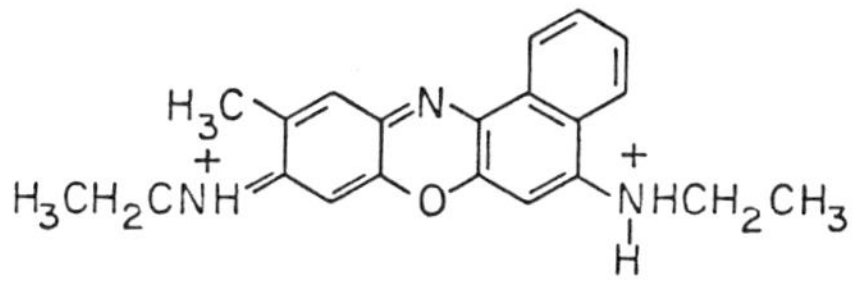

Tri-cationic IV

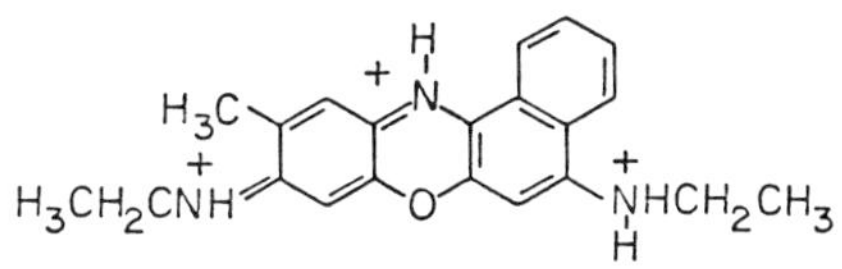

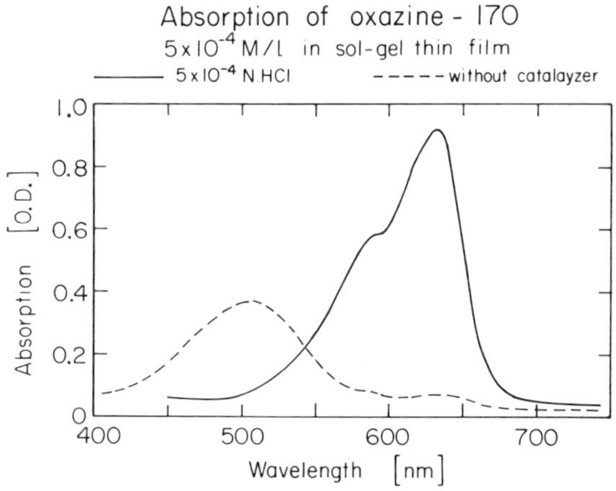

Fig. 7. Absorption spectra of oxazine-170 in sol-gel glass prepared at different pH conditions.

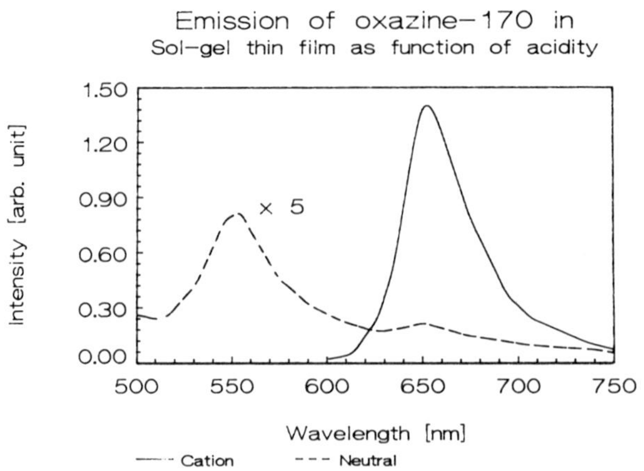

Fig. 8. Emission spectra of oxazine-170 in sol-gel glass prepared under different pH conditions.

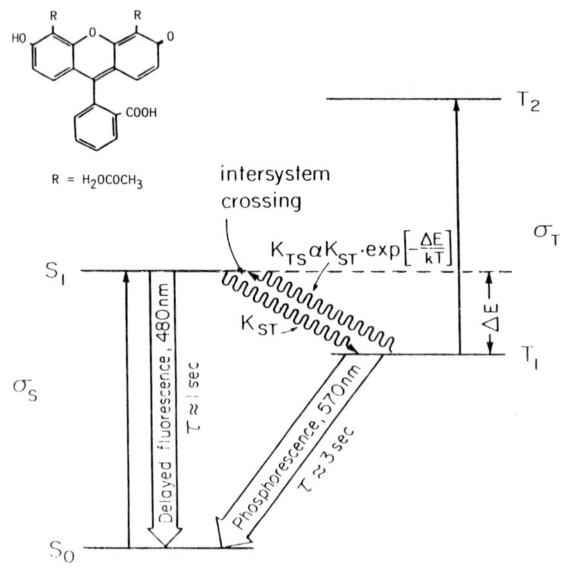

Fig. 9. Non-linear susceptibilty of organic molecules at room temperature

V. NONRADIATIVE SINGLET-TRIPLET TRANSITIONS AND NONRADIATIVE LOSSES FROM TRIPLET STATES

The origin of the large non-linear susceptibilty at room temperature of organic molecules having good singlet-triplet transfer incorporated in a glass is illustrated in Fig.9. Following optical excitation from the ground single state of the molecule S_0 an intersystem crossing transfers population to the lowest lying triplet state T_1. Provided ΔE, the energy difference between the first excited singlet S_1 and the triplet state T_1 is of appropriate value, delayed fluorescence from the excited singlet state occurs after a back intersystem crossing that returns the population to the excited singlet state from vibrational excited states of the lowest lying triplet manifold. Many of the relaxation mechanisms that can quench the triplet state do not exist when the dye is held rigidly in a solid matrix, and hence the lowest-lying triplet state has a very long lifetime. Because of the longevity of the triplet state, significant population can be trapped in that state even by weak optical fields. The saturation intensity I_s is therefore quite small less than $100mWcm^{-2}$, and since the non-linear susceptibilty χ^3 varies inversely with I_s, the non-linear χ^3 susceptibility is extremely large, ~0.1esu as compared with ~10^{-12} for the liquid CS_2.

First fluorescein [61] and its derivatives [51,62] have been studied as materials for non-linear optics. Fluorescein appears in different forms which depend on the pH of the surroundings. The absorption spectra reveal the various species. At pH 0 the cationic form is obtained, at pH 3 the neutral form, at pH 5.5 the mono-anion and at pH 11 or 12 the di-anion. Addition of a heavy ion of mercury in the form of mercury acetate [51,61], see Fig.9, increases the singlet S_1 -> T transition rate followed by the decrease of saturation intensity I_s.

Fig.10 presents the absorption spectra of fluorescein in sol-gel glass (curve 1), in lead-tin-fluoride glass (heavy glass) [63] (curve 2) and in boric acid (curve 3). The absorption spectrum of fluorescein in the heavy glass is red shifted in respect to the spectrum in boric acid and to the spectrum in sol-gel. Bathochromism of fluorescein in liquids was

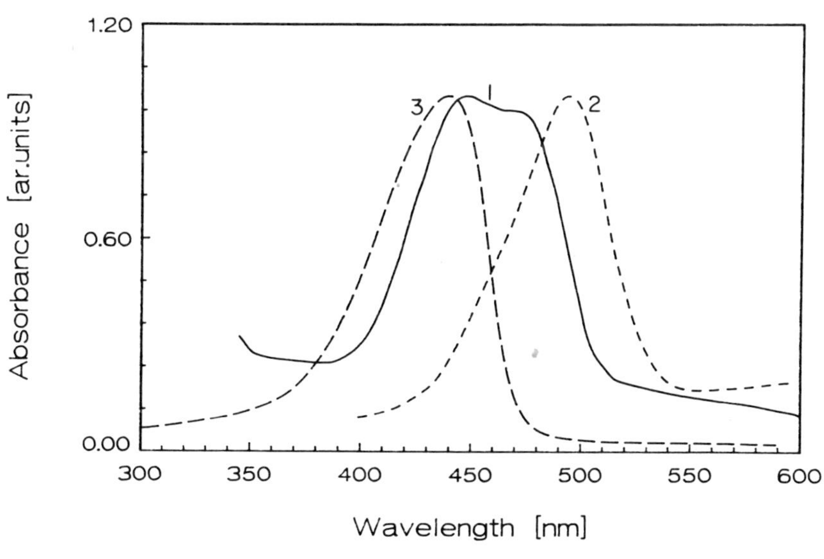

Fig. 10. Absorption spectra of fluorescein in (1) sol-gel glass; (2) lead-tin-fluoride glass; (3) boric acid glass.

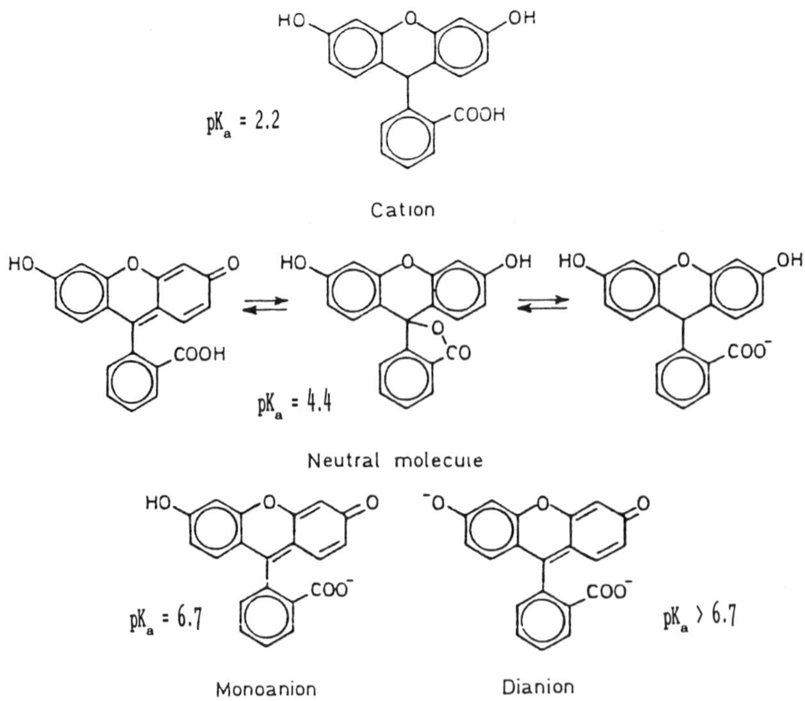

Fig. 11. Various forms of fluorescein.

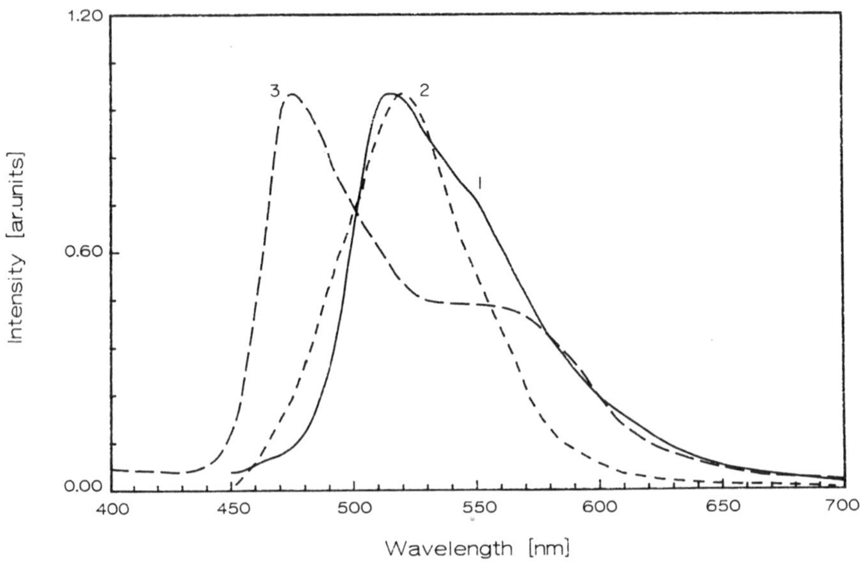

Fig. 12. Emission spectra of fluorescein in (1) sol-gel glass; (2) heavy glass; (30 boric acid glass.

investigated in the past and it was found that in neutral or basic surroundings (pH above 7) fluorescein exists in its most stable anionic form and its absorption spectrum peaks at about 480 nm, while at acidity of around 0 or less the spectrum shifts to about 435 nm, which is known to be due to the creation of a cationic fluorescein [64]. Various forms of fluorescein are shown in Fig.11. In solids we observe the same phenomena, however interpretation of these spectra must be in more general terms of

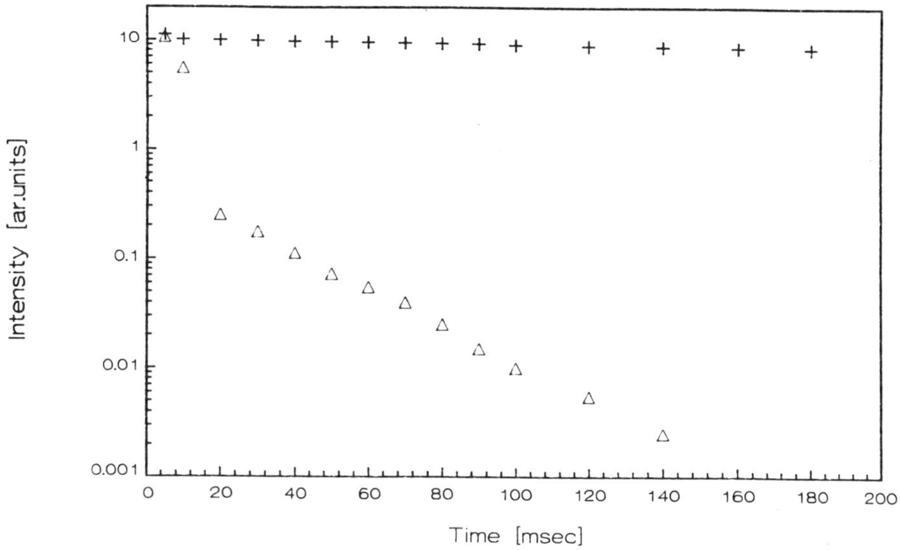

Fig. 13. Semi-log plots of emission intensities of fluorescein +++++ in boric acid; ΔΔΔΔ in heavy glass.

Lewis acids. In a boric acid glass the cationic form exists because of the action of boron as a strong Lewis acid [65]. The behaviour of fluorescein in sol-gel glass suggests a slightly basic character of the glass [51]. In the heavy glass the absorption spectrum is predominantly anionic and shows an additional red shift. In this glass we expect also a strong singlet-triplet coupling because of the presence of heavy metals.

Fig.12 presents the emission spectra of fluorescein in sol-gel glass (curve 1), in heavy glass (curve 2) and in boric acid (curve 3). While in boric acid the emission peaks at about 480 nm, in the heavy glass and the sol-gel glass the emissions are characteristic of a stable anionic form of fluorescein peaking at about 515 nm. These observations are consistent with the strong acidic character of the boric acid and with the basic character of the heavy and sol-gel glasses.

Fig.13 presents the semi-log plots of emission intensities of fluorescein in boric acid (++++) and in heavy glass (ΔΔΔΔ) monitored at 580 nm and excited by a nitrogen laser. Lifetimes calculated from these plots are 1.15±0.1 sec and 25±2 msec respectively. The former is in fair agreement with literature values [62,63]. In sol-gel glass the approximate lifetime of the emission at 580 nm is about 200 μsec, much too short to be presented on this figure. The lifetime of the excited singlet was found to be 3.5 nsec. The exceptionally long lifetime of fluorescein in boric acid can be attributed to the blue shift of the first singlet and to a weak triplet quenching in this material. The first increases the energy gap between the singlet and triplet. This process decreases the rate of back-transfer of the energy from triplet to singlet which is determined by the product of the singlet-triplet rate and the Boltzmann population of upper levels of the triplet. On the other hand, due to a weak triplet quenching, the triplet population decays mainly via the back-transfer channel. These conditions are rarely met in other dyes; either the singlet-triplet coupling is too strong, which results in a strong coupling to the ground state and relatively short lifetime, or the energy gap between singlet and triplet is too small, which increases the rate of back-transfer, the result of which is again decreased lifetime of delayed

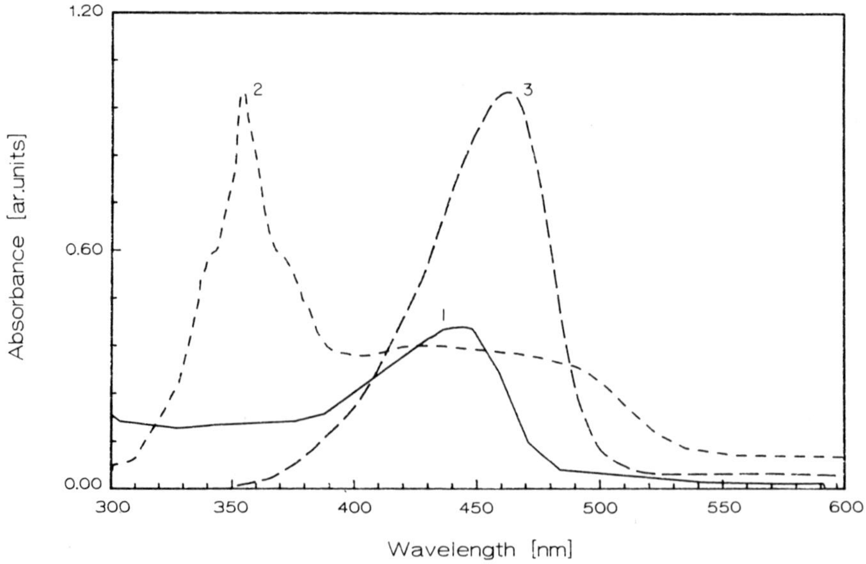

Fig. 14. Absorption spectra of Acridine Yellow in (1) sol-gel glass; (2) boric acid glass; (3) heavy glass.

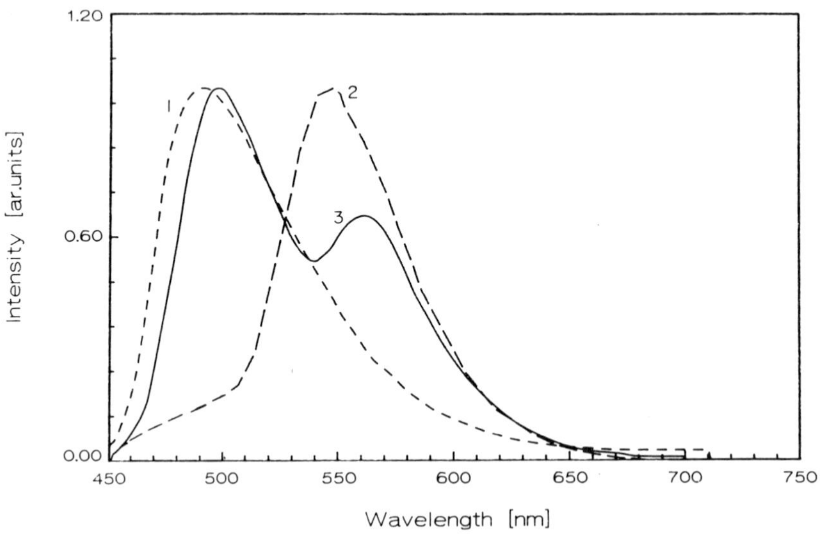

Fig. 15. Emission spectra of Acridine Yellow in (1) sol-gel glass; (2) boric acid glass; (3) heavy glass.

luminescence. This is best illustrated by the lifetime of fluorescein in the heavy glass, which, in our opinion is due to both increased singlet-triplet rate because of the presence of heavy atoms and to the smaller energy gap between singlet and triplet.

Fig.14 presents the absorption spectra of Acridine Yellow in sol-gel glass (curve 1) in boric acid (curve 2) and in heavy glass (curve 3). Only in boric acid the strong absorption peak at about 350 nm can be observed distinctly. In heavy glass and sol-gel glass this peak is smeared out to a

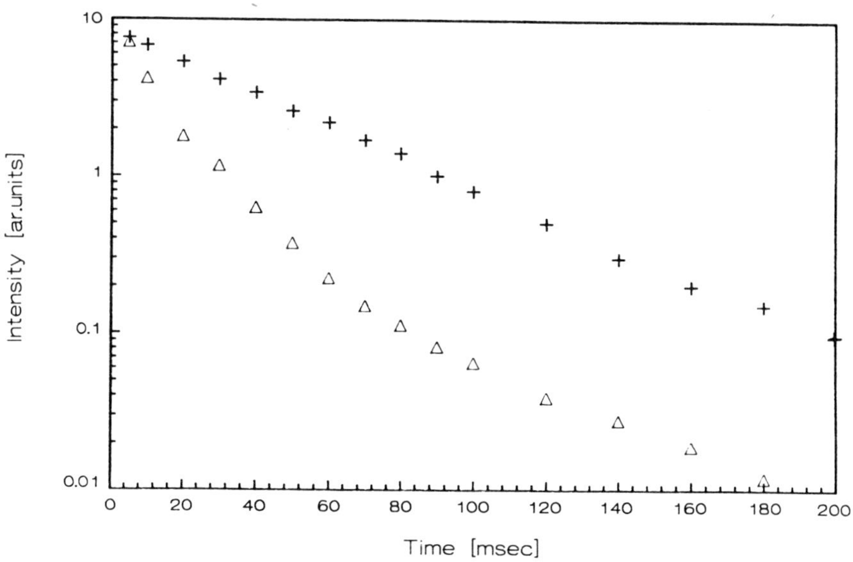

Fig. 16. Semi-log plots of emission intensities of Acridine Yellow +++++ in boric acid; ΔΔΔΔ in heavy glass.

considerable width. The spectra in boric acid and in heavy glass are red shifted in respect to the absorption band in the sol-gel glass. Fig.15 presents the respective emission spectra, curves 1,2 and 3. The emission spectrum in the heavy glass is red shifted in respect to the emission spectrum in the sol-gel glass. In addition a phosphorescence band can be seen at about 570 nm. The emission spectrum in the boric acid is red shifted in respect to both the former spectra. It is not clear whether it consists of a pure phosphorescence or is a mixture of delayed fluorescence and phosphorescence.

Fig.16 presents the semi-log plots of luminescence intensities of Acridine Yellow, collected at 570 nm, in boric acid and in heavy glass (crosses and triangles, respectively). As in the case of fluorescein, the lifetime of the dye in boric acid is longer than in the heavy glass, 46±5 msec and 27±2 msec respectively. However the difference between the two lifetimes is much smaller than for fluorescein. The singlet-triplet rate of Acridine Yellow is about 80% of the total decay rate of the singlet [67], hence it appears that the presence of heavy atoms accelerate this rate by at most a factor of two, the result of which is also a twofold increase in the decay rate of the triplet to the ground state. This is in a contrast to fluorescein for which the singlet-triplet rate is about 3% for the anionic form [64] and not more than 10% for the cationic form [51].

For Acridine Orange the absorption spectra reveal essentially the same features discussed in connection with absorption spectra of Acridine Yellow: red shift of the absorption spectra in boric acid and heavy glass in respect to the absorption spectrum in the sol-gel, very broad absorption band in boric acid and a smeared out UV absorption band in sol-gel and heavy glass. In the emission spectra of Acridine Orange in heavy glass, sol-gel and boric acid we observe the red shift of the luminescence in the heavy glass with a phosphoresence bump at about 610 nm and a strong red shift of the luminescence in boric acid, which consists probably mainly of phosphorescence. Lifetimes calculated from the semi-log plots im boric acid and heavy glass are 25±1 msec and 13±1 msec respectively. Here again we observe the same phenomenon as discussed in connection with Acridine Yellow; Although the lifetime in heavy glass is shorter than in boric acid, the presence of heavy atoms increases the S-T rate by at most a factor of two because that rate is already fast enough without the heavy atom. In sol-gel glass the delayed emission/ phosphorescence could not be detected at a timescale longer than 200 µseconds.

Self-diffraction in Acridine Orange incorporated in a thin film of boric acid was studied in the laboratory of Prof. C. Klingshirn in Kaiserslautern: The material was irradiated by two coherent laser beams of Ar^+ laser at 514.5nm (about 1mW/ cm^2). The two beams were crossed at an angle of 5 degrees and the spots of the diffracted light could be observed on a screen due to four-wave mixing.

Saturation intensities are presented in Table V.

Table V. Lifetimes and saturation intensities of organic dyes in boric acid and heavy glass [mW/cm^2].

Dye	Matrix	Lifetime [msec]	Saturation Intensity [mW/cm^2]	χ^3
acridine yellow	boric acid heavy glass	46±5 27±2	493 200	0.005 0.06
acridine orange	boric acid heavy glass	25±1 13±1	652 273	0.006 0.06
fluorescein	boric acid heavy glass	1115±50 25±2	2.7 91	7.6 0.3

VI. DOPANTS AS PROBES FOR SOL-GEL TRANSFORMATION

The addition of an ion or a molecule the optical properties of which, absorption and fluorescence, change during ther sol-gel transition, is a sensitive probe for studying this transition. Trivalent europium was used as an indicator for dehydration [68]. The total fluorescence of Eu(III) from the $^5D_0$ electronic state exhibits a gradual increase and a relative increase of intensity of the hypersensitive transition $^5D_0 \rightarrow {}^7F_2$ as function of time and temperature of dehydration of the gel prepared by hydrolysis of TMOS. This behaviour is explained by formation of a chemical bond between Eu(III) and the oxygens of the silicate group whereby Eu(III) acquires a symmetry similar to that in a regular glass obtained by melting the appropriate oxides. The degeneracy of the $^7F_1$ and $^7F_2$ states is removed in low-symmetry sites in the glass as reflected by the half-width of the emission bands.

The influence of the disappearance of pyrene excimers was used in order to compare the quality of the glasses obtained from tetramethoxysilane (TMOS) and tetraethoxysilane (TEOS) [69]. The relative concentrations of pyrene monomer to excimer indicate the velocity of gel formation which is much higher in methoxysilane than in its ethoxy homolog. This is concluded from a study of time-dependent absorption and emission behaviour.

Finally cobalt has been used as an indicator of transformation of the gels into a final glass during heating [32]. Here divalent cobalt was used as an optical probe for changes occurring during the glass formation. Cobalt initially present in a octahedrally coordinated position was transformed gradually into a tetrahedral Co(II) coordination as a result of elimination of water and an increase of density of the glass (Fig. 3).

VII. SILVER ISLANDS

In a recent paper [58] we have reported a study of interaction of the luminescent organic dye, Rh6G, incorporated in sol-gel silica films with silver aggregates. These glasses are manufactured by the relatively simple technique of hydrolysis of alkoxysilanes at temperatures much lower than required for plain glasses, which makes possible incorporation of organic

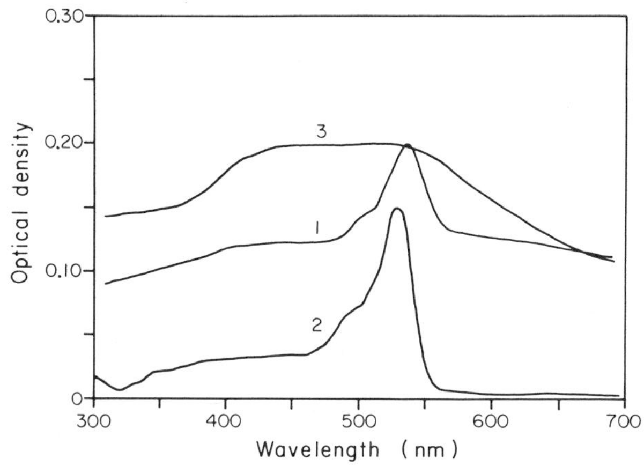

Fig. 17. Absorption spectra of sol-gel films on both sides of sample. (1) containing Rh6G and silver; (2) Rh6G; (3) Silver.

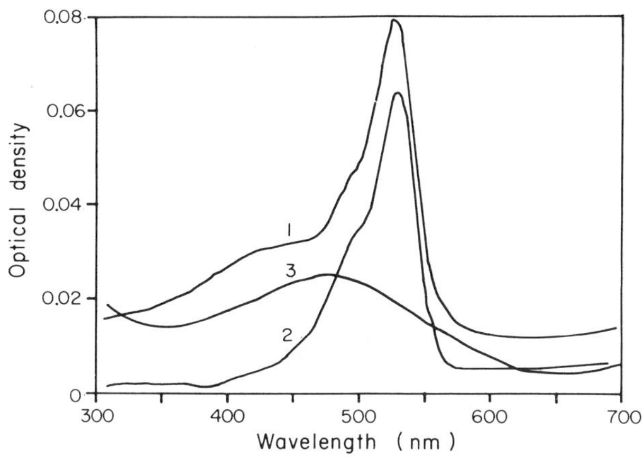

Fig. 18. Absorption spectra of sol-gel films on one side of sample only. (1) containing Rh6G and silver; (2) Rh6G; (3) Silver.

as well as inorganic materials in the sol-gel glass. The silver aggregates were formed shortly before and at the time of solidification of the sol-gel glass by means of reduction of ionic silver by organic reagents or by a stannous ion. Spectroscopic properties and lifetimes of a luminescent dye, Rh6G, incorporated in films of sol-gel glasses have been studied. Intensity of luminescence is dependent on conditions of preparation and up to 2-fold enhancement of steady-state luminescence has been observed. In such a way bulk samples doped by metallic silver and by a dye, and thin doped films containing each of the constituents separately or together, could be prepared.

Figure 17, curve 1, presents the absorption spectrum of the double film, i.e two films, one containing only silver and the second containing only Rh6G ($2x10^{-3}$ M/l in starting solution) deposited one on top of the another. Curve 2 presents the absorption spectrum of a film containing Rh6G deposited on a film containing silica only and curve 3 presents the absorption spectrum of a film containing only silver, deposited on top of a film containing only silica. These samples were dip-coated 1 hour after mixing of the appropriate solutions.

Here the absorption peak of Rh6G in the presence of silver is at 536 nm as compared to the absorption peak of Rh6G alone, which is at 528 nm. Since in this case (of the double film) we do not expect interaction between the film containing Rh6G and the film containing silver, we attribute the red shift to a combined effect of superposition of spectra of silver and spectra of Rh6G with the antireflecting action of Rh6G [70]. This assumption is reasonable, since in emission spectrum of such double film (not shown) this red shift does not appear. Moreover, due to the still high reflectivity of the double films, the intensities of emission spectra of Rh6G are weaker in the presence of silver than in its absence.

Figure 18, curves 1, 2, and 3 present the absorption spectra of a single film, i.e where all of the reagents are contained in a single sol-gel film. Thickness of a such single film is thus about half of the thickness of a double film. These samples were also dip-coated 1 hour after mixing of the appropriate solutions. Labels of the curves are the same as in Fig.17. Here we observe that the absorption spectrum of Rh6G in the presence of silver exhibits a blue shift of about 4 nm in respect to the absorption of Rh6G in the absence of silver.

By comparison of these two figures we see that one can obtain the absorption spectrum of Rh6G by removing the "baseline" due to light scattering on the silver aggregates from the curves 1 in Figs.17 and 18. In both cases, however, one cannot expect to reconstruct the absorption curve of silver by removing the absorption of Rh6G from these curves, because in the case of double films the layer of sol-gel film doped by Rh6G acts as an anti-reflective coating [70] and in the case of single films the interactions with the dye molecules change the absorption spectra.

Figure 19, curves 1 and 2, presents the emission spectra of Rh6G in a single film dip-coated 1 hour after mixing of the composite solutions, excited at 533 nm by unpolarized light. Curve 1 presents the emission spectrum of Rh6G in the presence of silver aggregates, while curve 2 presents the emission spectrum without the silver. Curves 3 and 4 represent the emission spectra excited at 350 nm, respectively. Both films were prepared at identical conditions. The enhancement of the emission is roughly proportional to enhancement of absorption (Fig.18) and is

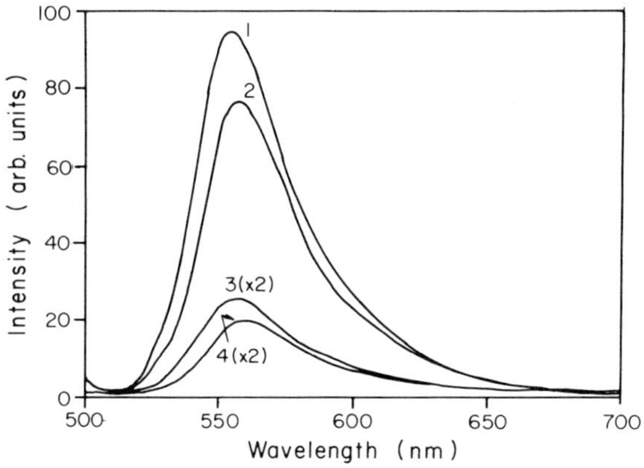

Fig. 19. Emission spectra of Rh6G of single thin film.
(1) with silver, excitation at 533 nm;
(2) without silver, excitation at 533 nm;
(3) with silver, excitation at 350 nm;
(4) without silver, excitation at 350 nm.

virtually independent of excitation wavelength. Both spectra 1 and 3 are blue shifted by about 4 nm, like the corresponding absorption spectra, Fig.18, and both are enhanced by about 30% in respect to the emission intensity of Rh6G alone. This phenomenon shows that although the dye is excited at a wavelength which is off-resonance with the silver aggregates, the fast (subpicosecond) relaxation of the absorbed energy toward the first singlet of Rh6G enables the dye to interact with the silver aggregates exactly as under direct excitation to the first singlet. The polarization ratios for curves 1 and 2 are slightly wavelength dependent and at 550 nm they are about 0.2 and about 0.3 respectively (excitation of the second singlet at 350 nm does not yield useful information since absorption dipole of this state is oriented at almost right angles to the emission dipole of the first singlet). The second number is in agreement with the values measured recently in our laboratory for various concentrations of Rh6G in sol-gel and plastic films [71].

Lifetimes of Rh6G in absence of silver is presented in Table IV. The lifetime of Rh6G in the presence of silver is 2.8 ±0.2 nsec and the first 0.2-0.3 nsec of its decay profile are nonexponential. This nonexponentiality indicates a distance dependent interaction between the dye and the silver aggregates; the dye molecules closest to the silver aggregates decay mainly by enhanced nonradiative routes, while those further away decay mainly by enhanced radiative routes.

In a systematic series of measurements of the enhancement and spectral shift as a function of concentration of Rh6G at constant concentration of silver we found an enhancement of about 100% at a concentration of 10^{-3} M/L. This is accompanied by a red shift of about 5 nm. On the other hand, at a concentration of $5x\ 10^{-3}$ M/L we did not observe any enhancement, or spectral shift. This phenomenon is in qualitative accord with both the theoretic and experimental results reported recently for sub-monomolecular layers of a dye on silver islands [72].

A large effort has recently been devoted to the study of relaxation dynamics of excited molecules embedded in porous materials [73-82]. Direct energy transfer experiments allow the gain of information about the underlying structure in which the molecules of the donor and acceptor ion are dispersed [25] or, the obtained fractal dimensions from dynamic measurements are sometimes attributed to the distribution of acceptors on the surface of the support around a donor [83]. A special issue of Chemical Physics has been devoted recently to relaxation and transport of excitation in solids, most of them porous solids [84].

Some Vycor glass and sol-gel glasses have a fractal-like character. The dynamics of excitation in these materials is expected to be influenced by the self-similar nature of these structures as proposed by Klafter and Blumen [85] who have demonstrated that the direct trapping process is determined by the fractal dimensionality $\bar{d}$. The survival probability P(t) of the excited state of the energy donor at time t is given by [85]

$$P(t) = \exp[-\gamma(t/\tau)^{\beta} - (t/\tau)] \qquad (1)$$

where

$$\beta = \bar{d}/s \qquad (2)$$

τ is the radiative decay lifetime of the donor, and s is the order of the multipolar electronic energy transfer (EET) rate $w(R) \propto R^{-s}$. The proportionality factor is

$$\gamma = \chi_A (d/\bar{d}) \Gamma(1 - \bar{d}/s)(R_0/a)^{\bar{d}} \qquad (3)$$

where χ_A is the fraction of the fractal sites occupied by the acceptors ($\chi_A \ll 1$), a is an (average) size of a unit cell, and R_0 is the "critical" radius for EET, corresponding to $w(R_0)\tau = 1$. As is apparent from equations (1) and (2), direct unistep EET on a fractal is determined by the fractal dimensionality.

It should be noted that the dynamic measurements which are a reflection of the processes occurring on a molecular scale do not correspond always to scales larger than 100 Å such as small-angle neutron scattering, SANS, and small-angle X-ray scattering, SAXS [86]. For instance, by measuring energy transfer between rhodamine B and malachite green on porous Vycor glass we have established a fractal dimensionality of 1.74±0.12 for the irregular structure over the correlation lengths of 1 # R_0 where R_0 is the critical radius for EET for the system studied [73]. On the other hand Schaefer et al [87] using the methods of SANS and SAXS on a scale larger than 200 Å did not reveal fractal geometry.

The studies of triplet annihilation of naphthalene in porous glasses by Kopelman et al [88] have also revealed that Vycor glass samples have demonstrated a fractal-like annihilation kinetics.

It should be noted that sometimes "regular" glasses exhibit also fractal-like behaviour as reflected by energy transfer studies [1], for example, recent studies on decay by cross-relaxation of $^1D_2 \rightarrow {}^3H_4$ emission of Pr(III) ions in fluorozirconate glasses [89] have shown that when the heating temperature of the studied samples approximates the glass transition temperature a dimensionality of less than 3 (namely 2.75) is obtained for the spinoidal decomposition of the glass. This is in opposition to the dimensionality of 3 in the same glass when the studies were performed at temperatures lower than the glass transition temperature. The decrease of dimensionality with heating seems to indicate a change in the Pr(III) ion distribution.

ACKNOWLEDGEMENTS

I am grateful to Mrs. Esther Greenberg for her valuable assistance in preparing the manuscript, to Professor C.K. Jorgensen and Dr. Marek Eyal for numerous discussions and to Dr. Valery Chernyak, David Brusilovsky and Raz Gvishi for providing me with experimental data. The financial support of the following institutions is acknowledged: the Krupp Foundation and the Volkswagen Foundation, the US Israel Binational Fund and the European Community.

REFERENCES

1. R. Reisfeld, Fractal behaviour of porous glasses. NATO Int'l. School on Atomic and Molecular Spectroscopy, 8th Course, Disordered Solids: Structures and Processes, Erice, Italy, June 15-29, 1987, being published by Plenum Press, Ed. B. DiBartolo.
2. J.D. Mackenzie, J. Non-Cryst. Solids 100, 162 (1988).
3. V.R. Kaufman, D. Avnir, D. Pines-Rojanski and D. Huppert, J. Non-Cryst. Solids 99,379 (1988).
4. K.C. Chen, T. Tsuchiya and J.D. Mackenzie, J. Non-Cryst. Solids 81, 227 (19860.
5. C.J. Brinker and G.W. Scherer, J. Non-Cryst. Solids, 70, 301 (1985).
6. C.J. Brinker, K.D. Keefer, D.W. Schaefer, R.A. Assink, B.D. Kay and C.S. Ashley, J. Non-Cryst. Solids 63, 45 (1984).
7. H. Schmidt, H. Scholze and A. Kaiser, J. Non-Cryst. Solids 63, 1 (1984).
8. S. Sakka, A. Ceram. Soc. Bull. 64, 1463 (1985).
9. K. Kamiya and T. Yoko, J. Mat. Sci. 21, 842 (1986).
10. M. Yamane, A. Aso and T. Sakaino, J. Mat. Sci. 13, 865 (1978).
11. M. Yamane, S. Aso, S. Okano and T. Sakaino, J. Mat. Sci. 14, 607 (1979).
12. G. Gowda, J. Mat. Sci. Lett. 5, 1029 (1986).
13. D. Kundu and D. Ganguli, J. Mat. Sci. Lett. 5, 293 (1986).
14. M. Nogami, J. Mat. Sci. 21, 3513 (1986).
15. C. Alquier, M.T. Vanderborre and M. Henry, J. Non-Cryst. Solids 79, 383 (1986).
16. S.P. Mukherjee,, J. Zarycki and T.P. Traverse, J. Mat. Sci. 11, 341 (1986).
17. B.E. Yoldas, J. Non-Cryst. Solids 38/38, 81 (1980).
18. H. Suzuki, H. Saito and T.J. Hayashi, J. Mat. Sci. 19, 396 (1984).
19. T.W. Zerda, I. Artaki and J. Jonas, J. Non-Cryst. Solids 81, 365 (1986).
20. I. Artaki, T.W. Zerda and J. Jonas, J. Non-Cryst. Solids 81, 381 (1986).
21. A. Osaka, M. Yuasa, Y. Miura and K. Takahashi. J. Non-Cryst. Solids 100, 409 (1988).
22. J.D. Mackenzie, in: Science of ceramic chemical processing, eds. L.L. Hench and D.R. Yulrich (Wiley, New York, 1986) ch.12.
23. E.J.A. Pope and J.D. Mackenzie, J. Non-Cryst. Solids 87, 185 (1986).
24. R.T. Morrison and R.N. Boyd, Organic Chemistry, (Allyn and Bacon, Inc., Boston, Third edition 1973).
25. P. Levitz, J.M. Drake and J. Klafter, J. Chem. Phys. 89, 5224 (1988).
26. S.G. Gregg and K.W. Sing, Adsorption, Surface Area and Porosity, 2nd ed. (Academic Press, London, 1982).
27. D. Avnir and P.J. Pfeifer, J. Chem. Phys. 79, 3558 (1983).

28. J.M. Drake, P. Levitz and A. Sinha, Mater. Res. Soc. Symp. Proc. 73, 305 (1986).
29. G. Porod, in Low Angle Scattering, eds. O. Glatter and O. Kratky (Academic, London, 1982), Chap.2.
30. H. Bale and P.W. Schmidt, Phys. Rev. Lett. 53, 596 (1984).
31. P.W. Schmidt, in Characterization of Porous Solids, eds. K.K. Unger, J. Rouquerol, K.S.W. Sing and H. Kral (Elsevier, Amsterdam, 1988).
32. R. Reisfeld, V. Chernyak, M. Eyal and C.K. Jørgensen, Spectroscopy of Co(II) in sol-gel films as a probe of kinetics of gelation. To be submitted to Chem. Phys. Lett.
33. R. Reisfeld and C.K. Jørgensen, Lasers and Excited States of Rare Earths, Springer-Verlag, Berlin, Heidelberg, New York, 1977.
34. R. Reisfeld and C.K. Jørgensen, Excited state phenomena in vitreous materials. In: Handbook on the Physics and Chemistry of Rare Earths, 9, chapter 58, eds. K.A. Gschneidner and L. Eyring (North-Holland, Amsterdam, 1987) 1-90.
35. L.A. Riseberg, Phys. Rev. A 7, 671 (1973).
36. M.J. Weber, in Laser Spectroscopy of Solids, eds.W.M. Yen and P.M. Selzer (Springer, Berlin, 1981) pp. 189-239.
37. M.J. Weber and S.A. Brawer, J. Phys. (Paris) 43, 291 (1982).
38. M.J. Weber, in Proceedings of the International Conference on Lasers '82 New Orleans, LA, Dec. 13-17, 1982.
39. R. Reisfeld, J. Electrochem. Soc. 131, 1360 (1984).
40. R. Reisfeld, in Radiationless Processes, eds. B. DiBartolo and V. Goldberg (Plenumn, New York, 1980) pp. 489-498.
41. R. Reisfeld and Y. Eckstein, J. Chem. Phys. 63, 4001 (1975).
42. R. Reisfeld, J. Hormadaly and A. Muranevich, Chem Phys,. Lett. 38, 188 (1976); J.. Hormadaly and R,. Reisfeld, ibid, 45, 436 (1977).
43. Y. Toyozawa, in Dynamical Processes in Solid State Optics, ed. R. Kubo, (1976) p 90.
44. R. Englman, Nonradiative Decay of Ions and Molecules in Solids, (North-Holland, Amsterdam, 1979).
45. H. Mack, R. Reisfeld and D. Avnir, Chem. Phys. Lett. 99, 238 (1983).
46. R. Reisfeld, N. Manor and D. Avnir, Solar Energy Materials 8, 399 (1983).
47. D. Avnir, V.R. Kaufman and R. Reisfeld, J. Non-Cryst. Solids 74, 395 (1985).
48. D. Brusilovsky, R. Reisfeld and M. Eyal, Spectroscopy of $Ru(bpy)_3^{2+}$ in sol-gel and fluoride glasses and polyvinyl alcohol. To be submitted to Chem. Phys. Lett.
49. A. Basu, H.D. Gafney, D.J. Perettle and J.B. Clark, J. Phys. Chem. 87, 4532 (1983).
50. D. Avnir, D. Levy and R. Reisfeld, J. Phys. Chem. 88, 5956 (1984).
51. R. Reisfeld, M. Eyal and R. Gvishi, Chem. Phys. Lett. 138, 377 (1987).
52. R. Reisfeld, R. Zusman, Y. Cohen and M. Eyal, 147, 142 (1988).
53. R. Reisfeld, M Eyal, V. Chernyak and R. Zusman, Solar Energy Materials 17, 439 (1988).
54. M. Eyal, R. Gvishi and R. Reisfeld, J. Physique, Colloque 48, C7 471 (1987).
55. R. Gvishi and R. Reisfeld, Chem. Phys. Lett. 156, 181 (1989).
56. R. Reisfeld, D. Brusilovsky, M. Eyal, E. Miron, Z. Burshtein and J. Ivri, New solid-state tunable laser in the visible. Chem. Phys. Lett. in publication.
57. E.J.A. Pope and J.D. Mackenzie, MRS Bulletin, March 17 - May 15, 29 (1987).
58. R. Reisfeld, M.Eyal and D. Brusilovsky, Chem. Phys. Lett. 153, 210 (1988).
59. S.J. Strickler and R.A. Berg, J. Chem. Phys. 37, 814 (1962).

60. R. Gvishi, Spectroscopic and physical properties of dyes in sol-gel films. M.Sc. Thesis. (1989).
61. M.A. Kramer, W.R. Tompkin and R.W. Boyd, Phys. Rev. A 34, 2026 (1986).
62. M. Eyal and R. Reisfeld, J. Luminescence 40-41, 539 (1988).
63. W.R. Tompkin, R.W. Boyd, D.W. Hall and P.A. Tick, J. Opt. Soc. Am. B 4, 1030 (1987).
64. M.M. Martin and L. Lindquist, J. Luminescence 10, 381 (1975).
65. G.N. Lewis, D. Lipkin and T.T. Magel, J. Am. Chem. Soc. 63, 3005 (1941).
66. T. Todorov, L. Nikolova, N. Tomova and V. Dragostinova, Opt. Quant. Electron. 13, 209 (1981).
67. J.G. Calvert and J.N. Pitts, Photochemistry, (John Wiley and Sons, New York, 1966).
68. D. Levy, R. Reisfeld and D. Avnir, Chem. Phys. Lett. 109, 593 (1984).
69. D. Brusilovsky and R. Reisfeld, Chem. Phys. Lett. 141, 119 (1987).
70. V.N. Rai, Appl. Opt. 26, 2395 (1987).
71. J.R. Lacovitch, Principles of Fluorescence Spectroscopy, (Plenum Press, New York, 1983).
72. H.R. Wilson, Solar Energy Materials 16, 223 (1987).
73. U. Even, K. Rademan, J. Jortner, N. Manor and R. Reisfeld, Phys. Rev. Lett. 52, 2164 (1984).
74. W.D. Dozier, J.M. Drake and J. Klafter, Phys. Rev. Lett. 56, 197 (1986).
75. R. Kopelman, in: Transport and relaxation in random materials, eds. J. Klafter, R.J. Rubin and M.F. Shlesinger (World Scientific, Singapore, 1986).
76. J. Warnock, D.D. Awschalom and M.W. Shafer, Phys. Rev. B 34, 475 (1986).
77. R. Kopelman, S. Parus and J. Prasad, Phys. Rev. Lett. 56, 686 (1987).
78. P. Levitz and J.M. Drake, Phys. Rev. Lett. 568, 686 (1987).
79. D. Rojanski, D. Huppert, H.D. Bale, Xie Dacai, P. Schmidt, D. Farin, A. Seri-Levy and D. Avnir, Phys. Rev. Lett. 56, 2505 (1986).
80. N.J. Turro, M. Gratzel and M. Braun, Angew. Chem. Intern. Ed. 19, 675 (1980).
81. J. Klafter, J.M. Drake and A. Blumen, in: Excited state spectroscopy in solids, eds. U.M. Grassano and N. Terzi (North-Holland, Amsterdam, 1986).
82. J.M. Drake, P. Levitz, S.K. Sinha and J. Klafter, Chem. Phys. 128, 199 (1988).
83. D. Pines-Rojanski, D. Huppert and D. Avnir, Chem. Phys. Lett. 139, 109 (1987).
84. Chemical Physics, 128, (1) December 1988, Special Issue on Relaxation and Transport of Excitations in Solids.
85. J. Klafter and A. Blumen, J. Chem. Phys. 80, 875 (1984).
86. U. Even, K. Rademann, J. Jortner, N. Manor and R. Reisfeld, Phys. Rev. Lett. 58, 285 (1987).
87. D.W. Schaefer, C. Laermans, R. Orbach and H.M. Rosenberg, Phys. Rev. B 28, 4615 (1983).
88. R. Kopelman, S. Parus and J. Prasad, Phys. Rev. Lett. 56, 1742 (1986).
89. M. Ferrari, E. Duval, A. Boyvirent and A. Boukenter, J. Non-Cryst. Solids 99, 210 (1988).

LASER SPECTROSCOPY TECHNIQUES APPLIED TO Cr^{3+} -DOPED MATERIALS

W. Nie, A. Monteil and G. Boulon

Laboratoire de Physico-Chimie des Matériaux
Luminescents
Université Claude Bernard - Lyon 1
Unité Associêe au CNRS No 442
Bât. 205 - 69622 Villeurbanne Cédex - France

ABSTRACT

Laser spectroscopy techniques have been applied to Cr^{3+} -doped YAG and GSGG crystals. Owing to a high resolution dye laser, ^2E and $^4A_2$ splittings, zero-phonon $^4A_2 \rightarrow {}^2E$. $^2T_1$, $^4T_2$ transitions and vibronic levels of ^2E doublets of Cr^{3+} non-equivalent centers are recorded. Energy transfers between Cr^{3+} multisites are revealed by site selective spectroscopy and fluorescence decays.

I. INTRODUCTION

Spectroscopic properties of Cr^{3+} in solid state materials are particularly interesting since Cr^{3+} can be used as either an activator to give laser actions or a sensitizer to improve the fluorescence efficiency of rare earch ions.

The fluorescence of Cr^{3+} is determined by the crystal field strength. The two examples given in the present paper have crystal field strengths at Cr^{3+} sites varying from strong in YAG to a weaker one in GSGG. This variation of the crystal field strength has been mainly connected with the Cr^{3+} - O^{2-} distances [1]. Owing to both the best optical quality of YAG and the high resolution of the dye laser, some important optical properties of Cr^{3+} in YAG, such as the presence of non-equivalent crystal field sites [2], the positions of zero-phonon $^4A_2 \rightarrow {}^2T_1$, $^4T_2$ transitions [3], vibronic levels of ^2E doublets [4] and energy transfer between Cr^{3+} non-equivalent centers [5], are observed. As for GSGG, the site selective spectroscopy technique has to be employed to

Advances in Nonradiative Processes in Solids
Edited by B. Di Bartolo, Plenum Press, New York, 1991

overcome to the mixture of $^4T_2$ and ^2E levels of non-equivalent Cr^{3+} centers [5.6].

II. EXPERIMENTAL

The YAG crystals were kindly donated by Dr. J. Mares. They were prepared by Monocrystaly Turnov, Research Institute for Single Crystals. Leninova 175, 51119 Turnov, Czechoslovakia. The measurements were carried out on the YAG sample containing 2% Cr^{3+} in weight concentration. The GSGG sample was kindly donated by Dr. L. Andrews at GTE Laboratory (Waltham, MA, USA).

The absorption spectrum at 7K was obtained on a Varian 2300 spectrophotometer in association with a liquid He cryostat. Emission and excitation spectra were made using a 10Hz Quantel Nd:YAG frequency doubled pulsed laser (time constant 15 ns) pumping a tunable dye-laser (spectral width 0.1 cm^{-1}). The emission was transmitted through a one meter Higer and Watts monochromator and detected by a RCA GaAs photomultiplier. The signal was then fed into a Ortec photon counter/boxcar integrator and further processed by a Tektronic 4051 computer.

III. RESULTS

III.A. Optical Parameters of Cr^{3+} - Doped YAG

As one of the most important and studied laser hosts, YAG has perhaps the best crystalline quality suitable for fine spectroscopic measurements. The spectroscopy of Cr^{3+} in this material is naturally interesting and has been widely studied by many researchers since more than twenty years. However, there are some important optical parameters such as the $^4A_2$ splitting, zero-phonon $^4A_2 \rightarrow {}^4T_2$, $^2T_1$ transitions, vibronic levels of excited states (^2E doublets), etc., which have never been directly and accurately obtained from the optical spectroscopy measurements [7]. In the present paper, we show some fine spectroscopic data from which the above optical parameters can directly be measured.

1. The $^4A_2$ splitting: $\Delta(^4A_2)$. Like the ^2E level, $^4A_2$ is also split by the trigonal field together with the spin-orbit interaction. A big difference between these two levels is that their splittings depend qualitatively on different trigonal field parameters (ν for ^2E and ν' for $^4A_2$), and are quantitatively incomparable. $\Delta(^4A_2)$ is rather small. It

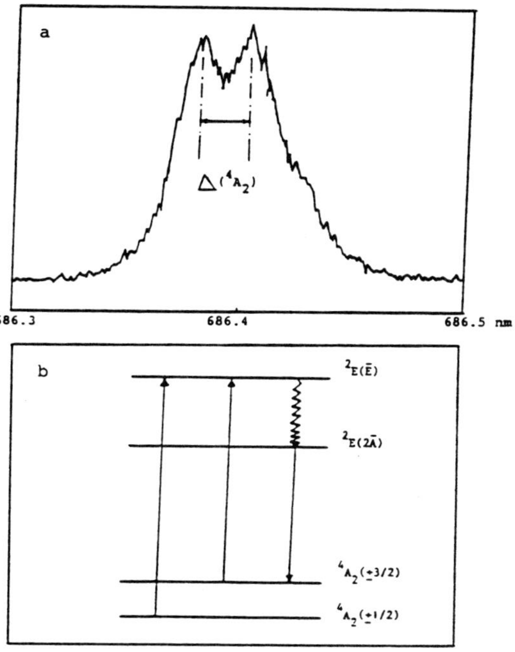

Fig. 1. a) Excitation spectrum of YAG:Cr^{3+} at 4.2 by monitoring the R_1 (687.3nm) emissionline:
b) principle of the $^4A_2$ splitting measurement.

has indirectly been obtained to be 0.53 cm^{-1} by the theoretical fitting of EPR results [8]. We have got it from an excitation spectrum at low temperature using a high resolution dye laser as shown in Fig. 1a. The principle is shown in Fig. 1b. By monitoring the $^2E(2\bar{A}) \rightarrow {}^4A_2(2\bar{A})$ or $^2E(2\bar{A}) \rightarrow {}^4A_2(2\bar{E})$ transition, two absorption lines corresponding to the $^4A_2(2\bar{E}) \rightarrow {}^2E(\bar{E})$ and $^4A_2(2\bar{A}) \rightarrow {}^2E(\bar{E})$ transitions are observed. $\Delta(^4A_2)$ is directly measured by the energy difference between the two absorption lines which is found to be 0.47 cm^{-1} for the R site. It should be noted that the method is only appropriate to the host crystals like YAG which have the best optical quality without continuously varied crystal field multisites available for Cr^{3+} ions. In the case of existing continuously varied crystal field multisites, another method can be employed to avoid the low resolving power caused by the energy transfer between multisites using the above one [9].

2. The ^2E Splitting: $\Delta(^2E)$. $\Delta(^2E)$ is much bigger than $\Delta(^4A_2)$. It can be measured directly by emission spectra at high temperatures in the following cases: i) without the overlap of emissions from non-equivalent Cr^{3+} centers; ii) the crystal field strength is strong enough to get emissions from $^2E(\bar{E})$ and $^2E(2\bar{A})$ levels. In these cases, a line pair

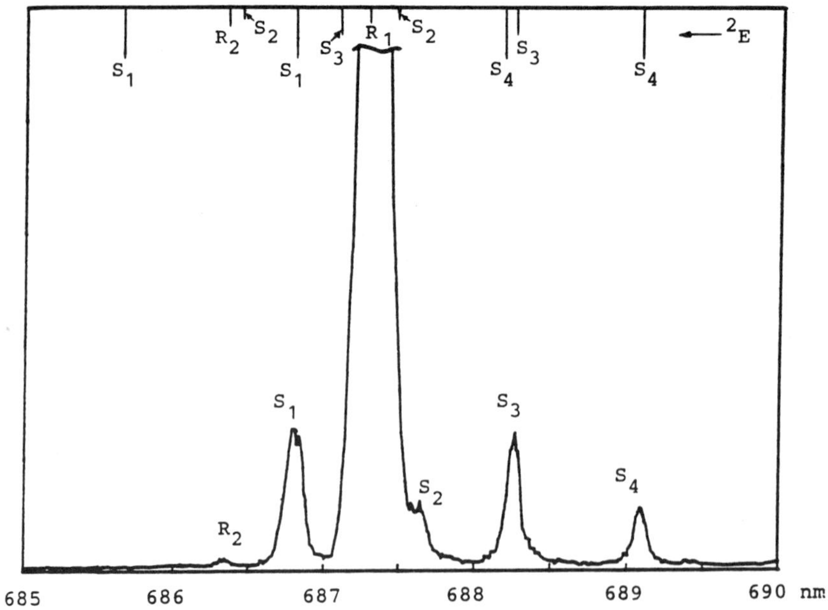

Fig. 2. Emission spectrum of YAG:Cr^{3+} at 4.2 K under excitation at 532 nm.

corresponds to the $^2E(E) \rightarrow {}^2A_2$ and $^2E(2\bar{A}) \rightarrow {}^2A_2$ transitions from one Cr^{3+} center may be distinguished by their possession of the same lifetime. This is usually observed for the R site of Cr^{3+} in all the crystals with strong crystal field strength.

Otherwise, due to either the overlap of $^2E \rightarrow {}^4A_2$ transitions from non-equivalent Cr^{3+} centers in the strong crystal field case or the disappearance of the $^2E \rightarrow {}^4A_2$ emission in the weak crystal field case, $\Delta(^2E)$ can only be measured by excitation spectra.

Figure 2 shows the emission spectrum of YAG:Cr^{3+} at 4.2K under excitation at 532 nm. According to reference 2, the weak emission lines are labelled "S-lines". To get $\Delta(^2E)$ of each Cr^{3+} center, excitation spectra are recorded by monitoring the R and S emission lines. The principle of the measurement is shown in Fig. 3f. At liquid helium temperature, the fluorescent transition occurs only from the lowest excited state $^2E(2A)$. So, by monitoring the $^2E(2\bar{A}) \rightarrow {}^4A_2$ transition of each center, two strong abosrption lines corresponding to the $^4A_2 \rightarrow {}^2E(\bar{E})$, $^2E(2\bar{A})$ transitions of the S_1 center appear to indicate the energy transfer from S_1 to S_4. Similarly, we can find the evidence of the following energy transfer: $S_1 \rightarrow S_2$, $R \rightarrow S_2$, $R \rightarrow S_3$ and $R \rightarrow S_4$. But two pairs of lines in the excitation spectrum of the S_2 center (Fig.3c) at

(685.85 nm, 687.0 nm) and (686.53 nm, 687.45 nm) which don't correspond to any one of the R and S centers must arise from the energy transfer from two new sites other than the R and S sites. Since they don't appear in the emission spectrum (Fig. 2) probably due to a small number of sites in the crystal, we name them X_1 and X_2 centers.

3. Zero-phonon $^4A_2 \rightarrow {}^2T_1$, $^4T_2$ Transitions. The zero-phonon $^4A_2 \rightarrow {}^2T_1$, $^4T_2$ transition lines of the Cr^{3+} R site in YAG can be observed by both absorption and excitation spectra at low temperature whereas those of the S centers can only be obtained by excitation spectra at low temperature.

Figure 4 shows the excitation spectrum of YAG:Cr^{3+} at 4.2 K by monitoring the R_1 emission line in comparison with the absorption spectrum at 7K. The two spectra coincide with each other. This spectral range is just between the $^4A_2 \rightarrow {}^4T_2$ broad band and the phonon sideband of ^2E doublets. The first intense absorption line located at the edge of the $^4A_2 \rightarrow {}^4T_2$ broad band can be assigned to the zero-phonon $^4A_2 \rightarrow {}^4T_2$ transition. The other three lines could be due to the $^4A_2 \rightarrow {}^4T_1$ transitions. We know that the "T" level is split into "A" and "E: levels by the trigonal field and both "A" and "E" levels are further split into "E" and "2A" levels in the presence of the additional spin-orbit interaction. The splitting of "A" is very small such as the splitting of $^4A_2$ (0.47 cm^{-1}). So the three weak lines in Fig. 4 should arise from the abosrption of split $^2T_1$ levels. From the line shapes, the two narrow lines clearly show "E" characteristics which correspond to the $^4A_2 \rightarrow {}^2T_1$ ($^2E(\bar{E})$), $^2T_1$($^2E(2\bar{A})$) transitions, and the other one can be assigned to the $^4A_2 \rightarrow {}^2T_1$ ($^2A_2$) transition.

Similarly, we get the $^4A_2 \rightarrow {}^2T_1$, $^4T_2$ transitions of the S centers by excitation spectra as shown in Fig. 5. All the spectral data are summarized in table 1.

If we pay attention in detail, a question is naturally raised: what are the splittings of $^4T_2$?

4. Vibronic Levels of ^2E Excited States. The vibronic levels of the ground state $^4A_2$ can be obtained by emission spectra at low temperature [10.11] whereas those of the first excited state ^2E doublets cannot be studied by fluorescence spectra neither at low temperatures since the emission occurs only from the lowest vibronic level of the lowest excited state $^2E(2\bar{A})$ nor at high temperatures due to the overlap

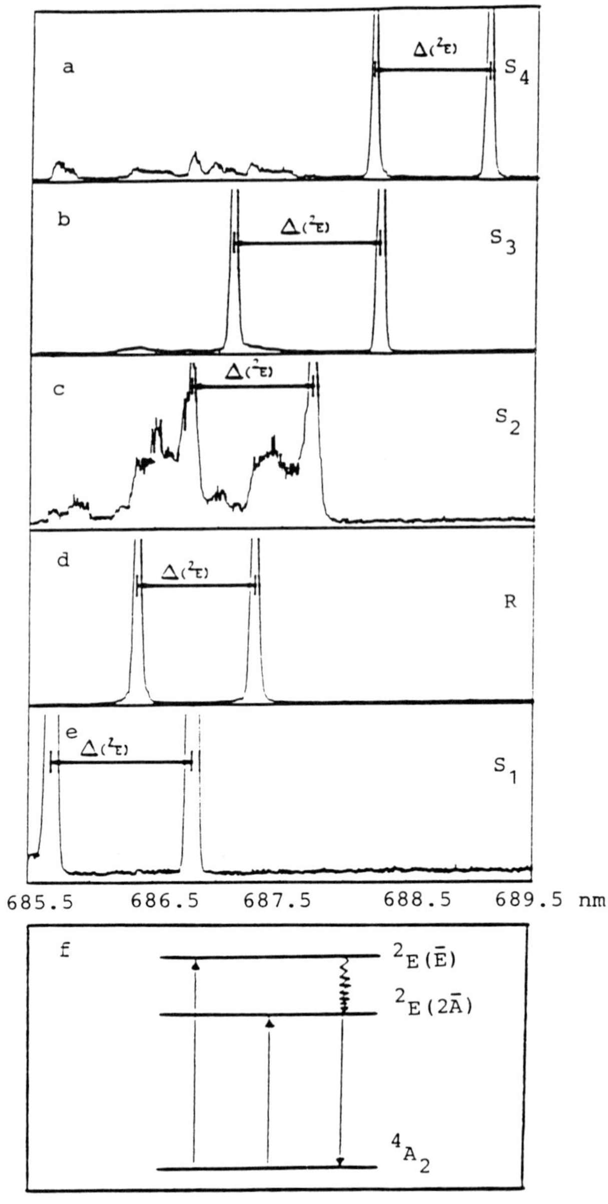

Fig. 3. Excitation spectra of YAG:Cr^{3+} at 4.2 K by monitoring the R_1 and S Lines. Monitored wavelengths: a) 689.1; b) 688.25; c) 687.75; d) 687.3 and e) 686.8 nm; f) principle of the ^2E splitting measurement.

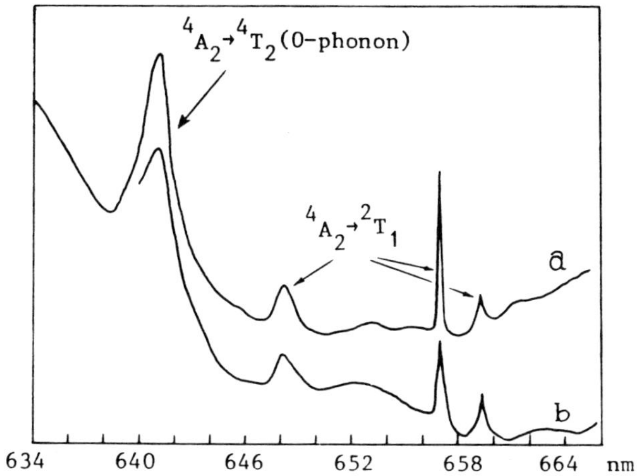

Fig. 4. a) Excitation spectrum of YAG:Cr^{3+} at 4.2 K by monitoring the R_1 line (687.3 nm): b) Absorption spectrum of YAG:Cr^{3+} at

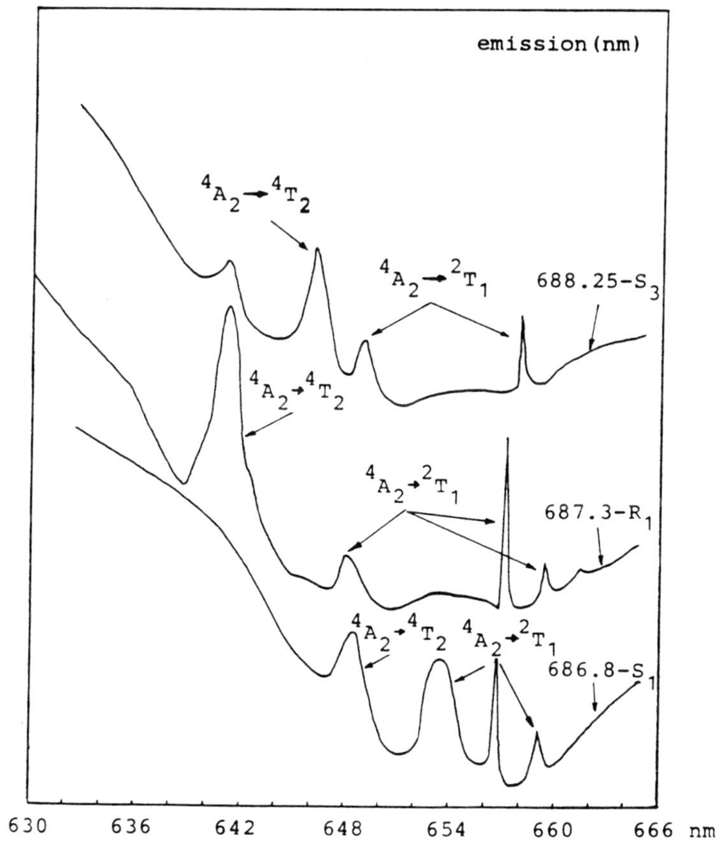

Fig. 5. Excitation spectra of YAG:Cr^{3+} at 4.2 K by monitoring the R and S lines.

Table 1: Optical parameters for Cr^{3+} multisites both in YAG:Cr and YAG:Cr, Nd at 4.2 K.

transition	$^4A_2 \to {}^2E(2\bar{A})$		$^4A_2 \to {}^2E(\bar{E})$		$\Delta(^2E)$	τ_1/τ_2	$\Delta(^4A_2)$
site	nm	cm^{-1}	nm	cm^{-1}	cm^{-1}	ms	cm^{-1}
S_1	686.8	14556	685.68	14580	24	5.2/3.0	
X_1	687.0	14552	685.85	14576	24		
R	687.3	14546	686.4	14565	19	9.2/5.7	0.47
X_2	687.45	14542	686.52	14562	20		
S_2	687.75	14536	686.78	14557	21	8.8/5.7	
S_3	688.25	14526	687.09	14550	24	8.6/5.7	
S_4	689.1	14508	688.19	14527	19	12.8/9.7	
Y_1	687.11	14550	686.19	14570	20		
Y_2	688.8	14514	687.99	14533	19		
Y_3	688.98	14510	687.46	14542	32		
Y_4	689.11	14507	687.59	14539	32		

transition	$^4A_2 \to {}^4T_2$		$^4A_2 \to {}^2T_1(^2A_2)$		$^4A_2 \to {}^2T_1(^2E(\bar{E}))$		$^4A_2 \to {}^2T_1(^2E(2\bar{A}))$		Dq
site	nm	cm^{-1}	nm	cm^{-1}	nm	cm^{-1}	nm	cm^{-1}	cm^{-1}
S_1	648.5	15416	653.5	15298	656.7	15224	659.1	15168	1542
S_3	646.1	15473	649.0	15404	658.0	15194	/	/	1547
R	641.3	15589	648.1	15425	657.1	15214	659.4	15161	1559
S_4	636.0	15719	652.4	15324	658.4	15184	/	/	1572

τ_1/τ_2: lifetimes of Cr^{3+} in YAG:Cr and YAG:Cr,Nd samples respectively.
"Y" centers are observed only in the YAG:Cr,Nd sample.

with those of the ground state $^4A_2$. The only possibilities are to record absorption and excitation spectra at low temperatures. However, the resolution of general spectrophotometers is not sufficient to resolve phonon-assisted absorption lines as shown in Fig. 6.

Figure 6 shows the absorption spectrum of YAG:Cr^{3+} at 7K. Apart from the zero-phonon $^4A_2 \rightarrow {}^4T_2$, $^2T_1$, ^2E transitions, the phonon sidebands of the ^2E excited states ($^2E(\bar{E})$ and $^2E(2\bar{A})$) are not resolved,due to the low resolution of the spectrophotometer.

Figure 7a shows the excitation spectrum of YAG:Cr^{3+} at 4.2 K by monitoring the R_1 emission line. A part of Fig. 7a is shown in Fig. 7b. The schematical interpretation is represented in Fig. 7c. Owing to the high resolution of the dye-laser and the monochromator, ^2E vibronic level absorption lines are well resolved.

The interpretation of these phonon assisted absorption lines is based on the assumption that the phonon energies, due to either different vibrational modes or multiphonon modes, coupling to ^2E excited states should not be very different to those coupling to the $^4A_2$ ground state, which seems to be reasonable since both ^2E and $^4A_2$ belong to the same electron configuration t_{2g}^3. Moreover, the phonon wavenumbers of $^2E(2\bar{E})$ should be practically identical with those of $^2E(2\bar{A})$. The detailed interpretation is given in Table 2. i and i' lines correspond to absorptions of the vibronic levels of $^2E(2\bar{A})$ and $^2E(\bar{E})$, respectively. The reproducible 19 cm^{-1} between i and i' lines that is equal to the ^2E splitting $\Delta(^2E)$ of the Cr^{3+} R center is indicative.

5. Energy Transfer Between Nonequivalent Cr^{3+} Centers. From Table 1, we can see that by the ^2E splitting, non-equivalent Cr^{3+} centers can be divided into two groups: i) R, S_2, S_4, X_2 with $\Delta(^2E) \simeq 20$ cm^{-1} and ii) S_1, S_3, X_1 with $\Delta(^2E) \simeq 24$ cm^{-1}. From Fig. 3, we can find that the energy transfer takes place from group 1 to group 2 sites. This shows us the quantitative site preference of energy transfer among non-equivalent Cr^{3+} centers. Different energy transfer mechanisms are revealed by site selective fluorescence decay studies as shown in Fig. 8.

Figure 8 shows the decay curves of the R_1, S_1, S_2 and S_4 emissions at 4.2K under excitation into the S_1 line. In the long time part, we find again the decay profils under excitation at 532 nm whereas in the short time part initial rising features are manifested. The R_1 line has an initial rise time about 5.2 ms which is equal to the lifetime of the

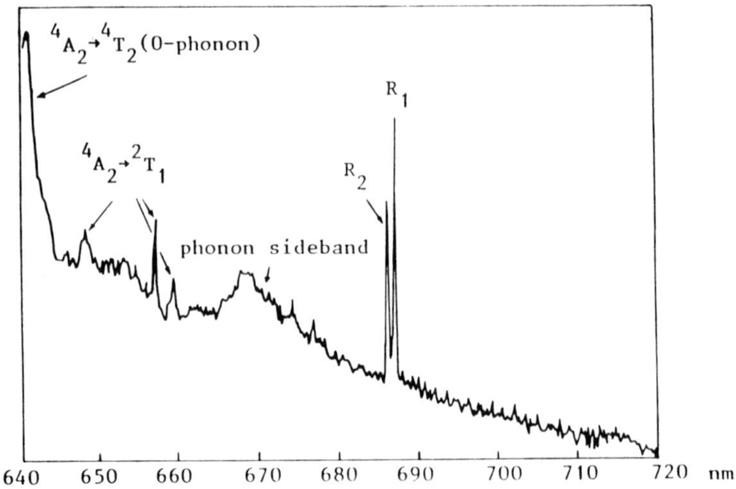

Fig. 6. Absorption spectrum of YAG:Cr^{3+} at 7 K.

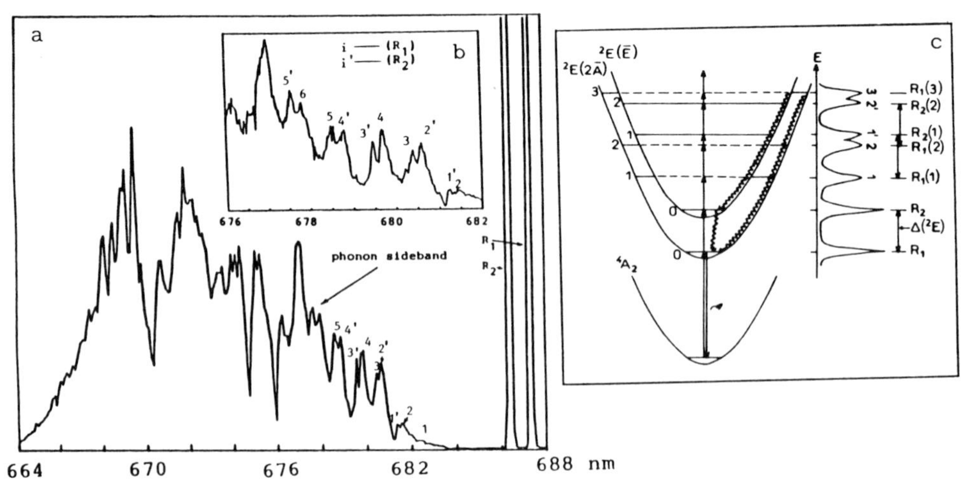

Fig. 7. a) Excitation spectrum of YAG:Cr^{3+} at 4.2 K by monitoring the R_1 emission Line (687.3 nm).

b) A magnification of a).

c) Schematical principle.

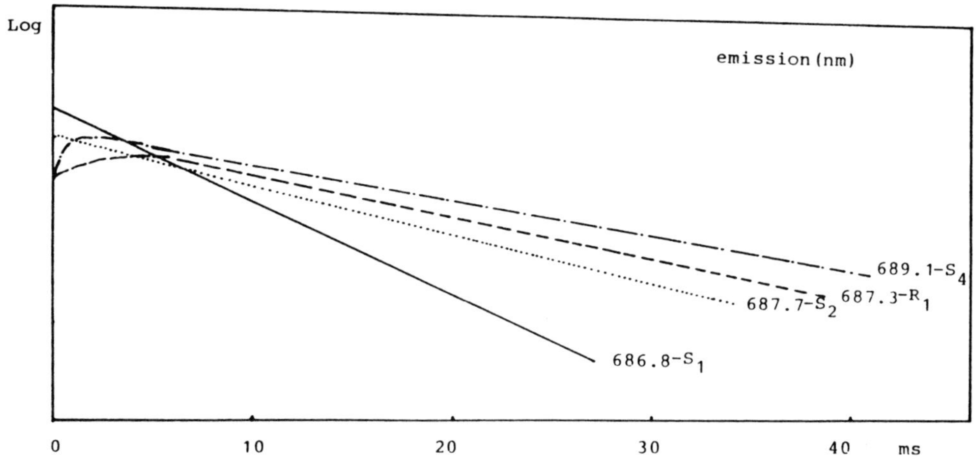

Fig. 8. Decay curves of the R_1 and S lines of YAG:Cr^{3+} at 4.2 K under excitation into the S_1 (686.8 nm) line.

S_1 line indicating that the energy transfer is 100% radiative. S_2 and S_4 have shorter initial rise times implying complicated energy transfer mechanisms.

III.B. Spectroscopic Data on GSGG:Cr^{3+}

The same types of experiments were also carried out on GSGG:Cr^{3+} crystals which have intermediate crystal field strength. Due to the existence of a strong mixture between $^4A_2$ and ^2E states, either for one given site or for the four different sites (α, β, ν and δ [1.5]), the interpretation of the spectroscopic data becomes a difficult task.

Emission and absorption spectra of GSGG:Cr^{3+} at 6K are shown in Fig. 9. Four emission lines that may be assigned to the $^2E(2\bar{A}) \rightarrow {}^4A_2$ transition of Cr^{3+} at non-equivalent crystal field sites are observed. Similarly, the ^2E splittings are measured by excitation spectra as shown in Fig. 10. An additional important feature in comparison with the excitation spectra in YAG is the appearance of a broad absorption line located closely to the $^4A_2 \rightarrow {}^2E(\bar{E})$ transition line (Fig. 10 α and β), which can reasonably be interpreted to be the $^4A_2 \rightarrow {}^4T_2$ zero-phonon line of the Cr^{3+} center according to its intermediate crystal field strength. For instance, for Cr^{3+} at α sites, the $^4A_2 \rightarrow {}^4T_2$ zero-phonon transition is at 694.5 nm. The experimental data are summarized in Table 3. The energy gap ε between $^2E(2\bar{A})$ and $^4T_2$ are found to be 38 cm^{-1} and 56 cm^{-1} for α and β centers, respectively.

Table 2. Phonon wavenumbers of the split ^2E excited states of Cr^{3+} in YAG.

i or i'	nm	cm^{-1}	assignment(cm^{-1})
2	681.49	14669.7	124 – $R_1(2)$–$^2E(2\bar{A})$
1'	681.25	14674.9	110.1 – $R_2(1)$–$^2E(\bar{E})$
2'	680.61	14688.6	123.8 – $R_2(2)$–$^2E(\bar{E})$
3	680.41	14693	147.3 – $R_1(3)$–$^2E(2\bar{A})$
4	679.71	14708.1	162.4 – $R_1(4)$–$^2E(2\bar{A})$
3'	679.47	14713	148.5 – $R_2(3)$–$^2E(\bar{E})$
4'	678.80	14727.8	163 – $R_2(4)$–$^2E(\bar{E})$
5	678.53	14733.6	187.9 – $R_1(5)$–$^2E(2\bar{A})$
6	677.80	14749.5	203.8 – $R_1(6)$–$^2E(2\bar{A})$
5'	677.56	14754.8	190 – $R_2(5)$–$^2E(\bar{E})$

$R_1(^2E(2\bar{A}))$	$R_2(^2E(\bar{E}))$	$^4A_2$
	110.1	
124	123.8	124
147.3	148.5	142
162.4	163	162
187.9	190	184

Table 3. Optical parameters of GSGG:Cr^{3+} at 6K.

Site	α	β	γ	δ
$^2E \rightarrow {^4A_2}$ (nm)	696.4	695.0	694.2	693.2
Lifetime (μs)	245	340	530	330
^2E splitting (cm^{-1})	20	28	38	30
$^4A_2 \rightarrow {^4T_2}$ 0-phonon (nm)	694.5	692.2	/	690.1

The overlap of energy levels of non-equivalent Cr^{3+} centers makes the analysis of spectra complicated. The adaptability of the site selective spectroscopy is very limited on this sample. For example, in Fig. 11, exciting into the $^2E(2\bar{A})$ levels of the four non-equivalent centers, we get very different emission spectra which is a result from

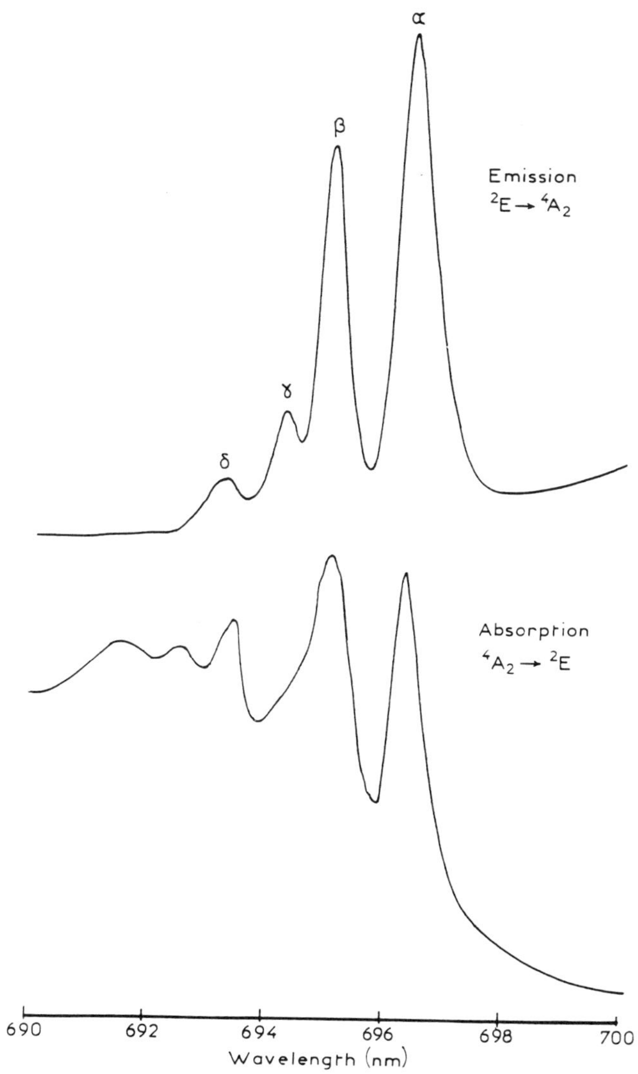

Fig. 9. Absorption and emission spectra of GSGG:Cr^{3+} at 6K (the emission spectrum is obtained by 580 nm excitation).

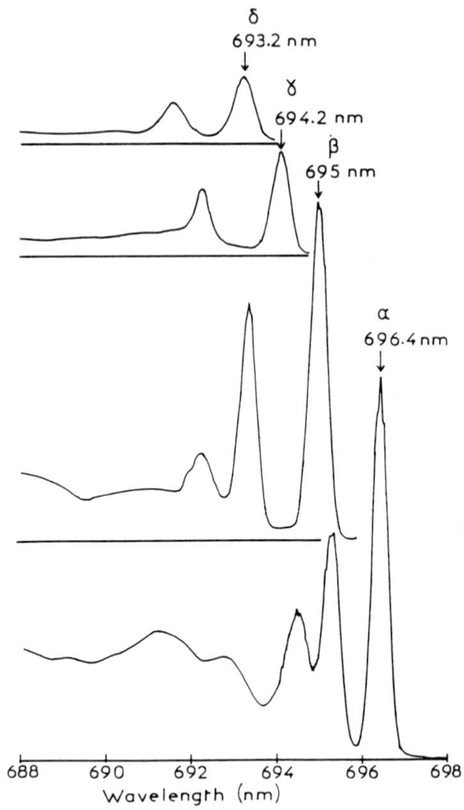

Fig. 10. Excitation spectra of the four sites of GSGG:Cr^{3+} at 6K.

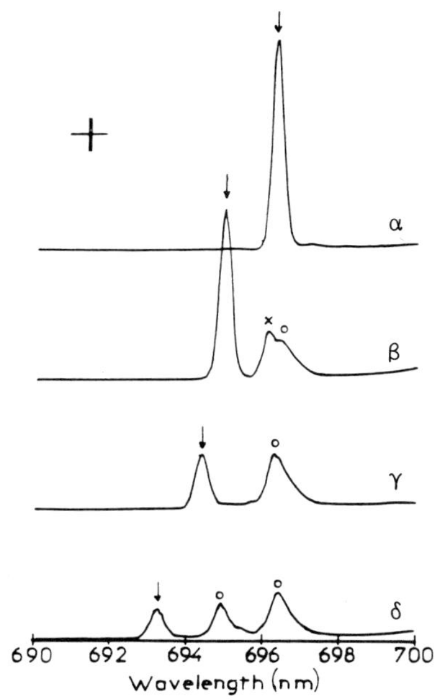

Fig. 11. FLN spectra of the four sites of GSGG:Cr^{3+} at 6K.

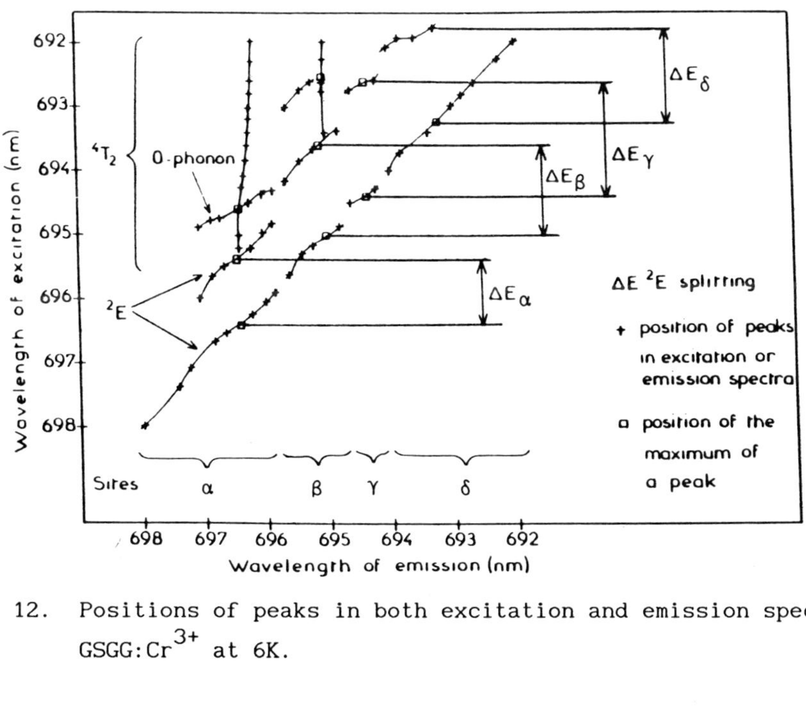

Fig. 12. Positions of peaks in both excitation and emission spectra of GSGG:Cr^{3+} at 6K.

energy level overlaps of non-equivalent Cr^{3+} centers at the excitation wavelengths. When exciting into $^2E(2\bar{A})$ of the α center, we observed only the $^2E(2\bar{A}) \rightarrow {}^4A_2$ resonant emission (Fig. 11a). Excitation into the $^2E(2\bar{A})$ of the β center, besides the resonant $^2E(2\bar{A}) \rightarrow {}^4A_2$ emission of the β center, the $^2E(2\bar{A}) \rightarrow {}^4A_2$ transitions of a selected part of Cr^{3+} at α sites (the X line) due to the overlap of $^2E(\bar{E})$ of the α center with $^2E(2\bar{A})$ of the β center and a non-selected α sites (the O line) relaxed from $^4T_2$ of the α center which overlaps also with the $^2E(2\bar{A})$ level of the β center; simiilarly, exciting into $^2E(2\bar{A})$ of the ν center, the zero-phonon $^4T_2$ level of the α center is also excited; exciting into $^2E(2\bar{A})$ of the δ center, both the zero-phonon $^4T_2$ level of the β center and vibronic levels of $^4T_2$ of the α center are excited.

In order to distinguish energy levels of the four non-equivalent centers, fluorescence line narrowing techniques are employed both on emission and excitation spectra. As each site is inhomogeneously broadened due to continuously varied crystal field strength, narrowed lines may be obtained either in emission spectra by monitoring selectively emission wavelengths. The positions of narrowed peaks are drawn in Fig. 12. From the positions of peak maxima, we can find that for α and β centers, a third line in the higher energy side of the two ^2E components which can be assigned to the $^4A_2 \rightarrow {}^4T_2$ zero-phonon transition appears. The vertical lines correspond to the continuous absorption of vibronic levels of the $^4T_2$ level.

IV. CONCLUSION

Using a high resolution dye-laser, important optical parameters such as $\Delta(^2E)$, $\Delta(^4A_2)$, lifetimes, zero-phonon $^4A_2 \rightarrow {}^2T_1$, $^4T_2$, ^2E transitions of non-equivalent Cr^{3+} centers in YAG are precisely obtained. Three components of $^2T_1$ split by the trigonal field together with the spin-orbit interaction for each center are also observed. Comparing optical parameters of non-equivalent Cr^{3+} centers in YAG, we can find that the Dq value varies in the same order with the lifetime whereas no relation with the position of the $^2E \rightarrow {}^4A_2$ transition. Optical parameters of non-equivalent Cr^{3+} centers in GSGG are obtained. A site selective spectroscopy technique ·is employed to distinguish energy levels of non-equivalent centers in GSGG. Each Cr^{3+} site in YAG has a well defined crystal field strength whereas in GSGG emission lines of non-equivalent centers are inhomogeneously broadened due to continuously varied crystal field strength.

REFERENCES

1. G. Boulon, C. Garapon and A. Monteil, Proceedings of the First International School on Excited States of Transition Elements. Ksioz Castle. Poland June 1988, p.96, World Scientific Singapore - London - Eds. B. Jezowska, J. Lengendziewicz, W. Strek.
2. W. Nie, G. Boulon and J. Mares, Chem. Phys. Lett. 160, 579 (1989).
3. W. Nie, G. Boulon and A. Monteil, Chem. Phys. Lett. 164, 106 (1989).
4. W. Nie, G. Boulon and A. Monteil, J. Physique 50, 3309 (1989).
5. G.F. Imbusch, T.J. Glynn and G.P. Morgan, Proceedings of the First International School on Excited States of Transition Elements. Ksioz Castle, Poland, June 1988, p.125, World Scientific - Singapore - London - Eds. B. Jezowska, J. Legendziewricz, W. Strek.
6. A. Monteil, W. Nie, C. Madej and G. Boulon, Optical and Quantum Electronics (to be published March 1990).
7. M.O. Henry, J.P. Larkin and G.F. Imbusch, Proc. R.I.A., Sect. A97 (1975).
8. J.W. Carson and R.L. White, J. appl. Phys., 32, 1787, (1961).
9. W. Nie, F.M. Michel-Calendini, C. Linares, G. Boulon and C. Daul, J. Lumin. (to be published 1989).
10. I.N. Douglas, Phys. Stat. Sol. (a) 9, 635 (1972).
11. W. Nie, A. Monteil and G. Boulon, Optical and Quantum Electronics (to be published March 1990).

HOW TO PLAY WITH SPRINGS AND PULSES

IN A CLASSICAL HARMONIC CRYSTAL

M. Dominoni and N. Terzi

Dipartimento di Fisica dell'Università
via Celoria 16, 20133 Milano ITALY

ABSTRACT

The classical dynamics of a harmonic lattice is presented. The aim of these lectures is not that of explaining once more the elements of the lattice dynamics. On the subject there is infact a large number of excellent treatises, either extended or short, starting from the milestone book of Born and Huang.

Here we examine how pulses, locally generated inside a crystal, propagate into the surrounding harmonic crystal, when the whole system of dispersion curves is taken into account. We focus attention on the process of propagation and not on that of pulse generation. A pulse is assumed to be already present at t=0 with a given shape and very localized. At t>0 it relaxes into the sourrounding threedimensional crystal with which it interacts through harmonic springs. This is the socalled regime of ballistic propagation when any source of scattering is disregarded.

We consider the propagation inside both a perfect and an imperfect crystal. For the last case we examine the effects due to a very localized change of force constants induced by a substitutional imperfection.

How the pulse is generated [17] is not considered here. Actually, it results from the interaction of the crystal with an external source, here assumed classical. Classical means that the source is able to perturbe the lattice by generating elastic pulses, but not to generate a single or few phonons, which are quantum objects. It describes, for instance, the semiclassical approach in the study of the electron-phonon interaction of bound electrons, where the electron is treated by quantum mechanics and the lattice classically. It describes also the behavior of the average values of position and momentum quantum variables or the atoms when the phonon field is in a coherent state (such a state is also called semiclassical just because of this property). When the source generates phonons in a coherent state, a first description of the phonon propagation can be done via the expectation values of position and momenta of the atoms, *i.e.* through the study of the dynamics of the corresponding classical values. This relatively easier problem is here treated. It is however a first approximation of the whole quantum problem. The complete quantum behavior of the phonon wavepacket propagation is recovered when also the fluctuations at all the orders are taken into account.

Advances in Nonradiative Processes in Solids
Edited by B. Di Bartolo, Plenum Press, New York, 1991

The plan of these lectures is the following. First, the lattice dynamics of a harmonic nonconducting crystal is briefly reviewed, mainly in order to define useful quantities. The equation of motion is discussed in terms of normal modes. Both the perfect and the imperfect crystal, containing a low concentration of substitutional impurities, are considered (Section I, II).

In order to follow the propagation of a pulse inside the crystal, a specific case must be studied, by solving numerically the equation of motion inside a specific crystal. We have chosen the diatomic KI and LiF ionic crystals as our model crystals, and the very reliable *Breathing Shell Model* as our lattice dynamics model.

We have considered a very localized initial pulse, as that generated, within the limits of the classical approximation, by excitating a bound electron in the Condon approximation. The relaxation of the lattice around the excited electron (*i.e.* the time dipendent dynamics of the n.n. ions), is numerically evaluated, for the cases of the thallium impurity in KI and the F_2^+ color center in LiF (Section III, IV). The relaxation time deduced for the last case is compared with the experiments reported in the literature (*Knox, Fork, Mollenauer '86*): a good agreement is found.

Here, we do not report and discuss the propagation of the pulse inside the crystal, described by the time dependent dynamics of the distant ions. From one hand it is a problem beyond the limits of these lectures. On the other hand, the linear and harmonic approximations, here adopted, correspond to the pulse dynamics in the socalled ballistic regime, when scattering and damping processes are still irrelevant, as in the first transient. Care must be taken in extending this approach to larger times and to further ions, reached by the pulse at delayed times, depending on its group velocity. However, we believe that, with all the necessary caution, this classical approach could be useful for a first undestanding of the socalled photoacoustics, *i.e.* the ways of propagation of lattice pulses induced by sudden optic transitions.

I. LATTICE DYNAMICS OF THE PERFECT CRYSTAL [1,2,4,16]

We begin with some elements of the general theory, that are useful to introduce the notation used.

I.A. The harmonic approximation

We denote the Bravais crystal lattice vectors by:

$$\mathbf{x}(l) = l_1\mathbf{a}_1 + l_2\mathbf{a}_2 + l_3\mathbf{a}_3 \tag{1}$$

where $\mathbf{a}_1, \mathbf{a}_2, \mathbf{a}_3$ are noncoplanar vectors, called lattice primitive vectors, and l_1, l_2, l_3 are three integers which we will refer collectively as l. The vector $\mathbf{x}(l)$ denotes the position of the lth unit cell of the lattice. We suppose that there are r atoms in the unit cell. The locations of these r atoms within the cell, at the equilibrium, are given by the vectors $\mathbf{x}(\kappa)$ $\kappa = 1,..,r$. In general the position vector of the κth atom in the lth unit cell is given by:

$$\mathbf{x}(l\kappa) = \mathbf{x}(l) + \mathbf{x}(\kappa) \tag{2}$$

We denote by $\mathbf{u}(l\kappa)$ the displacement from the equilibrium of the atom $(l\kappa)$. The total kinetic energy of the lattice is given by:

$$T = \frac{1}{2}\sum_{l\kappa\alpha} m_\kappa \dot{u}_\alpha^2(l\kappa) \tag{3}$$

where m_κ is the mass of the κth kind of atom and $u_\alpha(l\kappa)$ is the α-Cartesian component of $\mathbf{u}(l\kappa)$. The potential energy Φ of the crystal is assumed to be function of the istantaneous positions of all atoms. If Φ is expanded in a Taylor series in powers of the displacements $\mathbf{u}(l\kappa)$ of the atoms from their equilibrium position, we obtain:

$$\Phi = \Phi_0 + \Phi_1 + \Phi_2 + \cdots \tag{4}$$

where Φ_0 is the equilibrium potential energy,

$$\Phi_1 = \sum_{l\kappa\alpha} \phi_\alpha(l\kappa) u_\alpha(l\kappa) \tag{5}$$

$$\phi_\alpha(l\kappa) = \left(\frac{\partial \Phi}{\partial u_\alpha(l\kappa)} \right)_0 \tag{6}$$

$$\Phi_2 = \frac{1}{2} \sum_{(l\kappa\alpha)(l'\kappa'\beta)} \phi_{\alpha\beta}(l\kappa; l'\kappa') u_\alpha(l\kappa) u_\beta(l'\kappa') \tag{7}$$

$$\phi_{\alpha\beta}(l\kappa; l'\kappa') = \left(\frac{\partial^2 \Phi}{\partial u_\alpha(l\kappa) \partial u_\beta(l'\kappa')} \right)_0 \tag{8}$$

The coefficient $\phi_\alpha(l\kappa)$, the negative of the α component of the net force acting on the atom $(l\kappa)$, at the equilibrium is zero. The coefficients $\phi_{\alpha\beta}(l\kappa; l'\kappa')$ are called *atomic force constants*. Translational invariance requires that:

$$\sum_{l'\kappa'} \phi_{\alpha\beta}(l\kappa; l'\kappa') = 0 \tag{9}$$

while periodicity requires that the force constants depend on l and l' only through their difference:

$$\phi_{\alpha\beta}(l\kappa; l'\kappa') = \phi_{\alpha\beta}(l - l'\kappa; 0\kappa') = \phi_{\alpha\beta}(0\kappa : l' - l\kappa') \tag{10}$$

Neglecting the static energy Φ_0, the crystal hamiltonian in the *harmonic approximation* can be written as:

$$H = \frac{1}{2} \sum_{l\kappa\alpha} m_\kappa \dot{u}_\alpha^2(l\kappa) + \frac{1}{2} \sum_{(l\kappa\alpha)(l'\kappa'\beta)} \phi_{\alpha\beta}(l\kappa; l'\kappa') u_\alpha(l\kappa) u_\beta(l'\kappa') \tag{11}$$

The equations of motion of the lattice are thus given by:

$$m_\kappa \ddot{u}_\alpha(l\kappa) = -\frac{\partial \Phi}{\partial u_\alpha(l\kappa)} = -\sum_{l'\kappa'\beta} \phi_{\alpha\beta}(l\kappa; l'\kappa') u_\beta(l'\kappa') \tag{12}$$

I.B. About a convenient notation

We introduced the reduced displacements:

$$\overline{\mathbf{u}}(l\kappa) = m_\kappa^{1/2} \mathbf{u}(l\kappa) \quad , \tag{13}$$

and hereafter we denote $\overline{\mathbf{u}} = \mathbf{u}$. The dynamical matrix $\mathbf{D}$ of the crystal is defined as:

$$D_{\alpha\beta}(l\kappa; l'\kappa') = \frac{1}{(m_\kappa m_{\kappa'})^{1/2}} \phi_{\alpha\beta}(l\kappa; l'\kappa') \tag{14}$$

and then the equations of motion in the reduced displacements become:

$$\ddot{u}_\alpha(l\kappa) = -\sum_{l'\kappa'\beta} D_{\alpha\beta}(l\kappa; l'\kappa') u_\beta(l'\kappa') \tag{15}$$

The crystal hamiltonian, in the reduced displacements, can be written as:

$$H = \frac{1}{2}\sum_{l\kappa\alpha} \dot{u}_\alpha^2(l\kappa) + \frac{1}{2}\sum_{(l\kappa\alpha)(l'\kappa'\beta)} D_{\alpha\beta}(l\kappa; l'\kappa') u_\alpha(l\kappa) u_\beta(l'\kappa') \tag{16}$$

The (reduced) momentum of the atom $(l\kappa)$ is given by:

$$p_\alpha(l\kappa) = \frac{\partial L}{\partial \dot{u}_\alpha(l\kappa)} = \dot{u}_\alpha(l\kappa) \tag{17}$$

Thus we can write the hamiltonian in the form:

$$H = \frac{1}{2}\sum_{l\kappa\alpha} p_\alpha^2(l\kappa) + \frac{1}{2}\sum_{(l\kappa\alpha)(l'\kappa'\beta)} D_{\alpha\beta}(l\kappa; l'\kappa') u_\alpha(l\kappa) u_\beta(l'\kappa') \tag{18}$$

For future developements it is convenient to introduce the following abstract notation [16]. We suppose that the crystal is constituted by N unit cells, each containig r atoms . Thus the system has $3rN$ degrees of freedom, one for each component $u_\alpha(l\kappa)$ of the displacements of the rN crystal atoms. We consider the space constituted by the rN displacements $\mathbf{u}(l\kappa)$, hereafter called *crystal space*, of dimensions $3rN$. We denote by $|u\rangle$ the vector of this space whose components are:

$$\langle l\kappa\alpha|u\rangle = u_\alpha(l\kappa) \tag{19}$$

in the s.o.n.c. of vectors $\{|l\kappa\alpha\rangle\}$:

$$\begin{aligned} \langle l\kappa\alpha|l'\kappa'\beta\rangle &= \delta_{\alpha\beta}\delta_{ll'}\delta_{\kappa\kappa'} \\ \sum_{l\kappa\alpha} |l\kappa\alpha\rangle\langle l\kappa\alpha| &= \hat{I}_{3rN} \end{aligned} \tag{20}$$

The operator $\hat{I}_{3rN}$ is the identity operator in the crystal space. In the same way we can define a $3rN$ dimensional vector $|p\rangle$ whose components are:

$$\langle l\kappa\alpha|p\rangle = p_\alpha(l\kappa) \tag{21}$$

The vector $|u\rangle$, total field of the reduced displacements and the vector $|p\rangle$, total field of the reduced momenta, are connected as it's possible to show from (17), from the relation:

$$|p\rangle = \frac{\partial|u\rangle}{\partial t} \tag{22}$$

Thus we can write the crystal hamiltonian (18) in the form:

$$H = \frac{1}{2}\langle p|p\rangle + \frac{1}{2}\langle u|\hat{D}|u\rangle \tag{23}$$

and the equation of motion in the crystal space becomes:

$$-\hat{D}|u\rangle = \frac{\partial^2|u\rangle}{\partial t^2} \tag{24}$$

where the operator $\hat{D}$, called dynamical operator, is connected with the dynamical matrix $D_{\alpha\beta}(l\kappa;l'\kappa')$ by:

$$D_{\alpha\beta}(l\kappa;l'\kappa') = \langle l\kappa\alpha|\hat{D}|l'\kappa'\beta\rangle \quad , \tag{25}$$

i.e. the dynamical matrix $D_{\alpha\beta}(l\kappa;l'\kappa')$ is the matrix representation of the dynamical operator $\hat{D}$ with respect to the s.o.n.c. $\{|l\kappa\alpha\rangle\}$.

I.C. The secular equation

We choose as a solution of the equations of motion (15) a function of the form:

$$u_\alpha(l\kappa) = u_\alpha(\kappa)e^{i(\mathbf{kx}(l)-\omega t)} \tag{26}$$

Substituting (26) in (15) we obtain:

$$\omega^2 u_\alpha(\kappa) = \sum_{\kappa'\beta} D_{\alpha\beta}(\kappa\kappa'|\mathbf{k})u_\beta(\kappa') \tag{27}$$

where the matrix $\mathbf{D}(\mathbf{k})$ is the Fourier transform of the dynamical matrix given by:

$$D_{\alpha\beta}(\kappa\kappa'|\mathbf{k}) = \sum_{l'} D_{\alpha\beta}(l\kappa;l'\kappa')e^{-i\mathbf{k}[\mathbf{x}(l)-\mathbf{x}(l')]} \tag{28}$$

We observe that the problem of solving the infinite set of equations (15) is reduced to the problem of solving a linear homogeneous set of equations in the $3r$ unknowns $u_\alpha(\kappa)$. This reduction is possible because from (10) the $D_{\alpha\beta}(\kappa\kappa'|\mathbf{k})$ is indipendent from l. The wave vectors $\mathbf{k}$, determined by imposing on the displacements the cyclic boundary conditions of Born-von Karman, are given by:

$$\mathbf{k} = \sum_{i=1}^{3} \frac{m_i}{L}\mathbf{b_i} \tag{29}$$

where m_i are integers, $L^3 = N$ and $\mathbf{b_i}$ are the primitive vectors of the reciprocal lattice, defined by:

$$\begin{aligned} \mathbf{a_i}\mathbf{b_j} &= 2\pi\delta_{ij} \\ \mathbf{b_1} &= \frac{2\pi}{v_a}\mathbf{a_2}\times\mathbf{a_3} \qquad \textit{and cyclic permutations} \end{aligned} \tag{30}$$

being $v_a = \mathbf{a_1}(\mathbf{a_2}\times\mathbf{a_3})$ the volume of the unit cell.

The equation (27) is the eigenvalue equation of the matrix $\mathbf{D}(\mathbf{k})$. The eigenvalues are the solutions of the secular equation:

$$\det\left(D_{\alpha\beta}(\kappa\kappa'|\mathbf{k}) - \omega^2\delta_{\alpha\beta}\delta_{\kappa\kappa'}\right) = 0 \tag{31}$$

which is an algebraic equation of $3r$ degree in the unknowns ω^2. We denote by $\omega_j^2(\mathbf{k})$ $j = 1,\cdots,3r$ the eigenvalues for every value fixed of $\mathbf{k}$. The eigenvalues $\omega_j^2(\mathbf{k})$, the squares of the normal mode frequencies, must be positive owing to the condition of the macroscopical stability of the crystal . The $3r$ functions $\omega_j(\mathbf{k})$, for each value of $\mathbf{k}$, can be regarded as the branches of a multivalued function $\omega(\mathbf{k})$. It is possible to show that there are 3 acoustic branches verifing the relation:

$$\omega_j(\mathbf{k}) \to 0 \; as \; \mathbf{k} \to \mathbf{0} \; , \tag{32}$$

and $3(r-1)$ optical branches.

For each eigenvalue $\omega_j^2(\mathbf{k})$ exists an eigenvector $\mathbf{e}(\kappa|\mathbf{k}j)$, called polarization vector, which satisfies the equation:

$$\omega_j^2(\mathbf{k})e_\alpha(\kappa|\mathbf{k}j) = \sum_{\kappa'\beta} D_{\alpha\beta}(\kappa\kappa'|\mathbf{k})e_\beta(\kappa'|\mathbf{k}j) \tag{33}$$

As $\mathbf{D}(\mathbf{k})$ is a Hermitian matrix, we can choose the eigenvectors $\mathbf{e}(\kappa|\mathbf{k}j)$ so that they satisfy the orthonormality and closure relations:

$$\begin{aligned} \sum_{\kappa\alpha} e_\alpha^*(\kappa|\mathbf{k}j)e_\alpha(\kappa|\mathbf{k}j') &= \delta_{jj'} \\ \sum_{j} e_\beta^*(\kappa'|\mathbf{k}j)e_\alpha(\kappa|\mathbf{k}j) &= \delta_{\alpha\beta}\delta_{\kappa\kappa'} \end{aligned} \tag{34}$$

We note that:

$$D_{\alpha\beta}(\kappa\kappa'|-\mathbf{k}) = D_{\alpha\beta}^*(\kappa\kappa'|\mathbf{k}) \tag{35}$$

If we replace $\mathbf{k}$ by $-\mathbf{k}$ in the equation (32), take the complex conjugate of the resulting equation and make use of (35), we obtain:

$$\sum_{\kappa'\beta} D_{\alpha\beta}(\kappa\kappa'|\mathbf{k})e_\beta^*(\kappa'|-\mathbf{k}j) = \omega_j^2(-\mathbf{k})e_\alpha^*(\kappa|-\mathbf{k}j) \ : \tag{36}$$

$\omega_j^2(-\mathbf{k})$ are the eigenvalues of $\mathbf{D}(\mathbf{k})$, and we can take:

$$\omega_j^2(-\mathbf{k}) = \omega_j^2(\mathbf{k}) \tag{37}$$

As the vectors $\mathbf{e}^*(\kappa|-\mathbf{k}j)$ and $\mathbf{e}(\kappa|\mathbf{k}j)$ satisfy the same equation, we have that:

$$\mathbf{e}^*(\kappa|-\mathbf{k}j) = e^{i\varphi}\mathbf{e}(\kappa|\mathbf{k}j) \tag{38}$$

where $e^{i\varphi}$ is an arbitrary phase factor. Hereafter we use the Leibfried choice:

$$\mathbf{e}^*(\kappa|-\mathbf{k}j) = -\mathbf{e}(\kappa|\mathbf{k}j) \qquad \varphi = \pi \tag{39}$$

By using the definition (28), the Eq. (30), and by introducing the reciprocal lattice vectors:

$$\mathbf{K}(m) = \sum_{i=1}^{3} m_i\mathbf{b_i} \qquad m_i\, integers \quad , \tag{40}$$

we obtain:

$$\mathbf{D}(\mathbf{k}+\mathbf{K}(m)) = \mathbf{D}(\mathbf{k}) \tag{41}$$

This involve that:

$$\begin{aligned} \omega_j(\mathbf{k}+\mathbf{K}(m)) &= \omega_j(\mathbf{k}) \\ \mathbf{e}(\kappa|\mathbf{k}+\mathbf{K}(m)) &= \mathbf{e}(\kappa|\mathbf{k}) \end{aligned} \tag{42}$$

In this way we can limit the choice of the vector $\mathbf{k}$ to the first Brillouin zone. This is equivalent to take the integers m_i in (29) so that:

$$-\frac{L}{2} \le m_i \le \frac{L}{2} \tag{43}$$

We can see from (29) that any allowed vector $\mathbf{k}$ occupies a volume:

$$\Delta^3 k = \frac{v}{L^3} = \frac{v}{N} \tag{44}$$

being $v = (2\pi)^3/v_a$ the volume of the unit cell of the reciprocal lattice and thus of the first Brillouin zone. The number of the allowed wave vectors $\mathbf{k}$, which give distinct solutions is:

$$\frac{v}{\Delta^3 k} = N \tag{45}$$

Since for each $\mathbf{k}$ there are $3r$ distinct branches, the number of the solutions is equal to the number of the degrees of freedom of the crystal.

I.D. Diagonalization of the dynamical matrix

In the crystal space the squares of the normal mode frequencies are the solutions of the eigenvalue equation:

$$\hat{D}|\mathbf{k}j\rangle = \omega_j^2(\mathbf{k})|\mathbf{k}j\rangle \tag{46}$$

As $\hat{D}$ is Hermitian, the eigenvalues $\omega_j^2(\mathbf{k})$ are real and the eigenvectors $|\mathbf{k}j\rangle$ can be chosen in such a way to be a s.o.n.c. in the crystal space:

$$\begin{gathered}\langle\mathbf{k}j|\mathbf{k}'j'\rangle = \delta_{jj'}\Delta(\mathbf{k}-\mathbf{k}') \\ \sum_{\mathbf{k}j}|\mathbf{k}j\rangle\langle\mathbf{k}j| = \hat{I}_{3rN}\end{gathered} \tag{47}$$

where $\Delta(\mathbf{k}-\mathbf{k}')$ is equal to zero unless $\mathbf{k} = \mathbf{k}' + \mathbf{K}(m)$, in that case is equal to one. It is possible to show that the s.o.n.c. $\{|\mathbf{k}j\rangle\}$ and $\{|l\kappa\alpha\rangle\}$ are connected by the relation:

$$\langle l\kappa\alpha|\mathbf{k}j\rangle = \frac{1}{\sqrt{N}}e_\alpha(\kappa|\mathbf{k}j)e^{i\mathbf{k}\mathbf{x}(l)} \tag{48}$$

Infact:

$$\begin{aligned}\langle l\kappa\alpha|l'\kappa'\beta\rangle &= \sum_{\mathbf{k}j}\langle l\kappa\alpha|\mathbf{k}j\rangle\langle\mathbf{k}j|l'\kappa'\beta\rangle = \\ &= \frac{1}{N}\sum_{\mathbf{k}j} e_\alpha(\kappa|\mathbf{k}j)e_\beta^*(\kappa'|\mathbf{k}j)e^{i\mathbf{k}[\mathbf{x}(l)-\mathbf{x}(l')]} = \\ &= \delta_{\alpha\beta}\delta_{\kappa\kappa'}\frac{1}{N}\sum_{\mathbf{k}} e^{i\mathbf{k}[\mathbf{x}(l)-\mathbf{x}(l')]} = \delta_{\alpha\beta}\delta_{ll'}\delta_{\kappa\kappa'}\end{aligned} \tag{49}$$

where we use Eq. (33) and the relation:

$$\sum_{\mathbf{k}} e^{i\mathbf{k}\mathbf{x}(l)} = N\delta_{l0} \tag{50}$$

By using again Eqs. (48) and (34) and from the fact that:

$$\sum_{l} e^{i\mathbf{k}\mathbf{x}(l)} = N\Delta(\mathbf{k}) \tag{51}$$

it's possible to show that:

$$\langle \mathbf{k}j|\mathbf{k}'j'\rangle = \sum_{l\kappa\alpha} \langle \mathbf{k}j|l\kappa\alpha\rangle\langle l\kappa\alpha|\mathbf{k}'j'\rangle = \delta_{jj'}\Delta(\mathbf{k}-\mathbf{k}') \tag{52}$$

I.E. Equations of motion: a new form of the solutions

From what we have seen above, the displacements from the equilibrium can be written as:

$$u_\alpha(l\kappa) = \langle l\kappa\alpha|u\rangle = \sum_{\mathbf{k}j} \langle l\kappa\alpha|\mathbf{k}j\rangle\langle \mathbf{k}j|u\rangle \tag{53}$$

If we denote:

$$\langle \mathbf{k}j|u\rangle = Q(\mathbf{k}j) \tag{54}$$

we can write:

$$u_\alpha(l\kappa) = \frac{1}{\sqrt{N}} \sum_{\mathbf{k}j} e_\alpha(\kappa|\mathbf{k}j)Q(\mathbf{k}j)e^{i\mathbf{k}\mathbf{x}(l)} \tag{55}$$

The $Q(\mathbf{k}j)$ are the normal coordinates of the system. Infact the kinetic energy of the crystal can be written in the form:

$$T = \frac{1}{2}\left(\frac{\partial}{\partial t}\langle u|\right)\left(\frac{\partial}{\partial t}|u\rangle\right) \tag{56}$$

Introducing the s.o.n.c. $\{|\mathbf{k}j\rangle\}$ we have:

$$\begin{aligned} T &= \frac{1}{2}\sum_{\mathbf{k}j} \frac{\partial}{\partial t}\langle u|\mathbf{k}j\rangle \frac{\partial}{\partial t}\langle \mathbf{k}j|u\rangle \\ &= \frac{1}{2}\sum_{\mathbf{k}j} \dot{Q}^*(\mathbf{k}j)\dot{Q}(\mathbf{k}j) \end{aligned} \tag{57}$$

Similarly the potential energy of the crystal:

$$\Phi = \frac{1}{2}\langle u|\hat{D}|u\rangle \tag{58}$$

can be written in the form:

$$\begin{aligned} \Phi &= \frac{1}{2}\sum_{(\mathbf{k}j)(\mathbf{k}'j')} \langle u|\mathbf{k}j\rangle\langle \mathbf{k}j|\hat{D}|\mathbf{k}'j'\rangle\langle \mathbf{k}'j'|u\rangle = \\ &= \frac{1}{2}\sum_{\mathbf{k}j} \omega_j^2(\mathbf{k})Q^*(\mathbf{k}j)Q^*(\mathbf{k}j) \end{aligned} \tag{59}$$

Thus the crystal hamiltonian is given by:

$$H = \frac{1}{2}\sum_{\mathbf{k}j}\left\{\dot{Q}^*(\mathbf{k}j)\dot{Q}(\mathbf{k}j) + \omega_j^2(\mathbf{k})Q^*(\mathbf{k}j)Q(\mathbf{k}j)\right\} \tag{60}$$

The conjugated moment of $Q^*(\mathbf{k}j)$ is:

$$P(\mathbf{k}j) = \frac{\partial L}{\partial \dot{Q}^*(\mathbf{k}j)} = \dot{Q}(\mathbf{k}j) \tag{61}$$

Then the hamiltonian can be put in the form:

$$H = \frac{1}{2} \sum_{\mathbf{k}j} \{P^*(\mathbf{k}j)P(\mathbf{k}j) + \omega_j^2(\mathbf{k})Q^*(\mathbf{k}j)Q(\mathbf{k}j)\} \tag{62}$$

From the Hamilton's equations

$$\dot{Q}(\mathbf{k}j) = \frac{\partial H}{\partial P^*(\mathbf{k}j)} = P(\mathbf{k}j) \qquad -\dot{P}(\mathbf{k}j) = \frac{\partial H}{\partial Q^*(\mathbf{k}j)} = \omega_j^2(\mathbf{k})Q(\mathbf{k}j)\,, \tag{63}$$

we obtain the equations of motion in the normal coordinates $Q(\mathbf{k}j)$

$$\ddot{Q}(\mathbf{k}j) + \omega_j^2(\mathbf{k})Q(\mathbf{k}j) = 0 \;: \tag{64}$$

each normal coordinate describes an indipendent mode of vibration of the crystal. The $Q(\mathbf{k}j)$ solutions of particular physical problem are obtained imposing suitable initial conditions.

Owing to the request that the displacements $u_\alpha(l\kappa)$ are real and to the Leibfried option (39) one has:

$$Q(-\mathbf{k}j) = -Q^*(\mathbf{k}j) \tag{65}$$

I.F. <u>The Green's functions</u> [4,16]

Now we show some properties of the Green's functions of the perfect lattice, which will be useful in the study of the dynamics of the perturbed crystal. The equations of motion (27) can be written in the form:

$$\sum_{l'\kappa'\beta} \left(\omega^2 \delta_{\alpha\beta}\delta_{ll'}\delta_{\kappa\kappa'} - D_{\alpha\beta}(l\kappa;l'\kappa')\right) u_\beta(l'\kappa') = 0 \tag{66}$$

We introduce the matrix $\mathbf{L}(\omega^2)$ defined by:

$$L_{\alpha\beta}(l\kappa;l'\kappa';\omega^2) = \omega^2 \delta_{\alpha\beta}\delta_{ll'}\delta_{\kappa\kappa'} - D_{\alpha\beta}(l\kappa;l'\kappa') \tag{67}$$

It's immediate to verify that the equations of motion (66) can be written in matrix form as:

$$\mathbf{Lu} = \mathbf{0} \tag{68}$$

Let be $z = \omega^2 - i\epsilon$ a complex number where $\epsilon > 0$. The matrix $\mathbf{G}(z)$, whose elements are the Green's functions of the perfect lattice, is defined as the inverse of the matrix $\mathbf{L}(z)$:

$$G_{\alpha\beta}(l\kappa;l'\kappa';z) = L^{-1}_{\alpha\beta}(l\kappa;l'\kappa';z) \tag{69}$$

Then:

$$\mathbf{LG} = \mathbf{GL} = \mathbf{I} \tag{70}$$

A convenient way to write the Green's functionas explicitely, is to work in the crystal space. In this space the operator $\hat{L}$ associated to the matrix $\mathbf{L}$, is by (67):

$$\hat{L} = z\hat{I}_{3rN} - \hat{D} \tag{71}$$

and the operator $\hat{G}$, is by (70):

$$\hat{L}\hat{G} = \hat{G}\hat{L} = \hat{I}_{3rN} \tag{72}$$

If we use the s.o.n.c. $\{|\mathbf{k}j\rangle\}$, we can write:

$$\langle \mathbf{k}j|\hat{L}\hat{G}|\mathbf{k}'j'\rangle = \delta_{jj'}\Delta(\mathbf{k}-\mathbf{k}') \tag{73}$$

and then from the completness relation (47) we obtain:

$$\sum_{\mathbf{k}"j"} \langle \mathbf{k}j|\hat{L}|\mathbf{k}"j"\rangle\langle \mathbf{k}"j"|\hat{G}|\mathbf{k}'j'\rangle = \delta_{jj'}\Delta(\mathbf{k}-\mathbf{k}') \tag{74}$$

The definition (70) implies that:

$$\begin{aligned} \hat{L}|\mathbf{k}"j"\rangle &= (z\hat{I}_{3rN} - \hat{D})|\mathbf{k}"j"\rangle = \left(z - \omega^2_{j"}(\mathbf{k}")\right)|k"j"\rangle \\ \langle \mathbf{k}j|\hat{L}|\mathbf{k}"j"\rangle &= \left(z - \omega^2_{j"}(\mathbf{k}")\right)\delta_{jj"}\Delta(\mathbf{k}-\mathbf{k}") \end{aligned} \tag{75}$$

If we substitute (75) in (74) and sum over all the vectors $\mathbf{k}$ of the first Brillouin zone, we obtain:

$$G(\mathbf{k}j;\mathbf{k}'j';z) = \langle \mathbf{k}j|\hat{G}|\mathbf{k}'j'\rangle = \frac{\delta_{jj'}\Delta(\mathbf{k}-\mathbf{k}')}{z-\omega_j^2(\mathbf{k})} \tag{76}$$

$G_{\alpha\beta}(l\kappa;l'\kappa';z)$ is obtained in a similar way by using the s.o.n.c. $\{|l\kappa\alpha\rangle\}$:

$$\begin{aligned} G_{\alpha\beta}(l\kappa;l'\kappa';z) &= \langle l\kappa\alpha|\hat{G}|l'\kappa'\beta\rangle = \\ &= \sum_{(\mathbf{k}j)(\mathbf{k}'j')} \langle l\kappa\alpha|\mathbf{k}j\rangle\langle \mathbf{k}j|\hat{G}|\mathbf{k}'j'\rangle\langle \mathbf{k}'j'|l'\kappa'\beta\rangle = \\ &= \sum_{\mathbf{k}j}\langle l\kappa\alpha|\mathbf{k}j\rangle\langle \mathbf{k}j|l'\kappa'\beta\rangle\frac{1}{z-\omega_j^2(\mathbf{k})} \end{aligned} \tag{77}$$

Owing to (48), it becomes:

$$G_{\alpha\beta}(l\kappa;l'\kappa';z) = \frac{1}{N}\sum_{\mathbf{k}j}\frac{e_\alpha(\kappa|\mathbf{k}j)e^*_\beta(\kappa'|\mathbf{k}j)}{z-\omega_j^2(\mathbf{k})}e^{i\mathbf{k}[\mathbf{x}(l)-\mathbf{x}(l')]} \tag{78}$$

Let us consider now the relation:

$$\lim_{\epsilon\to 0^+}\frac{1}{x-i\epsilon} = P\frac{1}{x} + i\pi\delta(x) \tag{79}$$

where P denote the Cauchy principal value. We can write the $G_{\alpha\beta}$ in the form:

$$\begin{aligned} G_{\alpha\beta}(l\kappa;l'\kappa':\omega^2-i0^+) &= \sum_{\mathbf{k}j}\langle l\kappa\alpha|\mathbf{k}j\rangle\langle \mathbf{k}j|l'\kappa'\beta\rangle\left(P\frac{1}{\omega^2-\omega_j^2(\mathbf{k})} + i\pi\delta(\omega^2-\omega_j^2(\mathbf{k}))\right) = \\ &= Re\,G_{\alpha\beta}(l\kappa;l'\kappa';\omega^2-i0^+) + i\,Im\,G_{\alpha\beta}(l\kappa;l'\kappa';\omega^2-i0^+) \end{aligned} \tag{80}$$

The real part of $G_{\alpha\beta}$ is the Hilbert transform of the immaginary part. If ω_L is the maximum frequency of the normal modes of the crystal, we have:

$$\begin{aligned} Re\,G_{\alpha\beta}(l\kappa;l'\kappa';\omega^2-i0^+) &= \frac{1}{\pi}P\int_0^{\omega_L}\frac{Im\,G_{\alpha\beta}(l\kappa;l'\kappa';\lambda)}{\omega^2-\lambda}d\lambda = \\ &= \sum_{\mathbf{k}j}\langle l\kappa\alpha|\mathbf{k}j\rangle\langle \mathbf{k}j|l'\kappa'\beta\rangle P\int_0^{\omega_L}\frac{\delta(\lambda-\omega_j^2(\mathbf{k}))}{\omega^2-\lambda}d\lambda \\ &= \sum_{\mathbf{k}j}\langle l\kappa\alpha|\mathbf{k}j\rangle\langle \mathbf{k}j|l'\kappa'\beta\rangle P\frac{1}{\omega^2-\omega_j^2(\mathbf{k})} \end{aligned} \tag{81}$$

I.G. The density of states

We define $G(\omega^2)d\omega^2$ as the number of the squared frequencies in the interval $(\omega^2, \omega^2 + d\omega^2)$ in the limit as $d\omega^2 \to 0$. Similarly $g(\omega)d\omega$ is the number of the frequencies in the interval $(\omega, \omega + d\omega)$, related to $G(\omega^2)$ by:

$$g(\omega) = 2\omega G(\omega^2) \,, \tag{82}$$

since both the distributions are normalized:

$$\int_0^{\omega_L^2} G(\omega^2)d\omega^2 = \int_0^{\omega_L} g(\omega)d\omega = 1 \tag{83}$$

where ω_L is the maximum frequency of the crystal.

We derive now an explicit expression for $G(\omega^2)$. The number of normal modes whose frequency is less or equal to ω^2, is given by:

$$N(\omega^2) = \int_0^{\omega^2} \sum_{\mathbf{k}j} \delta(\lambda - \omega_j^2(\mathbf{k}))d\lambda \tag{84}$$

By definition, the distribution function $G(\omega^2)$ is related to $N(\omega^2)$ by:

$$G(\omega^2)d\omega^2 = \frac{1}{3rN}[N(\omega^2 + d\omega^2) - N(\omega^2)] \tag{85}$$

Then:

$$G(\omega^2) = \frac{1}{3rN}\frac{d}{d\omega^2}N(\omega^2) \tag{86}$$

and by (83) we obtain:

$$G(\omega^2) = \frac{1}{3rN}\sum_{\mathbf{k}j} \delta(\omega^2 - \omega_j^2(\mathbf{k})) \tag{87}$$

We observe that from (80) we have:

$$Im\, G_{\alpha\alpha}(l\kappa; l\kappa; \omega^2 - i0^+) = \pi \sum_{\mathbf{k}j} \langle l\kappa\alpha|\mathbf{k}j\rangle\langle\mathbf{k}j|l\kappa\alpha\rangle\delta(\omega^2 - \omega_j^2(\mathbf{k})) \tag{88}$$

As $\{|l\kappa\alpha\rangle\}$ is a s.o.n.c. we obtain:

$$\sum_{l\kappa\alpha} Im\, G_{\alpha\alpha}(l\kappa; l\kappa; \omega^2 - i0^+) = \pi \sum_{\mathbf{k}j} \delta(\omega^2 - \omega_j^2(\mathbf{k})) \tag{89}$$

From the relation (87) finally we have:

$$G(\omega^2) = \frac{1}{3\pi rN}\sum_{l\kappa\alpha} Im\, G_{\alpha\alpha}(l\kappa; l\kappa; \omega^2 - i0^+) \tag{90}$$

II. THE DYNAMICS OF THE PERTURBED CRYSTAL [4,7,8,9,15,16]

We investigate now the dynamical properties of crystals containing defects. We suppose that only a small concentration of non interacting impurities is present in the crystal. In this way the effect of the impurities on the vibrational properties of the crystal is simply proportional to the effect produced by a single impurity and the results we obtain are corrected to the first order of impurity concentration.

II.A. The equations of motion

We consider a crystal constituted by N unit cells, each of them containing r atoms. We assume that one of the lattice sites of the crystal is occupied by a substitutional impurity, whose mass and interaction with the atoms of the host crystal may differ from those of the atom it replaces. Let :

$$p_\alpha(l\kappa) = m_{l\kappa}\dot{u}_\alpha(l\kappa) \tag{91}$$

where $u_\alpha(l\kappa)$, $p_\alpha(l\kappa)$ and $m_{l\kappa}$ are the (not reduced) displacements, the momenta and the mass of the atom in the lattice site $(l\kappa)$, respectively. The crystal hamiltonian, in the harmonic approximation, is:

$$H = \sum_{l\kappa\alpha} \frac{p_\alpha^2(l\kappa)}{2m_{l\kappa}} + \frac{1}{2} \sum_{(l\kappa\alpha)(l'\kappa'\beta)} \phi_{\alpha\beta}(l\kappa;l'\kappa')u_\alpha(l\kappa)u_\beta(l'\kappa') \tag{92}$$

and the equations of motion are:

$$m_{l\kappa}\ddot{u}_\alpha(l\kappa) = -\sum_{l'\kappa'\beta} \phi_{\alpha\beta}(l\kappa;l'\kappa')u_\beta(l'\kappa') \tag{93}$$

These equations differ from those of the perfect crystal in that the atomic mass $m_{l\kappa}$ now depends on the cell index l.Moreover, because the impurity atom destroys the periodicity of the crystal, the force constants $\phi_{\alpha\beta}(l\kappa;l'\kappa')$ no longer depend on the indices l and l' only through their difference, as it was for $\phi^o_{\alpha\beta}(l\kappa;l'\kappa')$, Eq.(10).

We seek for solutions of the form:

$$\mathbf{u}(l\kappa;t) = \mathbf{u}(l\kappa)e^{-i\omega t} \tag{94}$$

If we substitute (94) in (93), we have:

$$\omega^2 m_{l\kappa}u_\alpha(l\kappa) = \sum_{l'\kappa'\beta} \phi_{\alpha\beta}u_\beta(l'\kappa') \tag{95}$$

We define, as before, the reduced displacements (notice the mass of the perfect crystal):

$$\bar{u}_\alpha(l\kappa) = (m_\kappa)^{1/2}u_\alpha(l\kappa) \tag{96}$$

and the dynamical matrix:

$$D_{\alpha\beta}(l\kappa;l'\kappa') = \frac{1}{(m_\kappa m_{\kappa'})^{1/2}}\phi_{\alpha\beta}(l\kappa;l'\kappa') \tag{97}$$

We set $\bar{u}_\alpha = u_\alpha$. From (95) we have:

$$\omega^2 \frac{m_{l\kappa}}{(m_\kappa)^{1/2}}u_\alpha(l\kappa) = \sum_{l'\kappa'\beta} D_{\alpha\beta}(l\kappa;l'\kappa')u_\beta(l'\kappa')(m_\kappa)^{1/2} \tag{98}$$

If we put:

$$\epsilon_{l\kappa} = \frac{m_{l\kappa} - m_\kappa}{m_\kappa} \tag{99}$$

we obtain:

$$\omega^2(1 + \epsilon_{l\kappa})u_\alpha(l\kappa) = \sum_{l'\kappa'\beta} D_{\alpha\beta}(l\kappa; l'\kappa')u_\beta(l'\kappa') \tag{100}$$

Let the dynamical matrix of the perfect crystal be:

$$D^o_{\alpha\beta}(l\kappa; l'\kappa') = \frac{1}{(m_\kappa m_{\kappa'})^{1/2}}\phi^o_{\alpha\beta}(l\kappa; l'\kappa') \tag{101}$$

If we put:

$$\begin{aligned} L^o_{\alpha\beta}(l\kappa; l'\kappa'; \omega^2) &= \omega^2\delta_{\alpha\beta}\delta_{ll'}\delta_{\kappa\kappa'} - D^o_{\alpha\beta}(l\kappa; l'\kappa') \\ \delta D_{\alpha\beta}(l\kappa; l'\kappa') &= D_{\alpha\beta}(l\kappa; l'\kappa') - D^o_{\alpha\beta}(l\kappa; l'\kappa') \\ \delta L_{\alpha\beta}(l\kappa; l'\kappa'; \omega^2) &= \omega^2\epsilon_{l\kappa}\delta_{\alpha\beta}\delta_{ll'}\delta_{\kappa\kappa'} - \delta D_{\alpha\beta}(l\kappa; l'\kappa') \end{aligned} \tag{102}$$

the Eq.(99) can be rewritten in the form:

$$\sum_{l'\kappa'\beta} \left(L^o_{\alpha\beta}(l\kappa; l'\kappa'; \omega^2) + \delta L_{\alpha\beta}(l\kappa; l'\kappa'; \omega^2)\right) u_\beta(l'\kappa') = 0 \tag{103}$$

or in matrix form as:

$$(\mathbf{L^o} + \delta\mathbf{L})\mathbf{u} = \mathbf{0} \tag{104}$$

The coefficients $\delta D_{\alpha\beta}(l\kappa; l'\kappa')$ satisfy the general conditions of symmetry and translational invariance

$$\begin{aligned} \delta D_{\alpha\beta}(l\kappa; l'\kappa') &= \delta D_{\beta\alpha}(l'\kappa'; l\kappa) \\ \sum_{l'\kappa'} \delta D_{\alpha\beta}(l\kappa; l'\kappa') &= 0 \end{aligned} \tag{105}$$

II.B. The effects of the impurity

The equation (105) can be rewritten in the form:

$$\mathbf{L^o}(\mathbf{I_{3rN}} + \mathbf{G}\delta\mathbf{L})\mathbf{u} = \mathbf{0} \tag{106}$$

where $\mathbf{G}$ is the Green's functions of the perfect crystal in matrix form (see Eq. (69)). The condition that the equation (105) has nontrivial solutions is given by:

$$\det(\mathbf{L^o} + \delta\mathbf{L}) = \det \mathbf{L^o} \det(\mathbf{I_{3rN}} + \mathbf{G}\delta\mathbf{L}) = 0 \tag{107}$$

This is an algebraic equation in ω^2 of $3rN$ degree, whose solutions give the normal mode frequencies of the perturbed crystal.

Examine now the matrix $\delta\mathbf{L}$. If we assume that the impurity interacts directly with $n-1$ atoms, then the matrix $\delta\mathbf{L}$ have at most $3n \times 3n$ non-vanishing elements. These elements $\delta L_{\alpha\beta}(l\kappa; l'\kappa')$ are those labeled by the indices $(l\kappa)$, $(l'\kappa')$ which refer either to the impurity site or to the sites of the host crystal with which the impurity atom interacts directly.
We can recast the matrix elements as follows:

$$\delta\mathbf{L} = \begin{pmatrix} \delta\mathbf{l} & \mathbf{0} \\ \mathbf{0} & \mathbf{0} \end{pmatrix} \tag{108}$$

where $\delta\mathbf{l}$ is a $3n \times 3n$ matrix. We partition in the same way the elements of $\mathbf{G}$:

$$\mathbf{G} = \begin{pmatrix} \mathbf{g} & \mathbf{G_{12}} \\ \mathbf{G_{21}} & \mathbf{G_{22}} \end{pmatrix} \tag{109}$$

where $\mathbf{g}$ is a $3n \times 3n$ matrix, $\mathbf{G_{12}} = \mathbf{G_{21}^T}$ is a $3n \times (3rN - 3n)$ matrix and $\mathbf{G_{22}}$ is a $(3rN - 3n) \times (3rN - 3n)$ matrix. Then:

$$\det(\mathbf{I_{3n}} + \mathbf{g}\delta\mathbf{l}) = \det(\mathbf{I_{3rN}} + \mathbf{G}\delta\mathbf{L}) \tag{110}$$

where the first is an algebraic equation of $3n$ degree while the second is an algebraic equation of $3rN$ degree..

We denote by ω_{os} the normal mode frequencies of the unperturbed crystal, solutions of the equation:

$$\det \mathbf{L^o} = \prod_{s=1}^{3rN} (\omega^2 - \omega_{os}^2) = 0 \tag{111}$$

Similarly the normal mode frequencies ω_s of the perturbed crystal are the solutions of the equation:

$$\det(\mathbf{L^o} + \delta\mathbf{L}) = \prod_{s=1}^{3rN} (\omega^2 - \omega_s^2) = 0 \tag{112}$$

If we substitute (110) into (107),

$$\det(\mathbf{L^o} + \delta\mathbf{L}) = \det \mathbf{L^o} \det(\mathbf{I_{3n}} + \mathbf{g}\delta\mathbf{l}) \tag{113}$$

and use (111) and (112), we arrive at the form:

$$\det(\mathbf{I_{3n}} + \mathbf{g}\delta\mathbf{l}) = \prod_{s=1}^{3rN} \frac{\omega^2 - \omega_s^2}{\omega^2 - \omega_{os}^2} = 0 \tag{114}$$

Since this is an equation of $3n$ degree, there are $3rN - 3n$ factors to the numerator which must be identical to an equal number of factors to the denominator. This shows that at most $3n$ normal mode frequencies are modified by the impurity, while the remaining frequencies are the same of the perfect crystal.

II.C. The impurity space

The subspace of the crystal space, constituted by the $(n-1)$ displacements of the crystal atoms directly interacting with the impurity atom and by the dispalcement of the impurity itself, is hereafter called *impurity space*. The dimension of this space is $3n$. The set of the vectors $\{|l\kappa\alpha\rangle\}$ such that the indices $(l\kappa)$ denote one of the atom whose displacements belong to the impurity space, constitutes a s.o.n.c. for the impurity space:

$$\langle l\kappa\alpha|l'\kappa'\beta\rangle = \delta_{\alpha\beta}\delta_{ll'}\delta_{\kappa\kappa'}$$
$$\sum_{l\kappa\alpha} |l\kappa\alpha\rangle\langle l\kappa\alpha| = \hat{I}_{3n} \tag{115}$$

We have:

$$\langle l\kappa\alpha|\delta\hat{L}|l'\kappa'\beta\rangle \neq 0 \tag{116}$$

only if $|l\kappa\alpha\rangle$ and $|l'\kappa'\beta\rangle$ belong to the impurity space.
The equations of motion (103) projected into the impurity space become:

$$\sum_{l'\kappa'\beta} \left(L^o_{\alpha\beta}(l\kappa;l'\kappa';\omega^2) + \delta L_{\alpha\beta}(l\kappa;l'\kappa';\omega^2)\right) u_\beta(l'\kappa') = 0 \qquad (117)$$

Out of the impurity space, the equations of motion are those of the perfect crystal.

The $3n$ normal mode frequencies modified by the defect, are obtained solving the secular equation:

$$\det\left(\langle l\kappa\alpha|(\hat{L}^o + \delta\hat{L})|l'\kappa'\beta\rangle\right) = 0 \qquad (118)$$

where $|l\kappa\alpha\rangle$ and $|l'\kappa'\beta\rangle$ belong to the impurity space.

The study of the crystal dynamical properties can be therefore restricted to that into the impurity space, greatly semplifing the problem. Indeed, the number of atoms involved in the equations (117) is much smaller of the number rN of the atoms of the crystal.

II.D. Normal modes

We want to analyze the normal modes of the perturbed crystal. For this purpose we proceed along the same line used for the perfect crystal. First we rewrite the dynamical matrix $\mathbf{D}$ in the form:

$$D_{\alpha\beta}(l\kappa;l'\kappa') = \frac{1}{(m_{l\kappa}m_{l'\kappa'})^{1/2}}\phi_{\alpha\beta}(l\kappa;l'\kappa') \qquad (119)$$

and introduce the reduced displacements

$$\bar{u}_\alpha(l\kappa) = (m_{l\kappa})^{1/2} u_\alpha(l\kappa) \qquad (120)$$

denoted hereafter by $u_\alpha(l\kappa)$. The equations of motion become:

$$\omega^2 u_\alpha(l\kappa) = \sum_{l'\kappa'\beta} D_{\alpha\beta}(l\kappa;l'\kappa')u_\beta(l'\kappa') \qquad (121)$$

Then, we consider the operator $\hat{D}$, defined in the crystal space, and connected to the dynamical matrix $\mathbf{D}$. The squares ω_s^2 of the normal mode frequencies of the perturbed crystal, are the eigenvalues of the operator $\hat{D}$:

$$\hat{D}|s\rangle = \omega_s^2|s\rangle \qquad (122)$$

As $\hat{D}$ is Hermitian, we can choose the eigenvectors $|s\rangle$ so that they are a s.o.n.c. in the crystal space:

$$\begin{aligned} \langle s'|s\rangle &= \delta_{ss'} \\ \sum_s |s\rangle\langle s| &= \hat{I}_{3rN} \end{aligned} \qquad (123)$$

As in the case of the perfect crystal, the normal mode coordinates of the perturbed crystal are:

$$u_s = \langle s|u\rangle \qquad (124)$$

Because of the completness of the set $\{|s\rangle\}$ we can write:

$$u_\alpha(l\kappa) = \sum_s \langle l\kappa\alpha|s\rangle\langle s|u\rangle = \sum_s b_\alpha(l\kappa|s)u_s \tag{125}$$

where $b_\alpha(l\kappa|s) = \langle l\kappa\alpha|s\rangle$. Because $u_\alpha(l\kappa)$ are real and $b_\alpha(l\kappa|s)$ can be chosen real, the normal mode coordinates u_s can be chosen real.

II.E. Use of the group theory [4,15]

The point group of a crystal is defined by the set of symmetry operations which send the crystal into itself, leaving a lattice site unmoved. The perfect crystal is also translationally invariant owing to its periodicity, while the perturbed crystal has only the point symmetry operations. To exploit them we put the origin at the impurity site.

We denote by R the symmetry operations of the crystal point group $\mathcal{G}$ and by $\mathbf{R}$ the 3×3 real orthogonal matrix that represent this group. We assume that the group has h distinct symmetry operations.

The effect of a symmetry operation R on a lattice position vector $\mathbf{x}(l\kappa)$ can be written as:

$$\mathbf{x}(LK) = \mathbf{R}\mathbf{x}(l\kappa) \tag{126}$$

We introduce the operator $\hat{O}_R$ defined as:

$$\hat{O}_R f(\mathbf{r}) = f(\mathbf{R}^{-1}\mathbf{r}) \tag{127}$$

We can see that $\hat{O}_R$ is a linear unitary operator:

$$\hat{O}_R^\dagger = \hat{O}_R^{-1} \tag{128}$$

It is possible to prove that the set of the operators $\{\hat{O}_R\}_{R\in\mathcal{G}}$ constitutes a group isomorphic to $\mathcal{G}$.

As $\mathcal{G}$ is the crystal point group, the potential energy:

$$\Phi = \langle u|\hat{D}|u\rangle \tag{129}$$

must be invariant under a symmetry operation $R \in \mathcal{G}$. Consequently the dynamical operator $\hat{D}$ must commute with all the operators $\hat{O}_R$:

$$\hat{O}_R\hat{D} = \hat{D}\hat{O}_R \qquad \forall R \in \mathcal{G} \tag{130}$$

If we introduce the s.o.n.c. $\{|l\kappa\alpha\rangle\}$ of Eqs. (19)(20), we obtain:

$$\begin{aligned} D_{\alpha\beta}(l\kappa; l'\kappa') &= \langle l\kappa\alpha|\hat{O}_R\hat{D}\hat{O}_R^\dagger|l'\kappa'\beta\rangle = \\ &= \sum_{(l_1\kappa_1\gamma)(l_2\kappa_2\mu)} \langle l\kappa\alpha|\hat{O}_R|l_1\kappa_1\gamma\rangle D_{\gamma\mu}(l_1\kappa_1; l_2\kappa_2)\langle l_2\kappa_2\mu|\hat{O}_R^\dagger|l'\kappa'\beta\rangle \end{aligned} \tag{131}$$

The $3rN \times 3rN$ matrix $\{\langle l\kappa\alpha|\hat{O}_R|l'\kappa'\beta\rangle\}$ is a real orthogonal matrix as it is possible to show by (128). We can prove that:

$$S_{\alpha\beta}(l\kappa; l'\kappa') = \langle l\kappa\alpha|\hat{O}_R|l'\kappa'\beta\rangle = R_{\alpha\beta}\delta\left(\mathbf{R}^{-1}\mathbf{x}(l\kappa), \mathbf{x}(l'\kappa')\right) \tag{132}$$

where the last factor is a consequence of the fact that the symmetry operation R sends the crystal into itself only if the atoms $(l\kappa)$ and $(l'\kappa')$ are the same, *i.e.* only if:

$$m_{l\kappa} = m_{l'\kappa'} \tag{133}$$

Therefore the h matrices $\{\mathcal{S}\}$ provide a $3rN$-dimensional unitary representation $\mathcal{S}$ of the point group $\mathcal{G}$.

From the relation (130) it follows that:

$$\hat{O}_R\hat{D}|s\rangle = \hat{D}\hat{O}_R|s\rangle = \omega_s^2 O_R|s\rangle \tag{134}$$

Thus if $|s\rangle$ is an eigenvector of $\hat{D}$ corresponding to the eigenvalue ω_s^2, then $\hat{O}_R|s\rangle$ is also an eigenvector of the dynamical operator corresponding to the same eigenvalue.

Now we suppose that the normal mode s is degenerate. We denote by ω_s^2 the eigenvalue f_s times degenerate and by $|sj\rangle$ $j = 1,\dots,f_s$ the distinct eigenvectors chosen so that they are orthonormal:

$$\begin{aligned} \hat{D}|sj\rangle &= \omega_s^2|sj\rangle \\ \langle sj|si\rangle &= \delta_{ij} \end{aligned} \tag{135}$$

The vector $\hat{O}_R|sj\rangle$ is evidently a linear combination of the f_s independent vectors $|si\rangle$. Thus we can write:

$$\hat{O}_R|sj\rangle = \sum_{i=1}^{f_s} \Gamma_{ij}^{(s)}(R)|si\rangle \qquad \forall R \in \mathcal{G} \tag{136}$$

We want to prove that the h matrices $\mathbf{\Gamma}^s(R)$ $R \in \mathcal{G}$, of dimension $f_s \times f_s$, provide a representation of $\mathcal{G}$. Let it be:

$$R_3 = R_1 \circ R_2 \qquad R_i \in \mathcal{G}\ i = 1,2,3 \tag{137}$$

$$\mathbf{R_3} = \mathbf{R_2R_1}$$

Then we have:

$$\hat{O}_{R_3}|sj\rangle = \sum_{i=1}^{f_s} \Gamma_{ij}^{(s)}(R_3)|si\rangle \tag{138}$$

$$\begin{aligned} \hat{O}_{R_2}\hat{O}_{R_1}|sj\rangle &= \hat{O}_{R_2}\sum_k \Gamma_{kj}^{(s)}(R_1)|sk\rangle = \\ &= \sum_k \Gamma_{kj}^{(s)}(R_1)\hat{O}_{R_2}|sk\rangle = \sum_{ik} \Gamma_{kj}^{(s)}(R_1)\Gamma_{ik}^{(s)}(R_2)|si\rangle \end{aligned} \tag{139}$$

From the fact that $\hat{O}_{R_3} = \hat{O}_{R_2}\hat{O}_{R_1}$ we have that:

$$\Gamma_{ij}^{(s)}(R_3) = \sum_{k=1}^{f_s} \Gamma_{ik}^{(s)}(R_2)\Gamma_{kj}^{(s)}(R_1) \tag{140}$$

The contrary is immediate to prove.

As the $|sj\rangle$ are orthonormals we can show that the matrices $\{\mathbf{\Gamma}^{(s)}(R)\}_{R\in\mathcal{G}}$ provide an unitary representation of $\mathcal{G}$. Infact:

$$\begin{aligned}\delta_{ij} = \langle si|sj\rangle &= \langle si|\hat{O}_R^\dagger \hat{O}_R|sj\rangle = \\ &= \sum_{kl} \Gamma_{ki}^{(s)}(R)^* \Gamma_{lj}^{(s)}(R)\langle sk|sl\rangle = \\ &= \sum_{k} \Gamma_{ki}^{(s)}(R)^* \Gamma_{kj}^{(s)}(R)\end{aligned} \tag{141}$$

It is a general result that in absence of accidental degeneracies the set of h matrices $\{\mathbf{\Gamma}^{(s)}(R)\}_{R\in\mathcal{G}}$ of dimensions $f_s\times f_s$, forms an f_s-dimensional irreducible representation of $\mathcal{G}$. The f_s vectors $\{|sj\rangle\}_{j=1,\dots,f_s}$ are then the basis vectors for an f_s-dimensional irreducible representation of $\mathcal{G}$. If we consider the space of the displacements, we have:

$$\langle l\kappa\alpha|\hat{O}_R|sj\rangle = \sum_{i=1}^{f_s} \Gamma_{ij}^{(s)}(R)\langle l\kappa\alpha|si\rangle \tag{142}$$

and then from (132) we can write:

$$\sum_{l'\kappa'\beta} \mathcal{S}_{\alpha\beta}(l\kappa;l'\kappa')\langle l'\kappa'\beta|sj\rangle = \sum_{i} \Gamma_{ij}^{(s)}(R)\langle l\kappa\alpha|si\rangle \tag{143}$$

We denote by $\mathcal{Z}$ the $3rN \times 3rN$ matrices $\{\langle l\kappa\alpha|sj\rangle\}$. The the relation (142) can be written in the form:

$$\mathcal{S}(R)\mathcal{Z} = \mathcal{Z}\mathbf{\Gamma}(R) \tag{144}$$

where the $3rN \times 3rN$ matrix $\mathbf{\Gamma}(R)$ has the block-diagonal form:

$$\mathbf{\Gamma}(R) = \begin{pmatrix} \mathbf{\Gamma}^{(1)}(R) & \cdots & \cdots \\ \vdots & \mathbf{\Gamma}^{(2)}(R) & \cdots \\ \vdots & \vdots & \ddots \end{pmatrix} \tag{145}$$

The matrices $\mathbf{\Gamma}^{(1)}(R)$, $\mathbf{\Gamma}^{(2)}(R), \cdots$ appearing along the diagonal are the irreducible representation matrices correspondig to the eigenvalues $\omega_1^2, \omega_2^2, \cdots$. Thus the matrix of the eigenvectors $\mathcal{Z}$ is the matrix which by a similarity transformation simultaneously blok diagonalizes each of the matrices $\mathcal{S}(R)$ comprising a $3rN$- dimensionals representation of $\mathcal{G}$:

$$\mathcal{Z}^{-1}\mathcal{S}(R)\mathcal{Z} = \mathbf{\Gamma}(R) \tag{146}$$

II.F. The symmetry coordinates

As we have seen in the previous sections, the normal mode frequencies modified by the presence of the impurity are the solutions of the equation:

$$\det\left(\delta_{\alpha\beta}\delta_{ll'}\delta_{\kappa\kappa'} - \sum_{l_1\kappa_1\gamma} G_{\alpha\gamma}(l\kappa;l_1\kappa_1;\omega^2)\delta L_{\gamma\beta}(l_1\kappa_1;l'\kappa';\omega^2)\right) = 0 \tag{147}$$

where the indices $(l\kappa)$, $(l'\kappa')$ and $(l_1\kappa_1)$ label the $n-1$ atoms, interacting directly with the impurity, and the impurity itself. We consider the

$3n$-dimensionals impurity space; hereafter $\{|l\kappa\alpha\rangle\}$ is the s.o.n.c. of this space.

The matrices $\mathbf{G}$ and $\delta\mathbf{L}$ satisfy the transformation laws:

$$G_{\alpha\beta}(l\kappa;l'\kappa';\omega^2) = \sum_{(l_1\kappa_1\gamma)(l_2\kappa_2\mu)} \langle l\kappa\alpha|\hat{O}_R|l_1\kappa_1\gamma\rangle\langle l_2\kappa_2\mu|\hat{O}_R^\dagger|l'\kappa'\beta\rangle G_{\gamma\mu}(l_1\kappa_1;l_2\kappa_2;\omega^2)$$
$$\delta L_{\alpha\beta}(l\kappa;l'\kappa';\omega^2) = \sum_{(l_1\kappa_1\gamma)(l_2\kappa_2\mu)} \langle l\kappa\alpha|\hat{O}_R|l_1\kappa_1\gamma\rangle\langle l_2\kappa_2\mu|\hat{O}_R^\dagger|l'\kappa'\beta\rangle \delta L_{\gamma\mu}(l_1\kappa_1;l_2\kappa_2;\omega^2) \qquad (148)$$

under a symmetry operation R of the point group $\mathcal{G}$. The h matrices of dimensions $3n \times 3n$:

$$\mathcal{R}_{\alpha\beta}(l\kappa;l'\kappa') = \langle l\kappa\alpha|\hat{O}_R|l'\kappa'\beta\rangle \qquad (149)$$

provide a $3n$-dimensional representation $\mathcal{L}$ of the group $\mathcal{G}$. This representation is in general irreducible. The number of times c_Γ that an irreducible representation Γ of $\mathcal{G}$ occours in $\mathcal{L}$ is given by:

$$c_\Gamma = \frac{1}{h}\sum_{R\in\mathcal{G}} \chi^\Gamma(R)^*\chi(R)$$
$$\chi^\Gamma(R) = Tr\,\mathbf{\Gamma}(R) \qquad \chi(R) = Tr\,\mathcal{R} \qquad (150)$$

Now we construct a s.o.n.c. in the impurity space with the property that the matrix elements of any operator which commutes with $\hat{O}_R \quad \forall R \in \mathcal{G}$, are diagonals in that s.o.n.c. . Let us consider an arbitrary vector $|\varphi\rangle$. According to the group theory the projector on the μth row of a unitary representation Γ of $\mathcal{G}$ can be written as:

$$\hat{P}_\mu^\Gamma = \frac{f_\Gamma}{h}\sum_{R\in\mathcal{G}} \Gamma_{\mu\nu}(R)^*\hat{O}_R \qquad (151)$$

for any fixed ν, where f_Γ is the dimension of the Γ representation. Thus the vector:

$$\hat{P}_\mu^\Gamma|\varphi\rangle = |\varphi,\Gamma\mu\rangle \qquad (152)$$

belongs to the μth row of the Γ representation. The vectors $|\varphi,\Gamma\mu\rangle$, $\mu = 1,\cdots,f_\Gamma$, constitute a basis for the Γ representation. Infact as Γ is unitary, we have:

$$\begin{aligned}
\Gamma_{\mu\nu}(R)^* &= \sum_\gamma \delta_{\gamma\mu}\Gamma_{\gamma\nu}(R)^* = \\
&= \sum_\gamma \left(\sum_\lambda \Gamma_{\lambda\mu}(S)\Gamma_{\lambda\gamma}(S)^*\right)\Gamma_{\gamma\nu}(R)^* = \\
&= \sum_\lambda \Gamma_{\lambda\mu}(S)\left(\sum_\gamma \Gamma_{\lambda\gamma}(S)^*\Gamma_{\gamma\nu}(R)^*\right) = \\
&= \sum_\lambda \Gamma_{\lambda\mu}(S)\Gamma_{\lambda\nu}(SR)^* \qquad \forall S,R \in \mathcal{G}
\end{aligned} \qquad (153)$$

Thus for $S \in \mathcal{G}$ we have:

$$\begin{aligned}
\hat{O}_S\hat{P}_\mu^\Gamma|\varphi\rangle &= \hat{O}_S\frac{f_\Gamma}{h}\sum_{R\in\mathcal{G}} \Gamma_{\mu\nu}(R)^*\hat{O}_R|\varphi\rangle = \\
&= \sum_\lambda \Gamma_{\lambda\mu}(S)\frac{f_\Gamma}{h}\sum_{R\in\mathcal{G}} \Gamma_{\lambda\nu}(SR)^*\hat{O}_{SR}|\varphi\rangle
\end{aligned} \qquad (154)$$

Then if we apply the definition (151) we obtain finally:

$$\hat{O}_S \hat{P}_\mu^\Gamma |\varphi\rangle = \sum_\lambda \Gamma_{\lambda\mu}(S) \hat{P}_\lambda^\Gamma |\varphi\rangle \tag{155}$$

It is possible to show that if Γ and Γ' are two unitary irreducible representation, then:

$$\langle \varphi, \Gamma\mu | \varphi, \Gamma'\mu' \rangle = d_\Gamma \delta_{\Gamma\Gamma'} \delta_{\mu\mu'} \tag{156}$$

For all we have said, the irreducible representation Γ may be contained c_Γ times in the representation $\mathcal{L}$. This imply that each vector $|\varphi, \Gamma\mu\rangle$ is a linear combination of c_Γ vectors $|\Gamma\gamma\mu\rangle \quad \gamma = 1, \cdots, c_\Gamma$, and belongs to the μth row of the Γ representation as it results from (155). If we choose these vectors so that they are orthonormal, we have that the set $\{|\Gamma\gamma\mu\rangle\}$ is a s.o.n.c. of the impurity space:

$$\begin{aligned} \langle \Gamma\gamma\mu | \Gamma'\gamma'\mu' \rangle &= \delta_{\Gamma\Gamma'} \delta_{\gamma\gamma'} \delta_{\mu\mu'} \\ \sum_{\Gamma\gamma\mu} |\Gamma\gamma\mu\rangle\langle\Gamma\gamma\mu| &= \hat{I}_{3n} \end{aligned} \tag{157}$$

If we consider an operator $\hat{A}$ such that:

$$\hat{A}\hat{O}_R = \hat{O}_R \hat{A} \qquad \forall R \in \mathcal{G}, \tag{158}$$

we have:

$$\begin{aligned} \langle \Gamma\gamma\mu | \hat{A} | \Gamma'\gamma'\mu' \rangle &= \langle \Gamma\gamma\mu | \hat{O}_R^\dagger \hat{A} \hat{O}_R | \Gamma'\gamma'\mu' \rangle \\ &= \sum_{\lambda\nu} \Gamma_{\lambda\mu}(R)^* \Gamma'_{\nu\mu'}(R) \langle \Gamma\gamma\lambda | \hat{A} | \Gamma'\gamma'\nu \rangle \end{aligned} \tag{159}$$

From the group theory, because Γ and Γ' are irreducible representations, we have:

$$\sum_{R\in\mathcal{G}} \Gamma_{\lambda\mu}(R)^* \Gamma'_{\nu\mu'}(R) = \frac{h}{f_\Gamma} \delta_{\Gamma\Gamma'} \delta_{\lambda\nu} \delta_{\mu\mu'} \tag{160}$$

If we sum (159) over R we obtain::

$$h \langle \Gamma\gamma\mu | \hat{A} | \Gamma'\gamma'\mu' \rangle = \frac{h}{f_\Gamma} \left(\sum_\lambda \langle \Gamma\gamma\lambda | \Gamma\gamma'\lambda \rangle \right) \delta_{\Gamma\Gamma'} \delta_{\gamma\gamma'} \tag{161}$$

This relation shows that:

$$\langle \Gamma\gamma\mu | \hat{A} | \Gamma'\gamma'\mu' \rangle = A(\Gamma, \mu, \gamma\gamma') \delta_{\Gamma\Gamma'} \delta_{\mu\mu'} \tag{162}$$

The s.o.n.c. $\{|\Gamma\gamma\mu\rangle\}$ is that we seek for.

The vectors $|\Gamma\gamma\mu\rangle$ are hereafter called *symmetry coordinates*.

II.G. The projected density on the symmetry coordinates

We consider the point group $\mathcal{G}$ of the substitutional impurity. $\mathcal{G}$ is in general a subgroup of the point group of the perfect crystal. Then if $\hat{L}^o$ is the operator of the perfect crystal introduced in (70), the relation:

$$\hat{L}^o \hat{O}_R = \hat{O}_R \hat{L}^o \tag{163}$$

is satisfied. Since

$$\hat{O}_R^{-1} = \hat{O}_{R^{-1}} \qquad \forall R \in \mathcal{G} \;, \tag{164}$$

if we take the inverse of the equation (163) we have:

$$\hat{G}\hat{O}_R = \hat{O}_R \hat{G} \qquad \forall R \in \mathcal{G} \;, \tag{165}$$

where $\hat{G}$ is the Green's operator of the crystal. If $\{|\Gamma\gamma\mu\rangle\}$ is the s.o.n.c. (157), we have from (162) and (165) that:

$$\langle \Gamma\gamma\mu|\hat{G}|\Gamma'\gamma'\mu'\rangle = G(\Gamma\gamma\gamma'\mu)\delta_{\Gamma\Gamma'}\delta_{\mu\mu'} \tag{166}$$

By inserting in (165) the s.o.n.c. $\{|\mathbf{k}j\rangle\}$, it becomes:

$$\begin{aligned} G(\Gamma\gamma\gamma'\mu; z) &= \sum_{(\mathbf{k}j)(\mathbf{k}'j')} \langle \Gamma\gamma\mu|\mathbf{k}j\rangle\langle \mathbf{k}j|\hat{G}|\mathbf{k}'j'\rangle\langle \mathbf{k}'j'|\Gamma\gamma'\mu\rangle = \\ &= \sum_{\mathbf{k}j} \langle \Gamma\gamma\mu|\mathbf{k}j\rangle\langle \mathbf{k}j|\Gamma\gamma'\mu\rangle \frac{1}{z-\omega_j^2(\mathbf{k})} \end{aligned} \tag{167}$$

We make the extension to the complex field, as in sect. I.F. and obtain:

$$Im\, G(\Gamma\gamma\gamma'\mu; \omega^2) = \pi \sum_{\mathbf{k}j} \langle \Gamma\gamma\mu|\mathbf{k}j\rangle\langle \mathbf{k}j|\Gamma\gamma'\mu\rangle\delta(\omega^2 - \omega_j(\mathbf{k})) \tag{168}$$

We define the *crystal density of states projected into the symmetry coordinates* as it follows:

$$\rho^{0\Gamma\mu}_{\gamma\gamma'}(\omega^2) = \frac{1}{\pi} G(\Gamma\gamma\gamma'\mu; \omega^2) \tag{169}$$

Because of the completness of the set $\{|\mathbf{k}j\rangle\}$ and from (168) we obtain:

$$\int_0^{\omega_L^2} \rho^{0\Gamma\mu}_{\gamma\gamma'}(\omega^2) d\omega^2 = \delta_{\gamma\gamma'} \tag{170}$$

The coefficients $\langle \Gamma\gamma\mu|\mathbf{k}j\rangle$ are hereafter called *overlap integrals*. We have:

$$\begin{aligned} \langle \Gamma\gamma\mu|\mathbf{k}j\rangle &= \langle \Gamma\gamma\mu|\hat{I}_{3n}|\mathbf{k}j\rangle = \\ &= \sum_{l\kappa\alpha} \langle \Gamma\gamma\mu|l\kappa\alpha\rangle\langle l\kappa\alpha|\mathbf{k}j\rangle \end{aligned} \tag{171}$$

where $\{|l\kappa\alpha\rangle\}$ is the s.o.n.c. (114) of the impurity space.

Now we consider the Green's operator $\hat{U}$ of the perturbed crystal. It is defined by the relation:

$$\hat{U} = (\hat{L}^o + \delta\hat{L})^{-1} \tag{172}$$

and thus we have:

$$(\hat{I} + \hat{G}\delta\hat{L})\hat{U} = \hat{G} \tag{173}$$

As $\mathcal{G}$ is the point group of the defect, we have:

$$\langle \Gamma\gamma\mu|\delta\hat{L}|\Gamma'\gamma'\mu'\rangle = \delta L(\Gamma\gamma\gamma'\mu; \omega^2)\delta_{\Gamma\Gamma'}\delta_{\mu\mu'} \tag{174}$$

If we project the equation (171) on the symmetry coordinates we have:

$$\langle\Gamma\gamma\mu|(\hat{I}+\hat{G}\delta\hat{L})\hat{U}|\Gamma\gamma'\mu\rangle = G(\Gamma\gamma\gamma'\mu;\omega^2)\delta_{\Gamma\Gamma'}\delta_{\mu\mu'} \tag{175}$$

Thus we obtain the result:

$$\sum_{\gamma_2}\left(\delta_{\gamma\gamma_2}+\sum_{\gamma_1}G(\Gamma\gamma\gamma_1\mu;\omega^2)\delta L(\Gamma\gamma_1\gamma_2\mu;\omega^2)\right)U(\Gamma\gamma_2\gamma'\mu;\omega^2) = G(\Gamma\gamma\gamma'\mu;\omega^2) \tag{176}$$

In particular if the representation Γ is contained only once ($\gamma = 1$) in the $3n$-dimensional representation $\mathcal{L}$, we have:

$$U(\Gamma\mu;\omega^2) = \frac{G(\Gamma\mu;\omega^2)}{1+G(\Gamma\mu;\omega^2)\delta L(\Gamma\mu;\omega^2)} \tag{177}$$

The *perturbed projected density of states of the Γ-symmetry* (γ, γ' component) is:

$$\rho^{\Gamma\mu}_{\gamma\gamma'}(\omega^2) = \frac{1}{\pi} Im\, U(\Gamma\gamma\gamma'\mu;\omega^2) \tag{178}$$

If we introduce the s.o.n.c. $\{|l\kappa\alpha\rangle\}$ of the impurity space, we can write:

$$\rho^{\Gamma\mu}_{\gamma\gamma'}(\omega^2) = \sum_{(l\kappa\alpha)(l'\kappa'\beta)} \langle\Gamma\gamma\mu|l\kappa\alpha\rangle\langle l'\kappa'\beta|\Gamma\gamma'\mu\rangle\rho^{l\kappa\alpha}_{l'\kappa'\beta}(\omega^2) \tag{179}$$

It is possible to show that:

$$\int_0^{\omega_L^2}\rho^{\Gamma\mu}_{\gamma\gamma'}(\omega^2)d\omega^2 = \sum_{l\kappa\alpha}\left(\frac{m_{l\kappa}}{m_\kappa}\right)\langle\Gamma\gamma\mu|l\kappa\alpha\rangle\langle l\kappa\alpha|\Gamma\gamma'\mu\rangle \tag{180}$$

III. LATTICE RELAXATION FOLLOWING ULTRAFAST EXCITATION

Now we apply the formalism introduced in the previous sections to the study of the light absorption by impurity centers in insulators, which generates localized electronic states inside the forbidden gap [17]. The absorption of a radiation of suitable frequency causes an electronic transition between the localized electronic levels. Consequently there is a variation in the electron-lattice interaction, which istantaneous on the scale of the ionic times (Condon approximation).

We work with a classical local model, *i.e.* we assume that the interaction acts only between the impurity and its n.n. and that the sudden change of the interaction means a sudden change of the equilibrium positions of the crystal atoms. Thus, after the electronic transition has taken place, the atoms begin to oscillate in a classical way around the new equilibrium position,

III.A. The new equilibrium configuration [21]

Owing the local variation of the electron-lattice (E-P) interaction the equilibrium configuration is modified. If we evaluate the Taylor's series to the second order of the atomic displacements around the old configuration, the potential energy becomes:

$$\Phi = \sum_{l\kappa\alpha} f_\alpha(l\kappa)u_\alpha(l\kappa) + \frac{1}{2}\sum_{(l\kappa\alpha)(l'\kappa'\beta)} D_{\alpha\beta}(l\kappa;l'\kappa')u_\alpha(l\kappa)u_\beta(l'\kappa') \tag{181}$$

The linear term, due to the change in the EP interaction, arises when the electron is suddenly excited. The reduced forces $f_\alpha(l\kappa)$ are the coefficients of such interaction. In the linear EP approximation the quadratic term is assumed unchanged by the absorption process, while the changes in the springs due to the presence of the impurity are considered in the lattice dynamics.

The new equilibrium positions are shifted of $\xi(l\kappa)$ from the old positions and the displacements $\tilde{\mathbf{u}}(l\kappa)$ around the new equilibrium configuration are given by:

$$\tilde{\mathbf{u}}(l\kappa) = \mathbf{u}(l\kappa) + \xi(l\kappa) \tag{182}$$

In the new equilibrium configuration the harmonic potential energy can be written:

$$\Phi = \frac{1}{2} \sum_{(l\kappa\alpha)(l'\kappa'\beta)} \tilde{D}_{\alpha\beta}(l\kappa; l'\kappa')\tilde{u}_\alpha(l\kappa)\tilde{u}_\beta(l'\kappa') \tag{183}$$

where:

$$\begin{aligned}
\tilde{D}_{\alpha\beta}(l\kappa; l'\kappa') &= \left(\frac{\partial^2\Phi}{\partial\tilde{u}_\alpha(l\kappa)\partial\tilde{u}_\beta(l'\kappa')}\right)_o = \\
&= \Big\{\frac{\partial^2}{\partial\tilde{u}_\alpha(l\kappa)\partial\tilde{u}_\beta(l'\kappa')}\big(\sum_{l\kappa\alpha} f_\alpha(l\kappa)(\tilde{u}_\alpha(l\kappa) - \xi_\alpha(l\kappa)) + \\
&+ \sum_{(l\kappa\alpha)(l'\kappa'\beta)} D_{\alpha\beta}(l\kappa; l'\kappa')(\tilde{u}_\alpha(l\kappa) - \xi_\alpha(l\kappa))(\tilde{u}_\beta(l'\kappa') - \xi_\beta(l'\kappa'))\big)\Big\}_o = \\
&= D_{\alpha\beta}(l\kappa; l'\kappa')
\end{aligned} \tag{184}$$

Because the new configuration is an equilibrium configuration, for what said in section I, we have:

$$\begin{aligned}
\left(\frac{\partial\Phi}{\partial\tilde{u}_\alpha(l\kappa)}\right)_o = 0 = &\Big\{\frac{\partial}{\partial\tilde{u}_\alpha(l\kappa)}\big(\sum_{l\kappa\alpha} f_\alpha(l\kappa)(\tilde{u}_\alpha(l\kappa) - \xi_\alpha(l\kappa)) + \\
&+ \sum_{(l\kappa\alpha)(l'\kappa'\beta)} D_{\alpha\beta}(l\kappa; l'\kappa')(\tilde{u}_\alpha(l\kappa) - \xi_\alpha(l\kappa))(\tilde{u}_\beta(l'\kappa') - \xi_\beta(l'\kappa'))\big)\Big\}_o =
\end{aligned} \tag{185}$$

Then:

$$f_\alpha(l\kappa) - \sum_{l'\kappa'\beta} D_{\alpha\beta}(l\kappa; l'\kappa')\xi_\beta(l'\kappa') = 0 \tag{186}$$

Once the forces $f_\alpha(l\kappa)$ are known, the shifts $\xi(l'\kappa')$ between the equilibrium positions can be evaluated. This is the static asymptotic value. The dynamics is the aim of the next section.

III.B. The dynamics of the relaxation

Now we study the dynamics of the displacement (from the old to the new equilibrium position of the atoms belonging to the impurity space. In other words, we evaluate the time dependent relaxation of the defect neighbours which follows the process of light absorption by the localized electron.

The equation (186) can be rewritten in the form:

$$|f\rangle - \hat{D}|\xi\rangle = 0\,, \tag{187}$$

which is diagonal respect to the s.o.n.c. $\{|s\rangle\}$ of the normal modes of the perturbed crystal:

$$f_s - \omega_s^2 \xi_s = 0 \tag{188}$$

Let be $|l\kappa\alpha\rangle$ a vector of the impurity space. If we consider the s.o.n.c. $\{|\Gamma\gamma\mu\rangle\}$ of the symmetry coordinates of the defect, we can write:

$$\begin{aligned} u_\alpha(l\kappa;t) &= \sum_{\Gamma\gamma\mu} \langle l\kappa\alpha|\Gamma\gamma\mu\rangle u_{\Gamma\gamma\mu}(t) = \\ &= \sum_{\Gamma\gamma\mu} \langle l\kappa\alpha|\Gamma\gamma\mu\rangle \sum_s \langle\Gamma\gamma\mu|s\rangle u_s(t) \end{aligned} \tag{189}$$

where now $u_\alpha(t)$ is the displacement from the new equilibrium position, and $u_s(t)$ is given by:

$$u_s(t) = u_s(0)\cos\omega_s t = \xi_s \cos\omega_s t \tag{190}$$

From (188) we obtain:

$$\xi_s = \omega_s^{-2} f_s = \omega_s \sum_{\Gamma\gamma\mu} \langle s|\Gamma\gamma\mu\rangle f_{\Gamma\gamma\mu} \tag{191}$$

If we substitute (190) in (189) we have:

$$\begin{aligned} u_\alpha(l\kappa;t) &= \sum_{\Gamma\gamma\mu} \langle l\kappa\alpha|\Gamma\gamma\mu\rangle \sum_s \sum_{\Gamma'\gamma'\mu'} \langle\Gamma\gamma\mu|s\rangle\langle s|\Gamma'\gamma'\mu'\rangle \omega_s^{-2}\cos\omega_s t f_{\Gamma'\gamma'\mu'} = \\ &= \sum_{(\Gamma\gamma\mu)(\Gamma'\gamma'\mu')} \langle l\kappa\alpha|\Gamma\gamma\mu\rangle f_{\Gamma'\gamma'\mu'} \\ &\quad \frac{1}{2}\int \sum_s \langle\Gamma\gamma\mu|s\rangle\langle s|\Gamma'\gamma'\mu'\rangle\omega_s^{-2}\cos\omega_s t(\delta(\omega-\omega_s)+\delta(\omega+\omega_s))d\omega \end{aligned} \tag{192}$$

In the previous sections we have defined the projected density of states on the symmetry coordinates of the defect as:

$$\rho_{\gamma\gamma'}^{\Gamma\mu}(\omega^2) = \sum_s \langle\Gamma\gamma\mu|s\rangle\langle s|\Gamma\gamma'\mu\rangle\delta(\omega^2-\omega_s^2) \tag{193}$$

From the relations introduced in section II.F. we obtain:

$$\sum_s \langle\Gamma\gamma\mu|s\rangle\langle s|\Gamma'\gamma'\mu'\rangle\delta(\omega^2-\omega_s^2) = \delta_{\Gamma\Gamma'}\delta_{\mu\mu'}\sum_s \langle\Gamma\gamma\mu|s\rangle\langle s|\Gamma\gamma'\mu\rangle\delta(\omega^2-\omega_s^2) \tag{194}$$

because of the relation:

$$Im\, G(\Gamma\gamma\gamma'\mu;\omega^2) = \pi\rho_{\gamma\gamma'}^{\Gamma\mu}(\omega^2) \tag{195}$$

Consequently we can write (192) in the form:

$$u_\alpha(l\kappa;t) = \sum_{\Gamma\gamma\gamma'\mu} \langle l\kappa\alpha|\Gamma\gamma\mu\rangle f_{\Gamma\gamma'\mu} \int_0^{\omega_L} \rho_{\gamma\gamma'}^{\Gamma\mu}(\omega^2)\frac{1}{\omega}\cos\omega t d\omega \tag{196}$$

This is the equation we use to evaluate the relaxation of the atoms from the old to the new equilibrium, in the semiclassical approach. Notice that the relaxation is a true nonradiative process during which the local deformation (the initial "elastic pulse") propagates classically into the surrounding lattice.

IV. SOME EXAMPLES

The model crystals with which to test our results are two alkali halides (KI and LiF), whose dynamics is here described by the *breathing shell model* [3,5,6,14]. As centers we have considered the substitutional impurity Tl^+ in KI and the color center F_2^+ in LiF.

IV.A. The center $\mathbf{KI : Tl^+}$ [7,11,12]

This choice is due to the fact that both the perturberd lattice dynamics and the EP coupling cefficients are known with good accuracy [11,12]. Therefore, the transient relaxation can be directly evaluated from Eq.(196).

The impurity Tl^+ substitutes a positive ion. so that the point group is the same of that of the perfect crystal, *i.e.* the cubic group O_h. In what follows we assume that the impurity space extends only to the negative n.n. of Tl^+. We discuss first the parameters used to evaluate the perturbed lattice dynamics (Eqs.(196)-(199)) and then the values of EP induced forces.

The effect of the ion Tl^+ on the even irreducible representations A_{1g}, E_g, T_{2g} density of states does not depend on the mass variation:

$$\epsilon = -\frac{\Delta m}{m} \tag{197}$$

where Δm is the mass difference between the ion Tl^+ and the ion that it substitutes. It depends instead on the impurity induced change central force constant λ between the ion Tl^+ and the halogen ions around it:

$$\frac{\lambda}{2} = \Delta \left\{ \frac{\partial^2}{\partial r^2} \phi(r) \right\} \tag{198}$$

The variation of the noncentral force constant λ':

$$\frac{\lambda'}{2} = \Delta \left\{ \frac{1}{r} \frac{\partial}{\partial r} \phi(r) \right\} \tag{199}$$

is neglected, because it is an order of magnitude smaller than λ. λ is evaluated [12] from the difference between the effective constant force $f^*(r)$ of the host lattice and the effective constant force $\bar{f}^*(r)$ between the ion Tl^+ and the halogen ions:

$$\lambda = \bar{f}^*(r_o + \xi) - f^*(r_o)\ , \tag{200}$$

where $r_o = 3.52$Å is the interionic equilibrium distance and $\xi = 0.1$Å is the n.n. static relaxation caused by the presence of the ion Tl^+. In our calculation, we have used $\lambda = 0$ (isotopic model) and $\lambda/f^* = -0.17$, where $f^* = 2.45 \times 10^4\, g\, sec^{-2}$ (n.n. defect model).

We have evaluated projected density of states on the symmetry coordinates A_{1g}, E_g and T_{2g} of the impurity space, both for the unperturbed

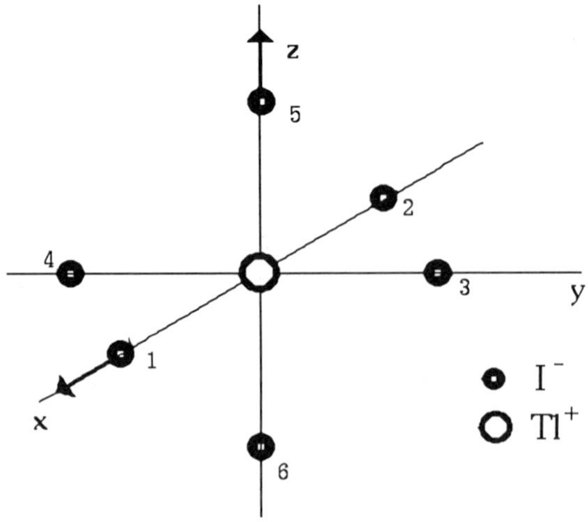

Fig. 1. The Tl^+ center in KI. The arrows on ions 1 and 5 indicate the directions of the displacements evaluated and reported.

and perturbed crystal. The figures reported (Figs. 2-7) are obtained considering 10^6 points in the first Brillouin zone. The frequency axis has been partitioned in intervals of 0.0094×10^{13} $rad\ sec^{-1}$.

Once the pertinent lattice dynamics is known, we need the EP coefficients in order to evaluate (196). We refer again to [12] for the estimate of these coefficients. We have considered the electronic transition ${}^1A_{1g} \rightarrow^3 T_{1u}$ (corresponding to the transition ${}^1S_o \rightarrow^3 P_1$ of the free ion) giving rise to the socalled A-band in the absorption coefficient. Notice that the ${}^3P_1$ level is threefold degenerate, so that a dynamical Jahn-Teller effect, here negleted, is present in the excited state.

For a transition polarized along the z axis and considering only the representations A_{1g} and E_g, the coefficients $f_{\Gamma\gamma\mu}$ are:

$$f_{A_1g} = \frac{a}{r_o + \xi} \qquad f_{Eg} = -2\frac{b}{r_o + \xi} \tag{201}$$

where $a = -0.698$ eV and $b = -0.363$eV for the n.n. defect model at T=0, and $a = -0.760$ eV $b = -0.406$ eV for the isotopic model.

Using the previous values, we are now able to evaluate the time dependence of the displacements, from the old to the new equilibrium positions, of the ions I^-, nearest neighbours to the impurity atom.

The ions perform damped oscillations whose amplitude depends on their position with respect to the polarization direction of the electronic transition. By referring to fig.1, the outward (radial) z-direction displacement of atoms 5 and 6 is parallel to the polarization direction, while the outward (radial) displacement of atoms 1-4 lies in xy plane and is then perpendicular to the polarization.

Notice that in our model the ionic motion has no transverse (shear) component, because we have neglected the T_{2g} EP coupling. From figs.8-11 we can deduce what follows. The radial motion in z-direction is different as whole from that in xy plane. Only the oscillation period is about the same in both the directions, being of about 5.1×10^{-13} *sec*. It corresponds to a frequency of

$$1.23 \times 10^{13} \; rad \; sec^{-1} \; ,$$

which is near to the frequency of the acustic longitudinal mode at the edge zone in the direction $\langle\zeta,\zeta,\zeta\rangle$:

$$\omega_{LA}(\langle\zeta,\zeta,\zeta\rangle) \approx 1,2 \times 10^{13} \; rad \; sec^{-1}$$

This is the frequency where the A_{1g}-projected density of states has a peak.

The time dependence of the amplitude is however the most interesting feature: there is a short time behavior showing an overall relaxation, followed by irreguliar oscillations of small amplitude.

Consider first the results obtained using the n.n. defect model (figs.8 9). From the short time behavior a relaxation time can be extimated which is of about $\tau_{xy} \sim 4 \; ps$ for the radial motion perpendicular to the polarization, and of abot $\tau_z \sim 0.9 \; ps$ for the radial motion parallel to the polarization.

Since the parallel initial amplitude (deduced from (188) with the values (201)) is larger than the perpendicular one, the n.n. relax in the due time from a cubic towards an elongated configuration along the z-direction. Such a configuration is however reached passing through a fase of irregular amplitude oscillations, sometimes even larger than the asymptotic relaxation (see fig.8). Finally let us examine the n.n. motion as obtained in the isotopic defect model. The perpendicular and parallel relaxation times, as extimated from figs.10 and 11, are $\tau_{xy} \sim 3 \; ps$ $\tau_z \sim 0.7 \; ps$, respectively. These times are shorter than those evaluated with the n.n. defect model. This seems a logical consequence of the fact that the stronger the n.n. forces constant the faster the pulse propagation and than the relaxation. With respect to the oscillation amplitudes, we can see how in the isotopic defect case the amplitudes are the half of those observed in the n.n. defect case. This fact can be again explained as the effect of the weakening of the bonds between the n.n. ions I^- and Tl^+.

Until now we have discussed the behavior of a very localized initial pulse, extended to the Tl^+ n.n.. Consider now the opposite limit, that of an acoustic wave as large as 10^3-10^4 lattice spacings (continous limit). Its propagation velocity can be obtained by using the Debye's model, via the relations:

$$\omega_D = c k_D \qquad n = \frac{k_D^3}{6\pi^2}$$

and the fact that:

$$n = \frac{2}{(2a)^3} \qquad a = 3.43\AA$$

$$\theta_D = 131^o K \qquad \hbar\omega_D = k_B \theta_D$$

where k_B is the Boltzmann's constant. We obtain for the KI crystal:

$$c \approx 2.390 \times 10^3 \; m \; sec^{-1}$$

This extended pulse takes about 10^{-10}-10^{-9} *sec* to pass as a whole. This time is much greater than the relaxation time we have obtained by the model adopted.

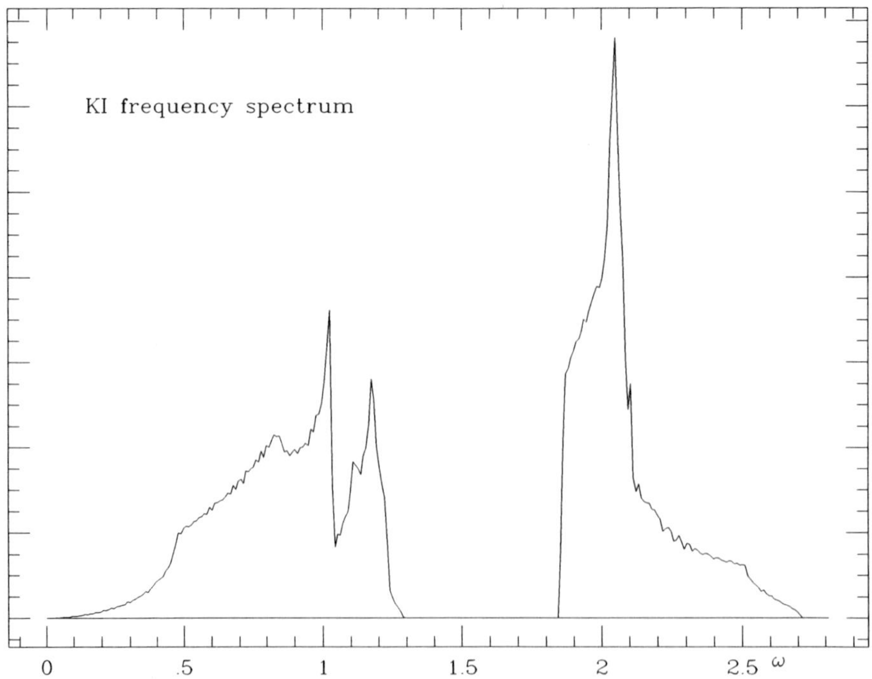

Fig. 2. KI **frequency spectrum. The** ω **axis is in units of** $10^{13}\ sec^{-1}$**.**

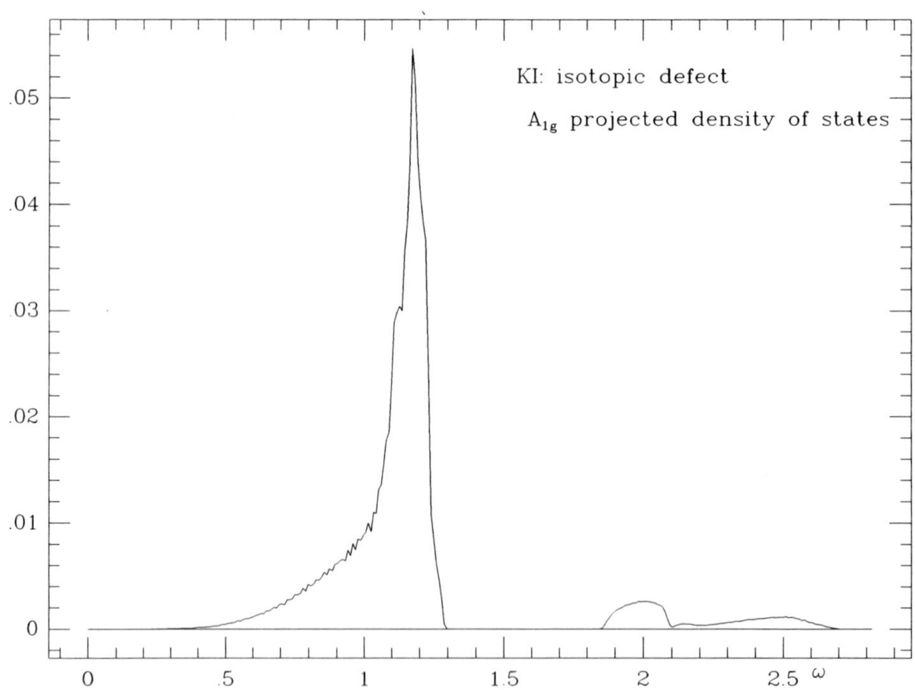

Fig. 3. Calculated A_{1g} **projected density of states of** KI **using the isotopic defect model. The** ω **axis is in units of** $10^{13}\ sec^{-1}$**.**

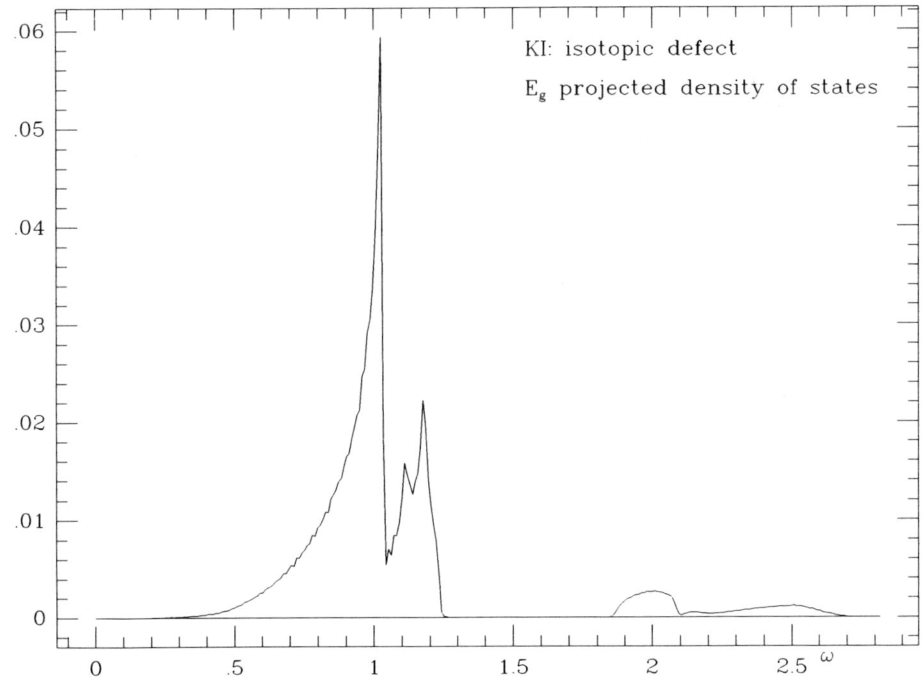

Fig. 4. Calculated E_g projected density of states of KI using the isotopic defect model. The ω axis is in units of 10^{13} sec^{-1}.

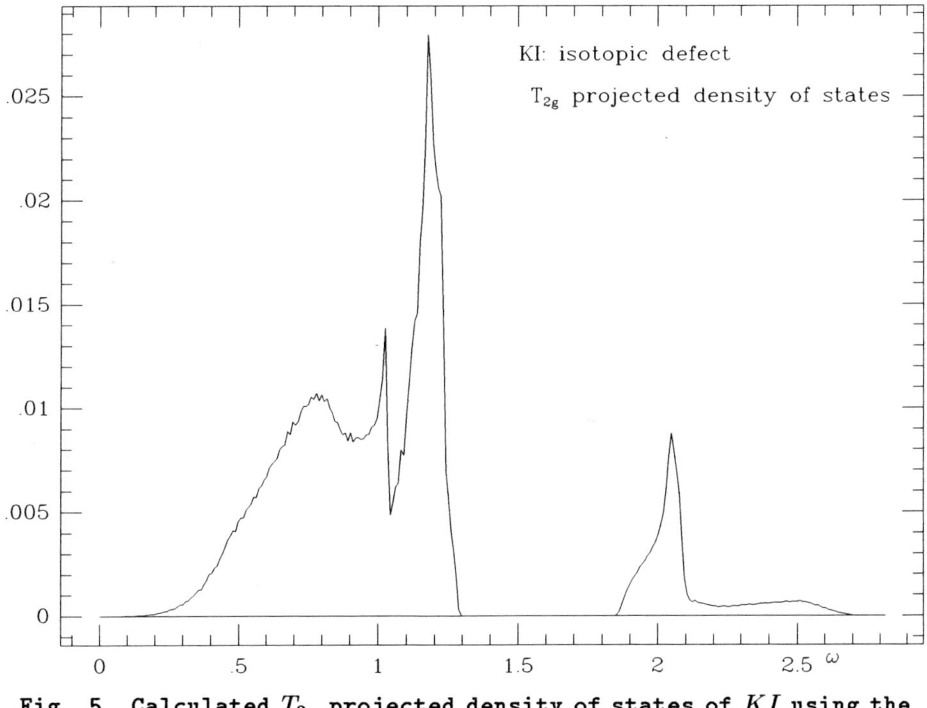

Fig. 5. Calculated T_{2g} projected density of states of KI using the isotopic defect model. The same results are obtained using the n.n. defect model. The ω axis is in units of 10^{13} sec^{-1}.

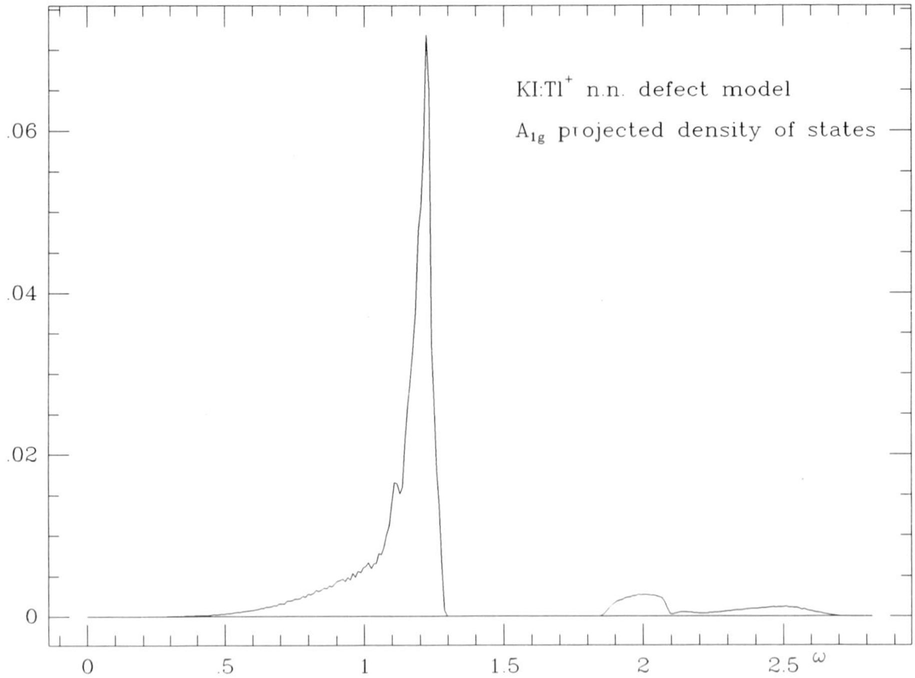

Fig. 6. Calculated A_{1g} projected density of states of KI using the n.n. defect model. The ω axis is in units of 10^{13} sec^{-1}.

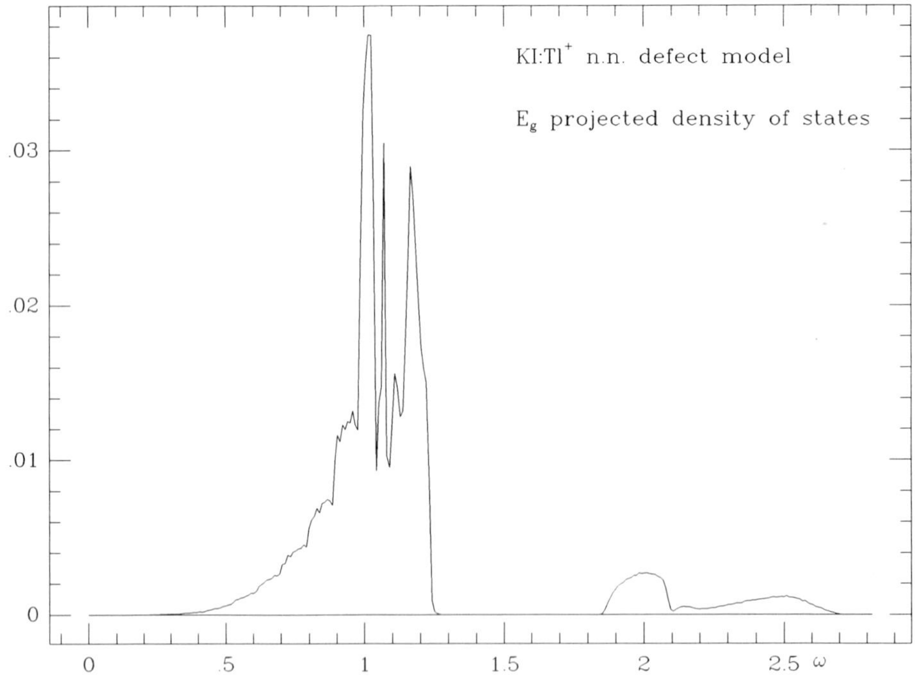

Fig. 7. Calculated E_g projected density of states of KI using the n.n. defect model. The ω axis is in units of 10^{13} sec^{-1}.

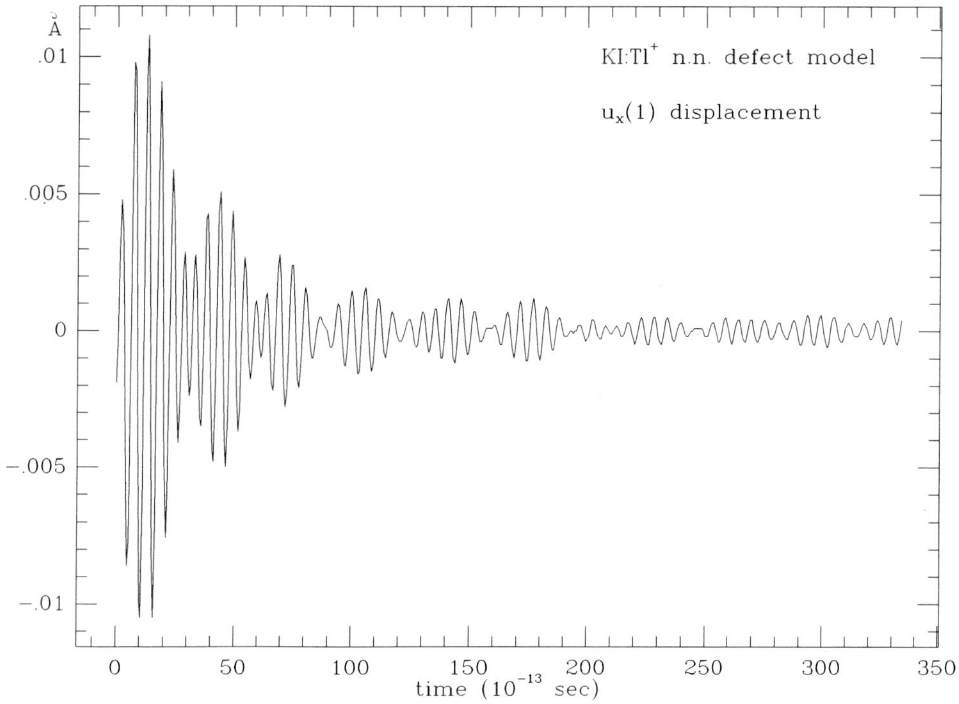

Fig. 8. The displacement of the I^- ion 1 in the x-direction (see Fig.1), obtained using the n.n. defect model and assuming that the electronic transition is polarized along the z-direction.

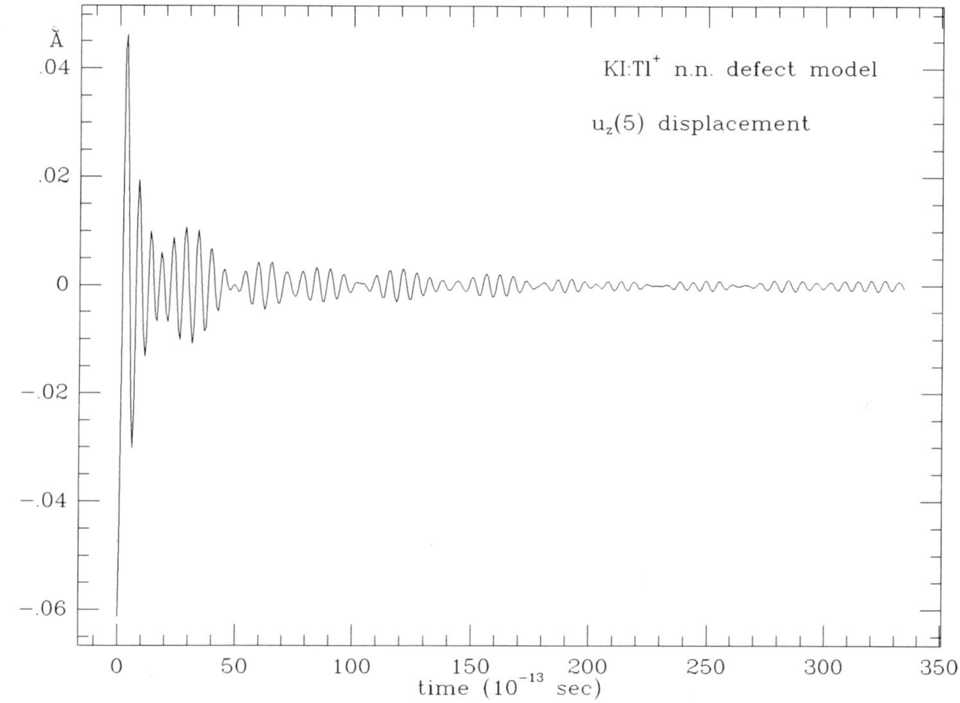

Fig. 9. The displacement of the I^- ion 5 in the z-direction (see Fig.1), obtained using the n.n. defect model and assuming that the electronic transition is polarized along the z-direction.

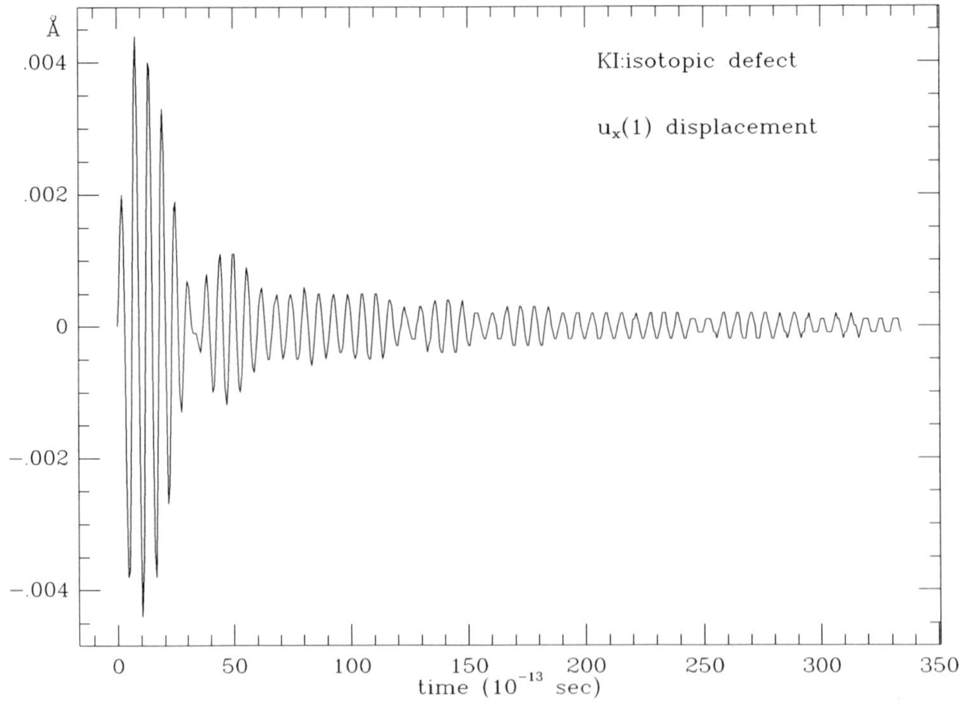

Fig. 10. The displacement of the I^- ion 1 in the x-direction (see Fig.1), obtained using the isotopic defect model and assuming that the electronic transition is polarized along the z-direction.

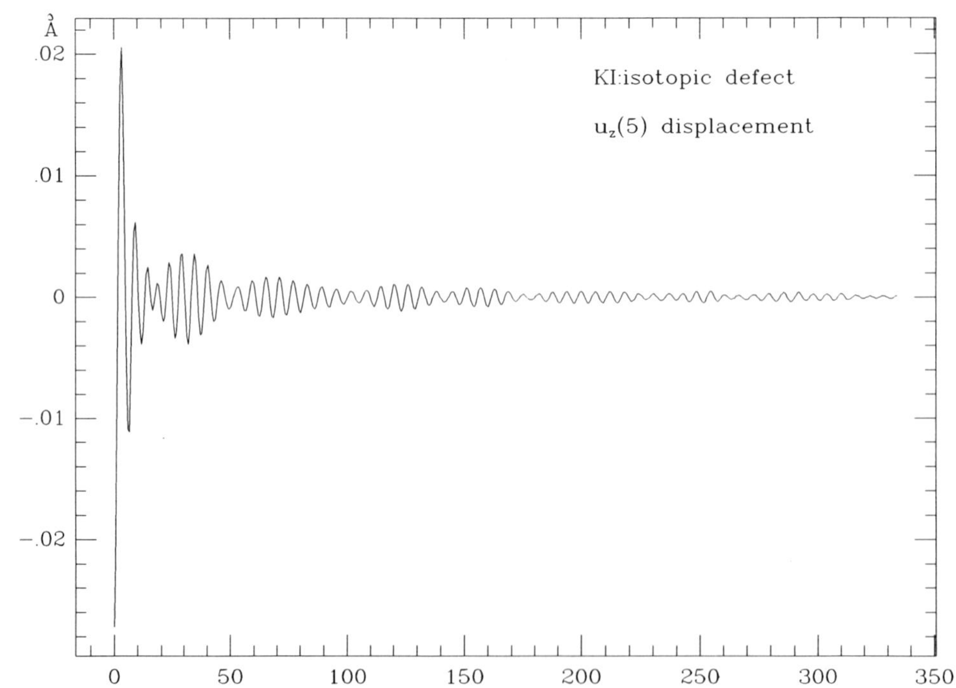

Fig. 11. The displacement of the I^- ion 5 in the z-direction (see Fig.1), obtained using the isotopic defect model and assuming that the electronic transition is polarized along the z-direction.

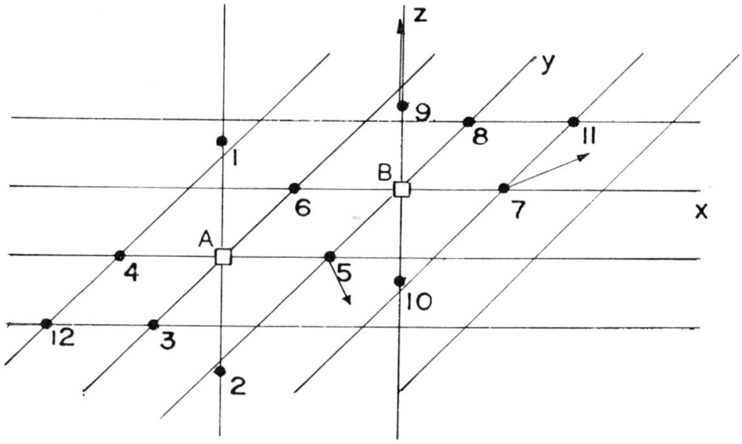

Fig. 12. The F_2^+ center, formed by two adiacent fluorine vacancies A and B and their neighbours: dots 1-10 indicate Li^+ ions, 11-12 F^- ions. The arrows on ions 5, 7 and 9 indicate the direction of the displacements evaluated and reported.

IV.B. The center $\mathbf{LiF : F_2^+}$ [13,18]

The center F_2^+ is constituted by a pair of anionic vacancies in the direction $\langle 1,1,0 \rangle$ binding a single electron. This center was proposed by Seitz (1954) and experimentally identified by Schneider and Rabin (1965) in the near i.r. absorption band of three alkali halides.

The experimental data differ in many aspects from the theoretical predictions. The model that seems reproduce the observed properties, at least in a semiquantitative way, is that introduced by Herman, Wallis and Wallis (1956). In this model the center is treated as a hydrogen ion molecule H_2^+, embedded in a dielectric medium of dielectric constant ϵ_o.

In LiF, the $1s\sigma_g \rightarrow 2p\sigma_u$ electronic transition of F_2^+ induces an absorption band at $645\ nm$, while the emission band is at $900\ nm$.

The defect space extends on the n.n. of both vacancies. The degrees of freedom are then 42. The point group of this defect is the group D_{2h}. In our calculations we have considered only the effect of the totally simmetrical N_1 irr. rep., which is contained six times in the impurity space.

For lack of information, we have modelled the center with the isotopic impurity, with equal EP coupling to all the modes. The results are reported for this reason in arbitrary units.

The densities of states of the perfect crystal are calculated considering 5.12×10^5 points in the first Brillouin zone and partitioning the frequency spectrum in intervals of $6.5 \times 10^{11}\ rad\,sec^{-1}$. The results are reported in Figs.18, 19 and 20. The n.n. ions perform quickly damped oscillations with a damping time of about:

$$1.4 \times 10^{-13}\ sec$$

This time is shorter than the damping time observed for $KI : Tl^+$. Notice however that Li and F ions are lighter than K and I ions of the other crystal. This is reflected also on the value of the sound velocity which is for the present crystal of:

$$7.760 \times 10^3 \; m\, sec^{-1}$$

This means that an acoustic perturbation extending over 10^3-10^4 lattice spacings takes about 10^{-11}-10^{-10} sec to pass a whole.

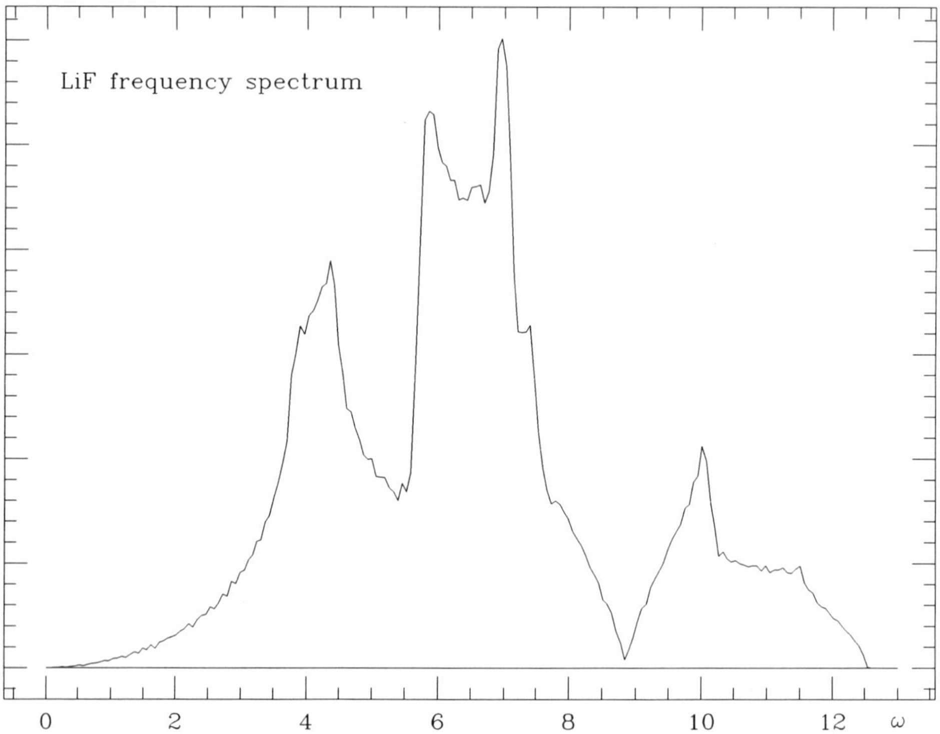

Fig. 13. LiF frequency spectrum. The ω axis is in units of 10^{13} sec^{-1}.

For $LiF : F_2^+$ we have some experimental results. Knox, Fork and Mollenauer [18] have deduced a relaxation time of less than 75 fs. Given the very simple approximations here done we can conclude that our results are in a fair agreement with the experimental ones.

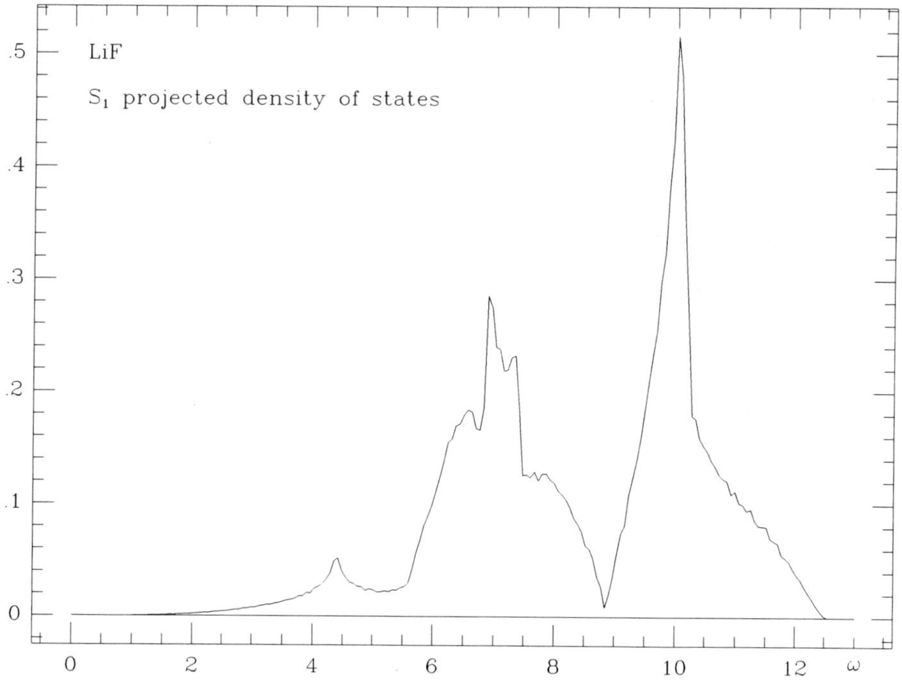

Fig. 14. Calculated S_1 projected density of states of LiF. The ω axis is in units of 10^{13} sec^{-1}.

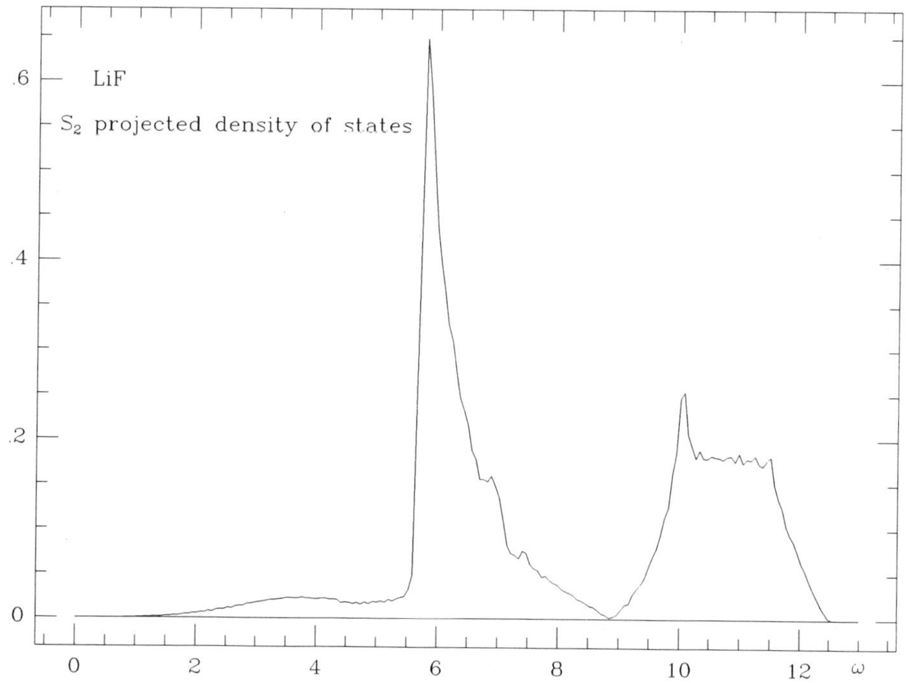

Fig. 15. Calculated S_2 projected density of states of LiF. The ω axis is in units of 10^{13} sec^{-1}.

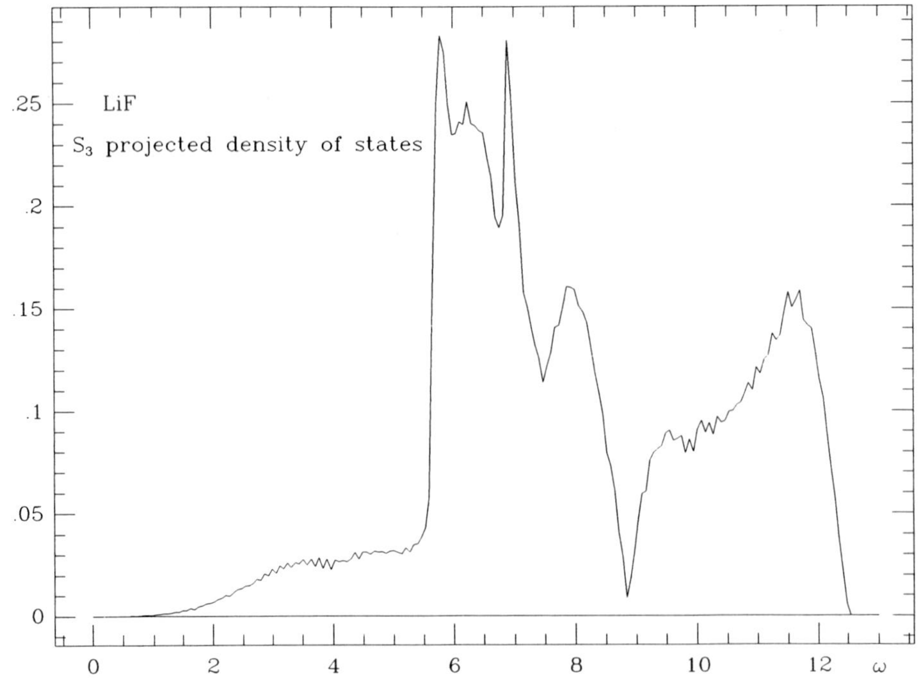

Fig. 16. Calculated S_3 projected density of states of LiF. The ω axis is in units of 10^{13} sec^{-1}.

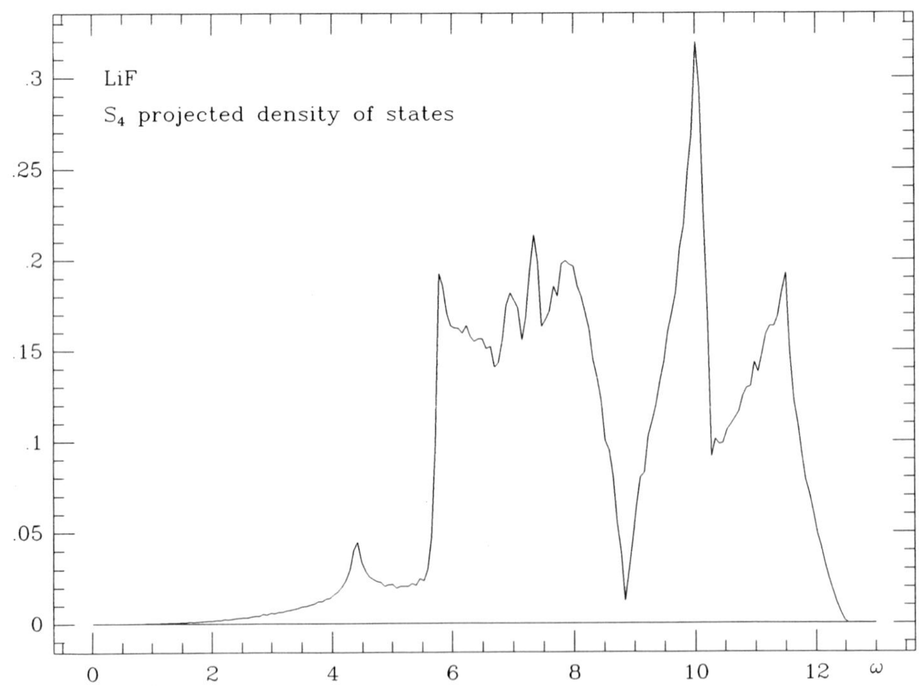

Fig. 17. Calculated S_4 projected density of states of LiF. The ω axis is in units of 10^{13} sec^{-1}.

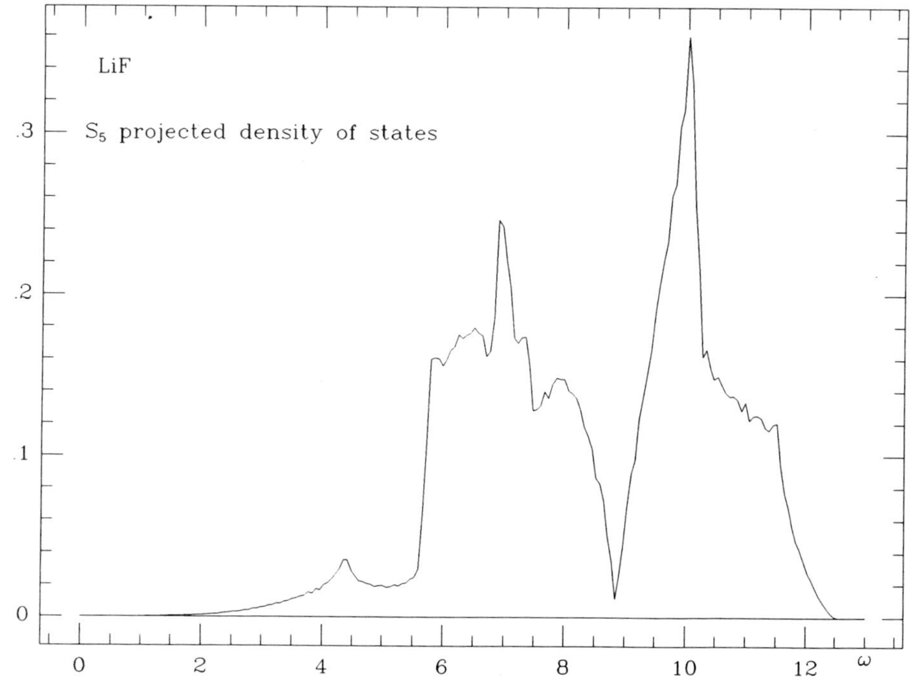

Fig. 18. Calculated S_5 projected density of states of LiF. The ω axis is in units of 10^{13} sec^{-1}.

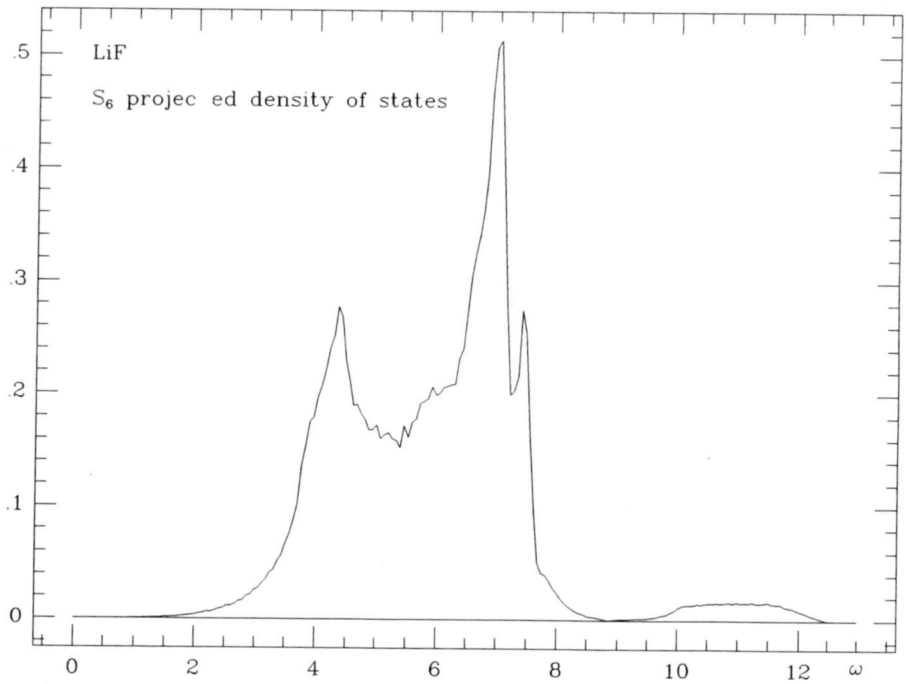

Fig. 19. Calculated S_6 projected density of states of LiF. The ω axis is in units of 10^{13} sec^{-1}.

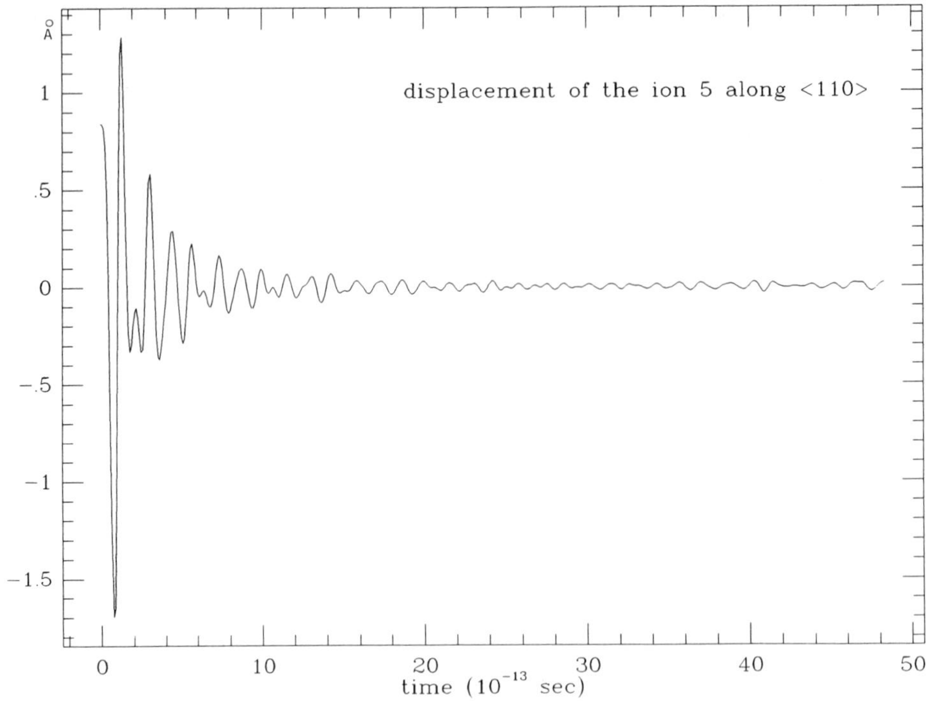

Fig. 20. The displacement of the Li^+ ion 5 in the direction drawn in Fig.10 in arbitrary units.

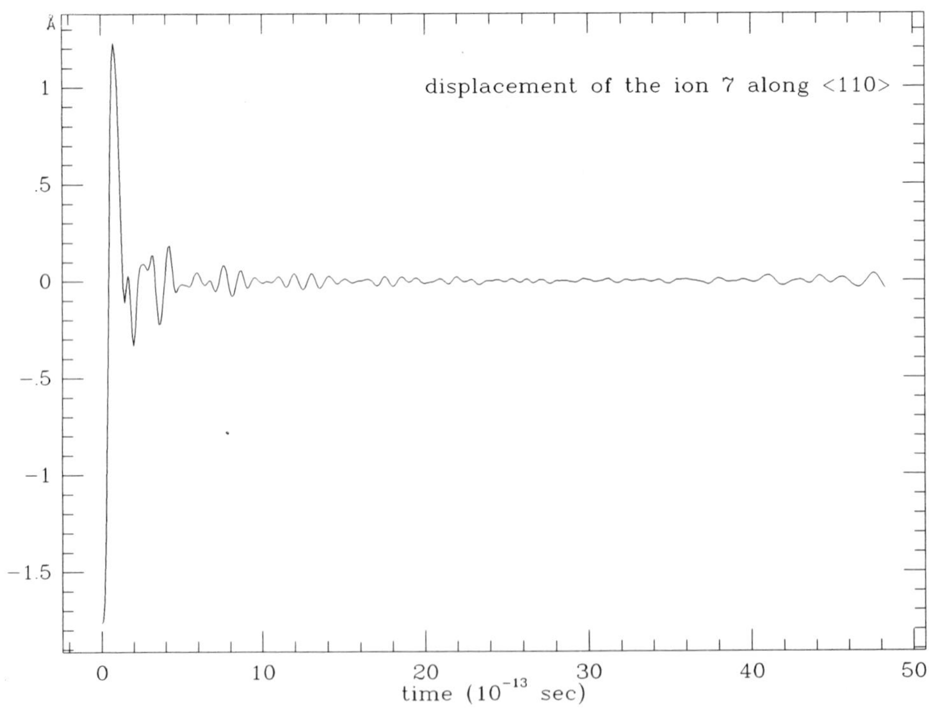

Fig. 21. The displacement of the Li^+ ion 7 in the direction drawn in Fig.10 in arbitrary units.

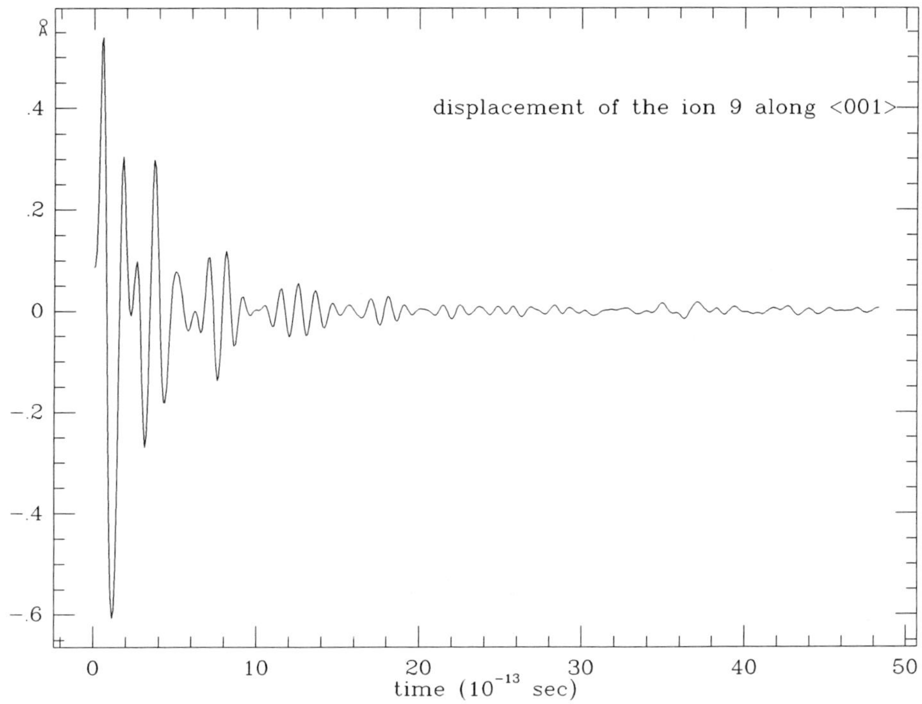

Fig. 22. The displacement of the Li^{+} ion 9 in the direction drawn in Fig.10 in arbitrary units.

REFERENCE

1. E. W. Kellermann, Trans. Royal Soc., 238, 513 (1940)

2. M. Born, K. Huang, "Dynamical Theory of crystal lattices," Oxford Univ. Press, London and New York (1954)

3. A. D. B. Woods, W. Cochran, B.N. Brockhouse, Phys.Rev., 119, 980 (1960)

4. A. A. Maradudin, E. W. Montroll, G.H. Weiss, I. P. Ipatova, "Theory of lattices dynamics in the harmonic approximation," Academic Press, New York and London (1971)

5. U. Schröeder, Solid St. Comm. , 4, 347 (1966)

6. V. Nüsslein, U. Schröeder, Phys. Rev., D27, 969 (1967)

7. G. Benedek ,G. F. Nardelli, Nat. Bureau of Standars, Misc. Publ., 287, 161 (1966)

8. G. Benedek ,G. F. Nardelli, Phys. Rev., 155, 1004 (1966)

9. R. Fieschi, G. F. Nardelli, N. Terzi, Phys. Rev. A, 138, 203 (1965)

10. G. Benedek, E. Mulazzi, Phys. Rev., 179, 906 (1969)

11. G. Benedek, N. Terzi, An evaluation of Raman spectra,phonon relaxation time, infrared and ultraviolet absorptions induced by Tl^{+} substitutional in KI, in: "Physics of impurity centres in crystals," Academy of sciences of Estonian, Tallin (1972)

12. G. Benedek, N. Terzi, Phys. Rev. B, 8, 1746 (1973)

13. M. A. Aegerter, F. Lüty, Phys. Stat. Sol. B, 43, 245 (1971)

14. B. Dochy,Phys. Stat. Sol. A, 59, 531 (1980)

15. G. F. Bassani, G. Pastori Parravicini, "Electronic states and optical transitions in solids," Oxford Pergamon Press (1975)

16. M. V. Klein, Localized modes and resonance states in alkali halides, in: "Physics of color centers," W. Beall Fowler, ed., Academic Press, New York and London (1968)

17. A. Giorgetti, N. Terzi, Solid. St. Comm., 39, 635 (1981)

18. W. H. Knox, L. F. Mollenauer, R. L. Fork, Chem. Phys. Series (Springer), 46, 277 (1986)

19. M. J. Rosker, F. W. Wise, C. L. Tany, Phys. Rev. Lett., 57, 32 (1986)

20. C. Sierens, W. Joosen, D. Schoemaker, Phys. Rev. B, 37, 3075 (1988)

21. N. Terzi, Phys. Rev. B, 23, 2769 (1981)

22. N. Terzi, Coherent wavepackets of phonons, in: "Collective Excitations in Solids," B. Di Bartolo, ed., Plenum Press, New York and London (1981)

THERMAL RESPONSE AND DECOUPLING OF EXCITATIONS AT LOW TEMPERATURES

M. J. Graf

Department of Physics
Boston College
Chestnut Hill, MA 02167

ABSTRACT

The manner in which the interactions ('coupling') between excitations in solids (for example, phonons and electrons) vary with parameters such as temperature or magnetic field not only gives information on the density of states, but can also provide for novel measurements of specific heat of a single set of excitations. The particular example of a two-dimensional electron gas confined to a GaAs-AlGaAs single quantum well is discussed, along with preliminary measurements on the decoupling of electrons and phonons in magnetic field at low temperatures.

I. INTRODUCTION

In solid systems at low temperatures the possible modes of energy excitation decouple into weakly interacting elementary excitations and quasiparticles - for example, phonons, electrons, plasmons, etc. One of the most fundamental properties of these systems is the density of states (DOS), which can be probed directly, as in superconducting tunnelling experiments, or by thermodynamic means, as in measuring the specific heat. For example, measurement of the specific heat of a Fermi system at low temperatures yields the DOS at the Fermi energy. Another indirect thermodynamic means would be to determine how heat is transferred between the sets of excitations. In this article I shall describe a recent effort [1] to measure the thermal conductance between two-dimensional electrons confined to a GaAs-AlGaAs quantum well and the crystal lattice

Advances in Nonradiative Processes in Solids
Edited by B. Di Bartolo, Plenum Press, New York, 1991

phonons. An extension of the method developed in Ref. 1 should allow for a direct measurement of the specific heat of the two-dimensional electron system (2DES) in a strong magnetic fields while avoiding the contributions of phonons, thermometers, etc. by using a modified AC calorimetric technique [2] at temperatures on the order of 0.1 K. [3]

Modern crystal growth techniques, particularly molecular beam epitaxy, have made the fabrication of extremely high quality layered crystal structures possible. 2DES such as doped GaAs-AlGaAs heterojunctions and quantum wells, where electrons are confined to the interface and thin layer of GaAs, respectively, are characterized by an extremely high electron mobility. The reduction of such extrinsic effects as impurity or surface roughness scattering has allowed for the observation of many new interesting phenomena, such as the integral and fractional quantum Hall effect. [4] By applying a strong magnetic field perpendicular to the plane of the 2DES, the density of states (which at low temperatures and zero magnetic field is constant up to the Fermi energy) is broken up into a series of Landau levels with spacing $\hbar\omega_c$, where $\hbar$ is Planck's constant and the cyclotron frequency $\omega_c = eB/m^*$. B is the applied magnetic field strength, e is the electronic charge, and m^* is the electronic effective mass (which equals 0.067 times the free electron mass in two-dimensional GaAs structures). Ideally these Landau levels are delta functions, but the effect of thermal agitation, impurity scattering, and electron-electron interactions is to broaden the levels. For very high quality samples at low temperatures, probing the shape of the DOS would then yield information on the electron-electron interactions. [5] This would be of particular use to help understand the fractional quantum Hall effect, where a many-body state is formed in strong magnetic fields by these interactions, and the quasiparticles have fractional electronic charge.

II. TECHNIQUE

The measurement to be described will be of σ_{e-ph}, the steady-state thermal conductance from electrons, confined to a GaAs single quantum well, to phonons in the crystal

$$\sigma_{e-ph} = \dot{Q}_{el} / (T_e - T_{ph}). \tag{1}$$

$\dot{Q}_{el}$ is the rate at which energy is transferred from the electrons to the phonons (which in the steady-state is equal to the rate at which the electrons absorb heat from an external source), and T_e and T_{ph} are the temperatures of the electrons and phonons, respectively. Qualitatively, one expects this conductance to depend on the available states into which both the electrons and phonons can scatter, and therefore on the density of states of each system. [6] Extracting detailed quantitative information would be quite difficult, but under the conditions whereby the DOS of 2DES only is modified in an extreme fashion -- that is, by application of a strong magnetic field -- changes in this DOS can be inferred from the date. Such measurements were made by Eisenstein et. al., [7] but from measurements of the phonon thermal conductivity variation with magnetic field. Our goal is to work in the other direction -- apply heat directly to and measure temperature of the electrons. While the data for σ_{e-ph} would not be any better than that of Ref. 7, our approach will allow us to directly measure the specific heat (and therefore gain detailed information of the DOS) of the 2DES.

The measurement consists of two basic parts: applying a known amount of heat directly to the electrons, and monitoring the difference between the electron temperature and the phonon temperature. the most common technique for accomplishing these would be the use of resistive devices. However, resistors complicate the experiment in two ways. First, a resistor placed on a sample couples to electrons in the sample via the sample phonons, and so also measures properties of the phonons. The variations in the 2DES are only observed to the extent that they modify the phonon behavior (this is the technique of Ref. 7). Second, the heat capacity of typical resistive devices is much larger than that of a typical 2DES. Since our long term goal is to measure the heat capacity of the 2DES, we do not wish to have the resistive device in thermal contact (via phonons) with the electrons and obscure their specific heat.

To heat the electrons we illuminate the sample with laser light at an energy well above the GaAs valence-to-conduction band energy gap, but less than the corresponding gap for AlGaAs, resulting in electron-hole pair creation. Three main problems must now be addressed. First, an appreciable amount of energy must be thermalized between the electrons and not all re-radiated by the electron-hole recombination. The ratio of radiated to total energy absorbed (the quantum efficiency) is actually quite high, usually over 0.90, for these devices. [8] However, most recombination events occur between electrons and holes at the k=0 band edges, so essentially all the photon energy in excess of the GaAs band gap energy is thermalized. A second possible problem is the magnetic field variation of the absorption coefficient of the 2DES at the laser energy. [9] We have measured this independently over the field ranges used here and find it to vary in a monotonic fashion, and by less than 4%. The third, and most troublesome, problem is that of changing the electron carrier density by the illumination. While the steady-state change due to excitation of electrons from the valence band is quite small due to the high rate of recombination, persistent photoconductivity effects, [10] where electrons are believed to be freed from trapping impurity centers can be non-negligible, and cause the free electron density to vary with illumination, field, etc. This effect has limited the range of magnetic fields where our technique is successful. We believe that by varying the growth parameters that this effect can be eliminated. This point will be discussed further with the results.

The measurement of the electron temperature has been accomplished by using the temperature dependence of the 2DES impedance as a 'self-thermometer'. In zero magnetic field, this temperature dependence is extremely small (for the samples we have studied), and so the thermometer is not very sensitive. In magnetic fields, however, the impedance becomes strongly temperature dependent. This is mostly true because near integral filling factors an energy gap opens up in the DOS (for extremely high quality samples this also occurs at some fractional filling factors -- the fractional quantum Hall effect). Here the conductivity is essentially thermally activated. To monitor this 'magneto-impedance', two NiAu electrodes were evaporated onto the surface. Each pad then forms one 'plate' of a capacitor, the other 'plate' being the 2DES itself. The effective circuit between the two pads is then 2 capacitors in series with the resistance of the 2DES. By putting an oscillating voltage across the pads and measuring the ac current through the loop, the circuit impedance can be measured as a function of magnetic field. The calculation of this field dependent impedance is quite complicated; however, for the purpose of this experiment, one only needs to measure the impedance versus field at several temperatures in order to calibrate this property and use it to determine the electron temperature as a function of magnetic field.

The use of capacitive coupling into the 2DES, rather than the more typical diffused metal contacts directly into the 2DES, results in the electrons in the electrodes, and wires connected onto them, being thermally isolated from the 2DES. The advantage of this is two-fold. First, this removes an alternative route for heat transfer from the electrons to the outside world (we wish to study the electron-to-phonon transfer rate). Second, the heat capacity of the electrons in the electrodes and wires is much greater than that of the 2DES, and so thermal isolation eliminates a large extraneous contribution to the heat capacity of the 2DES.

III. THERMAL CONDUCTANCE

We have studied GaAs - $Al_{0.3}Ga_{0.7}$ As single quantum well. The carrier density $n_e = 3.9\times10^{11} cm^{-2}$ and the mobility $\mu_e = 2.0\times10^5 cm^2 V$ -sec as determined from magnetoresistance and Hall measurements at 1.2K. A 100 kHz, 10 m V peak oscillating voltage was placed across the NiAu pads, and the current through the circuit (and therefore the impedance) was monitored with a current-sensitive preamplifier connected to a lock-in amplifier. Heating was of the 2DES was accomplished with a Krypton-pumped dye laser operating at 1640 meV. Optical fibers carried the light to the sample, which was immersed in a refrigerant of liquid ^{3}He. The temperature range of the measurements was 0.4K to 1.2K.

The ‘thermometer’ is calibrated by measuring the impedance of the circuit and sweeping magnetic field for several different temperatures of the ^{3}He (see Fig. 1a). The sample is illuminated with a very low-intensity beam to minimize the effects of transient electron carriers between the calibration and the actual heating data. In this calibration it is assumed that the electron and phonon temperatures are equal. The heating data is obtained by fixing the operating temperature of the refrigerator, which we assume is equal to the phonon temperature. This is equivalent to assuming the conductance from the phonons to the ^{3}He is much larger than the electron-phonon conductance. This will be justified later. The magnetic field is changed while illuminating the sample with higher intensity light (see Figure 1b). At higher intensity levels, heat is absorbed in the 2DES and so the electron temperature is offset from that of the phonons ($T_e = T_{ph} + \delta T$) due to a non-infinite thermal conductance σ_{e-ph}. Using the date from Figure 1a we can calculate the offset in electron temperature as a function of magnetic field, δ T (B) for several values of illumination intensity.

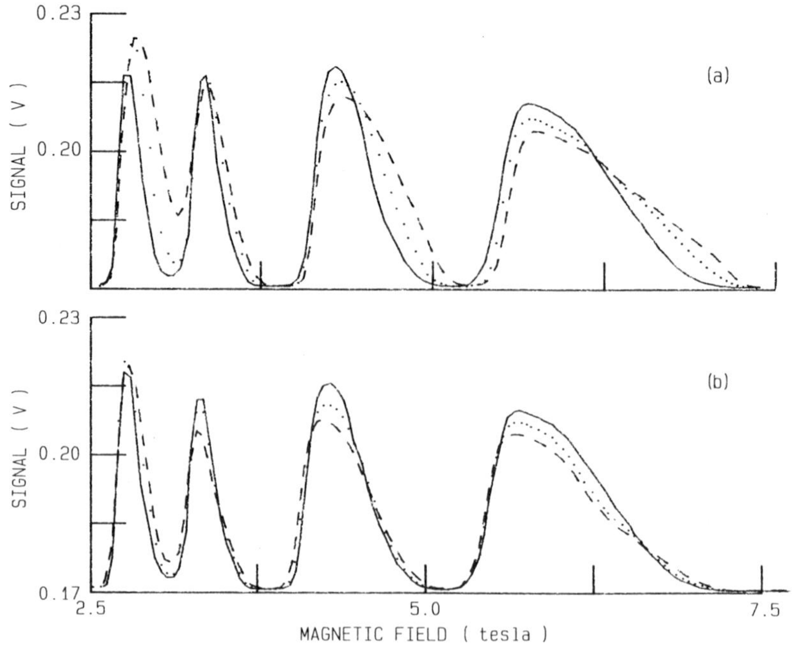

Figure 1. (a) The magnetocapacitance versus magnetic field at several different temperatures with a fixed, low-power illumination of 3 microwatts. The solid curve is taken at T = 0.4K, the dotted at 0.7K, and the dashed at 1.0K. (b) The magnetocapacitance versus magnetic field at a fixed bath temperature of 0.6K for different illumination intensities on the sample. The solid curve is taken with at 7 microwatts, the dotted at 23 microwatts, and the dashed at 53 microwatts.

This general procedure assumes the only effect of shining light on the sample is to heat the electrons in the 2DES. However, this is not completely true, as there are several areas along the magnetic field axis, particularly near the maxima in the signal, where the shape of the curves obtained by heating the sample and by eliminating it (Figures 1a and 1b) are qualitatively different. Based on previous experiences with other samples, this is probably caused by the particular details of the growth, possibly the existence of several thin buffer layers of GaAs located within a few hundred angstroms of the quantum well. However, near the minima, where the thermometer is most sensitive, it appears that indeed the predominant effect of illumination is to heat the 2DES. The focus of the analysis will be in the region of the minimum located at a field of 3.1 tesla (filling factor $\nu = 5$).

In Figure 2 the heat absorbed in the 2DES is plotted as a function of the calculated temperature offset of the electrons from the phonons. For this data set the phonon temperature is 0.5K and the applied magnetic field is 3.09 tesla. In the limit $\delta T/T << 1$ the data should be linear with a slope equal to the electron-phonon thermal conductance σ_{e-ph} [see equation (1)]. To estimate the absolute magnitude of the power absorbed and thermalized by the 2DES, the losses in the optical fibers have been measured, along with the absorption coefficient. Also, the quantum efficiency has been assumed to be 0.90; this may be somewhat small, but

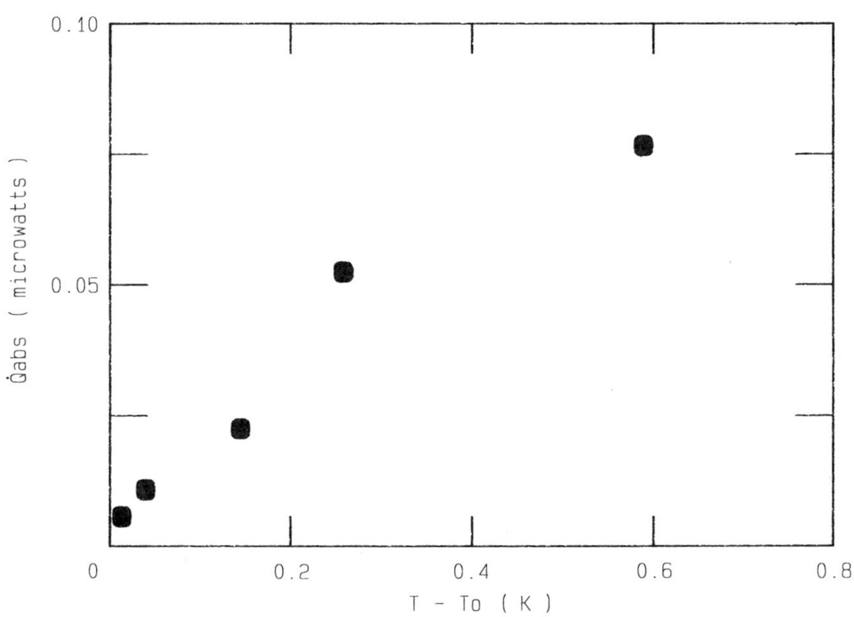

Figure 2. Estimated heat abosrbed by the two-dimensional electron gas versus the difference in electron temperature from the helium bath temperature at 0.5K and in an applied magnetic field of 3.09 Tesla.

leads to less than a 10% error in the estimate. One then finds from the figure

$$\sigma_{e-ph} \sim 0.2\times10^{-6} \text{Watts/K}.$$

This is within a factor of three of the measurements made by Wennberg et al. [11] from self-heating effects in a GaAs multiple quantum well. Previously, it was assumed that $\sigma_{e-ph} << \sigma_{ph-He}$, that is the phonon temperature equals that of the ^{3}He. Materials with densities and sound velocities similar to GaAs have an effective thermal conductance three orders of magnitude larger than this. [12]

In Figure 3 the percent change in electron temperature caused by absorption (at constant power) is plotted versus magnetic field for two different ^{3}He temperatures. The data at 0.6K shows a pronounced peak at the field corresponding to an integrally filled number of Landau levels (filling factor $v = 5$). At this field value an energy gap opens up in the density of states. This maximum in $\delta T/T$ indicates that the conductance is at a minimum. Thus the experimental results are consistent with the idea that as the number of available states for the electrons to scatter into reaches a minimum, so does the electron-phonon thermal conductance. The data taken at 1K have no such maximum, indicating that the thermal broadening of the Landau levels is sufficient to wash out the structure. This field variation is in agreement with the results of Reference 7.

IV. EXTENSION TO SPECIFIC HEAT

The AC calorimetric technique will now be reviewed before discussing how to adapt it to our particular system. In AC calorimetry, one has a sample placed in good thermal contact with a heater and thermometer, usually both resistors. This set-up, with a total heat capacity C_{total} comprised of the heat capacities of the sample plus heater and thermometer, is weakly linked thermally to a temperature bath via a thermal conductance σ_{link}. By applying a voltage across the heater which oscillates at frequency ω, one produces heat at the sample which goes as the square of the voltage divided by the heater resistance R_h. Thus, using the relation $\cos^2(\omega t) = 1/2 + 1/2\cos(2\omega t)$, the heat produced has a constant (time-independent) term plus one which oscillates at 2ω, which is given by

$$\delta T_{2\omega} = \frac{\sqrt{2}\, V_h^2}{2\omega R_h} C_{total}^{-1}, \tag{2}$$

where V_h is the amplitude of the heater voltage. Equation (2) is valid when heat is not transferred 'too rapidly' from the sample to the bath. More formally, one must have $2\omega\tau_b >> 1$. Here τ_b is the sample-to-bath relaxation time, given by

$$\tau_b = C_{total}/\sigma_{link} \tag{3}$$

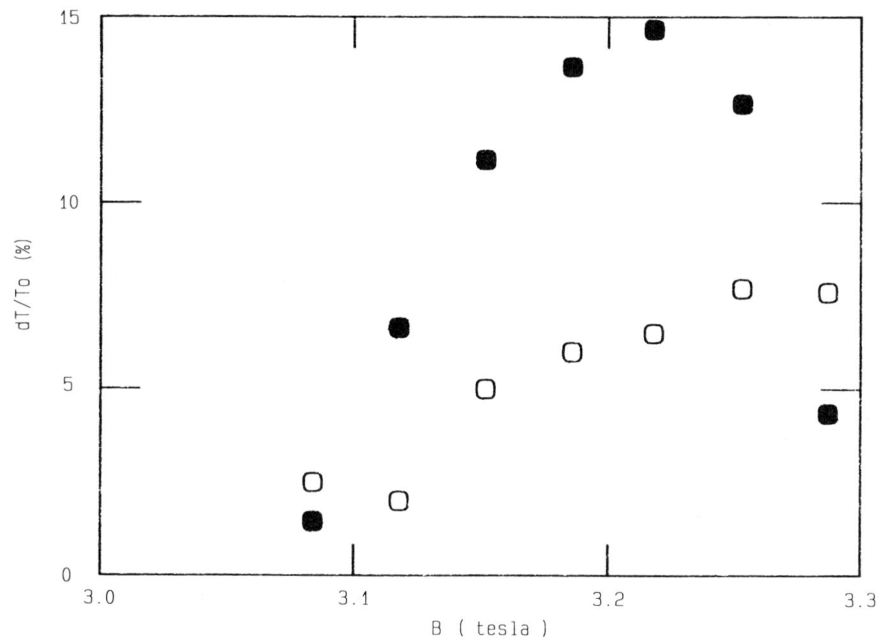

Figure 3. The percent change in the (small) offset of the electron temperature from the bath temperature versus magnetic field for two different bath temperatures. The filled circles are for data taken at 0.6K, and the open circles were taken at 1.0K.

Also, the heating must occur slowly enough so that the thermometer, heater, and sample are all in equilibrium. If the internal relaxation time is given by τ_{int} then the frequency must be in the region defined by

$$\tau_b^{-1} << 2\omega << \tau_{int}^{-1} \tag{4}$$

for the AC technique to work properly.

The major advantage of this technique is that the oscillating temperature component is easy to separate from noise with the use of a lock-in amplifier, and high sensitivity measurements can be made. However, one still has the problem that the contributions to the heat capacity of the GaAs phonons, heater, and thermometer will be much bigger than the contribution from the 2DES. One must go to great lengths to try and measure the contribution of the 2DES with any accuracy. [13,14]

In an AC measurement of the heat capacity using laser heating of the 2DES in GaAs, the main idea would be to have the role of the weak thermal link played by the electron-phonon thermal conductance. The advantages of the measurement technique developed in Reference 1 are obvious. First, both the heater (photons) and thermometer (the 2DES itself) make no extraneous contributions to the heat capacity. Further, the phonons now play the role of the thermal reservoir and also make no contribution to the measurement, allowing for very accurate determination of the heat capacity of the 2DES.

The main problem, apart from refinements in the heating/thermometry technique, would be to find a reasonable frequency window, as defined by equation (4), to operate in. The electrons thermalize very rapidly, on the order of picoseconds, and the upper bound would be 10^{10} - 10^{11}Hz. A greater concern is the lower limit. For a two-dimensional degenerate Fermi gas, the heat capacity (in zero magnetic field) is given by

$$C_{2DES} \sim \frac{\pi}{3} k_B^2 T g(E_F) A, \qquad (5)$$

where T is the temperature, k_B is Boltzmann's constant, $g(E_f)$ is the density of states at the Fermi energy, and A is the area. Using our estimate of σ_{e-ph} and equation (3) we find

$$\tau_b \sim 0.3 \times 10^{-6} \text{ seconds}$$

at 0.5K. This would restrict our measurements to above approximately 10 MHz, which can complicate the technique somewhat. However, the relaxation time has been found both theoretically [15] and experimentally [11] to vary at T^{-3}, so by running the experiment at a temperature of 50mK, the relaxation time would be of the order of milliseconds, and so the frequencies would be in the more tractable kHz range. Work is presently underway to make these low temperature measurements.

V. CONCLUSION

A new technique was described where direct heating of a two-dimensional electron gas was obtained by laser absorption, and the electron temperature was monitored via the magnetic-field dependence of the electron system impedance. This technique was used to measure the electron-phonon thermal conductance, which was found to be a minimum when the magnetic field was such that an integral number of Landau levels were completely full.

The technique may be most useful in possible specific heat measurements. By relying on the low-temperature decoupling of the electron quasiparticles from the phonon excitations, it should be possible to measure the extremely small electron heat capacity independent of the other excitations and quasiparticles in the system. The technique should be applicable to other small systems whose heat capacity is obscured by larger contributions from the surrounding solid.

The work was carried out in collaboration with Bennett Goldberg of Boston University. The experiments were performed at the Francis Bitter National Magnet Laboratory at MIT, which is sponsored by the National Science Foundation.

REFERENCES

1. M.J. Graf, B.B. Goldberg, J.S. Brooks, and M. Heiblum, Proceedings of the 19th International Conference on hte Physics of Semiconductors, p.185, W. Zawadzki, ed. (Polish Academy of Sciences, Warsaw, 1988).

2. P.F. Sullivan and G. Seidel, Physical Review 173, 679 (1968).

3. A similar technique is described by D.C. Glattli, E.Y. Andrei, and F.I.B. Williams, Phys. Rev. Lett. 60, 420 (1988).

4. For a review see The Quantum Hall Effect, R.E. Prange and S.M. Girvin eds. (Springer-Verlag, New York 1987).

5. A.H. MacDonald, H.C.A. Oji, and K.L. Liu, Phys. Rev. B34, (1986).

6. For example, see M. Prasad and M. Singh, Rev. B29, 4803 (1984).

7. J.P. Eisenstein, A.C. Gossard, and V. Narayanamurti, Phys. Rev. Lett. 59, 1341 (1987).

8. See for example, the work presented by D. Bimberg in this volume.

9. B.B. Goldberg, D. Heiman, M.J. Graf, D.A. Broido, A. Pinczuk, C.W. Tu, J.H. English, and A.C. Gossard, Phys, Rev. B38, 10131 (1988).

10. For example, see S. Prasad, H.P. Wei, D.C. Tsui, W. Schlapp, and G. Weimann, J. Appl. Phys. 63, 1793 (1988).

11. A.K.M. Wennberg, S.N. Ytterboe, C.M. Gould, H.M. Bozler, J. Klem, and H. Morkoc, Phys. Rev. B34 4409 (1986).

12. O.V. Lounasmaa, Experimental Principles and Methods Below 1K, (Academic Press, London, 1974).

13. E. Gornik, R. Lassnig, G. Strasser, H.L. Stormer, A.C. Gossard, and W. Weigmann, Phys. Rev. Lett. 54, 1820 (1985).

14. J.K. Wang, J.H. Campbell, D.C. Tsui, and A.Y. Cho, Phys. Rev. B38, 6174 (1988).

15. P.J. Price, J. Appl. Phys. 53, 6863 (1982).

NEW LASER CRYSTALS

A. Kaminskii

Institute of Crystallography
Academy of Sciences of the USSR
Lenisky pr. 59
Moscow 117333, USSR

The development of laser crystals is a complex problem, involving fundamental investigations in the areas of material science, solid state physics, and quantum electronics. We search for new lasing compounds, new operating laser schemes, and new channels of stimulated emission. This seminar described results obtained in the last two years in our laboratory. In the first part of the seminar we briefly described the laser properties of new laser crystals doped with Ln^{3+} ions, and

- with Pr^{3+} ions:

  a) $YAlO_3$(300 K, $^3P_0 \rightarrow {}^3F_3$, $^3F_4$, $^1G_4$ and $^1D_2 \rightarrow {}^3F_3$ channels),

  b) $LiYF_4$, $LiLuF_4$, BaY_2F_8, and LaF_3(300 K, $^3P_0 \rightarrow {}^1G_4$ channels),

  c) $LiYbF_4$ and $BaYb_2F_8$(110 K, $^1G_4 \rightarrow {}^3F_4$, $^3H_5$ channels);

- with Nd^{3+} ions:

  a) $NaBi(MoO_4)_2$ and $NaBi(WO_4)_2$(300 K, $^4G_{3/2} \rightarrow {}^4I_{11/2}$, $^4I_{13/2}$ channels);

- with Er^{3+} ions:

  a) BaY_2F_8(300 K, $^4S_{3/2} \rightarrow {}^4I_{9/2}$, $^4I_{11/2}$, $^4I_{13/2}$ and $^4I_{11/2} \rightarrow {}^4I_{13/2}$ channels).

In the second part of the seminar we considered the suitability of crystals doped with Ln^{3+} and TM activators as active media for tunable stimulated emission. In particular, we described the record broadband tunable room-temperature laser action (~0.8 - ~1.25 μm) of Cr^{3+} in trigonal compounds with Ca-gallogermanate disordered type structure ($La_3Ga_5SiO_{14}$, $Ca_3Ga_2Ge_4O_{14}$, $La_3Ga_{5.5}Nb_{0.5}O_{14}$ and other), as well as the low-temperature laser action (~1.32 - ~1.43μm) of Cr ions (we hope Cr^{4+}!!) in orthorhombic $LiNbGeO_5$ single crystals (space group D_{2h}^{16}-Pbnm) under laser pumping in the spectral range (~0.67 - ~1.04 μm).

DEGENERATE FOUR-WAVE MIXING TECHNIQUE FOR STUDYING VERY FAST RADIATIONLESS PROCESSES IN VERY SHORT TIME SCALE

P. EVESQUE

Laboratoire d'Optique de la Matière Condensée
Université P. et M. Curie, 4 place Jussieu
75252 PARIS Cedex 05, FRANCE

ABSTRACT

The standard transient grating experiment enables to measure relaxation times longer than the laser pulse duration. We enumerate the different mechanisms which can be investigated and show how powerful this method is. A counter part is that one has to know much about the physics of the system in order to interpret the experimental results.

A second unconventional method of performing a four-wave mixing (FWM) experiment is sensitive to the coherence induced between states. In a two-level system, it measures transverse relaxation time T_2 and allows a resolution time only limited by the coherence time of the laser, (which can be < 10fs). Our results obtained on a dye molecule, which is not a two-level system are not fully understood.

KEYWORDS: relaxation process, doped solids, depolarization, coherence

I. INTRODUCTION

Optical methods are efficient tools to investigate physical properties of solids. It is well known that absorption and emission spectra are of prime interest in determining the density of states and energy levels of semiconductors, doped crystals or glasses. Moreover, these techniques can also reveal information on the relaxation processes existing in the materials, especially when these techniques are associated with time resolved detection and pulsed excitations. In the same way, the invention of picosecond and femtosecond lasers which are characterized by very short resolution times and high power emission intensities has enabled to develop a new set of experimental techniques based on non-linear optics; for instance, they had allowed to get information on new parameters using either Raman or photon-echo experiments as well as hole burning techniques. But the higher the power of the laser is, the larger the order of the nonlinear response can be, so that one can also think to use the third order term in the electric field susceptibility and design a set-up which gives information on its value.

Among these last methods, the degenerate four-wave mixing (FWM) experiment, also called transient grating experiment [1-5], seems really powerful to us. It is based on a third order process in electric field, which lets three incoming electromagnetic beams of the same frequency interact in the sample, propagating in three different directions k_1, k_2 and k_3. This interaction creates an outcoming beam which propagates in a different direction and its intensity is detected and analyzed. The detected output signal follows the dynamical processes taking place in the medium so that it is through its temporal evolution that one can learn something about them. This signal does not only depend on the coherence properties between these three exciting waves, but also on the physical characteristics of the material and on its relaxation times. For instance, the system can keep in memory the phase of each incoming light beam during a rather large time so that it can make three beams interact which are not superimposed in time.

In the first part of the talk, we will pay attention to the classical way of performing this experiment; for instance, a typical transient grating [6] experiment is done by printing in the sample the grating at t=0 by letting two laser beams interfere; a third probe beam is delayed by time t compared to the two excited ones; it is diffracted by the grating and it then tests the amplitude of its fading. The physical nature of the different processes which can be investigated with this technique in doped solids and disordered materials are manyfold: it goes from coherent propagation, collision and diffusion of electronic excitations to energy transfer processes between donors and accepters. It also concerns heat diffusion and sound propagation. We do not intend to give a complete overview of the subject, but will focus our attention on a few examples which will illustrate the wide range of applications and the efficiency of the method. One of its more important application is to measure short life time T_1 of the electronic excited states, or reorientation or recombination processes. The resolution time of this experiment is the laser pulse duration.

When one wants to study some phenomenon related to an optical transition and characterized by a response time smaller than the picosecond, it becomes rather difficult to build a laser device with such a short emission duration. We will describe in the second part a second set-up and in some peculiar cases the resolution time of the experiment becomes the coherence time τ_c of the two laser beams; one can then study the coherence properties between the excited and ground states of the investigated sample and measure the transverse relaxation time T_2 of the medium [5]. However, this result is only true when τ_c is smaller than T_2, (so that it is better to use incoherent light such as a flash white lamp!); we will then discuss the different parameters which are measured with this device. We will even show that one has to be careful in interpreting such experimental results in the light of a two-level scheme for the medium, which is generally used to interpret the results of coherent degenerate FWM: our results, obtained at low temperature on a dye molecule doping a polymeric film, are incompatible with such a theory. We will argue that this disagreement comes from the complexity of the electronic structure of excited and ground states of the sample. In this second part, we will be interested in determining the validity domain of the theory of coherent degenerate FWM and give some ideas on how to improve it.

II. TRANSIENT GRATING AS A CLASSICAL DEVICE

II.A Set-up

We report on Figure 1 the classical set-up of a transient grating experiment. Two "write" (or "pump") pulsed laser beams of wave vector k_1 and k_2 interfere in a sample, excite this medium and create into it a periodic modulation of the real and of the imaginary part of the index of refraction. These two beams, in general, come from the same laser pulse, which has been split into two equal parts, and the optical paths of these two beams have been carefully kept equal in order to get a perfect phase matching between the two interfering electric fields and a stable grating. A third "probe" beam of wave vector k_3 is diffracted in the direction k_4. This can be treated theoretically as a third order nonlinear optical process or, alternatively as a Bragg scattering from a laser-induced refractive index grating [7]. The diffracted beam obeys the Bragg condition:

$$k_4 = k_3 \pm (k_1 - k_2) \tag{1}$$

This leads to an interfringe spacing Λ

$$\Lambda = \lambda/(2\sin(\Theta/2) \tag{2}$$

where Θ is the angle between the two pump beams of wavelength λ. In the linear response approximation, (i.e. when there is no absorption saturation), this prints a simple sinewave grating and the scattering efficiency at the Bragg angle is given by [4, 19]:

$$\eta = \exp(-2\alpha L)\ (\sin^2(d\pi\Delta n/2\lambda) + \sinh^2(d\Delta\alpha/4)) \tag{3}$$

where α and L are the absorption coefficient and the thickness of the sample respectively and d is the thickness of the grating. Eq. (3) can be expanded for typical experimental conditions so that the FWM signal efficiency is proportional to the square of their arguments. We will

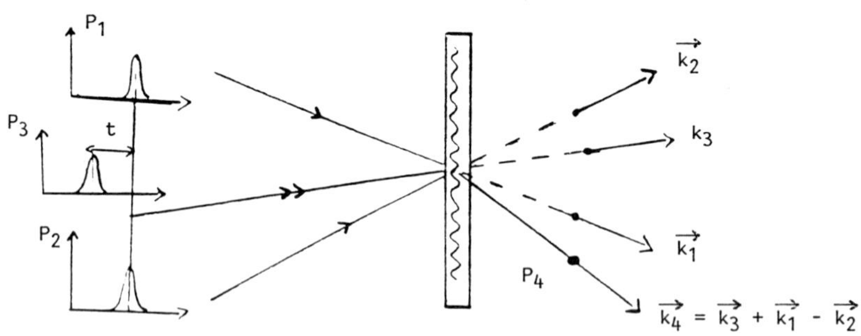

Figure 1. The experimental arrangement for a classical four-wave mixing (FWM) experiment: Beams 1 and 2 are the excitation beams, which come from the same pulsed laser beam, which interfere in the sample, printing in it an index of refraction grating. Beam 3 is either issued from the same pulsed laser as beams 1 and 2 but delayed by t compared to them, (or it is a continuous wave laser). One observes the intensity diffracted by the grating in the direction 4 as function of the delay t, (or as function of time).

only consider a thin grating and a not too large absorption coefficient in this paper, so that linear expansion of Eq. (3) is possible. When there is no absorption saturation, only the first order diffracted beam is produced by a sinewave grating so that higher order diffracted beams indicate nonlinear conditions.

Therefore, the diffracted intensity at a given time t will bring information on the value of the complex index of refraction and will allow us to follow the different relaxation processes. Some of these processes will lead to a fading characteristic time τ of the grating which is independent of the interfringe. Some other ones will depend on it, either linearly or quadratically, or even with a different powerlaw. We will describe these different behaviors and begin with the linear dependence of τ vs Λ since it is the simplest one which arises from a wave propagation due to a thermal excitation or a sound wave generation [8-14].

II.B Linear Dependence of the Relaxation Time vs. the Interfringe Spacing

Such a behavior can be obtained when the experiment deals with a non-absorbing medium; an electrostriction process [8] can occur which produces an emission of sound waves. In this case, the excitation light is directly coupled to the acoustic field of the material using a stimulated Brillouin scattering process. A photon of the high energy part of pump beam 1 (or 2) is destroyed, generating a photon at lower energy in pump beam 2 (or 1) and a phonon is emitted by the material, the energy and momentum of which is given by the conservation laws of energy and momenta. Two opposite directions of propagation of this phonon are possible: these two kinds of phonon are generated at t=0 and beat then in time. This leads to a diffracted beam intensity which oscillates in time with a period $T = \Lambda/2v$, where v is the sound velocity of the material [9].

However, similar sound waves can also be produced in a different way: when exciting an absorbing sample, the excited states can relax very fast leading to local heating of the sample and generating a pressure modulation [10] which creates two sound waves propagating in opposite directions. It can be demonstrated [10] that this process also leads to periodic oscillations of the diffracted beam intensity versus the delay between the probe and pump beams and that its period ($T = \Lambda/v$)

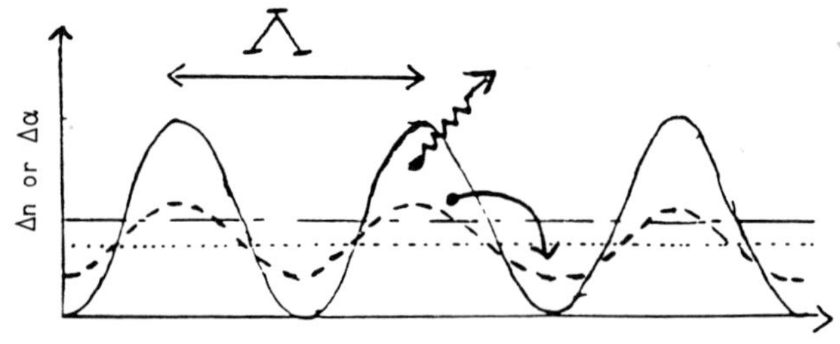

Figure 2. A transient sine-wave grating can relax either through local relaxation processes or through a diffusion process on a distance equal to the interfringe.

is twice that one obtained in the preceding paragraph. So FWM technique allows to measure the speed of sound.

When the medium absorbs the light, a periodic distribution of heat is in general generated leading to a heat conduction. In most cases, heat conduction obeys a diffusion equation which we will look at in the next section, but in some cases is does not. This last case is obtained when the momentum conservation law from collision between phonons is valid at a large length scale, which only occurs when umklapp [15] processes are unlikely to happen. But in order to be able to speak of heat propagation, one needs to get at a given time a local thermalization, which can be achieved through klapp processes (i.e. normal processes). So printing a grating with a wave number $k = 2\pi/\Lambda$ introduces a given momentum which is conserved. The diffracted beam intensity is periodically modulated in time and the experiment measures the speed v′ of the second sound, which is related the sound velocity v by [14]:

$$3v^2 = v'^2 \qquad (4)$$

Such second-sound waves have been detected at low temperature in superfluid helium [16], then in NaF [17] and on saphir [14]. The second sound has been detected in some materials at low temperature [16], but never at high temperature where heat transport occurs via diffusion.

A typical characteristic time of these oscillations ranges from .2ns to 30ns since the sound velocity lays around 1,000-5,000 m/s and the interfringe spacing between 1μm to 30μm.

II.C. Quadratic Dependence of the Fading Time vs. the Interfringe Spacing

In this section we shall focus our attention on the effect of a random walk multistep propagation on the fading time. Hence the transport will obey a master equation of the kind:

$$P_n = -\sum_m W_{mn} P_n(t) + \sum_m W_{nm} P_m(t) \tag{5}$$

where $P_n(t)$ is the occupation probability of site n at time t by an excitation and W_{nm} is the transfer rate from site m to site n. The transport can either concern a thermal or an electronic excitation. Eq. (5) can also be rewritten as a second order differential equation by expanding to second order P_n and assuming W_{nm} constant, independent of n and m:

$$\frac{\partial}{\partial t} p(r,t) = \Gamma *\sum_i \left(\frac{\partial^2}{\partial x_i^2} p(r,t) \right) \tag{6}$$

where $\Gamma = a^2 W_{nm}$. This is a diffusion equation governed by a Laplace operator. We will look for the solutions p(t,x) of the kind:

$$p(x,t) = p(t)*\exp(jKx) \tag{7}$$

where $K = 2\pi/\Lambda$ is the grating wave vector, since these solutions correspond to the symmetry of the set-up. In the small absorption limit, p is proportional to the amplitude of modulation of the index of refraction so that p is proportional to the diffracted electric field E. It is then the time dependence of p^2 which determines the time dependence of the first order diffraction intensity I according to $I = E^2$. Combining Eq. (6) and (7), one gets:

$$p^2(t) = \exp(-2\Gamma K^2 t) \tag{8}$$

The fading time is equal to $\tau = 1/(2\Gamma K^2)$ and varies as the square of the interfringe. Another demonstration of this behavior can be the following: Eq. (5) or (6) governs the random walk of a random walker; it is well known that the volume in which the random walker is localized at time t is defined by a radius of gyration R(t), the evolution of which varies as $\sqrt{t}$; the time τ at which the grating will have significantly faded due the diffusion process will correspond to $R(\tau) \approx \Lambda$ (see Fig.2), so that one gets approximately Eq. (8).

The random walkers which are governed by Eq. (6) may also spontaneously relax to a ground state and to dissipate their energy in the neighboring medium, so that a parallel fading process may occur, characterized by a lifetime T_1. In such a case, one has to add a $- p/T_1$ term in the right part of Eq. (6), which leads to the solution:

$$p^2(t) = \exp(-2t(\Gamma K^2 + 1/T_1) \qquad (9)$$

Eq. (5-9) describe fading by any kind of process governed by a random walk equation; this one can then concern an electronic state directly which has been first excited at time $t = 0$ by the two pump beams and which diffuses from now on, as in the case of a semiconductor; but it can also correspond to a heat diffusion process, or even to a molecule in solution which has been excited. In this last case, Γ is given by:

$$\Gamma = kT/(6\pi\ \eta\ R_H) \qquad (10)$$

where k is the Boltzman factor, T the temperature, η the viscosity of the liquid and R_H the hydrodynamic radius of the solvated molecule which migrates.

The order of magnitude of the fading time depends strongly on the material and on the nature of the random walker; for instance it takes about one year for the heat to diffuse on a 1-m length scale, so that one gets about $1\mu s$ fading time for a 1μ interfringe, according to Eq. (8) and to $K=2\pi/\Lambda$. For a fast semiconductor such as GaAs, the diffusion coefficient of the excited electrons is about $16cm^2/s$ [18], which leads to a typical erasing time of 100ps-1ns.

When the experiment deals with impurities in doped crystals, the diffusion coefficient may correspond to the migration of an optical excitation, which is first localized at time t=0 on an impurity, and which jumps from impurity to impurity after that time; in this case the diffusion coefficient will drastically depend on the impurity concentration. One expects that under most of the experimental range of concentration, this migration takes a long time; for instance such a migration has been observed between Cr^{3+} ions in emerald and alexandrite [19-20] for which the diffusion constant ranges around $1\mu^2/s$; no diffusion process has been found for the same ion in ruby. Such behaviors are important since it can influence strongly the luminescence

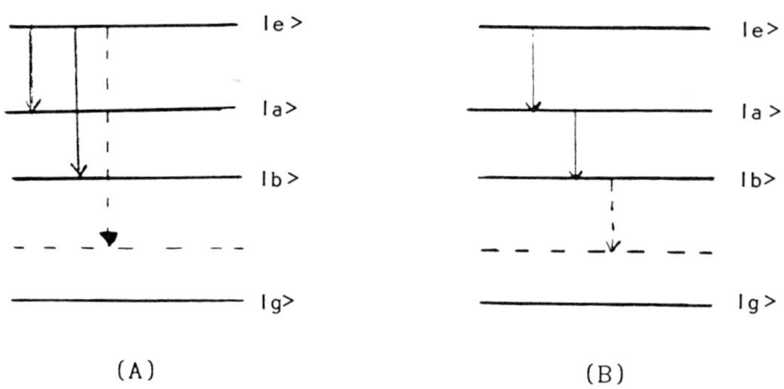

Figure 3. The two kinds of intramolecular relaxations of an excited state Ie>. In (A), the parallel scheme: direct transfers from level Ie> to levels Ia>, Ib>, ... are possible with a time constant T_a, T_b,..., leading to a pure exponential relaxation; $(1/T_e = (1/T_a + 1/T_b + \ldots))$. In (B), the sequential (or series) scheme: Ie> relaxes to Ia>, which relaxes to Ib> and so on; this sequential scheme leads to a multi-exponential decay.

efficiency of laser materials: when migration is efficient, trapping or fusion of excitations can occur efficiently.

In most cases, diffusion between donor states in doped crystal is fast enough, (when it occurs), so that an excitation migrates over a large number of impurities before having travelled over an interfringe distance; this means that averaging the jump time over the different geometrical configurations is possible and that the travel of the excitation is independent of the initial geometrical neighborhood of the excited ion. However this averaging process may not be possible in all cases; we will see in section II.E that this can bring some anomalous behaviors, but before looking at such behaviors we will first describe the different processes which can lead to a fading time which is independent of the interfringe spacing Λ and show that such a behavior may be due to a strong diffusion process.

II.D. Fading Time Independent of the Interfringe Spacing

1. Local parallel-sequential relaxation

A fading time independent of the interfringe means that deexcitation processes occur at a small length scale compared to the interfringe.

They can either come from an intramolecular process or from an intermolecular one. During an intramolecular relaxation process, the electronic excited state, excited at time t=0, relaxes to an intermediate state or to the ground state; this event is associated with an emission of phonon(s); it may also be induced by an intermediate absorption of a phonon so that these transition can strongly depend on temperature, (see other lectures). The change of population of the different electronic states modify the complex index of refraction of the medium which in turn modifies the intensity of the diffracted beam. Measuring the fading time τ allows to determine the relaxation time T_1; $\tau = T_1/2$ from Eq. (9). One also detects at longer time a tail superimposed to the transient grating decay, which is induced by heat diffusion. But interpretation of FWM results can be made difficult due to the simultaneous existence of more than one relaxation process; one gets in such a case a multiexponential decay rate. Moreover, the time dependence of the fading varies drastically with the way the deexcitation occurs when more than one process is involved; for instance two major schemes are possible, which are reported in Fig. 3: either the excited state |e> can be deexcited at the same time through different processes occurring in parallel to states |a>, |b>, ..., so that the relaxation probability $1/T_e$ at a given time is equal to the sum of the relaxation probabilities $\sum(1/T_a)$ and the FWM signal decays as a simple exponential; or the excited state |e> first relaxes to state |a>, which in turn is deexcited to state |b> and so on, (as in scheme B of Fig.3), so that the behavior of the fading is a multi-exponential; this is the sequential relaxation scheme. A general discussion of these two kinds of deexcitation process can be found in [21].

2. One-step energy transfer

We want to emphasize that a parallel scheme as in Fig.3A does not always lead to a simple exponential behavior, since it is also needed that all sites where the parallel processes occur are all identical in order to avoid the necessity of an averaging over the different configurations of the site. As a first example, suppose that one wants to study an excited state of an ion in a crystal under a polarized excitation and that the site of the doping ion has a local symmetry lower than the lattice one, so that there are different possible orientations for the lattice site; in such a case, in conjunction with the optical selection rules, the state of the site which is excited by the two pump laser beams might depend on the local orientation of the site as well as the relaxation process; so, an averaging over the different orientations seems necessary, which may bring a multiexponential decay of the grating. Another example of this effect can be found when studying intermolecular relaxation such as dipole-dipole one-step energy transfers from a donor to the neighboring acceptors. It is well known [21-25] that this leads to the stretched exponential behavior:

$$I(t) \approx p^2(t) = \exp -2\,(C_a V_d (t/\alpha)^{d/6}\,\Gamma(1-d/6) + t/T_1) \qquad (11)$$

where I(t) is the diffracted intensity, $C_a V_d$ is the average number of acceptor in the unit sphere in d dimension, (d = 3 in classical 3-D problems but in fractal space, it is the fractal dimension D); α is an adjustable parameter, which determines the strength of the donor-acceptor interaction and $\Gamma(x)$ is the Euler gamma function.

3. Chemical reaction formalism

When intermolecular processes are involved during the relaxation, this relaxation may correspond to a one-step energy transfer, as we have just mentioned, or to a multi-step one. In this last case, one expects a random walk process to occur characterized by a quadratic variation of the grating decay time τ as function of the interfringe $\Lambda = 2\pi/K$. However, in the presence of traps A, this excitation may not have time to travel over Λ and may be trapped before it. This is a sequential process, since excitation on site 1 jumps to site 2, 3 and so on until it reaches site n where the excitation meets an acceptor A and is then trapped. In most cases, one can take into account such a process by using a chemical reaction reaction formalism [21,24,26-27] and write:

$$p(t) = -kc_A p(t) - p(t)/T_1 \qquad (12)$$

where k is the constant of reaction, c_A is the trap concentration and T_1 is a lifetime due to other relaxation processes which occur in parallel. Solving Eq. (12) leads to a diffracted beam intensity $I(t) \approx p^2(t)$ decaying as

$$p^2 = \exp -\,(2t(kc_A + 1/T_1)) \qquad (13)$$

which is a simple exponential decay. Moreover, one can extend this chemical reaction formalism to the case of a high density of excited states; in this case collisions between two excitations are likely to occur so that fusion and annihilation become the most important deexcitation processes. The chemical reaction formalism leads to the differential equation:

$$\dot{p}(t) = -k'p^2(t) \qquad (14)$$

the solution of which is:

$$1/p(t) = 1/p(t=0) + k't \qquad (15)$$

In Eq. (14), we have omitted the p/T_1 term in order to emphasize the following effect: Eq. (14) depends on the square of p; so assuming that the fusion reaction occurs in a time scale much faster than the time needed by the random walker to travel over an interfringe, the fusion of excitations will concern excitations created in the same neighborhood. Thus the population variation will be equal to $k'tp^2(t=0)$ at short time t; this changes the simple sine-wave modulation of the index of refraction:

$$n \approx p\sin(Kx) \qquad (16)$$

and gives a second order term at short time:

$$n \approx p(0)\sin(Kx) - k'tp^2(0)\sin^2(Kx) \qquad (17)$$

This second order term will allow a diffraction in the second order of diffraction and will keep constant the diffraction intensity in the first order of diffraction. At longer times, other higher orders of diffraction will become observable. So looking at the evolution of the second or third order allows, in principle at least, to demonstrate the existence of nonlinear deexcitation processes and may prove the efficiency of fusion reaction.

4. Rotational relaxation, depolarization

The final effect due to diffusion that we want to comment on here, is the disorientation process: owing to optical selection rules, the non-spherical molecules or ions in a cubic environment (or in a liquid) which are excited by two polarized excitation beams have a preferential excitation so that the intensity of the diffracted signal depends on the probe beam polarization. After excitation, molecular orientational relaxation occurs due to the diffusion of either the molecules inside the liquid or of the excited states from molecules (ions) to molecules

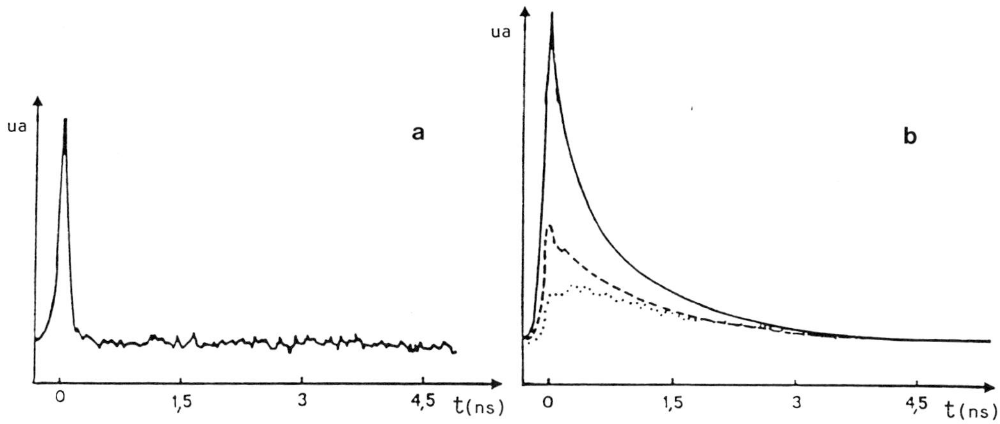

Figure 4. Short time response of a transient grating experiment performed on (a) a 8mM azobenzene bulk solution; (b) on a 8mM azobenzene solution imbedded in porous Vycor glass at three different probe beam polarizations (———, polarization parallel to the excitation beam; ....., polarization perpendicular to the excitation beam polarization; -----, polarization at the magic angle (see text)).

(ions). When the probe beam polarization is parallel to the pump beam polarization, molecular reorientation leads to a decrease in the diffracted signal; on the contrary, when the probe beam polarization is perpendicular to the pump beam polarization, reorientation leads to a rapid increase in the FWM signal. It can be proved that the diffracted signal is independent of these reorientation processes when the probe beam polarization is oriented at an orientation angle β_m of $54^{o}7'$ [9]. This angle, called the magic angle, does not depend on the intimate details of the relaxation processes (multi-exponentials, etc.). Therefore the behavior of the diffracted beam reflects the pure time dependence of the average excited- and ground-state populations. It can also be shown [28] that the time dependence of the molecular reorientation r(t) can be obtained from the experimental dependences of the intensities $I_{\parallel}(t)$ and $I_{\perp}(t)$ of the diffracted signals arising from probe beams polarized parallel and perpendicular, respectively, to the excitation beams. According to Tao [28],

$$r(t) = \left(I_{\parallel}(t) - I_{\perp}(t)\right) / \left(I_{\parallel}(t) + 2I_{\perp}(t)\right) \tag{18}$$

In the most general case, r(t) is the sum of three exponential decays since the principal axes of a molecule in a liquid are not equivalent and correspond to three different coefficients of rotational diffusion [29]. However, for a spherical molecule of hydrodynamics volume V_H in a liquid of viscosity η, r(t) turns out to be a pure exponential of time constant $\tau_r = V_H \eta kT$, (where T is the temperature and k the Boltzman constant); (one can compare Eq. (10) to the τ_r expression to get the relationship between rotational and translational diffusions).

So, following the works of Lord Rayleigh, Einstein, Langevin and Perrin on Brownian motion, one can consider that translational and rotational diffusions are related to each other and one may use a polarized experiment to determine the translational diffusion coefficient. We report on Figure 4 the result of such a polarized experiment performed on a dye molecule (the azobenzene, .008 mol) in solution (4.3mol methanol in toluene) embedded (Fig. 4b) or not (Fig. 4a) in a porous medium (the Corning 7930 Vycor glass).

As can be obtained by comparing the three curves of Fig. 4b, which correspond to the three different probe beam polarizations (parallel, perpendicular and at the magic angle), one may conclude:

- the excited state lifetime is T_1 = 2.2ns, from the time dependence of the experimental result performed at the magic angle;
- Using Eq. (18) together with the two other results of Fig. 4b leads to the rotational relaxation time τ_r = 450ps.

On the contrary, when the solution is not embedded in the porous medium, Fig. 4a exhibits only a sharp peak which corresponds to the coherent spike [30], since it is not observed when the probe beam is polarized perpendicular to excitation beam. Moreover, the constant value of the diffracted intensity in Fig. 4a means that the relaxation time of the excited state of the molecule in the bulk fluid much exceeds the observation time scale, $T_1 \gg$ 6ns.

These results mean then that:

- firstly, local diffusion is very fast in the pure liquid (τ_r < 100ps) whereas it is not in the porous (τ_r = 450ps).
- secondly, the relaxation of the excited state is much faster in the porous (T_1 = 2.2ns) than in the liquid alone ($T_1 \gg$ 6ns).

- Combining these two facts leads to the conclusion that the azobenzene molecule probably sticks on the internal Vycor porous surface so that its rotation is limited by the sticking-unsticking process; this one has a characteristic time of 450ps. Moreover, this sticking-unsticking process increases the speed of the excited state relaxation so that T_1 decreases from a value much longer than 6ns to 2.2ns.

A more complete discussion of these results can be found in [31].

II.E. About Anomalous Decay of the Transient Grating and About Fractals

For a complete discussion of the influence of disorder on diffusion processes, trapping, etc. on either Euclidean spaces or fractal ones, see review papers such as [21,24,32]. For instance it is well known that one-step energy transfer obeys a modified stretched exponential behavior when donors and acceptors lie on a fractal structure: in this case, Eq. (11) is still valid but d has to be replaced by the fractal dimension D [21,24].

We will focus our attention on grating decays due to diffusion in a disordered medium; moreover, the disorder, which will be considered here, is a topological one, which can be mapped using the fractal space formalism.

1. Diffusion on fractals

The fractal dimension D [21,24,34] of a fractal defines the scaling law of the number N(L) of sites contained in an Euclidean volume defined by its Euclidean length L:

$$N(l/a) \approx (L/a)^D \tag{19}$$

But in order to describe the random walk on a fractal, one also has to introduce the spectral dimension D_{sp} of the fractal [21,24,35,36]; it describes the number N(t) of different sites visited by the random walker between time t=0 and time t, when D_{sp} is smaller than 2:

$$N(t) \approx t^{Dsp/2} \tag{20}$$

(when D_{sp} is larger than 2, one gets $N(t) \approx t$, as one can find in [21,24,35,36]). A peculiar situation corresponds then to the case $D_{sp} < 2$, since the number of different sites visited increases slower than t in this case, which means in turn that the average time spent on each visited site increases continuously with time, which implies also that

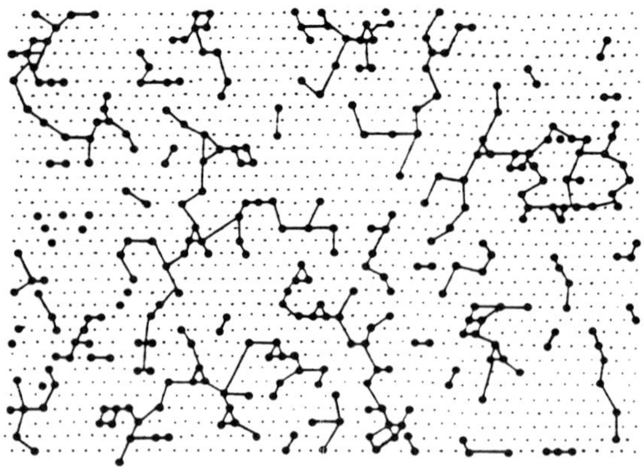

Figure 5. A percolation network near the percolation threshold: the system consists in a distribution of small, large and very large (infinite) fractal clusters on which the excitations can diffuse.

the random walker comes back very often to the same visited sites. Such a behavior is due to geometrical constraints, (a similar situation is got in the 1-D random walk, where $N(t) \approx \sqrt{t}$). When D_{sp} is < 2 and following de Gennes [37], one can then speak of a compact exploration since the time on each visited site increases with t, so that it becomes more and more unlikely that a site linked to a visited site will not have been explored as the time goes on. Using this approximation, (and this is an approximation since it takes the dead ends of the fractal structure too much into account), one can combine Eqs. (19) and (20) to get the time dependence of the mean volume (or Euclidean length R(t)) in which the random walker is localized at time t:

$$R(t) \approx t^{Dsp/2D} \tag{21}$$

In a FWM experiment the typical decay time τ of the gratings corresponds to the migration of the random walker over a distance equal to the interfringe Λ, so that using Eq. (21) leads to:

$$\tau \approx (\Lambda/a)^{2D/Dsp} \tag{22}$$

When the sample structure is well defined by a fractal between two length scales b and a (b>a) and that the interfringe grating Λ is larger than b, one has to combine this result with section II.C ones. One gets [24,38-40]:

$$\tau \approx (b/a)^{\left((2D/Dsp) - 2\right)} (\Lambda/b)^2 \tag{23}$$

It seems to me that only this last behavior could be detected on a pure fractal disordered sample, since a fractal disorder ranging over a length scale b larger than the interfringe will probably lead to a strong scattering of the light in the direction of the first order diffraction beam even in the absence of the two-pump-beam excitation. This scattering will hide the temporal dependence of the grating.

2. The percolation case

In some case, the disorder can be taken into account using a percolation model: for instance, take a perfect lattice and dope it randomly with an impurity; excite now an electronic states of few impurities and let them migrate by jumps from impurity to impurity; the diffusion law of this electronic state will strongly depend on the jump rules; (see for instance the review paper [32]). However, a simple jump rule could be that the only jumps which are allowed are those concerning two impurities which are separated by at most by a given length a; any allowed jump lasts the same time. In this case, and for a small impurity concentration range, one can get a diffusion problem similar to that one sketched in Figure 5, which is a percolation problem near the percolation threshold. We will now briefly look at this diffusion problem through a transient grating experiment in the case where the wave length of the laser beams is much larger than the jump length a. We will first make a few remarks.

a) One can see from Fig.5 that impurities are joined together in clusters of different sizes; there are small (single impurity, pairs, triads), large and very large clusters; this is a characteristic of the percolation model near the percolation threshold for which the size distribution obeys a power law (scaling law) [41]. However, there is a maximum size b of the finite cluster for a concentration c different than the critical concentration c_c. b will be called the correlation length of the disorder. b scales as:

$$b \approx a(c - c_c)^{-\nu} \qquad (24)$$

b) If $c > c_c$, an infinite cluster exists. The probability that an impurity pertains to this infinite cluster scales as:

$$P_\infty \approx (c - c_c)^{\beta} \qquad (25)$$

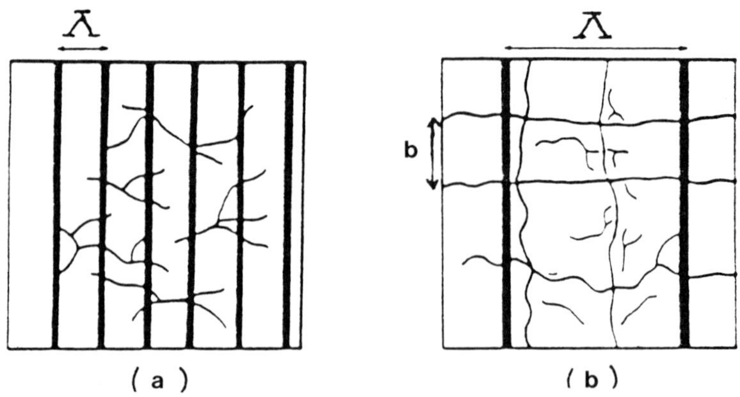

Figure 6. (a) Small-interfringe limit; (b) large-interfringe limit.

c) Moreover, each large finite cluster, (and we will see in the next paragraph this stands also for the infinite cluster at a length scale smaller than b), looks like a fractal, the spectral dimension D_{sp} of which always lies around 4/3 [35] whatever the lattice dimensionality is. Both, spectral and fractal dimensions are independent of the cluster size, since one cannot determine the size of a large cluster by only looking at a small part of it. The fractal dimension of the cluster varies from a 2-D percolation or a 3-D one, but does not depend on the intimate structure of the bonds.

d) The infinite cluster, when it exists, has the same fractal structure as the large finite ones (D & D_{sp}) at a length scale smaller than b; its structure turns out to be Euclidean for length scales larger than b.

e) Near the percolation threshold, the average distance between the impurities is a ($a \ll \Lambda$); as these impurities are randomly distributed, and as far as one is interested in a first order diffraction due to an interfringe Λ much larger than a, one can assume that the static scattering in the diffracted beam direction will be negligible compared to the diffracted intensity induced by the transient grating pattern. Thus it becomes possible to study diffusion on this structure.

f) So, FWM excitation will excite different clusters, either finite or infinite; some of them will be very small, so that their extension will be smaller than the interfringe Λ; an excitation on such clusters will never be able to diffuse over a distance equal to the interfringe and will look localized.

g) Moreover, by looking at Fig. 5, one can define two different behaviors for the FWM signal depending on whether the correlation length of the disorder b is larger than the interfringe or not; we have called these two behaviors the small- and large-interfringe limits; they are sketched in Fig. 6.

Let us consider now the large-interfringe limit. In the case of the large-interfringe limit, any finite cluster has an extension smaller than the interfringe, their excitations stand localized; on the contrary, an excitation on the infinite cluster will diffuse so that a part of the diffracted intensity will decrease within a characteristic time τ given by Eq. (23). Combining Eq. (23) & (24) leads to:

$$\tau \approx (\Lambda/a)^2\ (c - c_c)^{-2\nu((D/Dsp) - 1)} \tag{26}$$

Moreover, the probability that the infinite cluster has been excited at time t=0 is p_∞, which is given by Eq. (25); considering then that p_∞ is $\ll 1$, one can compute the diffracted intensities I_0 and I_∞ at time t=0 and t=∞ and get the grating contrast C:

$$C = (I_0 - I_\infty)/I_0 \approx (c - c_c)^{\beta} \tag{27}$$

Eq. (27) clearly indicates that only a small part of the grating will be washed out by diffusion. This part of the decay will obey a pure exponential decrease, except perhaps at short time, when the experiment is probing the fractal structure.

Let us consider now the small-interfringe limit. In the case of a small-interfringe limit, the system obeys scaling laws. The calculation has to take them into account Λ; this has been done in [39-40].

The characteristic fading time τ of the grating will correspond to an excitation which diffuses on a fractal over a length scale equal to the interfringe Λ; τ is then given by Eq. (22). In order to evaluate the grating contrast, one has to determine the probability that an excitation pertains either to the infinite cluster or to a finite one, the size of which is larger than the interfringe. This is simply performed using a renormalization group argument. When the correlation length b is larger than the interfringe Λ, the experiment does not test lengths equal to b and Λ. Any sample with a Λ correlation length b' larger than Λ looks like the studied sample or like a sample characterized by a correlation length b' equal to the interfringe Λ; (or $\approx$ the maximum size of finite cluster is b'). So, b' = Λ.

This new sample should exhibit a concentration c' given by Eq. (24); it corresponds also to the highest concentration possible for the large-interfringe limit case, the contrast of which is given by Eq. (27). Combining then Eq. (24) and (27) together with b' = Λ, one gets:

$$C = (1 - I_0/I_\infty) \approx (\Lambda/a)^{-\beta/\nu} \tag{28}$$

$$\tau \approx (\Lambda/a)^{2D/Dsp} \tag{29}$$

The decay of the grating is then characterized by τ; however, there is no theoretical reason that this decay is a simple exponential; moreover, the diffracted beam is diffracted by fractals the size of which is equal to the interfringe, so that more than one order of diffraction may be observed, and the intensities of the beams diffracted in these different orders most likely obey a power law distribution, (see other chapters in ref. [24] for more information on the scattering due to fractals).

II.F. Measure of the Dephasing Time T_2

In this section, we will briefly describe a FWM method which can be used to measure the dephasing time T_2 (also called transverse relaxation time) of a medium. A correct description of these properties would need to introduce the formalism of the density matrix, but we will omit it for sake of simplicity and give rough arguments which allow us to understand a bit of this topic and especially to see where the limitations of FWM techniques really lie. However, in order to tackle the theoretical problem the density matrix needs to be expanded to the third order of perturbation, which is a hard task especially when one wants to take into account the laser bandwidth, and the two densities of electronic states, (one for the ground state and the other for the excited one); so, one limits the analysis to a two-level model partly homogeneously broadened (width $1/t_2$) and partly inhomogeneously broadened. We will see in section III that this model is not always valid and will give a counter example. We report the reader to this next section III for references on the experimental studies and theoretical solutions.

Let us assume that we keep the delay t constant between the probe beam and the two pump ones; (we fix this delay t to a value approximately equal to the T_1 relaxation time of the sample). Let us now assume that we delay slightly a pump beam compared to the other, (delay t_D); the diffracted intensity will depend not only on the coherence time τ_c of the pump pulsed laser but also on a new parameter, the transverse relaxation time T_2 of the sample. Different cases are possible; this one which will be achieved will depend on the value of T_2/τ_c; we will also see that the pulse duration T_p and the relaxation time T_1 will become important when $T_2 << \tau_c$.

1. $\underline{T_2 >> \tau_c}$

This means that the coherence which has been induced in the sample by the laser pulse does not relax at once. This medium keeps in memory the characteristics of the laser during a time equal to T_2, even if they are not superimposed in time (i.e. even if $T_p << T_2$). The diffracted signal, as function of the delay t_D between the two pump laser beams, will exhibit a width of about T_2.

2. $\underline{T_2 << \tau_c}$

This means that the medium lost at once the memory of the exciting electric fields. The diffracted signal will then have a width characterized by τ_c. However two schemes are now possible which depend on whether T_1 is smaller or larger than τ_c.

- $T_1 << \tau_c$: The grating population relaxes at once and one measures the time duration T_p of the two exciting laser pulses. (The diffracted signal does not measure the coherence time of the laser but T_p).

- $T_1 >> \tau_c$: For delay longer than τ_c, one is printing a grating in the sample. The wave vector of the grating (i.e. the interfringe) is kept constant but the phase is fluctuating randomly as the time goes on, changing every τ_c. So the contrast of the accumulated gratings becomes negligible and the diffracted intensity tends to 0 after a delay t_D smaller than τ_c. One measures then the coherence time of the laser.

This crude analysis of FWM technique is very fruitful in many cases; however, it is based on many approximations, so that it can sometimes be an over-simplified situation in the next section. Moreover, it seems to us that the true theory of what we have observed still remains to be performed.

III. SELF-DIFFRACTION OR FWM EXPERIMENT WITH INCOHERENT LIGHT

Several experiments using degenerate FWM technique have been performed in order to bring information about optical dephasing times on a picosecond [42-43] or femtosecond [44-46] time scale. Most of them dealt with weakly bounded impurity ions or molecules imbedded in a host matrix, so that most of the obtained results can be satisfactorily described by using a two-level model scheme, considering both the homogeneous and inhomogeneous broadenings. However, some papers [42-45] report on similar experiments conducted on dye molecule (cresyl violet) doping polymer films. The dye molecule exhibits a large continuum of states which is induced through a strong coupling between electronic and vibrational states of the molecule. Therefore the system can hardly be considered as a two-level system.

We undertook this work with the goal of obtaining more detailed information about this problem: can we provide some experimental evidence that the unique optical dephasing time T_2 used in a two-level scheme is still valid in a multilevel system? We have used the so-called self-diffraction FWM method.

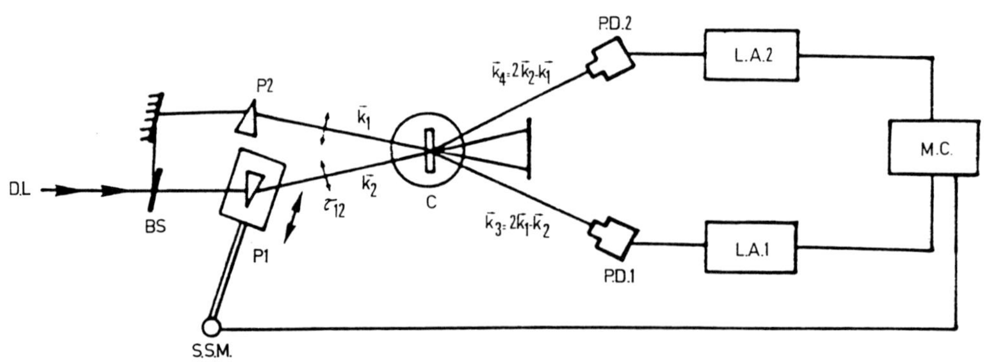

Figure 7. Experimental set-up for a self-diffraction FWM technique. Beam 1 (or beam 2) is diffracted by the interference of beams 1 and 2. This leads to beam 3 (or 4) of intensity S3 (or S4).

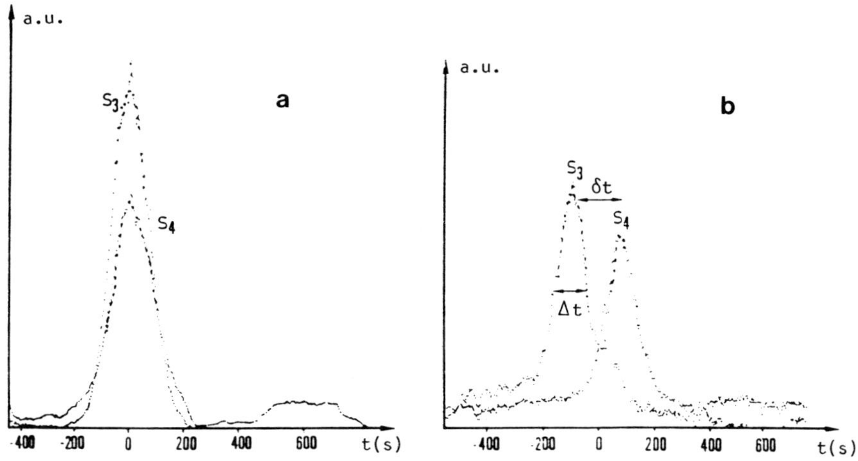

Figure 8. Typical diffracted signal intensities S3 and S4 vs. the delay t between beam 1 and 2. (a) 560-nm laser beams; (b) 625-nm laser beams and T=8K; a time shift δt appears between the two curves of width Δt.

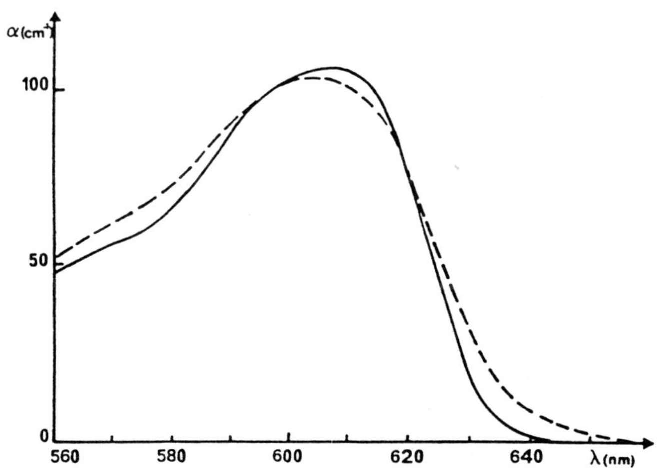

Figure 9. The absorption spectrum of cresyl violet at 10K (——) and 300 K (----).

III.A. Experimental Section

As can be seen in Figure 7, the basic idea is to spatially split a dye laser beam (coherence time τ_c) into two distinct beams (wave vectors k_1 and k_2) with equal intensity and same polarization. One of these two beams is then delayed by a variable time t with respect to the other one. These two beams are then focused at the same point in a .3mm thick sample, thereby giving rise to a FWM; (i.e. beam 1 (or 2) is diffracted by the grating due to beam 1 and 2 interference pattern). The two diffracted intensities are recorded as a function of the delay t; the very short variations of delay required by this study are achieved by moving a prism parallel to one of its bisectors by means of a stepping motor (one single step correspond to a 1.4 fs delay). The dye laser was pumped by a Q-switched-mode-locked picosecond Nd:YAG laser; the lasing dyes used were either a methanol solution of rhodamine B and DCM lasing at 625nm (10nm bandwidth and τ_c = 130fs) or a solution of rhodamine 6G which lases at 560nm (τ_c = 210fs). The samples were immersed in a Helium cryostat.

III.B. Results

Figure 8 reports a typical set of results obtained with the FWM technique on cresyl violet thin films. Figure 8a corresponds to an excitation wavelength at 560nm which lies around the top of the absorption band of the dye molecule, as one can see on Figure 9. In this case, the results do not depend on temperature: the two curves S3 and S4, corresponding respectively to $2k_1 - k_2$ and $2k_2 - k_1$ signals, are superimposed within the experimental error bar (20fs). Thus these curves do not show any splitting and their widths correspond to the coherence time of the laser.

Figure 8b is a typical result obtained with the 625nm laser excitation, lying on the steep low-energy side of the absorption band, (see Figure 9). In this case, the results crudely depend on temperature: at room temperature, S3 and S4 are superimposed although they are shifted by a time δt at low temperature. We will label δt the shift between S3 and S4 maxima and Δt the width of each S3 or S4 signals as function of t. (Due to the symmetry of the configuration, S3 and S4 signals shall be symmetric compared to t=0, so that they shall display the same width Δt).

We first notice that the Figure 8 results are quite similar to those previously published [45-46].

1. Spectral characteristics of diffracted and transmitted beams

According to the absorption spectrum (Fig. 9), the 625 nm emission is exactly located on the abrupt low-energy side of the cresyl violet absorption band, so that the film is acting not only as a nonlinear medium but also as a strong inhomogeneous filter which induces a distortion and/or a shift in the 625-nm laser emission band after passing through the sample. Due to the fact that the absorption band undergoes a slight bathochromic effect when the temperature is decreased, this may induce a temperature-dependent change of the coherence time τ_c of the transmitted and diffracted laser beams. We have then measured the spectral characteristics of these beams:

- In the high temperature range, diffracted and transmitted beams exhibit the same spectrum. Both are shifted by about 6-7nm towards the low-energy side; this shift is temperature-independent within the uncertainty (1nm).

- The width of the transmitted beam is 10nm at T=300K, whereas it reduces to 8nm at 10K; the incident beam bandwidth is 10nm.

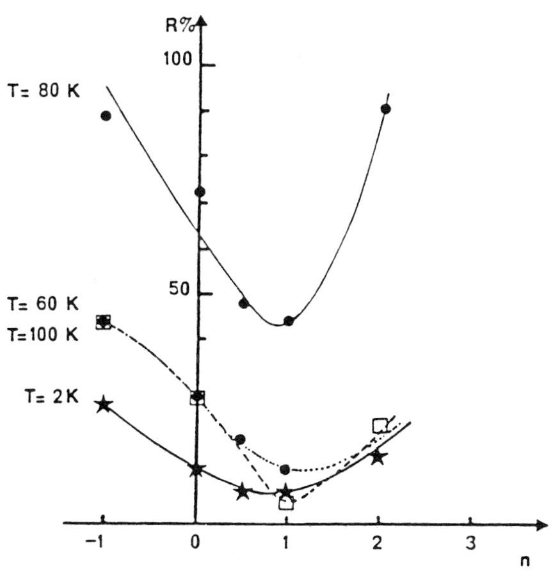

Figure 10. Experimental signal-to-noise ratio of the function $\delta t/\Delta t^n$ as a function of n at different temperatures, (625-nm excitation). The existence of the minimum for $n = N_o = 1$ proves a correlation between the noise of δt and of Δt and suggests a linear relationship between δt and Δt.

2. Temperature dependences of δt and Δt at 625-nm

We have determined the temperature dependence of the widths of the two S3 and S4 signals as function of temperature. A first series of experiments at 625nm have been performed and exhibited a steep dependence on at least one uncontrolled parameter since our results about δt and Δt where changing from one run to another (even when keeping the sample temperature constant). It was noticed (and this turned out to be crucial) that most of these signal fluctuations were to be attributed to the poor stabilization of the laser cavity, thereby giving rise to fluctuations of the wavelength of the laser emission on a time scale of 1 hour. After the laser cavity was stabilized, a second run of experiments has been performed leading to an experimental uncertainty of about 20 fs.

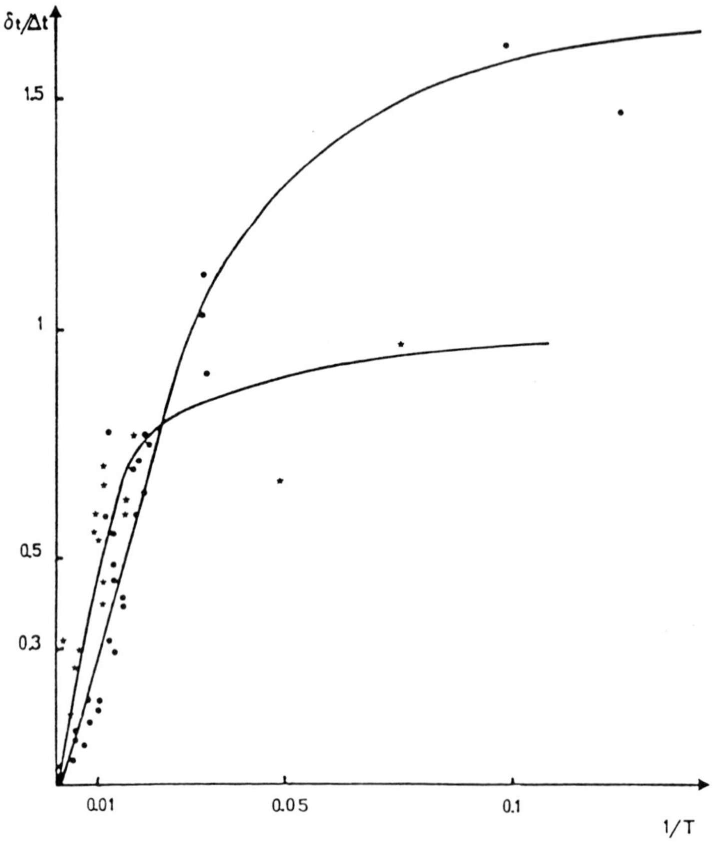

Figure 11. Temperature dependence of $\delta t/\Delta t$ for the 625 nm excitation. (+ first series of experiment; • second series of experiment).

In order to get information about the variations of the time shift δt as function of the width Δt, we have looked for correlation effects between the experimental values of δt and Δt. The procedure has been the following: during the first series of experiments (the results which exhibit very large fluctuations), several runs of S3 and S4 signals were performed at each temperature; then we could measure several values of the couples δt and Δt at the same temperature and compute the experimental values of $\delta t/\Delta t^n$. We have then experimentally determined the mean values and the standard deviations of quantities such as δt, Δt, $(\delta t/\Delta t^n)$, at each temperature.

Therefore, if one assumes the validity of a theoretical relationship between δt and Δt of the form

$$\delta t = \Delta t^{No} f(T,...) \tag{30}$$

where No is any number of f and any function of temperature T or any other hidden parameter, one expects the signal-to-noise ratio of $\delta t/\Delta t^n$ to be maximum for $n = No$. So, we have plotted in Fig. 10 the inverse of the signal-to-noise ratio as a function of the power n. For each temperature, the obtained results show up as a minimum for $n = No = 1\pm.5$, thereby proving the linear dependence of δt vs Δt.

We report the variations of $\delta t/\Delta t$ vs 1/T for the two different runs in Fig. 11. Both these curves exhibit a linear dependence at high temperature and a saturation at low one. Experimental data could be fitted with, at least, the following laws:

$$\delta t/\Delta t = B/(1 + T/T_s) \tag{31a}$$

$$\delta t/\Delta t = B\tanh(T_s/T) \tag{31b}$$

A different pair of B and T_s has been required to fit each curve. These ones are (B,T_s) about (.9,50K) for the first series and (1.7,20K) for the second one. Moreover, the product BT_s, which is the slope of $\delta t/\Delta t$ vs. 1/T at high temperature is different for each curve.

3. FWM and the failure of the two-level model at 625-nm

The results reported in Fig. 8 are quite similar to those previously observed by other authors [45-46] and have tentatively been

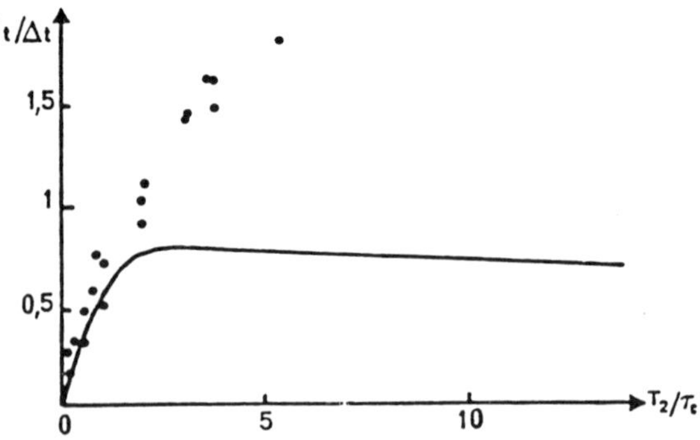

Figure 12: Ratio $\delta t/\Delta t$ as function of T_2/τ_c; (———) theoretical curve; • are the data points. T_2 is the dephasing time, τ_c is the laser coherence time. There is a misfit between theory and experiment.

interpreted within a simplified two-level model including a homogeneous and an inhomogeneous broadening. We will prove in this section that our results do not agree completely with this interpretation.

According to ref. [42], this model allows to relate the splitting δt between the two S3 and S4 curves reported in Fig. 8 to a T_2 time. It also predicts a T_2 dependence of the width Δt of the two signals. Using then Eq. (7) of ref. [42] and assuming a Gaussian-shaped autocorrelation function of the laser emission, we have first computed these widths Δt, the splitting δt and the ratio $\delta t/\Delta t$ as function of T_2 for a given τ_c, in order to compare these results with the experimental ones. The predicted theoretical variations of the ratio $\delta t/\Delta t$ are reported on Fig. 12, (solid line); it exhibits a maximum of about .8 for $T_2/\tau_c \approx 2.5$. We have also obtained: (i) when $T_2 \ll \tau_c$, $\delta t/\tau_c$ varies linearly as a function of T_2/τ_c and Δt is independent of T_2/τ_c; (ii) when $T_2 \gg \tau_c$, on the contrary, $\Delta t/\tau_c$ varies linearly against T_2/τ_c.

If now we compare the theoretical predictions with our experimental values, we find: (i) the results obtained with a 560-nm laser emission are correctly described by a $T_2 = 0$ dephasing time, whatever the temperature is, in accordance with published results [46-47]; (ii) on the contrary, these theoretical predictions cannot explain the linear relationship which exists between δt and Δt when one excites the sample

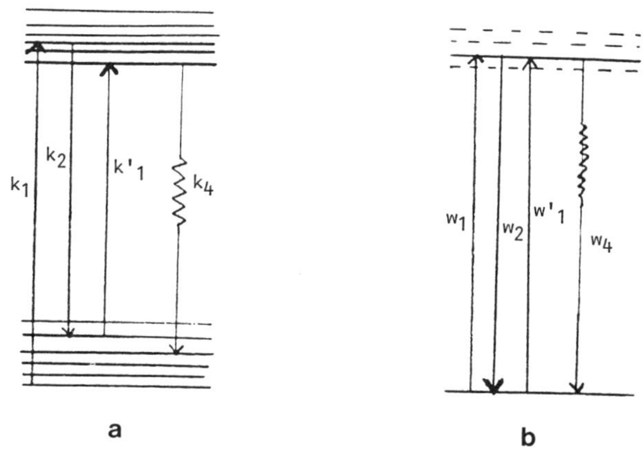

Figure 13: In a multi-level scheme, (a), the four FWM-optical transitions can occur at different frequencies inside the same molecule, even when $T_2 = \infty$; this is not the case in a two-level model with $T_2 = \infty$, as can be seen in scheme (b). The dashed lines in scheme (b) stand for the electronic levels of other molecules which are shifted due to the inhomogeneous broadening.

at 625-nm, as we have found in the last section. We will look forward to more details which demonstrate the inadequacy between this theory and the 625-nm experimental results.

In order to do so, we have computed the T_2 values using this model and the δt experimental values of the second run. Using these T_2 values we could theoretically determine Δt and predict the $\delta t/\Delta t$ values, which are given by the solid line of Fig. 12. We could also compute the experimental values $\delta t/\Delta t$ directly from the data; we also report these ones on Fig. 12 as a function of the computed T_2 values, (the points). If the theory was adequate we should get a perfect superimposition of the points on the curve; this is obviously not the case.

We will not go deeper into the discussion; more details on these experimental results can be found in [48]. These results are also not understood in the framework of Morita et al. theory for FWM [51] which takes into account the pulse duration, nor even on the approach of Mi et al. [52] who assume a multilevel system; however this last paper only takes into account a discrete series of levels and a very broad laser band, a situation which is not in accordance with our experimental conditions due to the drastic change of the optical absorption in the pulse bandwidth, so that T_2 and τ_c are comparable in our experiment. For more theoretical references on coherent effect using FWM technique, one can look at ref. [53-59].

A reason why the two-level model is inadequate for FWM on cresyl violet is probably schemed on Fig. 13: we see that on they dye molecule many different 4-photon absorption and emission processes are possible which are not taken into account in the two-level system. Moreover, if one takes account of the steep variation of the density of states and the laser bandwidth, this scheme (Fig. 13) predicts that there shall be a shift of the part of the spectral distribution of the laser which interacts with the medium when considering the first photon interaction or the third one, even in the case of a very long T_2; the two-level scheme is not able to take this shift into account.

REFERENCES

1. A.E. Siegman, J. Opt. Soc. Am. 67, 545 (1977).
2. K. Jarasiun, F. Vaitkus Phys. Stat. Sol. (a) 44, 793 (1977).
3. M.D. Fayer, Picosecond Phenomena III, Springer Series in Chemical Physics, 23 (Springer Verlag 1982).
4. D.R. Lutz, K.A. Nelson, C.R. Gochanour, M.D. Fayer, Chem. Phys. 58, 225 (1981).
5. Y.R. Shen, The Principles of Nonlinear Optics, edited by John Wiley & Sons, (New York 1984).
6. J.R. Salcedo, A.E. Siegmann, D.D. Dlott, M.D. Fayer, Phys. Rev. Lett., 41, 131 (1978).
7. H.J. Eichler, P. Gunter, D.W. Pohl, Laser Induced Dynamics Gratings, Springer Verlag, Berlin, (1986).
8. K.A. Nelson, D.R. Lutz, M.D. Fayer, L. Madison, Phys. Rev. B 24, 3261 (1981).
9. M.D. Fayer, IEEE J. Quant. Electron. QE22, 1437 (1986).
10. D.C. Auth, Appl. Phys. Lett., 16, 511 (1970).
11. H. Eichler, G.L. Salga, H. Stahl, J. Appl. Phys. 44, 5383 (1973).
12. H. Eichler, H. Stahl, J. Appl. Phys. 44, 3429 (1973).
13. R.I. Scarlett, Phys. Rev A, 6, 2281 (1972).
14. M. Chester, Phys. Rev., 131, 2013 (1963).
15. C. Kittel, Introduction to Solid State Physics, p.131, ed. John Wiley & Sons, 3rd edition.
16. L. Landau, J. Phys. USSR, 5, 71 (1941).
17. D.W. Pohl, S.E. Schwarz, Phys. Rev. B, 7, 2735 (1973).
18. R.J. Manning, A. Miller, D.W. Crust, K. Woodbridge, Optics Lett., 13, 868 (1988).
19. R.C. Powell, A. Suchocki, F.M. Durville, G.D. Gilliland, E.G. Behrens, G.J. Quarlesn, G. Boulon, J. de Phys. (Paris), colloque-C7, 48, C7-417 (1987).
20. A Suchocki, G.D. Gilliland, R.C. Powell, Phys. Rev. B, 35, 5830 (1987).
R.C. Powell, J. de Physique (Paris), 46, C7-403 (1985).
G.D. Gilliland, A. Suchocki, K.W. Ver Steg, R.C. Powell, Phys. Rev. B, 38, 6227 (1988).
21. A Blumen, J. Klafter, G. Zumofen, Optical Spectroscopy of Glasses, p. 199-265, ed. I. Zschokke, D. Reidel publsihing company, (1986).

22. D. Dexter, J. Chem. Phys. 21, 836 (1953).
23. M. Inokuti, F. Hirayama, J. Chem. Phys., 43, 1978 (1965).
24. P. Evesque, The Fractal Approach to Heterogeneous Chemistry, p. 81-103, ed. D. Avnir, John Wiley & Sons Ltd. (1989).
25. A. Blumen, J. Manz, J. Chem. Phys., 71, 4694 (1979).
26. P.G. de Gennes, J. Chem. Phys., 76, 3316 (1982).
27. P. Evesque, J. de Phys. (Paris), 44, 1217 (1983).
28. T. Tao, Biopolymers, 8, 609 (1969).
29. F. Perrin, J. Physique Radium VII, 5, 497 (1934)
30. B.S. Wherrett, A.L. Smirl, T.F. Bogges, IEEE J. Quant. Electron., QE-19, 680 (1983).
31. M.T. Portella, P. Montelmacher, A. Bourdon, P. Evesque, J. Duran, J. PHys., C-1, 981 (1989).
32. S. Havlin, D. Ben-Avraham, Adv. in Phys. 36, 695-798 (1987).
33. P. Evesque, J. Duran, A. Bourdon, J. Phys. C, 18, 2643 (1985).
34. B.B. Mandelbrot, Les Objects Fractals: Forme, Hasard et Dimension, Flammarion, Paris (1977).
35. S. Alexander, R. Orbach, J. de Physique (Paris) Lett., 43, L625 (1983).
36. R. Rammal, G. Toulouse, J. de Physique (Paris) Lett., 43, L625 (1983).
37. P.G. de Gennes, C.R. Acad. Sci. (Paris), 296, Ser. II, 881 (1983).
38. W.D. Dozier, J.M. Drake, J. Klafter, Phys. Rev. Lett., 56, 197 (1986).
39. P. Evesque, J. Duran, A. Bourdon, J. Phys. C, 18, 2643 (1985).
40. P. Evesque, J. Duran, A. Bourdon, J. de Physique (Paris), 46, C7-45 (1985).
41. D. Stauffer, Phys. Rep., 54, 1 (1979).
42. H. Nakatsuka, H. Tomita, M. Fujiwara, S. Asaka, Opt. Commun., 52, 150 (1984).
43. S. Asaka, H. Nakatsuka, M. Fujiwara, M. Matsuoka, Phys. Rev. A, 29, 2286 (1984).
44. A.M. Weiner, S. de Silvestri, E.P. Ippen, J. Opt. Soc. AM. B, 2, 654 (1985).
45. S. de Sivestri, A.M. Weiner, F.G. Fujimoto, E.P. Ippen, Chem. Phys. Lett., 112, 195 (1984).
46. H. Nakatsuka, M. Fujiwara, R. Kuroda, J. de Physique (Paris), 46, C7-511 (1985).
47. M. Fujiwara, R. Kuroda, H. Nakatsuka, J. Opt. Soc. Am. B, 2, 1634 (1985).
48. M.T. Portella, P. Montelmacher, A. Bourdon, P. Evesque, J. Duran, J. Phys. Chem., 91, 3715 (1987).
49. J.E. Golub, T.W. Mossberg, J. Opt. Soc. Am. B, 3, 554 (1986).
50. R. Zhang, X. Mi, H. Zhou, P. Ye, Opt. Comm., 67, 446 (1988).
51. N. Morita, T. Yajima, Phys. Rev. A, 30, 2525 (1984).
52. X. Mi, H. Zhou, R. Zhang, P. Ye., J. Opt. Soc. Am. B, 6, 184 (1989).
53. J.P. Lavoine, A. Boeglin, S.H. Lin, A.A. Villaeys, Phys. Rev. A, 38, 2896 (1988).
54. R. Trebino, Phys. Rev. A, 38, 2921 (1988).
55. P. Ye, Y. R. Shen, PHys. Rev. A, 25, 2183 (1982).
56. T. Yajima, H. Souma, Phys. Rev. A, 17, 309 (1978).
57. B. Hönerlage, M. Frindi, J. Miletic, Phys. Stat. Sol. (b), 149, 775 (1988).
58. M.A. Dupertuis, R.R.E. Salomaa, J. Opt. Soc. Am. B, 5, 130 (1988).

PHOTOTHERMAL OPTICAL NONLINEARITIES, OPTICAL BISTABILITY AND POSSIBLE APPLICATIONS

C. Klingshirn
Fachbereich Physik der Universität
Erwin Schrödinger Straße 1
D–6750 Kaiserslautern
Germany

ABSTRACT

The exciton–phonon interaction in semiconductors leads to a temperature dependence of the absorption edge described by the well known Urbach–Martienssen rule. Based on this dependence, strong illumination of semiconductors with photons in the absorption tail results via nonradiative recombination in sample heating and consequently in changes of the optical properties. These photo–thermal optical nonlinearities can be used to realize various types of optical bistabilities, of self–oscillations in (hybrid) ringresonators including deterministic chaotic behaviour. Applying additionally a dc electric field allows to build electrooptic bistable devices and modulators. They are model systems for possible future applications in digital optical data handling.

I. INTRODUCTION

The nonlinear optical properties of matter gained increasing interest in the last one or two decades both from the aspect of fundamental and applied research. This trend is underlined by an increasing number of national and international conferences and summer schools and special issues of journals devoted to this topic. Some of them are mentioned in [1–17].

Optical "nonlinearity" means all reversible changes of the optical properties like the spectra of transmission $T(\omega)$, reflection $R(\omega)$ or luminescence $L(\omega)$ induced by sufficiently strong illumination. In some cases dc electric fields are involved to create or to enhance the optical nonlinearity leading thus to nonlinear electrooptic devices.

The optical nonlinearities are either due to dense real or virtual excitations in the electronic system of matter leading to the so–called photo–electronic optical nonlinearities. In the other group, the absorbed part of the incident light field results via nonradiative recombination in a heating of the sample, which in turn gives rise to changes of the optical properties. This group of phenomena is known as photo–thermal optical nonlinearities. In this contribution we shall treat exclusively photo–thermal (electro–) optical nonlinearities.

The materials investigated concerning their nonlinear optical properties reach from gases and vapours over liquids to the solid state, including inorganic insulators and semiconductors or organic materials, with strong emphasis on semiconductors. We shall concentrate here consequently on the latter group of materials. In order to keep things simple, we even select only one semiconductor namely CdS, though this is not the only material investigated with respect to

Advances in Nonradiative Processes in Solids
Edited by B. Di Bartolo, Plenum Press, New York, 1991

photo–thermal optical nonlinearities. However, CdS is the most extensively studied one, and there are to our knowledge only few phenomena connected with photo–thermal optical nonlinearities, which have not been realized also or often exclusively in CdS.

II. PHOTOTHERMAL OPTICAL NONLINEARITIES

In this section we treat first the linear optical properties of CdS and their temperature dependence. In the second part we show how these dependences can be used to get photothermal optical nonlinearities.

II.A The temperature dependence of the absorption edge

The optical properties in the vicinity of the band–gap of semiconductors are determined by two–particle states, because an electron and a hole are involved in all transitions as indicated in connection with Fig 1a.

In absorption (recombination), a negatively charged electron in the conduction band and a positively charged hole in the valence band are generated (annihilated) simultaneously. These two particles interact due to their attractive Coulomb potential, forming a series of positronium–like pair states below the forbidden gap, called excitons. These exciton states are shown in Fig 1b. The optical properties of excitons have been reviewed already many times (see e.g. [18–21]) including previous schools of this series [20, 21]. So we shall be rather short in the description of their properties and refer the reader for more details to the above mentioned literature and references given therein. The binding energy of excitons is reduced from the value for positronium of 6.8 eV to values in the range from 1 to 200 meV depending on the dielectric constant and the reduced masses of the corresponding semiconductors. The Bohr radius of excitons varies correspondingly in semiconductors from about 0.8 nm to 20 nm and is thus considerably larger than the lattice constant (Wannier excitons). If the band–to–band transition is dipole–allowed, the exciton states with S type envelope function are strongly coupling to the electromagnetic field, forming a mixed state of electromagnetic field and the excitation in the electronic system, the quanta of which are called exciton–polaritons. The polaritons determine the optical properties of the solid state [21]. For semiconductors close to the absorption edge the relevant quasiparticles are the exciton polaritons mentioned above. For our purposes it is sufficient to adopt a more simplified point of view. Since photons have a rather small wavevector (= 10^5 cm^{-1}) as compared to the full width of the first Brillonin zone (= 10^8 cm^{-1}) we assume that exciton states with wavevector $\vec{K} = 0$ are only coupling to the radiation field. As a consequence, we expect a series of resonances in the absorption spectrum as shown schematically in Fig 1c. The oscillator strength decreases with increasing Bohr quantum number n_B as n_B^{-3}, but their mutual distance decreases. In the continuum states there is still some residual electron–hole pair correlation, leading to a rather constant absorption coefficient (Sommerfeld–factor) as compared to the square–root dependence $\alpha \simeq (\hbar\omega - E_g)^{1/2}$ expected from the density of states in an effective mass approximation for the model of Fig 1a which neglects the Coulomb–interaction between electron and hole. This square–root dependence is indicated in Fig 1c by a dotted line. Every resonance in the absorption spectrum is connected with a corresponding resonance like strukture in the real part of the refractive index $n(\omega)$ or in the reflection spectrum $R(\omega)$. With increasing lattice temperature the exciton resonances broaden due to exciton–phonon interaction as indicated schematically by the dashed line in Fig 1c. The exciton resonances may become even so broad that they do not show up any longer as individual structures, but the Coulomb–interaction between electron and hole with its influence on the oscillator–strength is still there. The degree to which the exciton resonance broadens depends on the ratios of exciton–binding energy E_x^b with the thermal energy $k_B T_L$ with the longitudinal optical phonon energies $\hbar\omega_{LO}$ and on the coupling strength α' of phonons with electrons and holes.

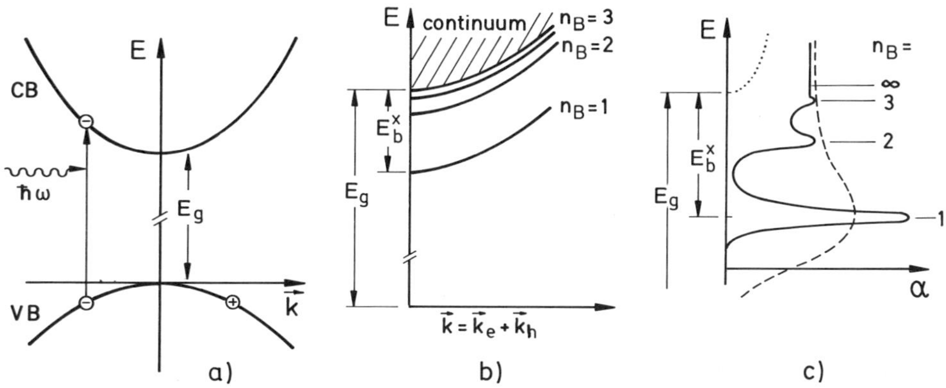

Fig 1 Schematic drawing of the bandstructure of an idealized, direct gap semiconductor (a), the corresponding dispersion relation of excitons (b) and the resulting absorption spectra for low (——) and high (– – –) temperatures and the squareroot dependence expected for simple band–to–band transitions (•••••) (c).

In the spectral range below the exciton resonance, the dependence of the absorption coefficient $\alpha(\hbar\omega, T_L)$ on photon–energy and temperature is described by the Urbach–Martienssen rule [22, 23]

$$\alpha(\hbar\omega, T_L) = \alpha_0 \exp\{-\sigma(T_L)(E_0-\hbar\omega)(k_B T_L)^{-1}\} \tag{1}$$

with

$$\sigma(T_L) = 2\,\sigma_0\, k_B T_L\, (\hbar\omega_{LO})^{-1} \tanh(\hbar\omega_{LO}/2k_B T_L) \tag{2}$$

where α_0, σ_0, E_0 and $\hbar\omega_{LO}$ are material parameters, k_B is Boltzmann's constant. E_0 corresponds roughly to the energy of the lowest free exciton at low temperatures. The parameter values for CdS used for the calculated curves in Fig 2 are $\alpha_0 = 8 \cdot 10^9$ cm^{-1}, $\hbar\omega_{LO} = 24$ meV; $\sigma_0 = 2.222$, $E_0 = 2.6275$ eV (for $\vec{E}\|\hat{c}$) [24]. CdS is a dichroitic material. For the polarisation $\vec{E}\perp\hat{c}$ the value of E_0 is somewhat lower. We use here the polarization $\vec{E}\|\hat{c}$ if not stated otherwise.

In Fig 2a we show calculated curves $\alpha(\hbar\omega)$ for various temperatures in a semi–logarithmic plot. The experimentally observed data follow the calculated ones over several orders of magnitude as indicated schematically by the dashed line but not up to α_0. Instead the measured curves bend over to form an exciton resonance peak. The heavy solid line in Fig 2b gives the transmission at 2.41 eV calculated for a 5 μm thick sample as a function of temperature neglecting reflection.

Though the Urbach–Martienssen rule is of rather general nature and can be observed in the absorption edges of crystalline semiconductors and insulators and even in amorphous materials, the theoretical description of it has been (and partly still is) a point of discussion over many years, since the simple models do not give agreement with experiment. A lifetime–broadening of the exciton resonance due to a decrease of the phase–relaxation time τ_2 induced by the scattering with phonons would result in a Lorentz–type line–shape, not in an exponential dependence. Simultaneous absorption of an incident photon and a (longitudinal) optical phonon would reproduce the square–root dependence of the exciton density of states, starting at $E_x-\hbar\omega_L$ because of the rather flat dispersion of optical phonons.

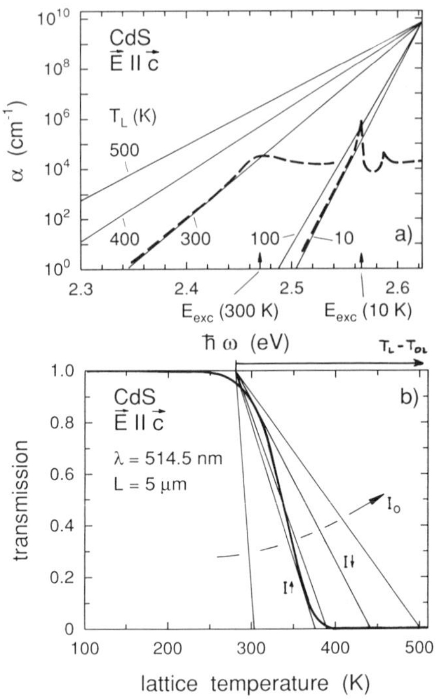

Fig 2 Calculated absorption spectrum of CdS for $\vec{E}\|\vec{c}$ for various lattice temperatures T_L. (——) and schematic drawing of the experimentally observed ones (– – –) (a), the transmission as function of T_L at $\lambda = 514.5$ nm (2.41 eV) (——) and various straight lines to construct the hysteresis loop of induced absorptive optical bistability (b).

The two theoretical models, which are considered as the most relevant ones, will be shortly outlined below. A phonon (optic or acoustic) propagating through the lattice of a semiconductor, modulates periodically the lattice constant and eventually produces a periodic electric field. Both effects influence the width of the forbidden gap e.g. via the deformation potential. At finite temperatures thermally excited phonons propagate randomly through the sample in all directions with average occupation probabilities of the various phonon modes given by the Bose–Einstein statistics. These random phonons result in random, temporarily variing fluctuations of the width of the forbidden gap. According to Sumi and Toyozawa [32], excitons are momentarily localized in this fluctuating, disordered potential relief. On the other hand, it is well known that disordered structures tend to produce exponential tails in the density of states and correspondingly exponential tails in the absorption spectra [26–31]. So, the Sumi–Toyozawa model attributes the Urbach–Martienssen rule to momentary localization of excitons caused by the random fluctuating potential produced by thermally excited phonons.

The other model by Dow and Redfield [25] is in principle based on the influence of electric fields on the absorption edge. It is well known, that a dc electric field applied to a semiconductor tilts the bands and leads to some tunneling of electron and hole wave–function into the forbidden gap. Consequently absorption becomes possible at lower photon energies exhibiting an exponential low energy tail in the absorption spectrum. This is the Franz–Keldysh effect. With respect to the Urbach–Martienssen rule, the electric field is produced by thermally excited optical phonons (or by acoustic ones via the piczo–electric effect) and the Urbach–Martienssen rule is correspondingly attributed to field ionisation or tunneling of excitons.

More recent experimental and theoretical investigations indicate that both effects contribute with a weight determined by the material parameters of the semiconductor or insulator under consideration [33].

II.B Photothermal Optical Nonlinearities

If we illuminate a semiconductor in the spectral region of the absorption edge, the first interaction between photons and matter takes place in the electronic system. We excite really or virtually excitons or more generally speaking electron–hole pairs. If created in sufficiently high density, these electron–hole pairs start to interact with each other, giving rise to a very rich variety of photo–electronic optical nonlinearities, which is, however, beyond the scope of this contribution. The reader is referred e.g. to [2, 5, 14, 16, 17, 34, 36] and the references therein. We consider in the following only really excited excitons and/or electron–hole pairs and neglect virtual excitation. These real excitations relax energetically and recombine with their respective time constants. In CdS and similar semiconductors the effective phase–relaxation times τ_2 are usually in the range below 10 ps, depending on the excitation conditions, the intraband relaxation–times τ_i are usually longer and the interband recombination–times τ are in the range 100 ps $\lesssim \tau \lesssim$ 10 ns. It has been found that the nonradiative recombination is in all semiconductors, including the direct gap ones, generally the dominant process. The yield η_r for radiative recombination is (except for some selected semiconductor lasers) usually considerably below 0.5 and often only around 10^{-2} or 10^{-3}. The absorbed incident optical energy is consequently transformed into other forms of energy which is the definition of absorption of light in contrast to extinction. This other form of energy is usual heat, i.e. an increase of the occupation probability of phonons. There might be an intermediate step like in Auger–type processes, where the energy, which is given away by a recombining electron–hole pair, is first transferred to another carrier as kinetic energy, which, however, is again converted to heat by intraband relaxation. The increase of the lattice temperature, which is a consequence of this heat–production, depends on the material via the specific heat and heat conduction and on the excitation conditions like pulse duration, size of illuminated volume, heat sinking etc. If the rise in sample temperature ΔT_L caused by the absorbed light is sufficiently large ($\Delta T_L \gtrsim 5$K or $\Delta T_L/T_L \gtrsim 10^{-2}$), the ΔT_L results in noticable changes of the optical

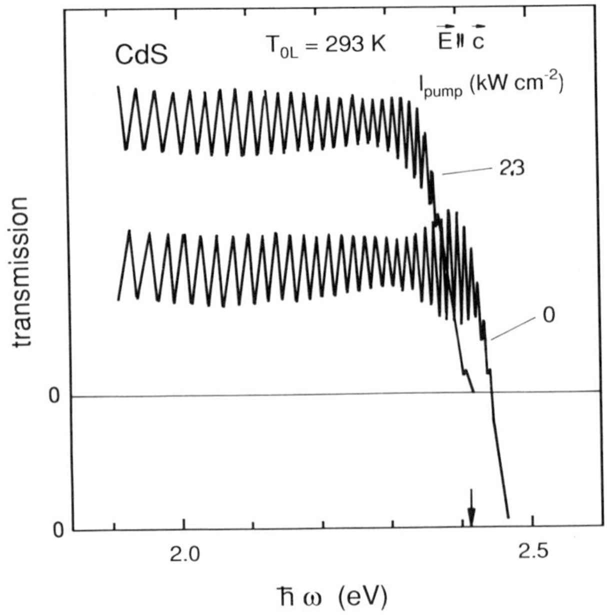

Fig 3 The transmission spectra of CdS at room–temperature without excitation and with $I_{exc} = 2.3$ kW cm^{-2}. The arrow gives the spectral position of the green Ar^+ laser–line at 514.5 nm.

properties. Below the exciton energy there is an increase of the absorption coefficient α ($\hbar\omega$, T_L) or a decrease of transmission as seen in Fig 2. In Fig 3 we present experimental data for CdS. The transmission spectra are taken with a weak, spectrally broad probe beam once without and once with illumination of the sample with the green line of the Ar^+ Laser (514,5 nm $\hat{=}$ 2,408 eV). In the transparent part, T ($\hbar\omega$) is modulated by Fabry–Perot (FP) modes caused by the as–grown, plan–parallel surfaces of the platelet type sample with a thickness of a few μm. It is obvious, that the illumination of the sample with the Ar^+laser leads to a red shift of the absorption edge, which gets also slightly flatter, in agreement with what we expect from eq (1).

From Fig 3 it becomes also obvious, that the combination of CdS and the green Ar^+ laser line forms an ideally matched pair for photothermal optical nonlinearities at room–temperature and for the polarization $\vec{E}\|\vec{c}$: the transmission at 514,5 nm decreases from values around T = 0.9 with increasing excitation intensity to values $T \lesssim 0.1$., i.e. we have a strong excitation induced increase of absorption at this wavelength. Furthermore, we observe a slight red–shift of the FP–modes in the transparent spectral region. This indicates an increase of the refractive index connected with the red–shift of the absorption edge due to Kramers–Kronig relations. This means we have two coupled effects for the coefficients of absorption α and diffraction n or for the complex index of diffraction $\tilde{n} = n + i\kappa$ with $\alpha = 2\,\omega\kappa(\omega)c^{-1}$: they depend on temperature and the increase of temperature above the surrounding– or bath temperature T_{0L} depends via nonradiative recombination on the excitation conditions summarized schematically in the following eq. (3) by the excitation intensity I_{exc}.

$$n = n(\omega, T_L);\ \alpha = \alpha(\omega, T_L);\ \kappa = \kappa(\omega, T_L) \tag{3}$$

with

$$T_L - T_{0L} = f\,(I_{exc}).$$

An excitation induced bleaching of absorption, as it occurs frequently in photoelectronic optical nonlinearities [37, 38], is rather rare in the photothermal case. It could occur directly in the resonances of free or bound excitons or for materials, where the exciton–energy increases with increasing temperature like in CuCl [39, 40], but we shall not follow this aspect further.

III. OPTICAL BISTABILITY

To summarize the preceeding section, we may state that photothermal optical nonlinearities generally lead to an increase of the absorption and of the refractive index under strong illumination in the spectral region below the exciton resonance. In this section, we describe first the principle of optical bistability and then various realizations due to the above mentioned photo–thermal optical nonlinearities.

III. A Principles of Optical Bistability

An optically bistable device has two different, reversible states of transmission and/or reflection for the same input parameters. One or the other of these states occurs, if one approaches the bistable region from low or high incident intensities. In other words, one has a stationary hysteresis loop in a plot of transmitted versus incident intensity. This is shown schematically in Fig 4.

In Fig 4a the sample is in a highly transmitting state if one starts for low values of the incident intensity I_0 and switches at a certain value $I\downarrow$ to a state of low transmission. If one reduces I_0, the system remains over a certain range in this low transmitting state and switches back only at a lower value $I\uparrow$. The difference between $I\downarrow$ and $I\uparrow$ defines the bistable range. The hysteresis–loop is revolved clockwise. Figure 4b shows a device with exactly the opposite behaviour. The hysteresis loop is revolved counter–clockwise. If there are more than two stable states over a certain range of I_0, one speaks about multistability.

A necessary but not sufficient precondition for optical bistability (ob) is an optical nonlinearity connected with a suitable feedback.

The optical nonlinearity can be an excitation induced increase of absorption (induced absorptive ob), a chance of the refractive index (dispersive ob) or a bleaching of absorption. The first mechanism leads to hysteresis loops of the type shown in Fig 4a, the second and third one to the one of Fig 4b. Induced absorptive ob has a built–in feedback, dispersive and bleaching ob need the feedback from an optical resonator, usually a Fabry Perot (FP) resonator.

We discuss in the following induced absorptive and dispersive ob.

In the case of absorptive ob, only a small fraction of I_0 is absorbed for small values of I_0. If, with increasing I_0, this small fraction is sufficient to increase α (in our case by an increase of temperature), a positive feedback sets in since an increasing α leads to an increase of $T_L - T_{0L}$ and so on. If the characteristics $T(T_L)$ is sufficiently steep, the system may switch in a state of low transmission T (i.e. high absorption A). Now almost all the incident light is absorbed and consequently I_0 can be lowered and the system remains on the lower branch until the switch back occurs for $I_0 = I\uparrow < I\downarrow$ resulting in a loop like Fig 4a. This type of induced absorptive ob has been found independently by various groups for photo–electronic, photo–thermal and electrooptic systems [38, 41, 42]. In the following we present a more quantitative approach according to [41]. We neglect here for simplicity reflection losses of the surfaces and

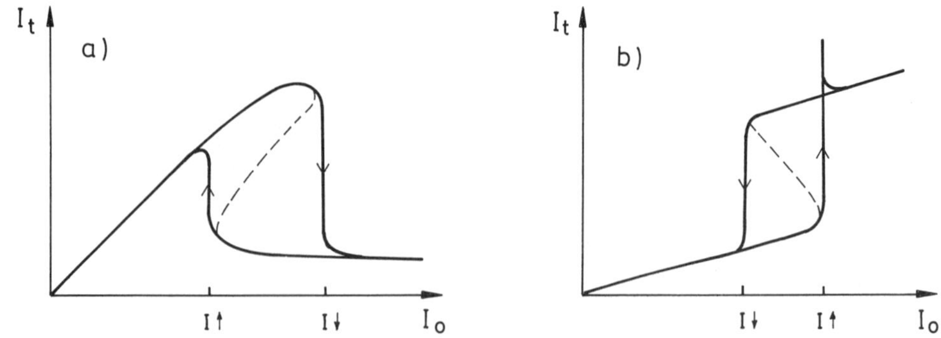

Fig 4 Schematic drawing of the input–output hysteresis–loops characteristic for optical bistability. The dashed lines give the third, unstable, usually not observable solution.

have to fulfill simultaneously two equations for the transmission T or the absorption A with $T(T_L) = 1 - A(T_L)$.

First eq (4)

$$T = \exp\{-\alpha(\hbar\omega, T_L)\,L\} = I_t/I_0 \tag{4}$$

holds. This relation can be measured by putting the sample in an oven, or calculated (Fig 2b) if the material parameters are known. For the other relation we assume that the increase of T_L is proportional to the absorbed intensity AI_0 with a proportionality factor C which contains the specific heat c', the sample thickness L, the heat conductivity and the cooling conditions described by a thermal relaxation time τ

$$\Delta T_L = T_L - T_{0L} = CA(T_L)I_0 = CI_0(1 - T(T_L)) \tag{5}$$

and

$$C = \tau(c'L)^{-1} \tag{6}$$

For homogenous laser excitation in a cylindrical part of the platelet type sample with radius r, only radial heatflow and a temperature bath T_{0L} at a distance R_0, τ reads

$$\tau = \frac{r^2}{4\kappa}\left(1 + 2\ln\frac{R_0}{r}\right) \tag{7}$$

Rearranging eq (5) leads to

$$T = 1 - \frac{\Delta T_L}{I_0 C} \tag{8}$$

This is a straight line with a negative slope proportional to I_0^{-1}. The solutions of eq (4) and (8) are the intersections of both curves (Fig 2b). If we start at low I_0, i.e. a steep slope, we get only one solution. With increasing I_0 a second solution appeares for $I = I\uparrow$ which splits for further increasing I_0 into two. The system remains, however, in the highly transmitting state until this solution disappeares for $I_0 = I\downarrow$. Then the system jumps to the solution on the lower branch, where it remains for further increasing I_0 and for decreasing one until $I\uparrow$. In Fig 4a we see all three solutions. It can be shown that the middle, dashed one is unstable, and can be observed experimentally only under special conditions as will be demonstrated later. It is obvious from the construction of Fig 2b that a certain steepness of $A(T_L)$ is necessary to obtain three solutions and thus optical bistability. The appearance of ob and the width of the loop can be also influenced strongly by the surrounding temperature T_{0L}. A decrease of T_{0L} increases the width of the loop $I\downarrow - I\uparrow$ and leads to higher temperatures on the low transmitting branch, which may eventually cause a thermal destruction of the device. An increase of T_{0L} leads to a decreasing width and finally to a strongly nonlinear but only single valued relation between I_t and I_0.

For dispersive ob, one starts in a transmission minimum of the FP resonator, i.e. preferentially destructive interference of the partial waves reflected from the front and rear partly transmitting mirrors of the FP. Consequently the intracavity intensity IFP and the transmitted one I_t are low. If IFP becomes with increasing I_0 sufficient to change the refractive index n of the material in the FP, the interference becomes more constructive, IFP increases, n too, and so forth. This positive feedback may result in an abrupt switching of T from a value close to the transmission minimum, to a highly transmitting state beyond the next transmission maximum at a value $I_0 = I\uparrow$. Since we have now a high IFP and consequently a large change of n, I_0 can be lowered below $I\uparrow$ and the system still remains in the highly transmitting state down to $I\downarrow < I\uparrow$. Note the inversion of $I\downarrow$ and $I\uparrow$ in Fig 4a and b.

One of the two equations, which we have to satisfy here simultaneously, is the Airy–function of the transmission of the FP.

It reads

$$T=(1-R)^2\{[\exp(\tfrac{\alpha L}{2})-R\exp(-\tfrac{\alpha L}{2})]^2+4R\sin^2\delta\}^{-1}=I_t/I_0 \qquad (9)$$

where R is the intensity reflection coefficient of one mirror. We assume for the moment that α is small and does not change much with the intracavity intensity IFP. Finally δ is the phaseshift of one half round trip in the FP. Eq. (9) is schematically shown in Fig 5.

For dispersive ob δ depends on the excitation conditions:

$$\delta = \delta_0 + \Delta\delta \qquad (10)$$

with

$$\delta_0 = n_0 \omega L c^{-1}$$

By choosing the material (n_0), the frequency ω and the thickness L of the FP, we define the initial phase δ_0. The weak absorption of IFP caused by α results in a sample heating and thus in a change of n and of δ. We assume a linear relation between Δn and ΔT_L for constant α and find

$$\Delta\delta = \Delta n\,(\Delta T_L)\,\omega L c^{-1} \sim \Delta T_L \sim \mathrm{IFP} \qquad (11)$$

The second equation is

$$T = \mathrm{IFP}\,\left(\frac{1-R}{1+R}\right)\frac{1}{I_0} \qquad (12)$$

We see that (12) describes again a straight line,this time with a positive slope decreasing with increasing I_0. The construction of the hysteresis loop and the appearance of the third unstable branch are analogous to the case of induced absorptive ob above. Here it is obvious that several dispersive loops may appear with increasing I_0. Multistability results in case of overlapping dispersive loops. It is also clear from Fig 5 that starting not in a transmission minimum by a suitable shift of δ_0 may significantly alter the shape of the $I_t = f\,(I_0)$ relation.

III. B Absorptive Optical Bistability

Room–temperature photothermal induced absorptive optical bistability has been observed for the first time in CdS [43] and has been investigated later on by various research groups [44–48].

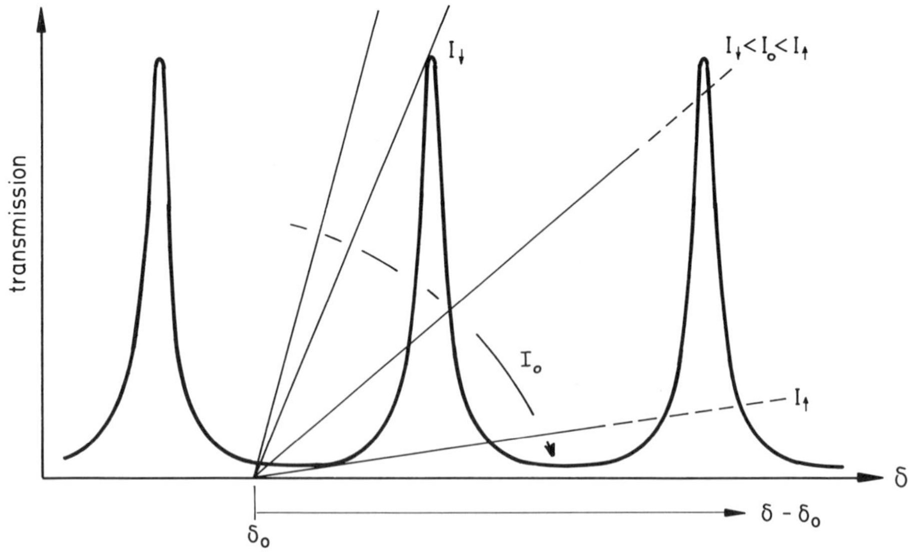

Fig 5 Schematic drawing of the transmission of a Fabry Perot Resonator as a function of the phase–shift δ (———). The straight lines help to construct the loop of dispersive optical bistability.

The experimental observation is rather simple. By some type of modulator, a smooth { (e.g. $I_0(t) = I_0 \sin^2 (2\pi t \tau_p^{-1})$ } pulse of duration τ_p, is cut out of the cw Ar+laser beam. With two photo–diodes one measures the temporal evolution of the incident and transmitted pulse $I_0(t)$ and $I_t(t)$, respectively. The time free plot $I_t = f(I_0)$ yields then the hysteresis loop in case of ob. We present in Fig 6 a measured hysteresis loops for various pulse duration τ_p. For the longest pulses (τ_p = 5s) one follows the stationary bistable hysteresis loop. With decreasing τ_p a dynamical blowing–up of the loop is found. This phenomenon is connected with the fact that a certain amount of energy must be deposited in the sample to heat it sufficiently to come from the state of low absorption to the one of high absorption. This process needs a finite time. If $I_0(t)$ changes significantly during this time, a dynamical overshooting of the switching intensities $I\downarrow$ (and of $I\uparrow$ for the switch back) occurs. For constant I_{max} this effect is more pronounced with decreasing τ_p.

Fig 6b shows a set of calculated hysteresis loops. They look very similar to the experimental ones, except for the fact that the given time– and intensity regimes deviate by several orders of magnitude. This is due to the fact that the curves have been calculated for a completely different optical nonlinearity, namely increasing absorption connected with the formation of an electron–hole plasma in CdS, i.e. a photo–electronic optical nonlinearity. For details of this phenomenon see e.g. [38, 49]. The fact that the curves look almost identical apart from scaling factors for time and intensity demonstrates that the dynamical behaviour of the hysteresis–loop is a universal phenomenon for this group of optical bistabilities.

Optical bistability is a first order phase transition in a driven system, as can be seen e.g. from the appearance of hysteresis [50]. In contrast to e.g. the gas–liquid transition of a real or van der Waals gas, the transition from one phase to the other does not follow a Maxwell–construction through the coexistence region. The transitions start from the points which separate the unstable from the metastable

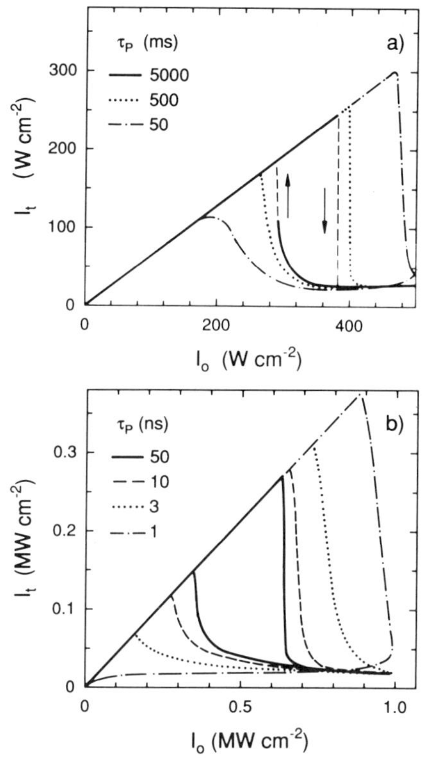

Fig 6 Measured hysteresis loops for photothermal induced absorptive optical bistability for various pulse lengths τ_p (a), calculated induced absorptive hysteresis loops due to the formation of an electron–hole plasma (b). From [38, 44, 49].

regime [50]. Consequently critical slowing down has to be expected in the transition from one branch to the other. This phenomenon has been investigated for the first time for induced absorptive ob independently by [44, 46] and subsequently by other authors [48]. The method is the following. $I_0(t)$ is switched on at $t = 0$ in a step–like function from a value below $I\downarrow$ to a value slightly above by a relative amount $\delta = (I_0 - I\downarrow)/I\downarrow$. Fig 7a shows an example for $I_0(t) = 0$ for $t < 0$. After a short transient feature, the system remains for a certain time τ_s on the upper branch until the switching sets in. This delay time τ_s diverges for $\delta \rightarrow 0$, while the actual switching time is rather independent of δ.

Fig 7b shows τ_s as a function of δ and a calculated curve exhibiting the logarithmic singularity expected for critical slowing down. The agreement between experiment and theory is good, but it must be stated that the absolute values of the calculated curve have to be scaled by a thermal relaxation–time constant τ which is almost two orders of magnitude larger than the one given by eq. (7). Two attempts have been made until now to explain this discrepancy. In one case the heat–flow in the sample during the switching process has been taken explicitely into account [51], in the other case the dependence of the absorption A (T,t) [48]. Recently the critical slowing down has been investigated as a function of the incident intensity for $t<0$ [52]. The delay depends only weakly on this quantity. The influence of heat sinking by attaching the samples to a glass or saphire substrate can change the delay–times by an order of magnitude [52]. If noise is applied to the input signal $I_0(t)$, the hysteresis loops get narrower if the frequency spectrum of the noise has sufficiently strong components below the inverse thermal relaxation time τ [52]. For a theoretical investigation of the influence of noise see [50].

Another challenge in ob is to measure the unstable branch excplicitely. The first successful attempt for induced absorptive ob has been reported in [44]. The bistable element has been placed in a hybrid ringresonator as the one described in section IV, however, with a short round trip time. Measuring the intensity transmitted through the sample, as a function of the intensity in the resonator in front of the sample under variation of the intensity sent into the ringresonator gives the unstable branch.

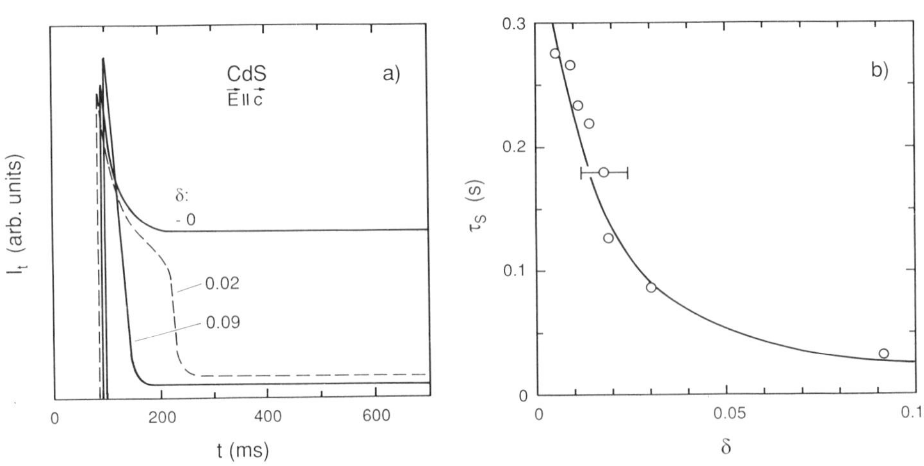

Fig 7 The transmitted intensity as a function of time, when the incident intensity I_0 is switched on from zero in a steplike function to $I\downarrow (1 + \delta)$ (a), the delay between the step in I_0 and the switching process τ_s as a function of δ; o experiment, theory (b). From [44].

Another technique, which has been used successfully both for dispersive [53] and induced absorptive ob [52, 54], is the following. The system is kept by a holding beam I_h on the upper branch in the bistable regime of induced absorptive ob. See Fig 8a. Then an additional pulse of a duration short as compared to the thermal relaxation time is applied to heat the sample further. This pulse can be e.g. a UV–N_2–laser pulse [54] or a pulse produced by a modulator in the Ar+ laser beam [52]. Fig. 8b.

This pulse heats the sample instantaneously in the illuminated area, consequently the transmission decreases. After some thermal relaxation time, a temperature profile and a transmission builds up, which are situated above or below the unstable branch. The unstable branch is the separatrix which separates the basins of attraction for the upper and lower branch, respectively.

If the energy of the additional pulse is properly chosen, the system can be prepared on the unstable branch which is actually a metastable state. Due to finite fluctuations of laser and due to the fact that a mathematical point on the unstable branch can be reached experimentally only with a relative precission $0<|\epsilon|<<1$, the system decays after some time either to the upper or lower stable branch as seen in Fig 8b. This time increases with decreasing $|\epsilon|$, indicating thus another type of slowing down. In the experiments described by [52], the longest time the system stood on the unstable branch was 40 ms. The point on the unstable branch is determined by the plateau in Fig 8b. A variation of the working point I_h allows to determine the whole unstable branch. A result of such a measurement is shown in Fig 8c. The stable

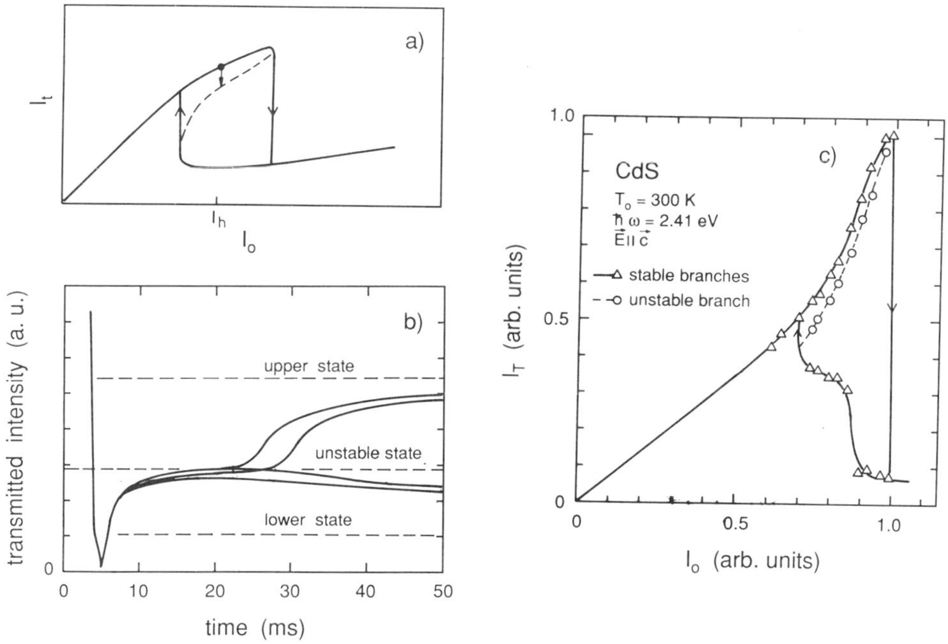

Fig 8 Principle of how to measure the unstable branch in optical bistability (a), temporal evolution of the transmission after application of a short additional pulse (b) measured unstable branch (c). From [52, 54].

branches have been determined by just measuring the stationary hysteresis loop. The fact that the relation $I_t = f(I_0)$ on the upper branch is slightly superlinear can be attributed to weak influences of a FP mode structure since the platelet type samples have more or less plan–parallel, as grown surfaces. The unstable branch, determined by the method outlined above, lies close to the upper branch. This occurs in induced absorptive ob, depending on sample thickness and steepness of the increase of absorption and is known also for photo–electronic optical nonlinearities (see e.g. Fig 72 in [50]).

Finally it should be mentioned that induced absorptive ob may occur due to its build–in feedback mechanism also repeatedly because of partial switching into the depth of the sample if the (heat–)diffusion is sufficiently small. For a theoretical description and experimental observations of this phenomenon, the reader is referred e.g. to [38, 50, 56, 57]. Lateral structures in the switching process have been described in [58, 59]. Induced absorptive ob in other materials or due to other effects are reviewed e.g. in [1, 3, 6, 7, 10, 11, 13, 14, 17].

III.C Dispersive and mixed types of optical bistability

Dispersive optical bistability due to photo–thermal optical nonlinearities has been investigated preferentially in interference filters [3, 7, 10, 13]. They consist of two evaporated stacks of quarter–wave layers of alternating high and low refractive index on both sides of a $m\lambda/2$ thick, usually also evaporated layer of the nonlinear material, which is often ZnSe or ZnS. The whole structure is produced on a glass substrate. Here it is desirable, that only a small fraction of the light is absorbed in this layer, both in the on and in the off state, in order not to deteriorate the finess of the FP resonator. As a consequence of this idea, the so called BEAT (bistable etalon by absorbed transmission) device has been developed [60]. In this case, the spacer layer in the interference filter is completely transparent, and the heating comes from a thin absorbing layer on the rear (output) side of the structure.

In order to be consistent with our previous sections, we present here again results obtained with CdS single crystal platelets.

In order to make dispersive ob more easily visible, the plan–parallel, as grown surfaces of the platelet type samples have been covered with dielectric reflecting coatings of an intensity reflectivity of about 0.6 as compared to 0.24 for an uncoated surface. The improvement of the finess by this coating can be clearly seen in the transmission spectra shown in Fig. 5 of [49].

In Fig 9a we show again the time–free relation between input and output intensities deduced from a measurement of $I_0(t)$ and $I_t(t)$ under stationary conditions. If we consider for the moment only the range of input intensities below 5 a.u. we find experimentally exactly the behaviour expected from the calculation of eq (9) to (12) when the initial phaseshift δ_0 in the FP is at or close to a transmission minimum. The spike at 4 a.u. in the switch up process is a transient feature when the system switches from a state close to the transmission minimum to a new state on the other side of the next transmission maximum. If we reduce now the intensity from 5 a.u., the system remains in the bistable regime on the upper branch and switches back from the transmission maximum of the FP resonator, which is indicated by a dashed line. This hysteresis loop is revolved counter–clockwise. If we increase I_0 beyond 5 a.u. we find a second hysteresis loop, which is revolved clockwise. This is the induced absorptive ob discussed before. Now the temperature dependence of α, which we neglected above, becomes important again. The sample switches at $I_0 = 6.5$ a.u. in the state of low transmission. The increase of absorption connected with this transition significantly reduces the influence of the FP resonator according to (9). When the incident intensity is reduced, starting from values around 7 a.u., we see the induced absorptive switch back at 5 a.u.. Now the FP acts again and indeed the system is still or again on the upper branch of dispersive ob and follows the route discussed just above.

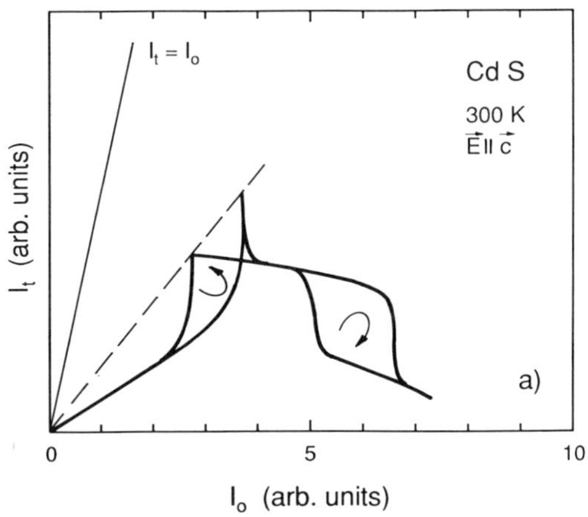

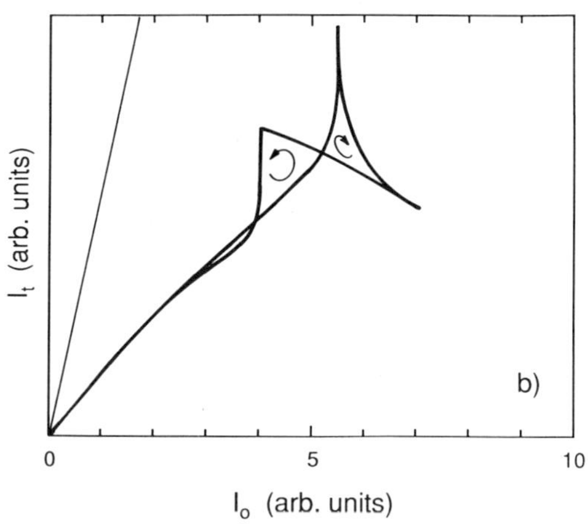

Fig 9 Mixed dispersive, induced absorptive optically bistable hysteresis loops for different original detuning. From [44].

Depending on the starting conditions, i.e. the phaseshift δ_0, the sample thickness etc., very different types of hysteresis loops can be obtained. In Fig 9b we show a butterfly loop where the absorptive and dispersive loops are mixed in a way that the system shows only two switch down processes. More information of this mixed type of absorptive and dispersive ob can be found concerning theory [44, 61] and experiment [44].

More recently another group was even able to observe photo–thermal dispersive ob and mixed absorptive and dispersive loops in samples with uncoated, as–grown surfaces [48]. In this case it is crucial to select the samples very carefully concerning their thickness and especially the plan–parallelity of the surfaces. An example for dispersive and mixed type of ob using photo–electronic nonlinearities in CdS is given in [38].

IV. SELF–OSCILLATIONS

In this section we want to investigate the scenarios which can be observed and calculated, if we bring an induced absorptive bistable or monostable device in a (hybrid) ringresonator. These investigations complement the ones for the Ikeda resonator, which is basically a dispersive nonlinear element in an optical ringresonator [62].

The system of an induced absorber in a FP or ringresonator has been treated first theoretically by [63]. The experiments for a hybrid ringresonator and further theoretical modelling has been done in much detail by M. Wegener [44, 45]. The results which we present here can be found in more detail in these references. First experimental findings concerning a purely optical resonator are given in [52].

As has been explained earlier, an excitation induced increase of absorption may lead to induced absorptive ob or only to a strongly nonlinear, single valued relation between $I_t = f(I_0)$, depending on the steepness of the increase of absorption with excitation and on the initial absorption.

In CdS both cases can be realized at room–temperature and for the green Ar+ laser line, due to the dichroism of the material which has already been mentioned. The transition from the upmost valenceband (called A–band, which has symmetry Γ_9) to the conduction band (Γ_7) is allowed only for a polarization $E \perp c$, while the transitions from the two lower valence–bands of symmetry Γ_7 are allowed both for $E \perp c$ and $E \| c$ [40].

First we assume that we use the polarization $\vec{E} \| \vec{c}$ for the ringresonator, i.e. photo–thermal induced absorptive ob with a hysteris–loop of the type shown in Fig 6 a or in Fig 8.

In Fig 10 we explain schematically what we expect by placing such a device in a ring–resonator, assuming that the round trip time τ_R is long compared to the characteristic switching time τ, which is according to Fig 7b some ms.

Fig 10a shows the ringresonator. The incident intensity I_0 is chosen, so that the part transmitted through the first beam–splitter TI_0 is below the switch–back intensity $I\uparrow$ of the bistable loop (Fig 10c), and the sample is supposed to be in the transparent state at the beginning. This means that the intensity falling on the sample I is almost completely transmitted (we neglect the sample reflectivity for the moment for reasons of simplicity). A small fraction of the transmitted intensity is coupled through the next mirror and detected by the detector D. After one round–trip time, the incident intensity and the transmitted one fall on the sample. Here we assume that we have to add simply the intensities. This approach is justified for the actual experiment as will be shown later. The above mentioned procedure repeats itself several times (Fig. 10d) with steplike increases of the intensity impinging on the sample and being transmitted. After some round trip times I exceeds eventually $I\downarrow$ and the crystal switches in the absorbing state. This means that almost no more light is transmitted. Consequently after another round trip time τ_R , only TI_0 falls on the sample, the sample switches back into the transparent state and the system starts again, giving rise to self oscillations with a period which is an integer multiple of τ_R .

The condition $\tau_R >> \tau$ would necessitate a ringresonator with a delay–time of at least 100 ms, this means approximately once around the globe. Since funds for such an experiment are rather unlikely, a corresponding electrooptic hybrid system was set up. Fig 10b. The incident light passes through an electrooptic modulator (EOM). A constant offset on the high–voltage amplifier (HVA), which drives the EOM models the constant intensity transmitted through the first mirror in Fig 10a. The nonlinearity is really the optical nonlinearity of the CdS sample in the Ar+ laser beam. The transmitted signal is detected (D), monitored by a scope and sent through an electronic delay line of duration τ_R . Then the signal goes back via HVA on the EOM which opens in proportion to the applied signal. The choice of the amplification of the electric signal in the delay–line and/or in the HVA allows to simulate various total reflectivities of the set of four mirrors and beamsplitters in Fig 10a. So the hybrid ringresonator is a rather good simulation of the purely optical one, with two minor exceptions. We loose the phase information of the light in the electronic branch of the hybrid resonator. On the other hand, a purely optical resonator of some 10^4 km length would also significantly surpass the coherence length of most laser systems, so that adding intensities is actually in agreement with the conditions in the optical resonator. The other approximation is that we neglect the formation of lateral structures which may appear due to partial switching of the device [64].

Our system can be described by two equations, which allow to describe all the experimental findings given below at least qualitatively, but in most cases even quantitatively.

$$\frac{d}{dt}\Delta T_L = \frac{\Delta T_L}{\tau} + A(\Delta T_L)\, I(t)\, \frac{1}{c'L} \tag{13}$$

with

$$A(\Delta T_L) = 1-\exp\{-\alpha(\Delta T_L)\, L\}$$

and

$$I(T)=I_0T+R^2I_t(t-\tau_R)=I_0T+R^2I(t-\tau_R)\exp\{-\alpha(\Delta T_L(t-\tau_R))\} \tag{14}$$

where ΔT_L is the increase of temperature above the surrounding temperature T_{0L} at time t

$$\Delta T_L(t) = T_L(t) - T_{0L}$$

c' and L are the heat capacity and the sample thickness, respectively and the other quantities have their usual meanings.

Equation (13) gives the rate equation for the heating and eq (14) is the delay equation of the (hybrid) ringresonator. The absorption coefficient $\alpha(\Delta T_L)$ is given by eq. (1) or deduced from experiment.

Now we want to present some experimental results observed with the above described hybrid ringresonator.

In Fig 11 we show various selfoscillation modes, if we change the input intensity I_0. We use the following nomenclature to describe the various modes: a (n/m) mode has a total period of $n\tau_R$ and m maxima. Correspondingly we see in Fig 11 a (2/1), a (3/1) and a (5/2) mode. The small spikes at the end and/or the beginning of the steps are transient effects and reflect the reaction–time of the crystal. The (3/1) and (2/1) modes are fundamental modes, the (5/2) one belongs to the next generation. According to theoretical predictions and experimental investigations [44, 45, 63], the various generations of modes can be constructed according to a Farey–tree structure i.e. by adding numerators and denominators independently. We expect thus between a (n_1/m_1) and a (n_2/m_2) mode a $(n_1 + n_2/m_1 + m_2)$ mode.

If the stability range of the various modes (n/m) is plotted as a function of I_0, theory predicts a complete devils staircase [44, 63]. This means that the range of I_0 over which an individual mode is stable, decreases with increasing generation as shown schematically in Fig 12. In the experiment the first two or three generations can be seen only, presumably due to slight fluctuations of the laser and the setup. So it is not yet possible to give statements e.g. about the fractional dimension of the complement of the "staircase".

If we reduce the round trip–time, the oscillations become smeared out, the strict locking of the oscillation periods into multiples of τ_R is relaxed, and finally the oscillations disappear and the system remains for constant I_0 in a temporarily constant state, i.e. we have a transition from a limit cycle to a fixpoint. Under these conditions it becomes possible to measure the unstable branch, if I_0 is varied and I_t is plotted as a function of I as indicated already in III. 2. Corresponding figures are given e.g. in [44].

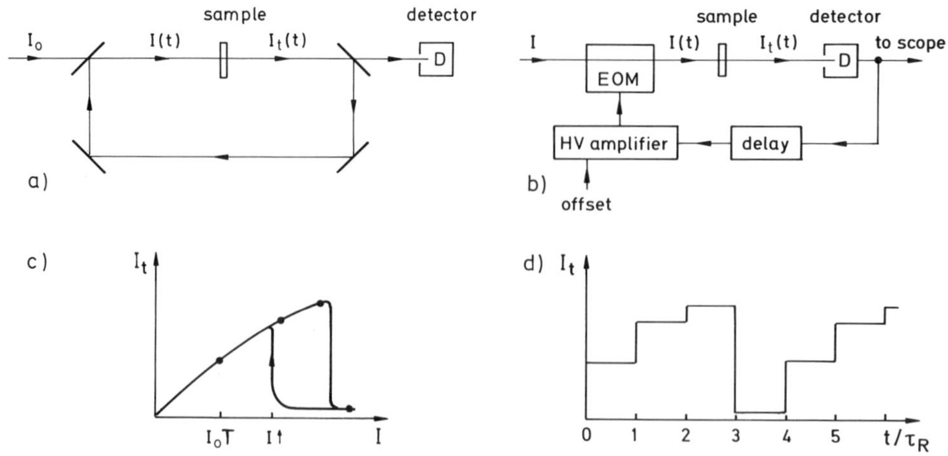

Fig 10 Basic scheme for an optical ring resonator containing an induced absorptive (bistable) sample (a), a realization by a hybrid opto–electronic ringresonator (b), the bistable hysteresis loop of the sample (c) and the temporal evolution of the selfoscillation (d).

If we interrupt the incident beam shortly, we can produce under the same system conditions various oscillation modes. An example is given in Fig 13. A small additional spike is introduced in the system, which has basically the same shape as the temporarily long structures before the perturbation. This coexistence of various modes at the same excitation conditions defines a new type of bi– or multistability. A (n/m) mode with i additional spikes is called a $(n/m)_i$ mode. So Fig. 13 shows a $(3/1)_0$, a $(3/1)_1$ and a $(3/1)_2$ mode. The basic reason that these higher modes appear is the rather large dimensionality of the phase space of our system. For constant I_0 we can choose independently the intensity falling on the sample and its temperature (or absorption coefficient). These are two independent variables, which can be chosen during one round trip τ_R/τ times. With $\tau \approx 50$ ms (see Fig. 76) and $\tau_R = 500$ ms we get a dimensionality $d \approx 2 \cdot \tau_R/\tau \approx 20$.

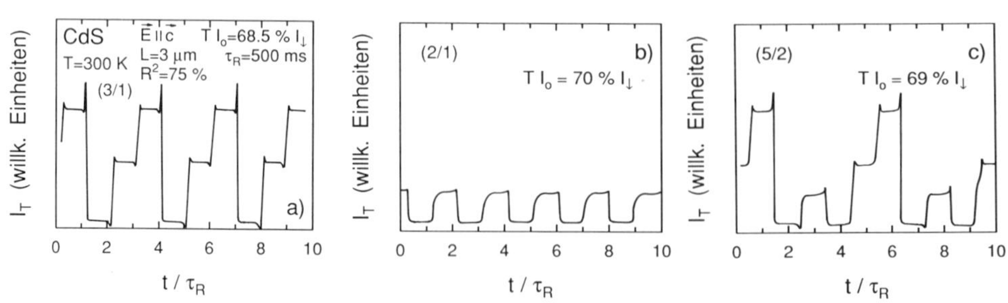

Fig 11 Various oscillation modes of the hybrid ring resonator of Fig 10 b. From [44].

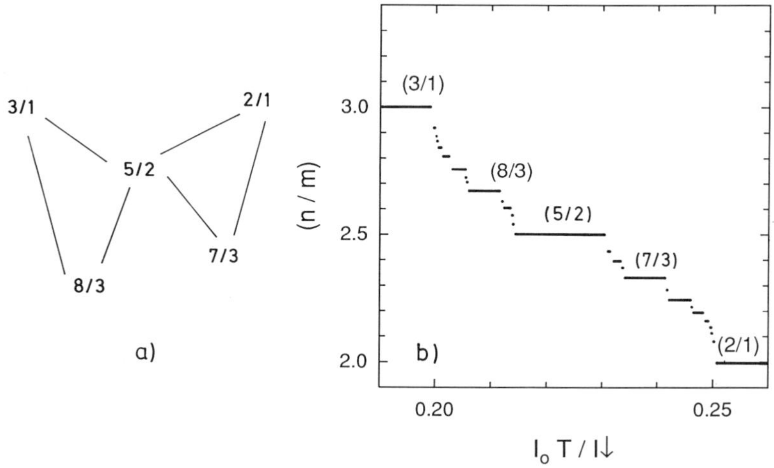

Fig 12 The principle of a Farey–tree (a), the stability range of various (n/m) modes as a function of the normalized input intensity (b). From [44].

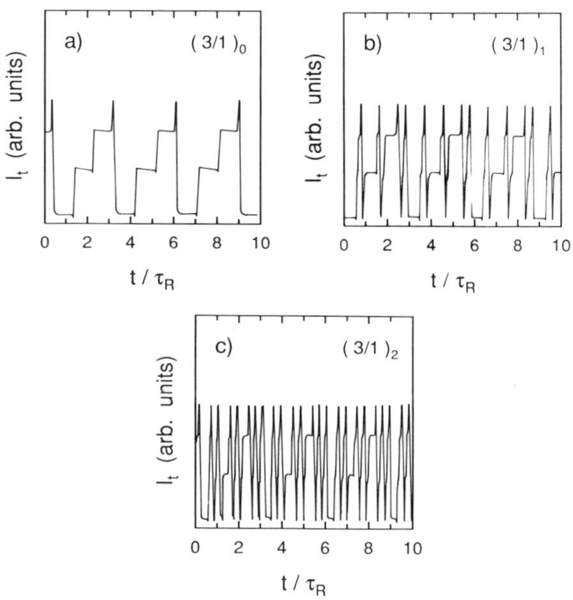

Fig 13 Various coexisting oscillation modes of the hybrid ring resonator. From [44].

If we interrupt I_0 for a time τ_S with $\tau < \tau_S < \tau_R$ during a $(n/m)_0$ oscillation, the crystal switches on the upper branch but no light is transmitted. After one round–trip, the incident intensity during τ_S is only TI_0 and so forth, producing the self–similarity of the additional spikes stated above.

Until now I_0 was kept constant. An external modulation of I_0 with a frequency f as

$$I_0(t) = I_0 \{ 1 + A \cos (2\pi ft) \} \qquad (15)$$

leads to new phenomena [45]. The most interesting regime is the one, where f is comparable to τ_R^{-1}. We find under all conditions periodic oscillations. The oscillation period locks either into integer multiples of τ_R for $A \leq 0.4$ or in multiples of f^{-1} for $A \geq 0.4$. No non–periodic oscillations have been found, even when τ_R and f^{-1} form a rather irrational relation close to the golden mean.

The periodic oscillations show again a Farey–tree structure if we consider $\tilde{T}f$, where $\tilde{T}$ is the total oscillation period devided by the number of various maxima m. The stability range of the various modes if investigated either as a function of τ_R for constant A or as a function of A for constant f yields again a devil's staircase [45].

To conclude this subsection, we want to demonstrate two cases where temporarily irregular oscillations appear.

In the first case, we use the same hybrid ringresonator as above, however, with a monostable optical device, i.e. we simply turn the polarisation from $E \| c$ to $E \perp c$, and get the $I_t = f(I_0)$ characteristic of Fig 14a. If we insert this monostable device in the hybrid ringresonator, we observe for constant I_0 again oscillations, which are for a certain range of I_0 periodic with a period corresponding to integer multiples of τ_R. For another range of I_0 values, we get temporarily irregular oscillations. Actually we see in Fig 14b to d a sequence of period doublings to deterministic chaotic behaviour. For a review of this topic see e.g. [65].

There are two criteria to determine whether a system shows deterministic chaotic behaviour. One is a positive Liapunov exponent λ or a positive Kolmagorov entropy K, which is the sum of the positive Liapunov exponents, which describe the temporal evolution of two initially closely lying points in phase–space. For positive λ the two points separate exponentially with time.

The other indication is the dimensionality of the trajectory of a system in phase space. A fix–point is evidently zero–dimensional, a limit cycle one dimensional etc. There are, however, trajectories with fractal dimension in phase space. In this case the system is said to move on a strange attractor. The fractal dimension can be defined in various ways. We chose here the so–called Haussdorff dimension D. We assume to cover the trajectors with d–dimensional hypercubes of length l i.e. lines for d = 1, squares for d = 2, cubes for d = 3 etc. For a given length l we need a certain number M(l). Decreasing l towards zero increases M(l) and one gets a relation

$$M(l) \approx l^{-D} \qquad (16)$$

with eq (16) we define D for the limit $l \to 0$. It is important to notice that D is independent on the embedding dimension d provided that $D < d$.

If we were able to measure the trajectory of our oscillator in the phase–space, we can try to find the supposedly fractal dimensionality of the strange attractor in Fig 14d. Actually we even do not know the precise dimensionality of the phase–space and we are just able to measure one variable e.g. $I_t(t)$.

Fortunately there is a procedure developed by Grassberger and Procaccia [66] to determine quantities, which are closely related to the Haussdorff dimension, to the Liapunov exponent and to the Kolmogorov entropy K, respectively, namely the

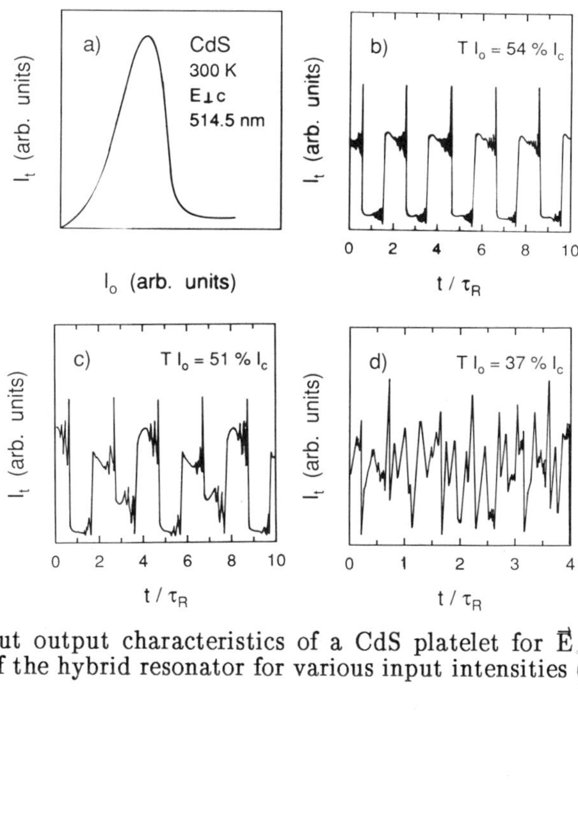

Fig 14 The input output characteristics of a CdS platelet for $\vec{E}.\vec{c}$ (a), oscillation modes of the hybrid resonator for various input intensities (b) to (d).

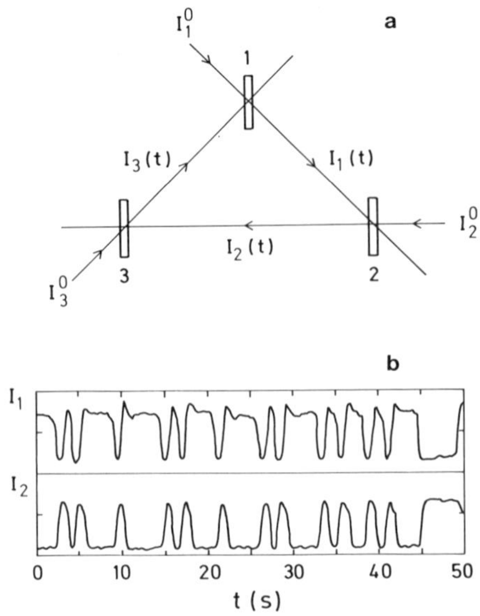

Fig 15 Setup of a purely optical ring resonator (a), temporal oscillations of two of the transmitted beams (b).

V. ELECTROOPTIC DEVICES

As will be discussed also in the last section below, the various types of mono– and bistable optical switches may have possible applications in digital optical data handling, i.e. bits are not stored and handled by electrical current or voltage levels and pulses, respectively, but as low and high light levels.

It soon became clear, however, that photons for optical switching are much more expensive than electrons for electronic switching. Consequently one tried to find some hybrid devices where only a small fraction of the necessary switching energy comes in as photons, the rest is contributed by an electrical battery via electrons. The first type of these elements consisted of multiple quantum wells imbedded as an intrinsic layer in a reversely based pin diode. The optical nonlinearity was the quantum confined Stark effect (QCSE) [41]. These devices can be used as electro–optic modulators, detectors and as bistable induced absorptive switches [41]. They have been called SEED (self electro–optic effect device) by the authors, hoping that they might be the seed for optical computing. Here we present a CdS photo–thermal SEED [67]. The principle of operation is quite simple. We evaporate Al or In contacts on the sample forming a narrow slit about 100 μm wide, which is (partly) illuminated by the Ar+ laser. If a voltage is applied to the contacts, both the dark– and the photo current dissipate energy in the sample and contribute thus to the sample heating.

The last experiment which we want to show in this section are selfoscillations in a purely optical resonator [52]. The concept is shown in Fig. 15a. We use three photo–thermal induced absorptive optically bistable devices. Each one has a holding beam, which is situated slightly below $I\uparrow$ and the transmission of one element is the switching beam of the next. One bistable element acts as an inverter i.e. it has high transmission T for low input intensity $I_0<I\uparrow$ and low transmission for high input intensity $I_0>I\downarrow$. Two of them in series restore to logic level, however, delayed by two switching times. Consequently we can understand the operation of the resonator in Fig. 15a in the following way. Two of the elements produce an optical delay while the third produces the oscillation in a similar way as the resonators shown in Fig 10a and b. It should be noted that the action of every of the three devices can be interchanged because they are basically identical, and that the system works not only with three bistable devices but in principle with all odd numbers larger than one. In Fig. 15b we show the temporal output of two of the three elements. The third one looks similar with a corresponding shift of the logic levels. It is obvious that the oscillation is not strictly periodic in time. Computer simulations lead to the contrary to a periodic oscillation. The origin of this discrepancy is not yet clear. It might be due to small temporal fluctuations of the laser beam, or the irregular behaviour might be an intrinsic effect of the system. As shown in connection with Fig 7, the switching time is not a constant, but is determined by the critical slowing down. Small differences in the characteristics of the three devices may add up to produce after several periods an irregularity in the oscillation. In this case, we would have again a type of deterministic chaotic behaviour.

correlation dimension D_2 and the - entropy K_2. The following relations hold

$$K \geq K_2 \text{ and } D_2 \approx D \tag{17}$$

In this procedure, the observed signal x(t) is taken at times t = O; $t = \tau_S$, $t = 2\tau_S$, ...$t = N\tau_S$. An d–dimensional phase–space can be created now by d points equidistant in time. With the help of several lengthy formula given e.g. in [66], it is possible to deduce K_2 and D_2.

For our resonator with oscillations of the type shown in Fig 14d one finds

$$D \approx D_2 = 2.6 \pm 0.3 \text{ and } K>K_2=(100\pm25)\tau^{-1} >0 \tag{18}$$

for imbedding dimensions d above 6, giving thus evidence for the deterministic chaotic behaviour of the system.

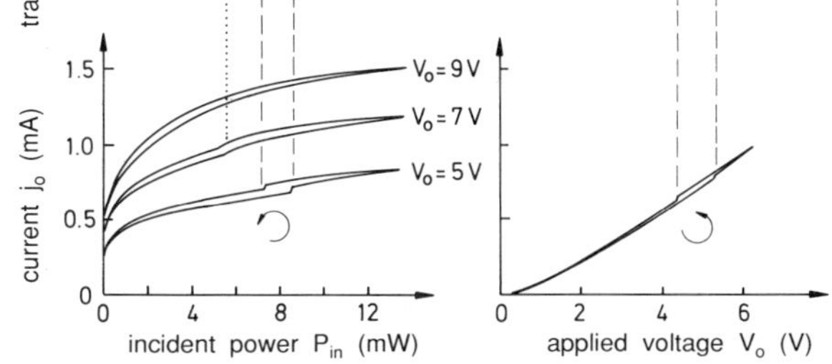

Fig 16 The four characteristics of a photo–thermal CdS SEED. From [67].

In Fig 16 we show the four characteristics of the photo–thermal SEED.

Fig 16a shows the optical input output characteristics for various applied constant voltages U. For U = O we get the usual induced absorptive bistability. Finite values of U shift the bistability to lower values of the optical input power due to the additional heating. At sufficiently high values of U, the ob disappeares, because the dark–current produces sufficient heat to rise the temperature T_{0L} of the sample without illumination to values where ob disappears. This process can be understood with the schematic diagram of Fig 2b.

In Fig 16b we keep I_0 constant at a value $I_0 < I\uparrow$, change U and measure the transmitted optical power. The device can be driven through a bistable loop by varying U. If I_0 is chosen in the bistable region without applied voltage $I1 < I_0 < I\downarrow$, we can switch I_t only down with increasing U but not back again [67]. Fig 17c and d correspond to Fig 17a and b respectively, with the difference that the electric current through the device is plotted on the Y axis. A switch in the absorbed power results also in a switch of the photocurrent. Usually the step–height in the current is only sublinearly dependent on the step height of the absorbed optical power, and even negative characteristics have been observed especially in photo–thermal ZnSe SEED devices [52]. The possible origin for this behaviour is manifold and includes among others: a sublinear increase of the carrier density with increasing generation rate, surface recombination, a decrease of mobility with increasing temperature and finally Schottky barrier effects, since it is difficult to produce simply ohmic, low resistivity contacts to CdS, ZnSe and some other II–VI compounds.

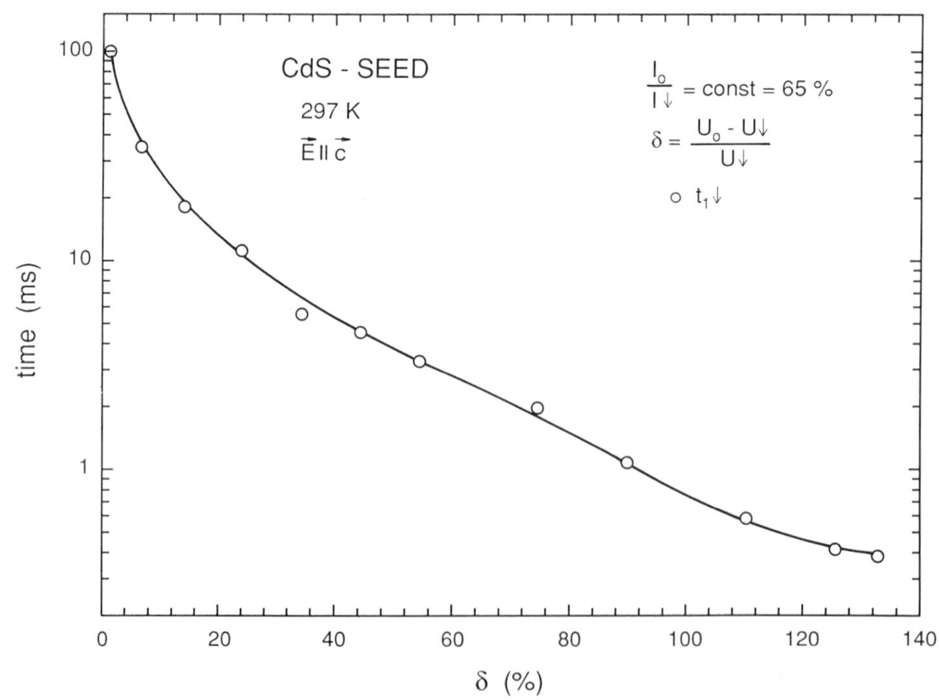

Fig 17 The critical slowing down of a photo–thermal CdS SEED as a function of the applied voltage step. From [68].

To demonstrate that the photo–thermal SEEDs have basically the same characteristics as the purely optical ones, we show in Fig 17 the critical slowing down of the switching time. Fig 17 corresponds to Fig 7b, with the only difference that the optical input power is kept constant $I_0 < I\downarrow$ and δ gives the relative amount, the applied voltage has been switched above the value for the switching threehold in a step–like function.

VI. SUMMARY AND OUTLOOK

We have shown that photo–thermal optical nonlinearities caused by exciton–phonon interaction and nonradiative recombination give rise to a rich variety of optical and electrooptical effects and devices such as mono– and bistable switches, memories, oscillators, etc.

Here we want to review several aspects under which the nonlinear optical properties of matter can be investigated.

Optical nonlinearities are always connected with manyparticle– and renormalization effects. In our case this was mainly the carrier–phonon interaction. In photo–electronic optical nonlinearities there are many additional effects of this group like elastic and inelastic scattering processes between excitons and/or carriers, exchange– and correlation effects, band– or phase–space filling etc. For recent reviews see e.g. [69–71]. Another aspect which can be studied in nonlinear optics are non–equilibrium phase–transitions, driven by external pump sources. The optical bistability is a first order phase transition of this type as explained in [50] and in the text, the formation of an electron–hole–plasma not treated here would be another example [69].

A third group of aspects which are strongly connected with the type of work presented here are nonlinear dynamics and synergetics [6, 14, 65, 72]. We observe the formation of temporal and/or spatial structures and of collective behaviour appearing in the systems. Optics are here especially valuable because it is often possible to study and to influence the nonlinearity and the feedback mechanism separately.

Finally we should mention again the aspect of optical computing. Shortly after the discovery of optical bistability, it became clear that memories and the basic logic connections (AND, OR, NOT) can be implemented optically. A rather optimistic phase followed and "optical computing" was expected to be rather close at hand. In the meantime it became obvious that it will not be possible to overcome electronic computing, which is the outcome of about five decades of research and development in a few years nor before the end of this century. This statement shall be made clear by a few numbers. The photo–thermal optical nonlinearities presented here can be optimized concerning switching speed and power. For a BEAT element the following data have been obtained with a laser beam focussed down to a diameter in the μm range [60]: A switching time for about 5 μs resulting in cycle times of 20 μs and input powers of 8 mW. Since these data are limited by some basic laws like energy conservation, the Dulong–Petit law for the specific heat, diffraction etc., a further significant improvement seems hardly possible. Indeed, optimal data for other photo–thermal devices are of the same order of magnitude or worse. On the other hand, it has been stated recently [73] what is necessary to overcome digital electronics by digital optics: 10^{12} simple operations per second. Due to the inherent parallelity of optical data handling, this figure can be achieved e.g. by a $10^3 \times 10^3$ array of pixels with a cycletime $\tau_c = 1$ μs or by a 30x30 array with $\tau_c = 1$ ns. Let us see what can be done with photothermal nonlinearities e.g. the BEAT–element mentioned above. A cycle time of 20 μs results in 2.10^7 pixels to get 10^{12} operations per second. With an input power of 5 mW and neglecting all losses of additional optics like (holographic) beam splitters, lenses etc., a total input power of the green Ar^+ laser line of 10^5 W is necessary, and this value is obviously beyond every realistic approach.

Optical switches and memories using photo–electronic optical nonlinearities and SEED elements are partly much faster and use lower (optical) input powers. Nevertheless they encounter also big problems concerning cascadability and fan–out values. The most promising candidates are socalled smart pixels, which consist in an integrated way of a photodetector as input device, some fast electronic switching circuitry and a luminescence or laser diode as output. If it becomes possible in the next decade to produce such smart pixels as one– and two–dimensional arrays of sufficient size and speed (see above), they may well bring a new range of data–handling and computing.

The partly pessimistic statement concerning presently available nonlinear optical devices does not make the research concerning optical computing obsolete. Systems like the one presented in this contribution serve as valuable model systems and allow to study basic properties and architectural concepts. When the first transistor was working, which is shown e.g. on the cover photo of [74], nobody was dreaming about VLSI and a Cray–machine. If we need now a 10W Ar+ laser and a whole table of optics to add optically 1+1 [7, 10, 13], we are already dreaming of optical computers. With all these dreams, scientists should nevertheless be careful what the possible use of the outcome of their research is, and to influence the use of their research to the extend they can for the benefit of mankind [75, 76, 77].

Acknowledgement: Most of the results presented here have been found in the research group of the author. He therefore thanks his coworkers for their excellent work in this field done e.g. by Dr. K. Bohnert, C. Dörnfeld, M. Graf Lambsdorff, A. Witt, A. Schmidt, J. Grohs and especially by Dr. M. Wegener. Thanks are also due to many colleagues all over the world for stimulating and fruitful discussions. Without being complete, I should like to mention especially Prof. Dr. H. Haug and Prof. Dr. W. Martienssen (Frankfurt), Prof. Dr. H.G. Schuster (Kiel), Prof. Dr. D. Jäger (Münster), Prof. Dr. D.S. Smith and Prof. Dr. B. Wherrett (Edinburgh), Prof. Dr. S.W. Koch and Prof. Dr. H.M. Gibbs (Tucson), Prof. Dr. J.B. Grun and Dr. B. Hönerlage (Strasbourg), Dr. D.S. Chemla and Dr. DAB Miller (Holmdel), Dr. S. Schmitt–Rink (Murray Hill), Dr. F. Henneberger and Dr. J. Gutowski (Berlin).

Financial support is acknowledged from various institutions, especially from the Deutsche Forschungsgemeinschaft through the Sonderforschungsbereich 185 Nichtlineare Dynamik, from the Materialforschungsschwerpunkt of the Land Rheinland–Pfalz at the University of Kaiserslautern and from the Commission of the European Communities through the Stimulation Action. The high quality CdS samples used in most of the experiments have been grown by Dr. G. Müller–Vogt at the Kristall– and Materiallabor of the University Karlsruhe.

References

[1] **Optical Bistability, Dynamical Nonlinearity and Photonic Logic,** B.S. Wherrett and D.S. Smith eds, Phil. Trans. R. Soc. London A 313, 191–451 (1984)

[2] **High Excitation and Short Pulse Phenomena,** M.H. Pilkuhn ed., J. Luminesc. 30 (1985)

[3] **Optical Bistability: Controlling Light with Light,** H.M. Gibbs, Acad. Press (1985)

[4] **Optical Bistability,** E.M. Garmire ed, special issue IEEE J.QE 21, 1339–1549 (1985)

[5] **Excitonic Optical Nonlinearities,** D.S. Chemla ed. special issue, JOSA B2, 1135–1243 (1985)

[6] **Optical Chaos,** J. Chrostowski and N.B. Abraham eds, SPIE Proc. 667 (1986)

[7] **Optical Bistability III,** H.M. Gibbs, P. Mandel, N. Peyghambarian and S.D. Smith eds, Springer Proceedings in Physics 8 (1986)

[8] **Laser–Induced Gratings,** H.J. Eichler, D. Pohland, P. Günter, Springer series in Opt. Sci. 50 (1986)

[9] **Dynamic Gratings and Four Wave Mixing,** H.J. Eichler ed., special issue, IEEE JQE 22, 1194–1542 (1986)

[10] **From Optical Bistability Towards Optical Computing,** P. Mandel, S.D. Smith, B. Wherrett eds., North Holland (1987)

[11] **Optical Computing 88,** J.W. Goodmann, P. Chavel and G. Roblin eds, SPIE Proc 963 (1988)

[12] **Nonlinear Optical Materials,** G. Roosen ed., SPIE Proc. 1017 (1988)

[13] **Optical Bistability IV,** W. Firth, N. Peyghambarian and A. Tallet eds., J. Physique, 49 C2 (1988)

[14] **Optical Nonlinearities and Instabilities in Semiconductors,** H. Haug ed, Acad. Press (1988)

[15] **Quantum Well Heterostructures and Superlattices,** J.J. Coleman ed. special issue IEEE JQE 24, 1579–1798 (1988)

[16] **Laser Spectroscopy of Solids II,** W.M. Yen ed., Topics in Applied Physics, 65, Springer (1989)

[17] **Optical Nonlinearity and Bistability of Semiconductors,** F. Henneberger ed. phys. stat. sol b 150, p 347–919 (1988)

[18] **Excitons,** K. Cho ed, Topics in Current Physics 14, Springer (1979)

[19] **Excitons,** E.I. Rashba and M.D. Sturge eds., Modern Problems in Condensed Matter Sciences 2, North Holland (1982)

[20] **Collective Excitations in Solids,** B. DiBartolo ed., NATO ASI series B 88, Plenum (1983)

[21] C. Klingshirn in Energy Transfer Processes in Cond. Matter, B. DiBartolo ed., NATO ASI series B 144, 285 (1984) Plenum

[22] F. Urbach, Phys. Rev. 92, 1324 (1953)

[23] W. Martienssen, J. Phys. Chem. Sol. 2, 257 (59) and ibid. 8, 294 (59)

[24] D. Dutton, Phys. Rev. 112, 785 (1958)
M. Kurik, phys. stat. sol. a 8, 9 (1971)
S. Spielberg, E. Gutsche and J. Voigt, phys. stat. sol. b 77, 233 (1976)

[25] J. D. Dow and D. Redfield, Phys. Rev. 85, 94 (1972)

[26] **Electronic Processes of Non Crystalline Solids,** N.F. Mott and E.A. Davies, Clarendon Press, Oxford (1971)

[27] **Models of Disorder,** J.M. Ziman, Cambridge University Press (1979)

[28] **Physics of Amorphous Solids,** R. Zallen, Wiley (1983)

[29] **Electronic Properties of Doped Semiconductors,** B.I. Shklovskii and A.L. Efros, Springer Series in Solid State Sciences 45 (1984)

[30] **Localization and Interaction,** D.M. Finlayson ed., Proc. Scottisch Universities, Summerschool in Physics 31 (1986)

[31] **Disordered Solids, Structures and Processes,** Intern. Summerschool, Erice (1987), B. Di Bartolo ed., in press

[32] H. Sumi and Y. Toyozawa, J. Phys. Soc. Japan 31, 342 (1971)

[33] J.G. Liebler, S. Schmitt–Rink and H. Haug, J. Luminesc. 34, 1 (1985)

[34] H. Haug and C. Klingshirn, Physics Reports 70, 315 (1981)

[35] B. Hönerlage, R. Levy, J.B. Grun, C. Klingshirn and K. Bohnert, Physics Report 124, 161 (1985)

[36] S. Schmitt–Rink, D.S. Chemla and D.A.B. Miller, Advances in Physics 38, 89 (1989)

[37] H.E. Swoboda, F.A. Majumder, V.G. Lyssenko, C. Klingshirn and L. Banyai, Z. Physik B 70, 341 (1988)

[38] M. Wegener, C. Klingshirn, S.W. Koch and L. Banyai, Semiconductor Science and Technology 1, 366 (1986)

[39] M. Dagenais and W.F. Sharfin, JOSA B 2, 1179 (1985)

[40] Landolt–Börnstein, New Series Vol. 17b, O. Madelung ed. Springer (1982)

[41] D.A.B. Miller, JOSA B 1, 857 (1984)

[42] D. Jäger, F. Forsmann and B. Wedding in IEEE JQE, p 1453 (1985)

[43] C. Klingshirn, M. Wegener, C. Dörnfeld, M. Lambsdorff, J.Y. Bigot and F. Fidorra in Ref. [7], p 129
M. Lambsdorff, C. Dörnfeld and C. Klingshirn, Z. Physik B 64, 409 (1986)

[44] M. Wegener and C. Klingshirn, Phys. Rev. A 35, 1740 and 4247 (1987)

[45] M. Wegener, C. Klingshirn, and G. Müller–Vogt, Z. Physik B 68, 519 (1987)

[46] F. Henneberger, phys. stat. sol. b 137, 371 (1986)

[47] I. Haddad, M. Kretzschmar, H. Rossmann and F. Henneberger, phys. stat. sol. b 138, 235 (1986)

[48] J. Gutowski, J. Hollandt and I. Broser, Z. Physik B (1989) in press

[49] F. Fidorra, M. Wegener, J.Y. Bigot, B. Hönerlage and C. Klingshirn, J. Luminesc. 35, 43 (1986)

[50] Dynamics of First–Order Phase Transitions in Equilibrium and Nonequilibrium Systems, S.W. Koch, Lecture Notes in Physics 207, Springer (1984)

[51] M. Kretzschmar, F. Henneberger, H. Rossmann and I. Hallad, phys. stat. sol. b 143, k71 (1987)

[52] J. Grohs, A. Schmidt, M. Kunz, C. Weber, A. Daunois, A. Rupp, W. Dötter, F. Werner and C. Klingshirn, Proc. Intern. Congr. of Optical Science and Engineering, Paris (1989) to be published by SPIE

[53] J.–Y. Bigot, A. Daunois, R. Leonelli, M. Sence, J.G. Mathew, S.D. Smith and A.C. Walker, Appl. Phys. Lett 49, 844 (1986)

[54] M. Wegener, C. Klingshirn, A. Daunois, J.–Y. Bigot, N. Cherkaoui Eddeqaqi and J.B. Grun, Appl. Phys. Lett. 52, 685 (1988)

[55] J.Y. Bigot, A. Daunois, N. Cherkaoui Eddeqaqi, J. Oberle, M. Wegener and C. Klingshirn in Ref [13], p 153

[56] G.R. Olbright, H.M. Gibbs, N. Peyghambarian, H.E. Schmidt, S.W. Koch and H. Haug in Ref [7], p 186

[57] C. Klingshirn, M. Wegener, C. Dörnfeld and M. Lambsdorff in Ref [10], p 252

[58] S.W. Koch and E.M. Wright, Phys. Rev. A 35, 2542 (1987)

[59] H.X. Nguyen, V.D. Egorov, A. Harendt and E.V. Nazanova, phys. stat. sol. b 148, 407 (1988)

[60] A.C. Walker, Opt. Comm. 59, 145 (1986) and H.J. Eichler V. Glaw, A. Kummrov, V. Penschke and S. Broede in Ref. [12], p 90

[61] S.W. Koch in Ref. [14], p 273

[62] K. Ikeda, H. Daido and O. Akimoto, Phys. Rev. Lett. 45, 709 (1980) and 48, 617 (1982)
J.V. Moloney, Phys. Rev. A 33, 4061 (1986)

[63] M. Lindberg, S.W. Koch and H. Haug, JOSA B 3, 751 (1986)

[64] I. Galbraith and H. Haug, JOSA B 4, 1116 (1987)

[65] **Deterministic Chaos,** H.G. Schuster, VCH (1988)

[66] P. Grassberger and I. Procaccia, Phys. Rev. Lett. 50, 346 (1983), Phys. Rev. A 28, 2591 (1983) and Physica D 9, 189 (1983)

[67] A. Witt, M. Wegener, V.G. Lyssenko, C. Klingshirn, G. Wingen, Y. Iyechika, D. Jäger, G. Müller–Vogt, H. Sitter, H. Heinrich and H.A. Mackenzie, IEEE JQE 24, 2500 (1988)

[68] W. Kazukauskas, J. Grohs, C. Klingshirn, G. Wingen and D. Jäger, to be published

[69] C. Klingshirn and H. Haug, Physics Reports 70, 315 (1981)

[70] B. Hönerlage, R. Levy, J.B. Grun, C. Klingshirn and K. Bohnert, Phys. Reports 124, 161 (1985)

[71] S. Schmitt–Rink, D.S. Chemla and D.A.B. Miller, Adv. in Physics 38, 89 (1989)

[72] **Synergetics, an Introduction,** H. Haken Springer Series in Synergetics 1

[73] A. Lohmann private communication

[74] **Physics of Semiconductor Devices,** 2nd edition, S.M. Sze, J. Wiley and Sons (1981)

[75] The Erice Statement, by P.A.M. Dirac, P. Kapitza and A. Zichichi, Erice (1982)

[76] Resolution der Deutschen Physikalischen Gesellschaft zur Abrüstung, Bonn (1989)

[77] **Das Prinzip Verantwortung,** H. Jonas, Insel Verlag (1979)

CRYSTALS: GROWTH METHODS AND CHARACTERIZATION

A. Scacco

Dipartimento di Fisica
Università La Sapienza
P.le A. Moro 2
00185 Roma, Italy

ABSTRACT

In the frame of a general survey of the methods used for crystal growth, particular attention is devoted to the most important or most modern techniques employed to produce materials relevant to basic research or technological applications. The essential features of the most common methods for crystal characterization are also discussed.

I. INTRODUCTION

Crystal growth always needs the careful control of a phase change. As a consequence, the following main categories of crystal growth methods can be defined:

- growth from the solid: S → S processes involving solid-solid phase transitions

- growth from the liquid: L → S processes involving liquid-solid phase transitions { growth from the melt / growth from solution }

- growth from the vapour: V → S processes involving vapour-solid phase transitions.

Fig. 1 shows how these main categories can be related to various growth techniques: some of them (strain anneal, aqueous solution growth, sublimation) are well known and used since several years, while others have been recently developed and successfully applied to technological materials (liquid encapsulation pulling, liquid phase epitaxy, molecular beam epitaxy). Specialists may only be interested in a detailed treatment of all the above techniques, so consideration will be limited to the most used or most modern methods.

In general, if the crystal is in dynamical equilibrium with its parent phase, the free energy G is at a minimum and no growth will be possible. For growth to occur the equilibrium must be disturbed by a change of the correct sign in temperature, pressure, chemical potential,

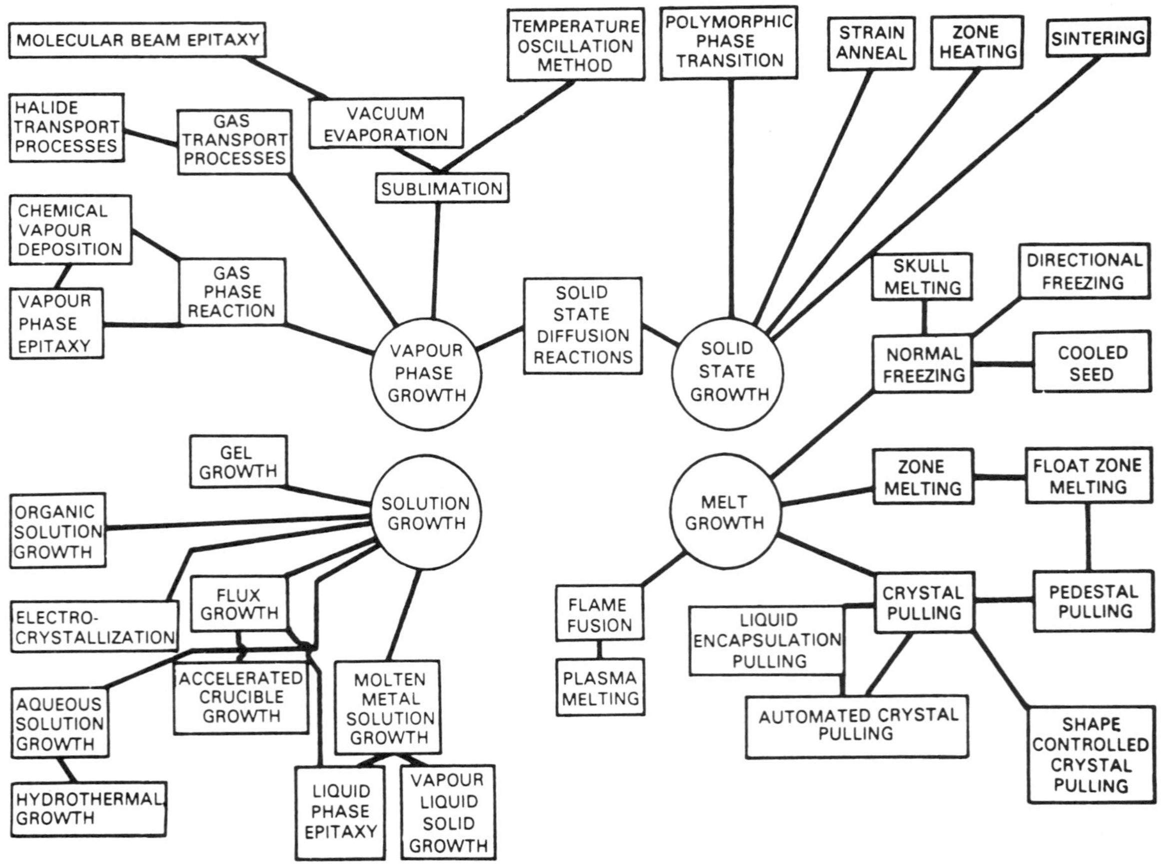

Figure 1. Crystal growth techniques (Ref. 8)

electrochemical potential or strain. In a well-designed growth process just one of these parameters is held minimally away from its equilibrium value to provide a driving force for growth. If the growth can be considered a reversible process, there is no change in the free energy of the system, given by

$$G = H - TS \qquad (1)$$

where H is the enthalpy (or heat content), T the absolute temperature, and S the entropy. For an isothermal process we have

$$\Delta G = 0 = \Delta H - T\Delta S \qquad (2)$$

where ΔH is the enthalpy variation (coincident with the appropriate latent heat in an isobaric growth) and ΔS the entropy variation related to the crystallization. It follows that

$$\Delta S = \Delta H/T \qquad (3)$$

and this change in the system entropy can be used to classify the crystal growth processes. Let us define the α factor as

$$\alpha = \Delta S/k = \Delta H/kT = L/kT_c \qquad (4)$$

where L is the molecular latent heat, k the Boltzmann constant and T_c the temperature of the phase change. Three main growth classes can be defined according to the value of α (that is, to the entropy change):

- low entropy range ($\alpha < 2$), typically growth of metals from the melt;

- middle entropy range ($2 < \alpha < 10$), for many cases of growth from the melt;

- high entropy range ($\alpha > 10$), typically growth from vapour phase.

A direct consequence of the entropy variation is the crystal habit. Crystals growing in the high entropy range usually undergo autonucleation processes, so that it is very difficult to obtain single crystals, due to simultaneous nucleations in the mass; moreover, every crystal exhibits faces on all its sides (typical example are ice crystals growing from vapour). Crystals growing in the middle entropy range may show faces in a particular crystallographic direction, or their habit can be faceless (typical example are semiconductors grown from the melt). Crystals growing in the low entropy range do not exhibit faces, the growth surface is rough and follows the isotherms in the growth medium.

II. SOLID GROWTH TECHNIQUES

Solid state growth, in general, requires atomic diffusion, which at normal temperatures is usually very slow. Thus techniques of growth in a solid phase are employed only when other methods cannot be used: typical example is the growth of materials showing a phase transition between the melting point and room temperature.

The annealing techniques are used mostly for metal crystals. If a polycrystalline metal rod is held at elevated temperature below the melting point for many hours, some grains grow at the expense of their neighbours. Since grain boundaries contain more free energy than bulk crystal, this process can be seen to lower the free energy of the rod. Such a secondary recrystallization can be induced by strain annealing

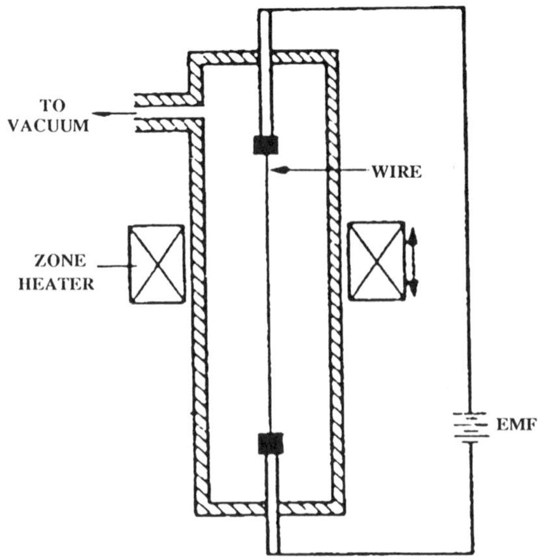

Figure 2. Pintsch process set-up.

under a temperature gradient (Al, Pb, Cu, Sn), or by zone heating (Fig. 2) applied to metal wires, as in the Pintsch process to produce single crystal wires for filament lamps (Mo, W). In metallurgy the annealing of a precompressed powder (sintering) and the annealing of a powder under pressure (hot pressing) are also used.

III. MELT GROWTH TECHNIQUES

In principle, all materials can be grown in single crystal form from the melt, provided they melt congruently (which involves no decomposition below or near the melting point), they do not undergo a phase transformation between the melting point and room temperature, and they have a manageable vapour pressure at the melting point. Melt growth is the fastest of the growth methods, its rate being independent of mass transport processes. For this reason it has been used for many metals, semiconductors, ionic crystals, and a few organic compounds. Often, especially with semiconductors and laser host crystals, some impurities can be deliberately added to the melt and dispersed in the grown crystals. Nucleation can be heterogeneous, if a crystal seed of the same material is used, or homogeneous, if a seed of another material is employed or the use of the seed can be eliminated.

It is convenient to divide melt growth into the following four main types of techniques.

III.A. Normal Freezing

In this inexpensive technique a molten mass is gradually frozen from one end to the other (Fig. 3). When this is achieved by means of a two-zone furnace, it is called the Bridgman-Stockbarger method, whose configuration is usually vertical (Fig. 4). Large metal crystals and

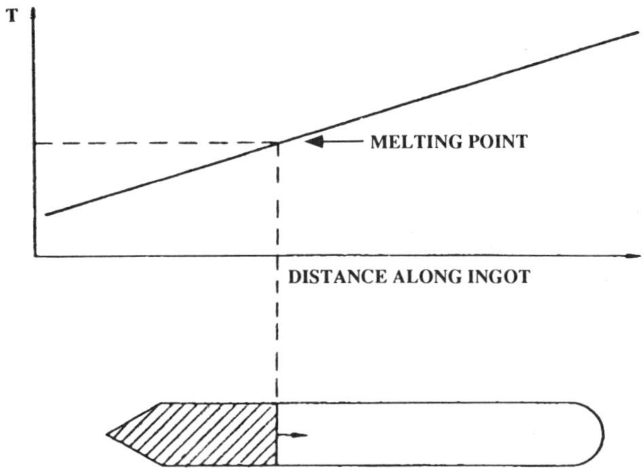

Figure 3. Normal (or directional) freezing

optical quality alkali halide single crystals for prisms and lenses are obtained in this way. The nucleation problem can be overcome by using a conical bottom of the crucible (Figures 3 and 4), so that the coldest region is essentially a point, or through the necking (Fig. 5) which contains the initial rapid growth and allows only one seed of fast-growing orientation to pass into the melt. If the material expands on freezing, horizontal normal freezing is preferred to minimize the container-pressure-effects. Three ways of moving the freezing interface are possible: moving ampoule (the melt is drawn through the temperature gradient), moving furnace (the melt is stationary), static freeze (furnace temperature is gradually reduced keeping the thermal gradient approximately linear, as in Fig. 3). If, in a normal freeze situation with constant cross section, there is an impurity with initial concentraction C_o and distribution coefficient k_o, then the composition of C of the material freezing when a fraction q of the melt has solidified is given by

$$C = k_o C_o (1-q)^{(k_o - 1)} \tag{5}$$

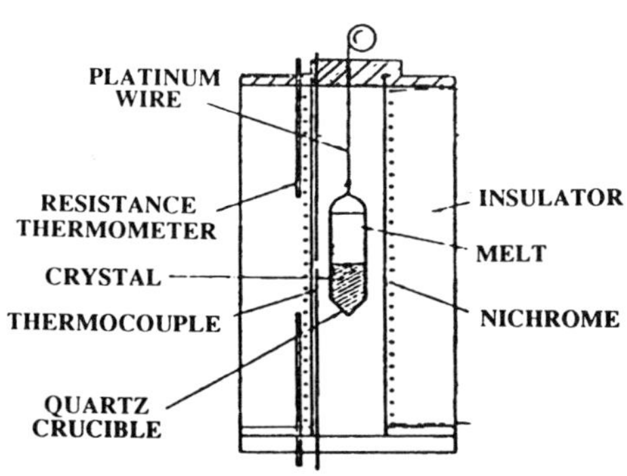

Figure 4. Bridgman set-up

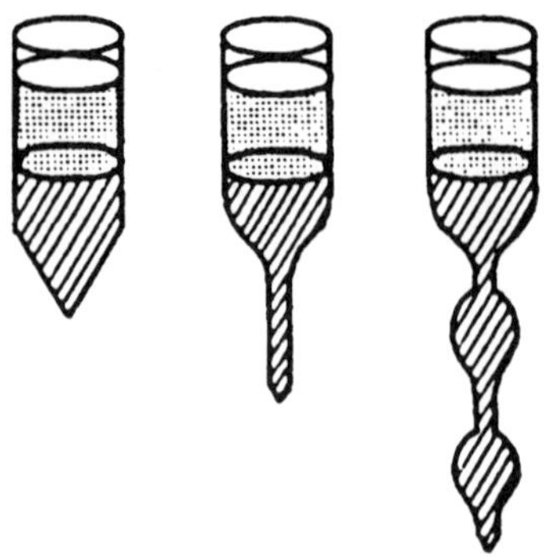

(a) (b) (c)

Figure 5. Single nucleation by: (a) conical bottom; (b) capillary; and (c) necking (Ref. 8).

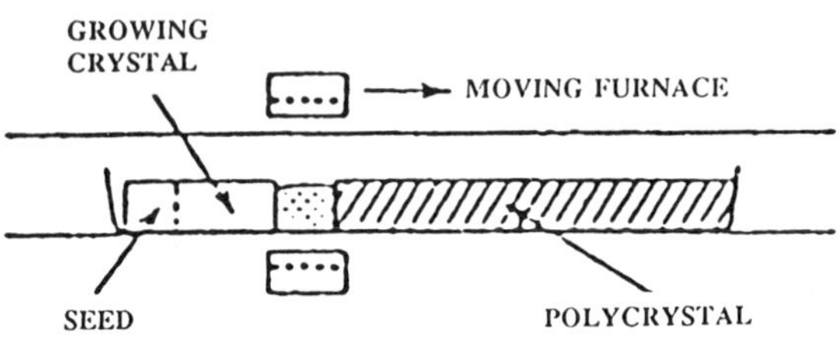

Figure 6. Zone-melting technique

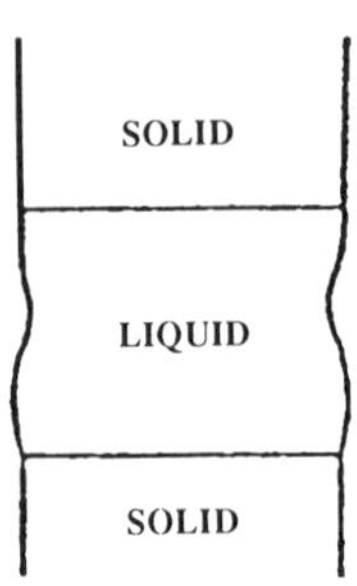

Figure 7. Float-zoning method

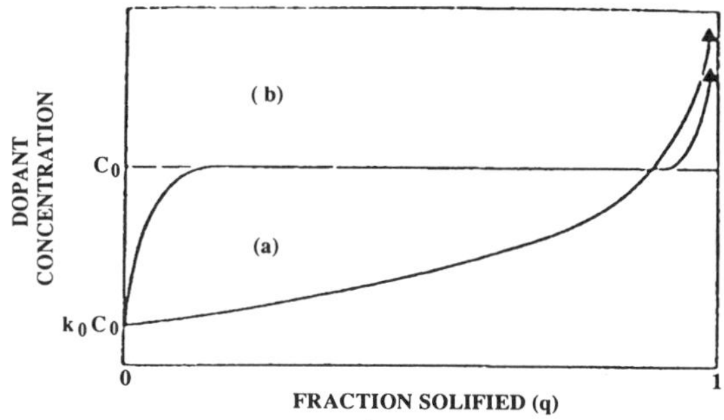

Figure 8. Impurity segregation after: (a) normal freezing and (b) zone melting.

This is the normal freeze equation and applies if there is complete mixing in the liquid phase, there is no diffusion in the solid phase and k_o is a constant. Since for many impurities k_o is appreciably different from unity, the middle region of the crystal is purified. Impurities, like P in GaAs, which have a preference for the solid phase, freeze out early while those, including most dopants in Si, which prefer the liquid phase, are swept to the end. This means that the crystals have a graded impurity content especially in the last part to freeze.

III.B. Zone Melting

In this case (Fig. 6) a molten zone is established at one end of the charge (or between the seed and the charge) and is advanced by moving either the container or the furnace. In the vertical configuration this method is called float zone refining (Fig. 7), and produces the purest single crystals known to man, since the molten zone, sustained by surface tension forces, does not require any container. Si single crystals with a very low oxygen concentration (unachievable in quartz containers) are produced extensively in this way. There is a distinct difference between impurities. In normal freezing, segregation is controlled by diffusion and convective mass transport of the impurity in the melt, whereas an effective distribution coefficient of one can be achieved in zone melting, due to the relatively small volume of the melt. As a

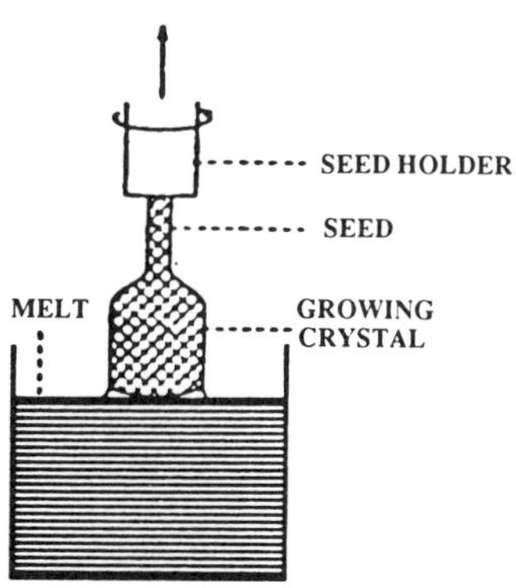

Figure 9. Czochralski apparatus

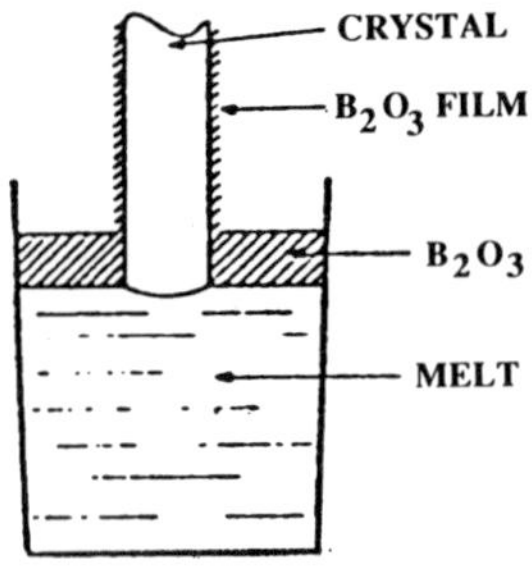

Figure 10. Liquid phase encapsulation

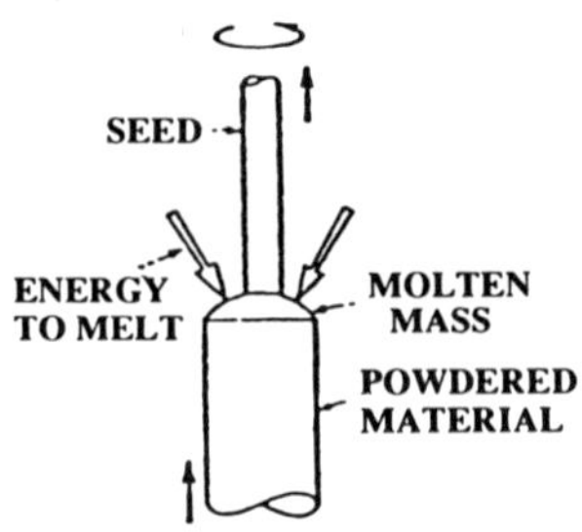

Figure 11. Flame fusion system

consequence, the distribution of a given impurity in the two cases is significantly different (Fig. 8), so that a homogeneous content of the desired impurity can be achieved by zone melting in a significant portion of the crystal.

III.C. Crystal Pulling

This technique is currently the most important method available for the growth of near perfect crystals of large size. A crucible is used to contain the melt, and a seed held by a rod is dipped partially into the liquid phase. After equilibration, the seed is pulled by the rod, which is at the same time rotated at an appropriate rate (Fig. 9). The rotation is employed to optimize compositional and thermal homogenization of the melt and to control the crystal shape. The growth apparatus can be exposed to the room atmosphere or most commonly enclosed so that growth can be carried out in vacuum or under controlled pressure. Numerous materials can be grown with this technique, usually known as the Czochralski method, including Si and refractory oxides. In case of compounds with volatile constituents, a layer of boric oxide B_2O_3 over the melt prevents the escape of the evaporating components (Fig. 10), originating the so-called liquid phase encapsulation technique. The crystal quality has been considerably improved in the recent years: for example the use of necking supplies dislocation free crystals, and automatic diameter control methods have been introduced.

III.D. Flame Fusion

Like float zone refining, this method has the great advantage of being crucibleless. The top of a seed crystal is molten with an oxygen-hydrogen flame or r.f. heating and powder of the material is fed to the liquid zone, so that the growth proceeds by pulling the seed (Fig. 11). This technique was introduced by Verneuil to grow refractory oxide crystals (Al_2O_3).

IV. SOLUTION GROWTH METHODS

Growth from solution is more widely used than the other techniques, mostly because it is simple and inexpensive. It requires lower temperatures and can lead to lower density of lattice defects, but crystal purity has defined limits due to the solvent inclusions. Supersaturation of the solution can be obtained in three different ways: by decreasing (or more rarely by increasing) temperature, or by evaporating the solvent, or by using a two-temperature system in which the solvent passes nutrient from source to seed (Fig. 12). At low temperature solution growth in aqueous medium is used for alums and Rochelle salt, at high temperatures oxides are grown from solution in molten salts.

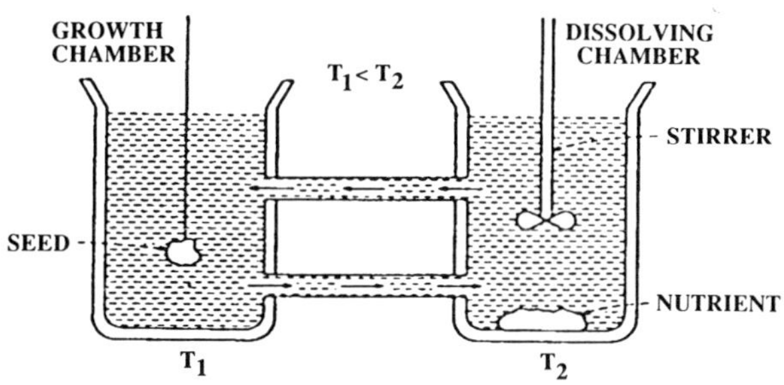

Figure 12. Solution growth set-up

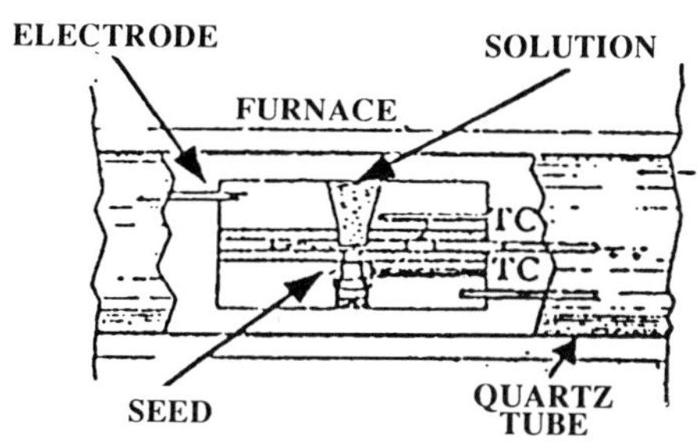

Figure 13. Liquid phase epitaxy

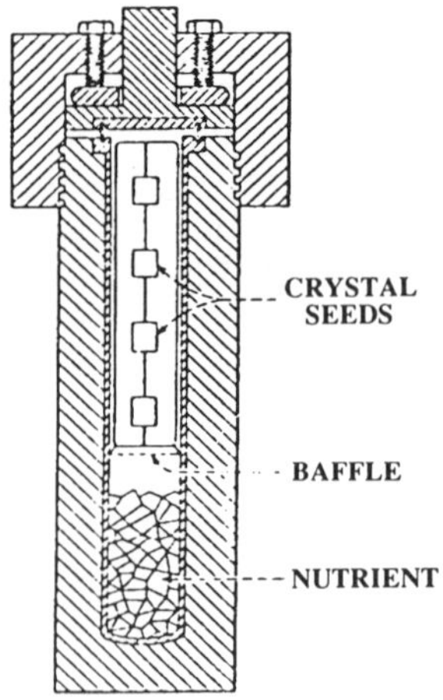

Figure 14. Hydrothermal growth

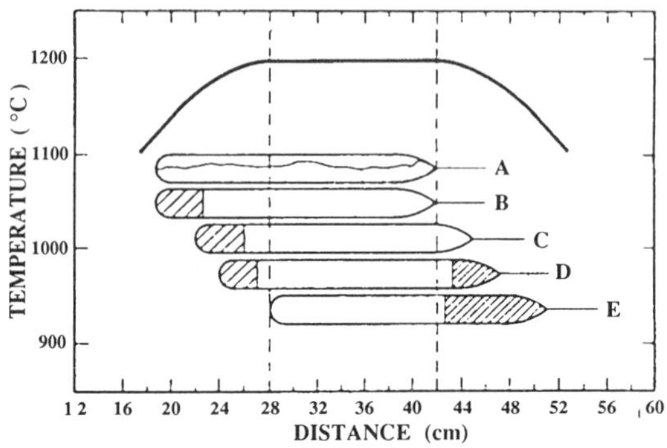

Figure 15. Growth by sublimation

Solution growth methods are at least an order of magnitude slower than melt growth. However, aqueous solution growth has produced the largest crystals known to man: the world's biggest artificial crystal is an octahedron of alum, weighing more than 100 kilograms, grown during three years and turned over in its bath every day to ensure the habit symmetry. Growth from metallic solutions is now widely used for the preparation of high quality epitaxial layers of III-V compounds and alloys (Fig. 13): all GaAs electronic devices are obtained from Ga-As solutions by controlled temperature decrease, and doping of the layers can be easily controlled.

Hydrothermal growth is essentially aqueous solution growth at elevated temperatures and pressures, and it can be useful for materials showing limited solubility under ordinary conditions. In the case of α-quartz (SiO_2) crystals (Fig. 14), the nutrient quartz originates at 400^oC and 2000 bars a saturated solution, which by convection is brought into contact with the seeds kept at different temperatures to achieve supersaturation. Of course, technological problems are involved by elevated temperatures and pressures operating for considerable periods of time.

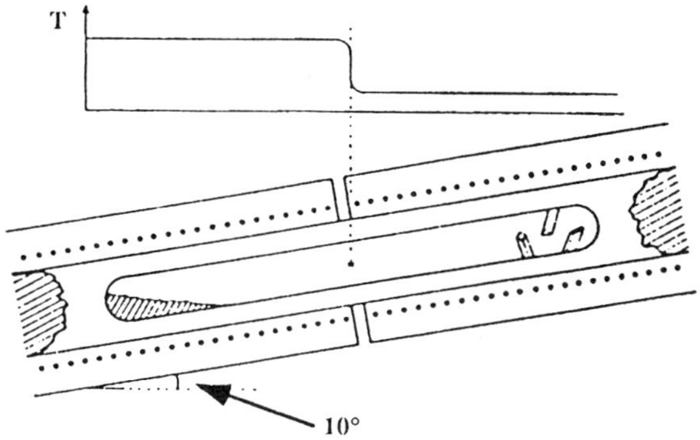

Figure 16. Growth by vapour transport

V. VAPOUR PHASE GROWTH

Bulk crystals, as well as thin layers, can be grown from the vapour phase. As the growth rate is quite slow, bulk crystals are prepared in this way when their growth by other methods is not avialable: this is the case of CdS, formed by volatile components (Fig. 15), which is grown by sublimation-condensation, and of the snow crystals. Besides these physical vapour depositions, various chemical methods are used, which can be divided into halogen transport techniques and gas phase reactions.

In the former class, a carrier gas (usually Cl_2 or I_2) is added to the powder of the metallic compound, which has to be grown, in a sealed quartz ampoule (Fig. 16): a suitable temperature gradient causes the transport, favoured by convective phenomena, of the non-volatile compound to the cold zone; the function of the carrier is to form a volatile salt with the metal, which after migration decomposes at the cold end of the tube originating metal crystals.

The gas phase reactions are now of considerable commercial importance, being widely used for growing epitaxial layers of semiconductors in the manufacturing of transistors and integrated circuits. Typical is the growth of a crystalline Si layer, pure or suitably doped, on a Si substrate by the following reduction of a volatile chloride:

$$SiHCl_3 + H_2 \longrightarrow Si + 3HCl$$

Vapour phase epitaxy is also relevant to obtain growth of GaAs layers, according to the reaction

$$2GaCl + As_2 + H_2 \longrightarrow 2GaAs + 2HCl$$

which occurs through various steps in different sections of the deposition arrangement.

Molecular beam epitaxy is a modern technique where the constituents of the resultant solid are simultaneously evaporated from separate sources onto the growth surface. It is used to grow a wide variety of semiconductor crystals in device structures as lasers, diodes, or microwave transistors. The process complexity requires sophisticated instrumentation controlling not only the growth, but also the quality of

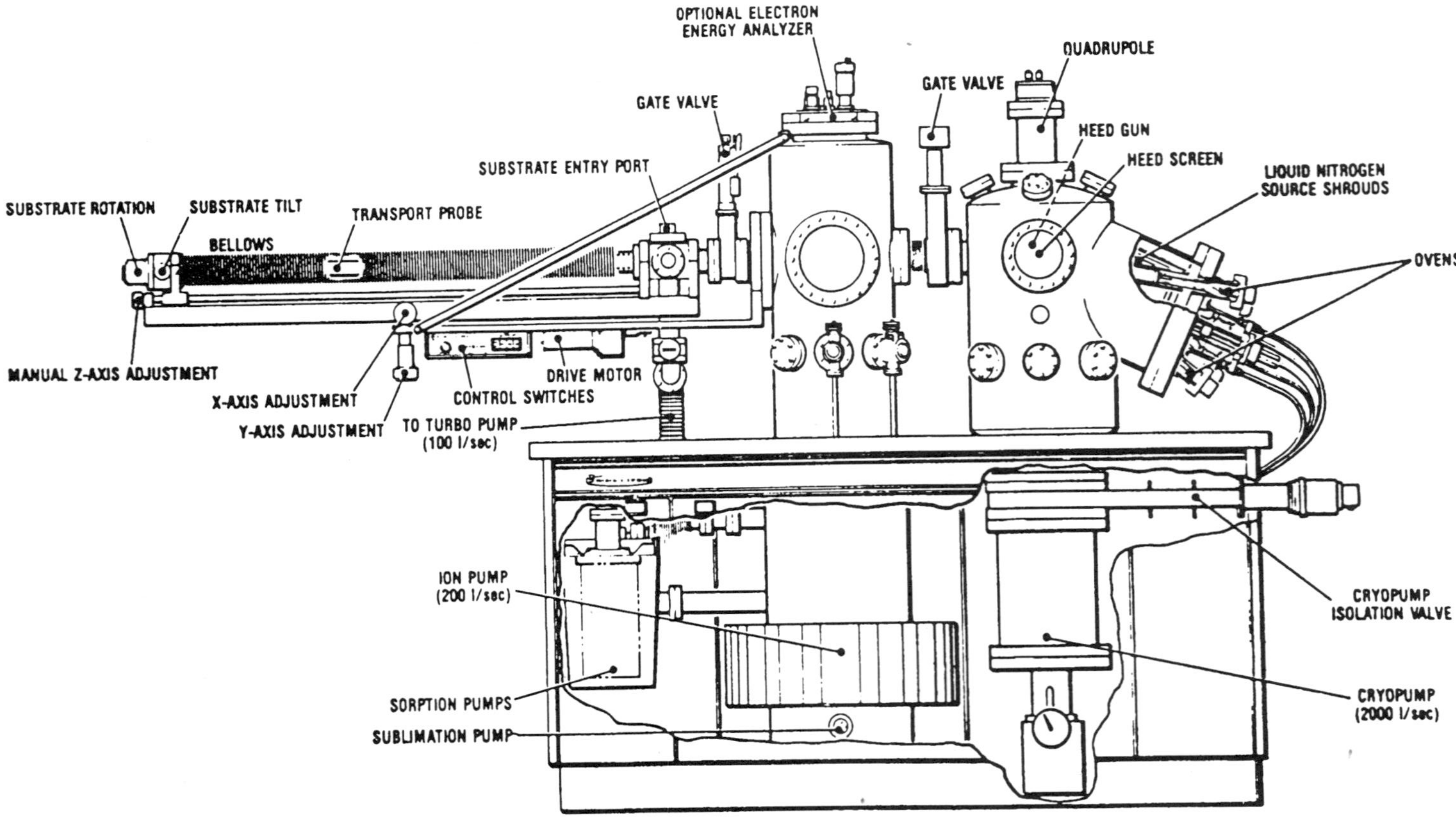

Figure 17. MBE system by Physical Electronics Inc. (Ref. 8)

the epitaxial layers. The system consists (Fig. 17) of a growth chamber which is cryopumped, an ion pumped analysis chamber, and a sample entry chamber. The sample sits on a single probe which accesses all chambers, each one of them being effectively isolated from the others. The evaporation sources are heated by resistive ovens or electron beam.

VI. CRYSTAL CHARACTERIZATION

The determination of the crystal quality is vital from both the consumers' and producers' points of view. Only proper assessment can determine that the crystals are satisfactory, or, if they are not, provide indication on the changes in the growth technique in order to improve the material quality. No one technique will supply all desired information on the crystal features, then the characterization of single crystals results from studies performed with various methods.

Imperfections in crystals are classified in diminishing order of size in terms of the number of dimensions in which they are large compared with interatomic distances in the material. Volume defects can differ from the rest of the crystal in orientation (grains), structure (second phases), or composition (precipitates). Area defects are given by free surfaces, interfaces, and stacking faults. Line defects are called dislocations, which so far could be avoided only in crystals of Si, Ge, GaAs. Point defects are usually divided into substitutional impurity atoms or ions, lattice vacancies, and interstitials, but larger aggregates are also possible. The threshold concentration of defects in a crystal is strongly dependent on the purpose of the material: for example, metal components of machines contain 10^8 - 10^{12} dislocations/cm^2, while silicon for integrated circuits must contain less than 10^4 dislocations/cm^2. Such levels of defect concentration are so low that sophisticated detection techniques become today more and more important.

The earliest methods for assessment of the structural perfection were etching and optical microscopy: they are not particularly fashionable at present, but their advantages (low cost, quick and easy use, simplicity of the result interpretation) make them still valuable.

Etching involves immersing the crystal in a special reagent, which dissolves the material with most rapid attack at the points where defects intersect the crystal surface. This produces etch pits at the points of emergence of dislocations, grooves along the lines of intersection with the surface of grain boundaries, and stacking faults (Fig. 18). Etch-pit counts quoted as dislocation densities are the standard method of specifying structural perfection in semiconductors.

There are a number of specialized optical microscopy techniques for crystal characterization. Phase contrast is suitable for observing regions of varying refractive index in transparent materials, while polarized light microscopy can be used for examining inhomogeneities in optically anisotropic materials and stress fields.

X-ray topography is another powerful source of information on crystal quality. In the Lang technique (Fig. 19) a line-by-line map of the intensity in the chosen beam, direct or diffracted, is recorded photographically as the crystal and the film are translated together. The resolution limit is a few microns, similar to those of optical techniques such as etching. The strain field around imperfections reduces the primary extinction of X-rays occurring in nearly perfect

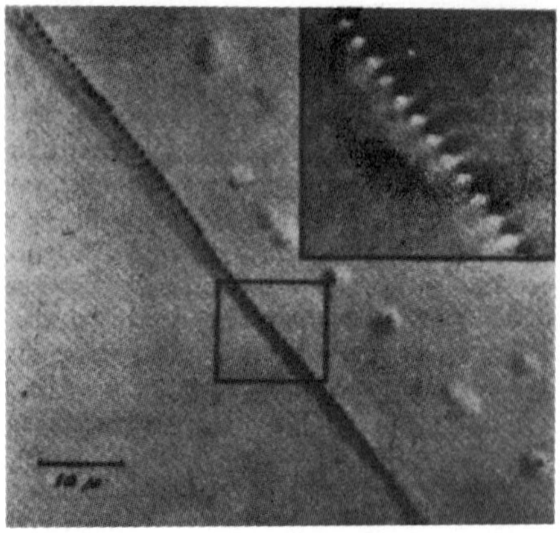

Figure 18. Etch-pits at the dislocations in Ge (Ref. 8)

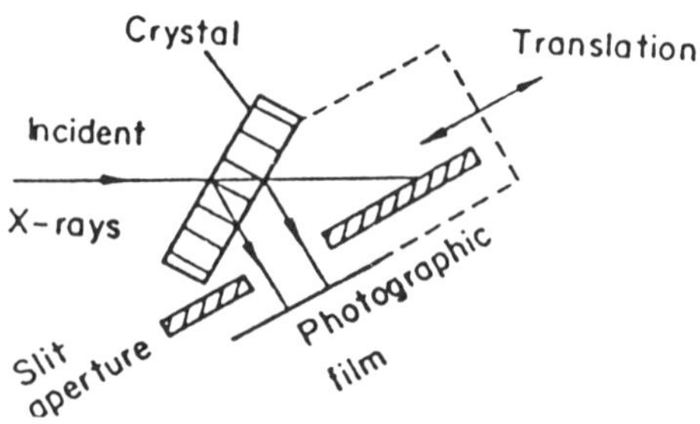

Figure 19. Lang X-ray topographic technique (Ref. 8)

Figure 20. X-ray Lang topography of a natural diamond (Ref. 8)

crystals, so that defects cause localized increases in the reflected intensity (Fig. 20). The technique is non-destructive and commercially available, but requires long times of exposure.

Transmission electron microscopy, whose resolution limit is about 1.5 Å, can easily produce both the micrograph and the electron diffraction pattern for the same area of the sample. The system, which in principle is working similarly to an optical transmission microscope (Fig. 21), is very expensive and complicated by the very small thickness of the specimens (about 2000 Å).

Scanning electron microscopy uses the forms of energy emitted by solid samples under electron bombardment (Fig. 22). The emitted X-ray photons have an energy which is characteristic of the chemical element of the emitting atom, and an intensity which is a measure of the element concentration in the specimen. The emitted electrons can be detected and different types of contrast observed, which gives information on physical properties. The cathodoluminescence can be spectrally analyzed to study activator impurities. Physical properties can also be investigated in semiconductors by means of the leakage current from the beam to earth via the specimen (absorbed electron current) and of the charge collection current (emitted current). In case of relatively thin specimens, transmitted electrons can be detected. In this way the five modes of scanning electron microscopy supply several contemporary types of results suitable for the sample characterization.

Classical optical experiments can give valuable information on the doping level in the crystals. Optical abosrption and emission (Fig. 23) are both successfully used to determine structure and concentration of impurities in a host lattice. Photostimulated and thermally stimulated luminescence, in particular, are able to reveal very low dopant concentrations (10 - 100 ppm). Segregation of various impurities,

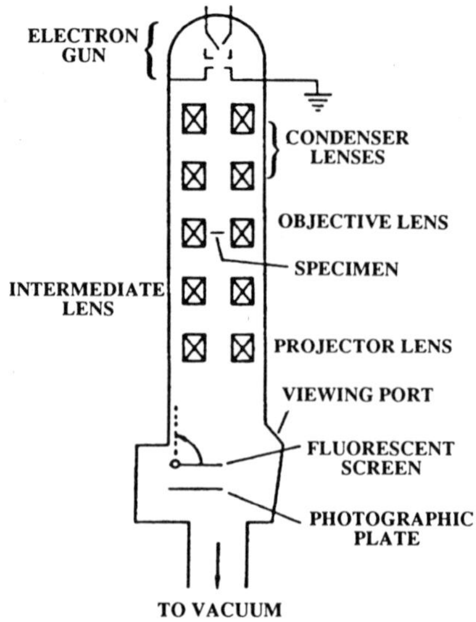

Figure 21. Scheme of a transmission electron microscope (Ref. 8)

occurring during the growth, can be easily measured with this technique (Fig. 24). These optical methods gained larger interest recently due to various technological developments: among them, the increasing availability of new and sophisticated light sources, such as Synchrotron radiation, and the wide diffusion of devices using crystals, such as solid state lasers.

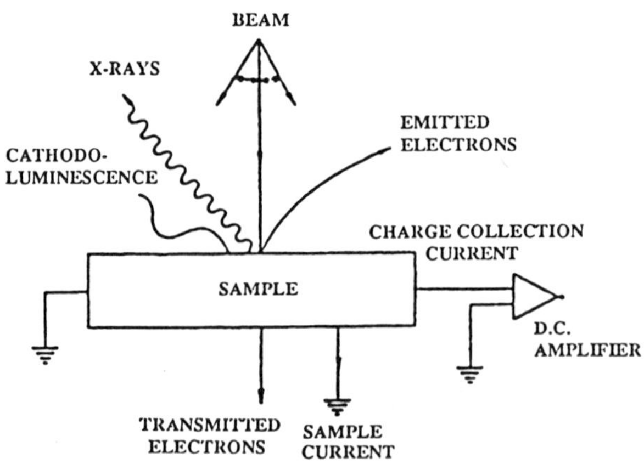

Figure 22. The types of energy dissipation in electron bombarded crystals which are employed as signals in scanning electron microscopy (Ref. 8).

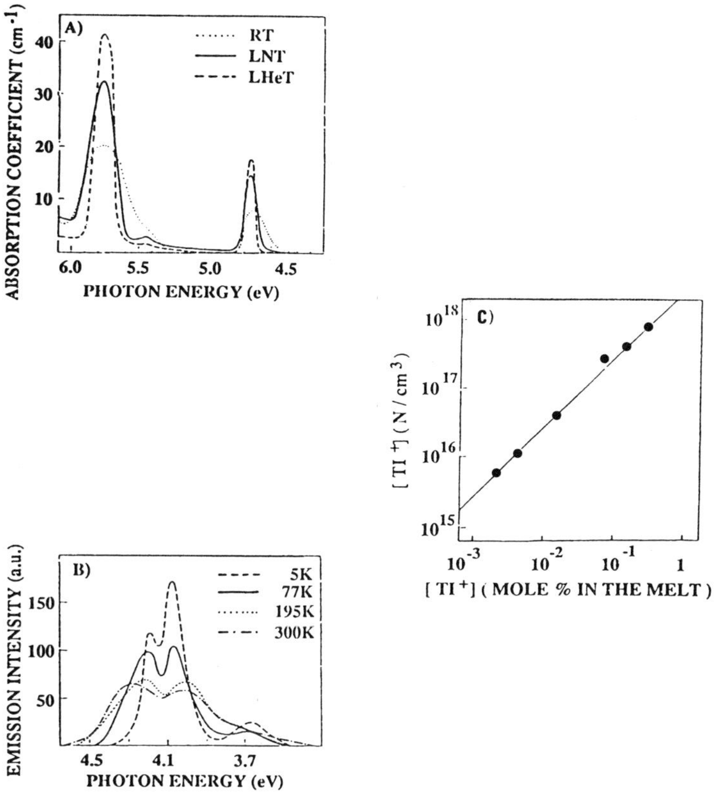

Figure 23. Absorption (A) and emission (B) spectra of Tl^+ impurity in NaBr crystals and linear dependence (C) of the crystal absorption at 4.65 eV on the dopant concentration in the melt. A) From Crystal Lattice Defects 7 (1977) 153. B) From physica status solidi (b) 84 (1977) 23.

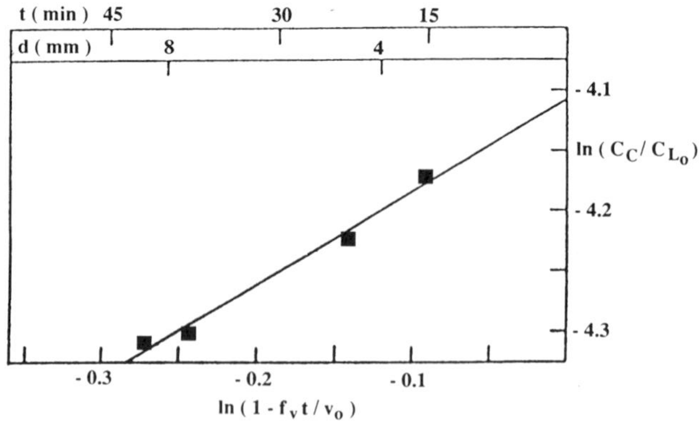

$$\ln \frac{C_c}{C_{L_o}} = a + b \ln \left(1 - \frac{f_v t}{V_o}\right)$$

C_c = dopant concentration in the crystal at the time t

C_{L_o} = initial doping of the melt

f_v = volume growth rate

V_o = initial volume of the melt

Figure 24. Segregation of Tl^+ impurity in the growth of NaBr single crystals. From Journal of Crystal Growth 41 (1977) 175.

REFERENCES

1. H.E. Buckley, Crystal Growth, John Wiley & Sons, 1950.
2. W.D. Lawson and S. Nielsen, Preparation of Single Crystals, Butterworths Scientific Publications, 1958.
3. J.J. Gilman, The Art and Science of Growing Crystals, John Wiley & Sons, 1963.
4. J.C. Brice, The Growth of Crystals from the Melt, North-Holland Publishing Company, 1965.
5. J.C. Brice, The Growth of Crystals from Liquids, North-Holland/ American Elsevier, 1973.
6. W. Bardsley, D.T.J. Hurle, and J.B. Mullin, Crystal Growth: A Tutorial Approach, North-Holland Publishing Company, 1979.
7. F. Rosenberger, Fundamentals of Crystal Growth, Springer-Verlag, 1979.
8. B.R. Ramplin, Crystal Growth, Pergamon Press, 1980.
9. Journal of Crystal Growth, North-Holland Publishing Company.
10. Progress in Crystal Growth and Characterization, Pergamon Press.

^{4}He AND ^{3}He CALORIMETRIC ABSORPTION SPECTROSCOPY: PRINCIPLES AND RESULTS ON InGaAs/AlInAs QUANTUM WELLS AND Fe IN InP AND GaAs

D. Bimberg, T. Wolf and J. Böhrer

Institut für Festkörperphysik, Technische Universität Berlin

Hardenbergstrasse 36, 1000 Berlin 12, Germany

ABSTRACT

A novel absorption technique, Calorimetric Absorption Spectroscopy (CAS), is compared to other photothermally based spectroscopic methods. CAS is more sensitive than any other low-temperature absorption technique reported hitherto and it is quantitative. The method is based on the integral detection of phonons emitted during nonradiative recombination processes. A low-temperature resistance thermometer acts as phonon detector. The sensitivity of CAS increases drastically with decreasing temperature. An increase in sensitivity by an order of magnitude to 10 pW is reached by decreasing the base temperature from 1.3 K (^{4}He-temperatures) to 0.5 K (^{3}He-temperatures). Depending on the excitation density, αd products down to 10^{-10} can be detected. The experimental set-up to perform CAS measurements at ^{3}He temperatures is described and modeled in detail. A direct determination of quantum efficiencies, obtained by combining CAS with calorimetric transmission spectroscopy (CTS), is demonstrated. Recent applications of CAS for the characterization of thin InGaAs/AlInAs quantum wells as well as for the study of the fine structure of deep traps like Fe centers in InP and GaAs are presented.

I. INTRODUCTION

Recent advances in semiconductor microstructures and optoelectronic devices are closely connected to research on nonradiative recombination processes in solids. Enhanced nonradiative recombination is very often the origin of catastrophic failure of optoelectronic devices, like semiconductor lasers. A key parameter which has to be determined in this context is the quantum efficiency of a device. Thus, techniques are required to study in particular the nonradiative recombination in these structures. On the other hand, with still improving material growth techniques like molecular beam epitaxy (MBE) or metalorganic vapour phase epitaxy (MOVPE), extremely thin and extremely pure semiconductor layers can now be fabricated. Quantum well structures often have a thickness of only a few monolayers. Epitaxial layers with residual background doping concentrations on the order of 10^{13} cm^{-3} have been grown by MOVPE [1]. A further advance of these methods depends on absorption techniques which are able to detect very small absorption signals originating from these ultrathin layers or from residual impurities in substrates and epitaxial layers. Many applications imply the detection of αd-products (α: absorption coefficient, d: layer thickness) of less than 10^{-4}.

A comparison of photothermal techniques will show that a novel absorption

Advances in Nonradiative Processes in Solids
Edited by B. Di Bartolo, Plenum Press, New York, 1991

technique, Calorimetric Absorption Spectroscopy (CAS) [2 - 5], is ideally suited to perform this task. During the last decade, a number of photothermal techniques were developed. Examples are photoacoustic absorption spectroscopy (PAS) and photothermal deflection spectroscopy (PDS). The heat produced in the sample upon nonradiative recombination of nonequilibrium charge carriers created by the absorbed photons is measured here more or less directly. CAS is more sensitive than any other low temperature absorption technique reported hitherto and it is quantitative. CAS at 1.5 K was introduced by Bubenzer et al. [2] already in 1980, and Bimberg et al. [3] in 1981 and extended to the mK range by Gruhl et al. [4]. A detailed description of an optimized CAS set-up has been given recently by Juhl and Bimberg [5].

CAS is based on the integral detection of phonons at low temperatures. Phonon emission is caused by nonradiative recombination processes during the relaxation to thermal equilibrium of an externally excited system. The phonon emission corresponds to a temperature increase in the sample and can be measured by a low-temperature thermometer. Such a detection system shows no wavelength dependence of its sensitivity in contrast to photon detectors. Therefore spectra can be taken in one continuous run over an extremely large energy range, which is not restricted by the detector system.

Theoretically, the sensitivity of CAS is given by the minimum heat power (ΔP) detectable by a low temperature calorimeter. ΔP is of the order of a few picowatts at 0.5 K. Using input powers (P) of 100 mW as produced by dye lasers, one would obtain a detectable αd-product of the order of $\alpha d = \Delta P/P = 10^{-10}$. The sensitivity of CAS is thus several orders of magnitude larger than that of other absorption techniques like Transmission, Photoacoustic or Photothermal Deflection Spectroscopy. This sensitivity is even much larger than one needs in most cases. Therefore the input power and the expense can be reduced by using cheap low-power and easily handleable illumination sources like tungsten lamps combined with monochromators of high spectral resolution ($<$ 0.1nm). CAS signals can be easily calibrated by an electrical heater simulating the absorbed light power which is converted into heat. The calibrated CAS spectra containing information on nonradiative recombination processes can be compared to transmission spectra containing information on both radiative and nonradiative transitions. This comparison yields quantum efficiencies otherwise difficult to determine.

This paper is organized as follows. First a comparison of photothermally based spectroscopic techniques is given in section II. In section III, Calorimetric Absorption Spectroscopy is presented and modeled in detail. A set-up for transmission measurements, called Calorimetric Transmission Sepctroscopy (CTS), is described in section IV. In addition, it is demonstrated that a combination of CAS and CTS leads to a direct determination of quantum efficiencies. Typical applications of CAS are presented in section V. In section VA we report CAS experiments on thin InGaAs/InAlAs quantum wells. In section VB an investigation of iron-related charge-transfer and internal transitions in semi-insulating GaAs and InP substrates is presented in order to demonstrate the potential of CAS for the analysis of low concentrations of impurities in semiconductors. Some remarks concerning the limitations of CAS from an experimental point of view are given in section VI.

II. COMPARISON OF PHOTOTHERMAL METHODS

A number of photothermally based spectroscopic techniques have been proposed and demonstrated during the last 10 years: Thermal Lensing (TL) [6], Photoacoustic Spectroscopy (PAS) [7,8], collinear and transverse Photothermal Deflection Spectroscopy (PDS) [9] and interferometric techniques [10,11]. A review has been given recently by Jackson et al. [9]. The following Table I is taken from this reference and completed by the relevant parameters of CAS. Column 4 contains a measure for the sensitivity of each technique: The minimum detectable αd - product multiplied with the illumination power. These values correspond to the smallest absorbed light power that can still be measured.

Table I. Comparison of photothermally based absorption techniques. The first lines are taken from Ref. 9 . $(\alpha d)_{min}$ x pump power designates the minimum detectable heat power.

Technique	Samples Already Studied	Experimental Set Up	Sensitivity $(\alpha \cdot d)_{min}$ x Pump Power (W)	Special Features
TL	Liquids	Difficult to align	10^{-7}... 10^{-8}	No mechanical contact. Permits hostile environment and *in situ* measurements.
Microphone PAS	Solids (bulk, powder), liquids, gases	Simple to align	10^{-4}... 10^{-5} (solids) 10^{-7} (gases)	Sensitive to mechanical and acoustical noise.
PZT PAS	Solids (bulk) liquids	Simple to align	10^{-5} (10^{-9}J pulsed)	Sample attachement difficult.
Collinear PDS	Optically clean transparent solids, liquids and gases	Difficult to align	10^{-7}... 10^{-8} (10^{-10}J pulsed)	No mechanical contact. Permits hostile environment and *in situ* measurements.
Transverse PDS	Condensed phase samples	Simple to align	10^{-5}	No mechanical contact. Permits hostile environment and *in situ* measurements.
Interfero-metry	Liquids and solids	Difficult to align	10^{-7}... 10^{-8}	No mechanical contact. Permits hostile environment and *in situ* measurements.
CAS	solids	Very simple to align	10^{-11} at 1.3K 10^{-14} at 50mK	Sample attachement simple including direct calibration

Obviously, CAS is much superior since it has the highest sensitivity of all methods by orders of magnitude. In addition, CAS is particularly suited for investigations at low temperatures where insight on electronic properties of solids is gained easily. Such investigations are in the center of interest here. We will therefore focus in the following chapters on a detailed description of an optimized CAS set-up and on typical applications of CAS. More details on other photothermal techniques can be found in this volume in the paper by U.M. Grassano.

III. CALORIMETRIC ABSORPTION SPECTROSCOPY (CAS)

III.A. The Calorimeter

In Fig. 1 the sample/thermometer arrangement is schematically depicted. The sample is suspended with nylon threads of low thermal conductivity from a sample holder and fixed in vacuum in a He-cryostat[12]. A sapphire chip serves as support for a low temperature thermometer. Sapphire is very suitable for this application due to its small heat capacity ($6 \cdot 10^{-9}$ J/gK at 0.5 K), high heat conductivity (0.3 W/mK at 0.5 K) and electrical insulation. The thermometer is thermally well coupled to the sample via a copper wire and connected to the He-bath by a thermal link producing a small temperature gradient between He-bath and sample/thermometer arrangement. A typical thermal link consists of a manganine wire, length 3 mm, Ø = 0.1 mm. The thermometer should be shielded and not be directly attached to the sample to prevent heating from absorption of light. An internal heater consisting of a gold film evaporated on the back side of the sapphire chip is installed to calibrate the complete set-up. The electric connections of the heater and of the thermometer are made of

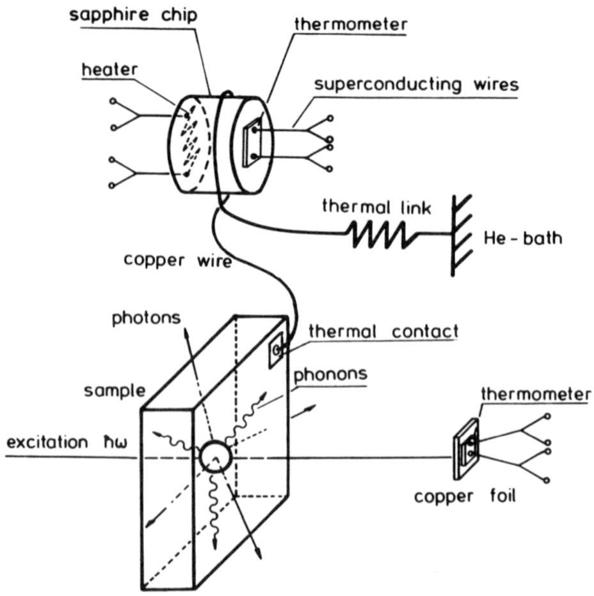

Fig. 1. Sample-thermometer arrangement for Calorimetric Absorption and Transmission Spectroscopy.

superconducting NbTi wires. This ensures that the heat can only flow through the copper and manganine wires to the He-bath, but not through electric connections of the thermometer and the heater. Otherwise a temperature gradient inside the thermometer would result.

III.B. The Thermometer

For base temperatures between 4.2 K and 1.3 K, a commercial Allan-Bradley carbon resistor serves as low-temperature thermometer. In order to reduce its heat capacity, the resistor is ground to approximately 0.3 mg. At 1.3 K its sensitivity is 100 Ω/mK. Typical thermometer rise times at 1.3 K are around 20 ms. Silver paint is used to mount the Allan-Bradley resistor on the sapphire chip and to connect it with two superconducting wires. Nevertheless, a four-point resistance measurement technique can be employed due to the vanishing electrical resistance of the superconducting material.

Still lower base temperatures (down to 0.4 K) can be obtained by employing ^{3}He instead of ^{4}He. We have recently installed a ^{3}He system. In this temperature range, a Ge-resistance thermometer turns out to be superior to the carbon element as its resistance drastically increases with decreasing temperature yielding a better temperature resolution. This can be seen from Fig. 2 where the resistance versus temperature characteristics of the Ge- and the Allan-Bradley thermometer are compared. We use an unencapsulated GR - 200 A - 100 element (Lake Shore Cryotronics, Inc.). Extra care must be taken when mounting the Ge-thermometer to avoid any strain which could alter the electrical characteristics of the resistor. Electrical connections are made by four gold wires bonded to the Ge-resistor. Again, each of them is continued with a superconducting NbTi wire to avoid any heat flow to the He bath other than through the thermal link.

III.C. Time Dependence of the Calorimetric Signal

In order to get information on the time dependence of the CAS signal ΔT and in particular on the temperature rise time τ at the thermal sensor we have to

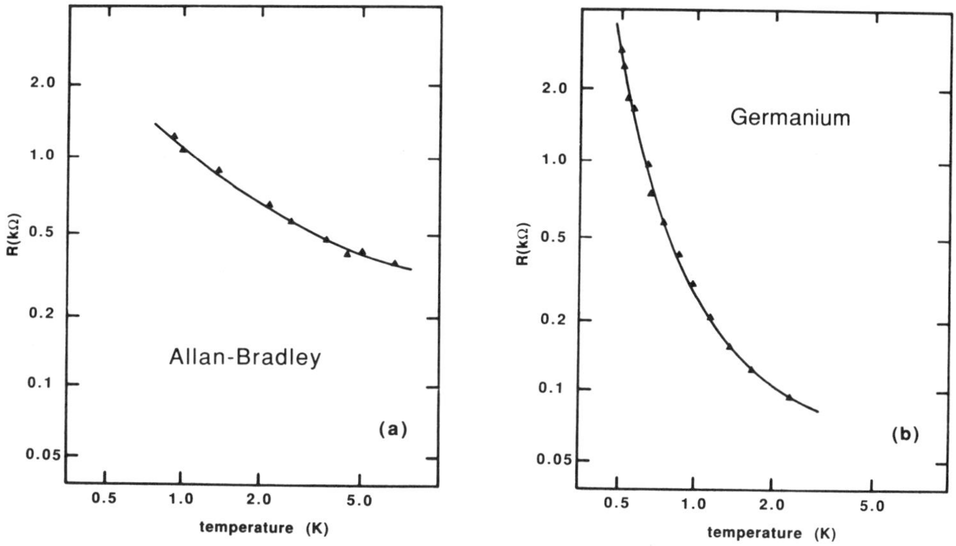

Fig. 2. Temperature versus resistance characteristic of an Allan-Bradley carbon resistor (a) and a n-Ge resistance thermometer (b).

investigate in detail the thermal circuit. Thermal capacities and thermal resistances are introduced in analogy to electric circuits. As shown in Fig. 3 the sample and the thermometer/heater arrangement represent thermal capacities C_S and C_T, respectively. The thermal link R_L (Fig. 1) as well as a resistor between sample and thermometer (R_S) -caused by acoustic mismatch at the sample-copper-wire interface - represent resistances. An additional resistor R_T exists between the sapphire-copper-wire interface. In contrast to R_L, the resistances R_T and R_S are undesirable. In the present case $C_S/C_T < 0.02$ and $C_S = 10^{-8}$ J/K. The thermal time constant of the sample (i.e. the time needed to reach a thermal equilibrium inside the sample) is estimated from the diffusivity to be approximately 20 μs [13], whereas the external thermalization time constant between He bath and thermometer is on the order of 1s. Hence C_S can be neglected with respect to C_T and a simple differential equation results:

$$\Delta T + C_T\,(R_T + R_L)\;\Delta\dot{T} = \Delta P \cdot R_L \qquad (1)$$

where ΔP is the light power absorbed by the sample and converted to heat by nonradiative recombination processes. The time dependence of the temperature rise ΔT at the thermal sensor, caused by illumination from $t = 0$ until $t = t_0$, is obtained by integrating Equ. (1):

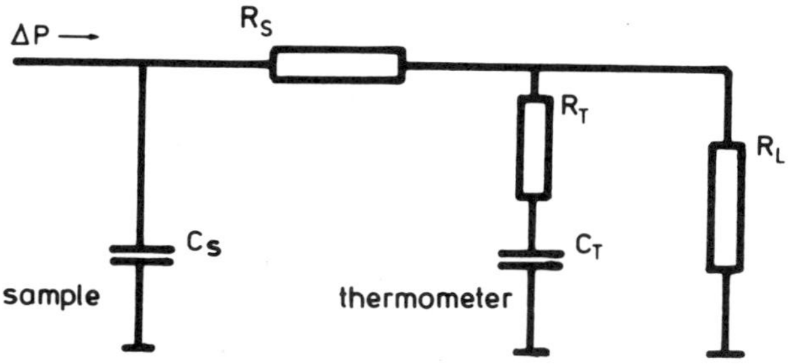

Fig. 3. Thermal circuit diagram for Fig. 1.

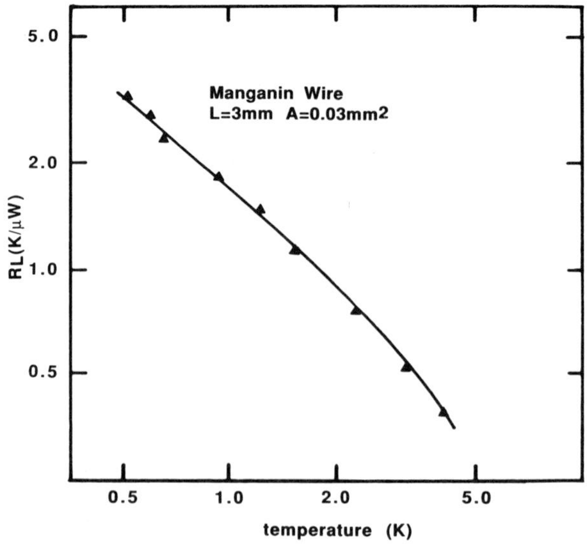

Fig. 4. Temperature dependence of the resistance of the thermal link (Manganine wire, diameter: 0.03 mm2, length: 3 mm).

$$0 < t \leq t_0: \Delta T = P \cdot R_L (1 - e^{-t/\tau}). \qquad (2)$$

The time constant τ is given by:

$$\tau = (R_T + R_L)\, C_T. \qquad (3)$$

After the illumination is switched off ΔT regains its steady state value with the same time constant τ for $t > t_0$. There exist two different possibilities to perform CAS experiments: pulsed and c.w. illumination.

1. Pulsed Illumination. If the sample is illuminated during a time t'_0 much smaller than the time constant τ, Equ. (2) simplifies to:

$$\Delta T\,(t'_0) = \Delta P\, R_L\, t'_0/\tau \qquad (4)$$

Using Equ.(3) and $\Delta Q = \Delta P \cdot t'_0$ for the heat created by the recombination process and $R_T \ll R_L$ (in our case $R_T = 0.03\, R_L$) we obtain the simple calorimetric relationship:

$$\Delta T\,(t'_0) = \Delta Q/C_T. \qquad (5)$$

The heat capacity of the thermometer C_T (including its appendages) determines the sensitivity of the experiment. Larger CAS signals are obtained in c.w. experiments.

2. c.w. experiments. If $t_0 \gg \tau$ Equ. (2) simplifies to:

$$\Delta T\,(t_0) = \Delta P \cdot R_L. \qquad (6)$$

Now the thermal resistor R_L determines the height of the temperature increase ΔT and consequently the sensitivity of the experiment. It is evident that the c.w. method ($t_0 \gg \tau$) has several advantages in comparison to the pulsed method: The calibration of the CAS signals is simpler, since heating time and illumination time do not need to be identical. Secondly, the CAS signal for a given absorbed light power is much larger, whereas the ΔT peak value found at pulsed illumination is reduced by t'_0/τ

as compared to a c.w. experiment. The measurement time is reduced by a factor of the same order. Furthermore, in c.w. measurements the maximum tolerable scanning speed of the spectrometer given by the ratio $\Delta\lambda/5\tau$, where $\Delta\lambda$ is the spectral bandwidth of excitation, is not much less than in a pulsed experiment. However, signal averaging can be done much more conveniently. Therefore, all spectra reported here are obtained under steady state conditions.

III.D. Temperature Dependence of the CAS Parameters.

The temperature rise ΔT of the thermometer is measured via a change of the electric resistance ΔR of the resistance thermometer. The relative resistance change $\Delta R/R$ is:

$$\frac{\Delta R}{R} = \frac{1}{R}\frac{\Delta R}{\Delta T}\Delta T = G\,\Delta T \qquad , \qquad (7)$$

where $G = 1/R\ (\Delta R/\Delta T)$ is the relative sensitivity of the thermometer. For c.w. measurements (Equ. 6) we obtain:

$$\frac{\Delta R}{R} = G\,R_L\Delta P \qquad . \qquad (8)$$

Equ. (8) shows that the thermal resistance R_L and the relative thermometer sensitivity G are the two key parameters governing the sensitivity of the entire CAS system.

In what follows we will show that the temperature at which the experiment is performed is of largest importance for the system performance (sensitivity and speed of data acquisition). It is obvious from Fig. 2 that the relative sensitivity of both the Allan Bradley and the Ge-resistor increases with decreasing temperature. In Fig. 4, the temperature dependence of the thermal resistor R_L for a fixed length and diameter of the manganine wire is depicted. R_L and G determine together the sensitivity of CAS under steady state illumination (see Equ. 8). R_L and G increase strongly with decreasing temperature.

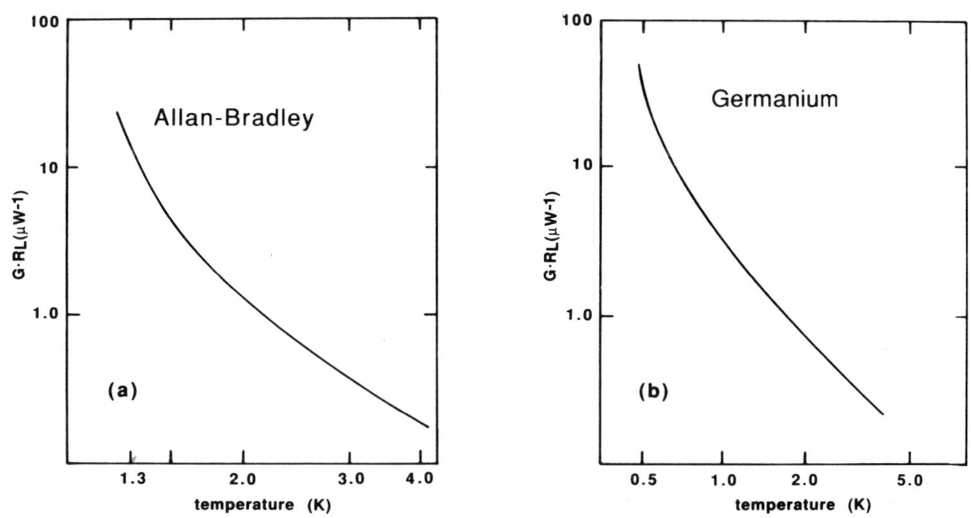

Fig. 5. Temperature dependence of the product of relative sensitivity of the thermometer and thermal resistance which determines the sensitivity of CAS under c.w. illumination. (a): Allan-Bradley carbon resistor, (b): n-Ge resistor.

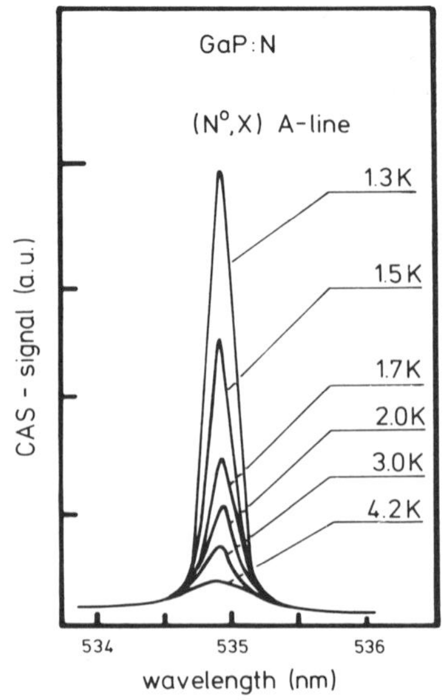

Fig. 6. CAS-signal of the A-line of GaP:N at several temperatures.

Between 4.2 K and 1.3 K the product $G \cdot R_L$, which controls the sensitivity of CAS (see Equ. 8), increases by a factor of 82 as shown in Fig.5. A direct measurement of the relative resistance changes $\Delta R/R$ at both temperatures is found to be in excellent agreement with this prediction. In order to illustrate the increase in sensitivity with decreasing temperature, Fig. 6 shows CAS spectra of the A-line of the N-bound exciton in GaP:N at several temperatures.

In principle, R_L can be varied by changing the length or diameter of the manganine wire. Thus, still higher sensitivities could be achieved by increasing R_L. The choice of R_L is limited, however, according to Equ. 3 as the time constant τ is given by the product $R_L \cdot C_T$. Increasing R_L increases the time constant. In order to have a tolerable speed of data acquisition, τ should not be much larger than 1s. Fig. 7 shows the temperature dependence of the time constant with the same thermal link as in Fig. 4. Although R_L increases with decreasing temperature, a decrease of the time constant is observed. Apparently, the increase in R_L is overcompensated by a very strong decrease of the thermal capacity C_T of the thermometer/heater arrangement with decreasing temperature. As shown in Fig. 7, τ reduces by about one order of magnitude between 4.2 K and 0.5 K. The data acquistion speed can in turn be increased by this factor.

It is obvious that the highest sensitivity of CAS is indeed obtained for minimum base temperature. Thus this temperature should be choosen as low as possible.

III.E. Sensitivity at ^{3}He and ^{4}He Temperatures

The sensitivity $\Delta R/R$ of the calorimeter depends on the thermal resistor R_L and the relative sensitivity G of the thermometer. The resistance resolution of the bridge (S.H.E., model PCB) is $\Delta R/R = 10^{-4}$. For a thermometer sensitivity $G(1.3K) = 5\ K^{-1}$ of the Allan-Bradley resistor, a minimum detectable temperature increase ΔT of 20μK is obtained from Equ. (7). For a typical thermal resistor R_L of $2 \cdot 10^6$K/W, Equ. (8) yields a minimum power of $\Delta P = 10$ pW which can be detected by CAS at 1.3 K.

Even higher sensitivities can be obtained by using ^{3}He instead of ^{4}He. Then the

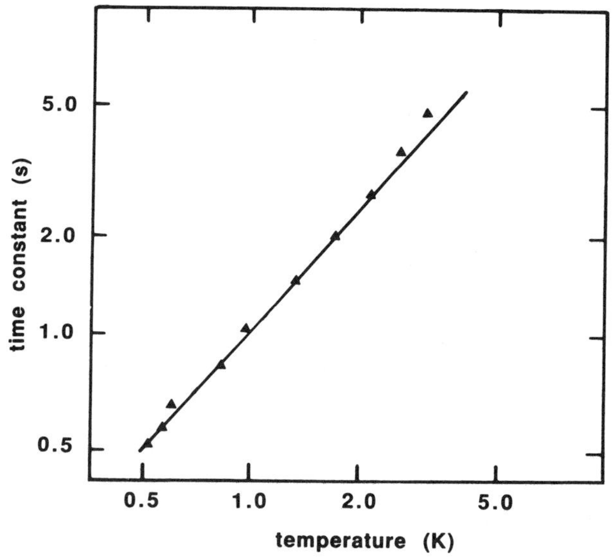

Fig. 7. Temperature dependence of the CAS time constant.

base temperature of the sample and the He bath drops from 1.3 K to a value of less than 0.5 K. The most important parameters of CAS at 0.5K and 1.3 K are compared in Table II. Using a germanium thermometer as CAS detector, a relative sensitivity G of 8.5 K^{-1} at 0.5 K is obtained which yields a minimum detectable temperature increase ΔT of 12 μK. With the length and diameter of thermal link unchanged, R_L increases from 1.9 K/(μW) at 1.3 K to 3 K/(μW) at 0.5 K. This results in an increase in sensitivity, as given by the product $G \cdot R_L$ or the minimum detectable power ΔP_{min},

Table II. Comparison of CAS parameters at 0.5 K (^{3}He temperatures) and 1.3 K (^{4}He temperatures).

	T=0.5K	T=1.3K
Thermometer	Germanium	Allan-Bradley
G (K^{-1})	8.5	5
ΔT_{min} (μK)	12	20
R_L (K/μW)	3	1.9
$G \cdot R_L$ (μW^{-1})	25.5	9.5
ΔP_{min} (pW)	4	10
τ (s)	0.5	1.5
NEP (pW)	10	250

of about a factor of 3. Simultaneously, the time constant τ decreases by the same factor from 1.5 s at 1.3 K to 0.5 s at 0.5 K. Theoretically, an even higher decrease of τ could be anticipated from Debye's law. The improvement of the time constant is limited, however, by the somewhat larger heat capacity of the germanium thermometer compared to the Allan-Bradley resistor. In total, if we define the quotient of sensitivity and time constant $\Delta P_{min}/\tau$ as a global figure of merit of the CAS system, the transition from ^4He to ^3He temperatures has lead to an improvement by about an order of magnitude.

As important as the sensitivity of the system, calculated from G, R_L and the resolution of the resistance bridge, is the noise equivalent power (NEP) which is given in Table II. Experimentally, we are dealing with a background-noise-limited detection where a number of sources contribute to the background noise. These include thermal radiation, electronic noise, temperature drift and external sources of heat generated inside the calorimeter/sample assembly like HF radiation and mechanical vibrations. Some of them will be discussed in more detail in section VI. With several improvements installed recently in order to further reduce the background noise in our set-up, we are now able to detect an absorbed power of as low as 10 pW at 0.5 K under standard measurement conditions. This represents an improvement by a factor 25 as compared to previous measurements at ^4He temperatures.

Dye lasers are the usual light sources for excitation spectroscopy. Their power in $\Delta\lambda$ =0.05nm surpasses 100 mW in the visible and near infrared. According to $\Delta P/P = \alpha d$, a dynamic range of 10^9 could be obtained and αd-products of less than 10^{-9} could be measured, if such a laser were employed for excitation. It is obvious from these numbers that such a high input power will not be needed for most practical cases. For a low power illumination source with P = 1 µW in $\Delta\lambda$ =0.1nm, the dynamic range is still 10^4 and αd-products of less than 10^{-4} can be measured. Operating with input powers of 1 µW allows using cheap and simple halogen tungsten lamps or infrared light sources combined with a monochromator of high spectral resolution covering a much larger wavelength range than a conventional dye laser.

Note that our detection system is completely independent of wavelength. It is only the intensity of the light source which determines the useful wavelength range for CAS measurements. Especially in the infrared regime CAS is by orders of magnitude superior to conventional transmission techniques. Typical applications for CAS at ^3He temperatures are problems where one has to deal with small absorption coefficients and small thicknesses. An important example is the characterization of transition metal impurities in thin epitaxial layers. Here the typical αd-product drops to 10^{-4}. Most absorption lines lie in the infrared region, and a high spectral resolution is required to distinguish the sharp lines characteristic for a certain impurity. Under these conditions, it becomes mandatory to work at ^3He temperatures.

III.F. The Experimental Set-Up

A block diagram of the complete experimental set-up for CAS and CTS is shown in Fig. 8. The sample is mounted in the vacuum system of a two-chamber cryostat having a ^4He filling volume of 700cm^3 and a holding time of 12 hours. A ^3He container is located inside the ^4He shield and connected to a closed cycle ^3He handling system. Approximately 10 l of ^3He gas corresponding to 13 cm^3 of liquid are sufficient to maintain a base temperature of less than 0.5 K for about 8 hours. A combined visible/infrared light source consisting of a 250 W tungsten lamp and a silicon carbide source contained in one housing is employed. A high-resolution 1 m f/10 monochromator is used to spectrally decompose the light incident on the sample.

The power of the source as a function of wavelength is measured calorimetrically using a black body in the place of the sample. A spectral bandwidth of typically 0.1 nm or less is used in our experiments. The optical path is flushed by dry nitrogen to remove atmospheric absorption. Spectra are presently being taken in the wavelength range between 5 µm and 400 nm. The resistance changes of the resistance thermometers are measured using a potentiometric conductance bridge (PCB). The

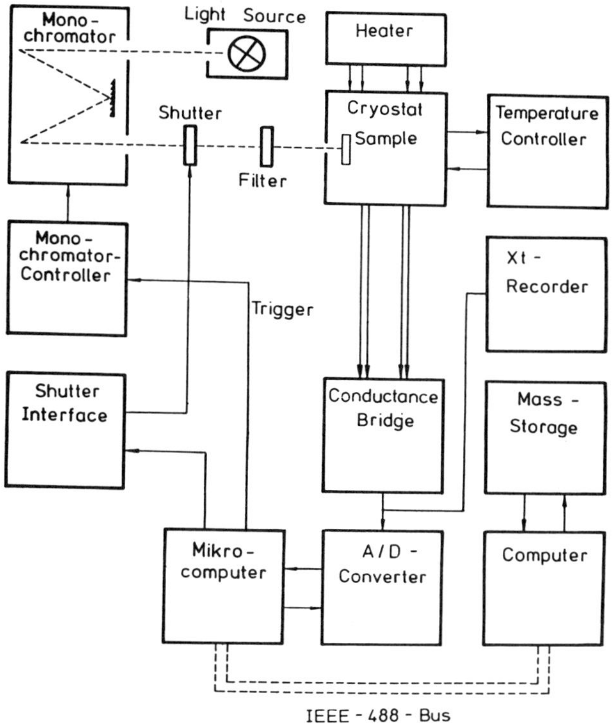

Fig. 8. Block diagram of the CAS set-up.

analog output signal of the PCB is digitized by an AD-converter. Data storage, processing and graphic output are performed by a microcomputer system. This computer also normalizes spectra to take care of the variations of the incident photon flux in the relevant wavelength range. In addition, the resistance changes of the thermometer are used to compute directly the absorbed power.

IV. CALORIMETRIC TRANSMISSION SPECTROSCOPY (CTS) FOR THE DETERMINATION OF QUANTUM EFFICIENCIES

A simple extension of the CAS system allows to perform transmission experiments with the same set-up. The basic idea is depicted in Fig. 1. A black painted copper foil, acting as wavelength-independent photon detector, is mounted directly behind the sample in order to absorb all light transmitted through the sample. The heat created by the absorption of photons can then be detected with a thermal sensor (e.g. a further Allan-Bradley resistor) mounted on the backside of the copper foil. All signal processing is done in a way identical to CAS.

A combination of CAS and CTS leads to a direct and absolute determination of quantum efficiencies. CTS measures the total absorption followed by dissipation via nonradiative and radiative recombination channels whereas the CAS signal is a measure of the nonradiative ones. The quantum efficiency η is given by [3]:

$$\eta = 1 - (\Delta P_{nr})/(\Delta P_{nr+r}) \qquad (9)$$

where ΔP_{nr+r} is the total absorbed light and ΔP_{nr} is the part of absorbed light which is dissipated via nonradiative recombination processes. If self-absorption mechanisms are neglected:

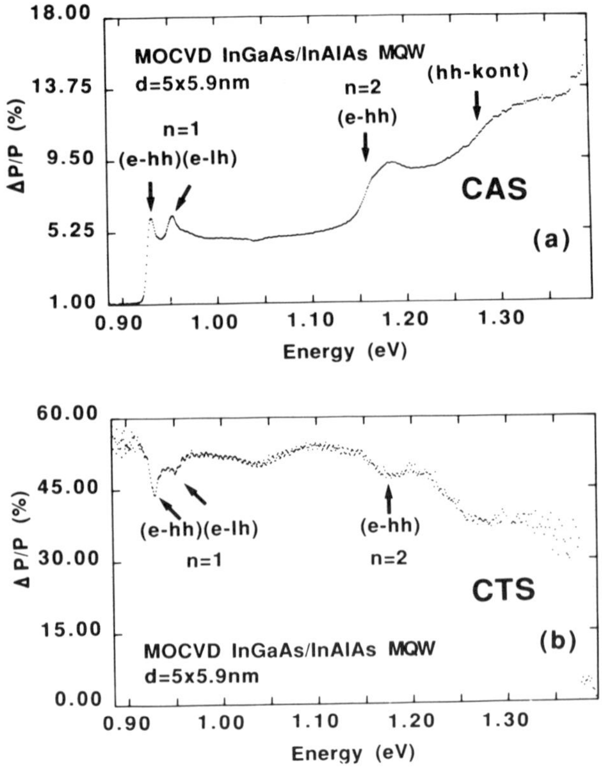

Fig. 9. CAS (a) and CTS (b) spectra of an InGaAs/AlInAs quantum well sample (5 wells with a layer thickness of 5.9 nm).

$$\Delta P_{nr} = \Delta P_{CAS} \cdot \tag{10}$$

The intensity ΔP_{CTS} transmitted by the sample, taking multiple reflections but not interference into account [14], is given by:

$$\Delta P_{CTS} = P\,(1 - R)^2 \frac{e^{-\alpha d}}{(1 - R^2 e^{-2\alpha d})} \quad . \tag{11}$$

The total absorbed light power ΔP_{nr+r} is:

$$\Delta P_{nr+r} = P\,(1-R)\,(1-e^{-\alpha d})/(1-Re^{-\alpha d}) \; . \tag{12}$$

From the transmission experiment the αd-product can be determined and inserted into Equ. (12) to obtain ΔP_{nr+r}. Insertion of Equs. (10) and (12) into Equ. (9) yields the quantum efficiency η .

V. APPLICATIONS OF CAS

V.A. Characterization of InGaAs/AlInAs Quantum Wells

A typical CAS measurement of a sample containing very thin layers is given in Fig. 9a. The spectrum of InGaAs/InAlAs quantum wells (5 wells, well thickness L_Z = 5.9 nm) grown by metalorganic vapour phase epitaxy (MOVPE) is depicted here. Fine structure resulting from excitonic absorption is clearly resolved. Note that the dots represent the original experimental results. No smoothing or averaging routines have been applied. CAS measurements of single quantum wells have already been reported by some of us earlier [5, 15].

In order to demonstrate the power of CAS in the detection of signals from very thin layers an estimate of the minimum layer thickness of a quantum well of which the absorption can still be detected by CAS shall be given here. A typical absorption coefficient for quantum wells of $\alpha = 10^5$ cm^{-1} is assumed. With a minimum detectable absorption ratio $\Delta P/P = \alpha d = 10^{-4}$ (corresponding to a wavelength resolution of better than 0.1 nm and an input power P below 1 µW) and with a quantum efficiency of 90 % a minimum "layer thickness" of $d = 10^{-3}$ nm is derived, which is more than two orders of magnitude less than the thickness of an atomic layer. Therefore, absorption contributions stemming from single monolayers can be easily detected even if the absorption coefficient of the layer is only 10^3 cm^{-1}.

From a detailed analysis of the spectrum shown in Fig. 9a, information on band discontinuities, oscillator strengths and specific properties of the sample can be derived. A detailed discussion of the various effects like interface roughness, clustering, alloy fluctuations and band filling, which contribute to the observed shape, is given elsewhere [15]. Here we wish to emphasize that CAS is indeed able to detect signals from very thin layers which in addition recombine predominantly radiatively.

In Fig. 9b the CTS spectrum of the same sample is depicted. Following the discussion in Sec. IV, a quantum efficiency can now be derived. At 0.932 eV, corresponding to the peak of the (e, hh) exitonic transition of the quantum well, a relative CAS signal of $\Delta P_{CAS}/P = 0.051$ is found (Fig. 9a). At the same photon energy, we obtain from the CTS experiment $\Delta P_{CTS}/P = 0.093$ (Fig. 9b). Using a reflection coefficient of $R = 0.278$ it follows from Equs. (11) and (12) $\Delta P_{nr+r}/P = 0.63$. The quantum efficiency of the exitonic (e,hh)-transition (electron heavy-hole transition) in our MOVPE InGaAs/InAlAs quantum wells is thus 0.92 according to Equ. (9). A more detailed evaluation, including self absorption effects, has been described by Juhl and Bimberg [5]. In our case, the quantum efficiency increases to 0.94 if this correction is taken into account.

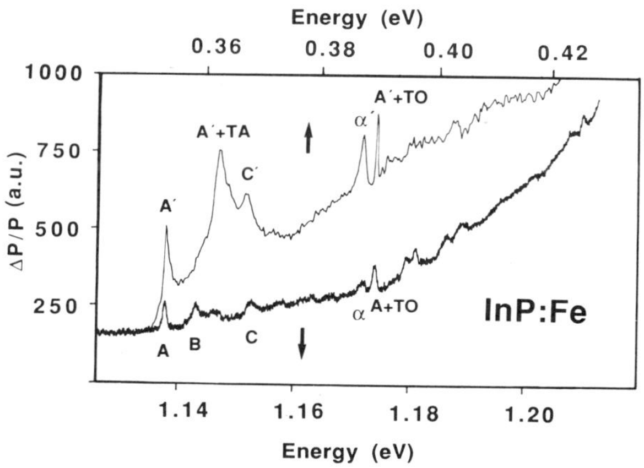

Fig. 10. CAS spectra of a semi-insulating InP:Fe substrate. Upper part: internal $^5E \rightarrow ^5T_2$ transition of Fe^{2+}, lower part: charge transfer process $Fe^{3+} + h\nu \rightarrow Fe^{2+}(^5T_2) + e^+$.

It should be pointed out that in spite of this predominantly radiative recombination process, the (e,hh) exitonic transition is clearly visible in the CAS spectrum, although only the nonradiative part (~6 %) of the recombination is detected.

V.B. Fe in InP and GaAs

Transition metals (TM's) in III-V semiconductors are at present the subject of extensive experimental studies since semi-insulating III-V substrate materials are commonly fabricated by TM doping, or by antisite defect formation in the case of GaAs. Especially InP:Fe is of great importance as a substrate in InP based device technology. CAS has been succesfully applied to study low concentrations of nonradiative recombination centers like TM impurities in semiconductors [16-18]. Recently, we have been able to reveal a bound-exiton related fine structure in the CAS spectra of InP:Fe [16,17]. So far, although a number of attempts have been undertaken to reproduce these results [19,20], only Fourier Transform Spectroscopy has been able to yield a similar resolution [20]. Here, we will present new results on semi-insulating InP:Fe and GaAs:Fe recently obtained in our laboratory [21].

In Fig. 10 two CAS spectra of a semi-insulating InP:Fe substrate are shown (sample thickness 300 µm, $n_{Fe} = 5 \times 10^{16}$ cm^{-3}). The spectrum in the lower part originates from the charge transfer process $Fe^{3+} + h\nu \rightarrow Fe^{2+}$ ($^5T_2$) + e^+ where an electron is transfered from the valence band to the exited $^5T_2$ level of Fe^{2+} [17]. Some of the fine structure observed here could be correlated with the known crystal field splitting of the $^5T_2$ level whereas the assignment of others, like the peak α, remained tentative. In the upper part of Fig. 10, a CAS spectrum of the internal $^5E \rightarrow {}^5T_2$ transition of Fe^{2+} is depicted. Note that this transition occurs much further in the infrared (around 3.5 µm). Thus, an extension of the useful spectral range of CAS and a significant increase in sensitivity preceeded this measurement. A striking similarity is observed between the two spectra. Apparently, the same final states are involved in both cases which can be only multiplet states of $^5T_2$. In particular the peak α in the charge transfer spectrum can now be unambiguously assigned to a transition from the valence band to the $^5T_2$ level of Fe^{2+}. A complete level diagram

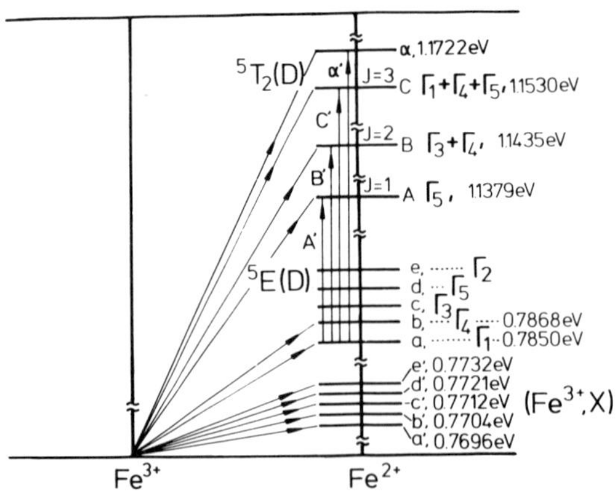

Fig. 11. Detailed energy level scheme of Fe in InP. All transitions observed by CAS are schematically indicated.

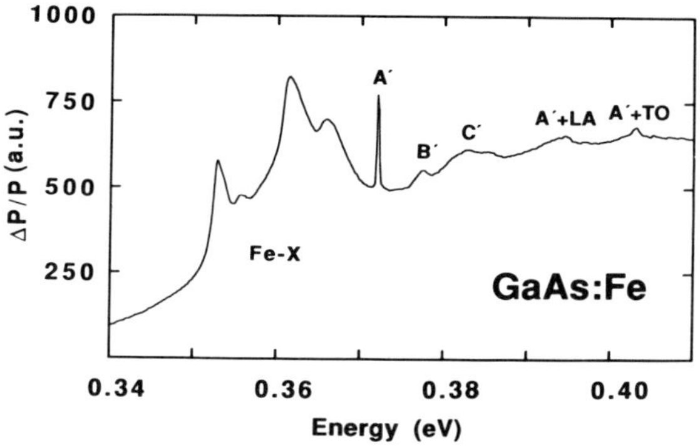

Fig. 12. CAS spectrum of the internal $^5E \rightarrow {}^5T_2$ transition of Fe^{2+} in GaAs. The fine structure originates from the same transitions as in Fig. 10 and Fig. 11. Broad maxima are due to iron-related complexes.

of InP:Fe including the excitonic, charge transfer, and internal transitions is shown in Fig. 11. The difference between the energy position of the peaks A and A' yields directly and with a much higher precision than available before the energy position of the iron acceptor level (Fe^{2+}/Fe^{3+}) in the band-gap of InP.

A CAS spectrum of a semi-insulating GaAs:Fe sample is shown in Fig. 12. Again the energy range corresponding to the internal $^5E \rightarrow {}^5T_2$ transition of Fe^{2+} is depicted. The peak labeled A' corresponds to the transition between the lowest ^5E and $^5T_2$ levels. As a result of the larger crystal field splitting in GaAs, this transition occurs at slightly higher energies than in InP (see Fig. 10). In addition, a number of broad peaks is observed at energies below the peak A'. A detailed analysis of their energy position and shape leads to assign them to Fe-X complexes. Possible candidates for X are either group IV elements or oxygen.

In conclusion, we have demonstrated that CAS is able to reveal a deep insight into the physical properties of TM impurities as in particular shown in the case of iron in InP. At the same time, CAS is sensitive enough to yield information on residual impurities in thin semiconductor samples which is extremely important for a further refinement of these materials.

VI. EXPERIMENTAL LIMITATIONS OF CAS

So far, we have demonstrated that CAS is extremely powerful in detecting very small absorption signals. Of course, a number of experimental problems have to be solved when dealing with such a highly sensitive technique like CAS. Some of them will be discussed in this section.

A very important point is the elimination of all unwanted sources of heat generated in the sample/calorimeter arrangement. Heat other than from the absorption of light would result in an unacceptable background in the CAS spectra, create noise and, by increasing the base temperature, lower at the same time the sensitivity of the system. Two major sources of heat are mechanical vibrations and HF-radiation. Thus, great care has to be taken to properly shield the set-up from HF-radiation. Vibrations originate essentially from building resonances and pumps.

The impact of both of them has to be reduced by an appropriate vibrational isolation of the cryostat from its support frame and by uncoupling the vibrations from the pumping lines.

Another source of heat which cannot be avoided in our set-up is the room temperature blackbody radiation impinging on the sample from the outermost window of the cryostat. For CAS measurement in the mK range, optical fibers have been successfully used to illuminate the sample. Then, however, a dye laser system is needed as an excitation source as otherwise the coupling loss would tremendously reduce the optical power available. Due to the very limited spectral range where dye lasers are available this alternative can be ruled out for most applications. As the impact of external sources of heat increases with decreasing base temperature (due to the decreasing heat capacity of the calorimeter/sample arrangement) a further decrease of our base temperature below 0.5 K would certainly be limited by this problem. Another point to be considered when dealing with ultrathin layers like quantum wells and epitaxial layers is the often disturbing background absorption originating from the substrate material. Especially when doped substrates are employed, the free carrier absorption can easily mask the small signals under consideration. In these cases the use of undoped or semi-insulating substrate materials is mandatory. Another solution consists of selectively etching the substrate away having only a thin well/barrier structure unattacked. For this reason, we have developed in our laboratory selective etching techniques for a number of material systems of current interest in III-V semiconductor physics.

Finally it should be mentioned that CAS is a typical low temperature method where it is by far superior to any other absorption technique. However, if temperature scans or evaluations at room temperature are required, other techniques have to be used.

VII. CONCLUSION

From an overview of photothermally based spectroscopic methods we conclude that Calorimetric Absorption Spectroscopy is the most sensitive one. In CAS, the temperature increase in the sample caused by nonradiative recombination processes is measured with a low-temperature thermometer. At 0.5 K the detectivity of CAS is 10 pW. Even when combining CAS with low power illumination sources covering a large wavelength range from 400 nm to 5 µm αd - products of less than 10^{-4} can be detected. From a theoretical model describing CAS, the key parameters for obtaining high sensitivities are derived. The sensitivity of CAS is found to be proportional to the product of the resistance of the thermal link between He bath and a thermometer and the sensitivity of the thermometer. Decreasing the operating temperature results in a much larger sensitivity and reduced measurement time. Between 1.3 K (^{4}He temperatures) and 0.5 K (^{3}He temperatures) an increase in sensitivity and speed of data acquisition by an order of magnitude is demonstrated. Typical applications of CAS result from its high sensitivity and the fact that exclusively nonradiation recombination processes are detected. In addition, the combination of CAS and calorimetric transmission spectroscopy allows the determination of quantum efficiencies. For typical absorption coefficients of quantum wells submonolayer sensitivity is reached. On the other hand, residual background impurities in semiconductor material down to concentrations of 10^{13} cm $^{-3}$ can be detected with CAS. Examples presented here include the recording of very small absorption signals from InGaAs/AlInAs quantum wells with high spectral resolution and the analysis of nonradiative recombination centers like Fe impurities in semi-insulating GaAs and InP substrate materials.

ACKNOWLEDGEMENTS

We are very grateful to W. Sander and S. Ewert for supplying the ^{3}He cryostat and many helpful discussions. This project has benefited a lot from many discussions with M. Meissner and initial work by A. Juhl and G. Jeiter. The quantum wells were kindly provided by N. Watanabe from Sony Corporation.

REFERENCES

[1] M. Razeghi, Ph. Maurel, M. Defour, F. Omnes, G. Neu, and A. Kozacki, Appl. Phys. Lett. 52, 117 (1988).

[2] A. Bubenzer, S. Hunklinger, K. Dransfeld, Journal of Non-Cryst. Solids 40, 605 (1980).

[3] D. Bimberg, A. Bubenzer, Appl. Phys. Lett. 38, 803 (1981).

[4] H. Gruhl, H.-P. Dorn and K. Winzer, Appl. Phys. B 38, 199 (1985).

[5] A.Juhl and D. Bimberg, J. Appl. Phys. 64, 303 (1988).

[6] R.L. Swofford, J.A. Morrel, J. Appl. Phys. 49, 3667 (1978).

[7] An overview of PAS is given in Y.H. Pao: Optoacoustic Spectroscopy and Detection, Academic, New York, 1977.

[8] For PAS at low temperatures see R. Kuhnert and R. Helbig, Appl. Opt. 20, 4149 (1981).

[9] W.B. Jackson, N.M. Amer, A.C. Boccara, D. Fournier, Appl. Optics 20, 1333 (1981).

[10] J.S. Yoon, S.S. Lee, Appl. Phys. Lett. 46, 913 (1985).

[11] M. Itoh, J. Appl. Phys. 53, 5140 (1982).

[12] K. Albert, H. v. Löhneisen, W. Sander, H.J. Schink, Cryogenics 22, 417 (1982).

[13] R. Krüger, M. Meissner, J. Mimkes, A. Tausend, Phys. Stat. Sol. (a) 17, 471 (1973).

[14] P.J. Dean, D.G.Thomas, Phys. Rev. 150, 690 (1966).

[15] A. Juhl, D. Oertel, R. Bauer, C. Maczey, and D. Bimberg, Acta Physica Polonica, A 69, 877 (1986).

[16] A. Juhl, D. Bimberg, Semi-Insulating III-V Materials, Hakone, Japan, edited by H. Kukimoto and S. Miyazawa, (North-Holland, Amsterdam, 1986), p. 477.

[17] A. Juhl, D. Bimberg, Appl. Phys. Lett. 50, 1292 (1987).

[18] S. Houbloss, A. Nakib, A. Vasson, A.-M. Vasson, C.A. Bates, J.L.Dunn, W. Ulrici, J. Phys. C: Solid State Phys. 20, L467 (1987).

[19] A. Görger, B.K. Meyer and J.-M. Spaeth, Semi-Insulating III-V Materials, Malmö, Sweden, edited by G. Grossmann and L. Ledebo, (North-Holland, Amsterdam, 1988), p. 331.

[20] B. Clerjaud, priv. communication.

[21] G. Jeiter and D. Bimberg, unpublished.

PARTICLES AND ELEMENTARY EXCITATIONS

G. Costa

Università degli Studi
Instituto di Fisica Galileo Galilei
Via F. Marzolo 8
35100 Padova, Italy

There are analogies in the formulation of quantum field theory (QFT) for elementary particles and of many-body theory for condensed matter systems. On one side photons, gravitons, etc. are identified with the quanta of radiation fields; on the other side phonons, magnons, etc. appear from quantization of collective waves in condensed matter.

Both in elementary particle physics and in solid state physics, a fundamental role is played by symmetry. We consider, specifically, the mechanism of spontaneous symmetry breaking, which takes place when the ground state of a system possesses a lower symmetry than the Hamiltonian and the equations of motion. In the case of continuous symmetry, and when only short-range interactions are relevant, the spontaneous breaking is accompanied by the appearance of Goldstone modes, which are massless bosons in QFT and elementary excitations without energy-gap in non-relativistic many-body theory. In particular, phonons and magnons can be interpreted as Goldstone modes. When also long-range interactions of the Coulomb type, such as those mediated by a massless vector field, are present, the would-be Goldstone mode acquires a non-vanishing mass or, correspondingly, a gap in the energy spectrum (like e.g. in the case of plasmons), and it could be interpreted as the longitudinal component of the vector field which becomes massive.

STATE OF THE ART OF "COLD FUSION"

F. Scaramuzzi

ENEA, Centro Ricerche Energia Frascati
Via E. Fermi, 27
00044 Frascati (Roma), Italy

Experiments and theories referring to the interaction of deuterium nuclei with the crystal lattice of some metals (typically palladium and titanium) - known with the name of "cold fusion" - were briefly reported, and an evaluation about the state of the art of this topic was presented.

The experiments are divided into two main categories: those producing a heat excesss with a poor evidence of the production of nuclear events (Fleishmann and Pons); and those that do not exhibit heat excess production, but show evidence of nuclear events, mainly neutron emission (Jones et al., Scaramuzzi et al).

After a description of these experiments, more details about those performed at ENEA (both in Frascati and in Casaccia), and a summary of the evidence produced at the recent "Workshop on Cold Fusion Phenomena" (Santa Fé, USA, 23-25 May 1989) was given. Further experimental evidence was then presented on other experiments not communicated at Santa Fé.

The flourishing of theories, mostly in Italy, was shortly illustrated. Two kinds of approaches are followed. A "fundamental" one, trying to explain how the Coulomb barrier between nuclei can be overcome and an "empirical" one, trying to justify fusion through solid state phenomena in the lattice, that could produce "hot" fusion on a microscopic scale.

In conclusion, the "heat excess" experiments, though repeated in other laboratories, are accepted with a diffuse skepticism on the nuclear origin of the heat production. On the other side, in spite of the contradictory and not reproducible character of the experimental evidence, there is a quite diffuse conviction that the experiments with nuclear events show the existence of nuclear phenomena, presumable connected with phase changes in the metal-deuterium system.

X-RAY ASTROPHYSICS: SATELLITE DATA

L. Scarsi

Istituto di Fisica Cosmica
Via M. Stabile, 172
Palermo, Italy

The use of satellites orbiting outside the earth atmosphere has allowed the observation of x-rays emission from celestial bodies.

This radiation is otherwise absorbed by the atmosphere.

By convention, x-ray emission is located in the energy range of 0.1 keV - 100 keV. The instrumentation aboard satellites typically consists of Grazing Incidence Telescopes and Collimated Detectors.

X-ray Astronomy has allowed for the observation of new phenomena in astrophysics. In the first decade of x-ray astronomy (1970-1980) roughly one satellite was launched per year; in the following years the number reduced to about half. For the decade going from 1990 to the year 2000 a series of international missions are planned, the so called "third generation" satellites.

Details about the phenomenology of interest were given.

EXCITED STATE ABSORPTION IN Tb ACTIVATED PHOSPHOR POWDERS

R. Raue

Philips GmbH
Forschungslaboratorium Aachen
Postfach 1980, Weisshausstrasse
D-5100 Aachen, F.R. of Germany

Optimal transitions between the $^5D_4$ excited state of the Tb^{3+} $(4f)^8$ configuration and the crystal field splitted components of the $(4f)^8 5d$ configuration have been observed using high-resolution laser excitation of Tb-doped powder samples. The Excited State Absorption (ESA) spectra of Tb^{3+} in YAG and YAGaG show broad excitation bands due to these transitions and superimposed relatively narrow lines due to transitions within the $(4f)^8$ configuration.

A theoretical model has been developed for the f-d transitions revealing a proportionality between the intensities of the $(4f)^8$ $^5D_4 \rightarrow (4f)^7$ (5d) and the $(4f)^8$ $^7F_6 \rightarrow (4f)^7$ (5d) transitions. This behavior is also observed in the experimental spectra. Although the final 4f-levels are positioned at the same energy as the 5d bands, the general features of the 4f-4f transitions can be successfully interpreted on the basis of the Judd-Ofelt hteory.

The collaboration of K.J.B.M. Nienivestey and W. Busselt is acknowledged.

ULTRAFAST, TIME RESOLVED PHONON SPECTROSCOPY

D.A. Young

Department of Electrical Engineering
Princeton University
Princeton, NJ 08544, USA

New techniques using ultrashort light pulses for the generation and detection of high frequency phonons have been developed. This has enabled measurements of the Kapitza Resistance (Thermal Boundary Resistance) between dissimilar materials at temperatures up to 300 K. These measurements were presented and the generation and detection of tunable, monochromatic phonons at frequencies up to 1THz(10^{12}Hz) were discussed.

SLOW EMISSION IN NATURAL DIAMOND

L.M. dos Santos

Universidade de Aveiro
Departamento de Fisica
3800 Aveiro, Portugal

Most brown diamonds show visible luminescence with decay times of the order of a millisecond. It corresponds to different centers with two phonon lines coupled to relatively low energy phonons and Huang-Rhys factors larger than 4.

The strongest bands originate at 2.57, 2.82, 2.496(S_3), 2.442(S_4), 2.082 and 2.052eV.

In all these centers the decay pattern shows a similar behavior: single exponential at low temperatures, a temperature region where it becomes two exponential, only to become single exponential again at higher temperatures.

In a few of these centers on increasing the temperature a new emission occurs at higher energies, with a decrease in decay time, and constant overall intensity. The S_3 center was analyzed in detail as it presents all these features.

The band is due to coupling to lower phonons with a Huang-Rhys factor of 5. Below 20 K the decay is 0.18 ms and above 40K is 0.35 ms, indicating thermalization with a nonradiative level. Above 120 K a band originating at 2.621eV starts to grow, the ratio of transition probabilities of the two levels estimated from the relative intensities of the bands accounting well for the decrease in decay time was observed above 120 K.

FLUORESCENT LAMP MAINTENANCE

H.C.G. Verhaar

Philips Lighting B.V.
Advanced Materials Development
P.O. Box 80020, Building EDW-629
5600 JM Eindhoven, The Netherlands

Fluorescent lighting is the major field in which phosphors are applied. As with other applications we are interested in which way the efficiency decreases during their lifetime. The maintenance behavior of fluorescent lamps is divided into two different processes. In the one process, called extrinsic phosphor degradation, the phosphor stays intact. The decrease in efficiency is caused by the formation of light-absorbing products, mainly consisting of carbon and mercury. This well-investigated process can be prevented (e.g.) by the application of the protective coatings that reduce the interactions leading to the Hg_xC_y formation.

In the other process, called intrinsic phosphor degradation, the phosphor itself is damaged and the loss in efficiency is caused by the formation of defects and color-centers. These can either absorb excitation radiation, absorb emission, interfere with energy-transfer, or reduce the phosphors quantum yield due to deviations from local symmetry. This process can be induced by UV-photons (185nm Hg-line) or by ionic bombardment (e.g. by Hg^+/e^-). Examples were given of the application of protective coatings absorbing 185nm (with or without a contribution to phosphor efficiency) and of coatings that prevent ionic bombardment from taking place. In addition the use of fluxing agents is able to drastically improve phosphor stability by preventing induced NR-transitions to occur.

NONRADIATIVE TRANSITIONS AT THE N3 CENTER IN DIAMOND

T. Monteiro

Universidade de Aveiro
Departamento de Fisica
3800 Aveiro, Portugal

The N3 optical center occurs in natural diamonds and is due to three substitutional nitrogen atoms bounded to a vacancy. It has trigonal symmetry, the ground state being a ^2E state. The zerophonon line (XPL) occurs at 2.985 eV and absorption and emission bands show a mirror image relationship. The N2 (2.596 eV) and N4 (3.603 eV) absorption bands are due to symmetry forbidden vibronic induced $^2A_2 \rightarrow {}^2A_1$ transitions. The quantum yield of luminescence is 0.29. ESA has been observed and four-wave mixing shows at room temperature the presence of two metastate states of 0.73 and 1.2 ms lifetimes.

We detected both delayed luminescence and luminescence from a lower energy when N3 is excited. From the temperature dependence of intensity and decay times of these slow luminescence we conclude that the metastate states must have different spin multiplicities and the processes by which they repopulate N3 or suffer further nonradiative decay were discussed. Namely at room temperature the decays agree with the results obtained by four-wave mixing.

For excitation into the N4 absorption band, beside an emission similar to the one observed for N3 excitation, N3 emission with a decay of 0.4 ms is found at all temperatures, indicating a further nonradiative path from N4 to N3.

OPTICAL PROPERTIES OF DIAMOND AND SOME APPLICATIONS

S. Albin

Department of Electrical and Computer Engineering
Old Dominion University
Norfolk, VA 23529-0246, USA

Diamond has many unique properties which are suitable for optical and electronic applications. Optical properties of synthetic bulk diamond films are investigated. Bulk diamond samples containing nitrogen impurities are irradiated with 1 MeV electron beam to generate color centers. The effect of irradiation on absorption and fluorescence spectra are studied. Polycrystalline diamond films are deposited on Si substrate using a microwave plasma enhanced chemical vapor deposition process. Raman spectra of these samples reveal the presence of sp^2 bonded graphitic carbon, in addition to sp^3 bonded diamond phase. These films exhibit interesting features in fluorescence and electroluminescence. Epitaxial diamond films can provide a means to study pure and doped diamond films. Some examples of the applications of diamond in optically controlled high power switches, color center lasers, electroluminescent displays and integrated optics were discussed.

PHOTOSTIMULABLE X-RAY STORAGE PHOSPHORS

H. von Seggern

Siemens AG
Dept. ZFE AMF 33
Paul Gossen Strasse 100
8520 Erlangen, F.R. of Germany

The ability of some ionic crystals to store X-ray information in form of locally trapped electron/hole pairs and the release of this information upon optical stimulation in form of emitted light has led to a new and promising concept of X-ray imaging. THis concept and the physical properties and requirements of applicable materials were discussed. Phosphors utilized so far commercially are BaFBr:Eu and RbBr:Tl. The nature of the electron trap, the luminescence center, as well as the charge transport between these centers were addressed. Discussion of the obtained results led to a suggestion for a formation process of photostimulated-luminescence-centers during X-ray irradiation.

SOME ASPECTS OF THE EFFICIENCY OF PHOSPHOR SCREENS

G.E. Giakoumakis

Applied Physics Lab - Solid State Div.
Physics Dept. - University of Ionnina
P.O. Box 1186
451 10 Ioannina Greece

We reported on the absolute efficiency of rare earth oxysulfide (Y_2O_2S:Tb) x-ray phosphor screens. The measurements have been carried out under fluoroscopy conditions both in transmission and in reflection mode observation. All the experimental results were discussed on the basis of a uniform screen model and analytical expressions were presented permitting the theoretical calculation of the absolute efficiency of the screens.

SOME CATHODOLUMINESCENT AND THERMOLUMINESCENT MATERIALS

R. Chapoulie

Université Bordeaux 3
Laboratoire de Physique du Solide
CRIAA (CNRS UA 1003) MSHA
Esplanade des Antilles
33405 Talence Cedex, France

Some natural minerals and synthetic materials are investigated using cathodoluminescence properties; materials such as quartz, feldspars, calcium carbonates, fluorites, gemstones, technical ceramics... Some thermoluminescent dosimeters are also being under study: their crystal growth, their sensitivity to very weak radioactive doses, the linearity of their responses to thermoluminescence. Thermo-stimulated luminescence (TSL) is used to detect the presence of defects in crystalline and non-crystalline solids, for instance in glass made for nuclear waste confinement, in optical fibers, in phosphors..., but also TSL is used as a dating method for archaeological and geological samples (ancient ceramics, burnt stones, some crystalline pigments, volcanic rocks...).

HOT WALL EPITAXY FOR REPAIRING II-VI COMPOUND THIN FILMS

M. Grün

Universität Kaiserslautern
Fachbereich Physik
Erwin Schrodinger Strasse
D-6750 Kaiserslautern, F.R. of Germany

After a short introduction to epitaxy the principles of the Hot Wall Technique was described. For comparison Molecular Beam Epitaxy was quoted.

The use of Hot Wall Epitaxy for growing thin films of II-VI semiconducting compounds was established. The technique is also suitable for preparing heterostructures, too.

The collaboration of U. Becker and C. Klingshirn is acknowledged.

THEORY OF ELECTRON-PHONON REPAIRING II-VI COMPOUND THIN FILMS

K.C. Mishra

GTE Products Corporation
100 Endicott Street
Danvers, MA 01923, USA

An electron-phonon coupling scheme was presented in order to explain the nonradiative relaxations among the spin-forbidden states. It was shown that the nonradiative relaxation will occur due to the Born-Oppeinheimer term in the total Hamiltonain, whereas the mixing among the different spin states will be due to the spin-orbit interaction.

OPTICAL PROPERTIES OF CdSSe DOPED GLASSES

A. Uhrig

Universität Kaiserslautern
Fachbereich Physik
Erwin Schrödinger Strasse
D-6750 Kaiserslautern, F.R. of Germany

We have investigated the optical properties of semiconductor microcrystallites embedded in a glass matrix. The basic semi-conductor materials CdS, sulphur and selenium are added to the glass melt. By annealing process the microcrystallites grow in the glass matrix to a size which depends on the annealing temperature and time.

By the confinement of the electrons and holes to a region of about one nm, which is the radius of the microcrystallites, the optical properties change considerably in comparison to the bulk material.

In linear absorption measurement we have seen on such samples a blue shift of the absorption edge with decreasing microcrystallite radius and the quantization of the conduction band states. This quantization shows up as two steps in the absorption spectrum.

In luminescence under high excitation we found a blue shift of the high energy side of the luminescence peak for large crystallites and even a second peak growing with increasing excitation intensities at smaller radii. This second peak is tentatively attributed to recombination process with two electron-hole pairs (i.e. a biexciton) present in one microcrystallite.

The collaboration of C. Klingshirn is acknowledged.

INFLUENCE OF LIGAND'S COVALENCY DEGREE TO CF-PARAMETERS OF Ln^{3+}-IONS IN DOPED CRYSTALS

S.E. Sarkisov

Institute of Crystallography
Academy of Sciences of the USSR
Lenisnky pr. 59
Moscow 117333, USSR

Using Raman scattering data the influence of covalency of next nearest neighbor coordinational groups (relative to the Ln^{3+}-ion) to the ligands forming a CF around a Ln^{3+}-ion in the host was considered within the frameworks of CF point charge model. Experimental results showing the CF-parameters changing as a function of lattice strength constants values in Nd^{3+}-doped crystals were presented.

SPECTROSCOPIC EFFECTS OF SMALL $^4T_2$-^2E ENERGY SEPARATIONS IN $3d^3$-ION SYSTEMS

C.J. Donnelly

Department of Physics
University College
Galway, Ireland

The energy of the $^4A_2 \rightarrow {}^4T_2$ transition of $3d^3$ ions in octahedral sites decreases with increasing temperature, with a consequent reduction in the energy separation, Δ, between $^4T_2$ and ^2E levels. This can significantly affect the spectroscopic properties of Cr^{3+}-doped materials with small Δ values, such as GSGG:Cr^{3+}. This material is of interest because of its use in tunable laser systems. The temperature dependence of the luminescence was investigated and can be understood if the mixing of ^2E and $^4T_2$ electronic-vibrational states is estimated and the value of Δ at low temperatures for the dominant Cr^{3+} site is taken ≈ 0.

THE β-ALUMINAS: OXIDE HOST MATERIALS FOR DIVALENT LANTHANIDE IONS

L.A. Momoda

Department of Materials Science and Engineering
University of California - Los Angeles
405 Hilgard Avenue
Los Angeles, CA 90024-1595, USA

The unique crystal structure of the β-alumina family of compounds permits the introduction of optically active ions into the material by ion exchange techniques. This low temperature synthesis route has allowed us to controllably dope divalent lanthanide ions into an oxide host material. Preliminary spectroscopic results as well as the future of the materials were discussed.

The collaboration of B. Dunn is acknowledged.

INVESTIGATIONS ON LOCALIZED EXCITONS IN CdS_xSe_{1-x} MIXED CRYSTALS

H. Schwab

Universität Kaiserslautern
Fachbereich Physik
Erwin Schrödinger Strasse 13
6750 Kaiserslautern, F.R. of Germany

The gradual exchange of sulfur by selenium in a CdS crystal leads to new physical effects. In these alloy semiconductors the density of states has an expenential tail on the low energy side of the square root dependence. The excitons show the same dependence in this region and they have a strongly reduced mobility, i.e. they are localized. It also shows up that the luminescence spectra of these crystals are dominated by the recombination of localized excitons.

We have investigated the intensity of the luminescence polarized parallel and perpendicular with respect to the polarization of the exciting beam for k‖c. From these spectra the degree of polarization was calculated, as a function of density, photon energy and temperature. This gives informations about the local anisotropy of the crystal caused by the compositional disorder and on the decay of the polarization memory.

The collaboration of V.G. Lyssenho, A. Reznitsky, and C. Klingshirm is acknowledged.

THE VIBRONIC INTENSITIES IN THE OPTICAL TRANSITIONS OF Gd^{3+}

J. Sytsma

Physical Laboratory, State University
P.O. Box 80 000
3508 TA Utrecht. The Netherlands

Vibronic intensities in optical transitions of Rare Earth ions are expected to be small. In contradiction with this we found a strong coupling in the emission of Gd^{3+} in YOCl. It is explained as a large Frank-Condon contribution due to a strong polar visibility.

STRONGLY COUPLED PLASMAS AND CRYSTALLIZATION: SIMPLE MODELS FOR THE DISPERSION RELATION

M. Minella

Department of Physics
Boston College
Chestnut Hill, MA 02167, USA

A system of quasi-localized charges was suggested as a model for the Classical Strongly Coupled Plasma near its crystallization limit. The charges oscillate harmonically around their equilibrium positions, but these are not fixed on a crystal lattice, as in the ordinary classical theory. They constitute a "random" base, whose ensemble average is fixed through the Equilibrium Pair Correlation Function (or functions, in the case of many species systems).

The Linear Response Functions of the system are then functionals of those Correlation Functions, which are becoming increasingly available from Montecarlo calculations.

The Dielectric Functions and Modes were shown for:

1) The One Component Classical Strongly Coupled Plasma. The dispersion is easily obtained at all wavelengths and for a very wide density range. When thermal effects are accounted for in a simple semiphenomenological way, remarkable agreement with accurate Molecular Dynamics Computer Simulations is obtained.

2) A Binary Ionic Mixture, compensated by a static background of electrons, and a Two Component (with opposite charges) Plasma; the effect of correlations on the plasmon frequency was studied, and the calculations of modes at all wavelengths was sketched.

THE EFFECT OF GLASS STRUCTURES ON THE FLUORESCENCE PARAMETERS OF TRIVALENT CHROMIUM

D.B. Hollis

Department of Physics
Paisley College of Technology
High Street
Pailey, Renfrewshire PA1 2BE
Scotland, UK

The effects of glass structure on the fluorescence parameters of trivalent chromium have been studied for a range of concentrations of chromium in a range of types of glass.

Ultraviolet-visible absorption spectra were only weakly sensitive to concentration and glass type. For a series of sodium aluminoscilicate glasses, electron paramagnetic spectra were strongly sensitive to glass type. The fluorescence decay data gave a concentration quenching parameter which was similar in the different glasses, and an ion-glass nonradiative loss parameter which was strongly sensitive to the type of glass for both the sodium aluminosilicate glasses and other types of glasses. From the decay data we calculated the quantum yields, and again a correspondence with glass type was noted.

Consideration of the phase diagrams relevant to the component oxides of the glasses indicated the phases incipient, but not yet formed, in the glasses. The quantum yields, the ion-glass nonradiative parameters, and the electron paramagnetic resonance results were all closely related to the symmetries of the chromium sites within these incipient phases.

Best quantum yield came from rhombically distorted octahedral sites, and lowest yield came from hexagonal sites.

COMPARISONS OF DIFFERENT MODELS FOR THE TEMPERATURE DEPENDENCES OF SPECTRAL LINE-WIDTH AND LINE-SHIFT

X.X. Zhang

Department of Physics
Boston College
Chestnut Hill, MA 02167, USA

The mechanisms for the thermal broadening and thermal shift of sharp lines were reviewed.

The temperature dependences are dependent on the phonon density of states. For comparison, different models based on the Einstein, Debye, and Vibronic effective density of states were used to fit the experimental data obtained for the $MgO:V^{2+}$ R line. Comparison with the specific heat fitting was also made.

The collaboration of B. Di Bartolo and P. Hong is acknowledged.

LUMINESCENCE AND ENERGY MIGRATION IN $(Sr,Eu)B_4O_7$, A SYSTEM WITH $4f^7$-$4f^6 5d$ CROSS-OVER IN THE EXCITED STATE

A. Meijerink

P.O. Box 80 000
3508 TA Utrecht, The Netherlands

The luminescence and energy migration in $Sr_{1-x}Eu_xB_4O_7$ have been studied. At 4.2 K line emission is observed due to the $4f^7(^6P_{7/2}) \rightarrow 4f^7 5d$ $(^8S_{7/2})$ broad band emission appears.

At temperatures below 20K energy migration occurs in EuB_4O_7 over the $^6P_{7/2}$ level of the Eu^{2+} ion. The decay curves of the Eu^{2+} emission in EuB_4O_7 were analyzed using a random-walk model for two-dimensional energy migration. From the analysis a value of $10^7 s^{-1}$ for the transfer, violability of the excitation energy between nearest Eu^{2+} neighbors was derived. Above 20 K the energy migration changes to three-dimensional energy migration over the thermally populated $4f^6 5d$ level of Eu^{2+}.

LUMINESCENCE AND ENERGY TRANSFER IN EUROPIUM TELLURITE ANTIGLASS PHASES

J.P.M. van Vliet

Physical Laboratory, State University
P.O. Box 80 000
3508 TA Utrecht, The Netherlands

The luminescence and energy transfer properties of two lower symmetrical Europium tellurite anti-glass phases were reported, viz. of $Eu_{1.79}TeO_x$, which has a pseudo-tetragonal structure, and of $Eu_{1.08}TeO_x$, which has a monoclinic, ordered crystal structure.

In the pseudo-tetragonal phase the Eu^{3+} ions occupy two major types of sites. In the monoclinic phase there are at least 7 different crystallographic sites available to the Eu^{3+} ions. At low temperatures transfer of excitation energy to the sites with the lowest $^5D_0$ levels occurs, and only these sites luminescence. Above 10 K backtransfer to higher $^5D_0$ levels takes place.

The energy migration process, which becomes important at temperatures above 20 K for the pseudo-tetragonal phase and above 10 K for the monoclinic phase, is two-phonon assisted. For both compounds the assistance mechanism up to about 50 K is two-site non-resonant, at higher temperatures one-site resonant.

The collaboration of G. Blasse is acknowledged.

ANOMALOUSLY SLOW RELAXATION AT THE PERCOLATION THRESHOLD

S. Jain

Department of Chemistry
University of Durham
Science Laboratories
South Road
Durham DHI 3LE, UK

We investigated the time dependence of the equilibrium spin-spin autocorrelation function of a two-dimensional Ising model at the bond percolation threshold. In the paramagnetic phase; the relaxational law of a stretched exponential with a temperature-dependent exponent. Our results indicated that recent suggestions concerning the relaxational dynamics of dilute magnets above the critical temperature needed to be modified for a percolating network. Kohlrausch relaxation was also exhibited in the Griffiths phase for short times. In the asymptotic limit ($t \to \infty$), the behavior was consistent with a recent exact result.

RECOMBINATION PROCESSES IN AlGaAs/GaAs MULTIPLE QUANTUM WELLS (MQW)

D. Oberhauser

Universität Kaiserslautern
Fachbereich Physik
Erwin Schrodinger Strasse
D-6750 Kaiserslautern, F.R. of Germany

A short view of the physical consistence and the optical properties of MQW was given.

The quantum confinement of these quasi two dimensional structures leads to an energetic splitting between light hole and heavy hole exciton at the Γ-point of GaAs.

Measurements of the recombination of light hole and heavy hole under resonant excitation were presented.

RESTRICTED GEOMETRIES IN COMPOSITE POLYMERIC MATERIALS

Ö. Pekcan

Technical University of Istanbul
Department of Physics
Maslak-Istanbul 80626, Turkey

A direct energy transfer method and Fractal analysis were used to study the internal morphology of Polymer Latex Particles produced by non-aqueous dispersion polymerization. These particles, with diameters ranging from 0.2 to 1mm and containing ca. 95 mol % of amorphous glassy polymer (PVAc) and 5 mol % of rubbery polymer (PEHMA) formed interconnecting networks within each particle. Fluorescence intensity decay of the naphthalene (N) donor in the presence of anthracene (A) acceptors was measured, and cylindrical and spherical restricted geometries were employed. Apparent dimension of $d \approx 2$ was observed for N and A within the PEHMA phase. Dimensionality was changed when the particles were dissolved and/or when they were swelled with hexadecane.

DOES ANDERSON TRANSITION EXIST IN RUBY? AN ABSOLUTE INTENSITY AND LIFETIME STUDY

X.X. Zhang

Department of Physics
Boston College
Chestnut Hill, MA 02167, USA

We have re-examined the energy transfer and Anderson transition issue in ruby by making a careful study of the temperature dependences of the absolute intensities and lifetimes of R lines and N lines. This study was carried out in the temperature range 15 K to 300 K for three samples of different Cr^{3+} concentrations (0.05, ~0.5, and 2.5%). In order to explain the experimental results, a new energy transfer picture was proposed. We assumed that the excitation energy is transferred from the R centers to N centers via the irregular paired ions called R^X. The excitation energy is localized at R^X at low temperatures due to the randomly Cr^{3+}pairing induced potentials and becomes delocalized at higher-temperatures. The temperature effect on the Anderson transition was discussed.

The collaboration of B. Di Bartolo is acknowledged.

SUMMARY OF THE MEETING

G.F. Imbusch

Department of Physics
University of College
Galway, Ireland

The theme of this school - nonradiative processes in solids -- was treated mainly from the viewpoint of optical spectroscopy, and was concerned primarily with the effects of nonradiative processes on the luminescence properties of solids. Many quite diverse topics were presented during the course of the school, and it seems helpful to classify the majority of the formal presentations into three materials classifications:

(i) single electronic centers in insulators,
(ii) extended states in insulators,
(iii) semiconductors.

These listings are shown in the table. By far the largest number of lecturers were in the first category.

Table: Materials classification of the topics covered by some of the lecturers.

SINGLE ELECTRONIC CENTERS IN INSULATORS	SEMI CONDUCTORS	EXTENDED STATES IN INSULATORS
Curie		
Di Bartolo	Ulbrich	Grill
Struck	Klingshirn	Terzi
Imbusch		
Auzel		
Blasse		
Reisfeld		
Kaminskii		
Smets		
Baldacchini		
Grassano		

I. LECTURES

Professor Curie, in the opening lectures, enumerated the different nonradiative pathways encountered under practical experimental conditions, and he listed some of the models employed to describe these nonradiative processes. His task, he felt, was to identify the problems which would be dealt with, and hopefully understood, before the end of the school.

As a theoretical background for the later discussions, Professor Di Bartolo gave a careful quantum description of how an electronic center in a solid interacts with the vibrations of the lattice in which the center is embedded. He showed how the adiabatic approximation, allied to the Condon approximation, allows one to decouple the electronic and vibrational subsystems, and at the end he showed the natural emergence of the configurational coordinate model. This model can be employed to describe radiative and nonradiative processes at a single electronic center in terms of the electronic states of the center and the vibrational modes of the local ionic complex. At this early point questions were raised about the validity of the adiabatic approximation and, in particular, of the Condon approximation. These questions surfaced regularly, at informal discussions among the participants in the school.

Dr. Struck took as a starting point the single configurational coordinate model (SCC model) in the harmonic approximation, and he showed how a consistent mathematical framework can be developed to treat nonradiative multiphonon processes. The expression for the nonradiative decay rate consists of an electronic prefactor multiplied by a vibrational overlap factor (W_p), the value of W_p depending critically on the arrangement of the configurational coordinate parabolae. This approach, developed by Struck and Fonger in a series of papers, uses the Manneback recurrence relations to derive the overlap factors for various interaction opperators. The emphasis in the struck and Fonger approach is in calculating the appropriate overlap integrals; the electronic prefactor is chosen for best fit to the measured nonradiative decay rates.

The talk of Professor Imbusch was concerned with the numberical predictions of the SCC model in the harmonic approximation, and for the intermediate coupling strengths ($S \approx 3$). The predictions of this model are not well described by the simpler activation model. He showed, by numerically solving the Schrödinger equation in some simple cases, that anharmonicities in the vibrational potential functions can drastically alter the value of the vibrational overlap factor. Similarly, changes in the vibrational frequencies for the two electronic states can greatly affect the value of the overlap factor.

Professor Grill described the relatively new field of phonon spectroscopy - the generation, transport, and detection of phonons, both broadband and monochromatic - and he introduced the intriguing idea of stimulated phonon amplification (as opposed to stimulated photon amplification). In a wide-ranging overview he covered such diverse topics as phonon dispersion relationships in practical crystalline materials, phonon focussing, and the specific techniques for phonon generationa and detection. One senses that, with the coming on-stream of free electron lasers with tuned emission in the infrared, the provision of high quality materials, and new methods of phonon detection, a wholly new technique for probing solids is being developed.

A classical harmonic lattice was the model used by Professor Terzi to describe how a pulsed ionic distortion, generated at a specific location in the crystal, propagated into the surrounding harmonic lattice in accordance with the laws of classical dynamics, and this purely classical analysis yields values for the relaxation time which are in reasonable agreement with experimental measurements for $KI:Tl^{o}$. The more difficult quantum problem of how to properly describe phonon propagation, when a coherent wavepacket of phonons is generated, was briefly referred to.

Professor Ulbrich in his lectures considered the nature of the one-electron states in a semiconductor crystal. He showed that the eigenstates are coherent states of the whole crystal, characterized by a k vector, and he described some of the complicated energy versus k relationships in different directions across the Brillouin zone for GaAs-type materials. He went on to relate these Bloch states to the localized exciton states formed by the optical generation of an electron-hole pair. He considered the efficiency of radiative recombination across the band gap, as well as the effect of increasing temperature on these transitions.

Staying on the subject of semiconductors, Professor Klingshirn gave a careful explanation of the band edge absorption transition in CdS, a II-VI direct band gap semiconductor with transitions in the visible part of the spectrum. He showed how photothermal optical nonlinearities can be created with very little power at this band edge transition. CdS, he showed, was an excellent model system to demonstrate the principles of optical bistability.

In a wide-ranging talk, Dr. Auzel pointed out the role played by nonradiative processes in determining the suitability of materials for operation as laser media; these processes can be both positive and negative. He showed how ion-phonon and ion-ion interactions (which can lead to efficient energy transfer and up-conversion processes) enter the optical pumping and laser cycle. He cited a number of examples of these processes.

In the course of his lectures Professor Blasse gave many examples of how intrinsic nonradiative and energy transfer processes can affect the luminescence properties of a material. He showed that, with a knowledge of the configurational coordinate picture of the electronic and charge transfer states, the radiative and nonradiative properties of a center can be predicted in many cases, and this allows one to design materials for a specific luminescence property. From the viewpoint of nonradiative processes he pointed out how the "softness" of a lattice site – its ability to relax when the optically-active dopant ion is excited – strongly influences the nonradiative rate. Of particular interest was his account of the luminescence properties of rare earth cryptates, where the cryptand ligand acts as a cage to hold the rare earth ion. The size of this cage can be changed, creating looser sites for the rare earth ion with a consequential enhanced nonradiative decay rate.

Nonradiative deexcitation of the excited states of rare earth and transition metal ions in glass was discussed by Professor Imbusch. Whereas the situation for rare earth ions in oxide and fluoride glasses was very similar to that in crystals -- both cases being covered by the same universal energy gap formula -- the nonradiative decay rates for transition metals in glass were universally found to be higher than for crystals, leading to disappointingly low quantum efficiency (QE) values, particularly so in the case of chromium-doped glass, which has been the subject of intense study. This low QE was attributed by him to the flexible character of glass at the local atomic level, which leads to a large configurational coordinate change for the excited Cr^{3+} state. That a large configurational coordinate change leads to an enhanced nonradiative decay rate was already established by George Blasse.

Professor Reisfeld described the sol-gel technique, which is a low temperature technique permitting the incorporation of organic dyes into inorganic glasses to give a new luminescent species with its own special radiative and nonradiative properties.

Professor Kaminiskii showed the impressive list of laser crystals covering the red and near infrared, many with wide tunable ranges, which have been fabricated in his laboratory.

In a change of emphasis, Dr. Smets reminded the school that the use of phosphor materials in X-ray intensifying screens, CR tubes, and fluorescent lamps is a huge industry and a major commercial motivation for continued spectroscopic research. In addition to high efficiency of conversion of electrical power to useable light, the industry is concerned with the cost of rare earths and with the degradation of the phosphor with lamp usage – matters which do not greatly interest the more academic spectroscopist. He emphasized that the development of the highly efficient tricolor lamps and of the more recent Super de Luxe lamps, with their excellent color rendition, can be regarded as deriving from the large amount of spectroscopic and materials research performed over many years.

Professor Baldacchini in his lectures described the physical properties of color centers in alkali halides. Not only is there a large variety of color centers, but the nonradiative processes can be quite complicated. He made much use of the configurational coordinate model in his analysis of these nonradiative processes. The absence of luminescence in some color centers was explained years ago by Dexter, Klick and Russell, and this explanation was put on a more quantitative basis by Bartram and Stoneham who give a criterion for the appearance of luminescence. This criterion is based on the idea that cross-over relaxation takes place before the rare earth ion can reach the relaxed excited state, and as a result no luminescence occurs. Baldacchini showed that all is not right with this generally-accepted picture. He presented some convincing evidence that all the excited centers reach the relaxed excited state and the absence of luminescence is attributed to efficient nonradiative effects from the relaxed excited state.

Professor Grassano described some of the techniques used to measure fast nonradiative rates. Of particular interest are the techniques to try to measure fast intraconfigurational relaxation in color centers with subpicosecond relaxation times, times comparable with the period of optical mode vibrations. Experiments to detect "hot luminescence" – luminescence from higher vibrational levels during the downward nonradiative relaxation of the color center – were described by him.

The principle of transient grating spectroscopy as well as some of the attendant experimental details were described by Dr. Evesque. This technique probes the spatial extent of the excited states and their dynamic evolution through diffusion and perfolation. As he showed, the interpretation of the results can be quite complicated if one is dealing with a complex physical system with a number of simultaneous relaxation processes. He introduced the fractal dimension involved in some of these dynamical processes.

The quantum efficiency (QE) of the photoluminescence process, defined as the ratio of luminescence photons out to pumping photons in, is the quantity which describes the competition between radiative and nonradiative processes. This is not an easy quantity to measure accurately; spectroscopists often only measure it indirectly using comparison methods or infer its value from measured decay rates. Methods which directly measure the heat generated by the released phonons permit a direct detection of the nonradiative process and of the quantum efficiency. A number of such techniques were described in detail by Professor Bimberg. He pointed out that photoacoustic spectroscopy -- which is based on the acoustical detection of the released heat -- was originally described by Alexander Graham Bell in 1880.

In his wide-ranging lecture Professor Scacco described the very sophisticated methods of crystal growth required to produce and parametrize the high-purity high-quality crystals with specific dopings which are part of today's high technology needs. Dr. Hollis organized an informal workshop on glass fabrication techniques which was of particular interest to a number of the participants. Interesting short seminars were given on ultrafast time-resolved phonon spectroscopy (Dr. Young) and on specific heat measurements at very low temperatures (Professor Graf).

II. SEMINARS

A feature of the school was the number of special seminars related to the main theme of nonradiative processes, as well as the large number of short seminars in which the participants describe their own research activity. These are briefly summarized below.

Technical problems relating to the commercial use of phosphors (Giakoumakis) and of X-ray storage phosphors (von Seggern) were described, as were the problems associated with fluorescence lamp maintenance (Vehaar). A number of participants (Momoda, Meijerink, Raue, Sarkisov, Sytsma, and van Vliet) described topics in rare earth ion spectroscopy. The luminescence properties of Cr^{3+} in laser-type crystals were the subject of a long seminar by Professor Boulon, while Dr. Buoncristiani described his experiments on $Cr^{3+} \rightarrow Tm^{3+}$ energy transfer. Short seminars on various aspects of Cr^{3+} luminescence were given by Donnelly and by Zhang. Dr. Hollis described his experiments employing Cr^{3+} ions as EPR probes in glasses, the relatively sharp signals in many of the glasses indicating that Cr^{3+} ions enter a small number of specific sites in these glasses.

In the semiconductor area, the optical properties of CdSSe crystals and glasses were analyzed by Schwab and Uhrig, respectively. Hot wall epitaxy for the preparation of II-VI compound thin films was described by M. Grün. D. Oberhauser spoke on the recombination processes in AlGaAs/GaAs multiple quantum wells.

The radiative and nonradiative properties of some luminescent centers in diamond were described by dos Santos and Monteiro, while S. Albin reminded the participants that diamond is a semiconductor with a large energy gap (≈5.4 eV) whose extrinsic semiconducting properties have many natural advantages. The study of the luminescence properties of purposely-doped diamond is now possible with the emergence of techniques for depositing diamond films. Diamond, with its large band gap, in principle permits the possibility of radiative transitions over the visible spectrum on selected dopants.

In his short seminar Dr. Mishra pointed to conceptual difficulties with the approximations and procedures used for calculating nonradiative decay rates, and he presented his own preliminary analysis of this problem.

S. Jaiin presented the results of his computer analysis of relaxation times in a two-dimensional Ising spin model. M. Minella presented an analysis of a system of quasi-localized charges near the crystallization limit, and Ö. Pekcan described experiments to probe the internal morphology of polymer latex particles produced by non-aqueous dispersion polymerization.

III. OTHER ACTIVITIES

A particular and much-appreciated feature of the school was a small number (3) of excellent lectures on topics which, though outside the main theme, were of broad interest to the participants.

In the first such lecture Professor Costa used the language and concepts of particle physics to describe solid state phenomena. The collective motion of the lattice (the phonons), as well as magnons and plasmons are the elementary excitations, conduction electrons in semiconductors are the "dressed" particles or quasiparticles, and phonons can be interpreted as Goldstone modes.

In a long lecture, noted for its balance and objectivity, Professor Scaramuzzi gave a comprehensive review of the recent experiments relating to the phenomenon of "cold fusion". He answered the many questions with patience and good humor.

Professor Scarsi spoke on X-ray astronomy, a new window on the universe, which promises to greatly extend our knowledge of stellar and galactic evolution, particularly when the space telescopes and satellite sensors, scheduled for launching in the next decade, begin transmitting.

Although the formal lecture and seminar classes were periods of strenuous intellectual activity for most of the participants, the hours outside the formal classes were marked by informality and open discussion and by the enjoyment of shared activities which are the unique experiences of the Ettore Majorana School of Spectroscopy.

The participants of a course on "Advances in Nonradiative Processes in Solids", held June 15-29, 1989 in Erice, Sicily.

PARTICIPANTS

	Names	Fields of Work
1.	Abramidou, Stavroula Department of Physics National and Capodistrian University of Athens Athens, Greece (30) 1 2236416	Semiconductors, and their applications in industry. Materials that have useful applications in electronics and microelectronics Laser materials spectroscopy.
2.	Albin, Sacharia Department of Electrical and Computer Engineering Old Dominion University Norfolk, VA 23529-0246, USA (1) 804 683 4967	Laser materials and photonics. Diamond.
3.	Auzel, François Centre National d'Etudes des Télécommunications 196, Av. H. Ravera 92220 Bagneux, France (33) 1 45 29 52 02	Laser materials (glasses and crystals). T.M. & R.E. spectroscopy; nonradiative transitions. Defects; electron-microscopy.
4.	Baldacchini, Guiseppe ENEA Centro Ricerche Energia Frascati C.P. 65 00044 Frascati (Roma), Italy (39) 9400 5365	Color Center Physics. Molecular Spectroscopy
5.	Balda de LaCruz, Rolindez Departamento de Fisica Aplicada I Escuela Technica Superior de Ingenieros Industriales Y de Telecommunicacion Alda, Urquijo, S/N 48013 - Bilbao, Spain (34) 4 4416400 Ext.355	Laser materials (glasses and crystals). T.M. & R.E. Spectroscopy.
6.	Beckwith, Clyfe Department of Physics Boston College Chestnut Hill, MA 02167, USA (1) 617 552 3575 BITNET:BECKWICL@BCVMS	Photoacoustics. Two dimensional semiconductor systems.

7.	Bimberg, Dieter Institut für Festkörperphysik Technische Universität Berlin Hardenbergstrasse 36 1000 Berlin 12 F.R. of Germany (49) 30 314 2783 (49) 30 314 2082	Physical properties of III-V compounds. Superlattices.
8.	Blasse, George Physical Laboratory State University P.O. Box 80.000 3508 TA Utrecht, The Netherlands (31) 30 532226	Luminescence materials.
9.	Boulon, Georges Université Claude Bernard-Lyon Laboratoire de Physico-Chimie des Matériaux Luminescents Unité Associée au CNRS No.442 Bat 205 43 Boulevard 11 Novembre 1918 69622 Villeurbanne Cedex, France (33) 78 89 81 24	Optical properties of inorganic materials (crystals, glasses, Glassceramics) - Laser Materials.
10.	Buoncristiani, Martin NASA Langley Research Ctr. MS468 Hampton, VA 23665-5225, USA (1) 804 864 1569	Optical Properties of Solid State Laser Materials.
11.	Chapoulie, Rèmy Université Bordeaux 3 Laboratoire de Physique du Solide, CRIAA (CNRS UA 1003) MSHA Esplanade des Antilles 33405 Talence Cedex, France (33) 56 80 84 43 p.68	Characterization of defect centers in solids (phosphors, ancient ceramics glasses, technical ceramics) using Cathodoluminescence, Thermo-Stimulated luminescence and Electron Paramagnetic Resonance.
12.	Chen, Yimin Changchun Institute of Physics Chinese Academy of Sciences Changchun, P.R. of China Fax: (86) 431 55378	Luminescence, electronic structures.
13.	Colombo, Mirco Universität Bern Institut für anorganische, analytische und physikalische Chemie Freiestrasse 3 CH-3000 Bern 9, Switzerland Fax: (41) 31 654499 Phone: (41) 31 654254	Luminescence Line Narrowing in Transition Metal Compounds.

14. Costa, Giovanni — Elementary Particles.
Università degli Studi
Instituto di Fisica Galileo Galilei
Via F. Marzolo 8
35100 Padova, Italy
(39) 49 844 232

15. Curie, Daniel — Luminescence in crystals, and specifically under high pressures.
Faculté des Sciences
Laboratoire de Physique de
la Matière Condensée
Tour 13-4e Etage
Université Paris VI
75252 Paris Cedex 05, France
(33) 1 4336 25 25 Ext.4457

16. Curie, G. — Luminescence in crystals.
Faculté des Sciences
Laboratoire de Physique de
la Matière Condensée
Tour 13-4e Etage
Université Paris VI
75252 Paris Cedex 05, France
(33) 1 4336 25 25 Ext.4457

17. Di Bartolo, Baldassare (Rino) — Luminescence and Molecular Spectroscopy. Flash Photolysis. Photoacoustic Spectroscopy.
Department of Physics
Boston College
Chestnut Hill, MA 02167 USA
(1) 617 552 3601

18. Di Bartolo, S. John — Mathematical Physics.
Department of Physics
Boston College
Chestnut Hill, MA 02167 USA
(1) 617 552 3575

19. Dinerman, Bradley — Applied laser technology, laser-tissue interactions, skiing, fishing.
Department of Physics
Boston College
Chestnut Hill, MA 02167 USA
(1) 617 631 9235

20. Djeu, Nicholas — Solid-state laser materials and devices, nonlinear optical crystals and devices.
Department of Physics
University of South Florida
Tampa, FL 33620-5700, USA
(1) 813 974 2121

21. Donnelly, Connor J.M. — Optical Spectroscopy of Solids.
Department of Physics
University College
Galway, Ireland
(353) 91 24411

22. dos Santos, Lucilia Maria
Universidade de Aveiro
Departamento de Fisica
3800 Aveiro, Portugal
Fax: (351) 34 24968
Phone: (351) 34 22224 Ext.3106

Deep centers in brown diamonds (luminescence and time-resolved spectroscopy).

23. Evesque, Pierre
Laboratoire d'Optique de
la Matiére Condensée
Tour 13, Université Paris VI
4 Place Jussieu
7531 Paris Cedex 05, France
(33) 1 43 25 33 88

Mechanics of Sandpile, Avalanche, Segregation.
Transient grating in disordered materials.
Energy Migration, Transfer Fractals, Percolation.

24. Fernandez, Joaquin
Departamento de Fisica
Aplicada I
Escuela Tecnica Superior de
Ingenieros Industrials Y
de Telecomunicacion
Alda, Urquijo, S/N
48013 - Bilbao, Spain
(34) 4 4416400 Ext.336

Luminescence materials (glasses and crystals).
Photoacoustics and TR Spectroscopy.

25. Garcia, Maria Jesus Elejalde
Departamento de Fisica
Aplicada I
Escuela Tecnica Superior de
Ingenieros Industrials Y
de Telecomunicacion
Alda, Urquijo, S/N
48013 - Bilbao, Spain
(34) 4 4416400 Ext.337

Laser materials (glasses and crystals).
T.R. Spectroscopy.

26. Giakoumakis, George E.
Applied Physics Lab -
Solid State Division
Physics Department
University of Ionnina
P.O. Box 1186
451 10 Ioannina Greece
Lab: (30) 651 91073
Home: (30) 651 34239

Luminescence of phosphor screens excited by x-rays or electron beam.
Semiconductor materials study and characterization.

27. Göcmen, Ayten
Istanbul Technical Univ.
Department of Chemistry
80626 Maslak - Istanbul, Turkey
(90) 1 1763156
(90) 1 3570702

Chemical kinetics and photochemistry.

28. Graf, Michael
Department of Physics
Boston College
Chestnut Hill, MA 02167 USA
(1) 617 552 4128

Experimental condensed matter at low temperatures and high magnetic fields; heavy fermion systems; quantum fluids and solids.

29. Grassano, Umberto M.
Dipartimento di Fisica
II Università degli Studi
di Roma, Via O. Raimondo
00173 Roma, Italy
(39) 24990 571

Color Center Physics.
Solid State.

30. Grill, Wolfgang
Physikalisches Institut
der Universität
Robert Mayer - Str. 2-4
D-6000 Frankfurt am Main
F.R. of Germany
(49) 69 798 2620
(49) 69 798 2345 (Secretary)

Phonon Spectroscopy, Stimulated Emission of Phonons, Ultrasonics Tunnel Microscopy, Atomic Force Microscopy.

31. Grün, Michael
Universität Kaiserslautern
Fachbereich Physik
Erwin Schrödinger Strasse
D-6750 Kaiserslautern
F.R. of Germany
(49) 631 2064092

Epitaxy, II-IV semiconductor compounds, Heterostructures.

32. Gülen, Demet
Middle East Technical Univ.
Department of Physics
06531 Ankara, Turkey
(90) 4 223 7100 Ext.3291

Energy transfer in photosynthetic systems and organic molecular crystals.

33. Herren, Daniel
Universität Bern
Institut für anorganische,
analytische und
physikalische Chemie
Freiestrasse 3
CH-3000 Bern 9, Switzerland
Phone: (41) 31 654234
Fax: (41) 31 654499
Mail: U30P@CBEBDA3T.EARN

Electron transfer reactions of transition metal complexes on zeolite surfaces.

34. Hollis, David B.
Department of Physics
Paisley College of Tech.
High Street
Paisley, Renfrewshire PA1 2BE
Scotland, UK
(44) 41 887 1241 Ext.238

Solid state optical and vibrational spectra of crystals and glasses for use in electro-optical and opto-electronic systems.

35. Illarramendi, Maria Asuncion
Departmento de Fisica
Aplicada I
Escuela Tecnica Superior de
Ingenieros Industrials Y
de Telecomunicacion
Alda, Urquijo, S/N
48013 - Bilbao, Spain
(34) 4 4416400 Ext.355

Laser materials (glasses and crystals).

36.	Imbusch, George F. Department of Physics University College Galway, Ireland (091) 24411 Ext.705	Solid state optical spectroscopy. Crystalline and glass materials.
37.	Jain, Sudhir Department of Chemistry University of Durham Science Laboratories South Road Durham DHI 3LE, UK (44) 91 374 3123	Spin-glasses, percolation, disordered systems, Monte Carlo simulations & Molecular Dynamics; Dynamics of floppy Van der Waals systems.
38.	Kamber, Ivo Universität Bern Institut für anorganische, analytische und physikalische Chemie Freiestrasse 3 CH-3000 Bern 9, Switzerland Phone: (41) 31 654234 Fax: (41) 31 654499 Mail: U30P@CBEBDA3T.EARN	Time-resolved luminescence & transient absorption of transition metal doped zolites; chemisorbed & physisorbed molecules. Diffuse Reflectance Laser Flash Photolysis.
39.	Kaminskii, Alexander Institute of Crystallography Academy of Sciences of the USSR Leninsky pr. 59 Moscow 117333, USSR (7095) 135 64 01	Physics and spectroscopy of laser crystals.
40.	Klingshirn, Claus Fachbereich Physik der Universität Erwin-Schrödinger-Str. D-6570 Kaiserslautern F.R. of Germany (49) 631 205 2319	Nonlinear optics, optical bistability, many particle effects in II-IV and III-V semiconductor, bulk material, quantum wells and quantum dots.
41.	Kökten, Hatice Department of Physics Middle East Technical Univ. 06531 Ankara, Turkey (90) 4 223 7100 Ext.3291	Chemisorption and Band Calculations.
42.	Köngeter, Andreas Universität Stuttgart Institut fur Theoretische Physik Pfaffenwaldring 57/V 7000 Stuttgart 80 F.R. of Germany (49) 711 685 5207	Theory: Exciton-phonon systems, nonradiative processes.

43. Kyrkos, Stamatios
Department of Physics
National and Capodistrian
University of Athens
Athens, Greece
(30) 651 53940
(30) 1 671 8345

Laboratory Techniques
(semiconductors, superconductivity)

44. Laasch, Wiebke
Universität Hamburg
II Institut für
Experimentalphysik
Luruper Chausee 149
2000 Hamburg 50
F.R. of Germany
(49) 40 8998 2379
(49) 40 8998 2305
Mail: F41LAA@DHHDESY3.BITNET

Time-resolved VUV-Spectroscopy of rare gas solids doped with rare gases.

45. Manaa, Hacène
Université Claude Bernard
-Lyon, Laboratoire de
Physico-Chimie
des Matèriaux Luminescents
Unité Associée au
CNRS No. 442, Bat. 205
43 Boulevard 11 Novembre 1918
69622 Villeurbanne Cedex, France
(33) 78 89 81 24

Physico-chemistry of luminescent materials (Optical properties of transition metal impurities introduced in crystals).

46. Mazurak, Zbigniew
Institute of Low Temperature
and Structure Research
Polish Academy of Sciences
P. O. Box 937
50-950 Wroclaw 2, Poland
(48) 71 221071 Ext.60

Luminescence, laser crystals.

47. Meijerink, Andries
P.O. Box 80 000
3508 TA Utrecht,
The Netherlands
(31) 30 532214

Storage Phosphors for X-ray Imaging and Luminescence of Eu^{2+} in Those Phosphors.

48. Minella, Massimo
Department of Physics
Boston College
Chestnut Hill, MA 02167 USA
Phone: (1) 617 552 4515
BITNET:MINELLA@BCVMS

Dynamical properties of Coulomb systems, strongly coupled plasmas. (Theory)

49. Mishra, Kailash C.
GTE Products Corporation
100 Endicott Street
Danvers, MA 01923, USA
(1) 508 750 2822

Electronic Structure calculation of solids, molecules and atoms.

50. Momoda, Leslie Ann
Dept. of Materials Science
and Engineering
University of California
405 Hilgard Avenue
Los Angeles, CA,
90024-1595, USA
(1) 213 206 1255

Synthesis & Characterization of rare earth doped oxides; particularly the β-alumina family of materials.

51. Monteiro, Teresa Maria
Universidade de Aveiro
Departamento de Fisica
3800 Aveiro, Portugal
Phone: (351) 34 25085 Ext.3101
Fax: (351) 34 24968

Deep centers in crystals (diamond and III-V). Luminescence and time-resolved spectroscopy.

52. Oberhauser, Detlev
Universität Kaiserslautern
Fachbereich Physik
Erwin Schrödinger Strasse
D-6750 Kaiserslautern
F.R. of Germany
(49) 631 205 3112

Phase- and Energy- Relaxation processes in MQW.

53. Pekcan, Önder
Technical Univ, of Istanbul
Department of Physics
Maslak-Istanbul 80626, Turkey
(90) 1 176 3213
(90) 1 362 8608

Restricted geometries and/or Fractal structure in Polymeric composite materials.

54. Pereira, M. Estela
Universidade de Aveiro
Departamento de Fisica
3800 Aveiro, Portugal
Phone: (351) 34 25085 Ext.3108
Fax: (351) 34 24968

Deep centers in crystals (Diamond and III-V materials. Time-resolved spectroscopy.) Eventually for laser material.

55. Petersen, Ronald O.
Tektronix Inc.
Tektronix Industrial Park
P.O. Box 500; M/S 46-567
Beaverton, OR 97077, USA
Phone: (1) 503 627 5475
Fax: (1) 503 627 2670

Luminescent Materials.

56. Pode, Ramchandra Balaji
Department of Physics
Nagpur University
Nagpur 44010, India
(91) 712 31946

Color Centers, Color Center lasers in alkali halides.

57. Raue, Ralf
Philips GmbH
Forschungslaboratorium Aachen
Postfach 1980, Weisshausstrasse
D-5100 Aachen,
F.R. of Germany
(49) 241 6003 240

Luminescent Materials.

58. Reisfeld, Renata
Department of Inorganic and Analytical Chemistry
The Hebrew University of Jerusalem
Jerusalem, Israel
(972) 2 585 323

Energy Tansfer. Nonradiative and Radiative Transitions. Fractals, lasers, fiber optics, phase conjugation, fast processes in amorphous materials.

59. Sarkisov, Stephan
Intitute of Crystallography
Academy of Sciences of the USSR
Leninsky pr. 59
Moscow 117333, USSR
(7095) 135 64 01

Lattice dynamics.
Physics and spectroscopy of laser crystals.

60. Scacco, Augusto
Dipartimento di Fisica
Università di Roma - La Sapienza
Piazza A. Moro, 2
00185 Roma, Italy
(39) 6 49914 394

Optical properties of Defects and Impurities in Insulating Crystals.
Crystal grown techniques.

61. Scaramuzzi, Francesco
ENEA, Centro Ricerche Energia Frascati
00044 Frascati (Roma), Italy
(39) 6 9400 5300

Low Temperature Physics, Basic & Applied.
D_2-Ti system.

62. Scarsi, L.
Instituto di Fisica Cosmica
Via M. Stabile, 172
Palermo, Italy
(39) 91 332 898

X-ray Astronomy, with satellite data.

63. Schriever, Robert
Freie Universität Berlin
Fachbereich Physik
Institut für Atom - und Festkörperphysik (WE 1)
1000 Berlin 33
F.R. of Germany
(49) 30 838 6233

Dissociation and Recombination processes of small molecules in rare gas crystals.

64. Schwab, Holger
Universität Kaiserslautern
Fachbereich Physik
Erwin Schrödinger Strasse 13
6750 Kaiserslautern
F.R. of Germany
(49) 631 205 3112

Luminescence, Relaxation and Recombination of Localized and free Excitons in Semiconductor Mixed-crystals, Laser Induced Gratings, fast optical processes and spectroscopy.

65. Seidl, Michael
Universität Regensburg
Institut für Physik III
Angewandte Physik
Universitätstrasse 31
Postfach 397
8400 Regensburg
F.R. of Germany
(49) 941 943 2113

Laser spectroscopy of solid oxygen.

66. Smets, Bruno M.J.
Philips Lighting B.V.
Advanced Materials
Development, P.O. Box 80020
Building EDW-6
Eindhoven, The Netherlands
(31) 40 75697

Lamp phosphors; their optical properties, preparation and application in fluorescent lamps.

67. Struck, Charles W.
GTE Laboratories
40 Sylvan Road
Waltham, MA 02254, USA
(1) 617 466 2326

Luminescence and Thermodynamics.

68. Sytsma, Joost
Physical Laboratory,
State University
P.O. Box 80 000
3508 TA Utrecht,
The Netherlands
(31) 30 532202

Luminescence of Gd^{3+} in different host lattices:
Vibronic intensities in optical transitions.
Two photon excitation.

69. Terzi, Nice
Università degli Studi
Via Celoria 16
20133 Milano, Italy
(39) 2 2392 402
(39) 2 2392 352 (Secretary)
Fax: (39) 2 2366583
E.Mail: TERZI VAXM1.INFN.IT

Phonon field (coherent states).
Lattice dynamics (Perfect and imperfect lattice) and relaxation.

70. Uhrig, Aiga
Universität Kaiserslautern
Fachbereich Physik
Erwin Schrödinger Strasse
D-6750 Kaiserslautern
F.R. of Germany
(49) 631 205 3112

Low dimensional systems.

71. Ulbrich, Rainer G.
IV. Physikalisches Institut
Bunsenstr. 11
34 Göttingen
F.R. of Germany
(49) 551 394542

Semiconductor Physics.
Optical Properties of III-IV Compounds.

72. van Vliet, Jacques P.M.
Physical Laboratory,
State University
P.O. Box 80 000
3508 TA Utrecht
The Netherlands
(31) 30 532203

Luminescence, energy transfer and energy migration in R.E. inorganic systems.

73. Verhaar, Harry C.G.
Philips Lighting B.V.
Advanced Materials Division
P.O. Box 80020,
Building EDW-629
5600 JM Eindhoven
The Netherlands
(31) 40 755046

Fluorescent Lamp Phosphors and the reasons for induced NR-transitions.

74. Verweij, Hans
Physical Laboratory
State University
P.O. Box 80 000
3508 TA Utrecht
The Netherlands
(31) 30 532321

Solid-state Optical Spectroscopy Rare Earth Ions, Glasses.

75. von Seggern, Heinz
Siemens AG
Dept. ZFE AMF 33
Paul Gossen Strasse 100
8520 Erlangen
F.R. of Germany
(49) 9131 731923

Luminescence, photostimulated luminescence, X-ray storage phosphors.

76. Wesner, Joachim
Physikalisches Institut
der Universität
Postfach 11 19 32
Robert Mayer Str. 2-4
D-6000 Frankfurt am Main 1
F.R. of Germany
(49) 69 798 2925

Phonon propagation at high excitations in ruby (and Alexandrite). Sideband-spectroscopy, Ultrasonics, Data Processing, Computer Simulations.

77. Wittl, Franz
Universität Regensburg
Facultät für Physik
Universitätstrasse 31
Postfach 397
8400 Regensburg
F.R. of Germany
(49) 941 2065

UV- and VUV-spectroscopy of small molecules and atoms in solid rare gas materials.

78. Young, Daniel A.
Dept.of Electrical Eng.
Princeton University
Princeton, NJ 08544, USA
(1) 609 452 3710

Femtosecond time-resolved spectroscopy of electronic excitations in semiconductors. Ultrafast measurements of phonon propagation across interfaces and in dielectrics.

79. Zakrzewski, Adam
Institute of Physics
Polish Academy of Sciences
Al.Lotników 32/46
02-668 Warsaw, Poland
(48) 43 70 01 Ext.142

Quenching of luminescence of by transition metal impurities in A_{II}-B_{VI} semiconductors. Application of photo-electron spin resonance for studying these phenomena. Spatial and spectral aspects of the energy transfer processes.

80. Zhang, Xin Xiong
Department of Physics
Boston College
Chestnut Hill, MA 02167, USA
(1) 617 552 3291

Luminescence Spectroscopy.

INDEX